A HISTORY OF
TECHNOLOGY

A HISTORY OF
TECHNOLOGY

EDITED BY

CHARLES SINGER · E. J. HOLMYARD
A. R. HALL and TREVOR I. WILLIAMS

ASSISTED BY

E. JAFFÉ · NAN CLOW
and R. H. G. THOMSON

VOLUME II

THE MEDITERRANEAN
CIVILIZATIONS AND THE
MIDDLE AGES
c. 700 B.C. TO *c.* A.D. 1500

OXFORD
AT THE CLARENDON PRESS

Oxford University Press, Walton Street, Oxford OX2 6DP

OXFORD LONDON GLASGOW

NEW YORK TORONTO MELBOURNE WELLINGTON

KUALA LUMPUR SINGAPORE HONG KONG TOKYO

DELHI BOMBAY CALCUTTA MADRAS KARACHI

NAIROBI DAR ES SALAAM CAPE TOWN

ISBN 0 19 858106 8

First published 1956
Reprinted (with corrections) 1957, 1962, 1967, 1972, 1979

Printed in Great Britain
at the University Press, Oxford
by Eric Buckley
Printer to the University

PREFACE

FOR the first volume of this work the authors had necessarily to rely almost entirely on evidence derived from material recovered by excavation. This provides practically the only available information on the use of tools and on the sources and nature of the substances employed in the period treated. Though the Ancient Empires possessed their literatures, the fragments of them as yet available contain very few accounts of technical methods.

This second volume deals with the classical civilizations of the Mediterranean together with the cultures that arose in barbarian Europe, and especially in its north-west. The techniques of the latter developed from those of the Mediterranean, much as the Mediterranean techniques had drawn on those of the Ancient Empires. For the period and area with which this volume deals there is much more contemporary documentary evidence than was available for Volume I, but one must still rely largely, though to a diminishing extent as the centuries advance, on archaeological findings. Nevertheless for Volume II—and even more for Volume III—the literary evidence is of considerable bulk, though works of a specifically technological character remain comparatively few. Elements that we can reasonably call scientific begin to be traceable in both the literature and the methods of technology, though science does not yet occupy a recognized and independent position in them. For that we must wait until the sixteenth century. A much longer period was to elapse before technology gained formal academic or educational acceptance. Technological instruction, so far as Volumes II and III are concerned, was essentially and almost wholly a matter of apprenticeship.

Although the task of the historian becomes progressively less intricate as the literary evidence grows in bulk and completeness, it is at the same time rendered more arduous by the increasing wealth of surviving material objects from which he has to make his choice. This problem of choice will increase as the work advances: our five volumes are not nearly large enough, nor are the available authors numerous enough, to compose a world history of technology. We therefore attempt only the technology of Europe, which—with its parallel growth in the New World in the nineteenth century—for the period covered by Volumes III, IV, and V was more advanced and progressive than any other. For the earlier part of the period considered in Volume II, however, the technology of the Far East, and perhaps that of the Near East, was in advance of that of the west. Despite this, we are forced almost to omit these eastern achievements for lack both of space and of writers. Our work thus far is best regarded as a survey preliminary to the account of that rising technical supremacy of western Europe which will be exhibited in later volumes.

We must draw the attention of our readers to the looseness of the chronological framework of these volumes. There is no way by which the different branches of technology can be fitted into a chronological series of the type familiar in the political histories of nations and empires. Each technique develops in relation to specific economic needs, social conditions, or local opportunities. A technique may in part determine such needs, conditions, and opportunities, or it may be determined by them. The situation is often further complicated by the interrelations of the techniques themselves, and most techniques have been repeatedly directed or diverted by the varying alternate demands of peace and war.

Our work is essentially an attempt to apply an artificial, but not illogical, method of selection to illustrate those activities that provide material amenities, and to show how these arise, develop, and depend on each other. It is thus concerned mainly with the nature and evolution of processes, techniques, and devices—in fact with technology proper and not with its social and economic repercussions. Such selective treatment is, at least temporarily, a necessity, and though one object of this work is to provide material for the assistance of social historians, it makes no attempt to assess the effect of technological progress on society. Thus regarded, these volumes are perhaps more in the nature of annals than of history proper, but true history cannot be written until the relevant events have been adequately marshalled. When completed, this *History* will provide a moderately comprehensive survey of the development of western technology.

A word must be said about maps in this work. Volume I was concerned with the finds in a large number of excavations, widely scattered through the Near East and Europe. Most of these sites, though well known to archaeologists, do not appear in ordinary atlases and could therefore not be found by some readers. We consequently felt it necessary to add a series of maps indicating the positions of these sites. Comparable difficulty seldom arises in this volume and will not arise at all in the later volumes. Special maps are, however, still needed to indicate such matters as trade-routes, mining-areas, and so on.

There have been some changes in our editorial staff. Mr R. H. G. Thomson and Mrs J. M. Donaldson have both left us, and we take this opportunity to thank them for their services. Mrs A. Clow has become a member of our staff. We are glad to acknowledge the assistance of Dr H. Heimann and Miss Judith Moore, who were both with us for short periods.

The editors' thanks must again be expressed to the other members of the staff—Mrs E. Harrison, Mrs D. A. Peel, and Miss J. R. Petty. Their work, and especially their close attention to points of detail, has greatly lightened the task of the editors, whose thanks are due also to the staff of the Clarendon Press for their patient consideration of the many production problems inseparable from a complex venture of this kind, and particularly to Mr B. G. Gosling, who has taken a personal interest in this *History*.

We have suffered an irreparable loss by the death of Sir Wallace Akers in 1954. He took the greatest personal interest in the work from its inception and followed its progress in every detail, including a careful reading of the proofs. Sir Lindsay Scott, for whose further help we had hoped, also died before Volume I was published. Of those who have contributed to this volume, Mr Bromehead of the Geological Museum and Survey died just after the completion of his manuscript on mining; he had therefore no opportunity to correct the proofs. Dr Sherwood Taylor, Director of the Science Museum, collaborated with one of the editors in an article which here appears, and was able to approve the galleys and illustrations before his death in January 1956.

In the composition of this volume we have had much helpful guidance and advice. Our special thanks are due to Mr W. J. Worboys, a director of Imperial Chemical Industries Limited, who, with Sir Wallace Akers and Dr C. J. T. Cronshaw, was one of the original sponsors of this project. The enlightened and continued generosity of I.C.I. is making the production of these volumes possible and is bringing their completion well within sight. The understanding and enthusiasm of Mr Worboys have been a stimulus and encouragement to us all.

On the editorial side we have had so many helpful suggestions that it is difficult to select names for mention here, but we should not like this volume to appear without recording our thanks to Professor V. Gordon Childe, Dr O. G. S. Crawford, Professor R. J. Forbes, Mr Mahmud A. Ghul, Dr Joseph Needham, Professor M. M. Postan, Mr F. I. G. Rawlins, Professor D. Talbot Rice, Professor A. W. Skempton, Professor E. G. R. Taylor, and Sir Mortimer Wheeler.

As with Volume I, the officials of many libraries have given us invaluable help. We should like particularly to mention those of the British Museum, the London Library, the Patent Office, the Science Museum Library, and the Warburg Institute. In the preparation of the illustrations we have again relied to a great extent upon Mr D. E. Woodall: among other artists who have assisted us we would like to mention here Mr Thomas A. Greeves and Mr K. F. Rowland. In this volume the maps are mostly the work of Mr J. F. Horrabin. The indexes have again been made by Mr P. G. Burbidge.

CHARLES SINGER
E. J. HOLMYARD
A. R. HALL
TREVOR I. WILLIAMS

CONTENTS

ILLUSTRATIONS AND TABLES xi

LIST OF ABBREVIATIONS xliv

HISTORICAL NOTES by A. R. HALL li

PART I. PRIMARY PRODUCTION

1. MINING AND QUARRYING TO THE SEVENTEENTH CENTURY by the late C. N. BROMEHEAD, District Geologist for South-Eastern England on the staff of the Geological Survey and Museum, London 1

2. METALLURGY by R. J. FORBES, Professor of Pure and Applied Sciences in Antiquity, University of Amsterdam 41

3. AGRICULTURAL IMPLEMENTS by E. M. JOPE, Reader in Archaeology, The Queen's University of Belfast 81

PART II MANUFACTURE

4. FOOD AND DRINK by R. J. FORBES 103

5. LEATHER by JOHN W. WATERER, Honorary Secretary of the Museum of Leathercraft, London 147

with *A Note on Parchment* by H. SAXL, Nuffield Gerontological Unit, University of Leeds 187

6. SPINNING AND WEAVING by R. PATTERSON, Curator, Castle Museum, York 191

7. PART I. FURNITURE: TO THE END OF THE ROMAN EMPIRE by CYRIL ALDRED, Assistant Keeper in the Royal Scottish Museum, Edinburgh 221

PART II. FURNITURE: POST-ROMAN by R. W. SYMONDS 240

8. PART I. CERAMICS: FROM *c* 700 B.C. TO THE FALL OF THE ROMAN EMPIRE by GISELA M. A. RICHTER, Honorary Curator, Department of Greek and Roman Art, Metropolitan Museum of Art, New York 259

PART II. CERAMICS: MEDIEVAL by E. M. JOPE 284

9. GLASS AND GLAZES by D. B. HARDEN, O.B.E., Keeper, Department of Antiquities, Ashmolean Museum, Oxford 311

10. PRE-SCIENTIFIC INDUSTRIAL CHEMISTRY by the late F. SHERWOOD TAYLOR, formerly Director of the Science Museum, London, and CHARLES SINGER 347

with *A Note on Military Pyrotechnics* by A. R. HALL 374

PART III. MATERIAL CIVILIZATION

11. THE MEDIEVAL ARTISAN by R. H. G. THOMSON, Scientific Laboratory, National Gallery, London 383

12. BUILDING CONSTRUCTION by MARTIN S. BRIGGS 397

13. FINE METAL-WORK by HERBERT MARYON, O.B.E., Technical Attaché, Research Laboratory, British Museum, London 449

with *A Note on Stamping of Coins and Other Objects* by PHILIP GRIERSON, Lecturer in History, University of Cambridge 485

PART IV. TRANSPORT

14. ROADS AND LAND TRAVEL, WITH A SECTION ON HARBOURS, DOCKS, AND LIGHTHOUSES by R. G. GOODCHILD, Controller of Antiquities, Government of Cyrenaica (III, IV, and VI), and R. J. FORBES (I, II, V, VII, and VIII) 493

15. VEHICLES AND HARNESS by E. M. JOPE 537

16. SHIPBUILDING by T. C. LETHBRIDGE 563

PART V. PRACTICAL MECHANICS AND CHEMISTRY

17. POWER by R. J. FORBES 589

with *A Note on Windmills* by REX WAILES 623

18. MACHINES by BERTRAND GILLE, Archivist, Archives Nationales, Paris 629

with *A Note on Ancient Cranes* by A. G. DRACHMANN, Librarian, University Library, Copenhagen 658

19. HYDRAULIC ENGINEERING AND SANITATION by R. J. FORBES 663

20. MILITARY TECHNOLOGY by A. R. HALL 695

21. ALCHEMICAL EQUIPMENT by E. J. HOLMYARD 731

EPILOGUE

EAST AND WEST IN RETROSPECT by CHARLES SINGER 753

INDEXES 777

PLATES *at end*

ILLUSTRATIONS AND TABLES

Oriental bottle of enamelled glass. The Arabic inscription says: 'Glory to our Lord the Sultan, the wise, just, religious warrior-king.' c *1330*. London, Victoria and Albert Museum. From E. DILLON. 'Glass', Pl. XXIII. London, Methuen, 1907 FRONTISPIECE

TEXT-FIGURES*

I. MINING AND QUARRYING TO THE SEVENTEENTH CENTURY *by* C. N. BROMEHEAD

1 *Greek miners.* After P. BRANDT 'Schaffende Arbeit und bildende Kunst im Altertum und Mittelalter', fig. 75. Leipzig, Kröner, 1927. R. Steed 2

2 *Drainage of a Japanese mine.* After C. N. BROMEHEAD. *Antiquity* **16,** Pl. I, facing p. 200, 1942. Newbury, Edwards. D. E. Woodall 4

3 *Two coins from Damastion.* After J. M. F. MAY. 'Coinage of Damastion', Pl. XI, nos 10, 14b. London Oxford University Press, 1939. D. E. Woodall 5

4 *Map showing Roman mines in Europe.* Based on O. DAVIES. 'Roman Mines in Europe', Maps. Oxford, Clarendon Press, 1935. E. Norman 6

5 (Left) *A series of water-raising wheels*; (Right) *A single wheel.* From D. E. HUDSON. *Metallurgia* **35,** 161, figs 7, 8, 1947. Manchester, Kennedy Press 7

6 *Roman miners' iron picks.* After H. SANDARS. *Archaeologia* **59,** 328, fig. 12, 1905. London, Society of Antiquaries. D. E. Woodall 8

7 *Typical Roman miner's iron tools.* After H. SANDARS. *Ibid.*, Pl. LXX, facing p. 328, nos 1, 4, 6, 7. D. E. Woodall 9

8 *Stone relief of Roman miners.* After H. SANDARS. *Ibid.*, Pl. LXIX. D. E. Woodall 9

9 *Roman iron picks.* After W. PAGE (Ed.). 'The Victoria History of the County of Somerset', Vol. I, fig. 95. London, Constable, 1906. D. E. Woodall 10

10 *Illustrations from a Castilian manuscript.* After 'Lapidario del Rey Don Alfonso X', ed. by J. FERNANDEZ MONTANA, fols. 18ᵛ, 21, 33, 46. Madrid, Blasco, 1881. D. E. Woodall 13

11 *First stages in the excavation of a mine.* From AGRICOLA. 'Vom Bergkwerck', p. lxxvi. Basel, Froben and Bischoff, 1557 14

12 *Iron tools for splitting rock.* From AGRICOLA. *Ibid.*, p. cxii 15

13 *Hammers.* From AGRICOLA. *De re metallica*, p. 209 [109]. Basel, Froben, 1556 15

14 *Digging-tools.* From AGRICOLA. *Ibid.*, pp. 209 [109], 110 16

15 *Containers for hauling.* (Left) *Basket and wooden buckets*; (Right) *Ox-hide buckets.* From AGRICOLA. 'Vom Bergkwerck', pp. cxv, cxvi. Basel, Froben and Bischoff, 1557 16

16 (Left) *Wheelbarrows*; (Right) *Truck.* From AGRICOLA. *Ibid.*, pp. cxvii, cxviii 17

17 *Horse-whim for raising large loads.* From AGRICOLA. *Ibid.*, p. cxxx 18

18 *Horse-drawn sledge.* From AGRICOLA. *De re metallica*, p. 126. Basel, Froben, 1556 19

19 *Mine-drainage: chain of dippers.* From AGRICOLA. *Ibid.*, p. 133 20

20 *Mine-drainage: triple-lift suction-pump.* From AGRICOLA. *Ibid.*, p. 145 21

21 *Mine-drainage: rag-and-chain pump.* From AGRICOLA. *Ibid.*, p. 154 22

22 *Mine-ventilation.* From AGRICOLA. *Ibid.*, pp. 164, 165 23

23 *Miners with tools.* From E. BROWNE. 'Travels in Europe' (2nd ed.), Pl. on p. 170. London, 1685 24

24 *Ancient Greek limestone quarries.* After J. A. HAMMERTON. 'Wonders of the Past', Vol. 2, p. 1208. London, Amalgamated Press, 1934. D. Wyeth 26

* The names at the end of entries are those of the artists who drew the illustrations.

25 *Modern tools used in Oxfordshire.* (Right) and (Left) from W. J. ARKELL. 'Oxford Stone', figs 16, 18. London, Faber and Faber, 1947. (Centre) *Plug and feathers.* Based on W. J. ARKELL. *Ibid.*, fig. 9. D. E. Woodall 36

26 *Mason's axe and bolster.* From W. J. ARKELL. *Ibid.*, fig. 10 37

2. METALLURGY *by* R. J. FORBES

27 *A graded series of needles.* From AGRICOLA. 'Vom Bergkwerck', p. ccvii. Basel, Froben and Bischoff, 1557 45

28 *Red figured bowl, showing casting of a statue.* From T. SCHREIBER. 'Atlas of Classical Antiquities' (ed. by W. C. F. ANDERSON), Pl. VIII, fig. 6. London, Macmillan, 1895 51

29 *Black figured vase, showing a smithy.* From T. SCHREIBER. *Ibid.*, Pl. LXIX, fig. 6 58

30 *Black figured vase, showing the forge of Hephaistos.* London, British Museum, Greek vase B 507. Adapted from photograph. By courtesy of the Trustees. R. Steed 59

31 *Roman forge.* After P. BRANDT. 'Schaffende Arbeit und bildende Kunst im Altertum und Mittelalter', fig. 139. Leipzig, Kröner, 1927. Steffie Schaefer 60

32 *Roman locksmith.* Aquileia, Archaeological Museum. After U. E. PAOLI. 'Das Leben im alten Rom', Pl. LXI, 2. Bern, Francke, 1948. D. E. Woodall 60

33 *Anvil and tongs.* Reading, Silchester Collection. After photograph by courtesy of Reading Museum and Art Gallery. D. E. Woodall 61

34 *The smelting of argentiferous copper.* After H. T. BOSSERT and W. F. STORCK. 'Das mittelalterliche Hausbuch im Besitz des Fürsten von Waldburg-Wolfegg-Waldsee', Pl. XXXIX (fol. 36[a]). Leipzig, Seemann, 1912. D. E. Woodall 66

35 *In the centre is a cupellation-furnace.* After H. T. BOSSERT and W. F. STORCK. *Ibid.*, Pl. XXXVIII (fol. 35[b]). D. E. Woodall 67

36 *Hydraulic bellows.* Munich, Bayerische Staatsbibliothek, MS. lat. 197, fol. 30[v]. After photograph by courtesy of the Director. D. E. Woodall 68

37 *A medieval blast-furnace.* Kropa, Yugoslavia. After photograph by courtesy of Director C. Rekar, Ljubljana, Metalurškega Inštituta. D. E. Woodall 70

38 *Evolution of the iron-smelting furnace.* (A, top; B, C, D) After R. J. FORBES. 'Metallurgy in Antiquity', figs 81, 82. Leiden, Brill, 1950. (A, below) After L. BECK. 'Die Geschichte des Eisens', Part I, fig. 250 b. Braunschweig, Vieweg, 1884. D. E. Woodall 72

39 *Drawing heavy iron wire.* From BIRINGUCCIO. 'De la Pirotechnia', fol. 140[v]. Venice, Roffinello, 1540 74

40 *German nail-maker.* Nuremberg, Stadtbibliothek, Hausbuch der Mendelschen Zwölfbrüderstiftung. After F. BOCK (Ed.). 'Deutsches Handwerk im Mittelalter', Pl. VIII. Leipzig, Insel-Verlag, 1935. D. E. Woodall 76

41 *Italian forge of c 1492.* From A. P. V. E. N. MASSENA, PRINCE D'ESSLING. 'Études sur l'art de la gravure sur bois à Venise', Vol. 2, p. 173. Florence, Olschki, 1908 76

42 *A south German wood-carving.* Vienna, Kunsthistorisches Museum. After P. BRANDT. 'Schaffende Arbeit und bildende Kunst im Altertum und Mittelalter', fig. 415. Leipzig, Kröner, 1927. D. E. Woodall 77

3. AGRICULTURAL IMPLEMENTS *by* E. M. JOPE

43 *Diagram of Greek plough.* By the author 82

44 *Diagrammatic section of a plough.* Based on P. V. GLOB. *Acta Archaeologica* **16**, 95, fig. 2; 96, fig. 3; 97, fig. 4, 1945. Copenhagen, Munksgaard. E. M. Jope and T. A. Greeves 83

45 *Greek plough from Nikosthenes cup.* Berlin, Staatl. Museen. From P. BRANDT. 'Schaffende Arbeit und bildende Kunst im Altertum und Mittelalter', fig. 82. Leipzig, Kröner, 1927 84

46 *Reconstruction of a plough.* Diagram based on A. STEENSBERG. *Acta Archaeologica* **16**, 64, fig. 3, 1945. Copenhagen, Munksgaard. E. M. Jope and T. A. Greeves 84

47 *Diagrammatic reconstruction of a plough.* Adapted from A. STEENSBERG. *Ibid.* **7**, 255, fig. 4, 1936. T. A. Greeves 85

48 *Iron socketed plough-share.* Based on W. M. F. PETRIE. 'Gerar', Pl. XXVI, no. 2. London, British School of Egyptian Archaeology, 1928. T. A. Greeves 85

49 *Plough-model from Cologne.* From P. V. GLOB. *Acta Archaeologica* **16**, 111, fig. 16, 1945. Copenhagen, Munksgaard 86

50 (A) *Romano-British plough-model.* London, British Museum, No. 54.12-2776. (B) *Modern ploughing in the Himalayas.* Both after F. G. PAYNE. *Archaeol. J.* **104**, Pl. VIII, facing p. 97; Pl. VII, facing p. 96, 1947. London, Royal Archaeological Institute of Great Britain and Ireland. T. A. Greeves 87

51 *Foot-plough.* Paris, Bibliothèque Nationale, MS. lat. 8846, fol. 62ᵛ. After H. OMONT. 'Psautier illustré' [facs. ed.], Pl. XLV. Paris, Bibliothèque Nationale, 1906. T. A. Greeves 88

52 *Wheeled one-way plough.* London, British Museum, Cotton MS. Tib. B.V, fol. 3. After photograph by courtesy of the Trustees. D. E. Woodall 88

53 *Wheeled plough.* From J. WALTER (Ed.). HERRAD DE LANDSPERG, *Hortus deliciarum*, p. 84. Strasbourg, Le Roux, 1952 89

54 *Swing-plough.* London, British Museum, Add. MS. 42130, fol. 170. After 'The Luttrell Psalter'. Facs. edition by E. G. MILLAR, Pl. XCII. London, British Museum, 1932. By courtesy of the Trustees. D. E. Woodall 89

55 *Plough with mould-board.* London, British Museum, Add. MS. 47682 (Holkham Hall MS. 666, formerly in the possession of Lord Leicester), fol. 6. After photograph by courtesy of the Courtauld Institute of Art, University of London. T. A. Greeves 90

56 *Sketch of a plough.* Oxford, Bodleian Library, MS. Top. Lincs. d.1, fol. 53. After photograph by courtesy of the Curators and H. M. COLVIN. *Antiquity* **27**, 166, fig. 1; 167, 1953. Newbury, Edwards. T. A. Greeves 90

57 *Ploughing and harrowing scene.* Bayeux, Episcopal Palace. After J. VERRIER. 'La broderie de Bayeux', p. 6. Paris, Éditions Tel, 1946. T. A. Greeves 91

58 *Flanders plough.* Chantilly, Musée Condé. After P. DURRIEU. 'Les Très Riches Heures du Duc de Berry', Pl. III. Paris, Plon-Nourrit, 1904. D. E. Woodall 92

59 *Horse-drawn plough.* Holkham Hall MS. 311. After British Museum photograph. By courtesy of Lord Leicester and the Trustees of the British Museum. D. E. Woodall 94

60 *Sickle-handle.* From J. G. D. CLARK. 'Prehistoric Europe', fig. 54. London, Methuen, 1952 95

61 *Short-handled scythes.* Utrecht University Library, MS. 488, fol. 74. From E. T. DE WALD. 'The Illustrations of the Utrecht Psalter', Pl. CXIII. Princeton, N.J., Princeton University Press, 1933 95

62 *Eleventh-century long-handled scythes.* London, British Museum, Cotton MS. Jul. A. VI, fol. 6. From J. R. GREEN. 'A Short History of the English People', Vol. I, p. 157. London, 1892 96

63 *Scythe with a semicircle.* From R. VAN BASTELAER. 'Les Éstampes de Peter Bruegel l'Ancien', p. 202. Brussels, Van Oest, 1908 96

64 *Reconstruction of a Roman machine-reaper.* From R. BILLIARD. 'L'Agriculture dans l'antiquité d'après les Géorgiques de Virgile', fig. 10. Paris, Boccard, 1928 97

65 *Hafted fork, rake, flail.* New York, Pierpont Morgan Library, MS. 638, fols. 12ᵇ, 17ᵇ. After S. C. COCKERELL and M. R. JAMES. 'A Book of Old Testament Illustrations of the Middle of the Thirteenth Century', figs. 84, 118. Cambridge, The Roxburghe Club, 1927. By permission of the Director of the Pierpont Morgan Library. K. F. Rowland 98

66 *Spades* (right to left). [i, ii] Oxford, Bodleian Library, MS. Jun. 11, fol. 49. [iii–v] London, British Museum MSS.: Cotton Claud. B. IV, fol. 22ᵛ; Add. 18750, fol. 3; Stowe 17, fol 4ᵛ. After photographs by courtesy of the Curators of the Bodleian Library and the Trustees of the British Museum. K. F. Rowland 98

67 *Fork-headed pick.* After G. BRETT. *J. Warburg Courtauld Insts* **5**, Pl. XII, facing p. 38, 1942. By courtesy of the Warburg Institute, University of London. D. E. Woodall 99

68 *Sowing scene.* London, British Museum, Add. MS. 42130, fol. 170ᵛ. After 'The Luttrell Psalter'. Facs. edition by E. G. MILLAR, Pl. XCIII. London, British Museum, 1932. By courtesy of the Trustees. D. E. Woodall 100

TAILPIECE. *Threshing in a barn.* London, British Museum, MS. Kings 9, fol. 9ᵛ. After photograph by courtesy of the Trustees. D. E. Woodall 102

4. FOOD AND DRINK *by* R. J. FORBES

69 *Threshing with oxen.* After original photograph by courtesy of Dr O. G. S. Crawford, C.B.E. D. E. Woodall 105

70 *Underside of a Cypriot threshing-sled.* After original photograph by courtesy of Dr O. G. S. Crawford, C.B.E. D. E. Woodall 106

71 *Japanese pivoted pestle.* From F. PERZYNSKI. 'Hokusai', fig. 16. [H. KNACKFUSS (Ed.). Künstler-Monographien No. 68]. Leipzig, Velhagen and Klasing, 1904 107

72 *Upper stone of a pushing-mill.* From D. M. ROBINSON and J. W. GRAHAM. 'Excavations at Olynthus', Part VIII, Pl. LXXX, 1. Johns Hopkins University Studies in Archaeology no. 25. Baltimore, Johns Hopkins Press, 1938 108

73 *Mounting of a pushing-mill.* From D. M. ROBINSON and J. W. GRAHAM. *Ibid.*, fig. 34 108

74 *Early English quern types.* From E. C. CURWEN. *Antiquity* **15**, 15–32, figs 24 a, 25–27, 1941. Newbury, Edwards 109

75 *Diagram of the Iberian type of quern.* From V. G. CHILDE. *Ibid.*, **17**, 20, fig. 1, 1943 110

76 *Diagram of a donkey-mill.* From E. C. CURWEN. *Ibid.*, **11**, 139, fig. 2, 1937 110

77 *Set of four donkey-mills.* After A. NEUBURGER. 'The Technical Arts and Sciences of the Ancients' (trans. by H. L. BROSE), fig. 156. London, Methuen, 1930. D. E. Woodall 110

78 *Horse harnessed to a donkey-mill.* Rome, Museo Vaticano. After copyright photograph Alinari, Florence. D. E. Woodall 111

79 *Reconstruction of* mola olearia. From A. G. DRACHMANN. *Archaeol.-kunsthist. Medd.* **1**, 143, fig. 9, 1932–5. Copenhagen, Det Kongelige Danske Videnskabernes Selskab 111

80 *Reconstruction of Greek* trapetum. From D. M. ROBINSON and J. W. GRAHAM. 'Excavations at Olynthus', Part VIII, fig. 35. Johns Hopkins University Studies in Archaeology no. 25. Baltimore, Johns Hopkins Press, 1938 112

81 *Simple beam-press.* From C. DAREMBERG and E. SAGLIO. 'Dictionnaire des antiquités grecques et romaines', Vol. IV, Part I, fig. 5388. Paris, Hachette, 1904–7 113

82 *Press described by Cato.* From D. M. ROBINSON and J. W. GRAHAM. 'Excavations at Olynthus', Part VIII, fig. 36. Johns Hopkins University Studies in Archaeology no. 25. Baltimore, Johns Hopkins Press, 1938

83 *Press described by Hero.* From A. G. DRACHMANN. *Archaeol.-kunsthist. Medd.* **1**, 151, fig. 20, 1932–5. Copenhagen, Det Kongelige Danske Videnskabernes Selskab 114

84 *Christ in the mystical wine-press.* From J. WALTER (Ed.). HERRAD DE LANDSPERG, *Hortus deliciarum*, Pl. XLI. Strasbourg, Le Roux, 1952 115

85 *Pliny's first lever-and-screw press.* From A. G. DRACHMANN. *Archaeol.-kunsthist. Medd.* **1**, 146, fig. 14, 1932–5. Copenhagen, Det Kongelige Danske Videnskabernes Selskab 116

86 *Pliny's second lever-and-screw press.* From A. G. DRACHMANN. *Ibid.*, 148, fig. 16 116

87 *Screw-presses described by Hero.* From A. G. DRACHMANN. *Ibid.*, 158, figs 25, 27 117

88 *Roman baker's oven.* After P. BRANDT. 'Schaffende Arbeit und bildende Kunst im Altertum und Mittelalter', fig. 151. Leipzig, Kröner, 1927. D. E. Woodall 118

89 *Roman kitchen.* Igel, near Treves, Monument of the *Secundinii.* After P. BRANDT. *Ibid.*, fig. 156. D. E. Woodall 119

90 *Olive harvest.* London, British Museum, Greek vase B 226. From photograph by courtesy of the Trustees 122

91 *Medieval kitchen.* London, British Museum, Add. MS. 42130, fol. 207. After 'The Luttrell Psalter'. Facs. edition by E. G. MILLAR, Pl. CLXVI. London, British Museum, 1932. By courtesy of the Trustees. D. E. Woodall 124

92 *Roasting on the spit.* *Ibid.*, fol. 206b. After E. G. MILLAR. *Ibid.*, Pl. CLXV. D. E. Woodall 124

93 *Preparing the roast.* *Ibid.*, fol. 207b. After E. G. MILLAR. *Ibid.*, Pl. CLXVII. D. E. Woodall 125

94 *Medieval meal.* *Ibid.*, fol. 208. After E. G. MILLAR. *Ibid.*, Pl. CLXVIII. D. E. Woodall 125

95 *Meal with knife and fork.* Monte Cassino, Monastic Library, MS. 132 (Rabanus Maurus), p. 511. After photograph by courtesy of the Most Reverend Abbot. D. E. Woodall 126

96 *Small mill.* After 'La Sculpture romane' (photographies de Jean Roubier), Pl. XIX. Paris, Alpina, 1937. D. E. Woodall 127

97 and 98 *Duke William's feast.* Bayeux, Episcopal Palace. After J. VERRIER. 'La broderie de Bayeux', p. 26. Paris, Éditions Tel, 1946. D. E. Woodall 128, 129

99 *The journey of Dionysus.* Munich, Staatl. Antikensammlungen, Inv. No. 2044. After copyright photograph by Prof. Max Hirmer (Gesellschaft für wissenschaftliches Lichtbild), Munich. D. E. Woodall 130

100 *Greek and Italian amphorae.* With acknowledgements to the *National Geographic Magazine*, Washington. D. E. Woodall 132

101 *Packing pottery wine-jars in straw.* Treves, Rheinisches Landesmuseum. After P. BRANDT. 'Schaffende Arbeit und bildende Kunst im Altertum und Mittelalter', fig. 159. Leipzig, Kröner, 1927. D. E. Woodall 136

102 *Towing a cargo-boat.* After P. BRANDT. *Ibid.*, fig. 158. D. E. Woodall 136

103 *Galley on the Rhine.* Treves, Rheinisches Landesmuseum. After P. BRANDT. *Ibid.*, fig. 164. D. E. Woodall 137

104 *The vine-grower in Europe.* Nuremberg, Stadtbibliothek, Hausbuch der Mendelschen Zwölfbrüderstiftung. After F. BOCK (Ed.). 'Deutsches Handwerk im Mittelalter', Pl. I. Leipzig, Insel-Verlag, 1935. D. E. Woodall 138

105 *Botanical garden in 1500.* From H. BRAUNSCHWEIG. 'Das Buch der Cirurgia.' Facs. edition by G. KLEIN, Pl. X (from *De arte distillandi*, Strasbourg, 1500). Munich, Kuhn, 1911 143

5. LEATHER *by* JOHN W. WATERER

106 *'Trousers' and 'skirts'.* From H. OBERMAIER and A. GARCIA Y BELLIDO. 'El hombre prehistórico y los orígines de la humanidad' (3rd ed.), fig. 21 c, e. Madrid, Revista de Occidente, 1944 148

107 *Scrapers.* (A) *Of bone.* After L. PFEIFFER. 'Die steinzeitliche Technik', fig. 13. Jena, Fischer, 1912. (B) *Of iron.* After H. BLÜMNER. 'Technologie und Terminologie der Gewerbe und Künste bei Griechen und Römern' (2nd rev. ed.), Vol. 1, fig. 266. Leipzig, Teubner, 1912. (C) *Modern.* D. E. Woodall 148

108 *'Plane' scrapers.* (A) *Aurignacean.* From L. PFEIFFER. 'Die steinzeitliche Technik', fig. 71. Jena, Fischer, 1912. (B) *Modern Eskimo tool.* After L. PFEIFFER. *Ibid.*, fig. 13, 1 [2]. D. E. Woodall 149

109 *Egyptian harp.* London, British Museum, No. 38170. After photograph by courtesy of the Trustees. D. E. Woodall 149

110 *Skin pot.* After 'The Archaeological Survey of Nubia', Vol. 1, Pl. LXVI d. Cairo, Ministry of Finance, 1910. D. E. Woodall 150

111 *Egyptian leather-dressers.* From N. DE G. DAVIES. 'The Tomb of Rekh-mi-rē at Thebes', Vol. 2, Pl. LIV. New York, Metropolitan Museum of Art, 1943 150

112 *Outline sketch of a sandal.* Based on G. A. WAINWRIGHT. 'Balabish', Pl. IX, type 3. London, Egypt Exploration Society, 1920. Reconstruction by J. W. Waterer 151

113 *The stocks as used in chamoising leather.* From D. DIDEROT and J. L. R. D'ALEMBERT. 'Encyclopédie', Planches Vol. 2, Pl. IV. Paris, 1763 152

114 *De-hairing of cowhides.* From HANS SACHS. 'Eygentliche Beschreibung aller Stände . . . mit kunstreichen Figuren [by Jost Amman]', fol. P [IV]. Frankfurt a. M., Sigmund Feyerabend, 1568 152

115 *Eighteenth-century curriers' workshop.* From D. DIDEROT and J. L. R. D'ALEMBERT. 'Encyclopédie', Planches Vol. 2, Pl. I. Paris, 1763 153

116 *Shagreen spectacle case.* London, Museum of Leathercraft. After photograph. D. E. Woodall 157

117 *Detail of Arkwright's spinning-frame.* London, Science Museum, South Kensington. After photograph by courtesy of the Director. D. E. Woodall 157

118 *Fragments of skin clothing.* After G. A. WAINWRIGHT. 'Balabish', Pl. XI, B. 179. London, Egypt Exploration Society, 1920. D. E. Woodall 158

119 *Dagger-sheath.* Stade, Urgeschichtsmuseum. After photograph by courtesy of the Konrektor. D. E. Woodall 159

120 *Egyptian painting.* From N. DE G. DAVIES. 'The Tomb of Rekh-mi-rē at Thebes', Vol. 2, Pl. XIX. New York, Metropolitan Museum of Art, 1943 160

121 *Tuareg jewel-case.* Basel, Museum für Natur und Völkerkunde, III 2715. After A. GANSSER. *Ciba Review* no. 81, 2940, 1950. By courtesy of CIBA Limited, Basel. D. E. Woodall 161

122 *Silenus with wine-skin.* London, British Museum. From C. H. SMITH. 'Catalogue of the Greek and Etruscan Vases in the British Museum', Vol. 3, Pl. VI, E 24. London, British Museum, 1896. By courtesy of the Trustees 161

123 *Leather girdle.* After G. A. WAINWRIGHT. 'Balabish', Pl. X, no. 218. London, Egypt Exploration Society, 1920. D. E. Woodall 161

124 *Ivory statuette.* From J. E. QUIBELL. 'Hierakonpolis I', Pl. IX. London, British School of Egyptian Archaeology, 1900. *Patterned fragment of leather.* From W. M. F. PETRIE and J. E. QUIBELL. 'Naqada and Ballas', Pl. LXIV, no. 104. London, Quaritch, 1896 161

125 *Egyptian sandal-makers.* From N. DE G. DAVIES. 'The Tomb of Rekh-mi-rē at Thebes', Vol. 2, Pl. LIII. New York, Metropolitan Museum of Art, 1943 162

126 *Red-dyed sandals.* London, British Museum, No. 24708. After photograph by courtesy of the Trustees. D. E. Woodall 162

127 *Ay wearing red gloves.* After NORMAN DE G. DAVIES. 'The Rock Tombs of El Amarna', Pt VI, Pl. XXXI. London, Egypt Exploration Fund, 1908. Restored drawing by Nina de G. Davies 163

128 *An Egyptian ball.* After W. M. F. PETRIE. 'Objects of Daily Use', Pl. LI, no. 364. London, British School of Egyptian Archaeology, 1927. D. E. Woodall 164

129 *Neolithic bowl.* Schloss Gottorp, nr. Schleswig, Schleswig-Holsteinisches Landesmuseum. After photograph by courtesy of the Director. D. E. Woodall 165

130 *Picture on a Greek bowl.* London, British Museum, Greek vase E 86. After photograph by courtesy of the Trustees. D. E. Woodall. (Below) *Evolution of the half-moon knife.* Diagrams by J. W. Waterer 166

131 *Greek shoemakers.* Boston, Museum of Fine Arts, No. P 9024. After photograph by courtesy, Museum of Fine Arts, Boston, Mass. D. E. Woodall 167

132 *Soles of* caligae. After copyright photograph by J. W. Waterer. D. E. Woodall 168

133 *The binding of the Stonyhurst Gospel.* Stonyhurst College. After photograph in the Victoria and Albert Museum, London. Crown copyright reserved. D. E. Woodall 169

134 *A Viking shield* (reconstruction by J. W. WATERER). London, Museum of Leathercraft. D. E. Woodall 169

135 *Shoemaker's shop.* From HANS SACHS. 'Eygentliche Beschreibung aller Stände . . . mit kunstreichen Figuren [by Jost Amman]', fol. O II. Frankfurt a. M., Sigmund Feyerabend, 1568 170

136 *Moulded and riveted fire-bucket.* London, Barrow, Hepburn, and Gale Limited. After copyright photograph by J. W. Waterer. D. E. Woodall 171

137 *Casket covered in tooled leather.* London, Museum of Leathercraft. D. E. Woodall 172

138 Cuir bouilli *chalice-box.* Ipswich, Christchurch Mansions. After copyright photograph J. W. Waterer. D. E. Woodall 173

139 *Bombard, black-jack, and bottle.* London, Museum of Leathercraft. After. J. W. WATERER. 'Leather and Craftsmanship', Pl. XII B. London, Faber and Faber, 1950. D. E. Woodall 174

140 *Buff leather tunic.* London, Museum of Leathercraft. D. E. Woodall 175

141 *Pointed leather shoe.* London, Museum of Leathercraft. D. E. Woodall 176

142 *English riding-boot.* London, Victoria and Albert Museum, No. T. 425–1913. After photograph Crown copyright reserved. D. E. Woodall 176

143 *Methods of joining a shoe.* After sketches by J. W. Waterer. D. E. Woodall 176

144 *Gloves and mittens.* London, British Museum, Add. MS. 42130, fols 172, 173. After 'The Luttrell Psalter'. Facs. edition by E. G. MILLAR, Pls XCVI, XCVIII. London, British Museum, 1932. By courtesy of the Trustees. D. E. Woodall 177

145 *Left gauntlet of the Black Prince.* Canterbury Cathedral. After photograph by courtesy of the Dean and Chapter. D. E. Woodall 178

146 *Manufacture of gilt leather panels.* From 'Encyclopédie méthodique', Planches Vol. 2, Pl. IV. Paris; Liége, 1783 179

147 *Blind-stamped binding of the* Liber Winton. London, Society of Antiquaries. After photograph by courtesy of the Committee. D. E. Woodall 181

148 *Medieval riding-saddle.* Toronto, Royal Ontario Museum of Archaeology. After photograph by courtesy of the Director. D. E. Woodall 182

149 *Hide-covered coffer.* London, Museum of Leathercraft. D. E. Woodall 183

150 *Leather lanthorn.* London, Museum of Leathercraft. D. E. Woodall 184

151 *Norman stone-carving.* Oxfordshire, Hook Norton Church. After photograph of cast at the Museum of Leathercraft, London. D. E. Woodall 184

TAILPIECE. *Water-seller.* London, British Museum, Add. MS. 42130, fol. 201. After 'The Luttrell Psalter'. Facs. edition by E. G. MILLAR, Pl. CLIV. London, British Museum, 1932. By courtesy of the Trustees. D. E. Woodall 187

152 *St Luke copying his gospel.* Paris, Bibliothèque Nationale, MS. grec 189, fol. 206v. After H. OMONT. 'Miniatures des plus anciens manuscrits grecs de la Bibliothèque Nationale du VIe au XIVe siècle', Pl. LXXXIX, left. Paris, Librairie Ancienne Honoré Champion, 1929. D. E. Woodall 188

153 (Centre) *Stretching the parchment.* Bamberg, Staatl. Bibliothek, MS. Patr. 5, B. 2, 5, fol. IV. After photograph. D. E. Woodall; (Right) *Smoothing the finished sheet*; (Left) *Cutting the sheet.* Copenhagen, Kongelige Bibliotek, MS. 9.12, 5.4, fol. 2, 195r; fol. 3, 142r. After photograph by courtesy of the Director. D. E. Woodall 189

TAILPIECE. *A reader unrolling a scroll.* From E. PETERSEN. *Röm. Mitt.* **15**, fig. 5, 1900. Berlin, Deutsches Archäologisches Institut 190

6. SPINNING AND WEAVING *by* R. PATTERSON

154 *Shearing sheep.* After G. COGGIOLA. 'Le Bréviaire Grimani de la Bibliothèque S. Marc à Venise' (Facs. ed.), Vol. 2, Pl. XIII (fol. 7v). Leyden, Sijthoff, 1908. D. E. Woodall. (Below) *A pair of Roman shears.* From W. M. F. PETRIE. 'Tools and Weapons', Pl. LVIII, no. 19. London, British School of Egyptian Archaeology, 1917 192

155 *Fuller's teazel.* From L. H. BAILEY. 'Standard Cyclopedia of Horticulture', Vol. 1, fig. 1276. New York, Macmillan, 1925 193

156 *Hand-combing wool.* London, British Museum, Roy. MS. 10. E. IV, fol. 138. After photograph by courtesy of the Trustees. D. E. Woodall 194

157 *The bow as used to prepare wool.* Venice, Civici Musei d'Arte e di Storia. After photograph by courtesy of the Direttore Reggente. D. E. Woodall 194

158 *Preparation of flax-stalks.* London, British Museum, Print Room. VIRGILIUS SOLIS. Activities of the Months: November. From photograph by courtesy of the Trustees 196

159 *Combing flax.* Milan, Biblioteca Ambrosiana, MS. E. 24. inf., fol. 193v. After P. TOESCA. *L'Arte* **10**, 189, fig. 4, 1907. Milan, Industrie Grafiche Italiane Stucchi. D. E. Woodall 197

160 *Unwinding silk.* London, British Museum. *Kêng-chih t'u* [Forty-six processes of tillage and weaving compiled at the command of Emperor Kang-Hsi]. 1696 198

161 *Girl spinning.* London, British Museum, Greek vase D 13. From photograph by courtesy of the Trustees 200

162 *Rolling wool-fibres.* Berlin, Staatl. Museen. From F. HAUSER. *Jh. Öst. Archäol. Inst., Wien* **12**, Pl. I, 1909. Vienna, Hölder 200

163 *Use of the* epinetron. Athens, National Archaeological Museum, No. 2179. From A. XANTHUDIDES. *Athen. Mitt.* **35**, 324, 1910. Berlin, Deutsches Archäologisches Institut. (Below) *Sixth-century* epinetron. London, British Museum, No. B 598b. After photograph by courtesy of the Trustees. R. Steed 201

164 *Roman spindles and whorls.* London Museum. After 'London in Roman Times', Pl. XLVI, nos 1, 4, 6, 7. London Museum Catalogues No. 3, 1946. D. E. Woodall 202

165 *Spinning with a short distaff.* Monte Cassino, Monastic Library, MS. 132. After A. M. AMELLI. 'Miniature sacre e profane... illustranti l'Enciclopedia... di Rabano Mauro', Pl. XCVI. Tipo-Litografia di Montecassino, 1896. Steffie Schaefer 202

166 *Spinning with a long distaff.* Paris, Bibliothèque Nationale, MS. 9106. After H. WESCHER. *Ciba Review* no. 65, 2381, 1948. By courtesy of CIBA Limited, Basel. D. E. Woodall 203

167 *Spindle-wheel.* London, British Museum, Add. MS. 42130, fol. 193. After 'The Luttrell Psalter'. Facs. edition by E. G. MILLAR, Pl. CXXXVIII. London, British Museum, 1932. By courtesy of the Trustees. D. E. Woodall 203

168 *Spinning-wheel.* From 'Das mittelalterliche Hausbuch im Besitz des Fürsten von Waldburg-Wolfegg-Waldsee'. Facs. edition by H. T. BOSSERT and W. F. STORCK, Pl. XXXV (fol. 34[a]). Leipzig, Seemann, 1912 204

169 *Spinning-wheel.* Milan, Biblioteca Ambrosiana, Codice Atlantico, fol. 393[v]a. From photograph in the Science Museum, South Kensington, London. By courtesy of the Director 205

170 *Modern Siamese spoke-reel and spool-winder.* After A. SCHWARZ. *Z. Schweizerische Statistik* **64**, 313, Pl. I, figs 7, 10, 1928. Basel, Schweizerische Gesellschaft für Statistik und Volkswirtschaft. D. E. Woodall 206

171 *Fifteenth-century sketch of a silk-throwing mill.* Florence, Biblioteca Medicea-Laurenziana, Cod. Plut. 89. sup. 117. After photograph. D. E. Woodall 206

172 *Water-driven silk-throwing mill.* From V. ZONCA. 'Novo teatro di machine et edificii', facing p. 74. Padua, Bertelli, 1607 207

173 *The cross-reel.* Vienna, Öst. Nationalbibliothek, Einblattdruck. From P. BRANDT. 'Schaffende Arbeit und bildende Kunst vom Mittelalter bis zur Gegenwart', fig. 226. Leipzig, Kröner, 1928 208

174 *Hank-winding.* After A. L. GUTMANN. *Ciba Review* no. 14, 484, 1938. By courtesy of CIBA Limited, Basel. D. E. Woodall 208

175 *Nuns making cloth.* Milan, Biblioteca Ambrosiana, Cod. G. 301, fol. 3[r]. After photograph Laboratorio Fotografico della Biblioteca Ambrosiana. K. F. Rowland 209

176 *Warping a loom.* After nineteenth-century reproduction of the original and A. L. GUTMANN. *Ciba Review* no. 14, 484, 1938. By courtesy of the Librarian of the Stadsbibliotheek, Ypres, and of CIBA Limited, Basel. D. E. Woodall 209

177 *Winding threads for warp.* London, British Museum. From *Kêng-chih t'u* [Forty-six processes of tillage and weaving compiled at the command of Emperor Kang-Hsi]. 1696 210

178 *Odysseus and Circe.* Oxford, Ashmolean Museum, Cabirian vase. From E. PFUHL. 'Malerei und Zeichnung der Griechen', Vol. 3, fig. 615. Munich, Bruckmann 1923 211

179 *Late Roman vertical loom.* From *Virgilii picturae antiquae ex codicibus Vaticanis*, Pl. LII (Vat. lat. 3225, fol. 58[r]). Rome, 1835 211

180 *Roman weaving-comb.* Birmingham, City Museum, Acc. No. 285–52. After photograph by permission of the Museum and Art Gallery Committee of the Corporation of Birmingham. D. E. Woodall 212

181 *Horizontal loom.* Cambridge, Trinity College, MS. O. 9. 34, fol. 32[b]. After photograph by courtesy of the Master and Fellows. Steffie Schaefer 212

182 *Four-heddle loom.* Nuremberg, Stadtbibliothek, Hausbuch der Mendelschen Zwölfbrüderstiftung. After F. BOCK (Ed.). 'Deutsches Handwerk im Mittelalter', fig. 14. Leipzig, Insel-Verlag, 1935. D. E. Woodall 213

183 *Weavers at work.* After A. L. GUTMANN. *Ciba Review* no. 14, Cover, 1938. By courtesy of CIBA Limited, Basel. D. E. Woodall 213

184 *Cloth finishing.* Milan, Biblioteca Ambrosiana, Cod. G, 301, fol. 4[v]. After photograph Laboratorio Fotografico della Biblioteca Ambrosiana. K. F. Rowland 214

185 (Above) *Cloth shears*; (Below) *fulling.* After P. BRANDT. 'Schaffende Arbeit und bildende Kunst im Altertum und Mittelalter', fig. 167. Leipzig, Kröner, 1927. D. E. Woodall 215

186 *Fuller trampling cloth.* Semur-en-Auxois Cathedral. After copyright photograph Archives Photographiques, Paris. D. E. Woodall 216

187 *Water-driven fulling-machine.* From V. ZONCA. 'Novo teatro di machine et edificii', p. 42. Padua, Bertelli, 1607 216

188 *Fresco from Pompeii.* After P. HERMANN. 'Denkmäler der Malerei des Altertums', Pl. XXIV, lower strip. Munich, Bruckmann, 1904–31. D. E. Woodall 217

189 *Raising cloth.* Semur-en-Auxois Cathedral. After copyright photograph Archives Photographiques, Paris. D. E. Woodall 218

190 *Shearing* (*or cropping*) *cloth.* Semur-en-Auxois Cathedral. After copyright photograph Archives Photographiques, Paris. D. E. Woodall 218

191 *Roman napping-shears.* Cambridge, University Museum of Archaeology and Ethnology. After photograph by courtesy of the Curator. D. E. Woodall 218

192 *Bleacher with cage.* After P. MARCONI. 'La Pittura dei Romani', fig. 111. Rome, Biblioteca d'Arte Editrice, 1929. K. F. Rowland 219

193 *Cloth-press.* From A. MAU. 'Pompeji in Leben und Kunst' (2nd ed.), fig. 244. Leipzig, Engelmann, 1908 219

7. PART I. FURNITURE: TO THE END OF THE ROMAN EMPIRE *by* CYRIL ALDRED

194 *Throne from a Greek amphora.* Florence, Museo Archeologico. After G. M. A. RICHTER. 'Ancient Furniture', fig. 13. Oxford, Clarendon Press, 1926. D. E. Woodall 222

195 *Reconstruction of a Greek chair.* Based on G. M. A. RICHTER. *Ibid.*, figs 129, 135, 360, and sketch by C. Aldred. D. E. Woodall 222

196 *Bronze couch.* Berlin, Staatl. Museen. After G. M. A. RICHTER. *Ibid.*, fig. 308. T. A. Greeves 223

197 *A throne or arm-chair.* (A) *Terracotta model.* London, British Museum. (B) *Stone example.* Rome, Museo Nazionale Romano. After G. M. A. RICHTER. *Ibid.*, figs 247, 248. Steffie Schaefer 223

198 *Wooden table.* Brussels, Musée du Cinquantenaire, Inv. No. A 1857. After copyright photograph A. C. L. Brussels. D. E. Woodall 224

199 *Folding table.* After G. M. A. RICHTER. 'Ancient Furniture', fig. 322. Oxford, Clarendon Press, 1926. T. A. Greeves 224

200 *Sideboard.* Based on G. M. A. RICHTER. *Ibid.*, figs 332, 333, 336, and on A. MAIURI. 'Pompeii', fig. 59. Novara, Istituto Geografico de' Agostini, 1951. D. E. Woodall 225

201 *Scene from a fresco.* Naples, Museo Nazionale. From G. M. A. RICHTER. 'Ancient Furniture', fig. 343. Oxford, Clarendon Press, 1926 225

202 *Byzantine ivory panel.* After J. H. POLLEN. 'Ancient and Modern Furniture and Woodwork' (rev. ed.), Vol. 1, fig. 42. London, H.M. Stationery Office, 1908. By permission of the Controller. Steffie Schaefer 226

203 *Selection of Assyrian iron tools.* The Manchester Museum, Nos. 2452, 2453, 2454, 2460, 2461, 2464, 2466, 2470. Drawing based on sketches by the author and outline drawings from the originals by H. Spencer, made by courtesy of the Keeper. R. Steed 229

204 *Roman carpenters at work.* Rome, Vatican Library, Gold-glass. Adapted from A. KISA. 'Das Glas im Altertume', Pt III, fig. 357. Leipzig, Hiersemann, 1908. R. Steed 230

205 *Carpenter at work.* After A. MAIURI. 'Pompeii', fig. 83. Novara, Istituto Geografico de' Agostini, 1951. D. E. Woodall 230

206 *Selection of Roman carpenter's tools.* (A, O) London, British Museum, No. 1954 12–14.1. After photograph by courtesy of the Trustees. (B, D, E, G, R, S) After W. M. F. PETRIE. 'Tools and Weapons', Pl. XLIII, 33; Pl. XLV, 108, 111; Pl. XLVII, 41; Pl. LXXVIII, S 47, M 115. London, British School of Egyptian Archaeology, 1917. (C, P) Reading, Silchester Collection. After photographs by courtesy of Reading Museum and Art Gallery. (F, H–N, Q, T) After J. CURLE. 'A Roman Frontier Post and its People', Pl. LIX, 2, 3, 7–9, 12, 13, 14; Pl. LXVIII, 6; Pl. LXXXVIII, 3. Glasgow, Maclehose, 1911, for the Society of Antiquaries of Scotland. (A–T) T. A. Greeves 231

207 *Greek carpenter at work.* London, British Museum. From G. M. A. RICHTER. 'Ancient Furniture', fig. 345. Oxford, Clarendon Press, 1926 232

208 *Leg of a table.* Edinburgh, The Royal Scottish Museum. After drawing from the original by courtesy of the Museum. D. E. Woodall 235

209 *Selection of metal furniture-fittings.* (A, G, I–R) After W. M. F. PETRIE. 'Objects of Daily Use', Pl. XLIII, nos 51, 52; Pl. XLIV, nos 57, 62, 69; Pl. XLV, nos 81, 86, 87, 89, 90, 95; Pl. XLVI, nos 127, 129. London, British School of Egyptian Archaeology, 1927. (B–F, H) After G. M. A. RICHTER. 'Greek, Etruscan and Roman Bronzes', nos 432, 1230, 1231, 1232, 1234, 1235, 1262. New York, Metropolitan Museum of Art, 1915. (G) After J. CURLE 'A Roman Frontier Post and its People', Pl. LXVII, 23–25, 27. Glasgow, Maclehose, 1911, for the Society of Antiquaries of Scotland. (A–R) D. E. Woodall 236

7. PART II. FURNITURE: POST-ROMAN *by* R. W. SYMONDS

210 *Bookcase.* Florence, Laurentian Library, *Codex Amiatinus.* After 'The Lindisfarne Gospels', Pl. XXXVII. Facs. edition by E. G. MILLAR. London, British Museum, 1923. By courtesy of the Trustees. D. E. Woodall 243

211 *Carving in relief.* (Detail from Plate 15 B.) T. A. Greeves 245

212 *Carving with ground sunk.* T. A. Greeves 246

213 *Incised or scratch carving.* T. A. Greeves 246

214 *Gouge-carving* T. A. Greeves 247

215 *Chip-carving.* (Detail from Plate 15 A.) T. A. Greeves 247

216 *Turned legs on a bed.* London, British Museum, Sloane MS. 3983, fol. 11. After photograph by courtesy of the Trustees. D. E. Woodall 248

217 *Turned chair.* Paris, Louvre. Based on A. FEULNER. 'Kunstgeschichte des Möbels seit dem Altertum', fig. 5. Berlin, Propyläen-Verlag, 1927. D. E. Woodall 248

218 *Details of a turned table.* London, British Museum, Greek MS. 1928/229.494–28815, fol. 76v. After photograph by courtesy of the Trustees. D. E. Woodall 249

219 *Detail of turned stool.* Based on photograph by courtesy of the Trustees, The National Gallery, London. D. E. Woodall 249

220 *Carved panel.* Collection Dr Stent, Shere, Surrey. D. E. Woodall 250

221 *The great wheel.* From JOSEPH MOXON. 'Mechanick Exercises' (3rd ed.), Pl. XIV. London, 1703 250

222 *The treadle-wheel.* From JOSEPH MOXON. *Ibid.*, Pl. XVII 251

223 *Example of stick-furniture.* Collection S. W. Wolsey, London. After photograph. T. A. Greeves 251

224 *Detail of construction of stick-furniture.* Diagram by T. A. Greeves 252

225 *A selection of tools.* From JOSEPH MOXON. 'Mechanick Exercises' (3rd ed.), Pl. IV. London, 1703 252

226 *A selection of joiner's tools.* From JOSEPH MOXON. *Ibid.*, Pl. VIII 253

227 *Example of wicker-work chair.* After photograph by courtesy of the Museum of the History of Science, Oxford. D. E. Woodall 254

228 *Medieval meal-ark.* After photograph. T. A. Greeves 254

229 *The wedged joint.* (Detail from Plate 14 A.) T. A. Greeves 255

230 (Left) *Milk-churn.* After G. BERSU. *Ulster J. Archaeol.* 3rd ser. **10**, Pl. II a, facing p. 57, 1947. Belfast, Ulster Archaeological Society. K. F. Rowland. (Right) *Stave-built bucket.* From H. O'N. HENCKEN. *Proc. R. Irish Acad.* **43**, 141, fig. 13 A, 1936. Dublin, Royal Irish Academy 255

231 *The coffer-maker's X-chair.* Winchester Cathedral. After photograph by courtesy of the Dean and Chapter. T. A. Greeves 256

232 *Chair of coopered construction.* London, British Museum, Add. MS. 18851, fol. 1v. After photograph by courtesy of the Trustees. D. E. Woodall 257

8. PART I. CERAMICS: FROM *c* 700 B.C. TO THE FALL OF THE ROMAN EMPIRE *by* GISELA M. A. RICHTER

233 *Corinthian vase.* New York, Metropolitan Museum of Art, No. 25.78.46 a–b. After photograph by courtesy of the Metropolitan Museum. D. E. Woodall 260

234 *Moulded vase.* New York, Metropolitan Museum of Art, Acc. No. 39.11.7. After photograph by courtesy of the Metropolitan Museum. D. E. Woodall 260

235 *Athenian pottery-establishment.* Munich, Staatl. Antikensammlungen. After A. FURTWÄNGLER and C. R. REICHHOLD. 'Griechische Vasenmalerei', Vol. 1, p. 159. Munich, Bruckmann, 1900. D. E. Woodall 261

236 *Youth decorating a cup.* Boston, Museum of Fine Arts. From P. HARTWIG. *Jb. dtsch. archaeol. Inst.* **14**, Pl. IV, 1899 261

237 *Potter stoking the fire.* From O. RAYET. *Gaz. archéol.* **6**, 105, 1880 262

238 *Terracotta wheel-head.* London, British Museum. After photograph by courtesy of the Trustees. D. E. Woodall 262

239 *Foot of a vase showing string-marks.* New York, Metropolitan Museum of Art, Acc. No. 07.232.30. After photograph by courtesy of the Metropolitan Museum. D. E. Woodall 263

240 *Turned foot of a cup.* New York, Metropolitan Museum of Art, Acc. No. 12.234.2. After photograph by courtesy of the Metropolitan Museum. D. E. Woodall 263

241 *The return of Hephaistos.* (Above) *preliminary sketch*; (Below) *finished drawing.* Munich, Staatl. Antikensammlungen. From A. FURTWÄNGLER and C. REICHHOLD. 'Griechische Vasenmalerei', Tafeln Vol. 1, Pl. VII. Munich, Bruckmann, 1904 264

242 *Unfinished red-figured cup.* New York, Metropolitan Museum of Art, Acc. No. 11.212.9. After photograph by courtesy of the Metropolitan Museum. D. E. Woodall 265

243 *Black amphora.* New York, Metropolitan Museum of Art, Acc. No. G.R. 607. After photograph by courtesy of the Metropolitan Museum. D. E. Woodall 266

244 *Red bowl with black rim.* Athens, Agora Museum, No. P 16001. After photograph by courtesy of the American School of Classical Studies, Athens. D. E. Woodall 267

245 *West-slope ware.* Athens. After C. WATZINGER. *Athen. Mitt.* **26**, Pl. III, 1901. Athens, Deutsches Archäologisches Institut. D. E. Woodall 267

246 *Gnathian pot.* Toronto, R. Ontario Museum of Archaeology, No. 530–C. 783. After D. M. ROBINSON *et al.* 'A Catalogue of Greek Vases', Vol. 2, Pl. LXXXIX. Toronto, University of Toronto Press, 1930. D. E. Woodall 268

247 *Vase from Centuripe.* Catania, Private Collection. After G. LIBERTINI. 'Nuove ceramiche dipinte di Centuripe', Pl. 1, facing p. 188. *Atti e memorie della Società Magna Grecia 1932*, Rome, 1933. D. E. Woodall 268

248 *Megarian bowl.* Athens, Agora Museum, No. D 35. After H. A. THOMPSON. *Hesperia* **3,** 379, fig. 66 a. Athens, American School of Classical Studies, 1934. D. E. Woodall 269

249 Impasto *ware.* Boston, Museum of Fine Arts, No. 579–76.30(243). After A. FAIRBANKS. 'Catalogue of Greek and Etruscan Vases', Pl. LXXVIII. Cambridge, Mass., Harvard University Press, 1928. D. E. Woodall 269

250 Impasto *bowl on stand.* New York, Metropolitan Museum of Art. After photograph by courtesy of the Metropolitan Museum. D. E. Woodall 270

251 *Etruscan* bucchero *jug.* New York, Metropolitan Museum of Art. After G. M. A. RICHTER. *Studi Etruschi* **10,** Pl. XXIII, 1936. Florence, Istituto di Studi Etruschi. D. E. Woodall 271

252 *Etruscan* bucchero *jug.* Rome, Museo Etrusco Vaticano. After photograph copyright Anderson, Rome. D. E. Woodall 270

253 *Arretine vase.* New York, Metropolitan Museum of Art. After C. ALEXANDER. 'Arretine Relief Ware', Pl. XXXIII, 1a. *Corpus Vasorum Antiquorum*, Metropolitan Museum Fasc. 1. Cambridge, Mass., Harvard University Press for the American Council of Learned Societies, 1943. D. E. Woodall 272

254 *Arretine stamp.* New York, Metropolitan Museum of Art. After C. ALEXANDER. *Ibid.*, Pl. 1, 2. D. E. Woodall 273

255 *Arretine mould.* New York, Metropolitan Museum of Art. After C. ALEXANDER. *Ibid.*, Pl. 11, 1a. D. E. Woodall 273

256 *Modelling-tools.* After C. DAREMBERG and E. SAGLIO. 'Dictionnaire des antiquités grecques et romaines', Vol. 2, Pt II, fig. 3036. Paris, Hachette, 1904–7. D. E. Woodall 274

257 *Bowl of* terra sigillata *ware.* London, British Museum, Vase M 5. After photograph by courtesy of the Trustees. D. E. Woodall 274

258 *Cup with greenish glaze.* London, British Museum, Vase K 36. After photograph by courtesy of the Trustees. D. E. Woodall 275

259 *Stilt.* London, British Museum. After G. M. A. RICHTER. 'The Craft of Athenian Pottery', fig. 89. New Haven, Yale University Press, 1923. D. E. Woodall 275

260 *Greek terracotta head.* After photograph, Agora Excavations. Athens, American School of Classical Studies. D. E. Woodall 276

261 *Votive head.* Rome, Museo Vaticano. After G. KASCHNITZ-WEINBERG. *R. C. Pont. Accad. Archeol.* **3,** Pl. XXV, 1925. Rome, Pontificia Accademia Romana di Archeologia. D. E. Woodall 276

262 *Terracotta statuette.* New York, Metropolitan Museum of Art, Acc. No. C.P. 2748. After J. L. MYRES. 'Handbook of the Cesnola Collection', no. 2030. New York, Metropolitan Museum of Art, 1914. D. E. Woodall 277

263 *Terracotta statuette from Tanagra.* After photograph in the possession of Dr E. J. Holmyard. D. E. Woodall 277

264 *Terracotta mould.* London, British Museum, Terracotta E 67. From 'Catalogue of Terracottas', by H. B. WALTERS. London, Trustees of the British Museum, 1903. By courtesy of the Trustees 278

265 *Campana relief.* London, British Museum, No. D 640. After photograph by courtesy of the Trustees. D. E. Woodall 279

8. PART II. CERAMICS: MEDIEVAL *by* E. M. JOPE

266 *Plan of a potter's workshop.* After sketch by the author based on B. Hope-Taylor's excavation plans. D. E. Woodall 285

267 *Sequence of settling-tanks.* London, Victoria and Albert Museum, PICCOLPASSO MS., fol. 2. After photograph Crown copyright reserved. D. E. Woodall 285

268 *Rhenish salt-glazed stoneware jug.* London, Victoria and Albert Museum. After A. LANE. 'Style in Pottery', Pl. XXII. London, Oxford University Press, 1948. K. F. Rowland 287

269 *Plaque of polychrome ceramic.* Istanbul, Ottoman Museum. After D. T. RICE. 'Byzantine Glazed Pottery', Pl. IX b. Oxford, Clarendon Press, 1930. D. E. Woodall 288

270 *A potter at a turn-table.* Paris, Bibliothèque Nationale, MS. 11560, fol. 174. After A. DE LABORDE. 'La Bible Moralisée, Vol. 3, Pl. CCCXCVIII. Paris, Société Française de Reproduction des Manuscrits à Peintures, 1915. D. E. Woodall 288

271 *A potter's kick-wheel.* After E. HARTMANN. *Jb. kunsthist. Samml. Allerh. Kaiserhaus.* **2,** 105. Vienna, 1884. D. E. Woodall 289

272 *Potter at a kick-wheel.* From AGRICOLA. *De re metallica*, p. 217. Basel, Froben, 1556 290

273 *Italian potters.* London, Victoria and Albert Museum, PICCOLPASSO MS., fol. 16. From photograph Crown copyright reserved 291

274 *From Piccolpasso's diagram (potter's wheel).* London, Victoria and Albert Museum. *Ibid.*, fols 8, 9^v 291

275 *Some typical pottery-shapes.* After: (1) R. BLOMQVIST. *Medd. Lund Univ. Hist. Mus.* 1947–8, 104, no. 37, 1948. K. Humanistiska Vetenskapssamfundet i Lund. (2) E. M. JOPE. *Norfolk Archaeol.* **30,** 304, fig. 9, 1952. Norfolk and Norwich Archaeol. Society. (3) E. M. JOPE. *Proc. Somersetshire Archaeol. and Nat. Hist. Soc.* **96,** 138, fig. 3, i, 1951. (4, 9) E. M. JOPE *et al. Oxonensia* **15,** 49, fig. 16, l; 59, fig. 21, 7, 1950. Oxford Architectural and Historical Society. (5) E. M. JOPE. *Berks. Archaeol. J.* **50,** 57, fig. 4, 3, 1947. Berkshire Archaeological Society. (6) D. WATERMAN. *Antiq. J.* **33,** 212, fig. 1, 1953. Society of Antiquaries of London. (7, 8) E. M. JOPE. *Oriel Record.* Oxford, Oriel College. Diagrams by the author 292

276 *Puzzle jug.* After C. F. FOX *et al. Archaeologia* **83,** Pl. XXIX, 1933. Society of Antiquaries of London. D. E. Woodall 293

277 *A mould.* London, British Museum. After R. L. HOBSON. *Archaeol. J.* **59,** Pl. III, nos 10, 11, facing p. 8, 1902. London, Royal Archaeological Institute of Great Britain and Ireland. D. E. Woodall 293

278 *Tall baluster jug.* York, Yorkshire Museum. After B. RACKHAM. 'Medieval English Pottery', Pl. LXII. London, Faber and Faber, 1948. D. E. Woodall 293

279 *From Piccolpasso's diagram (screw-top).* London, Victoria and Albert Museum, PICCOLPASSO MS., fols 5, 6. From photograph Crown copyright reserved 294

280 *From Piccolpasso's diagram (kiln).* London, Victoria and Albert Museum, *Ibid.*, fol. 32. From photograph Crown copyright reserved 294

281 *Plan and section of a kiln.* After F. OELMANN. *Bonn. Jb.* **132,** 279, fig. 10; 280, fig. 11, 1927. Rheinisches Landesmuseum in Bonn. D. E. Woodall 295

282 *Romano-British pottery kiln.* After B. BUNCH and P. CORDER. *Antiq. J.* **34,** 219, fig. 1, 1954. Society of Antiquaries of London. D. E. Woodall 295

283 *Reconstruction of a potter's kiln.* After sketch by author. D. E. Woodall 296

284 *Plan of a double-ended pottery-kiln.* After sketch by author. D. E. Woodall 296

285 *Isometric reconstruction and site layout of a kiln.* Based on photograph of model at Science Museum, South Kensington, London, and on C. J. MARSHALL. *Surrey Archaeol. Coll.* **35,** 79, fig. 1; 81, fig. 3; 83, fig. 4, 1924. By courtesy of the Director of the Science Museum and of Surrey Archaeological Society. E. M. Jope 297

286 *Saggars* London, Victoria and Albert Museum, PICCOLPASSO MS., fols 63, 64. From photograph Crown copyright reserved 297

287 *Piccolpasso's picture of a pottery-kiln.* London, Victoria and Albert Museum, *Ibid.*, fol. 35. From photograph Crown copyright reserved 298

288 *Applying a lead-glaze.* From W. F. GRIMES. *Cymmrodor* **41**, frontispiece, 1930. London, Hon. Society of Cymmrodorion 300

289 *Polychrome painted bowl.* After *Ill. Lond. News*, 23 March 1940, p. 391, fig. 2. By courtesy of the Iranian Expedition of the Metropolitan Museum of Art. D. E. Woodall 302

290 Sgraffiato *ornamented bowl.* Collection D. Talbot Rice, Edinburgh. After D. T. RICE. 'Byzantine Glazed Pottery', Pl. XVIII a. Oxford, Clarendon Press, 1930. D. E. Woodall 303

291 *Isometric diagram of kilns.* From W. F. GRIMES. *Cymmrodor* **41**, 30, fig. 19 a, 1930. London, Hon. Society of Cymmrodorion 305

292 *Ornamental roof-finial.* After B. RACKHAM. 'Medieval English Pottery', Pl. XIV. London, Faber and Faber, 1948. D. E. Woodall 306

293 *Relief tile.* Formerly Paravicini Collection. After postcard by Medici Society Limited. D. E. Woodall 307

TAILPIECE. *A set of jugs.* After C. J. MARSHALL. *Surrey Archaeol. Coll.* **35**, 85, fig. 6, 1924. Guildford, Surrey Archaeological Society. E. M. Jope 310

9. GLASS AND GLAZES *by* D. B. HARDEN

294 *Sand-core bottle.* London, British Museum. After photograph by permission of the Trustees. K. F. Rowland 318

295 *Sand-core unguent-flask.* Stockholm, Nationalmuseum. After P. FOSSING. 'Glass Vessels before Glass-Blowing', fig. 18. Copenhagen, Munksgaard, 1940. D. E. Woodall 318

296 *Greenish moulded figurine.* Oxford, Ashmolean Museum. After photograph by permission of the Visitors. D. E. Woodall 319

297 *Head-rest of Tutankhamen.* Oxford, Ashmolean Museum, Griffith Institute. After photograph by permission of the Visitors. D. E. Woodall 319

298 *Inner and outer views of fragment.* Oxford, Ashmolean Museum, No. 1925. 563. After photograph by permission of the Visitors. D. E. Woodall 320

299 *View looking into a green carved alabastron.* London, British Museum. After photograph by permission of the Trustees. D. E. Woodall 321

300 *Amber jug.* Ray Winfield Smith Collection on permanent loan to the Metropolitan Museum of Art, New York. After R. W. SMITH. *Bull. Metrop. Mus.* **8**, 55, 1949. D. E. Woodall 322

301 *Painted bowl.* After D. SILVESTRINI in 'Raccolta di Scritti in Onore di Antonio Giussani', Pl. II. Como, Cavalleri, 1942–3. D. E. Woodall 323

302 *Colourless bowl.* Copenhagen, Nationalmuseet. After photograph by courtesy of the Museum. K. F. Rowland 324

303 *Mould-blown green hexagonal jug.* London, British Museum. After photograph by permission of the Trustees. K. F. Rowland 324

304 *Green unguent-flask.* London, British Museum. After photograph by permission of the Trustees. K. F. Rowland 325

305 *Group of glasses.* Mayen, Museum. After W. HABEREY. *Bonn. Jb.* **147**, Pl. XXX, fig. 2, 1942. Rheinisches Landesmuseum in Bonn. D. E. Woodall 326

306 *Openwork cage-cup.* After photograph by courtesy of the Treasury of St. Mark, Venice. D. E. Woodall 327

307 *Green bowl.* Berlin, Staatl. Museen. After photograph. K. F. Rowland 327

308 *Light-brown unguent-flask.* Toledo, Ohio. The Toledo Museum of Art. After photograph by courtesy of the Museum. K. F. Rowland 328

309 *Three-storeyed glass-furnace.* Monte Cassino, Monastic Library, MS. 132 (Rabanus Maurus), p. 427. After photograph by courtesy of the Most Reverend Abbot. K. F. Rowland 329

310 *Glassmakers and their furnace.* London, British Museum, Add. MS. 24189 (The Travels of Sir John Mandeville), fol. 16. After photograph by courtesy of the Trustees. D. E. Woodall 330

311 *Glass-making tools.* Drawing after F. HAUDICQUER DE BLANCOURT. 'The Art of Glass , Pl. facing p. 30. London, 1699 331

312 *Glass-making tools.* Drawing after J. M. GOOD *et al.* 'Pantologia', Vol. 5, Pl. LXXXI. London, 1819 332

313 *Pottery mould.* After C. J. LAMM. 'Mittelalterliche Gläser und Steinschnittarbeiten aus dem nahen Osten', Vol. 2, Pl. XIII, nos 1, 2. Berlin, Reimer and Vohsen, 1929. D. E. Woodall 332

314 (A) *Mould-blown jug.* (B) *Bottom of jug.* Oxford, Ashmolean Museum, No. 1948.36. After photograph by permission of the Visitors. D. E. Woodall 338

315 *Olive-green bell-beaker.* Oxford, Ashmolean Museum, No. 1927.258. After photograph by permission of the Visitors. D. E. Woodall 339

316 *Two colourless bottles.* (Left) Cambridge, Fitzwilliam Museum. By permission of the Syndics. (Right) Cologne, Wallraf-Richartz-Museum. After F. FREMERSDORF. 'Römische Gläser aus Köln', Pl. XVIII. Cologne, Völker-Verlag. D. E. Woodall 339

317 *Green claw beaker.* London, British Museum, No. 1947.10.9.1. After photograph by permission of the Trustees. D. E. Woodall 340

318 *Jug with ribbed decoration.* New York, Metropolitan Museum of Art. After R. W. SMITH. *Bull. Metrop. Mus.* **8**, 51, 1949. New York, Metropolitan Museum of Art. D. E. Woodall 340

319 *Greenish bowl with pincered decoration.* London, British Museum. After photograph by permission of the Trustees. D. E. Woodall 341

320 *Colourless carved glass.* After photograph by courtesy of the City Art Museum, St Louis. K. F. Rowland 341

321 *Colourless bowl with facet-cut.* Cologne, Römisch-Germanisches Museum, Inv. No. 295. After F. FREMERSDORF. 'Figürlich geschliffene Gläser', Pl. I, no. 1. Römisch-Germanische Forschungen, Vol. 19. Berlin, de Gruyter, 1951. K. F. Rowland 342

322 *Engraved flask.* Syracuse, Museo Archeologico Nazionale. After photograph by courtesy of the Soprintendente alle Antichità della Sicilia Orientale. D. E. Woodall 343

TAILPIECE. *Dark green boat-shaped saucer.* London, British Museum, No. S 153. After photograph by permission of the Trustees. K. F. Rowland 346

10. PRE-SCIENTIFIC INDUSTRIAL CHEMISTRY *by* F. SHERWOOD TAYLOR and CHARLES SINGER

323 *Lime-kiln reconstructed from painting.* London, Apsley House (Victoria and Albert Museum). After photograph Crown copyright reserved. T. A. Greeves 348

324 *The traditional alum cauldron.* From C. SINGER. 'The Earliest Chemical Industry', fig. 43. London, Folio Society, 1948. By courtesy of Mr. Derek Spence 348

325 *Preparation of woad.* After D. G. SCHREBER. 'Historische, physische und öconomische Beschreibung des Waidtes . . . in Thüringen', Frontispiece. Halle, 1752. T. A. Greeves 349

326 *Delineating drapery.* Gent, Universiteitsbibliotheek, MS. 10, fol. 69v. After P. DURRIEU. 'La Miniature flamande au temps de la Cour de Bourgogne', Pl. LXVIII. Brussels, Van Oest, 1921. D. E. Woodall 353

327 *Scraping tartar from wine cask.* From *Ortus sanitatis*, Vol. 2, Pt II, fol. 328. Strasbourg, [c 1497] 354

328 *Making fine charcoal.* Vienna, Öst. Nationalbibliothek MS. After O. GUTTMANN. *Monumenta Pulveris Pyrii*, fig. 38. London, Artists Press for the author, 1906. D. E. Woodall 360

329 *Dyers, from mural painting.* After P. MARCONI. 'La Pittura dei Romani', fig. 112. Rome, Biblioteca d'Arte Editrice, 1929. K. F. Rowland 364

330 *Dyers preparing figured fabric.* Cambridge, Trinity College Library, MS. O.9.34, fol. 32b. After photograph by courtesy of the Master and Fellows. Steffie Schaefer 364

331 *Bleaching and dyeing silk.* Florence, Biblioteca Medicea-Laurenziana, Cod. Plut. 89, sup. 117. After C. SINGER. 'The Earliest Chemical Industry', fig. 69. By courtesy of Mr Derek Spence. D. E. Woodall 365

332 *Merchants selling woad.* Amiens Cathedral. After photograph copyright Archives Photographiques, Paris. D. E. Woodall. 365

333 *Dyers stirring woad-vat.* London, British Museum Roy. MS. 15. E. II, III, fol. 269. After photograph by courtesy of the Trustees. D. E. Woodall 366

334 *A charcoal burner.* From BIRINGUCCIO. 'De la Pirotechnia', p. 62. Venice, Roffinello, 1540 368

335 *Nitre-gatherers.* Vienna, Öst. Nationalbibliothek MS. After O. GUTTMANN. *Monumenta Pulveris Pyrii*, fig. 38. London, Artists Press for the author, 1906. D. E. Woodall 368

336 *A saltpetre-works.* From LAZARUS ERCKER. 'Beschreibung allerfürnemsten mineralischen Ertzt', Pl. facing p. 134. Frankfurt a. M., Feyerabend, 1580 369

337 *Crystallizing saltpetre.* From LAZARUS ERCKER. *Ibid.*, p. 130 370

338 *Pharmacy.* After H. H. SCHAEDER. 'Ausbreitung und Staatengründungen des Islam vom 7.–15. Jahrhundert', p. 244. Propyläen-Weltgeschichte, Vol. 3. Berlin, Propyläen-Verlag, 1932. Steffie Schaefer 370

339 *Pharmacy.* Cambridge, Trinity College Library, MS. O. I. 20, fol. 239ᵛ. After photograph by courtesy of the Master and Fellows. Steffie Schaefer 371

340 *Sugar-production.* From STRADANUS (JAN VAN DER STRAET). *Nova Reperta*, Pl. XIII. Antwerp, [1600?] 372

341 *Greek fire.* After H. B. GOODRICH. *World Petroleum* **10**, 35 (bottom). New York, Palmer Publications, 1939. D. E. Woodall 376

342 *Reconstruction of a late-Roman force-pump.* From M. MERCIER. 'Le Feu grégeois', fig. 128. Paris, Geuthner, 1952 376

343 *A bomb or grenade.* After L. GOODRICH and F. CHIA-SHENG. *Isis* **36**, fig. 4, facing p. 118, 1946. History of Science Society, Cambridge, Mass. D. E. Woodall 378

344 *Manufacture of sulphur.* From AGRICOLA. 'Vom Bergkwerck', p. cccclxxx. Basel, Froben and Bischoff, 1557 379

345 *Grinding gunpowder.* After W. HASSENSTEIN (Ed.). 'Das Feuerwerkbuch von 1420', fig. 28. Munich, Verlag der deutschen Technik, 1941. D. E. Woodall 380

346 *A large water-driven gunpowder-mill.* From K. SIEMIENOWICZ and D. ELRICH. 'Volkommene Geschütz-, Feuerwerck-, und Büchsenmeisterey-Kunst', Pt II, fig. 4. Frankfurt a.M., 1676 381

11. THE MEDIEVAL ARTISAN *by* R. H. G. THOMSON

347 *An early fifteenth-century picture.* London, British Museum, Roy. MS. 15. D. III, fol. 15ᵛ. After photograph by courtesy of the Trustees. D. E. Woodall 384

348 *Cutting thin slabs of marble.* Monte Cassino, Monastic Library, MS. 132 (Rabanus Maurus), p. 418. From photograph by courtesy of the Most Reverend Abbot 385

349 *Scaffolding for a stone building.* Brussels, Bibliothèque Royale, MS. 9068, fol. 289. After J. VAN DEN GHEYN. 'Cronicques et conquestes de Charlemaine', Pl. CIV. Brussels, Vromant, 1909. D. E. Woodall 386

350 *The construction of a stone bridge.* Brussels, Bibliothèque Royale, MS. 9068, fol. 203. After J. VAN DEN GHEYN. *Ibid.*, Pl. XCV. D. E. Woodall 387

351 *Making bricks and tiles.* London, British Museum, Add. MS. 38122, fol. 78ᵛ. After photograph by courtesy of the Trustees. D. E. Woodall 388

352 *Squaring a piece of timber.* Öst. Nationalbibliothek, Cod. pal. 3033, fol. 140ᵇ. From A. SCHULTZ. *Jb. kunsthist. Samml. Allerh. Kaiserhaus.* **6**, 78, 1887. Vienna 389

353 *Ship-builders.* London, British Museum, Add. MS. 15268, fol. 105ᵛ. After photograph by courtesy of the Trustees. D. E. Woodall 390

354 *Carpenters making a bridge.* Brussels, Bibliothèque Royale, MS. 9066, fol. 85. After J. VAN DEN GHEYN. 'Cronicques et conquestes de Charlemaine', Pl. X. Brussels, Vromant, 1909. D. E. Woodall 390

355 *Woodcutters.* After P. BRANDT. 'Schaffende Arbeit und bildende Kunst vom Mittelalter bis zur Gegenwart', fig. 21. Leipzig, Kröner, 1928. D. E. Woodall 391

356 *The carpenter and his family.* Paris, Musée de l'École des Beaux Arts. After photograph copyright Giraudon, Paris. D. E. Woodall 391

357 *Noah building the ark.* Monreale, Duomo. After postcard copyright Edizione Giovanni Bucaro, Palermo. D. E. Woodall 392

358 *Carpenters working.* After copyright photograph Alinari, Florence. D. E. Woodall 393

359 *An auger- and gimlet-maker.* Nuremberg, Stadtbibliothek, Hausbuch der Landauerschen Zwölfbrüderstiftung, Vol. 1, fol. 12v. After Photo-Scheder by courtesy of the Director. D. E. Woodall 394

360 *A file-maker.* Nuremberg, Stadtbibliothek, Hausbuch der Mendelschen Zwölfbrüderstiftung. After A. UCCELLI. 'Storia della Tecnica dal medio evo ai nostri giorni', fig. 156. Milan, Hoepli, 1945. D. E. Woodall 394

361 *Ship-builders.* After G. A. DELL'ACQUA. 'La scultura romanica in Italia', Pl. XIX. Novara, Istituto Geografico de' Agostini, 1942. D. E. Woodall 395

362 *Blacksmiths.* London, British Museum, MS. Sloane 3983, fol. 5. After photograph by courtesy of the Trustees. D. E. Woodall 396

12. BUILDING CONSTRUCTION *by* MARTIN S. BRIGGS

363 *Temple of Hephaistos.* Drawing by M. S. Briggs 399

364 *Diagram showing use of metal.* (A) After F. A. CHOISY. 'Histoire de l'architecture', Vol. 1, p. 269, fig. 1. Paris, Gauthier-Villars, 1899: (B) According to W. B. DINSMOOR. 'Architecture of Ancient Greece' (3rd ed.). London, Batsford, 1950. (C) Based on C. R. COCKERELL. 'The Temples at Aegina and Bassae.' Bassae, Pl. VIII. London, 1860. M. S. Briggs 400

365 *Greek masonry details.* Based on various sources. Diagrams by M. S. Briggs 400

366 (A) *Greek lifting devices.* (B) *Method of rolling.* (C) *Ancones.* Based on various sources. Diagrams by M. S. Briggs 401

367 *Timber origins of Greek masonry.* Based on J. DURM. 'Die Baukunst der Griechen' (2nd ed.), fig. 87. Darmstadt, Bergsträßer, 1892. M. S. Briggs 402

368 *Greek masonry construction.* Based on J. DURM. *Ibid.*, fig. 114. M. S. Briggs 402

369 *Roof of the arsenal.* Based on F. A. CHOISY. 'Études épigraphiques sur l'architecture grecque', No. 1: 'L'Arsenal du Pirée', pp. 20–22, Pl. II. Paris, Société Anonyme de Publications périodiques, 1883. M. S. Briggs 403

370 *Greek roofing-tiles.* (A) After 'The Unedited Antiquities of Attica', ch. 6, Pl. VII. London, Society of Dilettanti, 1817. (B, C, D) After W. B. DINSMOOR. 'Architecture of Ancient Greece' (3rd ed.), fig. 16. London, Batsford, 1950. M. S. Briggs 403

371 *Roman aqueduct.* Drawing by M. S. Briggs 404

372 *Roman Temple.* Drawing by M. S. Briggs 405

373 *Roman marble facings.* After J. H. MIDDLETON. 'The Remains of Ancient Rome', Vol. I, figs 14, 15. Edinburgh, 1892. M. S. Briggs 406

374 *Types of Roman bricks.* After G. T. RIVOIRA. 'Roman Architecture', figs 16, 17. Oxford, Clarendon Press, 1925. M. S. Briggs 408

375 *The Roman wall of London.* Based on W. PAGE (Ed.). 'The Victoria County History: London', Vol. 1, fig. 13. London, Constable, 1909. M. S. Briggs 408

376 *Part of terrace of the Roman temple of Jupiter at Baalbek.* After reconstruction by T. WIEGAND. 'Baalbek', Vol. I, Pl. XXII. Berlin, Leipzig, de Gruyter, 1921. M. S. Briggs 409

377 *The 'Great Stone'.* Drawing by M. S. Briggs from a photograph 410

378 *Details of Roman brickwork.* (A, B, D) Based on J. H. MIDDLETON. 'The Remains of Ancient Rome', Vol. I, figs 3, 4, 5. Edinburgh, 1892. (C) Based on F. A. CHOISY. 'Histoire de l'architecture', Vol. I, fig. 10A. Paris, Gauthier-Villars, 1899. M. S. Briggs 411

379 *Roman vaulted construction.* Based on J. DURM. 'Die Baukunst der Römer' (2nd ed.), fig. 702. Stuttgart, Kröner, 1905. M. S. Briggs 412

380 *Roman timber roofs.* (A) After F. A. CHOISY. 'L'Art de bâtir chez les Romains', fig. 85. Paris, Ducher, 1873. (C) According to CARLO FONTANA. *Templum Vaticanum.* Rome, 1694. (D) After J. C. KRAFFT. 'L'Art de la charpente', Pl. XI Ć. Paris, Strasbourg, 1805. M. S. Briggs 414

381 *Romano-British roofing materials.* (A, C) After J. WARD. 'Romano-British Buildings and Earthworks', figs 76, 78. London, Methuen, 1911. M. S. Briggs 415

382 *Greek and Roman locks and keys.* (A, B) From H. DIELS. 'Antike Technik' (3rd ed.), figs 14, 7. Leipzig, Berlin, Teubner, 1924. (C) From BRITISH MUSEUM. 'Guide to the Antiquities of Roman Britain' by J. W. BRAILSFORD, fig. 41. London, British Museum, 1951. By courtesy of the Trustees 416

383 *Roman bronze doors.* After T. L. DONALDSON. 'Doorways from ancient buildings in Greece and Italy', Pls XIX, XX. London, 1833. M. S. Briggs 417

384 *Roman water-pipes.* After J. H. MIDDLETON. 'The remains of Ancient Rome', Vol. 2, figs 97 A, D, E, H, 98. Edinburgh, 1892. M. S. Briggs 419

385 *Roman hypocaust.* After J. WARD. 'Romano-British Buildings and Earthworks', fig. 83. London, Methuen, 1911. M. S. Briggs 420

386 *Romano-British flue-tiles.* After J. WARD. *Ibid.*, fig. 84. M. S. Briggs 421

387 *Church of San Vitale.* Diagram by M. S. Briggs 422

388 *Byzantine column-shafts.* After F. A. CHOISY. 'L'Art de bâtir chez les Byzantins', figs 10, 11. Paris, Société Anonyme de Publications périodiques, 1883. M. S. Briggs 423

389 *The mausoleum of Theodoric.* Based on C. E. ISABELLE. 'Les édifices circulaires et les dômes...' Paris, 1855. Diagram by M. S. Briggs 424

390 *Greenstead church.* Diagram by M. S. Briggs 425

391 *St John's chapel, Tower of London.* Drawing by M. S. Briggs 426

392 *Builder hammering shingles.* London, British Museum, Cotton MS. Claud. B. IV, fol. 19. After photograph by courtesy of the Trustees. D. E. Woodall 429

393 *Norman lock.* Paris, Musée des Thermes et de l'Hôtel de Cluny, No. 14203. After C. FRÉMONT. 'La Serrure', figs 79, 80. Études expérimentales de Technologie industrielle, 70e Mémoire. Paris, Frémont, 1924. E. Norman 430

394 *Salisbury cathedral.* Drawing by M. S. Briggs 431

395 *King's College chapel, Cambridge.* Drawing by M. S. Briggs 432

396 *Diagrams of vaulting.* By M. S. Briggs 433

397 *Section of nave of Amiens cathedral.* Based on E. E. VIOLLET-LE-DUC. 'Dictionnaire raisonné de l'architecture française', Vol. 1, p. 203. Paris, 1854, and on other sources. M. S. Briggs 435

398 *English Gothic tracery.* After F. BOND. 'An Introduction to English Church Architecture', Vol. 2, pp. 593, 675. London, Oxford University Press, 1913. M. S. Briggs 437

399 *French Gothic tracery.* After F. A. CHOISY. 'Histoire de l'architecture', Vol. 2, figs 33 A, 37. Paris, Gauthier-Villars, 1899. M. S. Briggs 438

400 *The 'Guildhall' at Blakeney.* Drawing by M. S. Briggs 439

401 *English Gothic technique.* (Left) After F. BOND. 'An Introduction to English Church Architecture', Vol. 1, p. 422. London, Oxford University Press, 1913. (Right) After N. LLOYD. 'History of English Brickwork', p. 415. London, Montgomery, 1925. M. S. Briggs 440

402 *Romanesque and Gothic timber roofs.* Diagrams by M. S. Briggs 441

403 *Late English Gothic doors.* Diagrams by M. S. Briggs 443

404 *Gothic roof drainage.* (A–C) Based on F. A. CHOISY. 'Histoire de l'architecture', Vol. 2, pp. 372–3, figs. 27–29. Paris, Gauthier-Villars, 1899. (D, E, G–J) Based on E. E. VIOLLET-LE-DUC. 'Dictionnaire raisonné de l'architecture française', Vol. 7, pp. 212–19. Paris, 1864. (F) After F. BOND. 'Gothic Architecture in England', p. 393. London, Batsford, 1905. Diagrams by M. S. Briggs 444

405 *Rustic building.* Monte Cassino, Monastic Library, MS. 132 (Rabanus Maurus), p. 386. After photograph by courtesy of the Most Reverend Abbot. D. E. Woodall 445

13. FINE METAL-WORK *by* HERBERT MARYON

406 *Hellenistic bronze cup.* London, British Museum, No. 82, 10–9, 23. After photograph by courtesy of the Trustees. D. E. Woodall 450

407 *Bronze crater.* Châtillon-sur-Seine (Côte-d'Or), Museum. After R. JOFFROY. *Ill. Lond. News*, 5 March 1955, p. 11. D. E. Woodall 451

408 *Detail of the frieze.* After R. JOFFROY. *Ibid.*, p. 16. D. E. Woodall 451

409 (Above) *Twelve-sided gold basket*; (below) *eight-sided bowl.* Bucharest, Museum. After A. I. ODOBESCU. 'Le Trésor de Petrossa', Vol. 1, Pl. XII and Pl. facing p. 90. Paris, 1889–1900. D. E. Woodall 452

410 *Reconstruction of gold hilt.* Paris, Cabinet des Médailles, Le Trésor de Childeric No. XLVIII. After photograph by courtesy of Bibliothèque Nationale, Paris. Reconstruction after H. ARBMANN. *Medd. Lund Univ. Hist. Mus. 1947–8*, 100, fig. 1, 1948. Kungl. Humanistiska Vetenskapssamfundet i Lund. D. E. Woodall 453

411 *Cup of Chosroes II.* Paris, Cabinet des Médailles, Le Trésor de St-Denis. After photograph by courtesy of the Bibliothèque Nationale, Paris. D. E. Woodall 454

412 *One of the votive crowns from Fuente de Guarrazar.* Paris, Musée des Thermes et de l'Hôtel de Cluny, No. 3113. After M. HAUTTMANN. 'Die Kunst des frühen Mittelalters', p. 268. Berlin, Propyläen-Verlag, [1929]. D. E. Woodall 455

413 *Anglo-Saxon brooch.* City of Liverpool Public Museums. After photograph by courtesy of the Trustees of the British Museum, London. D. E. Woodall 456

414 *Purse-mount.* London, British Museum. After 'The Sutton Hoo Ship Burial' (3rd impr.), Pl. XVIII. London, The Trustees of the British Museum, 1952. By courtesy of the Trustees. D. E. Woodall 456

415 St Cuthbert's cross. Durham Cathedral. After photograph by courtesy of the Dean and Chapter. D. E. Woodall 457

416 *Detail from baptistère.* Paris, Louvre. After D. S. RICE. 'Baptistère de St Louis', Pl. XIII. Paris, Éditions du Chêne, 1953. D. E. Woodall 457

417 *One of the shoulder-clasps.* London, British Museum. After 'The Sutton Hoo Ship Burial' (3rd impr.), Pl. XXIII. London, The Trustees of the British Museum, 1952. By courtesy of the Trustees. D. E. Woodall 458

418 *A crown formerly known as Charlemagne's.* After G. HAUPT. 'Die Reichsinsignien', Pl. VII. Leipzig, Seemann, [1929], details after photograph Kunsthistorisches Museum, Vienna (copyright reserved). D. E. Woodall 459

419 (Left) *A section of a pattern-welded sword*; (Right) *An experiment in pattern-welding.* After H. MARYON. *Proc. Cambridge Antiquarian Soc.* **41**, Pls. XXI, XXII (right), 1948. Cambridge, Cambridge Antiquarian Society. D. E. Woodall 460

420 *Mycenaean gold ring.* Cyprus Museum. After sketch from original by author. D. E. Woodall 460

421 *End of a gold sceptre.* London, British Museum, Roman Jewelry No. 2070. Drawing from original by D. E. Woodall 461

422 *Celtic bronze shield.* London, British Museum. After photograph by courtesy of the Trustees. D. E. Woodall 462

423 *Celtic mirror.* Gloucester, City and Folk Museums. After photograph by courtesy of the Director. D. E. Woodall 463

424 *Foot of hanging bowl.* London, British Museum, Sutton Hoo Treasure. After photograph by courtesy of the Trustees. D. E. Woodall 464

425 *The Royal Gold Cup.* London, British Museum. After photograph by courtesy of the Trustees. D. E. Woodall 465

426 *The Valence casket.* London, Victoria and Albert Museum. After photograph Crown copyright reserved. D. E. Woodall 466

427 *Cross-reliquary.* Rome, Museo Cristiano Lateranense. After copyright photograph Alinari, Florence. D. E. Woodall 467

428 *Roundel from a Byzantine paten.* Venice, St Mark's Treasury. After copyright photograph Giraudon, Paris. D. E. Woodall 467

429 *Roman silver bowl.* London, British Museum, Silver plate No. 169. After photograph by courtesy of the Trustees. D. E. Woodall 468

430 *Human figure.* After S. MARINATOS. *Bull. corr. hell.* **40**, Pl. LXIII, 1936. École Française d'Athènes. D. E. Woodall 469

431 *The silver vase from Chertomlyk.* From E. H. MINNS. 'Scythians and Greeks', fig. 46. Cambridge, University Press, 1913 470

432 (Above) *A golden bowcase*; (below) *Details*. Electrotype at Victoria and Albert Museum, London. After photograph by courtesy of the Trustees of the British Museum. D. E. Woodall 471

433 *A very fine example of Greek* repoussé *work*. London, British Museum, Bronzes No. 285. After photograph by courtesy of the Trustees. D. E. Woodall 472

434 *The Tara Brooch*. Dublin, National Museum of Ireland. After photograph by courtesy of the Director. D. E. Woodall 473

435 *Sword with silver* repoussé *hilt*. Oslo, Universitets Oldsaksamling. After photograph by courtesy of the Director. D. E. Woodall 474

436 *Shrine for St Patrick's bell*. Dublin, National Museum of Ireland. After photograph by courtesy of the Director. D. E. Woodall 475

437 *Life size head of Apollo*. Chatsworth Collection, Bakewell (Derbyshire). By permission of the Chatsworth Estates Company; after photograph by courtesy of the Trustees of the British Museum. D. E. Woodall 476

438 *Minoan cast bronze group*. Spencer Churchill Collection, Northwick Park, Blockley. After photograph by courtesy of Captain E. G. S. Churchill. D. E. Woodall 478

439 *Gilt bronze candlestick*. London, Victoria and Albert Museum, No. 6749–1861. After photograph Crown copyright reserved. D. E. Woodall 479

440 *Silversmith's workshop*. London, British Museum, Print Room, ÉTIENNE DELANNE No. D. 266. From photograph by courtesy of the Trustees 482

441 *Gold buckle*. London, British Museum. After photograph by courtesy of the Trustees. D. E. Woodall 483

442 *Electrum stater of Phocaea*. London, British Museum. After photograph by courtesy of the Trustees. K. F. Rowland 486

443 *Athenian tetradrachm*. K. F. Rowland 487

444 *Tetradrachm of King Eucratides*. K. F. Rowland 487

445 *Stater of the Bellovaci*. London, British Museum. After photograph by courtesy of the Trustees. K. F. Rowland 487

446 *A moneyer's equipment*. London, British Museum, Sculpture 1954 12–14.1. After photograph by courtesy of the Trustees. K. F. Rowland 488

447 *Upper and lower die*. After V. MILLER ZU AICHHOLZ *et al*. 'Österreichische Münzprägungen' (2nd ed.), Pl. L, no. 1. Vienna, Verlag Bundessammlung von Medaillen, Münzen und Goldzeichen, 1948. K. F. Rowland 488

448 *Sestertius of Vespasian*. K. F. Rowland 489

449 *Penny of Charlemagne*. K. F. Rowland 489

450 Carlino d'oro. K. F. Rowland 489

451 *A moneyer*. After J. A. DEVILLE. 'Essai historique et descriptif sur l'Église et l'Abbaye Saint-Georges-de-Bocherville, près Rouen', fig. 3. Rouen, 1827. K. F. Rowland 490

452 *The lettering on a French coin*. After A. DIEUDONNÉ. 'Les monnaies capétiennes ou royales françaises', 1re section, Pl. II, no. 47. Catalogue des monnaies françaises de la Bibliothèque Nationale. Paris, Presses Universitaires de France, 1923. K. F. Rowland 490

453 *Interior of a German mint*. Vienna, Staatsbibliothek, Cod. pal. 3033. From A. SCHULTZ. *Jb. Kunsts. Allerh. Kaiserhaus*. **6**, p. [85], fol. 149b, 1887 491

454 *Coin struck on one face only*. K. F. Rowland 491

455 *Corinthian die*. Oxford, Ashmolean Museum. After photograph by courtesy of the Visitors. D. E. Woodall 492

14. ROADS AND LAND TRAVEL *by* R. G. GOODCHILD and R. J. FORBES

456 *Map showing course of Persian royal road*. Based on W. M. CALDER. *Class. Rev*. **39,** 8, 1952. London, Classical Association. E. Norman 496

457 *Greek rut-road*. T. A. Greeves 499

458 (Map) *Italy: Roman Roads*. Based on C. DAREMBERG and E. SAGLIO (Eds). 'Dictionnaire des antiquités grecques et romaines', Vol. 5, fig. 7437. Paris, Hachette, 1911. J. F. Horrabin 501

459 (Map) *The main roads of Rome and the provinces.* Based on C. DAREMBERG and E. SAGLIO. *Ibid.*, figs 7434, 7439. J. F. Horrabin 502

460 (Above) *Diagrammatic cross-section of a principal highway in Italy*; (Below) *Diagrammatic cross-section of normal Roman highway.* From R. J. FORBES. 'Notes on the History of Ancient Roads and their Construction', figs 22, 23. Allard Pierson Stichting: Archaeol.-Historische Bijdragen, Vol. 3, Amsterdam, 1934 503

461 Via Praenestina. After copyright photograph by G. R. Swain, Kelsey Museum of Archaeology, University of Michigan, Ann Arbor. D. E. Woodall 504

462 (Map) *Britain: Roman roads.* Based on 'Map of Roman Britain' (2nd ed.). Chessington (Surrey) Ordnance Survey Office, 1931. By permission of the Director General. J. F. Horrabin 506

463 *Roman milestone.* London, British Museum. After V. E. NASH-WILLIAMS. 'The Roman Frontier in Wales', Pl. XLII, fig. 1. Cardiff, University of Wales Press, 1954. D. E. Woodall 507

464 *A reconstruction of the Roman Gate at Newgate.* Drawing by Alan Sorrell 509

465 *Roman bridge.* After copyright photograph by R. G. Goodchild. T. A. Greeves 510

466 *The* Ponte d'Augusto. After copyright photograph Anderson, Rome. T. A. Greeves 511

467 *Bridge built over the Danube.* After C. CICHORIUS. 'Die Reliefs der Trajanssäule', Plates Vol. 1, Pl. LXXII. Berlin, Reimer, 1896. D. E. Woodall 512

468 *The* Ponte Amato. After copyright photograph by R. G. Goodchild. T. A. Greeves 513

469 (Above) *Iron horse-sandal.* After F. MORTON. 'Hallstatt und die Hallstattzeit', Pl. XXXIII, below. Hallstatt, Musealverein, 1953. D. E. Woodall. (Below) *Iron horse-shoe.* From J. G. D. CLARK. 'Prehistoric Europe', fig. 178. London, Methuen, 1952 515

470 *Diagrammatic section of ship-sheds.* After sketch by R. G. Goodchild. T. A. Greeves 517

471 *Diagram of the port of Rome.* After sketch by R. G. Goodchild. T. A. Greeves 518

472 *Reconstruction of quays at* Leptis Magna. After copyright photograph by R. G. Goodchild. T. A. Greeves 519

473 *Roman harbour.* From R. T. GÜNTHER. *Archaeologia* **58**, 520, fig. 11, 1902. London, Society of Antiquaries 520

474 *Reconstruction of pharos.* From H. THIERSCH. 'Pharos. Antike, Islam und Occident', frontispiece. Berlin, Leipzig, Teubner, 1909 521

475 '*The Old Man*': *the pharos at Boulogne.* From R. E. M. WHEELER. *Archaeol. J.* second series, **36**, 38, fig. 5, 1929 (1930). London, Royal Archaeological Institute of Great Britain and Ireland 522

476 *Reconstruction of the Roman lighthouse.* Rome, Mostra Augustea della Romanità 1938. Model. After photograph by courtesy of the Director. T. A. Greeves 523

477 *Bavai–Tournai road.* Brussels, Bibliothèque Royale de Belgique, MS. 6419 (JACQUES DE GUISE. 'Chroniques de Hainaut'), Inv. No. 9242, fol. 270v. After photograph. D. E. Woodall 525

478 *German paviour.* Nuremberg, Stadtbibliothek, Hausbuch der Mendelschen Zwölfbrüderstiftung, fol. 77r. After photograph. D. E. Woodall 526

479 *Street in Pompeii.* After A. MAIURI. 'Pompeii', fig. 10. Novara, Istituto Geografico de' Agostini, 1951. D. E. Woodall 530

480 *A German ditch-digger.* Nuremberg, Stadtbibliothek. Hausbuch der Mendelschen Zwölfbrüderstiftung, fol. 77v. After photograph. D. E. Woodall 532

TAILPIECE. *A plan of the harbour of Calais.* London, British Museum, Cotton MS. Aug. I.II.70 (Detail). After photograph by courtesy of the Trustees. T. A. Greeves 536

15. VEHICLES AND HARNESS *by* E. M. JOPE

481 *Plan and front-elevation of the Dejbjerg wagon.* From O. KLINDT-JENSEN. *Acta Archaeologica* **20**, 91, fig. 59; 92, fig. 60, 1949. Copenhagen, Munksgaard 538

482 *Reconstruction of a Celtic chariot.* Adapted from C. F. FOX. 'A Find of the Early Iron Age from Llyn Cerrig Bach, Anglesey', fig. 13. Cardiff, National Museum of Wales, 1946 539

483 *Greek chariot.* London, British Museum, Greek vase No. B 364. From photograph by courtesy of the Trustees. Prepared by R. Steed 540

484 *Model of a Roman chariot*. London, British Museum, Bronzes No. 2695. After photograph by courtesy of the Trustees. K. F. Rowland 541

485 *Chariot on an Attic vase*. After H. L. LORIMER. 'Homer and the Monuments', Pl. XXV, 3. London, Macmillan, 1950. By permission of the Principal and Council of Somerville College, Oxford. Steffie Schaefer 541

486 *Front view of a chariot*. Paris, Cabinet des Médailles, Vase 202. From A. RUMPF. 'Chalkidische Vasen', Pl. IX, 3. Berlin, Leipzig, de Gruyter, 1927. Prepared by R. Steed 542

487 *Model of a chariot*. Berlin, Staatl. Museen. After E. MERCKLIN, *Amer. J. Archaeol.* **20**, 398, fig. 1, 1916. Cambridge, Mass. Archaeological Institute of America. K. F. Rowland 542

488 *Etruscan chariot*. Florence, Museo Archeologico. After photograph by courtesy of the Soprintendenza alle Antichità d'Etruria, Florence. K. F. Rowland 543

489 *Bas-relief of a boy's chariot*. Paris, Louvre. After R. J. E. C. LEFEBVRE DES NOËTTES. 'L'Attelage. Le cheval de selle à travers les âges', fig. 73. Paris, Picard, 1931. D. E. Woodall 544

490 *Scene on a cross*. After H. S. CRAWFORD. 'Handbook of Carved Ornament from Irish Monuments of the Christian Period', Pls XLIX, 155; L, 155. Dublin, Royal Society of Antiquaries of Ireland, 1926. D. E. Woodall 544

491 *Greek country-carts*: (Above) *Greek wedding procession*. From H. L. LORIMER. *J. Hell. Stud.* **23**, 137, fig. 3, 1903. London, Society for the Promotion of Hellenic Studies. (Middle) *A Greek cart*; (Below) *A Greek coster-cart*. From E. PFUHL. 'Malerei und Zeichnung der Griechen', Vol. 3, figs 169, 248. Munich, Brückmann, 1923 545

492 *Clay model of a wagon*. From E. H. MINNS. 'Scythians and Greeks', fig. 6, no. 2. Cambridge, University Press, 1913 545

493 *Relief of a baggage-wagon*. After C. CICHORIUS. 'Die Reliefs der Trajanssäule', Pl. XXIX, 96–97. Berlin, Reimer, 1896. D. E. Woodall 546

494 *Roman relief of a travelling-wagon*. Klagenfurt, Kirche Maria Saal. After U. E. PAOLI. 'Das Leben im alten Rom', Pl. LXXV. Bern, Francke, 1948. K. F. Rowland 546

495 *Late Saxon two-wheeled cart*. London, British Museum, Cotton MS. Tib. B. V, fol. 6. After photograph by courtesy of the Trustees. D. E. Woodall 547

496 *Travelling-carriage*. Paris, Bibliothèque Nationale, MS. franç. 2091, fol. 125 (detail). After H. MARTIN. 'Légende de Saint Denis', Pl. XLII. Paris, Champion, 1908. D. E. Woodall 547

497 *Baggage-wagon*. Brussels, Bibliothèque Royale de Belgique, MS. 9067, fol. 149v. After J. VAN DEN GHEYN. 'Cronicques et conquestes de Charlemaine', Pl. LXI. Brussels, Vromant, 1909. D. E. Woodall 547

498 *Cart with large wheels*. Brussels, Musées R. des Beaux-Arts, 'Dénombrement de Bethléem' (detail). After G. H. DE LOO. 'Peter Brueghel l'Ancien, son œuvre et son temps', Vol. 2 Planches. Brussels, Van Oest, 1905. D. E. Woodall 548

499 *Two-wheeled cart with hurdles*. London, British Museum, Add. MS. 42130, fol. 162. After 'The Luttrell Psalter'. Facs. edition by E. G. MILLAR, Pl. LXXVI. London, British Museum, 1932. By courtesy of the Trustees. D. E. Woodall 548

500 *Two-wheeled cart*. *Ibid.*, fol. 173b. After E. G. MILLAR. *Ibid.*, Pl. XCIX. D. E. Woodall 549

501 *German coach*. Paris, Bibliothèque d'Art et d'Archéologie de l'Université. After 'Kurtze gegrundte Beschreibung des Pfalzgraven bey Rhein hochzeitlichen Ehren Fests', fol. 29. Munich, Adam Berg, 1568. By courtesy of the Director. D. E. Woodall 549

502 *A turning-train*. After H. T. BOSSERT and W. F. STORCK. 'Das mittelalterliche Hausbuch im Besitz des Fürsten von Waldburg-Wolfegg-Waldsee', Pl. LVIII (fol. 52b). Leipzig, Seemann, 1912. D. E. Woodall 550

503 *A farm cart*. After B. PACE. 'Mosaici di Piazza Armerina', fig. 19. Rome, Casini, 1955. D. E. Woodall 550

504 (A) *Wheel with a felloe*. From J. CURLE. 'A Roman Frontier Post and its People', Pl. LXIX, 2. Glasgow, Maclehose, 1911, for the Society of Antiquaries of Scotland. (B) *Wheel found in a bog*. After G. BERG. 'Sledges and Wheeled Vehicles', Pl. XV, 4. Nordiska Museets Handlingar. No. 4. Stockholm, Levin and Munksgaard, 1935. (C) *Hub of a wheel of the Dejbjerg wagon*. After O. KLINDT-JENSEN. *Acta Archaeologica* **20**, 88, fig. 55, 1949. Copenhagen, Munksgaard. (B, C) D. E. Woodall 551

505 *Wheelwright's shop.* From HANS SACHS. 'Eygentliche Beschreibung aller Stände . . . mit kunstreichen Figuren [by Jost Amman]', fol. a. Frankfurt a. M., Sigmund Feyerabend, 1568 552

506 *Roman relief.* After J. SAUTEL. 'Vaison dans l'Antiquité', Vol. 3, Pl. LIX, fig. 2. Avignon, Lyons, Aubanel, 1926. D. E. Woodall 553

507 *War-chariots on a bone carving.* Paris, Musée des Thermes et de l'Hôtel de Cluny. After R. J. E. C. LEFEBVRE DES NOËTTES. 'L'Attelage. Le cheval de selle à travers les âges', fig. 90. Paris, Picard, 1931. D. E. Woodall 553

508 *Tenth-century harnessing.* Paris, Bibliothèque Nationale, MS. lat. 8085. After R. J. E. C. LEFEBVRE DES NOËTTES. *Ibid.*, fig. 142. D. E. Woodall 554

509 *Diagram of a Scythian saddle and stirrup.* From W. W. ARENDT. *Eurasia Septentrionalis Antiqua* **9**, 207, fig. 2, 1934. Archaeological Society of Finland 555

510 *Roman saddle.* From A. BANDURI. *Imperium Orientale sive Antiquitates Constantinopolitanae*, Vol. 2, Pl. VI, 13. Paris, 1711 556

511 *Saddled Chinese horse.* Collection Wannieck. After R. J. E. C. LEFEBVRE DES NOËTTES. 'L'Attelage. Le cheval de selle à travers les âges', fig. 287. Paris, Boccard, 1931. D. E. Woodall 557

512 *Stirrups*: (A) After N. FELTICH. *Archaeologia Hungarica* **1**, 14, fig. 12, no. 1, 1926. Budapest, Pest Magyar Nemzeti Museum. (B) After H. STOLPE and T. J. ARNE. 'La Nécropole de Vendel', Pl. XIV, fig. 1. K. Vitterhets Historie och Antikvitetsakademien, Monograph 17. Stockholm, 1927. (C, D) London Museum. After 'Medieval Catalogue', fig. 23, no. 4; fig. 27, no. 2. London Museum Catalogues No. 7. 1940. T. A. Greeves 559

513 *Spurs*: (A) After J. L. PÍČ. 'Le Hradischt de Stradonitz en Bohême' (trans. from the Czech by J. DÉCHELETTE), Pl. XXXI, fig. 2. Leipzig, Hiersemann, 1906. (B, C, D, E) London Museum. After 'Medieval Catalogue', fig. 29, no. 3; fig. 31, no. 2; fig. 34, no. 1; fig. 35, no. 7. London Museum Catalogues No. 7. 1940. T. A. Greeves 559

514 *Bits*: (A) London Museum. After 'Medieval Catalogue', fig. 21, no. 1. (B) After C. F. FOX. 'A Find of the Early Iron Age from Llyn Cerrig Bach, Anglesey', Pl. XXIV. Cardiff, National Museum of Wales, 1946. (C) After P. JACOBSTHAL. 'Early Celtic Art', Pl. CCLVIII, no. 4. Oxford, Clarendon Press, 1944. (D) Vienna, Heergeschichtliches Museum. After G. LAKING. 'A Record of European Armour and Arms', Vol. 3, fig. 1000. London, Bell, 1920. T. A. Greeves 559

TAILPIECE. *Mining-truck running on wooden rails.* After A. WOLF. 'History of Science, Technology, and Philosophy in the 16th and 17th Centuries', fig. 265. London, Allen and Unwin, 1935. D. E. Woodall 562

16. SHIPS AND SHIPBUILDING *by* T. C. LETHBRIDGE

515 *Greek sailing merchant-ship.* London, British Museum, Greek vase No. B 436. After photographs by courtesy of the Trustees. D. E. Woodall 564

516 *A 'Homeric' war-galley.* From C. DAREMBERG and E. SAGLIO. 'Dictionnaire des antiquités grecques et romaines', Vol. 4, Part 1, fig. 5265. Paris, Hachette, 1904–7 565

517 *Greek galley.* London, British Museum, Greek vase No. E 2. After photograph by courtesy of the Trustees. D. E. Woodall 565

518 *Greek trireme.* Athens, Acropolis Museum. After G. LA ROËRIE and L. VIVIELLE. 'Navires et marins de la rame à l'hélice', Vol. 1, p. 50. Paris, Duchartre and Van Buggenhoudt, 1930. K. F. Rowland 566

519 *'Winged Victory.'* London, British Museum, Coin of Demetrius Poliorcetes. After photograph by courtesy of the Trustees. K. F. Rowland 567

520 *Roman warship.* Rome, Vatican, Museo Egizio. After G. LA ROËRIE and L. VIVIELLE. 'Navires et marins de la rame à l'hélice', Vol. 1, p. 60. Paris, Duchartre and Van Buggenhoudt, 1930. K. F. Rowland 570

521 *Pictures of war-galleys.* (Above) After O. ELIA. 'La Pittura elenistico-romana'. Sezione terza. 'Pompei', fasc. III–IV, Pl. IV. Rome, Istituto Poligrafico dello Stato, 1941. D. E. Woodall. (Below) After G. LA ROËRIE and L. VIVIELLE. 'Navires et marins de la rame à l'hélice', Vol. 1, p. 57. Paris, Duchartre and Van Buggenhoudt, 1930. T. C. Lethbridge 571

522 *Roman pleasure-galley.* After G. UCELLI. 'Le navi di Nemi', fig. 96. Rome, R. Istituto d'Archeologia e Storia dell' Arte, 1940. D. E. Woodall 572

523 *Roman merchant ship.* After G. LA ROËRIE and L. VIVIELLE, 'Navires et marins de la rame à l'hélice', Vol. 1, p. 65. Paris, Duchartre and Van Buggenhoudt, 1930. K. F. Rowland 572

524 *Roman merchant ship in port*. From C. SINGER. 'The Earliest Chemical Industry', fig. 23. London, Folio Society, 1948. By courtesy of Mr Derek Spence 573

525 *Bronze model*. After H. SEYRIG. *Syria* **28**, Pl. IX, 1951. Paris, Institut Français d'Archéologie de Beyrouth. D. E. Woodall 575

526 *Remains of a Roman ship*. London Museum. After 'London in Roman Times', Pl. LX. London Museum Catalogues No. 3. 1946. T. A. Greeves 576

527 (A) *Model of Hjortspring boat*. Copenhagen, Nationalmuseet. After photograph by courtesy of the Director. (B) *Viking vessel found at Nydam*. Schloss Gottorp, Schleswig-Holsteinisches Landesmuseum. After A. W. BRØGGER and H. SHETELIG. 'The Viking Ships', pp. 48, 54. Oslo, Dreyer, 1951. (C) *Interior view of stern*. After 'The Sutton Hoo Ship-Burial' (3rd impr.), Pl. III (Photo C. W. Phillips). London, the Trustees of the British Museum, 1952. By courtesy of the Trustees. (A, B, C) T. A. Greeves 579

528 (A) *The restored Gokstad ship*; (B) *Interior*; (C) *Longitudinal section*. After A. W. BRØGGER and H. SHETELIG. 'The Viking Ships', pp. 111, 109, 120–1. Oslo, Dreyer, 1951. (A, B) T. A. Greeves, (C) prepared by R. Steed 580

529 *Gold model of a boat*. Dublin, National Museum of Ireland, Broighter Find. After photograph by courtesy of the Director. D. E. Woodall 582

530 (Above) *Two-masted ship*. After U. NEBBIA. 'Arte Navale Italiana', fig. 11. Bergamo, Istituto Italiano d'Arti Grafiche Editore, 1932. (Below) *Single-masted ship*. London, British Museum. Seals Egert. ch. 396. 5502. After photograph by courtesy of the Trustees. K. F. Rowland 582

531 *Single-masted ship with stern-post rudder*. After 'Elbing', p. 7. Ed. by MAGISTRAT ELBING. [Deutschlands Städtebau.] Berlin, Dari-Verlag, 1929. K. F. Rowland 583

532 *Pictures of boats with lateen sails*. Paris, Bibliothèque Nationale, MS. grec 510. After H. OMONT. 'Fac-similés des miniatures des plus anciens manuscrits grecs de la Bibliothèque Nationale', Pl. XX. Paris, Presses Universitaires de France, 1902. D. E. Woodall 583

533 *Castilian merchantman*. From C. SINGER. 'The Earliest Chemical Industry', fig. 49. London, Folio Society, 1948. By courtesy of Mr Derek Spence 584

534 *Sailing-galley*. From B. VON BREYDENBACH. 'Die heiligen Reyssen gen Jerusalem.' Mainz, Erhart Reuwich, 1486 584

535 *An early 'carrack'*. London, British Museum, Cotton MS. Dom. A. XVII, fol. 123r. After photograph by courtesy of the Trustees. K. F. Rowland 585

536 *A three-masted merchantman*. From M. LEHRS. 'Der Meister W. A.', Pl. XI. Leipzig, Hiersemann, 1895 586

537 *Section of planking*. From W. ABELL. 'The Shipwright's Trade', p. 21. Cambridge, University Press, 1948 587

17. POWER *by* R. J. FORBES

538 *A water-mill*. After M. M. BANKS. 'British Calendar Customs', Pl. I (copyright photograph J. D. Rattar, Lerwick). London, Glasgow, Orkney and Shetland Folk-Lore Society, 1946. T. A. Greeves 590

539 *A late illustration of a horizontal water-mill*. From J. DE STRADA. 'Kunstliche Abriss allerhand Wasser- Wind- Ross- und Handt Mühlen', fol. 6. Frankfurt a. M., Octavio de Strada, 1617 594

540 (A) *A 'horizontal' water-mill*; (B) *A 'vertical' mill*. From E. C. CURWEN. *Antiquity* **18**, 131, fig. 1. Newbury, Edwards, 1944 595

541 *Diagram of*: (A) *undershot wheel*; (B) *overshot wheel*; (C) *breast-wheel*. From E. C. CURWEN. *Ibid.*, p. 133, fig. 2 595

542 *A mill*. Manchester, John Rylands Library, The Buxheim St Christopher (detail). After photograph by courtesy of the Librarian. D. E. Woodall 596

543 *The stream*. London, British Museum, Add. MS. 42130, fol. 181. After 'The Luttrell Psalter'. Facs. edition by E. G. MILLAR, Pl. CXIV. London, British Museum, 1932. By courtesy of the Trustees. D. E. Woodall 596

544 *An ore-crushing machine*. From AGRICOLA. *De re metallica*, p. 223. Basel, Froben, 1556 597

545 *Model of undershot water-wheel*. After C. REINDL. *Wasserkraft, Münch.* **34**, 143, fig. 1, 1939. Munich, Berlin, Oldenbourg. T. A. Greeves 598

546 *Conjectural reconstruction of Barbegal mills.* Based on C. L. SAGUI. *Isis* **38**, 226, 1948. History of Science Society, Cambridge, Mass.; and F. BENOIT. *Rev. archéol.* sixième série **15**, 19–80, 1940. Paris, Presses Universitaires de France. T. A. Greeves 598

547 *Ground-plan.* After F. BENOIT. *Ibid.*, pp. 24–25, fig. 3. Paris, Presses Universitaires de France. T. A. Greeves 599

548 *Mosaic from the great palace.* After G. BRETT. *Antiquity* **13**, Pl. VII, preceding p. 345, 1939. Newbury, Edwards. D. E. Woodall 600

549 *Paddle-driven warship.* Oxford, Bodleian Library, MS. Canon. Misc. 378, fol. 75v. After photograph by courtesy of the Curators. K. F. Rowland 606

550 *A floating mill.* From VERANTIUS. *Machinae novae*, Pl. XVIII. Venice, 1620 (?) 607

551 *A corner of the harbour of Cologne.* From A. SCHRAMM. 'Bilderschmuck der Frühdrucke', Vol. 8, Pl. CLXIV, no. 761. Leipzig, Hiersemann, 1924 607

552 *Floating mills under a bridge.* Paris, Bibliothèque Nationale, MS. franç. 2092, fol. 37v. After H. MARTIN. 'Légende de Saint Denis', Pl. LXIV. Paris, Champion, 1908. D. E. Woodall 608

553 *An undershot water-wheel.* From J. DE STRADA. 'Kunstliche Abriss allerhand Wasser- Wind- Ross- und Handt Mühlen', fol. 5. Frankfurt a.M., Octavius de Strada, 1617 609

554 *Water-driven hammer-forge.* From J. DE STRADA. *Ibid.*, fol. 4 612

555 *Water-driven bellows.* From RAMELLI. 'Le diverse et artificiose machine', fig. 137. Paris, 1588 613

556 *The wind-organ.* From HERO OF ALEXANDRIA. 'Surviving Works' (ed. by W. SCHMIDT), Vol. I, fig. 44. Leipzig, Teubner, 1899 614

557 *The horizontal windmill.* From AL-DIMASHQI. 'Cosmographie' (ed. by A. F. MEHREN), p. 182. St Petersburg, 1866 615

558 *Modern Chinese windmill.* After A. WARWICK. *Nat. Geogr. Mag.* **51**, 483, 1927. Washington, National Geographic Society. D. E. Woodall 617

559 *Modern Aegean windmills.* After R. STILLWELL. *Ibid.* **85**, 610, Pl. X, 1944. Washington, National Geographic Society. T. A. Greeves 618

560 *Post-mills and tower-mills.* From STRADANUS (JAN VAN DER STRAET). *Nova reperta*, Pl. XI. Antwerp, [1600?] 619

561 *Post-mill.* New York, Pierpont Morgan Library, MS. 102, fol. 2 After C. R. MOREY, B. DA COSTA GREEN, *et al.* 'The Pierpont Morgan Library. Exhibition of Manuscripts held at the New York Public Library', fig. 7. By permission of the Director of the Pierpont Morgan Library. K. F. Rowland 623

562 *Post-mill.* London, British Museum, Royal MS. 10. E. IV, fol. 70. After photograph by courtesy of the Trustees. D. E. Woodall 623

563 *Post-mill.* London, British Museum, MS. Harley 3487, fol. 161. After photograph by courtesy of the Trustees. D. E. Woodall 624

564 *Tower-mill.* After copyright photograph by R. Wailes. D. E. Woodall 624

565 *Byzantine tower-mills.* Madrid Biblioteca Nacional, MS. 18246 (BUONDELMONTE. *De Insulis Archipelagi*). After photograph by courtesy of the Warburg Institute, University of London. D. E. Woodall 625

566 *Tower-mills on the Island of Rhodes.* From B. VON BREYDENBACH. 'Die heyligen reyssen gen Jerusalem.' Mainz, Erhart Reuwich, 1486 625

567 *A tower-mill.* From RAMELLI. 'Le diverse et artificiose machine', fig. 73. Paris, 1588 626

568 *A tower-mill with canvas sails.* From J. BESSON. 'Theatre des instrumens', Pl. L. Lyons, Vincent, 1578 626

569 *Hinged sails.* From VERANTIUS. *Machinae novae*, Pl. VIII. Venice, 1620 (?) 627

570 *A tower-mill with sails of wood.* From VERANTIUS. *Ibid.*, Pl. XI 627

TAILPIECE. *'The Mills of Babylon.'* London, British Museum, Royal MS. 15. E. VI, fol. 4b. After photograph by courtesy of the Trustees. K. F. Rowland 628

18. MACHINES *by* BERTRAND GILLE

571 *Greek cabinet-maker.* From L. D. CASKEY and J. D. BEAZLEY. 'Attic Vase Paintings in the Museum of Fine Arts, Boston', Pt II, Pl. XXXIV, no. 69. Boston, Museum of Fine Arts, 1954. Prepared by R. Steed 631

572 *Reconstruction of Hero's screw-cutter.* From A. G. DRACHMANN. *Archæol.-kunsthist. Medd.* **1**, 159, fig. 28, 1932–5. Copenhagen, Det Kongelige Danske Videnskabernes Selskab 632

573 *Hero's hydraulic organ.* From HERO OF ALEXANDRIA. 'The Pneumatics' (ed. and trans. by B. WOODCROFT), p. 105. London, 1851 634

574 *Philo's combined suction and force-pump.* From A. P. USHER. 'A History of Mechanical Inventions', fig. 17. New York, McGraw-Hill, 1927 634

575 *Hero's device for opening doors.* From HERO OF ALEXANDRIA. 'The Pneumatics' (ed. and trans. by B. WOODCROFT), p. 57. London, 1851 635

576 *Hero's cupping-glass.* From HERO OF ALEXANDRIA. *Ibid.*, p. 79 635

577 *Roman water-raising scoop-wheel.* After F. MAYENCE. *Bull. Mus. R. Art Hist.* troisième série **5**, 5, fig .5, 1933. Brussels, Musées Royaux d'Art et d'Histoire. D. E. Woodall 637

578 *Roman hoist.* After P. BRANDT. 'Schaffende Arbeit und bildende Kunst im Altertum und Mittelalter', fig. 175. Leipzig, Kröner, 1927. D. E. Woodall 637

579 *Carpenter using a plane.* Nuremberg, Stadtbibliothek, Hausbuch der Mendelschen Zwölfbrüderstiftung. After F. BOCK (Ed.). 'Deutsches Handwerk im Mittelalter', fig. 9. Leipzig, Insel-Verlag, 1935. D. E. Woodall 641

580 *Building operations.* Dublin, Trinity College Library, MS. E. i. 40, fols. 59[v], 60[r]. After photograph by courtesy of the Librarian. D. E. Woodall 641

581 *Water-driven stamp-mill.* After H. T. BOSSERT and W. F. STORCK. 'Das Mittelalterliche Hausbuch im Besitz des Fürsten von Waldburg-Wolfegg-Waldsee', Pl. XL (fol. 36[b]). Leipzig, Seemann, 1912. D. E. Woodall 642

582 *Bellows operated by cams.* After H. T. BOSSERT and W. F. STORCK. *Ibid.*, Pl. XLI (fol. 37[a]) 643

583 *Double-counterpoise sling-trebuchet.* From VALTURIO. *De re militari*, fol. not numbered. Verona, Johannes of Verona, 1472 644

584 *Villard de Honnecourt's sketch of saw.* Paris, Bibliothèque Nationale, MS. franç. 19093. After R. HAHNLOSER. 'Villard de Honnecourt. Kritische Gesamtausgabe des Bauhüttenbuchs', Pl. XLIV. Vienna, Schroll, 1935. D. E. Woodall 644

585 *Turning.* From G. ACLOCQUE. 'Les corporations, l'industrie et le commerce à Chartres du XIe siècle à la Révolution', Pl. III. Paris, Picard, 1917 645

586 *Turning a wooden bowl.* Paris, Bibliothèque Nationale, MS. lat. 11560, fol. 84. After A. DE LABORDE. 'La Bible Moralisée', Vol. 3, Pl. CCCVIII. Paris, Société française de Reproduction des Manuscrits à Peintures, 1915. D. E. Woodall 645

587 *Crank-rotated machine.* Paris, Bibliothèque Nationale, MS. franç. 616 (GASTON PHÉBUS, 'Livre de Chasse'), ch. XXV. Drawing after postcard Éditions Braun et Cie, Mulhouse (Ht-Rhin). D. E. Woodall 646

588 *Reconstruction of a winch.* After E. E. VIOLLET-LE-DUC. 'Dictionnaire raisonné de l'architecture française du XIe au XVIe siècle', Vol. 5, p. 215, fig. 3. Paris, 1861. D. E. Woodall 646

589 *Villard de Honnecourt's sketch of screw-jack.* Paris, Bibliothèque Nationale, MS. franç. 19093. After R. HAHNLOSER. 'Villard de Honnecourt. Kritische Gesamtausgabe des Bauhüttenbuchs', Pl. XLIV. Vienna, Schroll, 1935. D. E. Woodall 647

590 *Drawing of the interior of corn-mill.* From J. WALTER (Ed.). HERRAD DE LANDSPERG, *Hortus deliciarum*, p. 83. Strasbourg, Le Roux, 1952 648

591 *Water-mill.* Madrid, Palacio Episcopal, Arca de San Isidro. After W. W. COOK and J. F. RICART. *Ars Hispaniae*, Vol. 6, fig. 261. Madrid, Editorial Plus Ultra, 1950. D. E. Woodall 648

592 *Post-mill.* Munich, Bayerische Staatsbibliothek. After A. UCCELLI. 'Storia della Tecnica dal medio evo ai nostri giorni', fig. 20. Milan, Hoepli, 1945. D. E. Woodall 649

593 *Sharpening a sword.* Utrecht, University Library, MS. 488, fol. 35v. After E. T. DE WALD. 'The Illustrations of the Utrecht Psalter', Pl. LVIII. New York, Princeton University Press, 1932. D. E. Woodall 651

594 *Crank-propelled paddle-boat.* Paris, Bibliothèque Nationale, MS. lat. 11015 (GUIDO DE VIGEVANO), fol. 10v. After photograph by courtesy of the Director. D. E. Woodall 651

595 *Carpenter's tools.* Hamburg, Kunsthalle, Thomas-Altar. After B. MARTENS. 'Meister Francke', Tafelband, Pl. XXXIX (detail). Hamburg, Friederichsen and de Gruyter, 1929. D. E. Woodall 652

596 *Hand-mill worked by crank.* Munich, Bayerische Staatsbibliothek, Cod. lat. 197, fol. 18r. After copyright photograph by courtesy of the Director. Steffie Schaefer 653

597 *Mill operated by double-throw crank.* Munich, Bayerische Staatsbibliothek, Cod. lat. 197. After A. UCCELLI. 'Storia della Tecnica dal medio evo ai nostri giorni', fig. 238. Milan, Hoepli, 1945. D. E. Woodall 653

598 *Leonardo's screw-cutting machine.* Sketch at Bibliothèque de l'Institut de France, Paris, Leonardo MS. B, fol. 70r; model at the Science Museum, London. After copyright photographs by courtesy of the Director, Science Museum, South Kensington, London. T. A. Greeves 655

599 *Machine for boring.* Nuremberg, Stadtbibliothek. After F. M. FELDHAUS. 'Die Technik der Antike und des Mittelalters', fig. 371. Potsdam, Akademische Verlagsgesellschaft Athenaion, 1931. D. E. Woodall 655

600 (A) *Simple form of crane*; (B) *Early example of slewing crane.* Munich, Bayerische Staatsbibliothek, Cod. lat. 197, fols 1r, 38r. From photographs by courtesy of the Director. Prepared by R. Steed 656

601 *Double swivelling-crane.* Paris, Bibliothèque de l'Institut de France, Leonardo MS. B, fol. 49r. From C. RAVAISSON-MOLLIEN. 'Les Manuscrits de Léonard de Vinci.' Paris, Maison Quantin, 1883 657

602 *Reconstruction of Vitruvius's crane*, by the author and T. A. Greeves 659

603 *Part of a relief.* Rome, Museo Lateranense. After copyright photograph Alinari, Florence. D. E. Woodall 660

604 *Hero's 'hanger' for lifting a stone.* Diagram after sketch by the author. T. A. Greeves 661

605 *Hero's triple 'hanger'.* Diagram after sketch by the author. T. A. Greeves 661

19. HYDRAULIC ENGINEERING AND SANITATION *by* R. J. FORBES

606 *Medieval wooden-lined well.* From G. P. LARWOOD. *Norfolk Archaeology* **30**, iii, 227, fig. 2, 1951. Norfolk and Norwich Archaeological Society 664

607 *Conjectural reconstruction of Mycenaean spring-approach.* Based on O. BRONEER. *Hesperia* **8**, 338, fig. 16; 341, fig. 18, 1939. American School of Classical Studies in Athens. T. A. Greeves 665

608 *Cisterns at Carthage.* After J. A. HAMMERTON. 'Universal History', Vol. 4, p. 2042. London, Amalgamated Press, [1928]. D. E. Woodall 666

609 *Plan and section of spring.* After J. HEIERLI. *Anz. Schweiz. Altertumsk.* new series **9**, 268, fig. 58, 1907. Zürich, Antiquarische Gesellschaft. Steffie Schaefer 666

610 *Greek women.* Berlin, Staatl. Museen. After E. GERHARD. 'Etruskische und Kampanische Vasenbilder des königlichen Museums zu Berlin', Pl. XXX, fig. 3. Berlin, 1843. D. E. Woodall 667

611 *Diagrammatic section of the aqueduct of Samos.* Based on E. FABRICIUS. *Athen. Mitt.* **9**, Pl. VIII, 1884. D. E. Woodall 667

612 *Diagrammatic section of the aqueduct.* Based on F. GRÄBER. 'Die Wasserleitungen von Pergamon.' Abh. preuss. Akad. Wiss., phil.-hist. Kl., for 1887, Pl. II, 1888. D. E. Woodall 668

613 *Ruins of the* Claudia *and* Anio Novus *aqueducts.* After photograph. T. A. Greeves 668

614 *Map showing the routes of the Roman aqueducts.* Based on B. BUFFET and R. EVRARD. 'L'Eau potable à travers les âges', fig. 63. Liége, Solédi, 1950. J. F. Horrabin 669

615 *Modern* sāqiya. After A. NEUBURGER. 'The Technical Arts of the Ancients', fig. 284. London, Methuen, 1930. E. Norman 675

616 *Three types of water-lifting machine.* From A. UCCELLI. 'Storia della Tecnica dal medio evo ai nostri giorni', fig. 101. Milan, Hoepli, 1945 676

617 *Archimedean screw.* City of Liverpool Public Museums. After copyright photograph Liverpool University, School of Archaeology and Oriental Studies. D. E. Woodall 676

618 *Archimedean screw for water-raising.* After D. R. HUDSON. *Metallurgia* **35**, 157, fig. 1, 1947. Manchester, Kennedy Press. D. E. Woodall 677

619 *Negro slave.* London, British Museum, No. 37563. After photograph by courtesy of the Trustees. K. F. Rowland 677

620 *Shaduf.* From R. VAN BASTELAER. 'Les Estampes de Peter Bruegel l'Ancien', p. 23, No. 5. Brussels, Van Oest, 1908 680

621 *Sketch of simple polder.* According to ANDRIES VIERLINGH. 'Tractaet van Dyjckagie' (ed. by J. DE HULLU and A. G. VERHOEVEN), Plates, fig. 13. The Hague, Nijhoff, 1920. Adapted by E. Norman 682

622 *Typical dike-construction.* According to ANDRIES VIERLINGH. Based on *ibid.*, Plates, fig. 45. Adapted by E. Norman 684

623 *Another type of medieval dike.* According to ANDRIES VIERLINGH. Based on *ibid.*, Plates, fig. 47. Adapted by E. Norman 684

624 *A lift for transferring boats.* From ZONCA. *Novo teatro di machine et edificii*, p. 58. Padua, 1607 685

625 *A lock-chamber.* From ZONCA. *Ibid.*, p. 9 686

626 *A river divided into level 'pounds'.* Florence, Biblioteca Medicea-Laurenziana, Ashb. MS. 361 (Trattato dei Pondi). After W. B. PARSONS. 'Engineers and Engineering in the Renaissance', fig. 132. Baltimore, Williams and Wilkins, 1939. D. E. Woodall 687

627 *Diagram of three scoop-wheels.* Haarlem, Archief der Gemeente. After copyright photograph by courtesy of the Director. D. Wyeth 688

628 *Part of twelfth-century plan.* Adapted from R. WILLIS. *Archaeologia Cantiana* **7**, Pl. I, no. 2, 1868. 690

TAILPIECE. *Clearing out a water-course.* London, British Museum, MS. Egerton 1894, fol. 14[a]. After photograph by courtesy of the Trustees. K. F. Rowland 694

20. MILITARY TECHNOLOGY *by* A. R. HALL

629 *Early Greek armour.* From INSTITUTO DI CORRESPONDENZA ARCHEOLOGICA. 'Monumenti Inediti', Vol. 1, Pl. LI. Rome, Paris, 1829–33 696

630 *A stick-slinger.* From J. KROMAYER and G. VEITH. 'Heerwesen und Kriegführung der Griechen und Römer', Pl. XVIII, fig. 70. Munich, Beck, 1928 700

631 *East wall of Troy VI.* After C. W. BLEGEN *et al.* 'Troy, the Sixth Settlement', Vol. 3, Part II: Plates, No. 30. Princeton, Princeton University Press for University of Cincinnati, 1953. By courtesy of Professor Carl W. Blegen. D. E. Woodall 701

632 *Glacis and covered way.* After sketch by the author. D. E. Woodall 702

633 (A) *Early Roman* pilum. According to Polybius and based on P. COUISSIN. 'Les Armes romaines', fig. 126. Paris, Librairie Ancienne Honoré Champion, 1926. (B) *Various later* pila. After P. COUISSIN. *Ibid.*, figs 116, 118, 121. D. E. Woodall 703

634 *Roman armour*: (A, B, C, D) After P. COUISSIN. *Ibid.*, Pl. I, no. 3; Pl. II, no. 7; Pl. III, no. 14; Pl. V, no. 25. D. E. Woodall 704

635 *Cataphracts.* After C. CICHORIUS. 'Die Reliefs der Trajanssäule', Plates Vol. 1, Pl. XXVIII, nos 93–94. Berlin, Reimer, 1896. D. E. Woodall 706

636 *Auxiliary archers.* After C. CICHORIUS. *Ibid.*, Plates Vol. 2, Pl. LXXXVI, nos 309–10. Berlin, Reimer, 1900. D. E. Woodall 707

637 *Cross-bow and quiver.* After H. DIELS. 'Antike Technik', fig. 36. Leipzig, Berlin, Teubner, 1914. D. E. Woodall 707

638 *A* carroballista. After C. CICHORIUS. 'Die Reliefs der Trajanssäule', Plates Vol. 1, Pl. XLVI, nos 163–4. Berlin, Reimer, 1896. D. E. Woodall 708

639 *Hero's* gastraphetes. After sketch by author. E. Norman 708

640 Euthytonon. After J. KROMAYER and G. VEITH. 'Heerwesen und Kriegführung der Griechen und Römer', Pl. XVII, fig. 66. Munich, Beck, 1928. D. E. Woodall 709

641 Palintonon. After E. SCHRAMM. 'Die antiken Geschütze der Saalburg', Pl. IV. Berlin, Weidmann for Saalburgverwaltung, 1918. D. E. Woodall 710

642 *Another reconstruction of the* palintonon. After sketch by author. E. Norman 711

643 *Philo's use of wedges.* After sketch by author. E. Norman 712

644 *Philo's* chalcotonon. After sketch by author. E. Norman 712

645 *The onager.* After H. DIELS. 'Antike Technik', fig. 41. Leipzig, Berlin, Teubner, 1914. D. E. Woodall 713

646 *An Assyrian relief showing a battering-ram.* After P. E. BOTTA. 'Monument de Ninive', Vol. 1, Pl. LXXVII. Paris, 1849. D. E. Woodall 715

647 *The tortoise.* After C. CICHORIUS. 'Die Reliefs der Trajanssäule', Plates Vol. 1, Pl. LI, no. 181. Berlin, Reimer, 1896. D. E. Woodall 716

648 *A movable tower.* From J. KROMAYER and G. VEITH. 'Heerwesen und Kriegführung der Griechen und Römer', Pl. XXII, fig. 75. Munich, Beck, 1928 717

649 *The 'stepped' curtain-wall.* After sketch by author. D. E. Woodall 717

650 *Section through fortifications of Nicaea.* After C. DAREMBERG and E. SAGLIO. 'Dictionnaire des antiquités grecques et romaines', Vol. 3, Part II, fig. 5168. Paris, Hachette, 1904. D. E. Woodall 718

651 *Section through fortifications at Constantinople.* After C. DAREMBERG and E. SAGLIO. *Ibid.*, fig. 5167. D. E. Woodall 718

652 *Legionaries building a fort.* After C. CICHORIUS. 'Die Reliefs der Trajanssäule', Plates Vol. 1, Pl. XLII, nos 145–6. Berlin, Reimer, 1896. D. E. Woodall 718

653 *Map showing part of the German* limes. Based on E. FABRICIUS *et al.* (Eds). 'Der obergermanisch-raetische Limes', Abt. B, Vol. 1, Map. Berlin, Peters, 1937. E. Norman 719

654 *The fabric of chain-mail.* After sketch by author. D. E. Woodall 721

655 *Three different types of cross-bow:* (A, B, C) After R. W. F. PAYNE GALLWEY. 'The Crossbow', figs 45, 75, 87. London, Longmans, 1903. D. E. Woodall 722

656 *The cross-bow lock.* After R. W. F. PAYNE GALLWEY. *Ibid.*, fig. 55. D. E. Woodall 723

657 *Design for a siege cross-bow.* After RAMELLI. 'Le diverse et artificiose machine', fig. 193. Paris, 1588. D. E. Woodall 724

658 *Inventions in military engineering.* (A, B) Paris, Bibliothèque Nationale, MS. lat. 11015 (GUIDO DE VIGEVANO), fols 8ᵛ, 9ᵛ. After photograph by courtesy of the Director. D. E. Woodall 725

659 *Wind-driven fighting-car.* Paris, Bibliothèque Nationale. *Ibid.*, fol. 14ᵛ. After photograph by courtesy of the Director. D. E. Woodall 726

TAILPIECE. *Roman soldiers crossing a bridge.* After C. CICHORIUS. 'Die Reliefs der Trajanssäule', Plates Vol. 1, Pl. VII. Berlin, Reimer, 1896. D. E. Woodall 730

21. ALCHEMICAL EQUIPMENT *by* E. J. HOLMYARD

660 *Figures of Greek alchemical apparatus.* Adapted from F. SHERWOOD TAYLOR. *Ambix* 1, 40, figs 1, 2, 1937. London, Society for the Study of Alchemy and Early Chemistry. By courtesy of the Council of *Ambix* 733

661 (A) *A* tribikos. (B) *Reconstruction.* From F. SHERWOOD TAYLOR. *Annals of Science*, **5**, 3, 15 July 1945, figs. 4, 7. London, Taylor and Francis, 1947 734

662 *The* kerotakis. (A, B) From F. SHERWOOD TAYLOR. *J. Hell. Stud.*, **50**, fig. 4, 1930. London, Society for the Promotion of Hellenic Studies 735

663 *Distillation apparatus.* Oxford, Bodleian Library, MS. Bodley 645, fol. 11. After photograph by courtesy of the Curators. D. E. Woodall 739

664 *Athanor and apparatus for sublimation.* From GEBER. *Alchemiae Gebri Arabis philosophi solertissimi Libri*, p. 187. Bern, Petreius, 1545 739

665 *Further apparatus for sublimation.* From GEBER. *Ibid.*, p. 69 740

666 *A worker testing purity of roll-sulphur.* Munich, Bayerische Staatsbibliothek, Cod. germ. 600. After O. GUTTMANN. *Monumenta Pulveris Pyrii*, fig. 37. London, Artists Press for the author, 1906. D. E. Woodall 740

667 *Jean Béguin's apparatus for sublimation.* From E. J. HOLMYARD. 'Outlines of Organic Chemistry' (3rd. ed.), fig. 38 [reproduced from 'Élemens de Chymie', Lyons, 1658]. London, Arnold, 1953 741

668 *Part of Thomas Norton's laboratory.* London, British Museum, Add. MS. 10302, fol. 1. After photograph by courtesy of the Trustees. D. E. Woodall 741

669 *Apparatus for calcination.* From GEBER. *Alchemiae Gebri Arabis Libri*, p. 186. Bern, Petreius, 1545 742

670 *Apparatus for fusion.* From GEBER. *Ibid.*, p. 190 742

671 *Apparatus for crystallization.* From GEBER. *Ibid.*, p. 91 742

672 *Stills on furnace.* From HIERONYMUS BRAUNSCHWEIG. *Liber de arte distillandi de simplicibus*, fol. VI. Strasbourg, Grüninger, 1500 743

673 *Still of about 1500.* From H. PETERS. 'Aus pharmazeutischer Vorzeit', fig. 51. Berlin, Springer, 1889 743

674 *Alchemical furnace.* From LIBAVIUS. *Alchymia*, p. 166. Frankfurt a. M., 1606 744

675 *One of Glauber's furnaces.* From J. R. GLAUBER. '*Furni novi philosophici* oder Beschreibung der neu erfundenen Distillir-Kunst', Part IV, frontispiece. Amsterdam, Fabel, 1648 744

676 *Furnace and stills designed by Leonardo da Vinci.* Milan, Biblioteca Ambrosiana, Codice Atlantico, fol. 335^r b. From E. J. HOLMYARD. 'Chemistry to the Time of Dalton', p. 86. London, Oxford University Press, 1925 745

677 *Part of Thomas Norton's laboratory.* London, British Museum, Add. MS. 10302, fol. 37^v. After photograph by courtesy of the Trustees. D. E. Woodall 745

678 *Metallurgical balances.* From AGRICOLA. *De re metallica*, p. 267. Basel, Froben, 1556 746

679 *Libavius's 'chemical house'.* From LIBAVIUS. *Alchymia*, pp. 95, 97. Frankfurt a. M., 1606 748

680 *Alchemical stills.* Canterbury, Chapter Library, MS. Lit. B.8, fols 60^v, 61^r, 61^v. After photograph by permission of the Dean and Chapter. Steffie Schaefer 749

681 *Boyle's laboratory.* From E. MIDDLETON *et al.* 'The New Complete Dictionary of Arts and Science', Pl. XVII. London, 1778 750

TAILPIECE. *Woman laboratory assistant.* From M. PUFF VAN SCHRICK. 'Ein gutz nutzlichs büchlein von den ussgebrenten wassern', title-page. Strasbourg, Hupffuff, 1512 752

EPILOGUE: EAST AND WEST IN RETROSPECT *by* CHARLES SINGER

682 (Above) *Pattern on pallium of St Cuthbert*; (Below) *Its interpretation in Kufic-Arabic lettering.* From F. W. BUCKLER. 'Harunu'l-Rashid and Charles the Great', Appendix IV, p. 53. The Mediaeval Academy of America, Publication no. 7. Cambridge, Mass., 1931 756

683 (Map) *Muslim power.* Based on H. G. WELLS. 'The Outline of History', p. 615. London, Cassell, 1932. J. F. Horrabin 757

684 *Map showing the 'Old silk-route'.* Based on H. G. WELLS. *Ibid.*, p. 587; and *China reconstructs*, April 1955, p. 24. Peking, The China Welfare Institute. J. F. Horrabin 759

685 *'Circle of Pythagoras.'* From J. H. G. GRATTON and C. SINGER. 'Anglo-Saxon Magic and Medicine', fig. 23. Publ. of the Wellcome Historical Medical Museum, new series no. 8. London, Oxford University Press, 1952 760

686 *Part of the pattern and inscription of a Byzantine silk weave.* After O. VON FALKE. 'Kunstgeschichte der Seidenweberei', Vol. 2, figs 241, 242. Berlin, Wasmuth, 1913. D. E. Woodall 761

687 *Design of tenth-century silk.* Paris, Louvre. After copyright photograph Archives Photographiques Paris. D. E. Woodall 762

688 (Left) *Eagle-and-gazelle pattern*; (Right) *Copy of the pattern.* Adapted from O. VON FALKE. 'Kunstgeschichte der Seidenweberei', Vol. 1, Pl. XLVI, fig. 200; p. 124, fig. 201. Berlin, Wasmuth, 1913. Prepared by R. Steed 763

689 *Arabic lettering.* From A. H. CHRISTIE. *Burlington Magazine* **40,** 288, figs 2, 3, 1922. London, The Burlington Magazine 764

690 *Part of a reliquary.* Poitiers, Monastère de la Sainte Croix. After M. CONWAY. *Antiq. J.* **3,** 1, 1922. London, Society of Antiquaries of London. D. E. Woodall 765

691 *Byzantine cross.* London, Victoria and Albert Museum. After photograph. Crown copyright reserved. D. E. Woodall 765

692 *The 'Alfred Jewel'.* Oxford, Ashmolean Museum. After photographs by courtesy of the Visitors. D. E. Woodall 766

693 *Byzantine agate chalice.* Venice, San Marco, Tesoro. After copyright photograph Alinari, Florence. D. E. Woodall 767

694 *Essentials of the Roman abacus.* From C. SINGER. 'A Short History of Science', fig. 48. Oxford, Clarendon Press, 1941 768

695 (Map) *Some medieval trade routes.* Based on H. G. WELLS. 'The Outline of History', p. 761. London, Methuen, 1920; and C. SINGER. 'The Earliest Chemical Industry', fig. 50. London, Folio Society, 1948. By courtesy of Mr Derek Spence. J. F. Horrabin 769

PLATES

1 *Interior of a Japanese gold-mine.* London, British Museum, Print Room. HIROSHIGE II. Photograph by courtesy of the Trustees

2A *The big squared limestone block.* From T. WIEGAND. 'Baalbek', Tafeln Vol. I, Pl. VIII. Berlin, Leipzig, de Gruyter, 1921

2B *Marks of cross-ploughing.* Photograph by courtesy of the Nationalmuseet, Copenhagen

3A *Peasants' wedding-feast.* Vienna, Kunsthistorisches Museum. London, British Museum, Print Room no. C.67*. 1902–6–17–377

3B *Twelfth-century north Italian sculpture.* Relief at Parma Baptistery. ANTELAMI. Occupations of the Months: August. Copyright photograph I.E.I. da Libero Tosi, Parma

3C *A migrant philosopher.* 'Royal Commission on Historical Monuments—Westminster Abbey', Pl. 214. Photograph by permission of the Controller of H.M. Stationery Office

4 *Photomicrographs.* (A, B, C) Photographs by courtesy of British Leather Manufacturers' Research Association, Egham, Surrey

5A, B (A) *Egyptian loincloth*; (B) *detail.* Photograph by courtesy, Museum of Fine Arts, Boston, Mass.

5C *Section of the 'Book of the Dead'.* London, British Museum, No. 10281. Photograph by courtesy of the Trustees

5D *Chariot-wheels from the tomb of Tutankhamen.* Oxford. The Griffith Institute, Ashmolean Museum. Photograph by courtesy of the Visitors

6A *Portion of leather wall-hanging.* Owned by the Cooper Union Museum, New York, Acc. No. 1903–11–12. Photograph by courtesy of the Museum

6B *Leather panel.* Collection J. W. Waterer

7A *Red morocco binding.* Oxford, Bodleian Library, MS. Auct. F. 1. 14. Photograph by courtesy of the Curators

7B *Gold-tooled binding.* Melbourne, Public Library. By courtesy of the Trustees. Photograph copyright J. W. Waterer

8 *Reeling silk.* Florence, Gabinetto di Disegni, No. 1786F. (Drawing by Campagnola). Photograph by courtesy of the Director

9 *'The suitors surprising Penelope.'* London, National Gallery. Photograph reproduced by courtesy of the Trustees, The National Gallery, London

10A *Headboard of a bronze couch.* Rome, Museo Nuovo Capitolino. Copyright photograph Alinari, Florence

10B *Bronze table.* Naples, Museo Nazionale. Copyright photograph Alinari, Florence

11 *Chair of St Maximin.* Ravenna, Duomo. Copyright photograph Anderson, Rome

12 *St Joseph, the carpenter.* Collection of Comtesse Willy de Grunne, Brussels. Copyright photograph A.C.L. Brussels

13A *An ambry.* Collection Mrs Geoffrey Hart, London

13B *Court cupboard with open shelves.*

13C *A court cupboard.* Collection Mrs Geoffrey Hart, London

13D *An oak joined press.*

14A *Oak trestle table.*

14B *Long table with turned trestle supports.* Collection Mrs Geoffrey Hart, London

14C *Oak drawer-leaf table.*

14D *Walnut drawer-leaf table.*

15A *Small oak box.*

15B *Oak table-box.* Formerly in the Percival Griffiths Collection

15C *Chest constructed of six boards.* Collection Lord Henley, Watford Court, Rugby

15D *Oak chest with linenfold panels.* Collection Mrs Geoffrey Hart, London

16 *Athena and Victories crowning potters.* (A) *Unrolled painting*; (B, C) *Two views of the hydria.* Collection Dr Enrico Scaretti, Rome. Photograph by courtesy of the owner

17A *Potters at work.* Oxford, Ashmolean Museum, No. V 562. Photograph by courtesy of the Visitors

17B *White-ground Athenian krater.* Rome, Museo Etrusco Gregoriano. Copyright photograph Anderson, Rome

18A *Black-figured Athenian amphora.* New York, Metropolitan Museum of Art, No. 06.1021.69. Photograph by courtesy of the Metropolitan Museum of Art

18B *Black-figured Athenian stand.* New York, Metropolitan Museum of Art, No. 31.11.4. Photograph by courtesy of the Metropolitan Museum of Art

18C *Athenian cup.* New York, Metropolitan Museum of Art, No. 12.234.5. Photograph by courtesy of the Metropolitan Museum of Art

18D *Black-figured Etruscan hydria.* Photograph by courtesy of the Metropolitan Museum of Art, New York

19A, B *Greek terracotta group.* Olympia, Museum. Photographs Deutches Archäologisches Institut, Athens

20A *Head of a Greek terracotta statue.* Olympia, Museum. Photograph Deutches Archäologisches Institut, Athens

20B *Head of the terracotta statue of the Apollo from Veii.* Rome, Museo Nazionale di Villa Giulia. Copyright photograph Alinari, Florence

20C *Terracotta antefix.* Rome, Museo Nazionale di Villa Giulia. Copyright photograph Alinari, Florence

21A, B *Terracotta statuettes of cupids.* New York, Metropolitan Museum of Art, No. 06.1130 and 1131. Photograph by courtesy of the Metropolitan Museum of Art

21C *Melian relief.* London, British Museum, No. B 364. Photograph by courtesy of the Trustees

22A *Dish of* sgraffiato *ware.* New York, Metropolitan Museum of Art, No. 38.40.137. Photograph by courtesy of the Metropolitan Museum of Art

22B, C *Drug-jars.* From T. W. ARNOLD and A. GUILLAUME. 'The Legacy of Islam', Pl. VI, figs 27, 28. Oxford, Clarendon Press, 1931

22D *Painted screw-topped bottle.* London, Victoria and Albert Museum. Photograph Crown copyright reserved

22E *A tin-glazed plate.* From C. H. MORGAN. 'Corinth', Vol. 11, Pl. XXXIVb. Cambridge, Mass., Harvard University Press for American School of Classical Studies at Athens, 1942

23A *A large dish made at Valencia.* From T. W. ARNOLD and A. GUILLAUME. 'The Legacy of Islam', Pl. VI, fig. 29. Oxford, Clarendon Press, 1931

23B *Majolica plate from Caffagiolo.* London, Victoria and Albert Museum. Photograph Crown copyright reserved

23C *Majolica dish from Deruta.* London, Victoria and Albert Museum. Photograph Crown copyright reserved

24A *Goblet of blue-green faience.* Oxford, Ashmolean Museum. Photograph by permission of the Visitors

24B *Kohl-pot.* Oxford, Ashmolean Museum. Photograph by permission of the Visitors

24C *Green-glazed pottery jar.* Oxford, Ashmolean Museum. Photograph by permission of the Visitors

24D *Blue glass vase.* London, British Museum, No. 47620. Photograph by permission of the Trustees

24E *Cane inlay fragment.* London, British Museum, No. 29396. Photograph by permission of the Trustees

24F *Jar, green 'frit' ware.* Oxford, Ashmolean Museum. Photograph by permission of the Visitors

24G *Cup in lead-glazed pottery.* Oxford, Ashmolean Museum. Photograph by permission of the Visitors

25A *Greenish bowl.* Brussels, Musées Royaux d'Art et d'Histoire, No. 2282. Photograph by courtesy of the Museum

25B *Bowl with gilt decoration between two layers.* London, British Museum. Photograph by permission of the Trustees

25C *Colourless lid.* London, British Museum, No. 72.7.26.6. Photograph by permission of the Trustees

25D *Green bowl, from Atfik, Egypt.* London, British Museum, Oriental Department. Photograph by permission of the Trustees

25E *Greenish bowl, Frankish.* Oxford, Ashmolean Museum. Photograph by permission of the Visitors

26A *The 'Portland' cameo vase.* London, British Museum. Photograph by permission of the Trustees

26B *Colourless beaker.* London, British Museum, Slade 919. Photograph by permission of the Trustees

26C, D *Top and side views of Sargon II vase.* London, British Museum. Photographs by permission of the Trustees

26E *Colourless beaker with engraved figures, from Worringen.* Toledo, Museum of Art. Photograph by courtesy of the Museum

27A *Opaque white jug, with painted scene of Apollo and Daphne.* Ray Winfield Smith Collection, on permanent loan to the Corning Museum of Glass. Photograph by courtesy of Mr Ray W. Smith

27B *Bottom of a bowl with gilt decoration between two layers.* Oxford, Pusey House

27C *Ribbed bowl.* London, British Museum, Woodhouse Collection 66.5–4.43. Photograph by permission of the Trustees

27D *Mould-blown beaker.* London, British Museum, No. 73.5–2.209. Photograph by permission of the Trustees

27E *Colourless jug, with nipped ribbing, from Colchester.* London, British Museum. Photograph by permission of the Trustees

27F *Amber jug.* London, British Museum, Slade 85. Photograph by permission of the Trustees

28 (A–E) *Portions of bowls.* (F–I) *Portions of vessels.* A–D, F–I, London, British Museum, Print Room. By courtesy of the Trustees; E, photograph by courtesy of the Royal Museum, Pola

29A *'Street of Balconies.'* Copyright photograph by Christopher Brunel, Gerrards Cross, Bucks.

29B *A 'peristyle', or colonnaded courtyard.* From A. MAIURI. 'Pompeii', Pl. LXXIV. Novara, Istituto Geografico de' Agostini, 1951

30A *A medieval artist's conception of Noah building the Ark.* London, British Museum, Add. MS. 18820, fol. 12r. Photograph by courtesy of the Trustees

30B *A nobleman discusses the progress of builders.* London, British Museum, Add. MS. 19720, fol. 27. Photograph by courtesy of the Trustees

31A *Building the Tower of Babel.* London, British Museum, Add. MS. 18850, fol. 17v. Photograph by courtesy of the Trustees

31B *A remarkable picture of churches.* Vienna, Öst. Nationalbibliothek, MS. 2549, fol. 164r. From F. WINKLER. 'Die flämische Buchmalerei des XV. und XVI. Jahrhunderts', Pl. XIV. Leipzig, Seemann, 1925

32A *The Ardagh chalice.* Dublin, National Museum of Ireland. Photograph by courtesy of the Director

32B *The* baptistère *of St Louis.* Paris, Louvre. Photograph by courtesy of Dr D. S. Rice

33 *The King's Lynn cup.* King's Lynn Corporation. Photograph by courtesy of P. M. Goodchild and Sons, King's Lynn

34A *Roman bridge.* Copyright photograph Anderson, Rome

34B Via di Nola. From A. MAIURI. 'Pompeii', fig. 12a. Novara, Istituto Geografico de Agostini, 1951

35 *Winged Victory.* Paris, Louvre. Copyright photograph Archives Photographiques, Paris

36 *Richly ornamented harness.* Ghent, Altarpiece at St Bavo. From L. BALDASS. 'Jan Van Eyck', Pl. XLIV (right). London, Phaidon Press, 1952

37A *Remains of one of the Dejbjerg wagons.* Copenhagen, Nationalmuseet. Photograph by courtesy of the Director

37B *Galatea rides a wheel-propelled shell.* London, Wallace Collection, Cistern No. C. III. 100. Reproduced by permission of the Trustees of the Wallace Collection

38A *Street scene.* Munich, Bayerische Staatsbibliothek. Cod. gall. 369, fol. 10. Photograph by courtesy of the Director

38B *The large crane at Bruges.* Munich, Bayerische Staatsbibliothek, Cod. lat. 23638, fol. 11v. Photograph by courtesy of the Director

39 *Intersection of five aqueducts.* Munich, Deutsches Museum, Painting by Zeno Diemer. Photograph Deutsches Museum, München

40A *Section of polygonal masonry wall at Alatri.* Copyright photograph Alinari, Florence

40B *Aerial photograph of a section of Hadrian's wall.* Photograph by J. K. St Joseph. Crown copyright reserved—by permission of Air Ministry

41A *The* Porta Palatina *at Turin.* Copyright photograph Alinari, Florence

41B *The Roman* Porta Nigra *at Treves*

42A *Alchemical laboratory.* London, British Museum, Print Room 1870-6-25-660. Photograph by courtesy of the Trustees

42B *Picture by Stradanus.* London, British Museum, Print Room 157* a. 38, fol. 193. Photograph by courtesy of the Trustees

43A *A stylized representation by Teniers.* London, British Museum, Print Room 1856-7-12-375. Photograph by courtesy of the Trustees

43B *Priestley's laboratory.* From J. PRIESTLEY. 'Experiments and Observations on different kinds of Air', Vol. 1, Pl. facing last page. London, 1774

44 *Mosaic of the figure of an angel.* Germigny-des-Prés (Loiret), nr Orleans. Copyright photograph Archives Photographiques, Paris

TABLES

Chronological I	*1400–1 B.C.*	lv
Chronological II	*Survey of Roman Imperial History 31 B.C.–A.D. 476*	lvi
Chronological III	*A.D. 200–1500*	lviii
Metallurgy I	*Comparison of medieval Iron-smelting Techniques*	71
Metallurgy II	*The Smelting-works of the Siegen Area*	74
Glass Analyses		313
Transmission of certain Techniques from China to the West		770

ABBREVIATIONS OF PERIODICAL TITLES

(AS SUGGESTED BY THE WORLD LIST OF SCIENTIFIC PERIODICALS)

Abh. Heidelberg. Akad. Wiss., phil.-hist. Kl.	Abhandlungen der Heidelberger Akademie der Wissenschaften, philosophisch-historische Klasse
Abh. preuss. Akad. Wiss., phil.-hist. Kl.	Abhandlungen der preussischen Akademie der Wissenschaften, philosophisch-historische Klasse. Berlin
Ambix	Ambix. Journal of the Society for the Study of Alchemy and Early Chemistry. London
Amer. J. Archaeol.	American Journal of Archaeology. Archaeological Institute of America. Cambridge, Mass.
Amer. Nat.	American Naturalist. Lancaster, Pa.
Amtl. Ber. preuss. Kunstsamml.	Amtliche Berichte aus den Königlichen Kunstsammlungen *in* Jahrbuch der preussischen Kunstsammlungen. Berlin
Ancient Egypt	Ancient Egypt and the East. British School of Egyptian Archaeology. London
Angew. Chem.	Angewandte Chemie. See *Z. angew. Chem.*
Ann. Archaeol. Anthrop.	Annals of Archaeology and Anthropology. University of Liverpool, Institute of Archaeology. Liverpool
Ann. Inst. Étud. orient.	Annales de l'Institut d'Études orientales. Faculté des Lettres de l'Université d'Alger. Paris
Ann. Lavori pubbl.	Annali dei Lavori pubblici. [Continuation of *Giornale del genio civile.*] Rome
Ann. Serv. Antiq. Égypte	Annales du Service des Antiquités de l'Égypte. Cairo. (Presses Universitaires de France, Paris)
Annu. Brit. Sch. Athens	Annual of the British School in Athens. London
Antiq. J.	Antiquaries Journal. The Journal of the Society of Antiquaries of London. London
Antiquity	Antiquity. A Quarterly Review of Archaeology. Edwards, Newbury, Berks.
Anz. Schweiz. Altertumsk.	Anzeiger für Schweizerische Altertumskunde. [Continued as *Zeitschrift für Schweizerische Archäologie und Kunstgeschichte.*] Schweizerisches Landesmuseum. Zürich
Arch. Eisenhüttenw.	Archiv für das Eisenhüttenwesen. Fachberichte des Vereins Deutscher Eisenhüttenleute und des Max-Plank Instituts für Eisenforschung. (Ergänzung zu 'Stahl und Eisen'.) Düsseldorf
Arch. Hyg., Berl.	Archiv für Hygiene (und Bakteriologie). Munich and Berlin
Arch. int. Hist. Sci.	Archives internationales d'Histoire des Sciences. Publication trimestrielle de l'Union Internationale d'Histoire des Sciences. Paris

Archäol. Anz.	Archäologischer Anzeiger. Beiblatt zum Jahrbuch des Deutschen Archäologischen Instituts. Berlin
Archæol. J.	The Archæological Journal. Royal Archaeological Institute of Great Britain and Ireland. London
Archaeol.-kunsthist. Medd.	Archaeologisk-Kunsthistoriske Meddelelser. Kongelige Danske Videnskabernes Selskab. Copenhagen
Archaeologia	Archaeologia or Miscellaneous Tracts relating to Antiquity. Society of Antiquaries. London
Athen. Mitt.	Mitteilungen des Deutschen Archäologischen Instituts in Athen. Berlin
Atti Accad. Archeol. Lett. Arti Napoli	Atti della Accademia di Archeologia, Lettere e Belle Arti. Società Nazionale di Scienze, Lettere ed Arti in Napoli. Naples
Atti Mem. Soc. Magna Grecia	Atti e Memorie della Società Magna Grecia. Rome
Ber. dtsch. Keram. Ges.	Berichte der Deutschen Keramischen Gesellschaft. Berlin
Ber. Sächs. Ges. (Akad.) Wiss., phil.-hist. Kl.	Bericht über die Verhandlungen der Königlich Sächsischen Gesellschaft (Akademie) der Wissenschaften zu Leipzig. Philosophisch-historische Klasse
Berks. Archaeol. J.	Berkshire Archaeological Journal. Berkshire Archaeological Society. Reading
Bgham archaeol. Soc. Trans. Proc.	Birmingham Archaeological Society Transactions and Proceedings. Oxford University Press, London
Bibl. Éc. franç. Athènes Rome	Bibliothèque des Écoles françaises d'Athènes et de Rome
Bibl. Éc. haut. Étud.	Bibliothèque de l'École des hautes Études. Paris
Boll. Staz. sper. Pelli Mat. conc.	Bollettino ufficiale. Reale Stazione sperimentale per l'Industria delle Pelli e delle Materie concianti. Naples
Bonn. Jb.	Bonner Jahrbücher des Rheinischen Landesmuseums und des Vereins von Altertumsfreunden im Rheinlande
Brit. numism. J.	British Numismatic Journal and Proceedings of the British Numismatic Society. London
Bull. corr. hell.	Bulletin de Correspondănce Hellénique. École Française d'Athènes. Paris
Bull. Hist. Med.	Bulletin of the History of Medicine. Johns Hopkins University. Baltimore, Md.
Bull. Metrop. Mus.	Bulletin of the Metropolitan Museum of Art, New York
Bull. Mus. Far East Antiq.	Bulletin of the Museum of Far Eastern Antiquities. Stockholm
Bull. Mus. R. Art Hist.	Bulletin des Musées Royaux d'Art et d'Histoire. Brussels
Bull. N.Y. Acad. Med.	Bulletin of the New York Academy of Medicine. New York
Cah. archéol.	Cahiers archéologiques: Fin de l'Antiquité et Moyen Âge. Paris
Chymia	Chymia. Annual Studies in the History of Chemistry. Philadelphia
Ciba Rev.	Ciba Review. CIBA Limited, Basel

Ciba-Rdsch.	Ciba-Rundschau. CIBA Limited, Basel
Ciba-Z.	Ciba-Zeitschrift. CIBA Limited, Basel
Class. J.	The Classical Journal. Classical Association of the Middle West and South. Chicago
Class. Philol.	Classical Philology: a quarterly journal devoted to research in the languages, literature, history and life of classical antiquity. University of Chicago, Chicago, Ill.
Class. Rev.	Classical Review. Classical Association. Great Britain
Class. Wkly	The Classical Weekly. The Classical Association of the Atlantic States. Pittsburgh, Pa.
Collegium, Haltingen	Collegium. Zentralorgan des Internationalen Vereins der Lederindustrie-Chemiker. Haltingen
C. R. Acad. Sci., Paris	Comptes Rendus hebdomadaires des Séances de l'Académie des Sciences. Paris
Cymmrodor	Y Cymmrodor. Hon. Society of Cymmrodorion. London
Econ. Geol.	Economic Geology and the Bulletin of the Society of Economic Geologists. A semi-quarterly journal. New Haven, Conn.
Egypt. Res. Acc. and Brit. Sch. Archaeol. Egypt	Egyptian Research Account and British School of Archaeology in Egypt. British School of Egyptian Archaeology. London
Eurasia Septentrionalis Antiqua	Eurasia Septentrionalis Antiqua. Journal for East European and North-Asiatic archaeology and ethnography. Archaeological Society of Finland. Helsinki
Forsch. Fortschr. dtsch. Wiss.	Forschungen und Fortschritte. Korrespondenzblatt der Deutschen Wissenschaft und Technik. Berlin
Gas- u. Wasserfach	Gas- und Wasserfach. Journal für Gasbeleuchtung und Wasserversorgung. Munich
Geogr. J.	Geographical Journal. Royal Geographical Society. London
Germania	Germania. Korrespondenzblatt der Römisch-Germanischen Kommission. Deutsches Archäologisches Institut. Berlin
Gesnerus	Gesnerus. Vierteljahrsschrift für Geschichte der Medizin und der Naturwissenschaften. Schweizerische Gesellschaft für Geschichte der Medizin und Naturwissenschaften. Zürich
Glass Ind.	Glass (and Ceramic) Industry. Moscow
Glastech. Ber.	Glastechnische Berichte. Frankfurt a. M.
Hammaburg	Hammaburg. Vor- und frühgeschichtliche Forschungen aus dem niederelbischen Raum. Museum für Hamburgische Geschichte und Hamburger Vorgeschichtsverein. Hamburg
Handb. Altertumswiss.	Handbuch der Klassischen Altertumswissenschaft begründet von Iwan von Müller. [Continued as Handbuch der Altertumswissenschaft.]
Hermeneus	Hermeneus. Maandblad voor de antieke Cultuur. Zwolle

Hermes	Hermes. Zeitschrift für Classische Philologie. Berlin
Hesperia	Hesperia. Journal of the American School of Classical Studies at Athens
History	History. Journal of the Historical Association. London
Ill. Stud. social Sci.	Illinois Studies in the Social Sciences. Illinois University. Urbana
Iraq	Iraq. British School of Archaeology in Iraq
Isis	Isis. An International Review of the History of Science. History of Science Society, Washington University, U.S.A.
Istanbuler Mitt.	Istanbuler Mitteilungen. Deutsches Archäologisches Institut, Abteilung Istanbul
J. asiat.	Journal asiatique. Paris
J. chem. Educ.	Journal of Chemical Education. Division of Chemical Education of the American Chemical Society. Easton, Pa.
J. Cuneiform Stud.	Journal of Cuneiform Studies. American Schools of Oriental Research. New Haven, Conn.
J. econ. Hist.	The Journal of Economic History. Economic History Association. New York
J. Egypt. Archaeol.	Journal of Egyptian Archaeology. Egypt Exploration Society. London
J. Hell. Stud.	The Journal of Hellenic Studies. Journal of the Society for the Promotion of Hellenic Studies. London
J. Hist. Ideas	Journal of the History of Ideas. A quarterly devoted to cultural and intellectual history. New York
J. R. Anthrop. Inst.	Journal of the Royal Anthropological Institute of Great Britain and Ireland. London
J. R. cent. Asian Soc.	Journal of the Royal Central Asian Society. London
J. R. micr. Soc.	Journal of the Royal Microscopical Society. London
J. R. Soc. Antiq. Ireland	Journal of the Royal Society of Antiquaries of Ireland. Dublin
J. Rom. Stud.	Journal of Roman Studies. Society for the Promotion of Roman Studies. London
J. Soc. Glass Tech.	Journal of the Society of Glass Technology comprising Proceedings and Reports, Transactions and Abstracts of Papers from other Journals. Sheffield
Jb. dtsch. Archäol. Inst.	Jahrbuch des Deutschen Archäologischen Instituts. Berlin
Jb. kunsthist. Samml. Allerh. Kaiserhaus.	Jahrbuch der Kunsthistorischen Sammlungen des Allerhöchsten Kaiserhauses. Vienna
Jh. Öst. Archäol. Inst., Wien	Jahreshefte des Österreichischen Archäologischen Instituts. Vienna
K. Vitterhets Hist. Antik. Handl.	Kungl. Vitterhets Historie och Antikvitets Akademiens Handlingar. Stockholm
Library	The Library. A quarterly review of bibliography. Transactions of the Bibliographical Society, second series. London

Man	Man. A monthly Record of Anthropological Science. Royal Anthropological Institute. London
Med. J. Rec.	Medical Journal and Record. New York
Medd. Lund Univ. Hist.	Meddelanden från Lunds Universitets Historiska Museet *in* Arsberättelse. Kungliga Humanistiska Vetenskapssamfundet i Lund
Meded. vlaamsche Acad., Kl.Wet.	Mededeelingen van de Koninklijke vlaamsche Academie, Klasse der wetenschappen. Brussels
Mem. Asiat. Soc. Beng.	Memoirs of the (Royal) Asiatic Society of Bengal. Calcutta
Métaux et Civil.	Métaux et Civilisations. Les métaux dans l'histoire, les techniques, les arts. Éditions Métaux, Saint-Germain-en-Laye
Metrop. Mus. Occas. Pap.	Occasional Papers of the Metropolitan Museum of Art. New York
Min. & Metall., N.Y.	Mining and Metallurgy. American Institute of Mining and Metallurgical Engineers. New York
Mitt. Gesch. Med. Naturw.	Mitteilungen zur Geschichte der Medizin und der Naturwissenschaften (und der Technik). Hamburg and Leipzig
Mon. ant.	Monumenti Antichi. Pubblicati per cura della Accademia Nazionale dei Lincei. Rome
Mon. Germ. hist.	*Monumenta Germaniae historica.* Preussische Akademie der Wissenschaften
Nat. geogr. Mag.	The National Geographic Magazine. National Geographic Society. Washington
Nature	Nature. London
Nature, Paris	La Nature. Revue des sciences et de leurs applications à l'art et à l'industrie. Paris
Norfolk Archaeol.	Norfolk Archaeology. Norfolk and Norwich Archaeological Society. Norwich
Numism. Chron.	The Numismatic Chronicle and Journal of the Royal Numismatic Society. London
Osiris	Osiris. Studies on the history and philosophy of science, and on the history of learning and culture. Bruges
Pat. lat.	*Patrologia Latina. Cursus completus patrologiae*, ed. by J. P. MIGNE. Series Latina, 221 vols. Paris, 1857–79
Pays Gaumais	Pays Gaumais. Musée Gaumais. Virton, Luxembourg
Philol. Wschr.	Philologische Wochenschrift. Leipzig
Proc. Brit. Acad.	Proceedings of the British Academy. London
Proc. Geol. Ass., Lond.	Proceedings of the Geologists' Association. London
Proc. R. Irish Acad.	Proceedings of the Royal Irish Academy. Dublin
Proc. Soc. Antiq. Lond.	Proceedings of the Society of Antiquaries of London. London
Proc. Soc. Antiq. Scotld	Proceedings of the Society of Antiquaries of Scotland. Edinburgh

Quart. Dept. Antiq. Palest.	Quarterly of the Department of Antiquities in Palestine. Government of Palestine. Oxford University Press, London
Quell. Gesch. Naturw.	Quellen und Studien zur Geschichte der Naturwissenschaften und der Medizin
Quell. Kunstgesch. Kunsttechn.	Quellenschriften für Kunstgeschichte und Kunsttechnik des Mittelalters und der Renaissance. Ed. by R. EITELBERGER VON EDELBERG, continued by A. ILG *et al.* Vienna
R. C. Ist. Lombardo Sci. Lett.	Rendiconti dell' Istituto Lombardo di Scienze e Lettere. Milan
R. C. Pont. Accad. Archeol.	Atti della Pontificia Accademia Romana di Archeologia. Serie 3: Rendiconti. Rome
Rec. Bucks.	Records of Buckinghamshire. The Journal of the Architectural and Archaeological Society for the County of Buckingham. London and Bedford
Res. Rep. Soc. Antiq.	Reports of the Research Committee of the Society of Antiquaries of London
Rev. archéol.	Revue archéologique fondée en 1844. Presses Universitaires, Paris
Rev. d'Assyriol.	Revue d'Assyriologie et d'Archéologie orientale. Presses Universitaires, Paris
Rev. belge Archéol. Hist. Art	Revue belge d'Archéologie et d'Histoire de l'Art. Académie Royale d'Archéologie de Belgique. Brussels
Rev. Philol.	Revue de Philologie, de Littérature et d'Histoire anciennes. Klincksieck, Paris
Rheinisches Museum.	Rheinisches Museum für Philologie. Frankfurt a. M.
Röm. Mitt.	Mitteilungen des Deutschen Archäologischen Instituts in Rom. Berlin
Saalburgjb.	Saalburgjahrbuch. Bericht des Saalburgmuseums
S. B. heidelberg. Akad. Wiss., Phil.-hist. Kl.	Sitzungsberichte der heidelberger Akademie der Wissenschaften, Philosophisch-historische Klasse
Skr. svenska Inst. Rom	Skrifter utgivna av Svenska Institutet i Rom. Lund
Speculum	Speculum. A journal of mediaeval studies. Mediaeval Academy of America. Cambridge, Mass.
Stahl u. Eisen, Düsseldorf	Stahl und Eisen. Zeitschrift für das deutsche Eisenhüttenwesen. Verein Deutscher Eisenhüttenleute. Düsseldorf
Studi Etruschi	Studi Etruschi. Istituto di Studi Etruschi ed Italici. Florence
Studi Sardi	Studi Sardi. Reale Università. Istituto per gli Studi Sardi. Cagliari
Surrey archaeol. Coll.	Surrey Archaeological Collections relating to the History and Antiquities of the County. Surrey Archaeological Society. Guildford
Syria	Syria. Revue d'art oriental et d'archéologie. Institut Français d'Archéologie de Beyrouth. Paris

Tech. Stud. fine Arts	Technical Studies in the Field of the Fine Arts. Fogg Art Museum, Harvard University. Cambridge, Mass.
Tech. et Civil.	Techniques et Civilisations. Éditions Métaux, Saint-Germain-en-Laye
Technikgeschichte	Technikgeschichte. Jahrbuch des Vereins Deutscher Ingenieure. Berlin
Trans. Bristol archaeol. Soc.	Transactions of the Bristol and Gloucestershire Archaeological Society. Bristol
Trans. Cumberland Antiq. Soc.	Transactions of the Cumberland and Westmorland Antiquarian and Archaeological Society. Carlisle
Trans. Newcomen Soc.	Transactions. Newcomen Society for the Study of the History of Engineering and Technology. London
Trans. Proc. Amer. philol. Ass.	Transactions and Proceedings of the American Philological Association. University of Wisconsin. Madison, Wis.
Trans. St. Albans and Herts. archit. and archaeol. Soc.	Transactions of the St. Albans and Herts. Architectural and Archaeological Society. St. Albans, Municipal Library
Trans. Soc. Glass Tech.	See *J. Soc. Glass Tech.*
Trierer Z.	Trierer Zeitschrift für Geschichte und Kunst des Trierer Landes und seiner Nachbargebiete. Rheinisches Landesmuseum. Treves
Umschau	Die Umschau. Halbmonatsschrift über die Fortschritte in Wissenschaft und Technik. Frankfurt a. M.
Wasserkraft, Münch.	Wasserkraft (und Wasserwirtschaft). Munich
Wheat Stud. Stanf. Univ.	Wheat Studies. Food Research Institute, Stanford University. Palo Alto, California
Z. angew. Chem.	Zeitschrift für angewandte Chemie und Zentralblatt für technische Chemie (continued as *Angewandte Chemie*). Gesellschaft Deutscher Chemiker. Leipzig, *now* Grünberg, Hesse
Z. Assyriol.	Zeitschrift für Assyriologie und verwandte Gebiete. De Gruyter, Berlin and Leipzig
Z. Ver. dtsch. Ing.	Zeitschrift des Vereins Deutscher Ingenieure. Düsseldorf

HISTORICAL NOTES

A. R. HALL

In this volume the history of technology is traced from a period in which the Greek city-state was rising from the ruins of the Mycenaean culture, to another in which the modern European nation-state was attaining self consciousness. In the intervening 2000 years civilization and political dominance moved westwards, while the technical superiority possessed by the ancient empires of the Near East was similarly transferred, through the Graeco-Roman world, to the formerly barbarous peoples of the west. This great technical tradition, whose continuity is perhaps most clearly manifest in the more delicate techniques (chs 8, 9, 13) was further enriched in four important ways. By about the tenth century, when the restoration of European civilization after the collapse of the Roman Empire was well begun, it already incorporated elements derived from Greek, and to a less extent Roman, craftsmanship, to which were added a number of improvements introduced by the Germanic tribes. Mainly after the tenth century many important innovations came from the Far East, and medieval Europe itself entered upon a period of almost unbroken technical invention.

It is impossible within the limits of this volume to describe the historical processes by which Celtic and Germanic peoples became the indirect heirs of Hellas, by which communication was fitfully maintained and often broken between the alien civilizations of the Far East and that of Christendom, and by which in the Mediterranean world techniques, goods, and ideas flowed across the barrier created by irreconcilable creeds. To trace adequately the evolution of the European cultural unity at the close of the Middle Ages, now detached from the ancient foci of learning and craft-skill in Egypt, Mesopotamia, and Greece, it would be necessary to trespass into religious, political, and economic history. The chronological tables on pp lv–lix seek to provide some indication of the time-relations of events in these other branches of history, and they may thus help to elucidate allusions to matters that cannot be fully explained in the text. A more complete historical background may be sought in well-known works easily available in public libraries.

If we view two millennia of history from a European perspective, the principal phases emerge clearly. The first is pre-eminently Greek (table I). While the tribes of western Europe—even the Latins—were slowly advancing through the Iron Age, the Greeks, heirs of the Minoan civilization, profited greatly from their close links with yet older cultures in the Fertile Crescent and Egypt. Repulsing the onslaught of the Persian Empire, they developed their own political and social organization in the city-state, which reached maturity in the fifth and fourth centuries B.C. Athens at this time shows the highest appreciation of art and thought, a full and conscious political experience, and the dependence of a rich and sophisticated society upon trade and manufacture.

Knowledge of Greek technology, of Hellenism in all its aspects, was spread in three chief ways. Alexander's empire, replacing the state of endless war among a multitude of small states, carried Greek culture to Persia and the Indus. Barbarian love of ostentation drew Greek products into the Danubian basin and even to the Baltic. And the Greeks themselves, under constant economic pressure in their mountainous homeland, migrated to colonies established along the whole shore of the western Mediterranean, thus emulating the Phoenicians in outlining trade-routes that have lasted until recent times.

Yet, to most of Europe, Greek civilization came not directly but through the conquests of Rome. Here again the history of technology is an important factor in the rise and fall of the great empire that was born in the west and died in the east (table I and II). Rome flourished upon the strength of its army, and declined when that army could no longer defend an immense frontier. Until the second century B.C. the aggressive spirit of the Latin tribes was largely confined to the Italian peninsula, where they absorbed the culture of the Greek and Etruscan cities that they later conquered. Then simultaneously, about 140 B.C., Rome extended its power to Greece and to the Carthaginian dominion in north Africa and Spain. Thenceforward the legions marched undefeated through Egypt and the Near East, Gaul, and Britain, and even into Germany beyond the Rhine and the Danube.

Thus was created a single vast area in which men and goods moved fairly freely, having a certain cultural lowest common denominator for its inhabitants whether they lived by the Thames, the Tiber, or the Tigris. It was so vast and rich that its influence was far wider than its political limits, reaching from Scandinavia to southern India. Under Rome these remote regions were better known to western scholars than at any time till comparatively recently (Epilogue). The originally frugal Roman citizen enjoyed the fruits of his valour; uninventive in the arts of civilization himself, he borrowed from the Greeks whom he patronized, often as slaves. Through the political ubiquity of Rome a measure of Hellenism spread, with varying degrees of dilution and debasement, over a large part of the civilized world and over much of Europe that was formerly counted barbarous. Even the tribes beyond the Roman frontiers were partially hellenized.

Despite great opportunities for trade, within and without the Empire, and the huge consumption of Rome itself and other great cities, besides that of the imperial armies, there was no 'industrial revolution'. The reasons for this circumstance were social rather than technical. In the end, technological inadequacy was at least partly responsible for the economic collapse of the third century A.D.; for Roman life was ever supported by the remorseless labour of the many and the costly manual dexterity of the few.

By the third century the Roman Empire had failed economically, politically, and militarily. The virile, extroverted Latin had succumbed to Greek elegance and oriental religion. Neither Diocletian nor Constantine could restore the ancient grandeurs. The third phase of history begins with the inrush of the first barbarians—the Goths and Vandals—and ends with that of the last—the Northmen. Yet however great the ethnographic displacement, vestiges of Hellenism remained. Latin was still the language of religion and learning in the west, and the spiritual leadership of the Roman Church was the greatest unifying factor there. With Charlemagne's coronation on Christmas Day 800 the surviving ideal of a western Christian empire, of political and cultural unity throughout Europe, was fulfilled. In the east the Greeks at Byzantium held a considerable but diminishing fragment of the Roman state intact until 1453, a fragment divided by self-interest, by religion, and by language from the west, but nevertheless an important intermediary. For it was partly through Byzantium that the technical and learned traditions of the past returned to Europe.

The third phase (table III), the immediate post-Roman period, is devoid of great material achievements. Its majestic political experiment, Charlemagne's empire, failed within half a century. It saw the meteoric rise of Islam, with its conquests in the Near East, north Africa, and Spain dividing the 'new world' of Europe from the old world of the eastern Mediterranean. Yet the early Middle Ages also witnessed the Christianization of Europe, the restoration of some measure of political and economic stability, and the beginnings of a new social order. All this, particularly the disappearance of slavery as a vital economic institution, was of importance to the development of technology. In medieval Europe the great majority of producers were free or partly so. Their initiative in part contributed to the great extension of the area of cultivated land continuing throughout the Middle Ages.

In the fourth phase, from about the tenth to the fifteenth century, medieval civilization reached its highest level. It was not a period of tranquillity, political or intellectual, yet the tensions of the age were productive of great achievements. Europe was disturbed and divided by conflicts between emperors and popes, by international wars, by the efforts of kings and princes to establish effective central governments in the face of feudal opposition. There were revolts of the peasant masses against the economic and social inequalities imposed upon them, and cruel suppressions of religious heterodoxy. For two centuries the west threw its military strength intermittently upon the Islamic world, and at last upon Byzantium, in the crusades. Moreover the 'new learning'—Greek science, medicine, and philosophy, derived at first from Islamic sources—seemed for a time likely to challenge ecclesiastical dominance in the realm of thought before it was assimilated into the medieval Christian tradition. But, despite poverty, plague, and war, populations slowly increased in number, and urban life advanced in security and amenity. Craftsmen especially enjoyed an increasing measure of privilege and liberty, jealously protected by their guilds. With the revival of town life there arose a form of mercantile autonomy as a by-product of the local particularism innate in feudalism, flourishing strongly in the Hanse towns, the German Free Cities, and the north Italian city-states, all of which played major roles in the economic life of the later Middle Ages. Growing international commerce lessened the need for local self-sufficiency. The release of a tremendous creative energy is manifest in the building of churches and the thronging of students to the universities, in great works of literature and philosophy, and in the beginnings of technical invention.

Here we may pause, with the failure of the medieval ideal that marks the fifteenth century. The Holy Roman Empire was now a political nonentity, and the Church had but recently surmounted a long-drawn crisis in which its spiritual prestige was gravely lessened. France and England were prostrated by the last stages of the Hundred Years War. Byzantium was an indefensible enclave in a Turkish empire stretching through the Balkans and obstructing Europe's route to the Near and Far East. Portuguese sailors were still groping down the west coast of Africa and towards the Azores with results as yet unspectacular. In Italy, however, there were clear signs of the emergence of a new social order and a new learning. The wealthy merchant-class, assuming leadership in its city-states, laid the economic foundations of a cultivated and secular society, encouraging that emulation of Greek achievements in art and literature which culminated in the sixteenth century. In Italy, too, was most fully manifest the possession of that technological skill whose development is traced in this volume. Europe now excelled in the use of power, in the crafts of metal-working, weaving, building, and ship-construction, in chemical arts and agricultural methods, and in an economic organization offering to technological skill a full measure of activity and reward. With these material advantages its modern history begins.

TABLE I. 1400–1 B.C.

DATE B.C.	GREECE AND THE AEGEAN	THE NEAR EAST	ITALY AND W. EUROPE	DATE B.C.
1400	Destruction of Minoan culture			1400
1300	Settlement of Achaean Greeks		Late Bronze Age 'Terramare' settlements in Italy	1300
1200	Trojan War [?]	Expulsion of Egyptians from Asia		1200
1100	*Proto-Geometric Period* (c. 1100–900): Dorian influence begins			1100
1000	Beginnings of Greek settlement in Asia Minor and Aegean Islands		Iron metallurgy begins in Austria; *'Villanova' Culture in Italy* and *Iron Age I: Hallstatt Period* (c. 1000–600)	1000
900	*Geometric Period* (c. 900–650)	Hegemony of Tyre in Phoenicia; Disruption of the Egyptian Kingdom and series of Civil Wars	Arrival of Etruscans in Italy	900
800		Carthage founded		800
	Period of Homer and formation of Greek alphabet	Age of Assyrian Conquests (Shalmaneser III, 859–824; Ashurbanipal III, 668–625)	Rome founded 753 (mythical) Greek colonization of Sicily and S. Italy	
700	First colonizations of W. Mediterranean	Egypt under Assyrian rule		700
600	Beginnings of Greek philosophy and science. Solon's laws at Athens (594)	Neo-Babylonian Empire (Nebuchadrezzar 605–561); Saitic Renaissance in Egypt	Celtic invasions of N. Italy	600
500	*Zenith of Greek city-state* (c. 500–350)	Persian Empire universal (Cyrus the Great, 550–530)	Roman Republic founded	500
	Peloponnesian Wars end in ascendancy of Sparta. Art and philosophy flourish in Athens (Plato, 429–347, Aristotle, 384–321)	Persians defeated by Greeks at Marathon (490), Salamis (480)	Greek and Carthaginian colonization of S. France and E. Spain. Carthage establishes control of W. Mediterranean and Spain	
		Expedition of Cyrus fails at Cunaxa		
400			Iron Age begins in Britain; *Iron Age II: La Tène Period* (c. 400–1)	400
	Alexander's Empire, 337–323	Alexander destroys Persian Empire (334–331) occupies Egypt and marches to India	Samnite wars	
300		Ptolemy I (323); *Ptolemaic Period in Egypt* (323–30)	Rome supreme in central Italy	300
	Period of Successor states	Seleucid Empire extends from Bosphorus to Persian Gulf; Zenith of Library at Alexandria	Rome occupies Sicily except Syracuse; *Punic Wars*	
200	Rome intervenes in Greece		Roman conquest of Spain	200
	Greece subjected to Roman rule (146)		Destruction of Carthage	
100		Roman wars in Asia	La Tène culture in Britain	100
		Four provinces organized by Pompey (65–62); Cleopatra VII (47–30). Egypt becomes a Roman province	Crisis in Roman government. Caesar becomes sole ruler (48), having conquered Gaul (58–51) and visited Britain (54)	
1				1

TABLE II

CHRONOLOGICAL SURVEY OF ROMAN IMPERIAL HISTORY

31 B.C.–A.D. 476

B.C.

31	Octavian (AUGUSTUS), Emperor (27), reorganizes the government. Greek influences in literature, art, and science, extend in Roman life and thought. Time of Virgil (70–19 B.C.), Horace (65–8 B.C.), Ovid (43 B.C.–A.D. 17), Livy (59 B.C.–A.D. 17), and Vitruvius (first century B.C.).
4	Birth of Christ.

A.D.

4–6	Roman frontier advanced to the Danube.
9	The Rhine adopted as the frontier in Germany after failure to reach the Elbe.
43–51	Invasion of Britain under CLAUDIUS (41–54).
54–68	Christian persecution under NERO.
68–69	Emperors elected by the legions. VESPASIAN (69–79) victorious.
77–84	Conquest of Britain extended into Scotland.
79	Eruption of Vesuvius buries Pompeii and Herculaneum.
83	DOMITIAN (81–96) campaigns across the Rhine; begins construction of the *limes* (figure 653).
101–7	TRAJAN (98–117); Empire reaches its greatest extent (figure 459). Signs of breakdown of municipal administration; financial organization of the Empire severely strained.
132–5	Revolt of the Jews, followed by their dispersion.
161–80	MARCUS AURELIUS, the Stoic Emperor. Wars with the barbarians on Danubian frontier.
200	During the third century the government and economic life of the Empire collapse. The military basis of political poweris openly revealed, and Emperors become creatures of their armies. Ruinous inflation. Intellectual and artistic endeavour cease.
193–211	Under SEPTIMIUS SEVERUS the Senate is humiliated and permanent settlement of the legions recognized.
213–17	Wars on the German frontier.
229	Sassanian dynasty in Persia begins wars with Rome.
235–84	Military anarchy—some of the short-lived Emperors were Illyrians, not Romans. Christianity is severely persecuted. Worship of Mithras flourishes in the armies.
284–305	DIOCLETIAN, a great Illyrian administrator, reorganizes the Empire on autocratic lines with an effective bureaucracy but a heavy burden of taxation. The Empire divided along a north–south line running centrally through modern Jugoslavia, Diocletian ruling the eastern portion and establishing a colleague at Rome.
306–37	CONSTANTINE the Great wins the whole Empire by force from MAXENTIUS in 312, establishes Christianity as its official religion.
324–37	The Empire reunited with its capital at Constantinople (330).

TABLE II (*continued*)

A.D.	
330–79	Basil of Caesarea established a monastic rule popular in the East. Monasteries increasingly become centres of art and learning.
361–3	JULIAN the Apostate attempts to restore paganism.
376	Visigoths cross Danube, defeat VALENS at Adrianople.
340–420	St Jerome translates the Bible into Latin (Vulgate); encourages monasticism.
395	The Empire permanently divided between East and West.
395–408	Visigoths overrun Balkans, and Huns much of Asia.
c 400	Under HONORIUS (395–423) the western Empire controlled by the Vandal Stilicho, who repels the Visigoths but fails to prevent the barbarian occupation of Gaul.
407	Roman troops withdrawn from Britain.
410	Visigoths sack Rome and establish in Spain (419) the first stable barbarian kingdom.
430–1	Vandals establish a kingdom in north Africa.
438	Theodosian Code of Roman law.
441	Saxons settle in Britain.
451–2	Huns under Attila frustrated in invasions of Gaul and Italy.
455	Sack of Rome by the Vandals.
476	Deposition of the last Roman Emperor in the west by Odovacar.

TABLE III. A.D. 200–1500

DATE		WESTERN EUROPE	BYZANTIUM	PERSIA AND ISLAM		DATE
A.D.						A.D.
200				Disappearance of Hellenistic principalities. Wars with Rome		200
				Sassanian empire founded. Rise of Zoroastianism	Sassanian Period	
		Diocletian (284–305)				
300		Roman Empire re-united and christianized by **Constantine**	Byzantium (Constantinople) becomes the capital of the empire	Conquests of Roman territory		300
400		Visigothic Kingdom of Toulouse (419–507)				400
			Theodosius II (Theodosian Code, 438)			
		Burgundians settle Geneva–Lyons region	Under **Marcian** the Ostrogoths settle Pannonia	Armenia forcibly converted to Zoroastianism		
	Merovingian	Clovis unites the two branches of the Franks, advances to the Loire; Theodoric leads Ostrogoths into Italy		Christianity (Nestorian) tolerated. College of Edessa removed to Nisibis		
500		Conversion of Franks to Christianity				500
		Rule of St. Benedict (529) initiates Western monasticism	Intermittent warfare between Byzantium and Persia, usually favourable to the Persians who sack Antioch and exact tribute			
		Reconquest of Italy by Justinian from Ostrogoths	**Justinian** (527–65); codification of Roman Law N. Africa recovered from Vandals Barbarians raid Balkans			
		Lombard occupation of N. Italy		Arabia becomes a Persian province	Mohammed	
600		Pope Gregory the Great. Conversion of England	Most of Italy lost to the Lombards Under Chosroes II Persia overruns almost the whole of Byzantine Empire while the Avars occupy Thrace. Their joint siege of Constantinople fails (626)			600
		Dagobert (628–38)	Revival of Byzantine power with invasions of Mesopotamia and Assyria	Muhammad (570–632). Muslim era begins 622. Egypt (639–43) and Persia (635–50) conquered by the Muslims		
			Palestine and Syria lost to Islam (636–40) Administrative reorganization of the empire under **Constans II.** Armenia occupied by Arabs (653)	Islamic conquests extended to Morocco, the Indus, and the Aral Sea	Omayyad Caliphate	
		Decline of Merovingians and rise of Carolingian 'Mayors of the Palace'				
700		Spain conquered by Muslims	Settlement of Slavs and Bulgars in Balkans	Religious schism in Islam		700
		Defeat of Muslims at Poitiers	Leo III (717–41) strengthens the defences of the empire, resists Arab siege of Constantinople	Conquest of Spain and Georgia		
		Umayyad dynasty established in Spain	Successful campaigns against the Arabs and Bulgars	Civil war leads to downfall of the Caliphate		
		Charlemagne (768–814) absorbs Lombardy (774), Saxony (785), Bavaria (788), the Spanish March (795) into Frankish Kingdom, Austria (796)		Foundation of Baghdad		

Date	Western Europe	Byzantium	Islam	Date
800	**Charlemagne** crowned Emperor	Byzantium riven by the Iconoclastic controversy. Refugees arrive in the West	Harun al-Rashid (785–809)	800
	Division of Charlemagne's Empire		Ma'mun the Great (813–33). Extensive patronage of arts and sciences at Baghdad and elsewhere	
	Invasions of Europe by Northmen, Magyars, and Muslims	Conquest of Sicily by Muslims Bulgars christianized. Slavic alphabet invented	Conquest of Crete and Sicily Egypt becomes independent	
	Scandinavians settle in England and Normandy. Alfred (871–99) divides England with the Danes	Russians appear at Constantinople First schism with Rome	Revolt of eastern provinces (Persia); under Samanids (872–999), Bokhara an important cultural centre	
900	Feudalism develops in France Cluniac reforms of the Church (910) The Saxon Emperors begin northward and eastward expansion of Germany against Slavs and Magyars	Byzantine conquests extend to the Danube	Egypt restored to the Caliphate (908) Age of disintegration and rise of local dynasties, including the Umayyads in Spain (756–1031)	900
	Capetian dynasty established in France	Conversion o f Russians to Christianity begins Syria reconquered		
	Carolingian (800–1000)		*Abassid Caliphate* (800–1000)	
1000	England part of a Danish empire (Canute, 1017–35)	Final submission of Bulgars (1018)		1000
	Zenith of Holy Roman Empire under **Henry III** Christian reconquest of Spain begins Norman Conquest of England (1066) Normans settle in S. Italy and Sicily Beginnings of translation from Arabic into Latin N. Italian city-states develop	Total schism with Roman Church Byzantine rule in S. Italy ends (1068) Seljuks destroy Byzantine power in Asia Minor (1071)	Seljuk Turks enter Baghdad (1055)	
1100	First Crusade	Latin Kingdom of Jerusalem created Italian city-states win commercial privileges from Latins and Greeks	Omar Khayyam, mathematician and poet, d c 1123 Break-up of Seljuk empire. The provinces ruled by military chiefs	1100
	Second Crusade Renewed German territorial expansion	Zenith of Byzantine culture under Manuel Comnenus (1143–80)		
	Strengthening of French monarchy under Philip II Third Crusade	War with Venice, losses severe Loss of Bulgaria	Saladin unites Egypt and Syria, captures Jerusalem	
1200	University of Paris founded Zenith of medieval papacy (Innocent III, 1198–1216)			1200
	Fourth Crusade	Constantinople sacked (1204) Latin Empire founded	Mongols defeat Seljuks, occupy Anatolia	
	Mongols invade, take Cracow (1241) Reception of 'Latin Aristotle' in Europe (Aquinas, 1227–74), development of universities The Great Interregnum (1256–73) marks failure of Emperors' attempt to unite Germany	Michael Palaeologus recaptures Constantinople, expelling the Latins from Byzantium Wars with French, Catalans, Genoese, Venetians, in Greece and by sea	Ottoman Turks begin to occupy Byzantine territory	
1300	France leading Power before outbreak of Hundred Years War (1337)			1300
	Clement V removes Papacy to Avignon	Break-up of the Byzantine empire proceeds rapidly. Western provinces lost to Latins, eastern to the Turks	Great expansion of the Ottoman Kingdom begins. Byzantine provinces in Asia Minor conquered, Gallipoli occupied	
	Beginnings of Italian Renaissance (Dante, 1265–1321; Petrarch, 1304–74; Boccaccio, 1313–75) Black Death removes one-third of population Great Schism stimulates reform movement in Church. Rise of Lollards and Hussites	Civil war between rival emperors	Bulgaria annexed and Serbia subjected: Magyars defeated on Danube	
1400	Period of Great Councils (1409–49). End of Schism	Turks penetrate into Balkan provinces Futile union of Greek and Roman Churches	(Timurlane defeats Ottomans, breaks Mongol power, forms brief empire extending from Syria to India and China, 1395–1405)	1400
	End of Hundred Years War, English lose all France save Calais First printed books	Capture of Constantinople by the Turks: end of the empire	Ottoman Empire restored	
	Wars of the Roses (1455–85) Christian conquest of Spain completed Age of great voyages of discovery		With the fall of Byzantium, the Turks controlled the whole of the Balkans, much of Hungary, and penetrated into Bohemia and Poland	
1500				1500

I

MINING AND QUARRYING TO THE SEVENTEENTH CENTURY

C. N. BROMEHEAD

I. GREEK MINING

IN the discussion of early mining and quarrying a rigid adherence to chronological lines is undesirable and indeed impossible. For instance, while Greek mining technology was well advanced by 500 B.C., that of Britain remained in a primitive stage until Roman civilization made itself felt there at the end of the first century A.D. For the present purpose, technological stages are more to the point than regnal or political dates. Thus gunpowder, though in use at least as early as the eighth century in China and the thirteenth in Europe, was not recorded as used for mining until the seventeenth century. Steam as a source of power for pumping, first mentioned early in the seventeenth century, was not in effective use for mining till the eighteenth. It is therefore convenient in discussing the technology of mining to take a combination of explosives and steam as marking the dawn of the modern era, and to defer their consideration to a later volume.

We are fortunate in our knowledge of mining under the Greeks. The silver-lead mines of Laurion near Athens were famous for centuries, during which technique advanced greatly. Though technology was despised by Greek writers, the employees being mostly of the slave class and not citizens, sundry scraps of information have found their way into Greek literature. Furthermore, these very mines have been reopened in modern times and their ancient workings investigated (figure 1).

The Laurion mines were first worked by Mycenaeans in the second millennium B.C. and later abandoned. Athenian working began about 600 B.C. Silver was scarce in Athens in the time of the laws given by Solon (594 B.C.), but by 500 B.C. royalties on Laurion appear in the Athenian budget. The mineral worked was mostly galena (lead sulphide), of which the silver content was such that the mines were always spoken of as 'silver' mines. The ore yields an average of about 60 oz per ton. From the middle of the nineteenth century the mines, left untouched since about A.D. 100, were worked by a French company, mainly for zinc.

Geologically the ground at Laurion is made up of three beds of limestone or marble separated by two of mica-schist, sometimes traversed by dikes of granite and gabbro.[1] Galena occurs throughout, but is concentrated at the junctions, more especially at the tops of the two lower beds of limestone, where it is overlain by schist. The ascending liquids, which brought the ores, encountered obstruction from the schists and concentrated at the planes of contact. The top limestone is the highest bed of the district.

FIGURE 1—*Greek miners. Ore is dislodged with a pick and collected in baskets. What appears to be a lamp made of a wick protruding from an amphora hangs in the centre. Corinthian clay tablet of the sixth century* B.C.

The lead at Laurion is accompanied by zinc and iron. At the outcrop of the upper contact the galena, blende, and pyrites (sulphides of lead, zinc, and iron respectively) gave place to cerussite (lead carbonate), calamine (a silicate or carbonate of zinc), and haematite (an oxide of iron). The red colour due to the last substance makes the outcrop conspicuous; probably it was worked at an early date for iron, and this was used for the manufacture of tools with which to work the argentiferous galena. The earliest remains are open-cast, with short adits.[2] Later, more than 2000 shafts were sunk, connected by galleries. It is noteworthy that there is practically no outcrop of the lower and richer contact-zone, the discovery and working of which were a matter for nineteenth-century inductive geology.

The shafts are exceedingly regular; they are rectangular and usually about $1{\cdot}9 \times 1{\cdot}3$ m. The centre line is very accurately vertical, but at about every 10 m of depth the cross-section is turned through an angle of 8 to 10°, so that the rectangle at the bottom may be at right-angles to that at the top. These stages seem to depend on the method of access; a succession of ladders, or tree-trunks with steps cut in them, fixed to the sides, left the middle free for hoisting ore by rope and pulley. Parts of the pulley-wheels and the markings of the axles at the top have been found. The deepest shaft is 117·6 m, the limiting depth being the water-table, here approximately at sea-level.

At the contact-zones, galleries about 1 m high by 0·75 m wide were driven to

[1] Gabbro is a name applied to several types of rock, but specifically to an igneous plutonite consisting of plagioclase and clinopyroxene, often containing olivine as well.

[2] Horizontal tunnels for access or drainage.

follow the ore-shoots, but betray also a definite plan. Shafts were sunk in pairs and parallel galleries driven from them, with frequent cross-cuts between the galleries to aid ventilation. The galleries in some cases run roughly up a dip-slope with a rise of mine-level and surface-level of 50–60 m, thus further aiding ventilation. No doubt the parallel galleries also served as inby and outby roads. Where rich bodies of ore were found they were stoped either overhand or underhand, according to circumstance. Branch galleries were cut leading out of the main roads at the necessary slope to reach the desired spot. From some of the stopes as much as 100 000 tons has been taken. In these stopes 'pillar and stall' work was used, the poorer ore being left as pillars.[1] Sometimes all was taken and pillars to support the roof were built from waste. Under Lycurgus in 338 B.C. a law was passed against the removal of pillars, some of which are over 9 m high. Strabo, however, says that in his time (? 63 B.C.– ?A.D. 21) the slag-heaps were being re-smelted and the stope-fillings robbed [1].

Many of the Laurion galleries show grooves in the sides where doors could be fitted to direct the ventilating air. In some cases the draught was increased by lighting fires. A sloping shaft was sunk to intercept a vertical working shaft; just before the normal junction a horizontal cut was made and the fire was placed on this level platform. Probably fanning with cloths was a common method; not much later Pliny (d A.D. 79) mentions this as a usual way of ventilating [2].

The work was done with wrought iron tools: hammers of approximately modern Geological Survey pattern, picks with wooden handles, chisels, and wedges. Each miner had his lamp, and niches were cut to receive them in the approaches to the face. It has been calculated that in shaft-sinking a miner averaged $4\frac{1}{2}$ m per month. Timbering was seldom necessary, but the remains show well cut mortise-and-tenon joints between props and lintels. An immense amount of wood was used for fuel at the smelting-works on the surface. Laurion was originally well wooded, but by the time of Strabo it had become as bare and arid as it is now.

There was a close similarity between ancient Greek mining methods and those in use in Japan until the middle of the nineteenth century (figure 2). A delightful colour-print by Hiroshigi II, published about 1850, shows the interior of a gold-mine (plate 1). Almost every feature—the step-ladders, and the picks and lamps of the miners—can be matched exactly in the Laurion mines. Similar lamps were found at Cassandra (Greece) and in the Roman mines on Mendip.

[1] Extraction of ore by driving a series of passages, generally horizontal, into the vein is known as stoping. This may be overhand, when each new stope is above the previous one, or underhand, when work proceeds downwards. In the pillar and stall system the roof of the stope is supported by leaving pillars of unworked ore.

The mineral left the mine at Laurion in comparatively small pieces, to be put through stamps and mills on the surface. The resultant powder contained not only galena but varying proportions of quartz, blende, pyrites, fluorite, iron ores, and other substances. It had next to be washed. The tables for this washing were of various sizes and angles of slope. One measures 7·5 m in length, another about 21 × 12 m over-all. They are of masonry, faced with concrete and a thin

FIGURE 2—*Drainage of a Japanese mine in the seventeenth century. Hand-bailing, Archimedean screws, and rope-and-bucket are used in succession. The screws are turned by cranks.*

layer of fine cement. The water reservoir has conical openings, from which strong jets played on the mineral below. The heaviest fraction collected almost at once in a shallow channel; the rest was sorted by gravity over a long, wide surface, slightly inclined. By an ingenious arrangement of channels the water was collected in a sump at one corner just below the reservoir to be bailed back for re-use. The supply of water was one of the chief difficulties. To conserve the winter rainfall innumerable cisterns were constructed, each with a cover of planks as a protection against evaporation.

There are curious sidelights on Greek mining. Thus the Attic orator Demosthenes (384–322 B.C.) clearly distinguishes between the mining-works and the mills where the ores were ground and washed [3]. Because of the difficulty of water-supply, the two departments, though under a single owner, might be a considerable distance apart. Demosthenes relates that in the mills the ore was ground to the consistency of flour. The importance of mining in northern Greece is indicated in the first half of the fourth century B.C. by the coinage of Damastion in Epirus, the designs on which show a miner's pick and an ingot of metal (figure 3).

The Greek author Polybius (?205–?125 B.C.) gives interesting notes on the silver mines at New Carthage (Cartagena in south-east Spain) [4]. He mentions blacksmith's bellows [5], but whether these were of the type still prevalent in primitive works in the Far East, that is, with a piston of square section, is mere speculation. That tools were chosen with care is indicated by a description (fourth century B.C.) of the useful varieties of steel—the Chalybean, best for carpenters' tools; the Lacedaemonian, best for files, drills, gravers, and stone-chisels; and several other types (p 228).

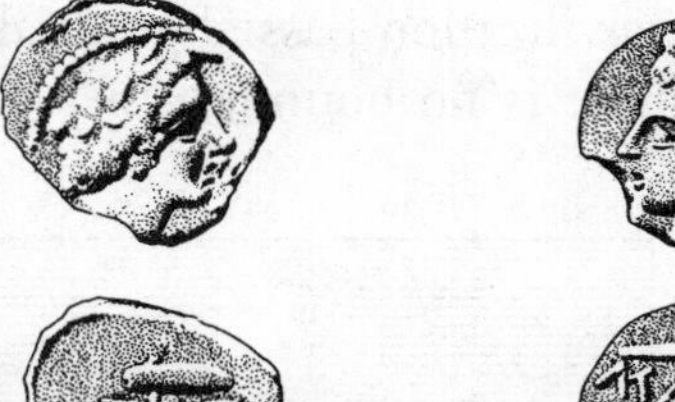

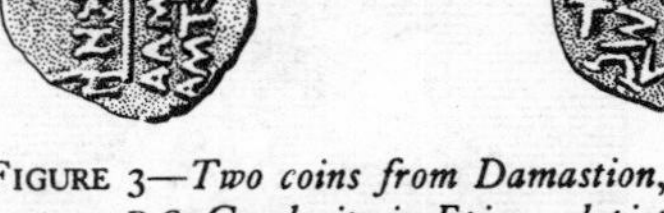

FIGURE 3—*Two coins from Damastion, a fourth-century B.C. Greek city in Epirus, depicting miners' picks. Diameters 15 mm.*

Our main literary source of information concerning mines is the 'Geography' of Strabo (d *c* A.D. 21), which was intended chiefly as a text-book for government officials. It treats products for export at some length but gives few technological details. Strabo relates that in his time more alluvial than mined gold was produced and adds that 'sieves woven after the fashion of baskets' were used for washing the material grubbed up with hoes or mattocks [6]. For silver-lead ores the washing was repeated up to five times. In working the alluvial deposits extensive trenching was carried out, the water being raised, where necessary, by means of Archimedean screws (figure 2).

A few materials were mined for purposes other than the direct production of metals; such was the pigment vermilion (p 361), of which the Spanish ores are described by Strabo as 'not inferior to that of Sinope' on the Black Sea [7]. This town was the place of export of the vermilion of Cappadocia, not its actual source. In the first century A.D. Dioscorides wrote as follows of *miltos synopike* (red ochre of Sinope, i.e. vermilion):

The best is dense and heavy, liver-coloured, free from stone, . . . of consistent colour. It is collected in Cappadocia in certain caves and is refined and taken to Sinope and sold. Hence its name. It has an astringent, drying, clogging faculty and so it is mixed in plasters for wounds, and in constipating, costive pastilles. Taken in an egg and administered as a clyster it stops the bowels. It is also given to those suffering from complaints of the liver [8].

As an army medical officer Dioscorides was mainly interested in drugs, but many minerals besides vermilion were mined as pigments. In the instance mentioned, vermilion was, of course, far superior to common red ochre, and much more profitable as an article of export trade.

II. ROMAN MINING

Greek civilization passed gradually into that of the Roman Empire. In technology there is no boundary between the two, though here and there, even in

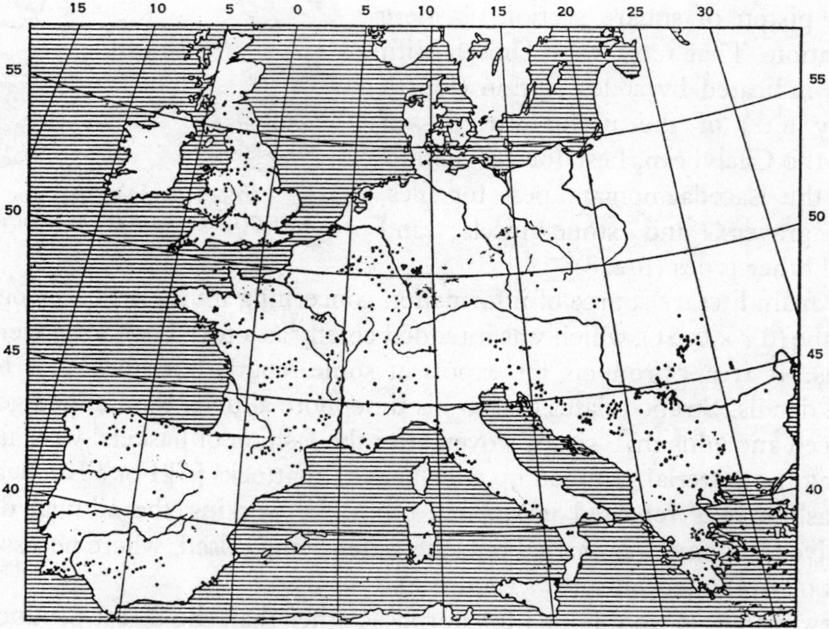

FIGURE 4. *Map showing the positions of Roman mines in Europe.*

mining and quarrying, the Greek love of science can be seen showing through the severely practical Roman aims (figure 4).

Greek furnaces occasionally, and doubtless at first accidentally, produced cast iron. Pausanias (second century A.D.) tells that one 'Theodorus of Samos was the first to discover how to pour [or melt] iron and make statues of it' [9]. Unfortunately he says nothing of the furnaces necessary to produce the high temperature required. Cast iron figures of animals, assigned to the seventh or sixth century B.C., have been recorded from Cyprus. Such statues were probably quite small and cast in one piece; they were toys, or were used for magical processes, and were popular in Alexandria and elsewhere. Late tales about them were gross exaggerations; such was the 5000 lb figure of Bellerophon on horseback mentioned in Saxon times by Bede [10]. The practical-minded Romans seldom refer to such curios, though a little cast iron was perhaps intentionally made in Britain.

With their magnificent aqueducts (ch 19), the Romans were able to provide adequate supplies of water for the washing of minerals. Even more to the advancement of mining, they drained the copper mines of Rio Tinto in south-

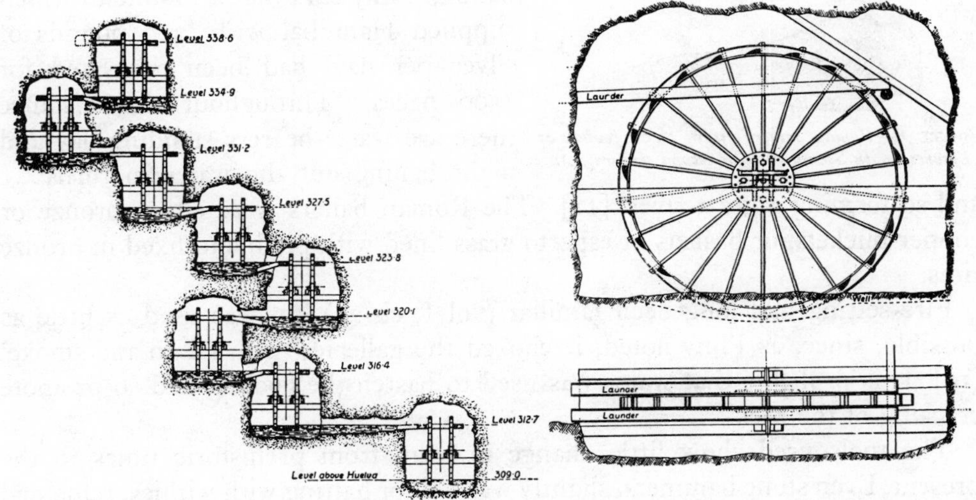

FIGURE 5—(Left) *A series of eight pairs of water-raising wheels used by the Romans to drain copper mines at Rio Tinto, Spain. The wheels were probably turned by treadmills.* (Right) *A single wheel, part of a Portuguese series, in elevation and plan. Diameter 5 m, giving a lift of 3·7 m. Second (?) century A.D.*

western Spain, and others as far afield as Britain. In fact it was in pumping-devices that Roman mining showed the greatest advance over that of the Greeks.

The most important of these devices was the water-wheel (pp 593 ff). At one site in a Rio Tinto mine a succession of eight pairs of such wheels raised the water 30 m. Many such wheels have been found in the mines of Spain and Portugal; they are usually about 4·5 m in diameter (figure 5). Another method of water-raising was the Archimedean screw, known to the Romans as a *cochlea* or 'water-snail' (pp 676 f).

For ordinary mining the adits of Roman mines were as small as possible, to economize labour. Sometimes the floor was just wide enough for one man to walk, but wider above to accommodate a pack on his back or a bucket in one hand. In some Roman mines in Portugal, however, there are adits for access to the workings measuring 5 m in height and width. Against one wall is a narrow water-channel a further 5 m deep. Every here and there is a chamber 10 m square, in the centre of which is a large granite block and round it a circular track worn by animals, all that remains of a 'chain of pots' machine for raising water (pp 637, 675).

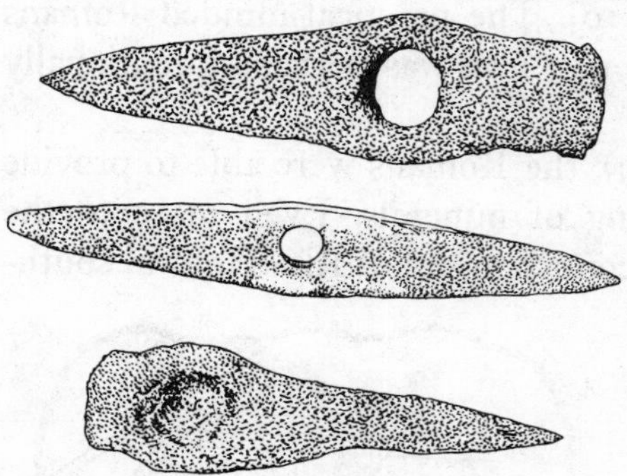

FIGURE 6—*Roman miners' iron picks found at Linares, south Spain. First or second century A.D.*

These various mechanical devices were auxiliary to the simple means of hand-bailing. Pliny says that a mountain which supplied Hannibal with '300 pounds of silver per day' had been excavated for 1500 paces. 'Throughout this distance there are water-bearers standing day and night bailing out the water in turns ... and so forming a great river' [11]. The Roman bailers were either bronze or copper buckets, or baskets of esparto grass lined with pitch and fixed in bronze rims.

Fire-setting had long been familiar (vol I, ch 20) but was used as little as possible, since, as Pliny noted, it choked the galleries 'with steam and smoke' [12]. It is probable that water was used to hasten the cooling and so promote cracking of the rock.[1]

The tools used show little change in form from prehistoric times to the present. Even stone hammers, slightly waisted for hafting with withies, remained long in use. There was naturally an increase in the use of iron in Roman times. Iron tools from the Roman province of Baetica (western Andalusia) can be dated with some approach to accuracy. They include single- or double-headed picks, straight or curved (figure 6), and wedges for hammering into the rock; the hammers might have two flat heads, or one flat head and a point. They usually weighed 5 to 10 lb and had short wooden handles (figure 7). Battering rams, with iron heads weighing as much as 150 lb, are mentioned by Pliny as used for breaking up quartz veins [12].

Gradual improvement was largely due to the replacement of slave-labour by skilled artificers, who were still supplemented by condemned criminals (figure 8).

[1] Though mentioned by both Pliny and Diodorus (first century B.C.) it is hardly likely that vinegar was ever used. It would have no effect except on limestone and little on that. The whole subject of this myth has been discussed in Agricola's *De re metallica*, trans. by H. C. and L. H. Hoover, p 118, note 414. London, 1912; New York, 1950.

By the time of Hadrian (A.D. 117–138) there were strictly regulated arrangements for the supply of hot water for the miners' baths, with times for their use by the women of the community when the men were below. The first British statute on this subject was passed in 1911! The unhealthy nature of

FIGURE 7—*Typical Roman miner's iron tools: pick, hammer, and spade. Parts of the wooden handles remain. From Linares, south Spain. First or second century A.D.*

employment in some mines, such as those whence ores of arsenic and mercury were obtained for pigments, was notorious even in the time of Lucretius (d 55 B.C.) [13].

The Roman invasion of Britain was no doubt mainly designed to tap the island's mineral wealth, though perhaps also to deprive refugees from Gaul of harbourage. The lead-ores of Mendip were being exploited by A.D. 49, within half a dozen years of the start of effective conquest. Besides dated pigs of lead, our evidence includes iron tools (figure 9) similar to those described above, and several leaden lamps, showing that some underground work was in progress [14]. The silver that accompanied the lead there and elsewhere, and the copper, iron, tin, and gold, contributed their share to the Empire's resources. But the Romans, as usual, to a large extent worked the mines in Britain by native labour and methods, supplying overseers and foremen as needed, and taking tribute. It is, therefore, not

FIGURE 8—*Stone relief of Roman miners. The foreman, on the left, carries a large pair of tongs; in front of him is a man carrying a pick like that shown in figure 7; and next a man carrying a lamp. All wear wide leather belts. It is not clear what the foreman carries in his left hand. From Linares, south Spain. First or second century A.D.*

surprising that technology in Roman Britain does not yield anything that is not better illustrated in lands less remote from the centre of government.

Definitely Roman work, as distinct from Romano-British, can be recognized at the gold-mine of Dolaucothy in south Wales; open-cast workings there are primitive in character, but a shaft and adit arc carried some 24 m below water-level [15]. From surviving fragments it has been recognized that a water-raising wheel was in use. An aqueduct on the site 7 miles long was of Roman design, though no doubt native labour was employed in making it. From the many engraved and partially engraved gems found here and elsewhere in connexion with Roman mines it would seem that gem-cutting was practised locally, perhaps by Roman officials as a hobby.

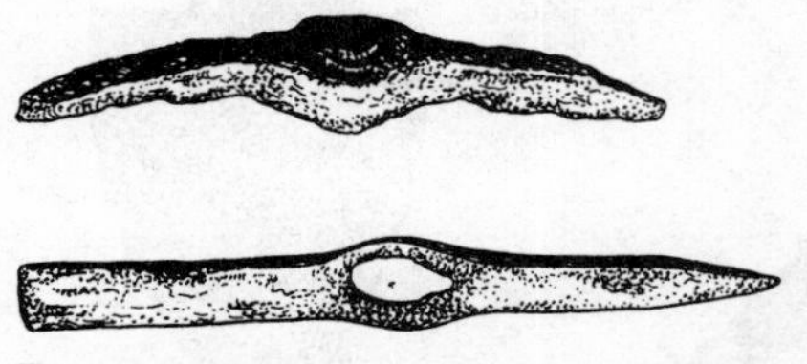

FIGURE 9—*Roman iron picks for lead-mining. From Mendip, England.* c *First century* A.D. *Scale 1/7.*

Tin was not much worked in Roman Britain till the middle of the third century A.D., when it appears to have passed under government control. The extraction (p 46) was evidently efficient, since a dish of about 350 is 99·18 per cent tin, and objects of nearly pure tin have been found as far north as Northumberland. But of the mining of British tin in Roman times, despite many legends, we know nothing.

Iron-working was practised in Britain from about 500 B.C., and was known to Julius Caesar as a local industry [16]. During the Roman occupation workings were very widespread, but mostly on a scale too small to be called mining. The more important centres were the Weald and the Forest of Dean. Besides the superficial limonitic iron ores, haematite (Cumberland and Wales), clay-band ironstone from the coal measures, the various Jurassic ores, and the Cretaceous ores were worked, with coal as fuel, various clays for furnaces, and limestone for a flux. Both wrought iron and cast iron of Roman age have been found in Lancashire, Warwickshire, and Sussex. But, again, we know nothing specific of the mining technology in these places.

III. MINING IN MEDIEVAL TIMES

Through the centuries after the fall of the Roman Empire tools and weapons of metal were in continued demand for agriculture and war, but there was little technological progress and the records of processes are few. Contemporary illuminations, tapestries, and bas-reliefs show implements and weapons: but where were these made, whence came the metals used, and how were they mined? The Anglo-Saxon and Carolingian kings issued coins of gold, silver, and bronze, but we do not know whether these coins were minted entirely from loot or tribute or survivals from the past ages, or from metals won by mining. That the period saw profound changes is obvious. The greatest activity was in central Europe. By the sixteenth century the words 'miner' and 'Saxon' had become almost equivalent. Saxon miners led the way in the Middle Ages, not only in their own country but throughout almost the whole of Europe. Mining was begun by them at Schemnitz in Czechoslovakia as early as 745, at Goslar in the Harz in 970, at Freiberg in Saxony in 1170, and at Joachimsthal in Bohemia in 1516 [17].

When the famous Spanish mine of Guadalcanal in the province of Seville was reopened in 1551 for the production of silver-lead, one of the administrators wrote to the government to ask for 200 or more Germans, skilled in mining and metallurgy. To take an example from the other extremity of Europe, the famous silver-mines of Kongsberg in Norway were first opened in 1623. 'Hereupon Christian the Fourth was pleased to give his name to the first groove and miners were sent for from Germany. These were the first inhabitants of the new built mine-town of Kongsberg, and the ancestors of the many thousands at present living there, who in process of time mixing with the Norwegians, each nation to this day performs divine service in its own language' [18].

The Germanic rebirth of mining contrasts with that of classical times, when Tacitus could write in A.D. 98 'Heaven has denied [to the Germans] gold and silver—shall I say in mercy or in wrath? I would not go so far as to assert that Germany has no lodes of silver and gold. But who has ever prospected for them?' [19]

We may notice one step in the change in Spain. When the famous mines there passed into Muslim hands during the Moorish occupation, the output of copper, lead, and other metals declined. The Moors then found that if water containing copper sulphate is allowed to run over iron, pure copper is deposited and the iron dissolved. As iron was cheap and abundant in Spain, this discovery yielded an efficient method of recovering copper from sulphide ore,[1] and direct mining of copper ores became less necessary. Many shafts and galleries went out of use, some being converted into aqueducts, which carried water containing copper sulphate from the mine shafts.

Medieval literature bearing on mining is misleading. It is true that there are many works showing some real, if small, knowledge of chemistry, but they are usually of Arabic origin, and are often intentionally obscure or ill translated, or shade into alchemy. There was also a big trade in artificial gems, for the most

[1] The sulphide ores, on exposure to air in the presence of water, are oxidized to soluble sulphates.

part glass coloured with metallic oxides, but we learn little of their manufacture from the texts. Theophilus, who was perhaps a monk of Helmarshausen near Paderborn (*c* 1110–40, pp 63 f), has a passage on the mining of gold: 'There are many kinds of gold, among which the best is produced in the land of Havilah, surrounded, according to Genesis, by the river Pison. When men, skilful in this art, have discovered the underground veins, they dig them up and, purifying the gold by fire and proving it in the furnace, they subject it to their use.' He adds that Arabian gold, which is pale, is often adulterated by the addition of one-fifth part of copper. His next chapter is on Spanish gold, 'which is composed from red copper, powder of basilisk, human blood and acid'. His description of how to prepare the powder of basilisk may conceal in symbolic language a method of extracting real gold from auriferous copper by suitable acids [20].

A less familiar ore, much used in medieval times, is stibnite (antimony sulphide, Sb_2S_3). In classical times antimony was largely used to form a bronze with copper. In medicine it was known as a powerful cathartic.

The *Lapidario* of King Alfonso X the Wise (1252–84) of Castile, a compilation that was finished in 1279, gives details of many stones. It is founded partly on classical work, partly on works in Arabic. It has important elements derived from alchemy, ancient chemistry, and astrology. The stone there called *ecce* is argentiferous stibnite, said to be worked at various localities in Spain and Portugal. Its main economic use was in the production of a 'beautiful gold-colour' on the surface of glass. The process described has recently been tested in practice; it may go back as far as the eighth century and have been employed in some of the famous mosque-lamps. The illuminations in the later manuscript of the *Lapidario* always include a well dressed expert, who remains on the surface to examine the specimens handed up to him by a labourer who searches for them at out-crops, digs in superficial deposits, or works underground. Shafts shown are circular, and the tools may have long or short handles fitted to iron heads; they include picks, hammers, wedges, and shovels or spades (figure 10). One shaft depicted has buckets; in another figure an expert weighs gems or ores in a balance.

Of the conditions of mining in England we gain some knowledge from documents on its legal aspect, which are numerous from about the year 1200. Most regions have their own peculiar regulations. Those of the Stannaries, that is, the tin-mining area of Cornwall and Devon, are very detailed. According to Carew's 'Survey of Cornwall' (1602) 'they *sincke a shaft*, or pit of five or six foote in length, two or three foote in breadth, and seven or eight foote in depth, to proove whether they may so meete with the Load [lode]'. He says that 'their ordinary tools are a pickaxe of iron, about sixteen inches long, sharpened at the one end to peck, and flat-headed at the other, to drive certain little iron wedges, wherewith they cleave the rocks. They have also a broad shovel the outer part of iron, the middle of timber, into which the staff is slopewise fastened' [21]. On Mendip, calamine (zinc carbonate or silicate) was mined as early as 1560. The laws of the High Peak of Derbyshire have a character of their own. Many of the terms are identical with those of Saxony; thus the head-man is a 'barmaster',

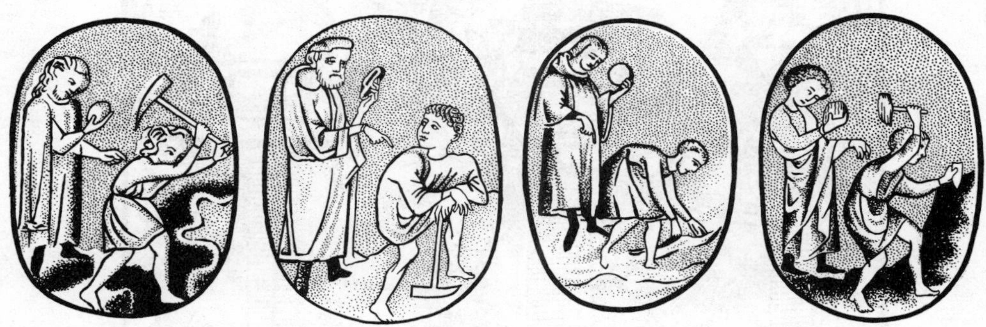

FIGURE 10—*Illustrations from the initials of a Castilian illuminated manuscript of the thirteenth century. From left to right: prospecting for a 'stone which attracts lead and resembles it'; quarrying marble; collecting gold from river sand; prospecting for iron with hammer and wedge.*

obviously the German *Bergmeister*. These various laws suggest some mining connexion between Roman, Anglo-Saxon, Norman, and medieval elements. In Cumberland copper was being mined by the Society of Mines Royal in the reign of Elizabeth I, but activities soon ceased from lack of fuel. Here also German experts were brought in. Copper was then imported from Sweden, till the use of coal for smelting initiated a revival.

The great textbook on every aspect of mining was the *De re metallica* (1556) of Georg Bauer (1494–1555), known as Agricola. He was a Saxon who took a medical degree in Italy and settled as a physician at the famous mining centre of Joachimsthal.[1] His first book was the excellent *Bermannus* (1530) [22], a remarkably clear dialogue on mineralogy; it was followed quickly by other works. His masterpiece, the *De re metallica* [23], in twelve books, treats the subject with great completeness. He writes:

> The first book contains the arguments which may be used against this art of mining, and against metals and the mines, and what can be said in their favour. The second book describes the miner, and branches into a discourse on the finding of veins. The third

[1] Coins minted from Joachimsthal silver were known as *thalers*, whence the Low German *dahler* and the modern 'dollar'.

FIGURE 11—*First stages in the excavation of a mine, showing both the surface of the hillside and the shafts beneath it. Note the ox-hide bucket (cf figure 15), crank-handled windlasses, ladder, and four-wheeled barrow. Figures 11 to 22 are from the* De re metallica *of Agricola, published in 1556.*

book deals with veins and stringers, and seams in the rocks. The fourth book explains the method of delimiting veins, and also describes the functions of the mining officials. The fifth book describes the digging of ore and the surveyor's art. The sixth book describes the miners' tools and machines. The seventh book is on the assaying of ore. The eighth book lays down the rules for the work of roasting, crushing, and washing the ore. The ninth book explains the methods of smelting ores. The tenth book instructs students of the metallic arts in the work of separating silver from gold, and lead from gold and silver. The eleventh book shows the way of separating silver from copper. The twelfth book gives rules for manufacturing salt, soda, alum, vitriol, sulphur, bitumen, and glass [24].

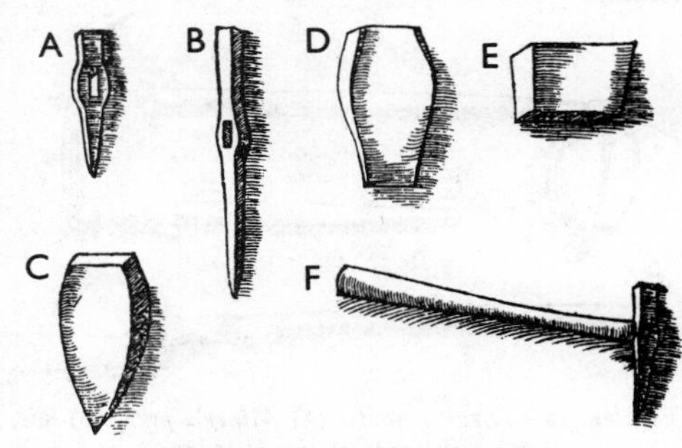

FIGURE 12—*Iron tools for splitting rock. All have flat heads to receive blows from a hammer.* (A), *says Agricola, 'is in daily use',* (B) *is 'to shatter the hardest veins'.* (C), (D) *and* (E) *are iron wedges.* (F) *shows the method of hafting* (A) *and* (B). (A) *is about 23 cm long. 1556.*

While our main concern here is with books five and six, all have some interest for the historian of mining.

The first necessity for a mine was the shaft, from the bottom of which, when a reasonably promising vein was reached, more or less horizontal passages were driven (figure 11). The section of the shaft was normally about 3 × 1 m. Most shafts were comparatively shallow, but in the *Bermannus* Agricola mentions that 'at Schneeberg from which so much wealth was taken within our memory one has reached 200 paces [in depth]'; when his interlocutor exclaims at this he replies that at Kuttenberg there are shafts more than 500 paces deep [25]. Such depths were, however, very exceptional. The 'pace' mentioned was probably the Roman double-pace of rather less than 6 ft (1·85 m). The shafts were as often sloping as vertical.

Agricola's illustrations and descriptions of the tools and equipment employed by sixteenth-century miners are very detailed (figures 12–16). After dealing with these matters, he continues:

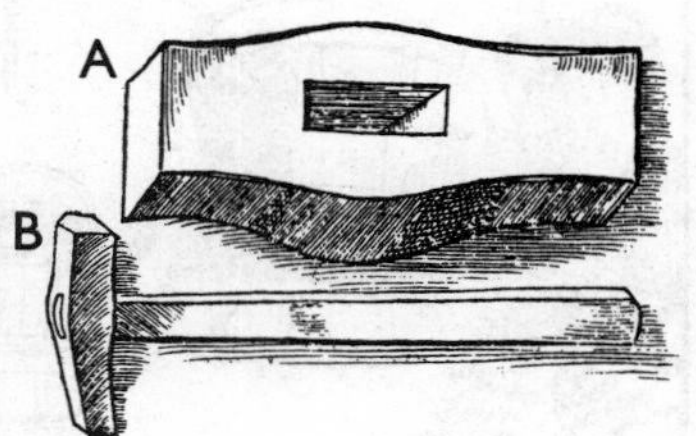

FIGURE 13—*Hammers.* (A), *the largest in a series, is about 30 cm long; it is used for driving wedges.* (B), *shown hafted, is the smallest. 1556.*

I will now explain their machines, which are of three kinds, that is hauling machines, ventilating machines and ladders. By means of the hauling machines loads are drawn out of the shafts; the ventilating machines receive air through their mouths and blow it into shafts or tunnels, for if this is not done, diggers cannot carry on their labour without great difficulty in breathing; by the steps of the ladders the miners go down into the shafts and come up again [26].

Windlasses were turned by one, two, or three men, the larger being furnished with a fly-wheel. A further step was the use of geared wheels, so that heavy weights could be raised or lowered by the use of horses (figure 17). An amusing illustration shows a horse-drawn sledge and a team of dogs with pack-saddles (figure 18); another shows a pair of goats turning a treadmill.

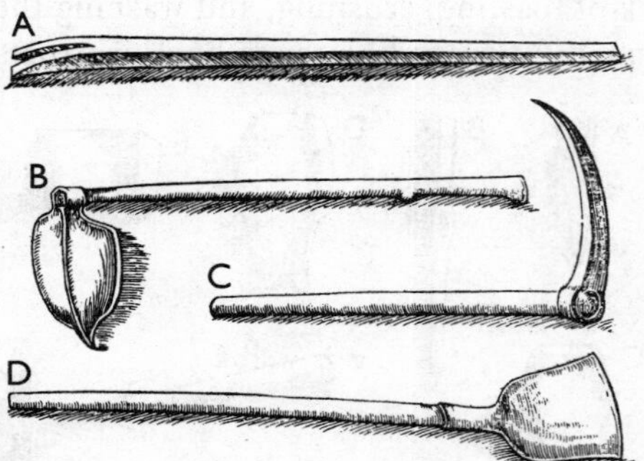

FIGURE 14—*Digging tools.* (A) *Miner's pike,* (B) *hoe,* (C) *pick, and* (D) *shovel. 1556.*

The most usual method of raising water was by chains of dippers turned by hand, by treadmill, by animals, or, when the position allowed, by a water-wheel; the larger kinds were seldom used, as the buckets or dippers were apt to get broken (figure 19). Of suction-pumps which drew by means of pistons, seven varieties were described, with details of all the necessary parts:

> The fifth kind of pump . . . is composed of two or three pumps whose pistons are raised by a machine turned by men, for each piston-rod has a tappet which is raised, each in succession, by two cams on a barrel; two or four strong men turn it. . . . Each of these three pumps is composed of two lengths of pipe fixed to the shaft timbers. This machine draws the water higher, as much as twenty-four feet. If the diameter of the pipes is large, only two pumps are made; if smaller, three, so that by either method, the volume of water is the same [27].

All these varieties go back in principle to classical times.

FIGURE 15—*Containers for hauling ore and spoil up shafts.* (Left) *Basket, and wooden buckets bound with iron.* (Right) *Ox-hide buckets. 1556.*

The seventh kind of pump, invented ten years ago, which is the most ingenious, durable, and useful of all, can be made without much expense. It is composed of several pumps, which do not, like those last described, go down the shaft together, but of which one is below the other, for if there are three, as is generally the case, the lower one lifts the water of the sump and pours it into the first tank; the second pump lifts it again from that tank into a second tank, and the third pump lifts it into the drain of the tunnel. A

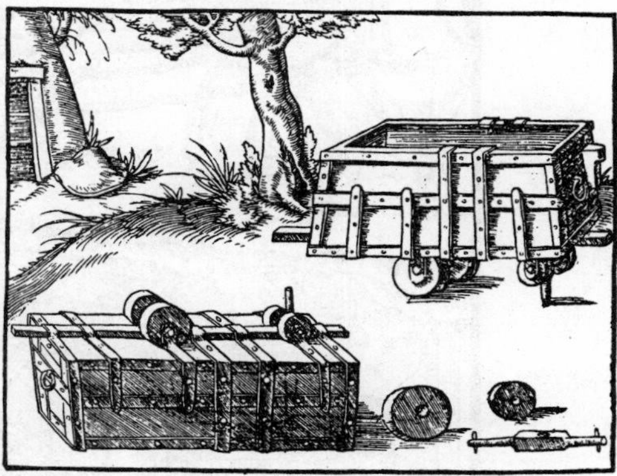

FIGURE 16—(Left) *Wheelbarrows.* (Right) *Truck. 'A large blunt pin fixed to the bottom of the truck runs in a groove of a plank in such a way that the truck does not leave the beaten track'* (*Agricola's description*). *Wheelbarrows and trucks were used to shift excavated material along tunnels. The trucks had the greater capacity, and were preferred for the larger and longer tunnels. 1556.*

wheel fifteen feet high raises the piston-rods of all these pumps at the same time and causes them to drop together. The wheel is made to revolve by paddles, turned by the force of a stream which has been diverted to the mountain (figure 20) [28].

Of rag-and-chain pumps Agricola also gave full details. There were six kinds, of which figure 21 illustrates one.

This kind of machine is employed at the foot of the Harz mountains and in the neighbourhood. Further, if necessity arises, several pumps of this kind are often built for the purpose of mining one vein, but arranged differently in different localities according to the depth. At Schemnitz, in the Carpathian mountains, there are three pumps, of which the lowest lifts water from the lowest sump to the first drain, through which it flows into the second sump; the intermediate one lifts from the second sump to the second drain, from which it flows into the third sump; and the upper one lifts it to the drains of the tunnel, through which it flows away. This system of three machines of this kind is turned by ninety-six horses; these horses go down to the machines by an inclined shaft, which slopes and twists like a screw and gradually descends. The lowest of these machines is set in a deep place, which is distant from the surface of the ground 660 feet [29].

I will now speak of ventilating machines. If a shaft is very deep and no tunnel reaches to it, or no drift from another shaft connects with it, or when a tunnel is of great

FIGURE 17—*Horse-whim for raising large loads. The man below operates a brake consisting of a beam that can be lifted to bear on a drum fixed to the driving shaft. The load may be held stationary by catching the chain on a suspended hook. 1556.*

length and no shaft reaches to it, then the air does not replenish itself. In such a case it weighs heavily on the miners, causing them to breathe with difficulty, and sometimes they are even suffocated, and burning lamps are also extinguished. There is, therefore a necessity for machines which . . . enable the miners to breathe easily and carry on their work [30].

FIGURE 18—*Horse-drawn sledge, and dogs with pack-saddles. 1556.*

These included sundry forms of revolving fans, single- or double-acting bellows worked by hand or by machines similar to those already described (figure 22), and, finally, flapping with cloths. The last had been described long ago by Pliny for driving out noxious fumes when sinking a well [2].

Agricola noted that mines were most often abandoned because they did not yield metal, or became barren beyond a certain depth. But another important cause for the abandonment of workings was difficulty in keeping them dry. The draining of mines was one of the greatest engineering problems of the times, and a sharp stimulus to invention.

Sometimes the miners can neither divert the water into the tunnels, since tunnels cannot be driven far enough into the mountains, nor can they draw it out with machines because the shafts are too deep; or if they could draw it out with machines, they do not use them, the reason undoubtedly being that the expenditure is greater than the profits of a moderately poor vein.

FIGURE 19—*Mine-drainage. Chain of dippers powered by water-wheel. 1556.*

Another problem was ventilation, which the owners could not always overcome 'either by skill or expenditure, for which reason the digging of shafts and tunnels is sometimes abandoned' [31].

Agricola added to these causes of the abandonment of mines the presence in them of poisonous 'damps' (German *Dampf*, vapour) or of 'fierce and murderous demons'. Sometimes also 'the underpinnings become loosened and collapse, and a fall of the mountain usually follows; the underpinnings are then restored only when the vein is very rich in metal'. Agricola advised that deserted workings should not be reopened,

unless we are quite certain of the reasons why the miners have deserted them, because

FIGURE 20—*Mine-drainage. Triple-lift suction-pump powered by undershot water-wheel. 1556*

FIGURE 21—*Mine-drainage. Rag-and-chain pump, manually operated. The balls, which are stuffed with horsehair, are spaced at intervals along the chain and act as one-way pistons when the wheel revolves. 1556.*

we ought not to believe that our ancestors were so indolent and spiritless as to desert mines which could have been carried on with profit. . . . Therefore it is advisable to set down in writing the reason why the digging of each shaft or tunnel has been abandoned [32].

Such is Agricola's account of the mining practice of his day. We may well take to heart his final warning as to the necessity of keeping account of abandoned mines, a practice not made compulsory in England until recent years and still by no means complete.

A century and a quarter later, the Englishman Edward Browne[1] travelled in central Europe and published in London his 'Brief Account of some Travels

[1] Son of Sir Thomas Browne, author of *Religio Medici*.

FIGURE 22—*Mine-ventilation. Fan driven by overshot water-wheel. The large drum contains a four-bladed fan* (inset) *to which long feathers are attached. The drum is connected to the mine by a hollow trunk. 1556.*

in . . . Hungaria, Serbia, Bulgaria, Macedonia, Thessaly, Austria, Styria, Carinthia, Carniola, and Friuli, through a great part of Germany . . . with some observations on the Gold, Silver, Copper, Quick-silver Mines . . . in those parts'. It shows that there had been no appreciable change in methods of mining since Agricola (figure 23). Browne relates that:

At *Freiberg* they have many ways to open the *Ore* whereby it may be melted; as by *lead* and a sort of *Silver Ore* which holds *lead* in it. They have also *Sulphur Ore*, which after it is burned, doth help much towards the fusion of *Metals*. . . . Their *Treibshearth*, or *driving Furnace*, where the *Litharge* is driven off, agrees better with the Figure of it in *Agricola*, than those of *Hungary*. . . . Much of their *Ore* is washed, especially the poorest, and that which is mixed with *stones*, *quarts*, or *sparrs*.

This is peculiar in their working, that they burn the pounded and washed *Ore* in the *Roasthearth*, before they smelt it in the *Smeltzoven* or *melting Furnace*.

At those *Mines* of *Hungary* where I was, they used not the *Virgula divina*, or forked Hazel, to find out *Silver Ore* or hidden Treasure in the Earth; and I should little depend thereon. . . .

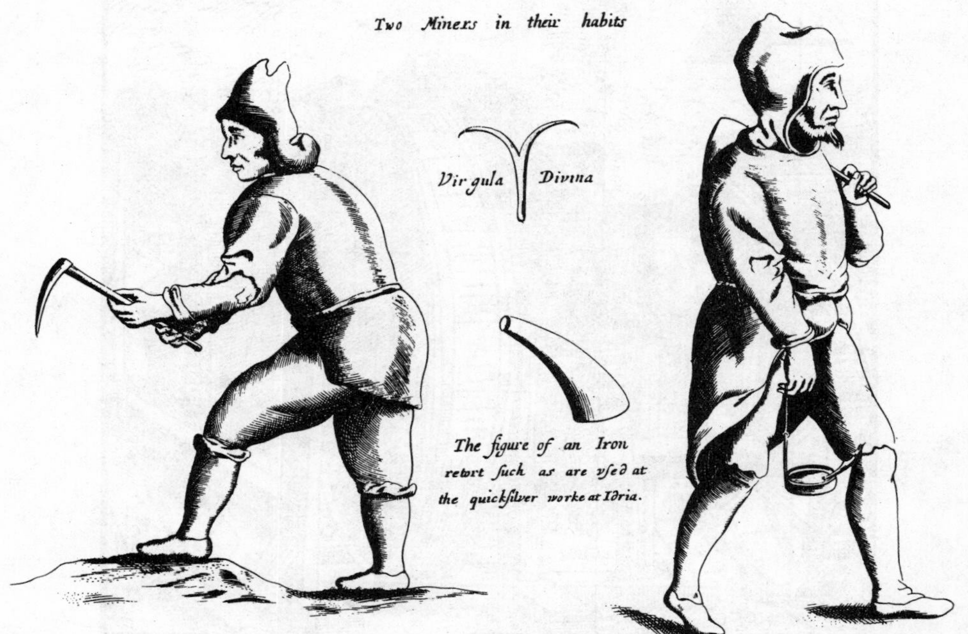

FIGURE 23—*Miners with tools, lamp, and divining-rod; also an iron retort for the extraction of mercury. Seventeenth century.*

I saw also another *Mine*, about eighty of our *Fathoms* deep, and much worked: They have divers sorts of *Ore* . . . *Silver* and *Copper*, *Silver* and *Lead*, or all three; but they work them only for *silver*. They have divers *damps* in these *Mines*, where it is deep. . . . The greatest trouble they have is by dust, which spoils their *Lungs*[1] and *Stomachs*, and frets their *Skins*. But they are not so much troubled with water; and have very good *Engines* to draw the water out [33].

IV. GREEK QUARRYING

Many Greek and Roman quarries bear well known names. The Isle of Paros in the Aegean and Mount Pentelicon in Attica yielded beautiful Greek marbles; from Roman Egypt came porphyry and yellow marble; and the mines of Carrara

[1] Though it has a new name, pneumoconiosis is obviously not a new disease.

were worked extensively by the Romans, though they became even more famous later. Despite all this we know little on the practical side. At the Athenian quarries all the processes connected with the work were grouped together in charge of a *technitēs*, who was not a citizen and therefore was largely ignored in contemporary literature. A *technitēs* was a skilled worker in almost any trade—mason, carpenter, and so forth—though a metal-worker was distinguished by a special title (*chalkeus* or *sidereus*).

Greek architecture reveals its origin from wooden buildings in a post-and-lintel technique (ch 12). The marble or other stone columns were formed from a series of drums, involving accurate mason's work but not very large blocks. It appears also that the masons were responsible for a large quantity of figure carving, the statues being merely roughed out by their designers.

The best account of quarrying is in the commentary on the 'Description of Greece' of Pausanias by Sir James Frazer (1854–1951). Pausanias wrote in the second century A.D., and recorded whatever might be expected to interest the intelligent tourist. The principal Greek quarries, he says, were on Mount Pentelicon, whence was hewn the stone for the Parthenon and other famous fifth-century buildings in Athens; Paros and Naxos in the islands yielded respectively white and grey statuary marble, and Mount Hymettos sandstone for general use in Athens.

The following account of the Pentelicon quarries is adapted from Frazer:

Pentelicus is a range of hills extending north-west to south-east for about four and a half miles. The ancient quarries lie on the south-western side of the highest peaks. Five-and-twenty of them may be counted, one above the other; the highest at over 3300 feet above the sea. An ancient road, very steep and rugged, leads up the eastern side of the principal gully and is roughly paved; the blocks of marble were probably brought down it on wooden slides. . . . The road appears to end at the principal quarry, 2300 feet above the sea. Here the rock has been quarried away so as to leave a smooth perpendicular wall. . . . The marks, delicate and regular, of the ancient chisels may be seen in horizontal rows on the face. The marks show that the ancients regularly quarried the marble in rectangular blocks, first running a groove round each block with the chisel and then forcing it out with wedges. The effect of this has been to leave huge rectangular cuttings in the side of the mountain.

The stone is a white marble of a close, fine grain, readily distinguished from Parian marble—the other white marble commonly used by Greek sculptors and architects—by its finer grain and opaque, milky whiteness; whereas the Parian is of large transparent crystals, of a glittering snowy whiteness. . . . Pentelic marble, alone among all Greek marbles, contains a slight tincture of iron; hence its surface, when long exposed to the weather, acquires that rich golden-brown patina which is so much admired on the columns of the Parthenon and other buildings. The Parian marble, on the other hand,

FIGURE 24—*Ancient Greek limestone quarries in Syracuse, the* Latomia dei Cappuccini.

though it weathers more easily than the Pentelic on account of its coarser grain, always remains dazzlingly white [34].

The marble from Hymettus, which is still quarried in large quantities, is inferior to the Pentelic and was apparently not in common use before the third century B.C. It was employed for tombstones, inscriptions, and the casing of buildings. The principal quarries are on the western side of the mountain. Vestiges of the ancient road by which the blocks were brought down from the quarries may be seen leading in serpentine curves down the slope.

Plato (d 347 B.C.) in his *Critias* records that the bare rocks round Athens used for quarrying were once covered with rich soil and supported timber suitable for the roofs of large buildings, some of the rafters surviving to his own time. Now, he says, the trees having been cut down, the rainfall is lost and the land will only support bees; that is, it was covered with heather [35]. The honey of Hymettus is still famous.

Perhaps the most impressive of Greek quarries is not in Greece itself but in Sicily, where the great limestone quarries of Syracuse were specially used under Hiero II (269–216 B.C.). One rock-face there is more than 27 m in height and 2 km long (figure 24). It has been estimated that, including stone worked in galleries, over 40 million cu m have been extracted. Some 7000 Athenian prisoners of war were employed here in 413 B.C.

V. ROMAN QUARRYING

The city of Rome was not to any great extent built of marble, though in Augustan and even earlier times it was largely dressed with it. Stone, brick, and concrete were common structural substances (ch 10). Both Seneca (d A.D. 65) and Pliny (d A.D. 79) pour scorn on the extravagance of those not content with these excellent materials. Seneca writes:

> I found the famous villa of Scipio Africanus built of squared stone [peperino]. The bath was narrow and rather dark. Under this low and sordid roof stood Scipio; he disdained not to tread so mean a floor. But who in our time would condescend to have such a bath? Men think themselves poor and mean unless the walls are decorated with large and precious roundels, unless Alexandrian marble is pointed and inlaid with Numidian mosaics, with borders faced with complex patterns arranged in many colours like paintings, unless their swimming-pools are lined with Thasian white marble, once a rare and wonderful sight in any temple [36].

Calcareous travertine was the stone earliest used in Rome. It formed the ancient walls of the city and can be seen *in situ* on the Tarpeian Rock. The Catacombs and much of the Colosseum are built of it. The travertine formation is about 150 m thick in all. The quarry most extensively worked during the Empire was near Tivoli, situated on the right bank of the river Aniene (*Anio*). 'The travertine of Tivoli is very porous, whitish or cream-coloured, hardens under exposure, is easily sawn into slabs, is submissive to the chisel and an excellent stone in every respect' [37].

A good example of the skill of the Roman quarryman from late Republican times to the present is the use of hard and durable basaltic rocks for paving. It was known to Livy (59 B.C.–A.D. 17), Vitruvius (first century B.C.), and others as *silex* or *lapis siliceus*. In modern language it is a basic, not a siliceous, rock. All the roads leading out of the capital, in some cases to great distances, 'were paved with *Lapis Silex* hewn in polygonal pieces' [38]. The measurements of these pieces were evidently exceedingly precise, for they were fitted together with great accuracy and skill. Remains of the ancient pavements are still visible in Rome and the neighbourhood, as on the *Via Praenestina* (figure 461) and the Appian Way, on which was one of the ancient quarries about 3 miles from the city. The *Via Sacra* was paved with this stone, from which the cloacae or sewers were also constructed.

Procopius (sixth century A.D.) alludes to the excellence of this paving:

> The breadth of this [Appian Way] is such that two wagons going in opposite directions can pass one another. The paving, which extends as far as Capua, was of naturally hard millstone. After working these stones until smooth and flat and cutting them to a polygonal shape, Appius fitted them together without mortar so securely and closed the joints so firmly that they look as though grown together. After sustaining so much traffic of all sorts they have neither separated at the joints, nor been worn and reduced in thickness [39].

Appius, however, had actually used gravel for his road, which was replaced by the basalt blocks within the Republican era (ch 12).

The most famous quarries in the whole world are those of the Carrara district, which embraces some half-dozen communes. In classical times they were usually referred to as of Luna, the nearest city and harbour. The marbles are mostly pure white,[1] though some beds are marked with green and some are pearly and tinged with pink. The quarries were described by Strabo (d *c* A.D. 21) [40] and, though traces of earlier working may be found, they became popular throughout the Empire under Augustus. Famous examples of the use of Carrara marble are the Apollo Belvedere, the Arch of Constantine, and various columns throughout Rome. Tibullus notes that the streets of Rome were crowded with wagons loaded with marble [41]. Modern working has obliterated the evidence of ancient technique.

After the fall of Rome little is known of quarrying for some centuries. Where building in stone took place, as in Rome, the material was often pillaged from temples, circuses, and town-buildings to be re-dressed for its new purpose. A little Carrara marble was wrought in the mid-fourteenth century. Michelangelo (1475–1564) reopened some quarries to acquire material for his famous statues, his 'David' being cut from the Polvaccio quarries. Numerous coloured marbles and breccias came from the same district.

[1] Pure white Carrara marble is often, but wrongly, called 'Sicilian' in the trade.

A few Roman quarries outside Italy require brief notice—especially those of Egypt, where conditions favour preservation of details. In the eastern desert are the great *Mons Claudianus* quarries of hornblende-granite, which may have been started by the Emperor Claudius (A.D. 41–54), employing Jewish prisoners. There is a pillar still *in situ* measuring 18×2·6 m. On account of its extensive use in Trajan's forum in Rome the rock is often called the 'granite of the forum'. It seems to have been first used for making mortars for grinding pigments, drugs, and the like. The large blocks were cut by wedge-holes; if saws were used, they were plain copper blades fed with sand and emery.

The great quarries of imperial porphyry were also founded under Claudius. Inscriptions show that at one time the granite and the porphyry workings were under the same overseer or procurator. These quarries were on the summits of the Dukhan range in the eastern desert of Egypt at the latitude of Asyut; they were always imperial property, perhaps because nowhere else was stone of the right colour for 'imperial purple' to be found. The main quarry is about 25 m high, and thousands of blocks, including the great sarcophagus of the Empress Helena (d *c* 330), which is a monolith 4 m high by 2·4 m long, were carried thence for some 155 km to the Nile.

It has been conjectured that the Egyptian sculptors worked both porphyry and basalt with stone balls, bronze tools, and emery (vol I, p 478). A copper blade set with square emery teeth about 1·6 mm long has been found in the Greek palace of Tiryns. It had been used for cutting limestone. For hard rocks a toothless blade was used; the hard grains of emery soon became embedded in the copper. It was inferior only to diamond for the purpose.

The word porphyry means purple[stone]. It is unfortunate that geologists have applied this name to a kind of stone different from that of these quarries. In modern nomenclature the word porphyry is a general term for igneous rocks with relatively large crystals in a fine-grained ground mass. This description would apply to Pliny's 'green Lacedaemonian marble' quarried between Sparta and Marathon; it consists of crystals, light green or brownish, in a dark-green felspathic base.

The Egyptian imperial porphyry was so popular in Rome that in Constantinople it became known as 'Roman stone'. Many pillars and blocks of it were taken from Rome during the later Empire. Constantine erected a column 30 m high, of 8 drums each 3·4 m long, a tribute to the careful work of the mason, although he found it necessary to cover the joints with laurel wreaths in gilded bronze. He also built in the palace the famous porphyry room lined with slabs of this rock. To that room later empresses had to go for the birth of their children;

hence the title *Porphyrogenitus*—'born in the purple'—for members of the royal family. The room is fully described, as her own birthplace, by the learned princess Anna Comnena (1083–?1148).

Of Roman quarries outside Italy among the most noteworthy are those of Baalbek near Damascus, where Antoninus Pius (A.D. 138–161) built a temple to Jupiter. Three blocks of limestone in the temple are 63 × 13 × 10 ft and are the largest known to have been used at any time in building (figure 376). Still in the quarry is one blocked out but not detached at the base, 69 × 14½ × 12 ft, estimated to weigh about 1500 tons (figure 377 and plate 2 A).

A contrast to the above is the green serpentine quarried around Gytheion in Laconia near the head of the gulf. The rock there, according to Pausanias, 'is not one continuous mass, but the stones are dug as pebbles. They are hard to work, but once wrought they might grace sanctuaries of the gods, and they are especially fitted to adorn swimming-baths and fountains' [42].

There is a well preserved Roman granite quarry in the Odenwald (Hesse, Germany), where the Romans obtained building material for such towns as Oppenheim, Mainz, Mannheim, Treves, Wiesbaden, and Aix-la-Chapelle. Fragments of cut stone still remain on the slopes. Surviving blocks have been named for descriptive purposes and are found in all stages of preparation, revealing the Roman mason's methods. The 'pyramid', for example, is split into three huge pieces by two horizontal rows of holes for the insertion of wedges. From the 'altar-stone' two blocks, for pillars, have already been removed [43]. The altar-stone, 3–5 yds long and almost 6 ft high, is technically the more interesting. Deep incisions accurately cut with a saw indicate the intention to split off blocks 20–24 inches in thickness. Holes indicate that wedges would have been used to help in this work.

> In this process the surface of fracture itself assumed a somewhat curved form which could be used to advantage when greater curvature was required. The saw-blade used must have had a length of at least 15 feet, and have produced cuts only one-sixth of an inch wide, that is, not wider than the most modern frame-saw. . . . The so-called 'Giant Pillar' . . . is also very striking. Its length measures 30 ft., its thickness at the lower end 5 ft. 1 in. and at the upper, 4 ft. 1 in.; the volume is therefore well over 9 cubic yards and the weight about 15 tons. In breaking off the pillars, the procedure consisted in marking the length of the pillar by deep incisions. Then a half of the pillar was worked to a state of completion. Along the sides of this half-column a deep furrow was chiselled in the block, and numerous wedge-holes were made in the furrow. After inserting and saturating the wedges with water, the back part of the pillar broke off in a convex shape, caused by the semicircular course of the lines of pressure due to the swollen wedges. . . . When the blocks had been broken off and prepared the fine work was begun; that is, they

were cut down to the right size, smoothed off and polished by exactly the same methods and with the same tools as we still use [44].

From the position of these quarries, it is clear that the saws employed in them were necessarily hand-worked, but by the end of the fourth century, and perhaps earlier, power-saws were in use. Ausonius in his poem on the Moselle (*c* A.D. 370) speaks of one stream which 'turns his millstones in furious revolutions and drives the shrieking saws through smooth blocks of marble [limestone], and so hears from either bank a ceaseless din' [45]. This stream enters the Moselle just below Treves, so that its limestone may have replaced the hand-sawn and harder granite of the Odenwald. Incidentally, the mill-stones mentioned by Ausonius were probably made from the lavas of Niedermendig near Andernach on the Rhine, which were wrought from very early times; many were imported into Britain, where some 30 Roman examples have been recorded. This stone is still being worked.

Much of the excellence of Roman building-stone may be attributed to scrupulous care in selection and to preliminary weathering at the quarry, as the practical Vitruvius suggests [46]. The Roman blocks were carefully marked at the quarries with a number, to record origin, and in many cases with the date.

In Britain much quarrying was done in Roman times. The most obvious source of information is the wall, known as 'Hadrian's', that extends from coast to coast from Wallsend in Northumberland to Bowness in Cumberland (plate 40B). Many quarries have been recognized there, mostly in grit-stone. Several have contemporary inscriptions. The best known is the 'written rock of Gelt', near Brampton, of A.D. 207. It states that a company of the Second Legion was employed there.

The stones used for the Wall and forts were very carefully chosen:

> In some parts of the line, in Cumberland especially, they were brought seven or eight miles. A quartzose grit was generally selected, because it was hard and its roughness gave a good key for the mortar. Behind the ashlar of the wall the core is of rubble firmly imbedded in mortar. The part of the stone exposed to the weather is cut across the 'bait' [bate, grain], so as to avoid the scaling off by the lines of stratification; the stone tapers towards the end which is set into the Wall. Bonding tiles, so characteristic of the Roman Wall of London, are altogether dispensed with. Stones of the shape and size described were just those which could be most easily wrought in the quarry, most conveniently carried on the backs of legionaries, and most easily fitted [47].

The care exercised by the Romans in choosing the beds and maturing the stone in the quarry before use is well seen at York. Magnesian limestone was brought from the Huddleston quarries near Tadcaster, probably by water down the Wharfe and up the Ouse, which was tidal to York. In the multangular tower on the walls the Roman stones are as fresh as when first laid, but the badly weathered stones, though larger, are a thousand and more years later. In the church of St Mary, Bishop Hill, much of the masonry is of Saxon age; it consists of re-used Roman stones. In a Roman monumental-mason's yard at Cirencester most of the stone was great oolite from a quarry close by, but one altar-pedestal was of Bath stone; the owner probably used Bath stone for his best work, as he set up an altar with his name in that city. The Roman quarries at Bath itself adjoined the Fosse Way near the present Bloomfield Crescent, south of the city, and may have reached as far as English Combe. There is evidence that the stone was mined as well as worked open-cast. Bath stone was used as far as Silchester, near Reading; a carving from Colchester has also been identified as of Bath stone.

It is hardly exaggerating to say that the Romans used nearly every one of the English building-stones. Tufa, so familiar in Rome, was naturally a favourite and was seized upon when it could be found; it was used near Maidstone (Kent), Dursley (Gloucestershire), and Malton (Yorkshire). Roofing was usually of tiles, but any fissile stone was welcomed. Cumberland, Welsh, and Charnwood (Leicestershire) slates have been found in Roman remains. Stonesfield (Oxfordshire) slates reached as far east as Buckinghamshire, and Pennant stone (from south Wales) to the Wiltshire Downs (p 415).

VI. POST-ROMAN QUARRYING IN ENGLAND

When the Romans departed the technique of stone-cutting fell at once into decay. At the very beginning of the fifth century the Britons sent for help to Rome, and one legion was sent back. 'Even so,' says Bede (673–735), 'the islanders raised the [Antonine] Wall, not, as they had been directed, of stone, having no artist capable of such work, but of sods' [48]. Bede relates that St Benedict of Wearmouth crossed into Gaul *c* 650 and brought back masons to build a church. For it he also imported glaziers, who could not be found in England [49].

The oolitic rocks of Bath and the Cotswolds were worked in Saxon times from about 675. Legend has it that the abbeys of Malmesbury and Sherborne were originally of wood. After a journey to Rome, St Aldhelm (d 709) went into the fields with his monks, flung down his glove, and said, 'Dig here and you shall find great treasure.' The treasure turned out to be the stone of which the abbeys were then rebuilt. In commemoration of this event the trademark of the present-day Bath and Portland stone firms is St Aldhelm's glove.

In the north of England the Saxons always used grit-stone and neglected the magnesian limestone, which was reintroduced only under Norman influence. Thus even on the east Yorkshire wolds 'Saxon' church architecture is of grit brought across the whole outcrop of the limestone and the Vale of York. At Ledsham, near Leeds, the tower and a few other features of Saxon date are of grit brought at least 8 miles; other parts of the church, of the thirteenth, fourteenth, and fifteenth centuries are of local magnesian limestone. Similarly, the world-famous northern carved and inscribed Saxon crosses are of grit. At Collingham there is such a cross of about 650, but in the same churchyard there is one of the twelfth century of the local limestone. In York Minster the mutilated statue called 'Our Lady of York' may mark the change; it appears to be copied from a Saxon illuminated manuscript in the Norman technique. The pre-Norman masons were apparently incapable of cutting the limestone, which needed saw, hammer, and chisel, whereas the grit could be carved elaborately with a bit of stick and a lot of patience. Perhaps the Danish subjugation of the north from about 850 may have a bearing on the subject.

From the Norman Conquest onwards there are frequent allusions to quarries in Britain, in documents giving the right to dig stone for building. Since the cost of stone lies mainly in transport it was natural that the nearest source should be used, and this sometimes determined the site of the building. Thus the discovery of a bed of suitable stone close to the site selected for the Conqueror's votive abbey of Battle was so opportune as to be deemed a miracle. Yet sometimes stone was brought from afar. The famous limestone of Caen in Normandy was much used in Britain, especially in the south-east; but it was soon found that, while it was unrivalled for detailed interior carving, many common English stones were preferable for exterior work. When a monastery was to be founded in an area where there was no suitable stone the material could often be brought only by water, which thus fixed the site (p 428).

For York Minster stone was brought from the quarries of Thevesdale, Huddeston, and Tadcaster down the Wharfe, and from Stapleton down the Aire into the Ouse, and so up to St Leonard's wharf, whence it was carried on sleds to the mason's yard. Westminster and London were mainly supplied from Surrey . . . and Kent. . . . The tough 'Kentish rag', used by the Romans for the walls of London, was much in demand for rougher masonry, and in a contract for building a wharf by the Tower in 1389, it was stipulated that the core of the walls should be of 'raggs' and the facing of 'assheler de Kent'. The Reigate stone, on the other hand, was of superior quality and more suited for fine work, and we find it constantly used for images, carved niches and window tracery [50].

Certain quarries at an early date acquired renown, such as those of Beer in Devon, from whose labyrinthine galleries stone was carried to St Stephen's, Westminster, in 1362, to Rochester in 1367, and elsewhere. Bath stone, mostly quarried near Box, Wiltshire, was sent in 1221 to the royal palace at Winchester.

Portland stone was already appreciated in the fourteenth century, when it was used at Exeter Cathedral and at Westminster Abbey. A list of stones bought in 1367 at Rochester is of interest as showing the various sources from which it was derived. There were bought 55 tons of Beer freestone, 62 tons of Caen stone, 45 tons of Stapleton freestone, 44 of Reigate stone, 195 of freestone from Fairlight, Sussex, 1850 tons of rag from Maidstone, and a large quantity from Boughton Mounthelsea, Kent. The prices paid for these stones are known. Portland stone attained its greatest fame at the hands of Wren after the Great Fire of London in 1666, and has been the chief stone of London ever since.

The manufacture of balls for artillery, a use for cut stone that might not occur to many technologists, was mainly centred in Kent. At first these were for the various forms of catapult (ch 20), but for long they were used in cannon before being replaced by cast iron. The Maidstone quarries were turning out stone shot during the early years of Henry VIII.

True slates were worked in Cornwall and Devon, not only for roofs but for facing walls of granite, the joints of which are apt to admit damp. The use of Cornish slates can be traced back to the thirteenth century, but, as in Wales and Cumberland, the use was for long a regional one. For roofing, the use of fissile limestones of the Jurassic rocks, such as occur at Collyweston, Northamptonshire, and Stonesfield, goes back to Roman times and remained popular throughout the Middle Ages.

Chalk, though in building now almost confined to cement-making, had three main uses. (*a*) The harder bands in the Lower Cretaceous were used, usually under the name of 'clunch', as a building-stone in some church architecture, for example in parts of Ely Cathedral, St Albans Abbey, Windsor Castle, other churches in the home counties, and some of the Cambridge colleges. (*b*) A huge amount of chalk has of course also been used for burning to lime. Thus at Chislehurst in Kent there are extensive galleries, up to 3 miles in total length. (*c*) In Essex and Kent much chalk was formerly obtained from narrow shafts sunk through the overlying layers for direct marling of the fields without burning. These shafts open out in the chalk and are known as dene-holes. There is an early reference to this practice in Agricola's *De natura fossilium* (1546). He there describes chalk as a form of fuller's-earth. It is, he says, 'the rock which forms headlands in Britain and France and is used in the former for fertilizing the fields' [51].

With chalk must be mentioned flint, which is found as nodules in it. These were worked in Neolithic times and perhaps earlier (vol I, p 558). For building-purposes the flints have sometimes been carefully worked into approximate cubes and set in mortar. Occasionally such squared flints alternate with chalk, with Caen stone, or with other whitish material, producing a chequered or magpie effect; examples are frequent in East Anglia and include Norwich guildhall. The setting of unbroken flints in mortar is a curious feature of buildings in East Anglia and elsewhere. The material was satisfactory when wrought stone could be obtained for the corners. When, in these stoneless regions, even that amount of stone was too expensive, the builders dispensed with corners, and the church towers are circular.

Strange to say, one of the hardest rocks used in English architecture belongs to the Tertiary beds of the Thames basin, otherwise entirely soft and useless. The sarsen (without pebbles), and the Hertfordshire pudding-stone (with pebbles), found only in isolated lumps up to 12 ft long and 50 tons in weight, consist of quartz-sand with a strong siliceous cement. These blocks are scattered over Buckinghamshire, south Oxfordshire, and Hertfordshire. They were largely used in the buildings of Windsor Castle and often for the foundations of the local churches.

By far the most important of English building-stones are those from the Jurassic oolitic limestones. Most of them are included in buildings at Oxford, which is near the centre of the outcrop. These have been specially investigated and the quarries exploited, the dates at which they were worked, the weathering-capacity of the stone, and the tools employed are known.

Lower in the geological column came the Taynton stones and those of several other quarries in the Windrush valley in Oxfordshire. These are mentioned in Domesday Book (1086), but there is evidence that they were in use for a thousand years before that. They were mostly cut with a 'jad' (deep groove), and the characteristic chevron marks can still be seen on many of the old faces. Plot relates in his 'Natural History of Oxfordshire' (1677) that single stones weighing 100 to 300 tons, of 16 cubic ft to the ton, were produced in his time; one of 30 cubic ft has been recorded [52]. More or less similar, though variable, stones from the same district were used not only in Oxford but at Windsor Castle. In recent times much repairing of old structures and some new work have been carried out in the Lower Oolite rocks of Clipsham, Rutland, where the Holywell quarry was worked from medieval times. Clipsham stone was used in Windsor Castle in

1363–8. Many stones used at Oxford and elsewhere have been spoilt by the use of iron cramps, which have rusted and consequently swollen and so caused the dressed stone to flake off.

Few ancient quarryman's tools are known, for they are constantly wearing out and are seldom preserved; but so little have the tools varied through the ages

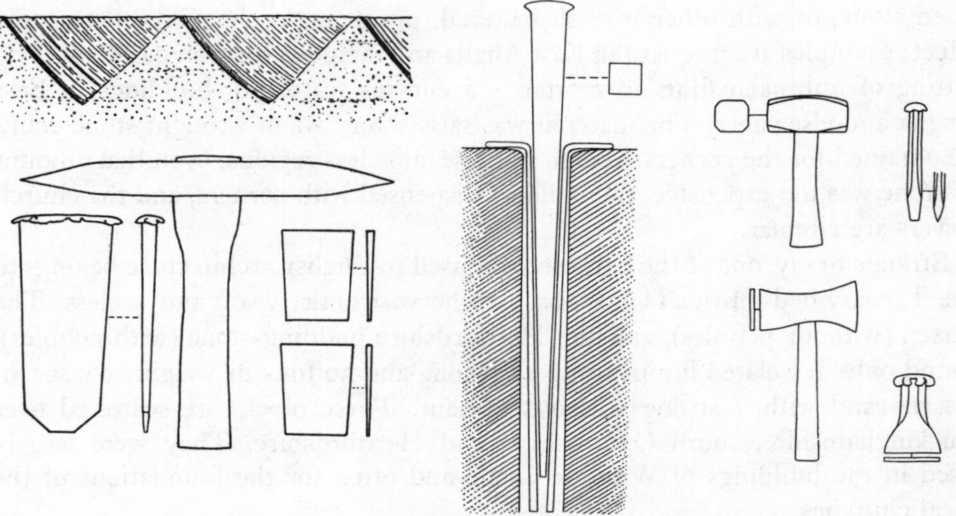

FIGURE 25—*Modern tools used in Oxfordshire for quarrying and shaping building stone.* (Left group) *Double-edged adze with a wedge and two slats, for splitting stone at the quarry. A series of triangular slots, as shown above, is cut along the line chosen for splitting. In each are placed two slats, and between them the big steel wedge is hammered down.* (Centre) *Also for splitting stone at the quarry, a plug and two feathers are placed in each of a series of holes 18 in deep, and the plug is hammered home.* (Right group) *Masons' tools. Above, hammer and punch; below, walling-hammer and pitching-tool. Work with the blunt edge of the walling-hammer or with the hammer and pitching-tool gives rough convex surfaces. These may be flattened with the hammer and punch, and further smoothed with the sharp right edge of the walling-hammer. Scale 1/8.*

that the forms chosen for illustration here are practically independent of date (figures 25–26).

Many types of stone have been used for ornamental purposes at various times and places. In England these are mainly represented by two 'marbles' and alabaster. Purbeck 'marble' is really a dark grey or brownish muddy limestone crammed with the white fossils of a fresh-water snail, *Paludina*. It takes a good polish and occurs near the top of the Jurassic. A similar rock, the Sussex 'marble', is also of fresh-water or estuarine origin and is found near the base of the succeeding Cretaceous. It is known also, from the locality where it is best developed, as Petworth 'marble'. The Purbeck industry is centred round Corfe Castle, Dorset, and was once a royal prerogative. It seems to have taken the place of the dark Tournai 'marble' from the other side of the channel. On the other hand Purbeck columns were used in Mont St Michel in Brittany, but were replaced by restorers who failed to identify their provenance. Purbeck stone is also found in the cathedral at Lisieux.

Many fragments of Purbeck stone, datable to the first century A.D., have been found in a Roman settlement at Studland, Dorset. It was used in 1205 in Chichester Cathedral, somewhat earlier in Durham, and even in Dublin. Its main use was for single thin columns in clustered piers, as at Canterbury (1175), London, Westminster, Exeter, Ely, Lincoln, and Durham, and also for recumbent effigies.

FIGURE 26—*Mason's axe and bolster. The bolster is a wide-bladed chisel used by the stone-carver on plane surfaces. From a brass of 1643 at Idbury church, Oxfordshire.*

The 'marble' was not only quarried at Purbeck but mostly worked into columns or carved into figures on the spot. These Purbeck effigies are distributed by the hundred all over the country. But however much we may admire the execution, we must not assume that they bear any resemblance to the persons whom they commemorate; the great majority of them are merely conventional. Thus in 1253 we find Henry III ordering the sheriff of Dorset to cause 'an image of a queen' to be cut and carried to the nunnery of Tarrant Keynston, there to be placed over the tomb of his sister, the late Queen of Scots. Heads only are frequently inserted as ornaments, and the carver, whether in Purbeck, other stone, or wood has sometimes seized the opportunity to introduce a representation of himself or a contemporary, or even a caricature.

Alabaster is the usual name for massive sulphate of calcium, as opposed to the wholly crystallized form selenite: both are covered by the term gypsum. Alabaster occurs in abundance in the red marl of the Upper Keuper formation. The chief quarries were at Tutbury, Staffordshire, and Chellaston, Derbyshire, but most of the stone, after winning, was worked at Nottingham. There was also a school of alabaster-carving at York, accessible by water down the Trent and up the Ouse, and workshops were to be found at London, Lincoln, Bristol, and other towns. It is a very soft material but capable of taking a polish and well adapted for finely detailed carving. Its general suitability is indicated by the name 'satin spar' given to a pure white, shiny, fibrous, and glossy variety. Its earliest known use is at Tutbury in the moulding of a Norman doorway of about 1160, but the great period for this work is from the middle of the fourteenth century onwards, when it became popular for effigies and also for small sculptured scenes for the reredos of altars. The latter are commonly known as retables. A substantial export trade in English alabaster sprang up, examples being sent to Iceland, Spain, Italy, and all intermediate countries. Many of the best specimens extant are in foreign countries, where they have to a large extent escaped iconoclastic activities.

Of alabaster effigies the most famous include those of King Edward II (d 1327) at Gloucester, which has a particularly beautiful head, and of his son, John of Eltham (d 1336), at Westminster. Of the retables, a complete set of scenes of the Passion is preserved at Naples. It consists of a central Crucifixion flanked on each side by three subsidiary scenes, attendant angels, and so on. These scenes were highly coloured with paint and gilded, only bare flesh being shown by bare alabaster. Much of the colour has now disappeared. The reredos of the Chapel of the Garter at Windsor Castle cost £200—a great deal of money at the time. It is of alabaster and was brought from Nottingham in ten carts, each with eight horses, which took 17 days on the journey. In several cases the actual carvers were considered sufficiently important to be mentioned by name.

REFERENCES

[1] STRABO IX, C 399. (Loeb ed. Vol. 4, p. 274, 1927.)
[2] PLINY *Nat. hist.*, XXXI, xxviii, 49. (Bohn ed. Vol. 5, p. 490, 1856.)
[3] DEMOSTHENES *Contra Pantaenetum*, 26–29 (XXXVII, 974). (Loeb ed. 'Private Orations', Vol. 1, p. 392, 1936.)
[4] POLYBIUS XXXIV, ix, 8–11. (Loeb ed. Vol. 6, p. 318, 1927.)
[5] POLYBIUS XXI, xxviii, 15. (Loeb ed. Vol. 5, p. 300, 1926.)
[6] STRABO III, C 146–8. (Loeb ed. Vol. 2, pp. 38, 44, 46, 1923.)
[7] *Idem* III, C 144. (Loeb ed. Vol. 2, p. 32, 1923.)
[8] DIOSCORIDES V, 96, i–ii. (Ed. Wellmann, Vol. 3, p. 67, 1914.)
[9] PAUSANIAS III, xii, 10. (Loeb ed. Vol. 2, p. 76, 1926.)
[10] BEDE, the Venerable. "The Seven Wonders of the World" in 'Complete Works', Latin and English ed. by J. A. GILES, Vol. 4, p. 13. Whittaker, London. 1844.
[11] PLINY *Nat. hist.*, XXXIII, xxxi, 97. (Loeb ed. Vol. 9, p. 74, 1952.)
[12] *Idem Ibid.*, XXXIII, xxi, 71. (Loeb ed. Vol. 9, p. 54, 1952.)
[13] LUCRETIUS *De rerum natura*, VI, 806 ff. (Loeb ed., p. 500, 1924.)
[14] HAVERFIELD, F. J. "Romano-British Somerset" in PAGE, W. (Ed.) 'The Victoria History of Somerset', pp. 334 ff. Constable, London. 1906.
[15] DAVIES, O. 'Roman Mines in Europe', pp. 13 ff. Clarendon Press, Oxford. 1935.
[16] CAESAR *De bello gallico*, V, xii. (Loeb ed., p. 250, 1937.)
[17] AGRICOLA, GEORGIUS. *De re metallica*, English trans. by H. C. HOOVER and LOU H. HOOVER, p. 5 and footnote 11. Mining Magazine, London. 1912.
[18] PONTOPPIDAN, ERIK (the Younger). 'The Natural History of Norway', part 1, pp. 183 ff. Linde, London. 1755.

[19] TACITUS *Germania*, v. (Loeb ed., p. 270, 1914.)
[20] THEOPHILUS PRESBYTER *Schedula diversarum artium*, III, 46–48. Ed. and trans. by A. ILG, p. 219. Quell. Kunstgesch. Kunsttechn., Vol. 7, 1874. The text in most manuscripts gives the river as Gihon, but see Gen. II. 11. On Theobald's more recent edition of the work and the latest theory on life and date of its author see p 351 and ref. [6].
[21] CAREW, RICHARD. 'The Survey of Cornwall', fol. 8v. Stafford for John Jaqqard, London. 1602.
[22] AGRICOLA, GEORGIUS. *Bermannus*. Froben, Basle. 1530.
[23] *Idem. De re metallica Libri XII*. Froben, Basle. 1556.
[24] *Idem Ibid*., trans. by H. C. HOOVER and LOU H. HOOVER, pp. xxix ff. Mining Magazine, London. 1912.
[25] *Idem. Bermannus* (in a collection of treatises by AGRICOLA), p. 432. Froben, Basle. 1546.
[26] *Idem. De re metallica*, trans. by H. C. HOOVER and LOU H. HOOVER, p. 160. Mining Magazine, London. 1912.
[27] *Idem Ibid*., pp. 181, 184.
[28] *Idem Ibid*., p. 184.
[29] *Idem Ibid*., pp. 194–5.
[30] *Idem Ibid*., p. 200.
[31] *Idem Ibid*., p. 217.
[32] *Idem Ibid*., p. 218.
[33] BROWNE, EDWARD. 'A Brief Account of Some Travels in Hungaria . . . and also Some Observations on the Gold, Silver . . . Mines . . . [and] An Account of Several Travels through . . . Germany', pp. 135 ff. Printed for B. Tooke, London. 1673, 1677.
[34] FRAZER, SIR JAMES G. 'Pausanias's Description of Greece', trans. with commentary, Vol. 2, pp. 424 ff. (Commentary). Macmillan, London. 1898.
[35] PLATO *Critias* in 'The Dialogues of Plato', trans. by B. JOWETT (4th ed. rev.), Vol. 3, p. 794. Clarendon Press, Oxford. 1953.
[36] SENECA *Epistulae morales*, LXXXVI. (Loeb ed. Vol. 2, p. 310, 1920.)
[37] PORTER, MARY W. 'What Rome Was Built With', p. 18. Frowde, London. 1907.
[38] LIVY XLI, xxvii, 5. (Loeb ed. Vol. 12, p. 276, 1938.)
[39] PROCOPIUS V, xiv, 7–11. (Loeb ed. Vol. 3, p. 142, 1919.)
[40] STRABO V, C 222. (Loeb ed. Vol. 2, p. 350, 1923.)
[41] TIBULLUS II, iii, 43–45. (Loeb ed., p. 265, 1912.)
[42] PAUSANIAS III, xxi, 4. (Loeb ed. Vol. 2, p. 132, 1926.)
[43] NEUBURGER, A. 'The Technical Arts and Sciences of the Ancients', trans. by H. L. BROSE, figs. 543–5. Methuen, London. 1930.
[44] *Idem. Ibid*., pp. 402 ff.
[45] AUSONIUS *Mosella*, lines 362–4. (Loeb ed. Vol. 1, p. 252, 1919.)
[46] VITRUVIUS II, vii, 5. (Loeb ed. Vol. 1, p. 110, 1931.)
[47] BRUCE, J. C. 'The Roman Wall' (2nd ed.), pp. 63, 65–66. J. R. Smith, London. 1853.
[48] BEDE, the Venerable. "Ecclesiastical History of England" in 'Complete Works', Latin and English ed. by J. A. GILES, Vol. 2, p. 63. Whittaker, London. 1843.
[49] *Idem*. 'Lives of the Holy Abbots of Wearmouth and Jarrow.' *Ibid*., Vol. 4, p. 367, 1844.
[50] SALZMAN, L. F. 'English Industries of the Middle Ages' (enl. ed.), pp. 85 ff. Clarendon Press, Oxford. 1923.
[51] AGRICOLA, GEORGIUS. *De natura fossilium* (in a collection of treatises by AGRICOLA), pp. 201–2. Froben, Basle. 1546.
[52] PLOT, ROBERT. 'Natural History of Oxfordshire', pp. 76 ff. Oxford, London. 1677.

BIBLIOGRAPHY

ADAMS, F. D. 'The Birth and Development of the Geological Sciences.' Baillière, London. 1938.

ALFONSO X. 'Lapidario del Rey D. Alphonso X. Codice original.' [A photolitho. reprod.] Ed. by J. FERNÁNDEZ MONTANA. Blasco, Madrid. 1881.

ARDAILLON, E. 'Les mines du Laurion dans l'antiquité.' Bibl. Éc. franç. Athènes Rome-fasc. 77. Thorin, Paris. 1897.

ARKELL, W. J. 'Oxford Stone.' Faber, London. 1947.

BROMEHEAD, C. E. N. "Geology in Embryo (up to 1600 A.D.)." *Proc. Geol. Ass., Lond.*, **56**, 89–134, 1945.

Idem. "Practical Geology in Ancient Britain." *Ibid.*, **58**, 345–67, 1947; **59**, 65–76, 1948.

COLLINGWOOD, R. G. and MYRES, J. N. L. 'Roman Britain and the English Settlements' (2nd ed.). Clarendon Press, Oxford. 1937.

EVANS, JOAN. 'Magical Jewels of the Middle Ages and the Renaissance, particularly in England.' Clarendon Press, Oxford. 1922.

HOPE, SIR WILLIAM H. St. J. and PRIOR, E. S. 'Illustrated Catalogue of the Exhibition of English Medieval Alabaster Work . . . June, 1910.' Society of Antiquaries, London. 1913.

HUDSON, D. R. "Some Archaic Mining Apparatus." *Metallurgia*, **35**, 157–64, 1947.

HULL, E. 'A Treatise on the Building and Ornamental Stones of Great Britain and Foreign Countries.' Macmillan, London. 1872.

KING, C. W. 'The Natural History of Gems or Decorative Stones.' Bell and Daldy, London. 1867.

LUCAS, A. 'Ancient Egyptian Materials and Industries' (3rd ed. rev.). Arnold, London. 1948.

NASH, W. G. 'The Rio Tinto Mine: its history and romance.' Simpkin Marshall, London. 1904.

PORTER, MARY W. 'What Rome Was Built With.' Frowde, London. 1907.

RICKARD, T. A. 'Man and Metals' (2 Vols.). McGraw-Hill, London. 1932.

SAGUI, C. L. "Economic Geology and Allied Sciences in Ancient Times." *J. econ. Geol.*, **25**, 65–86, 1930.

SANDARS, H. "The Linares Bas-Relief and the Roman Mining Operations in Baetica." *Archaeologia*, **59**, 311–32, 1905.

WEST, L. C. 'Roman Britain. The Objects of Trade.' Blackwell, Oxford. 1931.

2

METALLURGY

R. J. FORBES

I. METALS IN THE GRECO-ROMAN WORLD

THE Greco-Roman world depended far less on metallurgy than we do today. Thus metals played a very small part in its architecture, its aqueducts, and its ships. Above all it must be remembered that the machinery of the time was mainly of timber. Greco-Roman metallurgy was a continuation of that of the ancient Near East, and was essentially a phase of the Iron Age wherein copper and bronze were being only slowly displaced by the newcomers, iron and steel. Greeks and Romans added but two fundamentally important discoveries to the knowledge of earlier smiths, namely (*a*) the production of mercury and its application to the extraction of gold, and (*b*) the manufacture of the copper–zinc alloy, brass. Metallurgy remained essentially a charcoal-smelting technique, with all the limitations thus implied. Coal is scarce in the Mediterranean region, and its use in metallurgy was first attempted in western and central Europe by native smiths in their forges and smelting-furnaces. In the Mediterranean, deforestation had become serious by classical times, so that the prices of timber and charcoal were rising steadily and ominously. In this still agricultural world the mines and smelting-sites were mere 'islands in a sea of fields and meadows', as Rostovtzeff well put it. There were no true industrial areas.

Despite all this, the fact remains that the Romans left few surface deposits of ores for later generations to discover (figure 4). Their attempts to exploit deeper strata of ores were, however, limited by the cost of sinking shafts, hauling ores, and draining mines, and by shortage of labour. The last became acute in the second century A.D. The Romans took over Hellenistic mining and smelting enterprises and enlarged their scale, stimulating subdivision of labour and the rise of mining specialists. They drew up mining laws, and were organizers and administrators of mining enterprises rather than metallurgical innovators. Their lack of suitable prime movers defeated their attempts to exploit the deeper bodies of ore or to bring large quantities of metallurgical products into common use.

II. GOLD

During classical times Egypt lost its virtual monopoly of gold-production (vol

I, pp 579–82). Alluvial supplies were discovered in the Mediterranean region and in Europe, but the few important gold deposits in western Asia Minor, notably that on the banks of the Pactolus [1] and that near Astyra in the Troad [2], said to have been the source of the wealth of Troy, were becoming exhausted. The gold alluvial deposits of Thrace and Macedonia, exploited since the Bronze Age, were now but poor producers, as were those in the Ægean islands, though much native gold came to Greece and Rome by trade with central and western Europe. In classical tradition Britain was believed to produce much gold; this was probably from Ireland. The Romans organized large-scale gold-production in the Basque area of Spain by constructing big engineering works to tap and conduct huge quantities of water to the gold-bearing strata (p 670), which were in this way broken up [3]. Sufficient experience had been gained to locate gold-bearing ores by observation of certain white gravels [4], and it had become known that various copper ores, especially the oxidized top strata, often contained a profitable proportion of gold.

There were few changes in the actual refining processes, which consist mainly in the separation of metallic gold from its matrix. Assaying had long ago been developed[1] and there were no important changes in the methods of refining by cupellation, or in the salt and sulphur processes for separating gold from baser metals.[2] To these older processes was now added amalgamation. Because of the mobility of mercury, the extraction of the metal from its ores (chiefly the sulphide, HgS) presents little difficulty. On roasting, mercury distils over, so that a furnace with facilities for condensation is required. The Romans started to produce mercury in Spain about the turn of our era. The crushed gold ores were treated with mercury and the resulting amalgam separated from gangue by pressing it through leather. Finally the mercury was distilled off. This process is mentioned in a garbled form by Pliny [5] and was common practice in the early Middle Ages [6].

In Roman times liquation[3] was applied in Spain for the extraction of gold from copper, thus adding materially to the gold-production of the ancient world,

[1] The 'trial by fire' used for testing gold would, of course, remove valuable silver, if present, along with the other metallic impurities. Gold–silver alloys were rubbed on a touchstone, and the colour of the mark was judged either in the light of experience or against the mark of an alloy of known composition. Nor should we forget the more sophisticated correlation of composition with density, which was, according to legend, discovered by Archimedes in his bath (third century B.C.).

[2] Cupellation: impure gold and silver were melted with lead; in the furnace the molten lead was oxidized to a litharge containing all the base-metal oxides, leaving a button of silver–gold alloy. To separate silver from gold, use was made of sulphur, antimony sulphide, or a variety of salts including sodium chloride, with the general purpose of compounding the silver (e.g. as sulphide, chloride) and leaving a button of refined gold. The mineral acids were not yet in use.

[3] Liquation: the separation of metals by graded fusion of the metals themselves or their eutectic mixtures.

and recovering a valuable by-product from copper smelting. This new method of extracting gold came into use in the first century B.C., and had increasing importance. Copper and lead, when cast together, do not dissolve in each other to any appreciable extent, and silver is very much more soluble in lead than in copper; the liquation process made use of these facts. Copper containing silver and gold was melted with three or four times its weight of lead, and cast into cakes. During this operation the alloy of silver and gold was taken up by the lead. The lead was then slowly melted out of the cakes, taking with it the gold and silver and leaving behind a porous cake of copper. The gold and silver were recovered from the lead by cupellation.

The production of pure gold and silver by these various processes allowed the ancients to alloy them in different proportions for artistic and other uses; a famous alloy was 'Corinthian bronze', a gold–silver–copper combination [7]. The production of gold leaf was greatly improved by the Romans, who were able to make a leaf about twice as thick as our modern lower limit, one ounce of gold being beaten out into some 750 leaves 4 inches square [8]. Gold was also rolled. The amalgamation process was used to recover gold wire from gold cloth [9], and was one of the basic operations of the early alchemists (ch 21). Indeed, gold-refining was closely linked with early alchemy, and with the belief that metals attained perfection in nature and grew in the mines [10]—a belief probably encouraged by the dendritic forms of some native gold and silver deposits. Mines were sometimes closed down for a period, in the hope that the ores would replenish themselves.

III. SILVER AND LEAD

Lead and silver were largely obtained by smelting galena (lead sulphide), which generally contains small quantities of silver. The roasting and reduction techniques developed for this purpose in the ancient Near East (vol I, p 584) spread westwards in Mycenaean times from Asia Minor to Crete, the Ægean, and the Greek mainland. Most of the small galena deposits in the islands, such as those of Thasos, were exhausted by classical times, or even flooded by the sea, like those of Siphnos [11]. They were completely eclipsed by the discovery of rich galena strata at Laurion, near Athens, in the sixth century B.C. (p 2); these mines became a main support of the Athenian economy. Further mines were worked in Thrace and Macedonia, while galena and silver were obtained from Hungary, the Tyrol, the Harz mountains, Britain, and France, where Near Eastern production-techniques were adopted generally during the La Tène period in place of the more primitive local smelting-techniques of the later

Bronze Age. The silver mines of Sardinia, and above all those of Spain, were important in the Roman economy. Hannibal collected funds for his invasion of Italy (218–203 B.C.) from the rich silver mines of Cartagena in the south of Spain. The author of the first book of the Maccabees (*c* 100 B.C.) had no doubt of 'what the Romans had done in Spain for the winning of the silver and gold' (1 Macc. viii. 3). By classical times ores had become an important factor in world politics.

At Laurion an argentiferous lead was extracted from galena by a combination of roasting and smelting.[1] The lead-silver alloy thus obtained was then concentrated. The roasting converted the galena partly to litharge and partly to lead sulphate, and the smelting, carried out by raising the temperature when the correct stage of desulphurization was reached, yielded lead. The process was conducted in primitive furnaces built of clay and stones, which were re-used when the furnace broke down—after a relatively short life. In several parts of the Roman Empire, such as Britain, the furnaces were built on the slopes of hills, using the prevailing wind for draught. The crude lead, called *stagnum* [12], contained 45–180 oz, or even more, of silver per ton.

It seems that the ancients knew something of the principle of the nineteenth-century Pattinson process. If lead rich in silver is melted, the crystals first formed on cooling consist of almost pure lead. The liquid thus becomes richer in silver, and the process may be repeated until the silver content reaches 1–2 per cent. This alloy can then be cupelled to obtain the silver, and the litharge that is formed can be carried back to the smelting-furnace. Modern analyses have proved that this is what happened at Laurion. It is established that the ancient Greeks were able to desilver their lead so efficiently that only 0·02 per cent of silver was left in it. The Roman smelters improved even on this, their limit being about 0·01 per cent. This remained the limit even in the sixteenth century, when Agricola quotes a figure equivalent to 0·008 per cent. The modern process leaves no more than 0·0002 per cent of the silver.

The excellent early results were obtained only with great loss during the preliminary stages of the process. The complex of techniques—roasting, smelting, liquation, and cupellation—could not then be controlled by temperature-measurements and chemical analysis. Though the ancient metallurgist could desilver the crude lead quite efficiently, its extraction from the ore was very wasteful, and thus much of the silver was lost with the lead in the slag of the

[1] The processes may be simply represented as follows:

Roasting: $2PbS+3O_2 \rightarrow 2PbO+2SO_2$; $PbS+2O_2 \rightarrow PbSO_4$.

Smelting: $PbS+2PbO \rightarrow 3Pb+SO_2$; $PbS+PbSO_4 \rightarrow 2Pb+2SO_2$.

initial roasting and smelting. At Laurion, for example, to obtain a fairly pure lead a high temperature had to be used to ensure complete slagging of the impurities, which included zinc and iron. Thus over 10 per cent of the lead and more than 33 per cent of the silver remained in the slag.[1]

The Romans were well aware of such losses, and with their larger experience managed to smelt more efficiently. Thus Strabo relates that 'the silver mines of Attica [that is, at Laurion] were originally valuable but they failed. Moreover, those who worked them, when the mining yielded only meagre results, melted again the old refuse, or dross, and were still able to extract from it pure silver, since the workmen of earlier times had been unskilful in heating the ore in the furnaces' [13]. When production stopped at Laurion in the second century A.D., after the Romans had extracted further quantities of silver, something of the order of 0·005–0·01 per cent silver remained in over two million tons of slag, and a final extraction was undertaken in 1864 by a smelting firm.

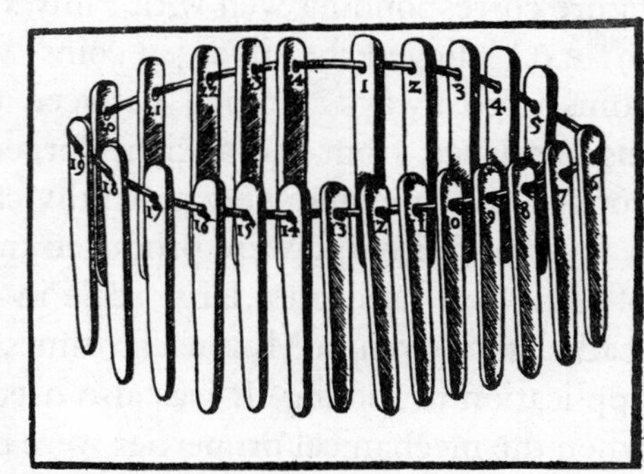

FIGURE 27—*A graded series of needles for rubbing on the touchstone. The colour of the streak left is compared with a test piece and indicates how much silver is present. Agricola gives lengthy details. Agricola, 1556.*

We know little of the construction of the furnaces used. It seems that the furnaces at Laurion were fairly tall, the upper portion serving for smelting, the lower for oxidation, before the cupellation. We are told that the lead-smelting furnaces of Spain were provided with chimneys 'so that the gas [sulphur dioxide] from the ore may be carried high into the air, for it is heavy and deadly'. Representations of such furnaces seem to have been found in Sardinia [14]. The Romans often used bowl-furnaces, excavated on the spot and having *tuyères* blowing on the surface to oxidize the lead and blow away the dross when cupelling. The cupellation was, however, usually carried out in small crucibles lined with bone-ash to absorb part of the dross.

The touchstone[2] (figure 27) was available for a first assay of silver, and,

[1] The efficient smelting of metals depends on an adequate control of the slag. The object of forming a slag is to flux the gangue, or earthy material mixed with the ore, and, no less important, to absorb impurities such as the oxides of unwanted metals. An acidic slag, e.g. silica, will remove basic oxides, and a basic slag, commonly lime, will take up acidic impurities. Unfortunately the important metallic oxides (e.g. of Pb, Fe, Sn) rarely fall sharply into either of these categories, and consequently too great an emphasis on the purity of the smelted metal may involve much loss of it in the slag.

[2] The touchstone was usually of the black, non-splintering, siliceous mineral known as lydite or Lydian stone.

alternatively, cautious heating in the hands of an expert could yield much information on impurities [15]. Probably a careful assay would involve gravimetric cupellation—in effect, refining on a small scale.

Silver was important not only in jewelry but in the coinage, which was commonly of silver. These coins usually contained some 10 per cent of copper, a figure corresponding well with Pliny's statement that the tribune Livius Drusus (91 B.C.) 'alloyed the silver [in coins] with one eighth part of bronze' [16]. Such coins were always stamped between two dies (figure 446); they can easily be distinguished from the ancient forgeries, which usually contained too much copper and zinc and were generally cast (p 485).

Lead and litharge were produced in considerable quantities. Sheets and slabs of lead were used on a large scale in buildings. Stone walls were bonded with lead clamps even in Mycenaean times, and sheet lead found its most frequent application in roofing. It was also used in cheap bronzes, as a substitute for tin, when the mechanical properties were unimportant and low casting-temperatures were in demand. Lead pipes played a prominent part in Greek and Roman water supply (ch 19), and lead solders were common; the recipes given by Pliny [17] for solders made of tin and lead compare well with those of today. The Romans knew that lead will draw black lines on parchment and our word 'lead-pencil' is derived from this use, though modern pencils contain not lead but graphite.

IV. TIN, ANTIMONY, AND ARSENIC

The Greek and Roman metallurgists found supplies of tin inadequate and difficult to get. The deposits of stream-tin and vein-ore in the Near East had long given out, so that the metal had to be imported from central and western Europe. The tin from northern Spain, Brittany, and Cornwall was indeed indispensable to classical metallurgy and was widely traded. 'The British metal', as it was sometimes called, was re-exported from Egypt to Somaliland and even to India. The mines of Spain had been exploited since the Early Bronze Age, and were worked continuously during Roman times until they became exhausted about A.D. 250. The mines of Brittany were worked from 500 B.C., but were probably soon abandoned because of Spanish competition. Cornish tin-production seems to have entered international trade about the same time; it continued at least until the days of Caesar. There is evidence that the production of tin declined during the early Empire, perhaps because supplies of stream-tin were becoming exhausted, but when extraction from vein-ore began in the third century A.D. the Cornish mines regained their old prominence, and broke the monopoly that Portugal and Spanish Galicia had enjoyed during the first and second centuries A.D.

The method of extracting tin from its ore (cassiterite, SnO_2) underwent little change. While stream-tin was still available, a simple smelting with charcoal yielded the metal. When vein-ore began to be worked, an initial calcination at 600–700° C, to remove volatile constituents such as sulphur and arsenic, preceded the smelting, but this operation was well understood by the ancient metallurgist. Lime was generally added as a flux when smelting.

The Romans constructed large aqueducts in Spain for the washing of stream-tin, and produced many millions of tons of it. The tin imported from Cornwall was cast into H-shaped slabs of a weight suitable for slinging one on either flank of a pack-animal. The remains of the old furnaces reveal that a surprisingly pure metal was obtained, but only at the cost of a heavy loss in the slag and by volatilization. Analysis of Cornish and Spanish tin and lead shows that the purity usually reached over 99·9 per cent.

The main use of the tin was for the manufacture of bronze, but its alloy with lead was almost as important. Tin-lead alloys were used as solders for many purposes [17], and also, from Roman times, as pewter for household vessels and other objects. For pewter, a proportion of about 70 per cent of tin to 30 of lead was common. During the Middle Ages only 5 to 15 per cent of tin was used, whereas in later times 80 to 85 per cent of tin was the general proportion. We are also told that 'a method discovered in the Gallic provinces is to plate bronze articles with white lead [i.e. tin] so as to make them almost indistinguishable from silver' [18]. This was effected by dipping the objects in molten alloys such as one containing brass, lead, and tin. Ancient nomenclature dubbed tin 'white lead' and lead 'black lead', and the confusion is made worse by the fact that antimony too is often called 'lead', though the more correct term is *plumbum cinereum*. The ancients also referred to antimony as *stimmi*, a name used both for the metal and for its ore, stibnite (Sb_2S_3).

Antimony was occasionally produced in pre-classical times by simply smelting stibnite with charcoal, and there was no change in this process during classical times. Antimony bronzes were often used in prehistoric Europe (Hungary, Gaul), and the metal found some use during classical times as a substitute for tin in bronzes. With arsenic, it played a large part in the experiments of the early alchemists.

Arsenic, as sulphide (realgar, As_2S_2, and orpiment, As_2S_3), is so commonly associated with the natural sulphides of other metals that it has appeared as a constituent of alloys from the earliest times. This was seldom advantageous, for arsenic has very little metallurgical value and confers its characteristic brittleness on its alloys. Arsenic compounds were, however, well known as pigments, ingredients of medical recipes, and poisons.

V. COPPER AND ITS ALLOYS

(*a*) *Copper and Bronze*. By classical times copper metallurgy had advanced far beyond the simple smelting of such surface ores as oxides and carbonates that had been practised earlier. Local pockets of such ores were worked, but had generally been exhausted by the sixth century B.C.

Most copper mines mentioned in classical literature yielded sulphide ores. The best and largest sources in classical antiquity were Cyprus, Asia Minor (with deposits near the two famous metallurgical centres Samos and Pergamum), Macedonia, Tuscany (near Volterra), Tyrol, Styria, Carinthia, Cornwall, Devon, Anglesey, and finally Spain and Portugal, where the deposits eclipsed all others in Roman times. Strabo waxes enthusiastic over the quality of these Spanish ores, saying: 'Up to the present time, neither gold, nor silver, nor copper, nor iron has been found anywhere in the world in a natural state either in such quantity or of such good quality. Mining is profitable beyond measure there, because one-fourth of the ore brought out by their copper-workers is pure copper' [19]. This refers to the Rio Tinto body of ores on the southern flanks of the Sierra Morena.

The evolution of copper-smelting in classical times involved no new principles. Locally, very old smelting-techniques served for small deposits of ores, but the larger production from the complex sulphide ores demanded much skill and attention for the removal of impurities, including other metals, from the copper. Hence in classical times, and indeed for many centuries to come, both primitive and more sophisticated smelting-methods were in use side by side.

While copper may be extracted from its oxide and carbonate ores by heating with charcoal, the processes necessary to deal with sulphide ores are a great deal more complicated. They comprise three stages. (*a*) The ore is roasted to drive off excess of sulphur and such volatile oxides as that of arsenic, and partially to oxidize iron sulphides; the roasting also enables the ore to be more easily broken up. (*b*) The ore is smelted after the addition of fluxes. The molten material separates into two layers in the fore-hearth, slag above and 'matte' below, the purpose of the operation being to obtain a matte of as nearly pure cuprous sulphide as possible. Inadequate control at this stage leads to loss of copper in the slag and a quantity of impurities in the matte. (*c*) The last stage of the operation is one of partial oxidation. The matte is melted, and, with charcoal floating on its surface, air is forced over the agitated liquid. (Nowadays air is blown through the liquid in a converter.) The remaining iron, lead, and other metals with negative electrode potentials, are thus oxidized, and oxidation of part of the copper sulphide follows. The interaction of the remaining copper sulphide and its various oxidation products yields copper and sulphur dioxide, the main reactions being:

$$2Cu_2S + 3O_2 \rightarrow 2Cu_2O + 2SO_2$$
$$2Cu_2O + Cu_2S \rightarrow 6Cu + SO_2.$$

Stage (*c*) must be carefully concluded so as to eliminate all impurities and yet to leave the metal more or less free from cuprous oxide, which induces brittleness. With skill, it was apparently possible to obtain very pure copper without further refining, for up to Greek times no further refining processes were known; yet specimens of ancient copper are often very pure.[1] The peculiar refining operation known as 'poling', which holds such a prominent place in elementary chemical text-books, may, however, have been known to some Roman metallurgists [20]. Wooden poles are forced under the surface of the molten copper and the volatile carbon compounds given off at that temperature reduce any oxide present. The first reference to it, however, is in a much later text by Theophilus on the purification of copper (*c* 950):

> Take an iron dish of the size you wish, and line it inside and out with clay strongly beaten and mixed, and it is carefully dried. Then place it before a forge upon the coals, so that when the bellows act upon it the wind may issue partly within and partly above it, and not below it. And very small coals being placed round it, place the copper in it equally, and add over it a heap of coals. When by blowing a long time this has become melted, uncover it and cast immediately fine ashes of coals over it, and stir it with a thin and dry piece of wood as if mixing it, and you will directly see the burnt 'lead' adhere to these ashes like a glue, which being cast out again superpose coals, and blowing for a long time, as at first, again uncover it, and then do as you did before. You do this until at length by cooking it you can withdraw the 'lead' entirely. Then pour it over the mould which you have prepared for this, and you will thus prove if it be pure. Hold it with the pincers, glowing as it is, before it has become cold, and strike it with a large hammer strongly over the anvil, and if it be broken or split you must liquefy it anew as before. If, however, it should remain sound, you will cool it in water, and you cook other [copper] in the same manner [21].

The reference to lead is not to be taken literally, since at this time the different metals and their oxides were not clearly differentiated. Theophilus is pointing

[1] Up to 99 per cent in Roman times. Modern processes yield by oxidation a refinement about 99·3 per cent, and by electrolytic methods over 99·9 per cent.

out a practical test that will show when the process has been completed. The final, important remark about brittleness is worthy of emphasis.

Naturally, in smelting sulphide ores the possibilities of returning intermediate products to some earlier stage were not always fully realized in antiquity; nor was such economy always necessary, for much depends on the impurities of the ore. Classical texts seldom reveal much about operations of this kind, and we still have insufficient analyses of ancient metals and slags to enable us to form an adequate picture of smelting-techniques at every ancient copper-smelting site. Roasting the ore, smelting the roasted ore, a second and possibly a third re-smelting, and the final refining were not always separate operations; some were often carried out simultaneously and even in the same furnace.

In Roman times, sulphide ores were reduced on a large scale to 'black' copper (about 95–96 per cent pure) almost everywhere in the smelting centres, and this semi-crude metal was refined in the manufacturing centres. In a few regions like Roman Britain, however, ingots of pure copper are fairly common.

The furnaces varied from centre to centre. Primitive bloomeries and forges such as pre-classical smiths used were still built, but shaft-furnaces came to be preferred. Most of these were 'two-storeyed', the upper and lower parts being connected by a fairly narrow neck [22]; the lower part, which was narrower than the upper part, was filled with charcoal, the upper with ore. They permit continuous smelting, as their long use has proved [23]. These furnaces were used at Mitterberg in the Tyrol (1600–800 B.C.), and in most Greek regions, including Cyprus. The Roman copper smelters of Populonia, Italy, and Rio Tinto, Spain, seem to have preferred furnaces of the bloomery-fire and Catalan hearth type (figure 38 B). Many classical authors agree that the charcoal used for copper smelting should be prepared from certain trees and plants [24].

The fluxes used depended on the impurities of the ore; but no doubt use was made of whatever was conveniently at hand, as long as it would easily combine with the gangue, or earthy part of the ore, to form a fusible mixture. In Cyprus the Romans used a flux with a high manganese content; in Rio Tinto, different forms of quartz; in Thasos, lime. The matte recovered during the last two smeltings was also often returned to the first smelting, as an ancillary flux. Even the prehistoric smelters of Mitterberg seem to have understood this procedure. However, these smeltings were hardly ever efficiently carried out, for the slags always contain much copper. The lead content of certain slags indicates the use of liquation in some cases. Lead or galena might also have been added to the ore for smelting, with the purpose of taking up impurities, as described by Biringuccio in the sixteenth century.

The refining of copper was carried out in a cupola-furnace like the normal pottery kiln. This type has been found at Salzburg and in Lorraine and is far better for melting copper at high temperatures than the shaft-furnace. Different qualities of refined copper were recognized, *aes coronarium* being sheet-copper,

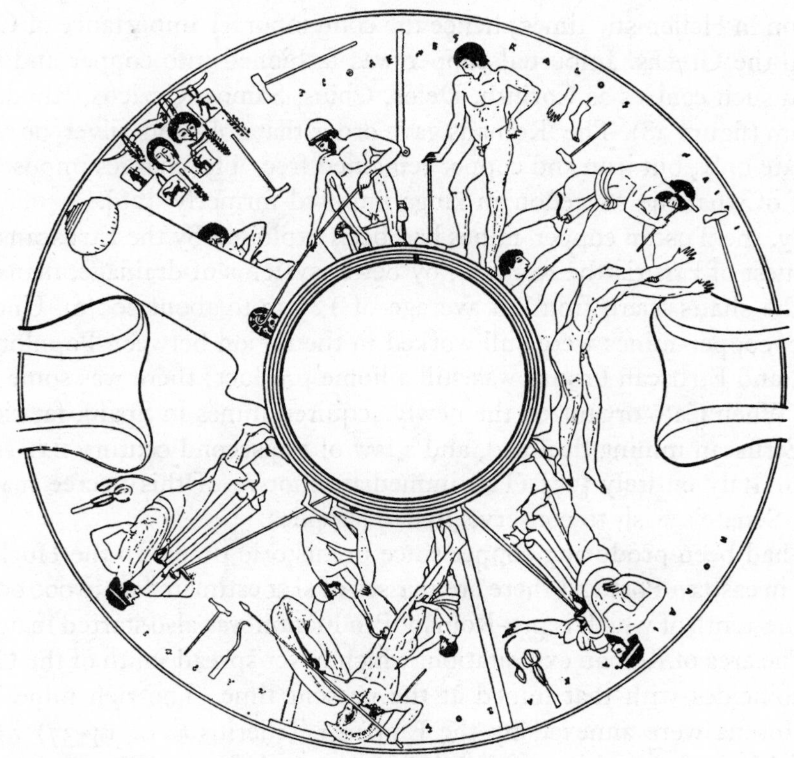

FIGURE 28—*Red-figure painting on a Greek bowl from Vulci, near Rome. It shows a foundry and the casting of a statue in bronze. Fifth century* B.C.

aes regulare malleable refined copper, and *aes caldàrium* the more impure black copper.

The production of copper in Cyprus partly supplied the demands of the mainland of Greece from Mycenaean times, when Greeks had founded trading-posts on the islands. However, the native kings, and later the Ptolemies, who conquered Cyprus, guarded their monopoly very strictly. The Roman emperors continued this policy, and granted the exploitation of the Cyprian mines to King Herod [25]. Though Cyprus as a source of copper was now eclipsed by Spain, its copper trade continued steadily until the fourth century A.D., and even through Byzantine rule. After an interlude of neglect, production was again taken up by the Frankish and Venetian rulers (1192–1573) and later by the Turks.

Local production in Greece ceased soon after Mycenaean times. By the time of Strabo, about the turn of the Christian era, the copper mines even on the island of Euboea were exhausted, and only the Othrys range had any significant production in Hellenistic times; hence the contemporary importance of Cyprian copper to the Greeks. Imported copper was fashioned into copper and bronze objects in such centres as Corinth, Delos, Chios, Samos, Cyzicus, Rhodes, and Pergamum (figure 28). The 'Romans gave order that gold and silver be worked by the state only, but iron and copper remained free. The charges imposed were only half of what the Macedonian kings imposed formerly' [26].

In Italy, the Tuscan copper-mines had been exploited by the Etruscans. After the conquest of Etruria the Romans, by better systems of drainage, managed to deepen the shafts there from an average of 125 m to about 200 m. Under the Republic, copper-mines were still worked in the region between Populonia and Volterra, and Etruscan bronze was still a home product: there was some tin ore near by. When Cato organized the newly acquired mines in Spain, far richer in metals, Etruscan mining declined, and a law of the second century B.C. stopped mining in Italy entirely [27]. The immediate motive of this decree may have been the Senate's wish to conserve Italian supplies.

Spain had been producing copper since prehistoric times. In the Huelva district and in eastern Portugal there are masses of slag estimated at 30 000 000 tons, at least one-tenth of which is pre-Roman. Production was also started in northern Spain. The area of Roman exploitation, which never spread south of the Guadalquivir, coincides with that mined at the present time. The rich mines of the Sierra Morena were annexed by the Emperor Tiberius (A.D. 14–37) when he executed Marius, proprietor of the *Aes Marianum* [28].

Roman Gaul depended largely on imports, though it used also small local deposits. Caesar claims that Britain had to import copper and bronze [29], but mines in Cheshire, north-west Wales, Anglesey, and Shropshire were at work well before the advent of the Romans, who continued the production. Many ingots—flat round cakes of copper weighing 30–50 lb—have been found bearing the abbreviated name of the contractors. This copper was mostly refined and often over 99 per cent pure. In Germany, copper ore was worked mainly outside the limits of Roman domination. The large copper-mines of the eastern Alps had declined during the Early Iron Age, and they virtually stopped production on a large scale after 400 B.C.—though some copper still came to Rome from these regions, as well as from Hungary.

Roman copper and bronze became an important item in world-wide trade, and copper coins were exported to Silesia, East Prussia, and the Baltic. There is a definite connexion between the copper alloys in use in those regions and that of Roman coinage. Copper exports to India by way of the Red Sea also developed—the home production of India even in Hellenistic times had been insufficient—and hoards of Roman coins have been found in trading-stations on the west coast of India [30]. We are told, too, that copper was exported to Oman and the Persian Gulf; this no doubt was surplus European metal shipped from Alexandria to Malabar and Barygaza (Broach) and thence re-shipped by Indian traders.

Roman large-scale mining and smelting required new forms of organization. Leases were given by the state to contractors, often Roman capitalists investing on a large scale. There was also a very far-reaching division of labour, but, contrary to common opinion, the mines and smelting centres were not served by slave-labour only. Though the unskilled work was largely done by convicts and slaves, many skilled experts were employed. We possess the mining-laws of the rich mine at Ajustrel in Spain. They contain tax-regulations, and divide the work at the mines into mining, cleaning, crushing, smelting, preparing, breaking-up, separating, and washing. Mining-dumps and piles of rock were taxed. The laws record many experts who had valuable privileges and who, like the copper- and bronze-smiths, formed powerful corporations.

Similarly, there was much specialization and division of labour in the manufacture of copper and bronze objects in Italy itself. At Capua by the end of Republican times there was a true factory system of manufacture, and large quantities of this copper and bronze ware, quite uniform in workmanship, are found throughout Italy and even in Germany, Sweden, and Finland. The metal was alloyed at Capua with tin or zinc, and cast, polished, carved, or forged. Plain kitchen utensils and farm implements required the service of many individual shops, and the copper-smiths were both craftsmen and salesmen. Obsolete metal articles were melted down and recast, and even articles in stock were liable to similar treatment for supplying immediate need. The copper-smiths were well organized; in Rome itself they formed a powerful guild, and one at Milan had 1200 members.

(*b*) *Brass*, an alloy of copper and zinc, was the most important newcomer amongst the alloys of classical times. The addition of zinc to copper forms alloys progressively stronger, harder, and less malleable than pure copper until the content of zinc reaches about 40 per cent. The most useful varieties of brass lie within this range, their general characteristic being the wide range of properties obtainable not only by varying composition but by cold-working. Their tensile strengths are comparable to those of the tin-bronzes (15 to 45 tons/in^2) and lower than those for steels (30 to 100 tons/in^2). A brass with 20 per cent of zinc simulates the colour of gold, while alloys rich in zinc are a dull white. The zinc ores used were the carbonate and silicate, both known as calamine.

Though brass appears to have been discovered in the first millennium B.C. (see below), neither the classical nor the medieval metallurgists were able to prepare metallic zinc, which indeed was not available until the sixteenth century A.D. All brass was prepared by a cementation process. A high temperature (about 1300° C) is needed to reduce calamine to zinc with charcoal, and since this is above the boiling-point of the metal its vapour distils off and, under ordinary circumstances, is reoxidized and condenses as zinc oxide. If, however, copper is heated in a mixture of powdered zinc ore and charcoal, a proportion of the zinc formed in the vicinity of the copper will diffuse into it and form a coating of brass. This was the old cementation process, in principle similar to the manufacture of steel by diffusion of carbon.

The inventors seem to have been the Mossynoeci, a people living in Pontus south of Trebizond, in whose territory lived many tribes or castes of smiths, such as the Chalybes. We are told that 'the bronze of the Mossynoeci excels because of its gloss and its extraordinary whiteness. They do not add tin but a special kind of earth, that is smelted with the copper. They say that the inventor did not disclose his secret, therefore the old bronzes of this region are remarkable for their excellent qualities and the later ones do not show them' [31]. These Mossynoeci, perhaps the Meshech of the Old Testament, seem to have discovered the manufacture of brass in the first millennium B.C.

Classical authors identified the natural zinc carbonate variety of calamine with the white zinc oxide which is frequently formed as a by-product in copper smelting, and which condenses against the furnace roof. They called both *cadmeia* or *cadmea* [32]. They considered the ore an impure product and preferred the purer condensate, which they sometimes called *pompholyx*. *Cadmeia* was identical with the later *tutiya* or *tuthy* of the Muslim alchemists and metallurgists, and with our tutty.

The earliest reference to brass is possibly the eighth-century inscription in the palace of the Assyrian King Sargon II (722–705 B.C.) at Khorsabad. It mentions the covering of wooden doors with a sheet of 'white bronze' from Musasir, the mountain region west of the Tigris. The Persians certainly used brass from the fifth century B.C. onwards, and Darius (521–486 B.C.) had a 'cup which looked like gold but had a disagreeable smell'. Persia was fairly rich in zinc ores, but brass-production on a large scale did not begin there before the sixth century

A.D., when this technique was exported to India; some two centuries later it passed on to China.

However, brass-manufacture spread only slowly to the west. Neither Homer (? eighth century B.C.) nor even Herodotus (fifth century B.C.) knew it. If we consider true brass to have an average of 30–38 per cent of zinc and not merely a little zinc due to the ore smelted, its manufacture was quite exceptional in Greece before the Augustan age (first century A.D.).

The same is true for Italy, where Republican coinage does sometimes contain up to 4 per cent of zinc. Veritable brass coinage was, however, issued by the Emperor Augustus, and the percentage of zinc in these coins, at first only some 17 per cent, gradually rose. Nevertheless, brass remained comparatively expensive, and even several centuries later, during the reign of Diocletian (284–305), its value was six to eight times that of copper. The important centres for its production were then Etruria and the Stolberg district near Aix-la-Chapelle, where deposits had been discovered between A.D. 74 and 77. This centre flourished notably between A.D. 150 and 300, but suffered a decline in the fourth century. In the fifth century production began again, and the district became prominent in the eleventh century, as did the then newly discovered deposits of zinc ores in the Meuse valley.

Roman brass objects were exported to Egypt to be re-exported to other parts of Africa, where the alloy was highly valued. Cosmas Indicopleustes (sixth century A.D.) informs us that in his days the natives of Abyssinia still considered it more valuable than silver. In Roman times it was a most important medium of exchange in the Red Sea region and the Abyssinian kingdom of Axum. Some of these brasses were manufactured in Cyprus from *cadmeia* recovered there during the smelting of copper ores.

VI. IRON AND STEEL

From 800 B.C. onwards iron came into use in central Europe on an increasing scale for weapons and tools. The earliest centre of iron metallurgy in Europe was Austria. Ancient Noricum comprised the greater part of Austria east of the Tyrol, and the Noric iron industry provided the munitions for many tribal wars and migrations in central Europe. By about 400 B.C., however, the centre of gravity had moved to the Celtic lands and to Spain. Classical civilization added little to the techniques of iron-manufacture, but the development of large-scale operations and the specialization of labour were prominent in the evolution of the industry.

In Roman times simple hearth-furnaces heated with charcoal still prevailed. After being roasted, the ore was mixed with charcoal and sometimes a flux. On heating, reduction took place and a tough, spongy bloom of iron collected at the bottom, covered by a liquid slag or scoria. The bloom was reheated and hammered into a compact mass, an operation that had to be repeated several times to drive out all the slag. Some bloomeries had a natural draught, like the Roman smelting-furnaces of Populonia, in Etruria, and of Wilderspool, near Warrington, and thus differed little from those of the prehistoric European smiths, except that coal was sometimes used in the process. Usually these bloomeries consisted of a kiln for roasting the ores, a smelting-furnace, and a smith's forge.

One of the oldest specialized iron-furnaces was the Catalan, employed by the early smiths of Catalonia (Spain) and France, and in use there up to the seventeenth century or even later (figure 38 B and p 71). Air was driven into the furnace by means of two bellows, working alternately to obtain a constant blast. Furnaces were still not permanent, and many were built anew after each smelting; they were fired with wood. These processes produced a malleable wrought iron directly from the ore.

Shaft-furnaces of the type described for copper-smelting (p 50) were also used; they were the forerunners of the later *Stückofen* (p 73) and the blast-furnace. They produced a more highly carbonized bloom, which was then worked up into malleable wrought iron or steel in a second operation. The temperatures obtained in such shaft-furnaces were higher but not high enough; and the iron did not remain long enough at such temperature to become sufficiently carbonized for the formation of liquid cast-iron. If such a thing happened it was by chance, and the ancient iron-smith would reject the product as poor and brittle. Thus lumps of cast iron found in Noricum are rejects.

Some prehistoric furnaces at Tarxdorf, Silesia, could produce a semi-fused bloom of about 50 lb in eight or ten hours with the help of 200 lb of charcoal; another 25 lb of charcoal would be needed for the subsequent forging and heating. The Roman furnaces in Noricum rarely produced blooms of more than 100 lb. The ore found there was an iron carbonate containing manganese, which made the manufacture of steel easier. If 'hard iron' or steel was desired, more and thicker charcoal was added, the smelting was continued longer, and the draught was reduced until the desired degree of carbonization had been obtained. But soft iron (wrought iron) was produced as of old, and larger pieces of iron were built up from smaller ones welded together. With their shaft-furnaces, hearths, and forges adapted to different operations like smelting, carbonizing, and welding, the Romans skilfully developed the processes derived from the smiths of the ancient Near East and those from Gaul.

The classical authors emphasize that direct production of steel from ore depends greatly on the character of the ore [33]. The smelters used no fluxes when smelting the more tractable type of ore. At Populonia, the very siliceous ore from the island of Elba was mixed with lime, as was the more clayey ore of Monte Valerio. At the Erzberg, slags of different fusibility have been found, the richest in iron being the oldest.

The possibility of producing steel by a direct process in Noricum was due to the spathic iron (iron carbonate) found there, and the use of well designed 3–6 ft shaft-furnaces. The Romans imported the very superior 'seric iron' from the east, believing it to come from China [34]. In fact it was crucible steel produced in the Hyderabad district of India and traded in round cakes 5 inches in diameter and $\frac{1}{2}$ in thick, weighing about 2 lb. In Islamic times the art of producing crucible steel came to Damascus and Toledo, and 'damask'[1] steel was manufactured in other parts of Europe from such cakes of steel. The cakes were flowed together in two or more directions by blows of the hammer, and the blades, thus shaped after prolonged forging, were quenched and drawn to the desired hardness, polished, and etched. Such blades have on their surface a characteristic watered or streaked appearance.

It seems that the Indian producers sent their steel to the Axumites of Abyssinia, who kept it a secret and allowed the Romans to attribute the steel to China. The Persian kings had received such steel from India, and we are told that Alexander the Great received three tons of it from an Indian king [35]. This imported steel was turned into fancy cutlery and weapons in places like Damascus and Irenopolis. Persian steel made by similar processes was regarded as second only to seric iron during the later Roman Empire.

Apart from this expensive importation there remained always the ancient cementation process, producing an outer layer of steel on wrought iron by carbonization in the forge. The Greek and Roman smiths handled this technique quite well, and knew how to control the final properties of the steel by quenching, tempering, and annealing. Roman swords of the fourth century A.D. show the following techniques for a good cutting-edge: (*a*) welding damask-strips on to both sides of a hard steel blade, (*b*) cutting-edges of hard steel welded on to iron, (*c*) case-hardening by extensive forging.

The iron industry of Cyprus was unimportant, and Crete, contrary to many

[1] Sometimes also called 'damascene', although this term is more usually applied to steel encrusted with gold, silver, or copper wire.

legends, had no smelting-sites. The main producer of iron in the eastern Mediterranean was Asia Minor, with its rich deposits of Pontus, the Taurus mountains, Phrygia, and Caria.

In the Greek homeland during the Mycenaean period iron was as costly as gold, but from the tenth century B.C. it was more generally smelted and gradually became common. In Homer *chalkeus* is a blacksmith, though the poet is describ-

FIGURE 29—*Black-figure painting on an Attic vase from Orvieto, showing a Greek smithy and selection of Greek tools. Of particular interest are the hinged tongs. Sixth century B.C.*

ing the Bronze Age. His heroes wield finely made bronze weapons, and yet Achilles offers prizes consisting of a bloom of crude iron [36]. Much iron was produced from small pockets of ore that had been depleted by the classical age of Greece, for instance in Samothrace and Euboea. Glaukos of Chios is reputed to have invented the welding of iron (*kollesis*) in the seventh century B.C.

Laconia, later famous for its steel, Lesbos, Siphnos, and a few other smaller islands started production only in the fifth century B.C. The iron from these smelting-sites was traded as bars, like the Spartan 'roasting-spits', which may have served for cooking but in trade represented both money and metal. In the Greek towns specialists forged iron into objects needed locally (figures 29 and 30). In the Athens of Pericles (d 429 B.C.) the blacksmiths were foreigners. Armour-factories were run by Pasion and Kephalos. The father of Sophocles was said to have been a blacksmith, and Demosthenes (d 322 B.C.) inherited a sword-factory. We are told that a Sicilian banker of the fourth century B.C. made something of a corner in Chalybean steel, '. . . recognizing that iron was an indispensable commodity, he once succeeded in buying up the produce of all the smelters of iron and made a profit of 200 per cent when a scarcity arose'. In Macedonia the mines were worked even in Roman times.

The Etruscans of northern Italy started the Elban and Tuscan iron mines about 900 B.C., and in these mines can be seen a powerful motive for the expan-

FIGURE 30—*Black-figure painting on an Attic vase, showing the forge of Hephaistos. Sixth century* B.C.

sion of early Rome. In the time of Varro (116–27 B.C.) travelling smiths supplied many villages in Italy [37]. In Rome itself blacksmiths appear in early records, but ritually the metal was a new-comer and certain religious orders were forbidden to use it in their ceremonies. The reader will remember similar taboos against iron in the Old Testament (Deut. xxvii. 5; Joshua viii. 31). The mines of Populonia furnished the arms for Scipio's invasion of Africa in 204 B.C., and indeed during the war period of 200–150 B.C. much of the iron still came from Italian mines, to be fashioned in the region between Rome and Capua. A senatorial decree of the second century B.C. forbidding mining in Italy had no effect on Elba, one of the earliest Greek colonies in Italy. The blooms, produced until well into the Imperial period, were carried thence to the mainland by way of Puteoli and other ports, to be shaped into weapons and tools. The depletion of the forests of Elba may have accounted for the fact that only the first stage was carried out on the island [38]. The Roman state had armour-factories of its own before 100 B.C. Military necessity led to the concentration of the industry in Puteoli, Syracuse, Reggio, Populonia, Volterra, and one or two other places, and this paved the way for later government control. Gradually the industry was transplanted to the provinces, in the neighbourhood of the mines. In general the Roman state was self-sufficient in metals, and only special products such as seric steel were imported. Apart from the state forges and factories, most provinces had their own smelting-sites. The local deposits in England and Gaul were worked for local needs; only in the case of abundant bodies of ore in massive formation, e.g. in Carinthia and in Aude, Gaul, did the Roman government confiscate the mines and entrust their exploitation to Imperial lessees.

FIGURE 31—*Roman forge of first century A.D. From a tombstone in the Domitilla catacomb.*

Specialization of products is often indicated in the texts. Many qualities of steel were recognized, even by the Greeks in the age of Alexander. Sinopic steel was used for carpenter's tools, Chalybic steel and Laconian steel for files and borers, Lydian steel for swords, and so on. In Roman times Spanish and Noric steel held first place for weapons, but even they yielded to seric and Parthian (Persian) steels from the orient when these were available.

The blacksmith always held a special place in classical society (figures 31 and 32). In the early Roman Empire there were large guilds of *fabri* at Milan and Brescia; both used Noric iron and steel. Como was famous in Pliny's days for its iron industry. Populonia still had a few iron-workers, but most of its iron trade had gone to Puteoli, where, however, no guilds of *fabri* seem to have existed—

FIGURE 32—*Roman locksmith with his tools and a specimen of his work. Second century A.D.*

possibly its smithies may have been manned by slave-labour. In the Po valley, on the contrary, the guilds of blacksmiths were strong enough to be a factor in Imperial policy.

In Spain the iron ore of the Basque country was not seriously exploited before the fifth century A.D., for it was the ore in the western part of Spain and Portugal that the Romans worked. Arms and cutlery were the articles usually exported. Their fame was due to particular manufacturing methods, such as burial and corrosion of the product, supposedly for the purpose of improving it, and the use of certain waters reputed to have special quenching properties. The importance of this industry is attested by a wealth of documents related to societies, officials, and private individuals connected with the extraction, smelting, and distribution of the iron. Certain alluvial beds in Catalonia, Alicante, and Toledo were worked only during the later Empire and the Middle Ages.

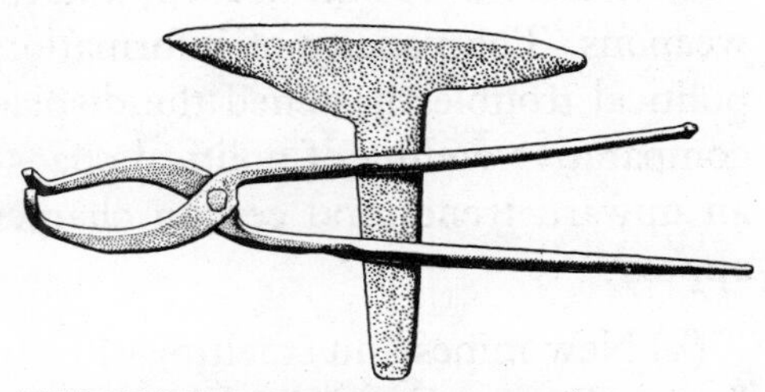

FIGURE 33—*Anvil and tongs of the fourth century A.D. Found at Silchester. Scale c 1/10.*

In central Europe the Romans controlled the iron deposits of Noricum and of the Sana valley in Bosnia. Slags of this period in Noricum amount to some 100 000 tons, which corresponds to some 30 000 tons of metal. The Romans came to prefer the white and yellow ores to the red spathic iron (crude ferrous carbonate) of the earlier smiths. Some of the ores they used contained titanium, an excellent component of steel. Roman exploitation of the Styrian Erzberg lasted until at least A.D. 316. To the north, iron was smelted in the central Jura, north-eastern Gaul, Luxembourg, northern Alsace, and the Eifel. At the same time the German tribes exploited much brown bog-iron ore (limonite, a hydrated iron oxide), and such deposits as those of the Siegen area were to acquire fame several centuries later. In Roman Britain the Weald and the Forest of Dean produced iron [39]. Other smelting and forging sites have been found, but they seem to have been worked for local needs rather than for export (figure 33). Some French towns may owe their rise to Celtic bloomeries.

VII. MEDIEVAL METALLURGY IN GENERAL

The fall of the Roman Empire was heralded by a gradual decline of central authority and consequently by a decline of metallurgy and mining. The drop in metallurgical output began in the third century A.D. [40], and it seems that, till the ninth or tenth century, surface ores or old workings were used by local smelters and smiths to meet a demand which diminished as disorder increased. Thus, in France, smelting continued in northern Burgundy, southern Champagne, Berry, Normandy, Brittany, Vendée, Périgord, Dauphiné, and the Pyrenees [41]. Again, in the Rhineland, Tuscany, Spain, and the eastern Alps iron was still produced. There is evidence that iron-production suffered less severely than that of silver, lead, copper, and tin. Axes, knives, spades, ploughshares, and similar tools were still needed, and much attention was naturally given to steel weapons. The paucity of information on copper and bronze suggests that the political troubles hastened the displacement of these metals by iron. With the comparative ending of political chaos by the ninth century, metallurgy showed an upward trend, and certain characteristics of medieval metallurgy began to appear:

(*a*) New mines and smelting-sites were established beyond the borders of the former Roman Empire.

(*b*) A literature of mining and metallurgy was gradually created, at first for the application of metals in art and decoration, later for the instruction of the army officer and engineer, the banker, the assayer, and the smelter.

(*c*) Coal was mined and came to be used on a fairly large scale for certain preliminary operations, though final smeltings were still carried out with charcoal. Thus metallurgy remains in essence a charcoal process. However, the shortage of timber became ever greater, and the general use of coal correspondingly increased.

(*d*) Water-power was gradually applied to metallurgy, and water-driven bellows, hammers, and stamping-mills displaced those operated by man-power and animal-power (figures 36, 346, 553, 554).

Whereas the metallurgy of other metals was virtually stationary until the sixteenth century, the large-scale production of iron and the use of water-power led to specialization of equipment, to achievement of higher temperatures, and to furnaces that made the production of cast-iron possible. By the end of this period real blast-furnaces were being constructed.

Though little is yet known from documents of the metallurgy of the early Middle Ages, much could be recovered by excavation. Modern analyses of the finds of the earlier periods have as yet received little attention. Where researches of this kind have been made, striking results have been obtained, as in those from Merovingian (481–752) France [42]. The older techniques of metallurgy were clearly not forgotten in the succeeding centuries. Syrian experts often worked in western Europe chasing silver, gilding helmets, making shields, and producing all kinds of metal inlay-work, like those mentioned in a decree of 374 in the Theodosian Code [43]. During the seventh century armourers worked in Burgundy, and spread out over northern and western France. They developed a very peculiar technique in making pattern-welded steel (p 457). Oriental steel weapons of this kind are completely homogeneous and develop the pattern by crystallization, but the western pattern-welded steel was built up by welding together strips of oriental steel, which were then cut, bent, and forged so as to develop the peculiar pattern. Objects of western pattern-welded steel came to be in demand even in Arab countries. Such facts throw a vivid light on the survival of very sophisticated techniques, and should make us wary before condemning the metallurgy of the so-called Dark Ages. We also have reason to believe that certain types of iron-furnaces, to be discussed later, were introduced at this period.

We become aware of the very fragmentary nature of our knowledge if we discuss the very few early handbooks dealing with metallurgical subjects. Oldest of these sources is an eighth-century book of recipes called *Compositiones ad tingenda musiva*, 'Recipes for colouring mosaics' [44]. This contains recipes for artists and craftsmen, mostly from earlier Greek and Byzantine books (p 351). The smelting of gold, silver, and lead ores is vaguely described and the working of the metals is given in more detail. The book also contains descriptions of how to make thin foils from such metals, the application of gold by way of its amalgam, and the manufacture of brass (*orichalcum*). The use of alloys containing lead indicates the scarcity of tin.

The *Mappae clavicula de efficiendo auro*, 'Key to the recipe of making gold' [45], also seems to date from the eighth century (p 351), though the oldest version now available is two centuries later. It is another manual for the manufacture of metal-work and alloys for the artist and craftsman, and contains many recipes for alloys to be used as substitutes for silver and gold. The bronzes mentioned again contain much lead.

The third manual, ascribed to Heraclius, *De coloribus et artibus Romanorum*, 'On the paints and arts of the Romans' [46], consists of three parts, the first two of which were probably composed by a citizen of tenth-century Rome; the third was added in France during the twelfth century. Here again we find descriptions of the refining of gold and silver, of amalgamation, of solders for different metals, of the manufacture of certain metals and alloys in thin sheets, and recipes for such alloys.

The fourth and most elaborate document is the *Diversarum artium schedula*, 'Essay upon various arrs' [47], of the priest Theophilus, who travelled widely through Europe, settled in a German monastery, and was possibly a Greek. This work may have been written during the tenth century but is usually given a rather later date. Theophilus was interested in gilding book-covers and illuminating manuscripts, in making stained-glass windows, in niello and other metal-craft, and in carving ivory. His discussion on methods of construction ranges from chalices, censers, and cymbals to church bells and organs. He does not forget the technology needed to make the materials of the artists, describing, for instance, the building of a glass-furnace and the preparation of glass of different colours. On the metallurgy of gold he discusses qualities, cupellation, and refining by amalgamation [48], together with the manufacture of gold foils, powdered gold, and alloys imitating gold. He treats of the solders for the lead strips in stained-glass windows, of the metallurgist's tools and how to make them, of the refining and casting of silver, of liquation [49], of the construction of cupellation-furnaces, and of the colouring of gold leaf. He gives details on smelting copper (p 49) from hand-picked surface ores such as malachite or oxidized sulphides, instructions for building the crucible-furnace for melting small quantities of copper [50], and methods of staining copper with different pigments. A description of bell-founding forms an important item in his book, which also discusses the bloomery-hearth [51], the soldering and welding of iron, the casting of tin and brass vessels [52], and the soldering of tin. Though this is no handbook on metallurgy, being intended for decorators and builders of churches, it gives a good picture of how completely the classical metallurgical techniques survived.

The casting of church bells was a new application of metallurgy. The first bells to be used in churches, in the fifth century, were of sheet-iron; their ancestors were the small table-bells used by the Romans. During the eighth century bells were first cast in bronze, and by the ninth such bells were common in western European churches. Their casting stimulated metallurgical skill, since the moulds were often much larger than any hitherto attempted. By 1250 there was a bell-founders' street in Lübeck, and certain individual bell-founders of the next century won international fame. Their experience was of great importance when, with the invention of gunpowder, firearms were introduced and the first bronze cannons and cannon-balls were cast (p 75); it was again of prime importance when cast iron came into more general demand.

After Theophilus we have few relevant documents to assist us. Though Albert the Great (1193?–1280) deals with metallurgy in one of his works his remarks are superficial [53]. There are scraps of information from medieval documents, but no real instructions or handbooks occur until the fifteenth century. Then the *Kriegsbücher*, 'Books on warfare', and *Rüstungsbücher*, 'Books on armament', mostly written in central Europe, deal with metallurgy in general. A quite unbroken literary tradition concerning metallurgical techniques begins by the end of that century with the *Bergbüchlein*, *Probierbüchlein* [54], and other books on metallurgy, mining, and assaying, which the new printing-presses began to produce on a larger scale than was possible with the *scriptoria* of the older monasteries. Information on medieval metallurgy derives mainly from finds at smelting-sites, from economic data, and from later traditions.

In certain parts of central Europe, notably in the Black Forest and the Harz, Saxony and Bohemia, Silesia and Hungary, resources tapped by primitive smiths waited for a more thorough exploitation. These regions still carried plenty of timber. They had also mountain streams and brooks to harness for power. Their gradual colonization by Flemish, German, and French settlers in the early Middle Ages made these regions the main source of medieval metallurgy. Within a few centuries the Germanic peoples became the mining and metallurgical experts of their age.

The older countries still played their part. Gold-washing continued in the Cévennes and in other parts of France and Spain. The silver mines of Devon were important enough to be leased by the Frescobaldi—one of the wealthiest Florentine firms of the Middle Ages—from Edward I in 1299; and the lead ores from Alston Moor were worked by Tillmann of Cologne (1359). In 1316 lead ores rich in silver were discovered at La Caunette, Languedoc.

Most gold and silver, both for trade and for church decoration, came from regions east of the Rhine. In 1136 traders carrying rock-salt from Halle to Bohemia by way of Meissen discovered the rich silver-bearing ores of Freiberg in Saxony, lodes of which had been bared by the spring floods. Samples carried back to the silver-producing centre of Goslar, Harz, were found to be very rich, and a 'silver-rush' took place. Mining and smelting at Freiberg were in full swing by 1170, and a metallurgical centre grew up, with some 30 000 inhabitants. Production rose steadily till 1348, when the Black Death caused an interruption of activities lasting for nearly a century. Nevertheless, in the middle of the fifteenth century Freiberg could boast of 52 smelting-works.

At Goslar silver was first produced from the copper lodes of the Rammelsberg, discovered in 968 and becoming important in the eleventh century. Later, the production of copper, zinc, and lead was developed there, and the rich Mansfeld lodes were discovered in 1215. By the fourteenth century, however, Stora Kopparberg in Sweden became a serious competitor. The ores of Hungary were smelted as far back as the days of Charlemagne (d 814), who manned the mines and smelting-works with captives. Here metals were produced until the Poles stopped the work by conquest in 1442. In the eastern Alps the silver mines of Trento were fairly important.

Most of these silver ores were very complex. Their treatment entailed liquation and other separative processes. By the twelfth century, hammers, stamps,

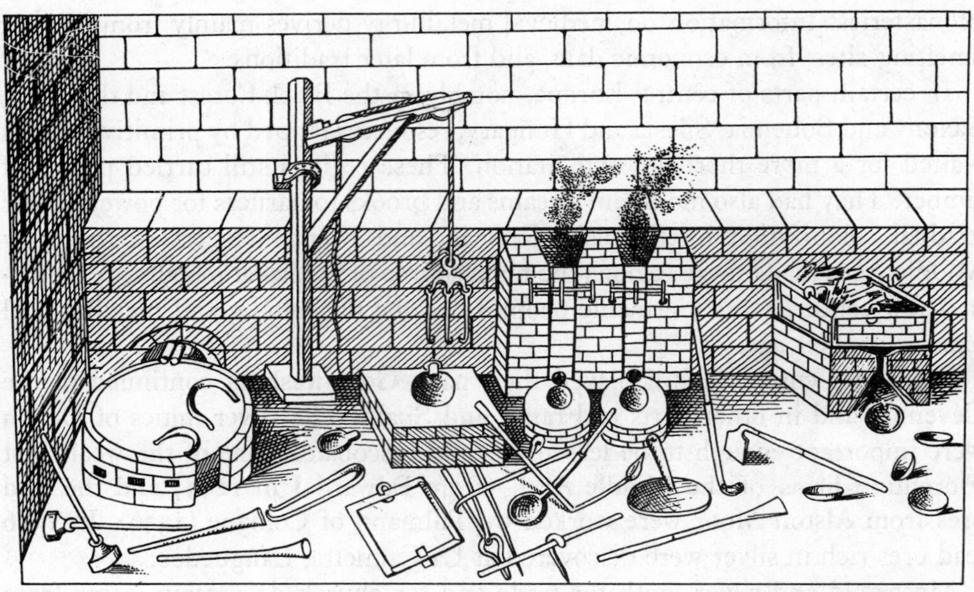

FIGURE 34—*The smelting of argentiferous copper in the fifteenth century. In this and the following illustration are shown the various stages in smelting copper to which lead or lead compounds have been added to extract the silver. In the centre are a pair of small blast-furnaces in the right-hand one of which the ore, which has probably first been roasted out-of-doors, is smelted with charcoal. The matte formed is then heated in the left-hand furnace with some form of lead, and the product, a mixture of copper and lead-silver alloy, ladled into small round moulds to form liquation cakes. The cakes are packed with wood in the liquation-furnace on the right of the illustration, and the lead dripping from these cakes forms the charge for the cupellation-furnace in figure 35. On the left of the illustration is a cupellation-furnace or refining-hearth for silver. The objects lying on it are scrapers for shaping the hearth, and tools for ramming the material into position are also shown. Next to it is another refining-hearth, probably used for copper. The* tuyères *of the bellows may be seen protruding into these furnaces. The artist has in all these drawings left out the upper parts of the furnaces. From the south German* Hausbuch, c *1480.*

and bellows driven by hand and foot were numerous in the mining towns. During the first decades of the thirteenth century water-wheels are mentioned as in use at the silver mines of Trento. By the next century, they had become fairly general in all Alpine regions. Such machines entailed heavier capital expenditure than formerly, and the rise of western capitalism was largely tied up with the development of mining and metallurgy. Mining laws and rights were formed on the Roman model, and industrial towns like Freiberg, Chemnitz, and Iglau arose as communities of specialists with their own laws and economy.

The feudal lords, like the bankers, promoted their rise, and slaves were no longer used.

In the wake of the new silver-metallurgy, new supplies of copper became available. Tin-mining in Cornwall had recovered in the tenth century and

FIGURE 35—*In the centre is a cupellation-furnace where the argentiferous lead for liquation is roasted to litharge, leaving purified silver. An operator can be seen drawing off the excess litharge with a hooked stick. Logs are used as fuel and the draught is by manually operated twin bellows. On the left are a set of stalls for 'drying' spent liquation cakes, i.e. for removing the remaining lead from the copper. The order of operation is: roasting (not shown), smelting and liquation (figure 34), cupellation of silver (figure 35), and refining (figure 34). From the south German* Hausbuch, *c 1480.*

production rose steadily. By the middle of the fourteenth century it averaged some 700 tons a year, double that of the preceding one. From the new zinc ores in the Meuse valley some brass was made, but this alloy did not come into general use until the fifteenth century, when further supplies of zinc ores became available from the Tyrol, Carinthia, and Moresnet (near Aix-la-Chapelle), thus stimulating the demand for copper. The lead production of Somerset, Durham, Cumberland, Shropshire, and Derbyshire was certainly equal to that of central Europe.

The smelting of these metals underwent no essential change. This becomes clear when we read a description of a complete copper-refinery in the finely illustrated *Mittelalterliche Hausbuch* of about 1480 (figures 34, 35) [55]. On the other hand, mechanization and the introduction of water-power had effected modifications, though their influence was much greater in the metallurgy of iron. As early as the tenth century the shortage of charcoal as a result of deforestation had become so serious that in certain regions its production was limited by law, and the efforts to adapt coal for metallurgical purposes were intensified.

FIGURE 36—*Hydraulic bellows of a furnace. They are operated by cams on the shaft of an overshot wheel (cf figure 582). From a fifteenth-century manuscript.*

Coal is usually said to have been first mined for forges. There are references in 1190–1230 to such mining for forges near Liége. In France also coal was mined for the same purpose during the twelfth and thirteenth centuries; a statute of Arles dated 1306 forbids its use in forges, but in 1345 it was mined for the forges of Marche-sur-Meuse. Coal was used even earlier in England, for Henry III in 1234 confirmed a privilege of King John to Newcastle-upon-Tyne. During that period 'sea-coal' was transported from Newcastle by barge to London, where there were soon (1273) complaints from the gentry about the smoke and smell. During the thirteenth century the coalfields of England, Scotland, the Saar, Liége, Mons, Aix-la-Chapelle, and Franche-Comté in the German Empire, and Lyonnais, Forez, Ales, and Anjou in France were being tapped for lime-burning and domestic fuel, and for the iron-smelters and blacksmiths. In England, in the Low Countries, and also in the industrial centres of the thirteenth century in Tuscany and northern Italy, smiths used much coal. Its sulphureous fumes made it unsuitable for anything but the primary stages of iron-smelting, but certain high-grade coals like anthracite were sometimes used in the later stages where today coke would be employed. Nevertheless it should be pointed out that the usual fuel for smelting up to the eighteenth century was still charcoal.

The advent of water-power was a much stronger force in the evolution of the modern metallurgy of iron (figure 36). Agricola dates the first application of water-driven bellows back to 1435, but this is certainly too late. In the eastern

Alps and Silesia water-driven bellows and hammers are mentioned in the eleventh and twelfth centuries [56]. In 1135 the Benedictine monastery of Admont in Styria had a water-mill at Leoben, to which in 1175 a stamping-mill was added, and we hear of others at Hradish in Moravia. From these eastern regions they moved into central Germany, where in the thirteenth century there were bloomeries in the river valleys of the Harz mining region; millponds are often mentioned in connexion with these works. Even in Denmark we find mention of the 'iron-mills' of Sorø (1197), built 'to manufacture iron from ore'.

The first application of water-power to French iron-metallurgy is also of the eleventh and twelfth centuries [57]. Again the monastic orders took the initiative in introducing it into metallurgy. In the Dauphiné, where the Grande Chartreuse was founded in 1084, water-mills were in use about 1200 [58]. The Cistercian engineer Villard de Honnecourt drew plans for all kinds of water-driven machinery, including a saw-mill, in his sketch book (1270) (figure 584).

The Templars owned the first iron-mill recorded in Champagne (1203) [59]. There, too, the sovereign ceded an iron-mill to a subject in 1249 [60]. About 1283 other iron-mills are mentioned as existing in Narbonne [61], the Pyrenees, and the Montagne Noire (southern France). Thus in the thirteenth century hammer-mills were not uncommon in France. Stamping-mills followed about a century later, when they appeared in the Saar. A contract on the water-forge of Briey (1323) also mentioned a furnace with water-driven bellows. Thus iron-works tended to move away from mines of ore and coal and towards streams and brooks, a trend reversed centuries later when the steam-engine arrived.

VIII. IRON AND STEEL IN THE MIDDLE AGES

As we have seen, the experience of Roman and early European smiths was not lost in the disorders preceding the tenth century. The metallurgy of iron even profited by the progress of skill and ingenuity, for iron tools and weapons were still in constant demand. The Celtic smiths of Gaul whom Caesar praised had made iron tires for their war-chariots and were said to be the inventors of chain-mail, but their swords still bent in battle and needed frequent straightening (p 456). The weapons of the new Norse rulers showed an improvement. The sagas of the early Middle Ages tell of the 'long swords' of the Vikings, the battle-axes of the Franks, and the dirks and swords of the Saxons. The smith maintained his status as a member of a strong guild.

Production continued in all the famous smelting-sites of earlier days, mostly in furnaces with natural draught built on hill-sides, and in some of the newer types equipped with bellows. The different tribes that invaded Europe often had good smiths among them, and new types of furnaces like the osmund[1] furnace; foreign workers seem to have introduced such improvements from the north and the east [62]. Most of the iron and steel was still directly produced from the ore by rather wasteful processes, but gradually new techniques for the production of pig iron were discovered. Such iron could have its carbon-content lowered to yield wrought iron, or raised to obtain cast iron.

FIGURE 37—*A medieval blast-furnace from Yugoslavia. The two air-ducts are clearly seen at the front. The sides of the furnace are coated with quartz corroded by the slag and give some idea of the high temperatures obtained. Scale 1/50.*

Styria and Carinthia gradually increased their production of iron and steel from the eighth century onwards, but they never reached a production of much over 2000 tons a year. In Franconia iron-production began in the tenth century; in Westphalia (the Siegen area) iron became important after 1228. Swabia and Hungary came to the fore in the same period. From the tenth century the Basque provinces became progressively more important both as producers and as exporters. The revival of the old forges of Lorraine, Burgundy, Dauphiné, the Cévennes, Poitou, Anjou, and Normandy dated from the same period. All over this area the number of forges and furnaces increased considerably during the twelfth century, and new smelting-sites such as those in the Piedmontese Alps and the Côte d'Or are now mentioned. In Britain the production of the furnaces of the Forest of Dean, the 'medieval Birmingham', rose to a new height during the twelfth century and provided Richard I with 50 000 horse-shoes to equip his cavalry for a crusade. The thirteenth century brought further developments in many regions. Thus there was considerable new activity in Lorraine, Champagne, and Dauphiné, and the town of Narbonne became an important centre for the export of iron and steel to Genoa and many parts of the Near East.

The coexistence of older and newer forms of furnaces and smelting-techniques during this period is very confusing, and the confusion is worse confounded by

[1] Osmund was a special type of Swedish iron used for making small objects such as fish-hooks and arrowheads. The origin of the name is unknown. In 1281 it is referred to as *ferrum Normannicum.*

the use of many local terms for the hearths and furnaces. Generally, medieval smelting-techniques were still very wasteful, as will be clear from table I.

TABLE I

Comparison of medieval iron-smelting techniques according to O. Johannsen [63]

Name	*Weight of iron per smelting operation in kilograms*	*Ratio of fuel consumption to ore, per cent*	*Yield on ore, per cent*
Bloomery hearth .	60–70	450 (*coal*)	12·5 (*on an iron ore containing 25 per cent iron*)
Corsican forge .	125	880 (*coal*)	38·5
Catalan forge .	150	360 (*charcoal*)	31·0
Stückofen . .	300–900	250 (*coal*)	39·0 (*and a slag containing 30 per cent iron*)

The ores required preliminary removal of the matrix by crushing and washing. This operation was thoroughly mechanized by the fifteenth century. Most ores were first roasted with green timber and then quenched with water (*löschen*), which dissolved part of the sulphur and copper compounds. Sometimes the ores were submitted to weathering before smelting.

In the bloomery-fire, the most primitive form of hearth, the ore was mixed with charcoal and then covered with fuel held together by a circle of stones. The bellows were directed towards the middle of the hearth. This wasteful process was in general applied only to very pure ores. The recovered bloom was reheated on a similar hearth, and the slag hammered out.

In the Corsican furnace iron ore was extracted in two stages: (*a*) roasting with partial reduction, (*b*) smelting, reduction, and carbonization. The permanent parts of the furnace (figure 38 A) are two masonry walls at right angles, constructed for the protection of the workers and the bellows respectively. Through the latter wall the *tuyère* supplying the blast protrudes into the centre of the furnace. Apart from these and a foundation, the rest of the furnace is only temporary, being built with charcoal, ore, and ashes to a special shape for each operation of roasting or smelting. Figure 38 A shows the horse-shoe shape that is built up for roasting with a blast not strong enough to cause extensive fusion. The roasted ore is broken up and, together with more charcoal, rebuilt in a V-shape. After heating the mass with a strong blast and continuous addition of materials, carbonized iron collects at the bottom.

The Catalan forge (figure 38 B) was primarily used for porous types of limonite (brown iron ore, approximately $2Fe_2O_3,2H_2O$). It was, however, also used for

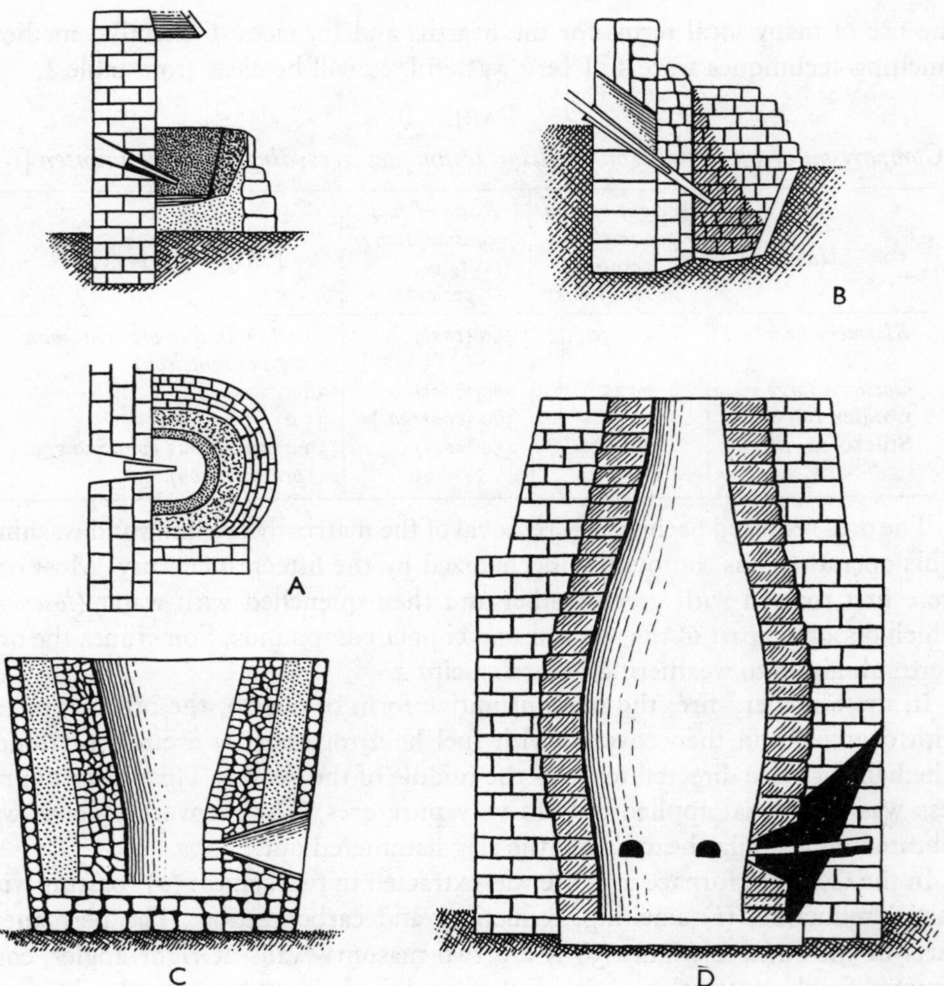

FIGURE 38—*Evolution of the iron-smelting furnace.* (A) *Sections in elevation and plan of the Corsican furnace. The only permanent parts were the two masonry walls with a* tuyère *for bellows. The crushed ore mixed with charcoal was built into a semicircle, supported on its inside by larger blocks of ore. The centre cavity was then filled with glowing charcoal and the temperature maintained until the ore had been sufficiently roasted. The semicircle was then broken up, and the temperature raised until a bloom of wrought iron, carbonized to a certain extent, was obtained, much as in the primitive bloomeries.* (B) *The Catalan furnace, first used in Spain, illustrates a more permanent type of construction and was popular in the Middle Ages.* (C) *The osmund furnace, which shows an attempt to heighten the furnace and insulate its contents.* (D) *The* Stückofen, *an improvement due to the medieval Rhineland iron-smelters. Here the ore undergoes continuous reduction by charcoal as it slowly falls through the height of the furnace, while the slag and metal are removed from the bottom. The greater height necessitates a stronger blast, which is supplied by water-driven bellows. Cast iron could be produced under favourable conditions in this type of furnace. Scale* c *1/60.*

haematite, the commonest ore, and other massive ores, which were first weathered and then roasted in large, wide furnaces containing some 20–35 *tonnes*[1] of ore. Balls of crushed pre-roasted ore were massed on to the inclined wall opposite the *tuyère*, and the other walls were stacked with charcoal. About 500 kg of ore were smelted at each operation. Crushed ore is more readily reduced but also more readily absorbed by the slag. If ores containing manganese are smelted, the absorption of impurities by the slag is improved, and it is then possible to produce an iron with about 0·3 per cent of carbon; such iron can be hardened by quenching and is a variety of steel.

Different types of shaft-furnaces were in use. The Romans and the early European smiths had used shaft-furnaces with natural draught, sometimes aided by bellows. They were mostly applied to ores containing manganese or low grade ores difficult to smelt. Some of them attained a height of up to 7 ft even in prehistoric times. As the demand for iron—and particularly for steel—rose and water-driven bellows were introduced, larger furnaces could be built and higher temperatures obtained. The first steps were thus taken on the road to the blast-furnace, the invention of which was, however, not properly developed before the sixteenth century [64].

The *Stückofen*, originating in Styria and popular by the Middle Ages, was a direct descendant of the Roman bloomery-furnace (figure 38 D). The first of these furnaces were some 10 ft high; later ones rose to 14 ft. In the fifteenth and sixteenth centuries they were still the best furnaces. When the smiths came to realize that pig iron could be efficiently cast, the blast of the *Stückofen* was handled in such a way that either pig iron or wrought iron could be produced at will. The best height was found to be 10–14 ft, and the furnace was provided with a special hole through which the slag was withdrawn. This improved *Stückofen* was called the *Blasofen* (figure 37). The survey of the Siegen area given in table II shows how this evolution and mechanization gradually took place, and how different types operated alongside each other.

The pieces of wrought iron made in the *Stückofen* weighed some 370 kg in 1430; they had increased to 400 kg by 1470, and had reached 500–600 kg by the end of the fifteenth century. The output of such furnaces was some 40–50 *tonnes* a year—about three times as much as that of the more primitive furnaces.

The introduction in the fourteenth century of water-driven bellows (figure 36), of which each furnace had at least two, initiated the production of big lumps of pig iron or 'salamanders'. Further water-power was necessary to operate the tilt-hammers that turned them into wrought iron (figure 554). Such

[1] 1 metric *tonne* = 1000 kilograms = 0·984 ton.

TABLE II

The smelting-works of the Siegen area according to Gillis [64]

Year	*1311*	*1417*	*1444*	*1463*	*1492*	*1505*
Smelting-works	?	19	26	25	18	16
Bloomeries	1	..	..	..	..	..
Pig iron works	..	..	2	4	11	13
Forges making steel in hearth-fires	?	1	8	11	9	15
Total	?	20	36	40	38	44
Total using water-power .	1	6	24	30	38	44
Disused	?	..	3	..	..	..

new furnaces appear near Namur in 1340, near Liége about 1400, and in the Nassau region about 1474. A new furnace for the manufacture of cast iron cannon-balls and guns was built at La Perche (north-west France) in 1486.

In the tall *Stückofen*, with more blast and harder charcoal, higher temperatures were reached and the slags contained less iron, while part of the iron contained sufficient carbon for liquefaction. The technique of further raising the carbon-content to produce a satisfactory cast iron was discovered in the thirteenth century, but it was mastered slowly and was not used on a large scale until the fifteenth century [65]. Formerly, wrought iron had been liquefied by adding stibnite, arsenic, antimony, and copper, but now several hundreds of kilograms of pig iron could be smelted with double the quantity of fuel to yield

FIGURE 39—*Drawing heavy iron wire. The workman is seated on a swing which moves with the crank as it is operated by the undershot wheel. He moves the tongs forward on the forward motion of the swing when the rope is slack, and a stirrup causes them to grip the wire when the rope is in tension. From Biringuccio*, Pirotechnia, *1540*.

cast iron. Filarete in his *Architettura* described such furnaces in the neighbourhood of Brescia about 1450.

Firearms were first made in Germany about 1325; they were of forged iron, but about 1350 cast bronze arms came into use. A generation later the first real cast iron cannons appeared, but bronze remained in use until the cast iron techniques were properly mastered in the fifteenth century. One Mercklen Gast of Frankfurt-am-Main, at the end of the fourteenth century, was the first to advertise his skill in making cast iron arms; and early in the fifteenth century casting directly from the furnace into the mould was achieved. Cast iron then gradually spread in western Germany, north-eastern France, and northern Italy. This coincided with the appearance of the term 'iron-founder', and did away, for example, with the technique of swaging cannon balls from wrought iron. Steel was still made by cementation in hotter hearth-fires, but it could also be produced in the *Stückofen* by a proper degree of decarbonization.

Mechanical hammers, actuated by 2-m water-wheels and weighing 500–1600 kg, were used to forge the blooms. For later stages in the working there were hammers of 300 kg, delivering 60–120 strokes a minute, and lighter ones of 70–80 kg, delivering 200 strokes a minute. Sheet-iron was produced from bars in such centres as Amberg, Bavaria.

Iron wire was made by forging until the tenth century, when the draw-plate was invented. The plate for wire-drawing has pierced in it a series of holes successively smaller in diameter (figure 39). It was made of fine cast iron from a kind of miniature crucible-furnace. Application of water-power to wire-drawing came early in the fourteenth century; it was in action in 1351 at Augsburg, and somewhat later at Nuremberg. The water-wheel turned a crank to which was attached a rope ending in a pair of tongs held by a stirrup (figure 39). The smith suspended himself on a swing so as to be able to grasp the wire issuing from the draw-plate with the tongs, and to move back with it as the crank drew it out. Such cumbersome methods were not, of course, necessary in the drawing of metals of lower strength, such as gold, silver, and copper.

The earliest iron needles had no eye but a closed hook. They were produced by the guild of needle-smiths of Nuremberg (1370) and that of Schwabach. The first needles with an eye were made in the Low Countries in the fifteenth century. Swords were the products of specialists at such towns as Milan, Brescia, and Passau. The art of forging them came to Solingen during the Italian wars of the Emperor Barbarossa (1152–90). Scythes were made in Styria, and from 1240 onwards in Cromberg and Plattenberg as well. These few examples go to show that during the Middle Ages specialization in the iron industry had gone

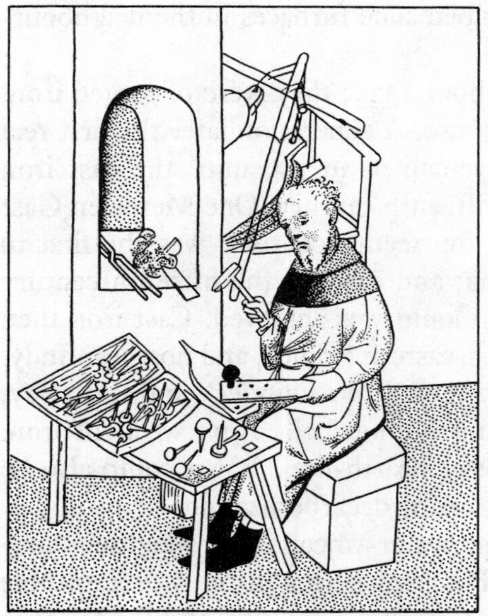

FIGURE 40—*German nail-maker of the late fifteenth century. By hammering the metal through a graded series of holes a primitive type of nail was formed.*

far (figure 40). Blacksmiths had already specialized on, for instance, anchors or horse-shoes (figures 41, 42).

Trade in iron and in different types of steel was spread over much of Europe. England produced sufficient iron but imported steel on a fairly large scale. France obtained her iron from Germany, Belgium, and Spain. Spanish iron was traded along the Atlantic coast. Up the Rhône valley iron came mainly from Italy. The steel of Styria and Carinthia travelled to Aquileia and Venice, and thence east as far as Turkey. As 'Hungarian ware' it reached western Germany and was traded down the Rhine and thence to England, where it was known as 'lymbrique stuff', a term derived from the names of the towns Leoben and Brücken. The steel from the Tyrol was sold at Innsbruck and reached England by the same route. The region between the Donau and the Main sold its sheet iron at Amberg. Central Germany obtained its iron mainly from the Harz region and Schmalkalden in Thuringia. The famous Hanseatic League traded in Spanish iron from the Basque provinces, Swedish osmund iron sold at Danzig and Lübeck, and Westphalian steel sold at Siegen and Cologne. In Belgium the iron trade centred at Liége with its strong *Corporation des Fèbvres*.

FIGURE 41—*Italian forge of* c *1492 representing Eloi at the smithy. From a Venetian title-page.*

During the latter half of the fourteenth century a severe decline in the production of metals set in, mainly owing to the Black Death (1348–50) and the ravages of the Hundred Years War (1338–1453). Severe shortage of labour on the land and elsewhere caused further decline of the smelting-works after the first wave of the pestilence had passed. In Cumberland alone the production of the iron-works fell by 50 per cent. Statutes of 1349 and 1351 attempted to control the labour situation and to prevent workers from leaving their places of employment. A general rise in prices followed, and those of iron and charcoal were doubled. Curiously enough, the price of steel rose by only 25 per cent, but at that period England imported

FIGURE 42—*A south German wood-carving of* c *1500 showing St Eloi at the smithy. The legend says that to simplify the business of shoeing he removed the leg, fixed on the shoe, and then replaced the leg on the horse.*

its steel, the home production being negligible, and used it only for armour, weapons, and the finer types of tools and instruments.

We find a similar situation elsewhere in Europe. In France, too, there was a tendency for labour to leave the smelting-works in favour of the farms; prices of steel and iron rose sharply and many forges and mines were destroyed in the wars. Franconia and Bohemia escaped the Black Death, but here the decline of mining and metallurgy set in as it did in Silesia, Hungary, the Rammelsberg, and Cornwall. When the industry recovered, after 1450, experienced metallurgists had become scarce in many countries, and German experts are mentioned frequently in contemporary documents in France and later in England.

Moreover, serious technical difficulties hampered the revival of mining and metallurgy. Many mines in the Harz region, in Saxony, Bohemia, Hungary, Alsace, Sweden, and Cornwall were now exhausted and deeper strata had to be tapped. This led to an inrush of sub-soil water, and the expenditure of more capital on drainage-machinery, driven by water-wheels or by horses in treadmills (figures 19, 20). Washing, breaking, and crushing were still often manual, outdoor operations; too many open-air hearths and tiny forges with hand- and foot-driven bellows worked uneconomically.

The remedy lay in concentration and mechanization. Concentration began to take place when the industry recovered. *Saigerhütten*, that is, large smelting-works consisting of small chambers, each containing special furnaces and hearths for each stage of the operation, and all built against a long thick central wall, were now erected in many places. The heavy capital expenditure required was possible only with the help of the bankers. Early western capitalism also provided the means for the mechanization of mines and forges. A new boom was heralded, to last from 1460 to 1530, when the flow of precious metals from the New World and a crisis in credit interrupted the adjustment of the industry to the new relations between capital and labour arising from this growing concentration and mechanization.

REFERENCES

[1] PLINY *Nat. hist.*, XXXIII, xxi, 66. (Loeb ed. Vol. 9, pp. 50 ff., 1952.)
HERODOTUS I, 93; V, 49. (Loeb ed. Vol. 1, p. 120, 1920; Vol. 3, pp. 50 ff., 1922.)
STRABO XIII, C 626; C 591. (Loeb ed. Vol. 6, p. 172; p. 44, 1929.)

[2] *Idem* XIII, C 591; XIV, C 680. (Loeb ed. Vol. 6, p. 44; pp. 368 ff., 1929.)

[3] PLINY *Nat. hist.*, XXXIII, xxi, 70–78. (Loeb ed. Vol. 9, pp. 54 ff., 1952.)
STRABO III, C 146. (Loeb ed. Vol. 2, pp. 38 ff., 1933.)

[4] AETHICUS ISTRICUS *Cosmographia*, II, 28, trans. by SAINT JEROME, ed. by H. WUTTKE (2nd ed.), p. 16. Dyk, Leipzig. 1854.

[5] PLINY *Nat. hist.*, XXXIII, xxxii, 99–100. (Loeb ed. Vol. 9, pp. 74 ff., 1952.)

[6] THEOPHILUS PRESBYTER *Diversarum artium schedula*, III, 36–37, ed. and trans. by R. HENDRIE. Murray, London. 1847. See also ref. [47].

[7] PLINY *Nat. hist.*, IX, lxv, 139; XXXIV, iii, 5–8 (Loeb ed. Vol. 3, p. 256, 1940; Vol. 9, pp. 128 ff., 1952); XXXVII, xii, 49.

[8] *Idem Ibid.*, XXXIII, xix, 61. (Loeb ed. Vol. 9, p. 48, 1952.)

[9] HUTIN, S. 'L'Alchimie', p. 72. Presses Universitaires de France, Paris. 1951.

[10] VITRUVIUS VII, viii. (Loeb ed. Vol. 1, p. 116, 1934.)

[11] HERODOTUS III, 57. (Loeb ed. Vol. 2, p. 72, 1921.)
PAUSANIAS X, xi, 2. (Loeb ed. Vol. 4, p. 426, 1935.)

[12] PLINY *Nat. hist.*, XXXIV, xlvii, 159. (Loeb ed. Vol. 9, p. 242, 1952.)

[13] STRABO IX, C 399. (Loeb ed. Vol. 4, p. 274, 1927.)

[14] *Idem* III, C 146. (Loeb ed. Vol. 2, p. 42, 1923.)
MINGAZZINI, P. *Studi Sardi*, **10**, 3, 1950.

[15] PLINY *Nat. hist.*, XXXIII, xliii–xliv, 126–27. (Loeb ed. Vol. 9, p. 94, 1952.)

[16] *Idem Ibid.*, XXXIII, xiii, 46. (Loeb ed. Vol. 9, p. 38, 1952.)

[17] *Idem Ibid.*, XXXIV, xlviii, 161. (Loeb ed. Vol. 9, p. 242, 1952.)

[18] *Idem Ibid.*, XXXIV, xlviii, 160, 162. (Loeb ed. Vol. 9, pp. 242 ff., 1952.)

[19] STRABO III, C 142. (Loeb ed. Vol. 2, p. 24, 1923.)
PLINY *Nat. hist.*, XXXIV, ii, 4. (Loeb ed. Vol. 9, p. 128, 1952.)

[20] *Idem Ibid.*, XXXIV, xx, 95. (Loeb ed. Vol. 9, pp. 196 ff., 1952.)

[21] THEOPHILUS PRESBYTER *Diversarum artium schedula*, ed. and trans. by R. HENDRIE, chap. 67, p. 313. Murray, London. 1847. See also ref. [47].

[22] DIOSCORIDES V, lxxv, 3 (ed. M. WELLMANN. Weidmann, Berlin. 1914; 'englished' by JOHN GOODYER A.D. 1655, ed. by R. T. GUNTHER, chap. 85, p. 625. Publ. by the author, Oxford. 1934.)

[23] TÄCKHOLM, U. 'Studien über den Bergbau der römischen Kaiserzeit.' Appelberg, Uppsala. 1937.

[24] THEOPHRASTUS *Hist. plant.*, V, ix, 3. (Loeb ed., 'Enquiry into Plants', Vol. 1, p. 468, 1916.)
PLINY *Nat. hist.*, XVI, viii, 23. (Loeb ed. Vol. 4, p. 402, 1945.)

[25] JOSEPHUS *Antiquitates Judaicae*, XVI, 4. ("Antiquities of the Jews", p. 474 in 'The Works of Flavius Josephus', trans. by W. WHISTON, newly ed. by D. S. MARGOLIOUTH. Routledge, London. 1906.)

[26] LIVY XLV, xxix, 4–14. (Loeb ed. Vol. 13, pp. 346 ff., 1951.)

[27] BESNIER, M. *Rev. archéol.*, cinquième série, **10,** 37, 1919.

[28] TACITUS *Annales*, VI, xix. (Loeb ed. Vol. 3, p. 184, 1937.)

[29] CAESAR *De bello gallico*, V, xii. (Loeb ed., p. 250, 1917.)

[30] WARMINGTON, E. H. 'The Commerce between the Roman Empire and India.' University Press, Cambridge. 1928.

[31] PSEUDO-ARISTOTLE *De mirabilibus auscultationibus*, lxii. (Loeb ed., p. 262, 1936.)

[32] DIOSCORIDES V, lxxiv–lxxv. (ed. M. WELLMAN, Weidmann, Berlin. 1914; 'englished' by JOHN GOODYER A.D. 1655, ed. by R. T. GUNTHER, chaps. 84 f., pp. 623 ff. Publ. by the author, Oxford. 1934.)

[33] PLINY *Nat. hist.*, XXXIV, xli, 143–46. (Loeb ed. Vol. 9, pp. 230 ff., 1952.)

[34] RICHARDSON, H. C. *Amer. J. Archaeol.*, **38,** 555, 1934.

[35] CURTIUS IX, viii, 1. (Loeb ed. Vol. 2, p. 432, 1946.)

[36] HOMER Iliad, XXIII, 825. (Loeb ed. Vol. 2, p. 556, 1925.)

[37] VARRO *De re rustica*, I, xvi, 4. (Loeb ed., pp. 220 ff., 1934.)

[38] DIODORUS V, 13. (Loeb ed. Vol. 3, p. 130, 1939.)
STRABO V, C 223. (Loeb ed. Vol. 2, pp. 354 ff., 1923.)

[39] See ref. [29].

[40] NEF, J. U. "Mining and Metallurgy in Medieval Civilisation" in 'The Cambridge Economic History', Vol. 2, p. 433. University Press, Cambridge. 1952.

[41] GILLE, B. 'Les origines de la grande industrie métallurgique en France', pp. 1–8. Collection d'histoire sociale, no. 2. Domat Montchastien, Paris. 1949.

[42] SALIN, E. and FRANCE-LANORD, A. 'Rhin et Orient', Vol. 2: 'Le fer à l'époque mérovingienne.' Geuthner, Paris. 1943.
FRANCE-LANORD, A. *Pays Gaumais*, **10,** 1, 1949.

[43] CODEX THEODOSIANUS X, xxii, 1. (*Theodosiani Libri XVI*, ed. by T. MOMMSEN and P. M. MEYER, Vol. 1, ii, p. 566. Weidmann, Berlin. 1905.)

[44] *Compositiones ad tingenda musiva*, ed., trans., and comm. by H. HEDFORS. Diss, Uppsala. 1932.

[45] WAY, A. (Ed.) *Archaeologia*, **32,** 187, 1847.

[46] HERACLIUS *De coloribus et artibus Romanorum*, ed., trans., and comm. by A. ILG. Quell. Kunstgesch. Kunsttechn., Vol. 4. Braumüller, Vienna. 1873.

[47] THEOPHILUS PRESBYTER *Diversarum artium schedula*, ed. in part and trans. with comm. by W. THEOBALD: 'Die Technik des Kunsthandwerks im zehnten Jahrhundert.' Verein Deutscher Ingenieure, Berlin. 1933; ed. and trans. by R. HENDRIE: 'An Essay upon various Arts by Theophilus called also Rogerus.' Murray, London. 1847.

The following references give Theobald's chapter numbers; where Hendrie's differ, these are given in parentheses.

[48] THEOPHILUS PRESBYTER *Diversarum artium schedula*, III, 46–49, 33–36.

[49] *Idem. Ibid.*, I, 23 (24), 30; II, 27, 35, 36; III, 4–22, 23.

[50] *Idem. Ibid.*, III, 62–64, 66 (63–65, 67.)

[51] *Idem. Ibid.*, III, 84 (85), 3.

[52] *Idem. Ibid.*, III, 91, 60 (61.)

[53] ALBERTUS MAGNUS *De Mineralibus* in *Opera omnia*, ed. by A. BORGNET, Vol. 5, pp. 1–116. Vivès, Paris. 1890.

[54] BERGWERK- UND PROBIERBÜCHLEIN, trans. and annot. by ANNELIESE G. SISCO and C. S. SMITH. Amer. Inst. Min. and Metallurg. Engrs, New York. 1949.

[55] DAS MITTELALTERLICHE HAUSBUCH, ed. by H. T. BOSSERT and W. F. STORCK, Pls 38, 39 (fols 35b, 36c). Seemann, Leipzig. 1912.

[56] JOHANNSEN, O. *Stahl u. Eisen, Düsseldorf*, **36**, 1226, 1916.

[57] GILLE, B. *Métaux et Civil.*, **1**, 89, 1945.

[58] BOUCHAYER, A. 'Les chartreux maîtres des forges.' Didier et Richard, Grenoble. 1927.

[59] ARCHIVES NATIONALES, PARIS: MS. S 4955, no. 10.

[60] *Ibid.*: MS. J 195, no. 34.

[61] *Ibid.*: MS. JJ 48, no. 160.

[62] RHODIN, J. G. A. *Engineer, Lond.*, **142**, 136, 1926.

[63] JOHANNSEN, O. 'Geschichte des Eisens' (3rd ed. rev.), pp. 120–26. Stahleisen, Düsseldorf. 1953.

[64] GILLIS, J. W. *Arch. Eisenhüttenw.*, **23**, 407, 1952.

[65] JOHANNSEN, O. *Stahl u. Eisen, Düsseldorf*, **30**, 1373, 1910.

BIBLIOGRAPHY

AGRICOLA, GEORGIUS. *De re metallica libri XII*. Froben, Basle. 1556. Eng. trans. and comm. by H. C. HOOVER and LOU H. HOOVER. Mining Magazine, London. 1912.

BAILEY, K. C. 'The Elder Pliny's Chapters on Chemical Subjects' (2 parts). Arnold, London. 1929, 1932.

FORBES, R. J. 'Metallurgy in Antiquity.' Brill, Leiden. 1950.

GOWLAND, W. "The Metals in Antiquity." *J. R. anthrop. Inst.*, **42**, 235–87, 1912.

JOHANNSEN, O. 'Geschichte des Eisens' (3rd ed. rev.). Stahleisen, Düsseldorf. 1953.

RICKARD, T. A. 'Man and Metals' (2 Vols). McGraw-Hill Book Company, London, New York. 1932.

STRAKER, E. 'Wealden Iron.' Bell, London. 1931.

For further references see:

FORBES, R. J. 'Bibliographia Antiqua, Philosophia Naturalis', Part 2: 'Metallurgy.' Nederlandsch Instituut vor het nabije Oosten, Leiden. 1940–50, and supplement I, 1952.

3

AGRICULTURAL IMPLEMENTS

E. M. JOPE

I. SOME GENERAL CONSIDERATIONS ON PLOUGHING

THE history of farm tools is part of the general history of agriculture. That subject cannot be considered as a whole in this work, though certain aspects of it are treated in other chapters (vol I, pp 539 ff., and vol II, pp 591 f., 670; figure 503).

The highest development of ancient Mediterranean agriculture was seen in the last centuries before Christ, especially in Italy and Ptolemaic Egypt, where great attention was devoted to the choice and improvement of agricultural tools and methods. The subsequent retraction from the large estates worked by slaves to an economy of smaller tenant farmers disseminated these improvements throughout the Roman Empire. Since then, Mediterranean agriculture has changed remarkably little.

In the colder and wetter conditions of northern and western Europe problems were very different. The bringing of these heavy, but ultimately more productive, lands under cultivation was begun in Roman times, though in some places outside the Roman frontiers. The process continued for more than a thousand years before it reached its widest extent. Herein lies the most significant agricultural development in the period covered by this volume.

The plough has been an implement of very great economic significance since the Bronze Age. Its importance was further enhanced when the heavier but more productive clay-lands were brought under cultivation for intensive cereal-growing. This process did not start in earnest until the Christian era. Before that, the plough-types of barbarian Europe were of a kind comparable with those current in the Mediterranean lands, and ultimately with those of the ancient civilizations of the Near East.

Mediterranean agriculture and that of the northern clay-lands had quite different problems. The former, working light permeable soils, has always been concerned with moisture-conservation, which is effected by keeping the surface layers continually powdered—the technique known as 'dry farming'. The latter, on heavier impermeable soils, is much more frequently concerned with efficient drainage. Nevertheless, the Mediterranean methods were not everywhere entirely unsuited to the northern lands, for the early agriculture of their barbarian peoples was largely confined to the lighter soils until the Iron Age. It is generally assumed that the climate of north-west Europe was warmer and drier in the Bronze Age, till about 500 B.C., than it has been since.[1] But intensive grain-production became increasingly urgent with the rising requirements of the Roman armies and towns. The Mediterranean methods of cultivation, which did little more than powder the surface of light soils, were inadequate for working the heavier ones which had then to be brought under cultivation. The change is reflected in radical alterations in the equipment and methods of ploughing.

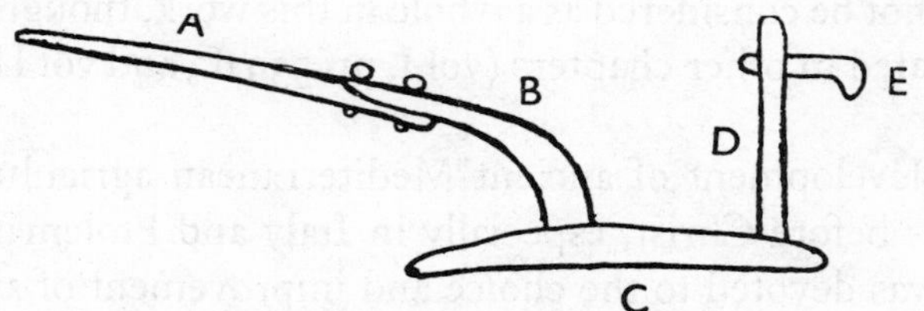

FIGURE 43—*Diagram of Greek plough:* (A) *draught-pole;* (B) *draught-beam;* (C) *stock;* (D) *stilt;* (E) *handle.*

Progress in exploitation of heavier lands, though suffering some decline with post-Roman unrest, continued for more than a millennium, and did not achieve its maximum extent until the thirteenth or fourteenth century. Grain-production then receded a little in some parts of Europe, for various economic and social reasons, but there was a rise in animal husbandry, particularly sheep-farming.

II. EARLY PLOUGHS

Neolithic farmers in Europe do not seem to have used any form of traction-plough. The tilling of their plots must have been done with hoe, digging-stick, and perhaps a precursor of the mattock. All early ploughs, alike in the ancient east, the classical world, and among the barbarians of northern and western Europe, were constructed according to the same general principle (figures 43–44). There was a stock, of which the leading point, lying nearly horizontal, did the actual breaking of the ground. The pull of the draught-team was transmitted to the stock through a beam and pole, and a handle was provided either by backward extension of the stock or on a separate stilt at the tail. In the simplest ploughs the stock and pole were formed in one piece from a tree-branch and its

[1] Possibly the assumed moister climate in the Late Bronze Age was more apparent than real. The observed moister soil-conditions may have been due to a redistribution of water over the surface of the land, as the accumulated effect of a millennium of forest-clearance and the abandonment of cleared areas when agriculturally exhausted.

junction with the trunk. Modifications of this type were merely attempted adaptations to local conditions of soil or terrain. Such an implement did little more than disturb the surface, pushing the soil and stones to either side.

The essential feature of ploughing is that it makes a continuous furrow. This is in contrast to hoeing or digging, where the blade is taken out of the soil in a series of discontinuous strokes. One type of ancient plough (vol I, figure 43) looks as though it had originated as a result of drawing a large Egyptian hoe

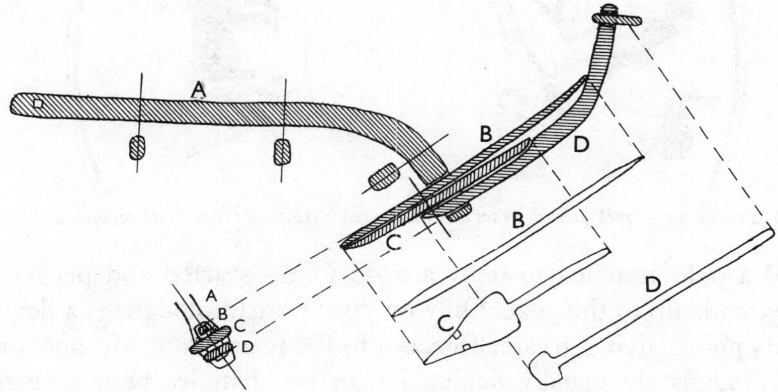

FIGURE 44—*Diagrammatic section of a plough from Donnerupland, central Jutland. Early Iron Age. 2·15 m from stilt to beam.* (A) *draught-beam of birch;* (B) *bar-shaped share of oak;* (C) *arrow-shaped share of oak;* (D) *stilt with handle of oak. Note the stock set at an angle to the pole.*

(vol I, figure 350) through the ground. The other well known type, with horizontal pole (vol I, figure 351), seems to have been a new idea, conceived from the start as a traction-plough. Such appearances may be illusory; the origin of the traction-plough remains uncertain, though we can trace the earliest introduction of some of the more significant features or changes of design in it.

It must have originated somewhere in the lands between Egypt and Persia, and spread gradually thence over much of Europe, north Africa, and the east, being introduced into China about the fourth century B.C.—an example of diffusion (vol I, ch 3) [1]. The earliest known example of it is on a Sumerian seal of the fourth millennium B.C. from the royal cemetery at Ur.

Details of construction of early ploughs. In a well designed example, the shape of the plough-beam and its angle of attachment to the stock would keep the plough working at an even depth with a minimum of effort by the ploughman. In many ploughs depicted from Greece and the Near East the draught-beam is

FIGURE 45—*Greek plough. From a black-figured Nikosthenes cup. Sixth century* B.C.

short, and a pole attached to it by a rigid joint—scarfed and pegged, or just lashed—extends up to the yoke. Only on post-Roman ploughs is a flexible joint seen at this point, giving greater freedom to the team. Both Mesopotamian and Egyptian ploughs are usually depicted with two handles branching from the stock (vol I, figures 43, 351). During the second millennium B.C. the handles may be seen equipped with an upright tube down which the seed-corn was dropped straight into the newly made furrow (vol I, figure 365). This was less wasteful than random scattering, though it was the latter method that survived. The forked double handle was easier and steadier to guide than the single stilt and inserted handle which the classical Greek world had in common with the barbarian peoples of northern Europe (figures 45–46, 52–53).

In surviving examples of this kind of plough from northern Europe of the Early Iron Age (*c* 500–100 B.C.), the pole is of one piece extending right back to the stock and branching from it. The handle-stilt is separate, and the stock nearly horizontal (figure 46). But there was also a type with the stock set at an angle and carried up to form a handle, as though it were a digging-stick or hoe-blade pulled along by the beam

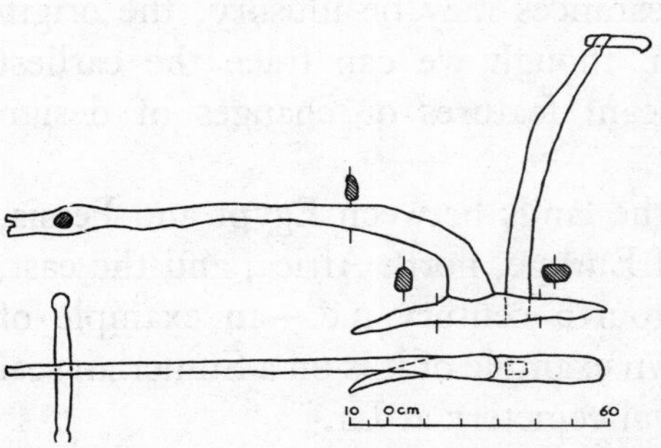

FIGURE 46—*Reconstruction of a plough from Vebbestrup, Jutland, Denmark. Early Iron Age. Note likeness to the Greek plough (figure 43) with horizontal stock, the draught-beam branching from it, and separate vertical handle-stilt.*

(figure 44). This type is known from Egypt (sixteenth century B.C.) and was in use in the Roman provinces.

Greek sources, such as the poet Hesiod (*c* 700 B.C.), enable us to appreciate

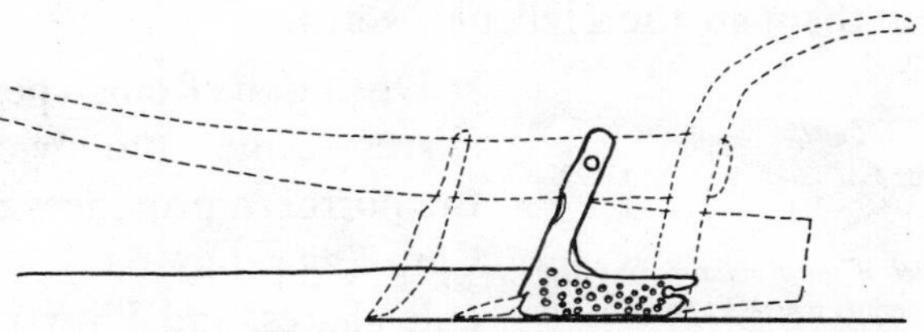

FIGURE 47—*Diagrammatic reconstruction of a plough from Tømmerby, Jutland. Possibly Iron Age. The surviving part is 52 cm high, made of beechwood, and shows pebbles inserted in the side.*

the careful choice of wood for the various parts of the plough. Holm-oak, noted for its toughness, was used for the beam, where the curved part took a particular strain; the stock was of oak and the pole was of elm or bay. Theophrastus (*c* 371–*c* 287 B.C.) says that oak will withstand rot in the earth and elm in the air [2]. The Early Iron Age ploughs from Denmark (figure 44) have beams of alder or birch, crooks of oak or hazel, plough-heads and stilts of oak, alder, birch, or lime, and shares of oak (or, in one case, apparently elder).[1] Particularly liable to wear are the faces of the stock rubbing against the sides of a furrow. They were therefore sometimes protected by numerous pebbles inserted in holes in the wood (figure 47). This practice has continued into modern times in some areas.

One of the most important and obvious improvements to a wooden plough, not always demonstrable in representations, is the protection of the cutting point with an iron shoe, the plough-share or plough-sock. There is no evidence for the earlier use of copper or bronze shares, and the Egyptians seem to have shod their ploughs with flint through much of the dynastic period. In the north, ploughs of the Iron Age (fifth century B.C. onwards) were frequently provided with separate long-shafted shares of hardwood, such as elder or oak, and some early ploughs in the Near East were probably also

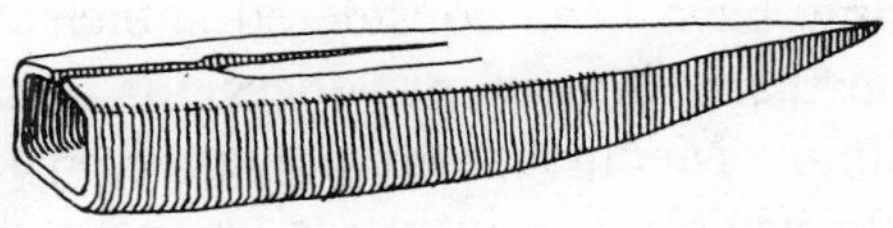

FIGURE 48—*Iron socketed plough-share from Gerar, Palestine. Twelfth to tenth century B.C. Scale* c *1/3.*

[1] The soft woods used for various parts of these Danish Iron Age ploughs suggest that they may be votive offerings in which had been included certain parts of working ploughs, such as the shares, made of harder woods.

thus equipped. As iron became generally available in the eastern Mediterranean area, plough-shares of it came to be regularly used. In Palestine iron shares were found at Gerar in a smith's workshop of the twelfth to the tenth century B.C. (figure 48), and others of before 926 B.C. were found at Megiddo. There is no evidence for them in the Hellenic world.

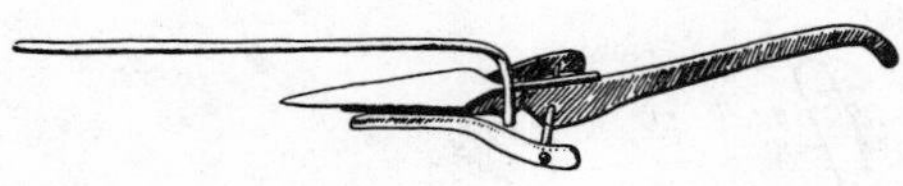

FIGURE 49—*Plough-model of the Roman period from Cologne, showing ground-wrests which scuffed aside the soil. Note tanged share.*

During the Roman period iron plough-shares came into widespread use in the northern provinces as well as in Italy. Cato (234–149 B.C.) mentions detachable shares, and Pliny refers to different types of share adapted to a variety of stocks [3]. Iron plough-shares were used among the Belgic tribes of Britain in the first century B.C. before the Roman invasion—possibly a Roman influence in advance of the conquering armies. These early iron shares are usually socketed, but occasionally of long slender tanged-spear form (figure 49).

III. PLOUGHS FOR HEAVY SOILS

During the Roman age, when Mediterranean life was making its first intensive contact with transalpine Europe, ploughs began to be designed to meet the problems of the heavier, more cohesive, soils which could not be so easily scuffed aside as sandy loam. To get a proper depth of cultivation the plough had both to cut and turn a sod. This was not in itself a new idea in Mediterranean practice, where the plough was sometimes used to heap the soil in ridges over the seed, but it was quite another matter to do this on sticky soils. Iron shares that gave horizontal undercutting had been in use for nearly a thousand years, though there were no coulters or vertical cutters until the Roman age; but a plough equipped with both vertically-cutting coulter and horizontally-cutting share still did not of itself turn over the sod.

Cutting the sod, shifting it to the side, and then breaking it up by cross ploughing, harrowing, beating with mallets, or leaving it to be weathered, may at one time have been considered sufficient. The type of plough mainly used in the northern Roman provinces for turning over the sod was developed from the simple Mediterranean plough. Roman examples with projecting bars or ground-wrests splaying outwards from the share were intended merely to push the soil aside (figure 49). These ground-wrests were then made as flat projecting wings, into which the sod was led by ears on the metal plough-share itself; the mass of loosened sod was then guided over to one side or the other,[1] with some measure

[1] Ploughs which can throw the sod to either side are misleadingly known as 'one-way' ploughs, because they enable the ploughman to travel up and down, turning the sod always to one side of the furrow, no matter which

of overturning, by keeping the plough tilted to that side as it moved. This tilting was helped by providing a keel on the sole of the plough (figure 50).

Many of the large iron coulters found on sites of Roman age are asymmetric. Their cutting-edges were forged by beating from one side only, and their subsequent wear shows that they were used to cut a sod that moved always to one side of the plough, usually the right. The iron shares, while not particularly broad, were used either as a protecting tip to an expanding wooden share, or set askew to the line of the furrow, so that furrows fairly large even by modern standards could be cleared at one stroke. The heavy iron coulters were wedged in place in a slot in what must have been a very robust plough-beam, capable of withstanding a considerable strain even though weakened by the slot.

These coulters and shares were mounted on a rectangular frame different from the Mediterranean type. It consisted of a nearly horizontal plough-beam and a share-beam, joined by a stilt at the back and a brace or 'sheet' just behind the share. This type of square frame is seen on many medieval ploughs. Such heavy irons imply something equally robust in Roman times. From a bog near Tømmerby in Denmark comes a portion of a plough-frame of this kind, the share-beam and sheet in one piece as a natural angled timber protected on the land side by inserted pebbles (figure 47).[1] It is perhaps of the pre-Roman or Roman Iron Age.

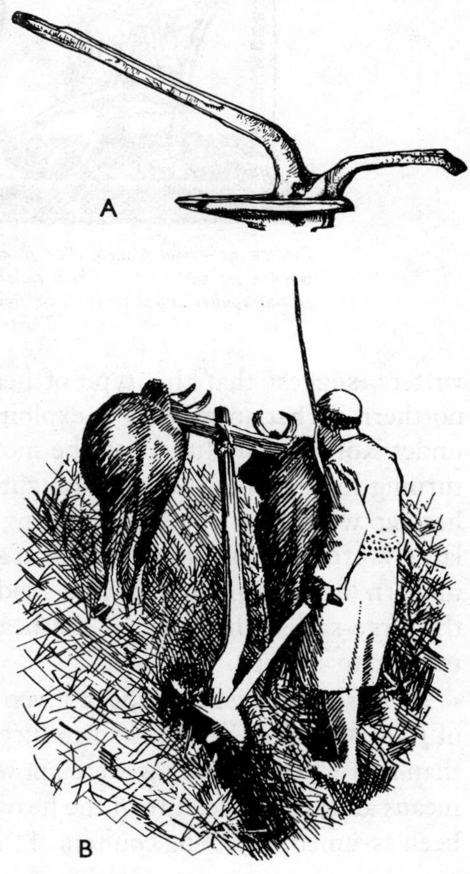

FIGURE 50—(A) *Romano-British plough-model from Sussex, showing a keel; note the socketed type of share.* (B) *Modern ploughing in the Himalayas, with a keeled plough tilted in action.*

The early date of the Tømmerby plough,[2] and scraps of evidence from Roman

way he is going. By their very symmetry these ploughs of the Roman age were evidently intended for 'one-way' work.

[1] In the standard reconstruction it is provided with wheels, as argued by Steensberg, but this is doubtful.

[2] As suggested by pollen-analysis on peat samples from crevices in the surviving part. It might, however, have been pushed down into the bog at a later date.

FIGURE 51—*Foot-plough. The front of the draught-beam is supported by a wooden leg with a foot which slides over the untilled surface in front of the cutting coulter and share (see also figure 56). From a French illuminated psalter. Thirteenth century.*

writers, suggest that this type of heavy plough was indeed developed by the northern barbarian peoples to exploit their own heavier lands, and not initiated under Roman stimulus. With the mould-board (p 89) as sod-turner, so that the turning force came from the draught-team, this type of plough could do much heavier work than the simple tilting type in which the sod-turning force was largely provided by the ploughman's wrists. It was the real precursor of the modern traction-plough, whose medieval intermediaries were responsible for the large-scale cultivation of the heavy lands that altered much of the face of temperate Europe.

Wheels on ploughs have been given undue prominence because of the remark of Pliny that they had been introduced in Bavaria. This really means little more than that wheeled ploughs were known to the Roman world. Wheels were by no means an essential feature of the heavier plough, though their presence has often been assumed. The long coulters of the Roman age in Britain, for instance, imply

FIGURE 52—*Wheeled one-way plough. Note coulter, movable to one side or the other, fixed in place with a peg and thong; also the detachable 'ear', which may be placed on either side of the stock. From a tenth-century calendar.*

FIGURE 53—*Wheeled plough. Note the flexible joint between draught-beam and pole. From Herrad of Landsperg's* Hortus deliciarum. c *1170*.

no more than a high beam. A pair of wheels was merely one way to support this. The beam could also have been supported on the yoke, as in more recent times, or, very simply, upon a sliding foot as seen in a French manuscript of about 1300 (figure 51). Insertion of the wheels under the beam in varying positions, according to a series of peg-holes, would have given, as in later ploughs, a valuable adjustment of ploughing-depth. There is little evidence for any widespread use of the wheeled plough until the eleventh century, when it begins to appear in north European illustrated manuscripts (figures 52–53, 57). It never gained universal acceptance, and wheel-less ploughs (figures 54–55) remained in use both for cheapness and for greater suitability in sticky or stony soils. On many farms both types were needed.

The mould-board is a characteristic part of the heavy plough of temperate Europe. It is a device for guiding the furrow-slice and so turning it over, as distinct from the ground-wrest, which cleared the bottom of the furrow. Mould-boards are not known before the eleventh century (figures 54–55), and there are no medieval documentary details on making them. In more recent times they have been made from smooth fine-grained woods such as apple, pear, or beech, which give a smoother passage of the sod than do coarser grained woods such as elm or oak.

FIGURE 54—*Swing-plough from the Luttrell Psalter. This shows the large mould-board, which turns over the sod after it has been cut by the coulter and share. Note the mallet for adjusting and tightening the pegs and wedges.* c *1338*.

FIGURE 55—*Plough with mould-board, from an English Bible. Note the donkey in the plough-team. Early fourteenth century.*

The introduction of this important sod-turning device was perhaps delayed by the inadequate design of its curvature. Experiment and the personal preference of plough-wrights no doubt led to some variety in the shape of mould-boards, still detectable in the eighteenth century, when some even had kinks and angles in their surfaces. The optimum shape of a mould-board was worked out systematically by Thomas Jefferson (1743–1826), third president of the United States.

As well as manuscript illuminations, there are a few medieval drawings of ploughs by accounting-clerks, in which the parts of the plough are usefully labelled. These are attempts by the clerk to locate for his own convenience the items constantly appearing in his accounts, and are not the product of first-hand

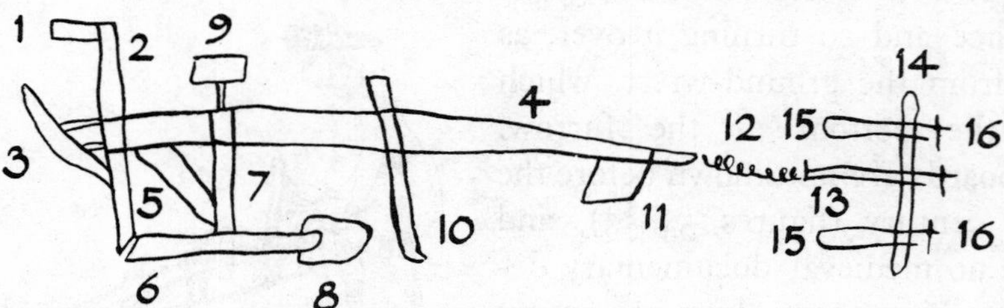

FIGURE 56—*Sketch of a plough by a thirteenth-century clerk. The names of the parts, which were written in Latin, are replaced by numbers:* (1) *the land-handle;* (2) *the plough-tail;* (3) *the furrow-handle or 'plough Stylte', attached to the plough-tail by two 'rough-staves';* (4) *the plough-beam;* (5) *the ear;* (6) *the share-beam or 'plough-head';* (7) *the sheath;* (8) *the plough-share;* (9) *the mallet or 'plough-beetle';* (10) *the coulter;* (11) *the plough-foot, an adjustable piece of wood or iron, attached to the front of the beam, regulating the depth of the ploughing;* (12) *the draught-chain;* (13) *the draught-pole;* (14) *the yoke;* (15) *the ox-bow or 'bonde'* (16) *a pin or peg.*

experience (figure 56). The mallet or plough-beetle often seen carried in a special hole in the beam on medieval ploughs was presumably used for tightening its pegs and wedges (figures 54, 55).

The draught-team in the early Near East and in the Greek world was usually a pair of oxen, though onagers were occasionally used (vol I, p 340). Yokes were placed across the horns or the necks, and sometimes traces were attached directly to the horns. Only in the earliest stages of agriculture was human plough-traction of any importance. In the Roman world also the usual draught-team was a pair of oxen, yoked to the pole. In some parts of the Empire the development of

FIGURE 57—*Ploughing and harrowing scene. Note the square harrow. From the Bayeux tapestry. Eleventh century.*

heavier ploughs may have needed four beasts, and it is possible that yet larger teams were used by the northern barbarians. In Denmark a simple plough of the Iron Age has a swingletree to which the draught-beasts were attached by traces, giving them greater freedom than by yoking to the pole (figure 46). The technique of traces and swingletree, though considerably used by Celtic peoples, often in conjunction with draught-pole and yoke, does not appear to have become general before the Middle Ages.

IV. METHODS OF PLOUGHING

By the end of the first millennium A.D. the four-ox plough-team was fairly common in temperate Europe, and it remained so through the Middle Ages, though smaller teams also continued to be used. The oxen were yoked either in pairs, in tandem, or sometimes four abreast on a long yoke. In the Domesday Book of 1086 the unit of assessment was usually a nominal plough-team of eight oxen, but it does not seem that the eight were commonly yoked at once to a single plough (though they were occasionally), and no illustration shows more than four. The suggested explanations, such as that they worked in two groups of four, morning and afternoon, or on alternate days, spending the rest of the time in grazing; or that one group may have been considered necessary for pulling a harrow after the plough (figure 57) are none of them satisfactory.

In medieval ploughing, horses or donkeys were occasionally used among the oxen. Horses became more effective when the padded horse-collar gradually found general use after the twelfth century (p 554), though even then oxen were preferred as more economical in fodder and harness and as not requiring shoeing. Not until the sixteenth century were horses widely chosen for ploughing on account of their greater speed.

Ploughing is usually depicted in both ancient and medieval sources as employing two men, a ploughman who guides the plough and a driver in charge of the team. Sometimes only one is shown, though guiding the plough and using the ox-goad at the same time must have been difficult (figures 45, 58). In post-Roman times a driver, especially if in charge of a team of four oxen yoked

FIGURE 58—*Flanders plough showing a superstructure for guiding the reins, really more appropriate to horse-traction. From the 'Book of Hours' of the Duc de Berry. Fifteenth century.*

abreast on a long yoke, walked backwards in front of his team—a 'caller', as his Welsh name (*hywell*) signifies. He himself may have encouraged his team by pulling with his hands on the yoke, but he often appears in front using his long goad-tipped rod, the origin of the measure of that name (figures 52, 503).

Ploughing and field-shape. The shape and appearance of fields depend on the plough, the draught-team, the soil, the contour of the land, and above all on the intentions of the ploughman in laying out his furrows. With cross-ploughing the fields naturally tend to be square, and this form is usually associated with small simple ploughs and a small draught-team. There is evidence for cross-ploughing in the Iron Age in Denmark and Holland (plate 2 B), and it was probably practised then and in the Late Bronze Age in Britain also. Larger ploughs and draught-teams, needed to make the heavier clay-lands worth working, are more troublesome and wasteful at turning. Hence with their use the ploughing was in strips, and cross-ploughing—both less necessary and less effective on heavy soils—was discarded, so that the fields tended to be long and narrow.

Ploughing in long strips does not, however, of itself imply ridges and furrows. Whether the field comes to show balks and linchets or ridges and furrows depends upon the type of plough and how the ploughman sets out his field. A plough that can turn the furrow-slice to either side ('one-way' plough, p 86), can leave a field quite level year after year even though it is ploughed in strips. This type of plough is most suitable along a slope, and its use produced the linchets or terraces so characteristic of many hill-sides in Britain.

In some parts of England land was ploughed in long strips in later Roman times, but without producing marked ridges and furrows. The striking ridge-and-furrow fields of the clay-lands of the English midland counties were produced by a swing- or wheel-plough, starting from a central furrow and working outwards to the full extent of the strip. The ridge-and-furrow pattern, however, becomes prominent only when the process is repeated year after year, as with the severally held strips in the open-field system. Ridge-and-furrow layout helped the drainage of fields on impermeable soils and was sometimes adopted for that reason. Thus though the presence of ridge-and-furrow systems may imply the presence at one time of open-field arable, the absence of open fields cannot be argued from the absence of ridges and furrows.

The history of the plough in Britain has been used as a key-index to the development of a land and a nation. Yet over-simplification and inadequately verified premisses have often led to misapprehensions, many of which are perpetuated in the standard histories. Uncritical acceptance of the Domesday nominal eight-ox plough-team as a working reality in the fields has been largely to blame. Thus a mythical history of Saxon Britain was evolved around a heavy wheeled plough, supposedly brought from their Germanic homelands by the earliest post-Roman invaders. No such thing occurred, for no such plough existed. The earliest Saxon farmers were alleged to have changed the landscape by bringing the heavy clay-lands under cultivation for the first time. While it is true that these lands were largely under cultivation in Domesday times (1086) the process did not reach full expansion until about 1300. There is almost no evidence concerning the agricultural methods and activities of the first Saxon settlers,[1] and a real break in the life of the country-side between the late Roman and early Saxon communities must not be always assumed. Archaeological material suggests that the process of bringing heavy lands to cultivation had started by Roman times and that it may even have suffered some set-back under later Roman recession, or as the Saxon intruders wrested lands from the native Romano-British population or intermingled with them. It is more likely that the great expansion of arable land took place in late Saxon times (eighth to eleventh centuries).

[1] There were no querns indicating the use of grain from the only pagan Saxon village site yet explored in England, at Sutton Courtenay in Berkshire; nor were there any from the neighbouring isolated Saxon huts. By contrast, rotary querns have recently been discovered in settlements and huts of this period in the Celtic west.

V. HARROWS AND RAKES

The harrow was known in the Roman world, where its chief use was for tearing out weeds. In the Middle Ages, and perhaps in Roman times as well, it was also used for covering the seed, which it did faster than the plough. The rolling-harrow, a cylinder of oak fitted with iron spikes, was also employed in the Middle Ages for breaking difficult ground and for levelling the threshing-floor. It was, however, only after the sixteenth century that it became general to roll the sown land to make it more nearly level for reaping and mowing.

FIGURE 59—*Horse-drawn plough and triangular harrow. From Virgil's 'Georgics', written in Flanders. 1473.*

Horse-draught, more rapid than ox-draught, was of particular service with the harrow, which to be effective in clod-breaking or seed-covering had to be moved fast. In medieval illustrations the harrow, pulled usually by one or two horses, is sometimes seen following the plough (figures 57, 59) to break down the soil yet further. It was a square wooden frame fitted with teeth and dragged over the soil. Its origin was a frame of thorn branches, such as may still be seen in seventeenth-century treatises on farming. The more easily manœuvred triangular-framed harrow (figure 59) was in general use in northern France by the thirteenth century, but not widely elsewhere until the sixteenth.

Wooden rakes (figure 65) were also used in the Middle Ages for such operations as covering seed and preparing ground in smaller plots.

VI. HARVESTING-IMPLEMENTS

In early times, corn was cut by small sickles with blades of flint or of bronze (vol I, figures 329, 356). Some flint sickles had serrated blades. Reaping in this way was slow and laborious. With the rise of grain-production, speedy harvesting became important, especially under the changeable weather conditions of central and north-west Europe. Continual use of the simple sickle required a handle specially shaped to the grip of the hand, which added to its cost (figure 60).

Significant changes in the equipment for harvesting grain and grass came in with the general use of iron. The balanced sickle, in which the blade is bent back at the handle-end and then curved forward in a long sweep, enabled grain to be cut with less strain on the wrist. It first appeared in Europe to the north of the Alps, in Switzerland and the Hungarian plain, and in the Iron Age from about the fourth century B.C. Its origin may be traced to the ancient Near East. Similarly, the short-handled scythe, worked with two hands, enabled grass to be reaped by a slicing action instead of direct cutting (figure 61). Leaf-reaping and lopping knives, of a pattern continuing into modern times, became widely used. The rising interest in the north in grass- and leaf-harvest was due to the need for cattle-fodder. Oxen were being increasingly stalled during the winter, and thus demanded more provision for feeding. Indoor feeding was also practised in the south, though there perhaps to get manure. Cato (second century B.C.) says that oxen should be fed on hay during the spring when at the plough, but only grazed during the winter [4].

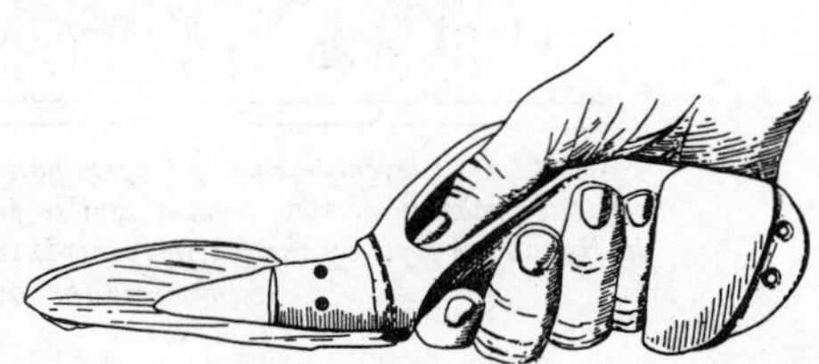

FIGURE 60—*Sickle-handle, elaborately shaped to fit the reaper's hand, from Mörigen, Switzerland. Late Bronze Age. Scale c 1/5.*

Both the balanced sickle, used with one hand to cut, and the balanced scythe, used with two hands to slice, with their often elegant curved blades, were disseminated widely through the Roman world. A tendency to change to open meadows from the leafy fields strewn with bushes, trees, and tussocks favoured the development of long swinging scythe-blades. The long-handled scythe already evolved in the Roman Empire became equipped, about the twelfth century, with its characteristic short bar-handle projecting from the long haft (figures 62–63). This pattern is still universal, just as the balanced sickle of the Roman age is effectively the modern form. By the late fifteenth

FIGURE 61—*Short-handled scythes. From the Utrecht Psalter. Late ninth century.*

FIGURE 62—*Eleventh-century long-handled scythes, one of which the reaper sharpens. The scene is probably derived from a similar picture in the ninth-century Utrecht Psalter, a work in Byzantine style by the Rheims school; the illustrator has changed the old type of plain long scythe into the new one with the bar-handle.*

century in Flanders, agriculturally the most advanced region of Europe, scythes were used with a small half-circle of bent withy attached near the base of the handle, to gather together the cut stems (figure 63). From the sixteenth century onwards the scythe supplanted the sickle for harvesting grain. The wide acceptance of this more efficient harvesting-tool was hampered during the Middle Ages by several factors, such as the legal status of the stubble as common property.

Several different techniques of reaping corn were used in the Roman world. The stems might be cut near the base, and the ears then cut off and collected. The ears might be held with one hand and the stems cut near the middle. Again, the ears might be cut from near the top of the standing stem by some form of composite blade. An extraordinary reaping-machine using the latter method and based on the old harvesting-combs is clearly described by Pliny and other writers [5]: it provides an illustration of Roman intensive agriculture (figure 64). The principle of the harvesting-comb went out of use after Roman times, though sickles with serrated blades have been used until recently. In the Middle Ages the ears were usually cut off near the top (figure 65), but if the corn was to be threshed in the sheaf the stems necessarily had to be cut low down (tailpiece).

FIGURE 63—*Scythe with a semicircle at the base for gathering the stalks. From an engraving after Pieter Brueghel's 'Summer'.* c *1566.*

These cutting-blades were sharpened in the fields as required, by sharpening-stones usually of fine-grained sandstone or of micaceous schist. In Roman and medieval times such stones were widely traded throughout northern Europe. There is some evidence that portable hammers and anvils were used in the fields, as they still are in some parts.

VII. THRESHING AND WINNOWING

The grain had to be extracted from the harvested sheaves by threshing, and

FIGURE 64—*Reconstruction of a Roman machine-reaper, from descriptions by Pliny and Palladius.*

the chaff blown away by winnowing. Medieval illustrations show sheaves being beaten with two-piece jointed flails (vol I, figure 39). The jointed flail (figure 65) was probably in use by late Roman times, for St Jerome (fourth century) seems to refer to it [6]. Gaul was perhaps its country of origin. But threshing was then largely done by the tread of oxen—later, of horses—or by the *tribulum*, a flint-studded board dragged over the threshing-floor by animals (figure 70).

In early times threshing was done with unjointed sticks, the use of which long survived in many areas. In temperate Europe, if threshing was to be done in the open, it had to be finished before the late autumn. Threshing under cover in Britain is mentioned by Strabo (d *c* A.D. 21), and from the thirteenth century onwards grain was being threshed far into the winter in the great barns, usually on monastic estates. Some medieval examples of such barns still survive (tailpiece).

The chaff was removed from the grain in a winnowing-fan. The fan, Latin *vannus*, was originally a specially shaped container—usually a basket (figure 65) —which was agitated in a draught to remove the chaff. Only later did the word

FIGURE 65—*Hafted fork; rake and jointed flail; winnowing basket. From an illuminated Bible.* c *1250.*

come to mean the vane that created the draught. Servius (fourth century A.D.) defines the *vannus* as the sieve of the threshing-floor, and says it is the equivalent of the Greek *liknon*, a wicker-cradle [7]. Presumably the draught was often artificially created by flapping, but among peasant communities winnowing must also have been done upon a windy hill-side, as until recent times in Ireland. The rotary winnowing-machine was a late introduction into the western world (sixteenth to seventeenth century), but it is said to have been known much earlier in China (p 770).

VIII. DIGGING-TOOLS

The soil was broken in early times with digging-sticks, hoes, and mattock-like implements, the latter stone-bladed like carpenters' adzes. The shoulder-blades of oxen, sometimes used as spades (vol I, figure 372), were more suitable for shifting soil than actually digging into it.

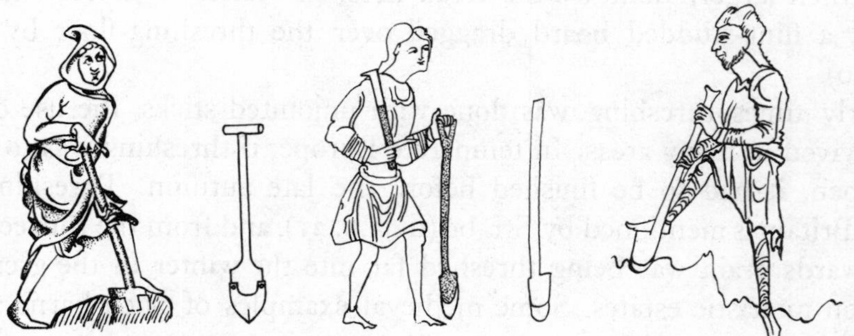

FIGURE 66—*Spades* (right to left)*: one-sided. From a manuscript of Anglo-Saxon biblica poetry,* c *1000. One-sided with handle. The object carried over the shoulder is probably a hoe. From Ælfric's Anglo-Saxon paraphrase of the Pentateuch, before 1050. Double-bladed spades with various haftings of the iron blade, from fifteenth-century manuscript illuminations. For a different type of spade see p 694, tailpiece.*

From Roman times wooden spades (vol I, figure 376), often shod with iron, have been extensively used. Strong iron mattocks and picks were turned out in smithies all over the Roman Empire, but these were more often the tools of the civil engineer than of the farmer. In medieval illustrations spades can be seen with both single- and double-sided blades. Their handles are finished with a cross-bar, fitted either direct on to the shaft-end or across a forked end (figure 66). Iron two-pronged heads were in general use from Roman times for digging and breaking the soil either as forks or picks (figure 67); lighter forks were used for handling hay- and corn-sheaves (figure 65).

IX. MISCELLANEOUS

Fencing. All cultivated land, whether for corn, meadow, olives, or vines, had to be protected against intruding animals. In the peasant pastoral life of classical times, sheep were brought at night into a fold of large upright stones or tree trunks [8]. In early times, as also in the Middle Ages, ditches, often with banks, were the usual protection. In the Iron Age and Roman period, and perhaps later also, stockades of timbers set in a foundation trench were made (vol I, figure 197). In the Middle Ages fencing was provided by quick-set hedges, wooden palisades, and, more usually, woven withies (figure 59). All these were fixtures, but temporary fencing was also required, and hurdle-making for this purpose was a considerable industry. Hurdles were further needed for sheep-folding on upland pastures. They seem usually to have been made of intertwined withies on a frame, just as today.

FIGURE 67—*Fork-headed pick or hoe in use. From a mosaic in the Great Palace, Constantinople. Early fifth century.*

Corn-drying kilns. With damp or imperfectly ripened corn, artificial drying is a necessary prelude to either threshing or milling. The drying of ears of corn before threshing has been regularly carried out since the Iron Age or earlier in the damp northern and western outer fringe of Europe, and the drying-kilns have undergone little modification during that time. They were either rectangular or circular chambers built of stone, or sometimes of clay, in which the ears or sheaves of corn were heaped on a hurdle. A flue led into this chamber, and at its mouth a fire was lit. An obstruction, such as a stone, prevented the flames from reaching the corn, the drying being done by the hot smoke. The winnowed grain was also dried before grinding, in smaller clay ovens. This procedure may have been derived from the other old tradition of burning off the husks instead of blowing away the chaff by winnowing.[1]

Axes with simple iron heads have been used since the Iron Age for such tasks as tree-felling, and in the Middle Ages, at any rate, for slaughtering beasts. The more complex varieties of axe-blade were probably carpenter's tools. *Shears* of simple unchanging pattern have been used since the Iron Age for sheep-shearing (figure 154). The introduction of scissors in Europe, about the sixth or seventh century, does not seem to have affected the use of shears for this operation, and they are so used today.

FIGURE 68—*Sowing-scene from the Luttrell Psalter.* c *1338.*

Wooden vessels were considerably used on the farm. From the Iron Age in temperate Europe stave-built buckets were used for milking, and well made stave-built churns for making butter or storing milk (figure 231). Pottery and metal vessels and turned wooden bowls were commonly employed for cooking and storage. The medieval sower carried his seed in a cooking-pot or bowl, or in his turned-up apron. Sometimes, however, he appears with a carefully made *seedlip* or *seedcot*, a wooden box-frame with wicker sides (figure 68), though the name *seedlip* implies a container of coiled straw.

Ladders, with rungs jointed into long beams, as nowadays, were used in Roman vineyards and during the Middle Ages.

There is a fourteenth-century illustration of a shepherd's movable plank-built hut on solid wheels, presumably to provide protection in exposed fields at lambing time.

[1] Even in early Roman Italy ears of emmer were roasted in an oven, as in the feast of *Fornacalia*, to propitiate the goddess Fornax.

REFERENCES

[1] A very different view of the origin and spread of agriculture is advanced by C. O. SAUER 'Agricultural Origins and Dispersals.' Bowman Memorial Lecture no. 2. American Geographical Society, New York. 1952.

[2] HESIOD *Opera et dies*, 429–30, 435–36. (Loeb ed. 'Homeric Hymns', p. 34, 1914.)
THEOPHRASTUS *Hist. plant.*, V, iv, 3. (Loeb ed. 'Enquiry into Plants', Vol. 1, p. 441, 1916.)

[3] CATO *De agri cultura*, cxxxv. (Loeb ed., p. 116, 1934.)
PLINY *Nat. hist.*, XVIII, xlviii, 171–4. (Loeb ed. Vol. 5, pp. 296 ff., 1950.)

[4] CATO *De agri cultura*, liv. (Loeb ed., pp. 68 ff., 1934.)
PALLADIUS *De re rustica*, VII, ii, 2.

[5] PLINY *Nat. hist.*, XVIII, lxxii, 296. (Loeb ed. Vol. 5, p. 374, 1950.)

[6] SAINT JEROME *Comm. in Isa.*, IX, 28. (*Pat. lat.*, Vol. 24, col. 326.)

[7] SERVIUS *Comm. Virg. Georg.*, I, 166.

[8] HOMER Odyssey, IX, 216 ff. (Loeb ed. Vol. 1, pp. 316 ff., 1919.)

BIBLIOGRAPHY

BISHOP, C. W. "Origin and Early Diffusion of the Traction-plough." *Antiquity*, **10**, 261–81, 1936.

CHILDE, V. GORDON. "The Balanced Sickle", in 'Aspects of Archaeology in Britain and beyond. Essays presented to O. G. S. CRAWFORD', ed. by W. F. GRIMES, pp. 39–48. Edwards, London. 1951.

CLARK, J. G. D. 'Prehistoric Europe', pp. 100–7. Methuen, London. 1952.

COLVIN, H. M. "A Medieval Drawing of a Plough." *Antiquity*, **27**, 165–7, 1953.

CORDER, P. "Roman Spade-Irons from Verulamium." *Archaeol J.*, **100**, 224–31, 1943.

CURWEN, E. C. 'Plough and Pasture. Past and Present.' Studies in the History of Civilisation no. 4. Cobbett Press, London. 1946.

CURWEN, E. C. and HATT, G. 'Plough and Pasture.' Schumann, New York. 1953.

DUIGNAN, M. V. "Irish Agriculture in Early Historic Times." *J. R. Soc. Antiq. Ireland*, **74**, 124–45, 1944.

GLOB, P. V. "Ploughs of the Døstrup type found in Denmark." *Acta Archaeologica*, **16**, 93–111, 1945.

Idem. 'Ard and Plough in Prehistoric Scandinavia.' Aarhus University Press, 1951.

GOW, A. S. F. "The Ancient Plough." *J. Hell. Stud.*, **34**, 249–75, 1914.

HARRISON, JANE E. "Mystica Vannus Iaechi. (The Winnowing Fan)." *Ibid.*, **23**, 292–324, 1903; **24**, 241–54, 1904.

MASSINGHAM, H. J. 'Country Relics', with drawings by T. HENNELL. University Press, Cambridge. 1939.

NIGHTINGALE, M. "Ploughing and Field-shape." *Antiquity*, **27**, 20–26, 1953.

PARAIN, C. "The Evolution of Agricultural Technique" in 'Cambridge Economic History', Vol. 1, chap. 3, pp. 118–68. University Press, Cambridge. 1941.

PAYNE, F. G. "The Plough in Ancient Britain." *Archaeol. J.*, **104**, 82–111, 1947.

RICHARDSON, H. G. "The Medieval Plough-team." *History*, **26**, 287–96, 1942.

SCOTT, SIR (WARWICK) LINDSAY. "Corn-drying kilns." *Antiquity*, **25**, 196–208, 1951.

STEENSBERG, A. "Northwest European Plough-types of Prehistoric times and the Middle Ages." *Acta Archaeologica*, **7**, 244–80, 1936.

Idem. "The Vebbestrup Plough." *Acta Archaeologica*, **16**, 57–82, 1945.

Idem. 'Ancient Harvesting Implements. A Study in Archaeology and Human Geography' (trans. by W. E. CALVERT with bibliography). Nationalmuseets Skrifter. Arkæologisk-historisk Række, no. 1, Copenhagen. 1943.

Idem. 'Farms and Watermills in Denmark through 2000 years.' Nationalmuseet, Tredje Afdeling. Arkæologiske Landsbyundersøgelser, no. 1. Copenhagen. 1952.

STEVENS, C. E. "Agriculture and Rural Life in the Later Roman Empire", in 'The Cambridge Economic History', Vol. 1, chap. ii, pp. 89–117. University Press, Cambridge. 1941.

Threshing in a barn. Flemish calendar picture for August. Fifteenth century.

4

FOOD AND DRINK

R. J. FORBES

I. PRODUCTION OF FLOUR AND BREAD

THE diet of the masses in ancient Greece and Rome consisted of little but bread and a porridge of wheat or barley supplemented by vegetable, fish, and spice. Cereals were relatively more important as a human food than they are today; most of the cereals we now produce are consumed by animals. Wheat and barley, the ancient staple crops of farmers of the Near East, ultimately spread over western Europe, mainly from the Mediterranean basin.[1] Climatic factors were of great importance in the choice of cereals in the colder regions of Europe (vol I, ch 14). Traditional methods of irrigation were also significant in that the elaborate systems so necessary in the Near East were applied only with reluctance.

A very important factor in determining the diet of the big cities of northern Italy was the cost of transport. It is estimated that the price of a low-value bulky commodity, such as wheat, was doubled after a haulage of 100 miles [2]. Large-scale production of wheat for Rome and the army was therefore impracticable; and, since it was both easier and more economical to drive cattle and pigs to market than to haul cereals over the Italian roads, farmers were diverted from wheat to fodder-grains such as barley.

The cheaper import by sea of wheat from Sicily, Egypt, and north Africa further favoured this policy, which explains the use of the artificial port of Rome at Ostia despite its stormy entrance (figure 458). The ever present threat of famine in Rome discouraged the use of the much better deep-sea harbour of Puteoli, because it entailed 150 miles of land-haulage. Seen in this light, Nero's plan to connect Ostia and Puteoli by a canal navigable by barges was economically sound.

Flour was of much poorer quality in Roman times than is commonly assumed, for the wheat was not thoroughly cleaned before the rather coarse grinding and

[1] The Latin word *triticum* (= that which is ground) in common usage implied wheat in general, of which two principal types were distinguished: *triticum* in a more restricted sense (macaroni- and rivet-wheats), and *siligo* (probably club- and bread-wheats). (See vol I, pp 363–7.) The former type was predominant throughout the Mediterranean in classical antiquity [1].

sifting. Our modern white flour is the product of a 70–75 per cent extraction, but most ancient flour was simply whole-grain meal, and even among the sifted flours an 80 per cent extraction was common [3].

Today, wheat before grinding passes through several stages of cleaning and is then 'tempered' or 'conditioned'. The latter operation, by adjusting the amount of water in the wheat, ensures the optimum elimination of husk from kernel during the first stages of grinding. In Pliny's time it was common practice in Babylonia, Palestine, and Egypt, but not in Rome, to temper before grinding.[1] But although it is recorded in classical times that in Egypt wheat for kings and nobles was first cleaned, and that in certain cities of Asia Minor a more refined milling-process which included preliminary washing was used, in general flour was simply ground and was consequently of an inferior colour and contained many impurities.

Of most of the articles in the miller's equipment—baskets, hand-sieves, and measures—the design was very ancient. Only the mill had been improved. The animal-mills used by Roman millers (figure 78) required more power and were less efficient than the hand-mills: in fact, the inefficient use of animal-power in Roman mills partly explains why human labour was so largely used for grinding.

Assuming that the flour was ground in a two-donkey mill, and that one man was required to control the animal and sift the flour ground, it has been calculated that the total energy needed to mill a bushel of flour amounted to about half a horsepower-hour, as compared with a total of 1½ to 2 horsepower-hours per bushel in a modern factory, where 60 per cent of the energy is used for cleaning and only 40 per cent for grinding. The pair of animals, moving in a small circle, could not have supplied more than about 2½ horsepower-hours each day: sufficient, with the energy supplied by the attendant, to grind and sift a maximum of 7 bushels. On this basis, at the equivalent of 20 lb of wheat per man (wages and subsistence) and 10 lb per donkey, which feeds mainly on grazing, it may be concluded that the cost of grinding came to about 40 lb of wheat for 7 bushels of flour, or about 10 per cent of the price of the flour [4].

Sieves had improved little. 'The Gallic provinces invented the horsehair sieve, while Spain made sieves and meal-sifters of flax and Egypt of papyrus and rush', says Pliny [5]. These sieves were mostly very coarse. That used to separate bran from flour was also used for *alica*, a kind of semolina with very large particles. Even the linen sieve for separating first-grade flour must have been coarse by our standards, for it was used also to separate chalk from *alica* [6].

[1] Evidently this difference was connected with the different types of wheat used. The later types of naked wheat needed grinding and sieving only, but the earlier husked cereals like emmer and spelt demanded a certain amount of cleaning, tempering, and pounding to separate the husk before grinding.

Bread always remained the staple food. The Greek *artos* means both bread and food in general, as in the Authorized Version of the Bible. In Bronze-Age Greece a coarse unleavened bread was eaten in the form of flat cakes, as in other parts of prehistoric Europe (vol I, pp 271–5). The lighter leavened bread was introduced from the east and was still a luxury in Greece in the days of Solon (sixth century B.C.) [7].

> Millet [says Pliny] is especially used for making leaven. If dipped in unfermented wine and kneaded it will keep a whole year. A similar leaven is obtained by kneading and drying in the sun the best fine bran of wheat itself, after steeping for three days in unfermented white wine. . . . At the present time leaven is made out of the flour itself, which is kneaded before salt is added to it and is then boiled down to a kind of porridge and left till it begins to go sour. Generally, however, they do not heat it up at all, but only use the dough kept over from the day before; manifestly it is natural for sourness to make the dough ferment, and likewise that people, who live on fermented bread, have weaker bodies, inasmuch as in old days outstanding wholesomeness was ascribed to wheat the heavier it was [8].

Pliny tells also that the Gauls and Iberians, when making beer from corn, used 'the foam that forms on the surface in the process for leaven; in consequence they have a lighter kind of bread than other peoples' [9].

FIGURE 69—*Threshing with oxen. Zlokutchene, near Sofia.*

II. EARLY EVOLUTION OF DISINTEGRATING TECHNIQUES

The production of flour, olive-oil, wine, and beer, the four mass-produced aliments of antiquity, required forms of crushing or disintegrating apparatus. Wheat-grains, extracted from the ears by threshing and winnowing, were dehusked by pounding, and then ground to flour. These operations were also necessary in the preparation of malt for beer. The fruit of the olive had first to be bruised, either by pounding or in an olive-mill, before its oil could be extracted by pressure. Presses have also been used since the earliest times to obtain the juice of grapes.

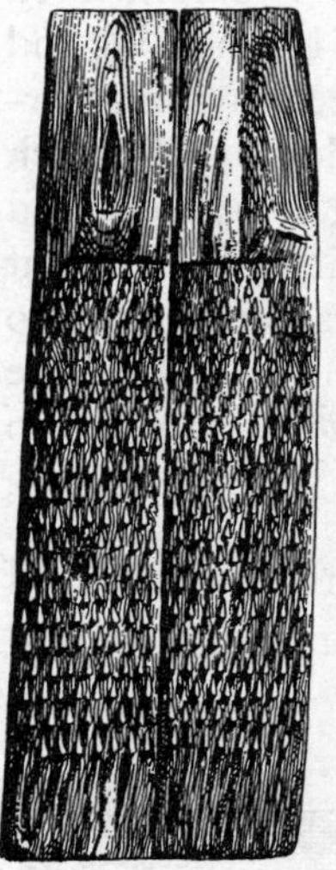

FIGURE 70—*Underside of a Cypriot threshing-sled studded with flints. Length about 2·5 m.*

(*a*) *Threshing* (figure 69). 'The ear itself when ripened is beaten out with threshing sledges on a threshing floor, in other parts by being trodden on by mares, and in other places threshed out with sticks' [10]. The second of these methods enumerated by Pliny has been finely illustrated by very early Egyptian artists (vol I, figure 221). In Palestine, Ruth, after working in the fields all day, 'beat out that which she had gleaned' (Ruth ii. 17).

A more elaborate instrument, the threshing-sled, was developed in Palestine and Mesopotamia. It consisted of a board or sled studded with flints or metal nails and driven over the threshing-floor by oxen (figure 70). It was sometimes weighted with stones and carried a driver. In the Bible it is referred to metaphorically as a symbol of utter destruction. This implement is still used in many parts of the Mediterranean region and Asia.

The hinged flail has its earliest reference in the commentary on Isaiah by St Jerome (d 420); it seems to have been invented in Gaul in the fourth century A.D. and during the Middle Ages slowly displaced the simple stick (p 97).

(*b*) *Pounding*. The pestle and mortar have evolved from forms used in Palaeolithic times to disintegrate foodstuffs. Their operation involves a combination of grinding and pounding. In Neolithic times these two processes became separated so that, for instance, grains were first dehusked by pounding, and then the kernels were ground to flour. The Roman bakers, although characteristically associated with milling, were known as 'pounders' (*pistores*) from their more ancient use of the pestle and mortar. Milling-implements such as the saddle-quern and later the rotary quern developed from early forms of mortar and pounding-stone. Dehusking mortars were often of wood and placed on a stand.

A simple mechanical aid to pounding is the elastic suspension of the pestle by rope to the branch of a tree or a springy lath of wood [11]. This method may be referred to in the only classical passage on working the pestle with a rope [12]. Attached at right angles to the end of a shaft the pestle may be worked like a hammer, a use known today in the Alps, and in the Far East for pounding rice and ores [13]. Further, the shaft may be pivoted and counter-balanced to be worked by hand or foot, as in Galicia, Poland, China, and Japan (figure 71). In mountainous countries there is a form which, instead of a counter-

FIGURE 71—*Japanese pivoted pestle. From a drawing by Hokusai. Nineteenth century.*

weight at the end of the shaft, has a bowl that can be made to fill with water so as to raise the pestle, spill the contents, and return to be refilled when the pestle falls. This simple prime-mover may well be the ancestor of the stamp-mill which, combined with a water-wheel, was elaborated in the Middle Ages (figure 544).

(*c*) *Milling: Development of the Rotary Quern.* For centuries after the beginning of public bake-houses in Rome (third or second century B.C.) bread-making remained a major task for the housewife. In these circumstances the saddle-quern (vol I, figures 175–6, 180) and various forms of grain-rubber[1] remained in common domestic use long after more effective forms such as the pushing-mill and the rotary quern had been introduced.

The pushing-mill was a development of the saddle-quern. In it both stones were flat and grooved, and the upper had a hopper with slit, so that the grinding-

[1] A type of shallow mortar suited to grinding rather than pounding, with squat bun-shaped pestles.

FIGURE 72—*Upper stone of a pushing-mill from Olynthus, seen from above. The depression forms a hopper from which the grain escapes through the slit. On either side are cuts across which a handle would have been placed, either to give the stone a motion towards and away from the grinder, or pivoted as in figure 73. Fifth century B.C. Scale 1/10.*

surfaces were continuously supplied with corn (figure 72). The upper stone might also be fitted with a stick running across as a handle (figure 73). This pushing-mill probably originated in Asia Minor and Syria [14]; it became familiar in Greece from the fifth century B.C. and in Rome in the first century B.C.

The rotary quern is important as the first major application of rotary motion since the invention of the potter's wheel and the lathe in the oriental Copper Age [15]. It seems probable that it evolved, not from the saddle-quern, but from grain-rubbers with spherical stones running in rimmed concave basins. Paired stones or grinders in which a projection on the upper stone pivots in a depression in the lower, or vice versa, were used in Palestine in the second millennium B.C. (vol I, p 200). With the addition of a handle-peg piercing the side of the upper stone, as in an example from Tel Halaf, Syria (ninth century B.C.?), these become rotary querns.

Archaeological evidence for the early use of this type of quern in Greece is not conclusive, and even in the third century B.C. its use was still somewhat rare.

From Greece the rotary quern spread as far north as Moldavia and the Alpine region, where we can trace it throughout the area of the Celtic La Tène Iron Age civilization (first century B.C.). Here it has a hopper in the upper stone. This type of mill reached southern England in the first century B.C. (figure 74 A).

By the time that the rotary quern had spread to Europe and the western

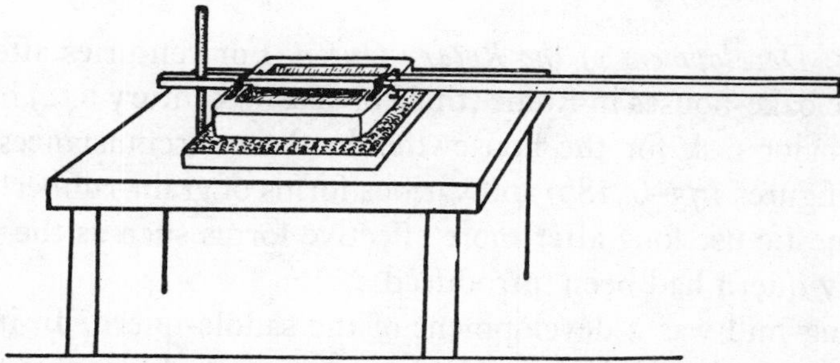

FIGURE 73—*Mounting of a pushing-mill to allow to-and-fro motion about a pivot. Diagram from a bowl of the third century B.C. from Megara near Athens.*

Mediterranean two types were distinguishable: the Iberian with two vertical handles (figure 75), and the central European with the horizontal radial handle of its eastern ancestors (figure 74) [15]. These querns appeared with the professional miller, and their evolution is related to the production of flour on

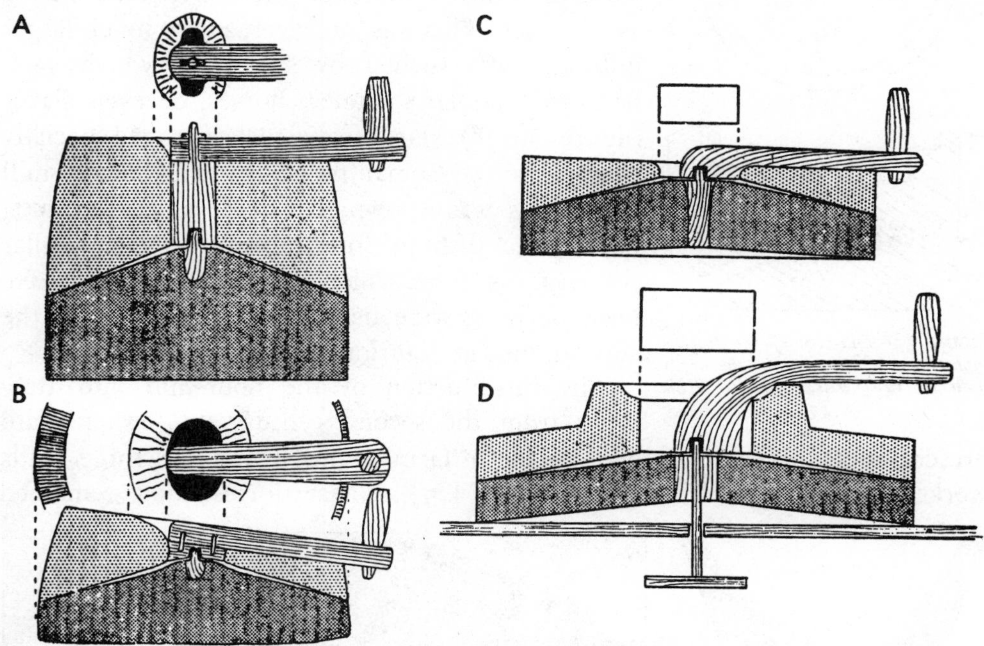

FIGURE 74—*Early English quern types (partly diagrammatic).* (A) *Before 50 B.C.;* (B) *first century A.D.;* (C) *second century A.D ;* (D) *fourth century A.D. Note increase in size of hopper and in ratio of grinding-area to weight.* (D) *has an upper stone adjustable for height. Scale* c *1/8.*

an industrial scale. Northern Germany and Scandinavia did not know the rotary quern until the third century A.D.

The spread of the classical portable rotary hand-mill with its two flat stones seems closely connected with the armies of Greece and Rome. The soldiers of classical armies ground their own corn, and their hand-mills were a common feature of Roman army-camps. One hand-mill was provided for each group of five to ten men, and each man carried enough flour for thirty days.

To prevent the mill-stones from contributing stone-grit to the flour, the developed rotary quern had a spindle protruding from the centre of the lower stone; the weight of the upper stone was transferred to this spindle through a 'rynd' or bridge of wood or iron, fixed across the perforation in the upper stone by which corn was fed to the mill. One of the simplest forms of this suspension

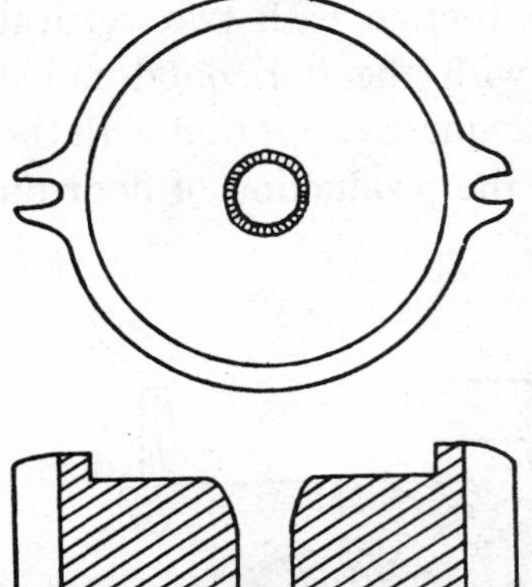

FIGURE 75—*Diagram of the Iberian type of quern with slots in the upper stone for two vertical handles. Scale 1/12.*

is shown in figure 74 A. The bearing was so adjusted that the stones touched only lightly, and most effectively at their edges.

The donkey-mill was thus mounted, the grinding-surfaces being grooved and the upper stone shaped as a hopper. This was, however, a very much larger mill and was turned by spokes to which were harnessed donkeys, mules, horses, or even slaves (figures 76–78). Hard Vesuvian lava proved specially suitable for the Roman mills [16]. The donkey-mill for cereals was known in Greece about 300 B.C., but in the fifth to fourth centuries B.C. similar contrivances, from which the grain-mill may have been derived, were used for crushing ore in the silver-mines at Laurion (p 44).

The introduction of the flour-mill into Italy dates from the second century B.C., when Cato prescribes as necessary apparatus for a farm a proportion of 'three mills worked by asses and one hand-mill' [17]. These forms long remained

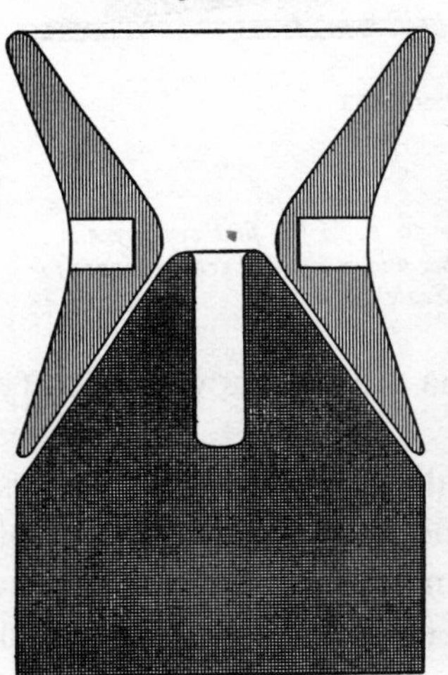

FIGURE 76—*Diagram of the two stones of a donkey-mill (the term is used to denote this large type of mill, whether turned by donkey, horse, or slaves). The depression in the lower stone is to receive a spindle which, by means of a rynd, supports the upper stone. The extension of the upper stone above the grinding-surfaces forms a hopper; the depressions in its side are for the frame that carries the harness.*

FIGURE 77—*Set of four donkey-mills and a baking-oven at Pompeii (A.D. 79). The picture shows what very small turning-circles the harnessed animals were forced to follow (figure 78). The flour collected on the circular platforms.*

the typical equipment of the millers of Rome, who resisted the introduction of the water-mill. The first certain literary reference to a rotary hand-mill in the Roman country-side is given by Virgil (70–19 B.C.) [18].

The important change in the corn-mill under the Empire concerns the methods used to turn it, rather than its construction.[1] The Romans of the Imperial period used fairly large fixed mills, about 75 cm in diameter, with the moving stone geared to a horizontal shaft to which was fixed a wheel operated by a slave [19].

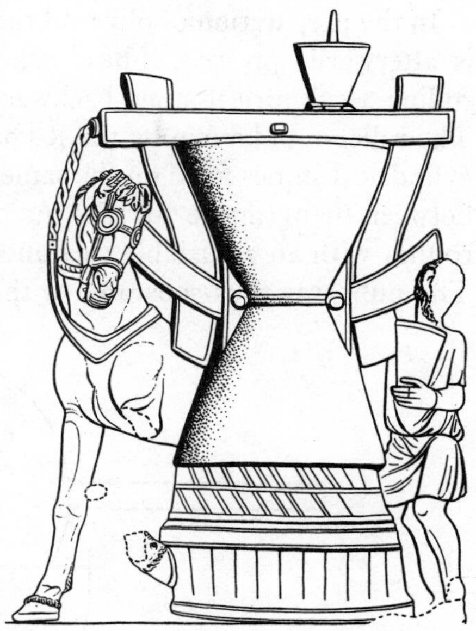

FIGURE 78—*Horse harnessed to a donkey-mill. The miller stands on the right. The grooves on the lower stone can be seen where it protrudes. Beneath this stone is an annular receptacle for the flour. Relief on a Roman sarcophagus of the second century* A.D.

At the present day, certain tribes of east and south-east Asia use cranks on their hand-mills, so that the operator has only to make a short to-and-fro movement. In Europe after 1430 there began to be used double cranks attached to the spindle of the hand-mill, and with a bearing above the operator (figure 597) [20].

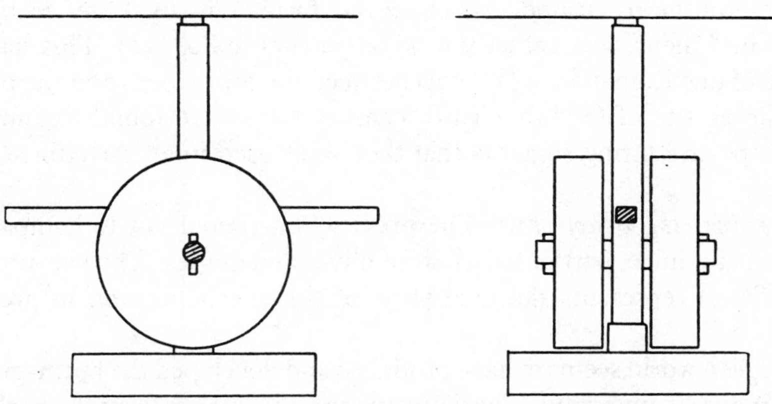

FIGURE 79—*Reconstruction of the* mola olearia *described by Columella, used for crushing olives. The fruit is spread on the flat circular trough and the mill-stones are turned round the vertical pivot. Note clearance between mill-stones and trough, which prevents the olive-stones from being crushed. The horizontal spokes are about waist-high. First century* A.D.

[1] See ch 17 for water-mills and windmills.

In the preparation of olive-oil the first stage is the crushing of the fruit, which is afterwards pressed. The crushing operation is most simply performed by rolling a cylindrical stone backwards and forwards over the olives in a trough. The roller-mill known by the Romans as *mola olearia* (oil-mill) consisted of two cylindrical stones fixed on the same horizontal axle, which was pivoted vertically between them (figure 79) [21]. As the centre pivot was turned, the rollers swept round, with an adjustable clearance, over the flat trough containing the olives. The pulp was thus separated without crushing the kernels. Forms of this mill

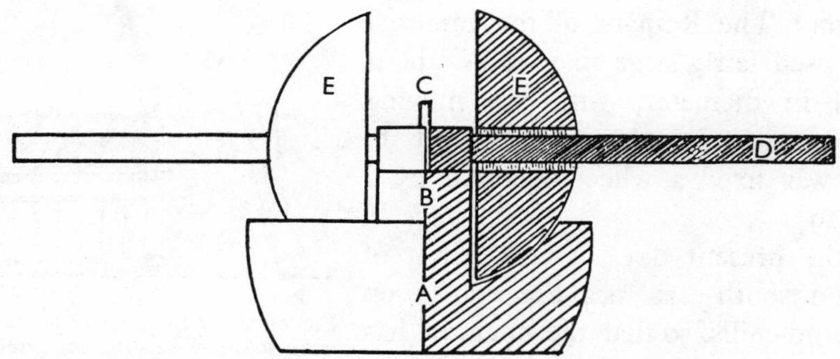

FIGURE 80—*Reconstruction of Greek* trapetum *from Olynthus, for crushing olives. In the middle of a large circular basin of lava* (A) *stands a solid column* (B). *On top of this column is a square hole holding an upright pin* (C) *fastened with lead. A wooden beam* (D) *fits over the pin and carries two heavy plano-convex mill-stones* (E). *As in the* mola olearia *the mill-stones turn about the centre pivot. These stones were of the order of 80 cm in diameter.*

survive in southern Europe. An improved form, said by Pliny to have been invented in Athens, was called the *trapetum* (figure 80) [22]. This had a fixed clearance of one Roman inch (1·8 cm) between the mill-stones and the basin [23]. At the Greek site of Olynthus (fifth century B.C.) were found five mill-stones whose shape and fitting suggests that they were used in an early form of *trapetum* [24].

(*d*) *Pressing: the Beam-press.* The press was invented not to compact but to express oil and juice, particularly that of olives and grapes. The bag-press (vol I, figures 186–7) represents the final stage of this mechanization in pre-classical times.

The Ægean world seems to have produced and developed the beam-press. How far back it goes is uncertain. The cultivation of olives dates from the Early Bronze Age in the southern Ægean, but there is no evidence of machinery for crushing them till later [25]. The earliest known remains of an olive-press and a trough for crushing olives were found in Crete, and are of Middle Minoan date (*c* 1800–

1500 B.C.). They are, however, insufficient for detailed reconstruction [26]. A Late Helladic (*c* 1600–1250 B.C.) beam-press for olives was found on one of the Cyclades. After about 1000 B.C. such presses became more frequent. Their construction is clearly shown on vases such as the black-figure ware of Athenian potters of the early sixth century B.C. (figure 81). The beam-press applies the principle of the lever. One end hinges in the recess in a wall or between two stone pillars, the other is drawn down and often loaded with heavy stones. The fruit, arranged in bags or between wooden planks, is crushed under the middle of the beam. Presses of this kind are still in use in viticulture in many regions.

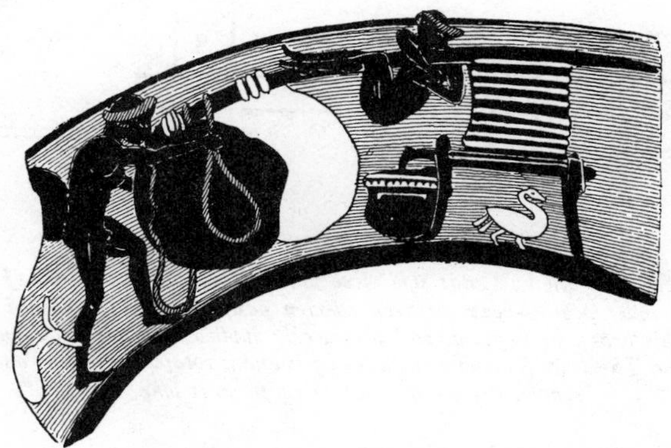

FIGURE 81—*Simple beam-press for extracting oil from olives and juice from grapes. The beam is held in a recess in the wall on the right, not shown. Loaded with weights, human and inanimate, it compresses the rope-bound fruit on the press-bed. The juice is collected in an urn beneath the press-bed. From a Greek vase of the sixth century.*

The further evolution of the beam-press can be traced in the works of classical authors such as Pliny, who writes:

> Some press the grapes with a single press-beam, but it pays better to use a pair, however large the single beams may be. It is length that matters in the case of the beam, not thickness, but those of ample width press better. In old days people used to drag down the press-beams with ropes and leather straps, and by hand-levers; but within the last hundred years the Greek pattern of press has been invented which has spars with furrows running round them in a spiral. Some put handles on the spar, others make the spar raise chests of stones with it, an arrangement which is highly approved. Within the last twenty years a plan has been invented to press with small presses and smaller press-houses, with a shorter upright spar running straight down into the middle, bearing down with the full weight from above on the lid laid on the grapes, and to build a superstructure above the press [27].

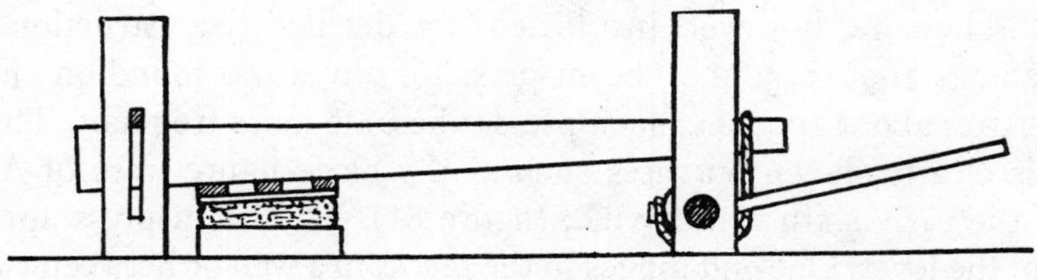

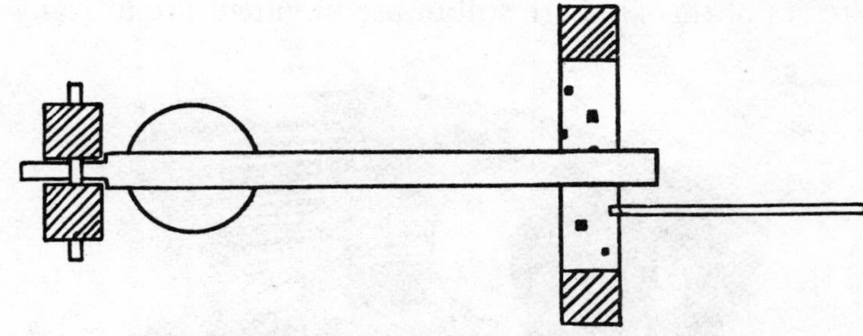

FIGURE 82—*Press described by Cato: side elevation and plan. The left end of the press-beam is caught under a cross-beam between wooden posts. Grapes or crushed olives are packed under this heavy press-beam, and pressure is applied with a rope wound round a drum on the right. The drum is mounted between posts and is rotated by pulling on detachable spokes. Press-beam may be up to 50 ft long.*

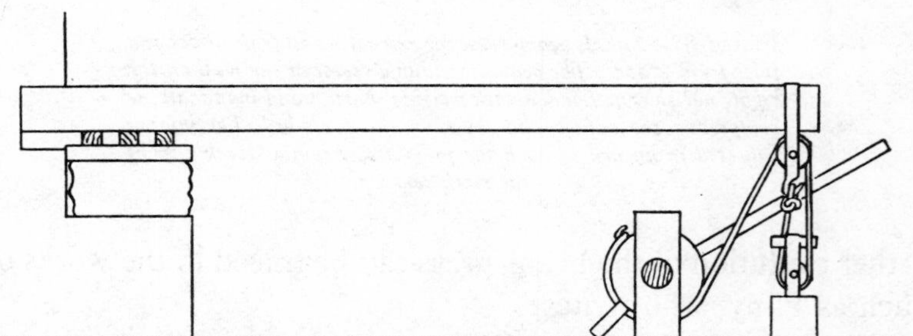

FIGURE 83—*Press described by Hero. The rope is taken in, as in figure 82, by turning the drum with detachable spokes. This raises the weight on the right, when a peg is put in the drum to prevent the rope unwinding. Hence a steady pressure is applied with very great leverage to the contents of the press-bed on the left. First century A.D. Press-beam may be up to 11 m long.*

Pliny clearly describes four types of presses. The first is the old beam-press of Cato (234–149 B.C.) slightly mechanized (figure 82) [28]. One end of the heavy press-beam, often up to 50 ft long, was caught under the cross-piece. The other was drawn down by a rope fastened to a horizontal drum 6–9 inches in diameter. This simple method of pulling down the press-beam, which sometimes itself weighed over 500 kg, was used until the end of the Roman Empire.

FIGURE 84—*Christ in the mystical wine-press. The artist has depicted a real screw-press. From the twelfth-century manuscript of Herrad of Landsperg.*

The second improvement, which allowed a steady prolonged pressure, was embodied in the press described by Hero (first century A.D.) though it was known much earlier (figure 83) [29]. It had a stone weight, press-beam, and winding drum. A rope from the press-beam ran under a pulley on the weight, and over a pulley on the press-beam, to the drum. The press-beam received the full weight of the stone when the rope was taken in on the drum. This form of the old stone-weighed press seems to have been invented in Greece.

The mass to be pressed was variously enclosed in rings of rope, woven esparto, or basketry. Alternatively, 'olives are crushed in wicker baskets or by enclosing the mash between thin laths, a method recently invented' [30]. Ropes wound around a mass of pulp under a press-beam are shown in a Greek picture mentioned above (figure 81). One or two gutters cut into the press-bed collect the juice and conduct it to the storage-vats.

We are informed that a team of four men could finish every day 25 pressings of 100 bushels of olives, each bushel yielding about six pounds of oil. Of grapes, every pressing yielded 22 gallons of must from 300 bushels of grapes.

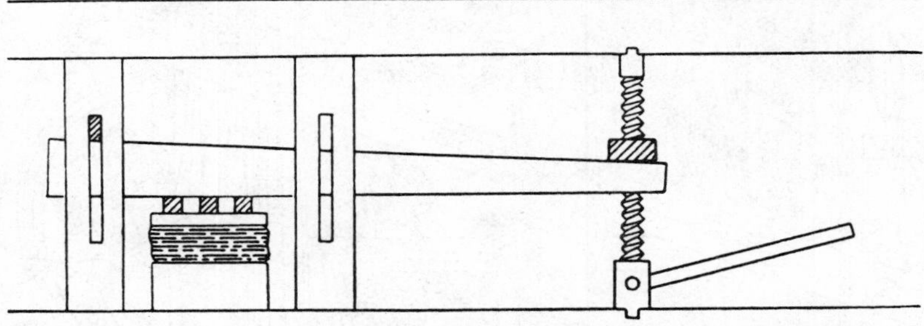

FIGURE 85—*Pliny's first lever-and-screw press. The screw has either one undercut bearing in the floor, or, as here, bearings in floor and roof. There is a second pair of slotted posts on the right of the press-bed. For filling, a beam is pushed through the slots in these posts and below the press-beam. The screw is then turned to bring down the end of the press-beam. First century* A.D. *(For Hero's method of cutting female threads see figure 572.)*

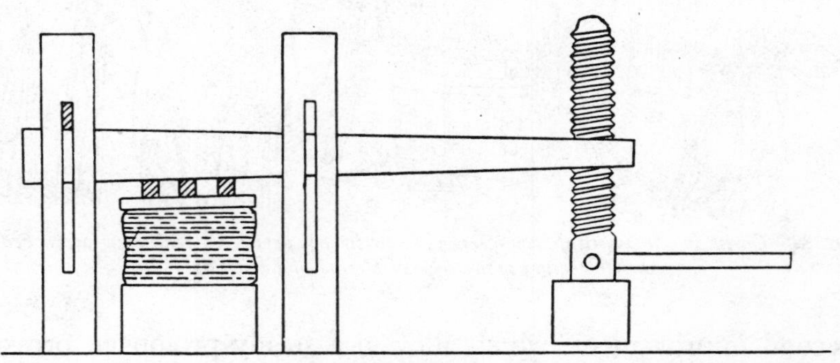

FIGURE 86—*Pliny's second lever-and-screw press. Bearings like those in figure 85 are here eliminated. First century* A.D.

Beam-presses were particularly suitable for large-scale operations, as with olives or grapes (figure 84). Where smaller quantities were involved, as for oil-seeds, herbs, or sheets of papyrus, two other forms, the *screw-press* and the *wedge-press*, were preferred.

The beginning of the screw is associated with the name of Archimedes (d 212 B.C.). By the time of Hero screw-cutters for inner screw-threads were known [31]. This agrees with the account of the screw-press by Pliny. Its introduction into Rome seems to have been at about the end of the first century B.C., and it was probably invented in Greece in the second or first century B.C.

The simplest application of the screw to a press was as a substitute for the spoked drum. The screw was fixed upright to a bearing in the floor, and its nut was placed over a fork at the end of the press-beam, which could thus be lowered by turning the screw (figure 85).

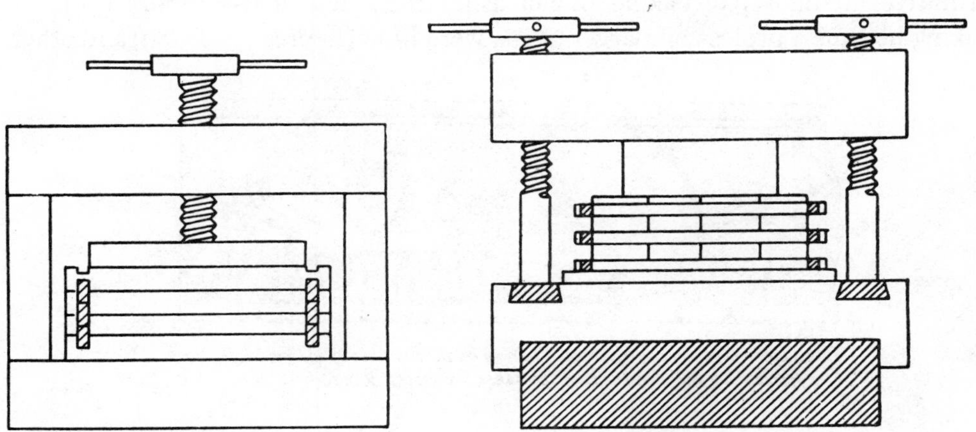

FIGURE 87—*Screw-presses described by Hero.* (Left) *Single-screw portable press. The screw is turned through the top beam, and its end bears down on the lid of the press. The fruit-container is constructed of interlocking wooden boards which can be built up to the height required.* (Right) *A heavier, double-screw press. Here the top beam bears down on the lid of the press. The feet of the screws are held in undercut bearings. The fruit-container is of vertical planks bound by three sets of cross-pieces. Width of press 140 cm. First century A.D.*

In an improvement of this type of press, described by both Hero and Pliny, the screw lifts a stone weight (figure 86). This type is still in use in Egypt, the Ægean, Italy, and the Alps. It is sometimes called the 'Greek' press, but it was common in Rome in the days of Vitruvius (first century B.C.). A press of this kind and of Roman date was found in Palestine [32], and examples are mentioned in the Talmud. The Egyptians also followed these developments.

Elimination of the lever action, so that the screw bore directly down from above, followed about A.D. 50, according to Pliny. Hero knew of two forms; the first operated by two screws, one at each end of a short beam, and the second by a single screw in the centre, rather like a modern copying-press (figure 87).

The wedge-press, for which there is no early literary reference, is depicted on monuments [33], and is described later in pharmaceutical literature. Above a depression in the stone base, which holds the pulp, are arranged alternate tiers of wooden wedges and planks, the uppermost bearing on the underside of a rigid frame. The wedges, which project on both sides, are driven in by hammering, thus compressing the pulp.

III. DIET OF THE GREEKS AND ROMANS

The prehistoric Greeks were familiar with the general methods of boiling, baking, and frying. Even after 600 B.C., with the spread of more efficient ovens, primitive methods like baking in hot ashes remained in use locally [34]. As baking became a profession, larger ovens were built (figures 77, 88–89) and other

FIGURE 88—*Roman baker's oven. The men on the right are kneading dough. Part of a frieze from the monument of the baker Eurysaces. Rome, first century B.C.*

bread-making processes, such as kneading, were mechanized—as, for example, in Pompeii.

In addition to flour, groats—that is, grain that has been hulled and crushed but not milled—were very popular, both in Greece and Rome. Their manufacture is described by Pliny:

> Groats are made from emmer. The grain is pounded in a wooden mortar to avoid the hardness of stone grating it up, the motive power for the pestle, as is well known, being supplied by the labour of convicts in chains; on the end of the pestle there is an iron cap. After the grain has been stripped of its coats, the bared kernel is again broken up with the same implements. The process produces three grades of groats—very small, seconds, and the largest kind. . . . Still these products have not yet the whiteness for which they are distinguished, though even at this stage they are preferable to the Alexandrian article. In a subsequent process, marvellous to relate, an admixture of chalk is added, which passes into the substance of the grain and contributes colour and fineness. . . .
>
> A variety of spurious groats is manufactured chiefly from an inferior kind of emmer growing in Africa. The ears . . . are mixed with sand and pounded, but even then there is a difficulty in rubbing off the husks, and only half the quantity of naked grain is produced; and afterwards a quarter of the amount of white lime is sprinkled into the grain, and when this has adhered they bolt it through a flour-sieve. The grain left behind in the sieve . . . is the coarsest. That which goes through, is sifted again in a finer sieve, and is

called seconds, and likewise the name of sieve-flour is given to that which in a similar manner stays behind in a third extremely fine sieve [35].

Much of the farinaceous food of classical antiquity was prepared by stirring the ingredients with boiling water to form a kind of porridge. Such were *maza*, compounded of flour, honey, salt, and oil; and *turon*, of flour, grated cheese, eggs, and honey. Broths were also commonly made.

The Homeric poems give the impression that meat (lamb, pork, and beef) was devoured in quantities by all, and that fish was consumed by the poor only. In

FIGURE 89—*Roman kitchen of the second century A.D. The oven is on the right, and ingredients are being stirred into a large bowl standing on it. From a monument at Igel, near Treves.*

fact, however, the main foods were farinaceous—porridge, bread, and groats made from wheat emmer, or barley, and a great variety of vegetables, particularly pulses (beans, peas, and lentils). To these were added fish and other sea-food, and spices were used for flavouring. Other ingredients of diet were mutton and beef, but the latter was costly. The upper classes could afford game; special preferences were the hare and the thrush. Few wild birds other than the thrush were eaten, and then only in times of distress. The rich indulged in the meat of ewes and kids and the more esteemed types of fish, such as eels. Nuts, acorns, chestnuts, almonds, and other fruits were very popular; onions were often eaten raw, even with wine. Salads were known in great variety. In general, meat was freely eaten only at sacrifices; it figured more prominently in the frequent religious celebrations than on the common bill of fare. The situation, in this respect, was very similar to that in Palestine during biblical times.

The Greeks had a curious abhorrence of cow's milk, which was considered unwholesome. From the time of Homer its consumption was denounced as intemperate, so that both milk and cheese were mainly from ewes and goats. Wine, drunk with one to three times its volume of water, was the main beverage. Barley-water was sometimes drunk, especially by invalids; there is a whole work on its preparation in the 'Hippocratic Collection' (*c* 400 B.C.).

With the expansion of Greek civilization in the sixth and fifth centuries B.C. oversea contacts brought dependence on imported grain, and diet changed. Wheat displaced barley, and by the fifth century barley-bread became generally regarded as fit only for the poor or for northern barbarians. Bakeries became more common in the ports and fish was eaten more abundantly. White wheat-bread and cakes compounded with oil, cheese, milk, wine, pepper, or anise were now normal. Pulses were no longer regularly eaten by the well-to-do. Salted fish, such as tunny, herring, and sardines now appeared more frequently, as well as many kinds of sauces, of which the most famous—*garon*—was prepared by leaving small fish in salt brine for some two to three months. Spices such as cummin, coriander, sesame, and silphium (from Cyrene) were grown or imported. A much greater variety of vegetables was cultivated, including cabbages, asparagus, gherkins, artichokes, spinach, celery, and native flavouring-herbs like fennel, marigold, and mint. Spices were imported from overseas, but indigenous aromatic or savoury herbs such as thyme, garlic, and cress were favoured in certain circles.

This rapid extension of the diet was accompanied by the publication of cookery-books, by the appearance of famous cooks working either as independent professionals or attached to a great household, and by the organization of dinner-parties or *symposia*. We also find the beginnings of a science of dietetics in the 'Hippocratic Collection' and other works of the fourth and third centuries B.C.

A similar evolution can be observed in ancient Rome at a somewhat later date. The standard dish of the peasants was a thick porridge made of emmer- or bean-flour, flavoured with salt. This was as a rule eaten with green vegetables, seldom with meat. It was supplemented by dishes made from groats. The ancient Roman bread was in the form of a flat, hard, unleavened cake baked in hot embers and dipped in milk before eating. It remained popular even after the introduction of leaven. Beans, peas, turnips, and onions completed the mainly vegetarian diet. Of the beet, the leaves were at first the only part eaten. Milk, cheese, honey, and olive-oil were generally used. Fruit, such as apples, plums, pears, pomegranates, and figs were well known; spices such as garlic, salt, and garden herbs completed the menu. The most ancient drink was milk diluted with water; wine came later. Fish was little valued by the ancient Romans; game was seldom eaten, and most of the meat consisted of goat, lamb, or pork.

A change of diet occurred about 300 B.C., with the spread of Greek civilization from the south of Italy. Wheat now began to displace the older emmer and einkorn (vol I, pp 363 ff) and a corresponding taste for white bread and pastries arose. During the second Punic war (218–201 B.C.) imports of wheat from Egypt, north Africa, and Sicily began. Egyptian grain soon displaced that of Italy in the Roman market. Professional baker-millers appeared about 170 B.C. Taste became more and more sumptuous and refined. Though cabbages, leeks, pulses, and peas were still common food, fruit-culture was now intensified and almonds, chestnuts, hazel-nuts, and walnuts were produced in much larger quantities. During the first century B.C. oriental fruits such as cherries, peaches, and apricots were introduced into the orchards of Italy. Citriculture did not succeed until the fourth century A.D.

Bigger supplies of meat now figured on the tables of the rich, and Italian farmers began to specialize in techniques for fattening geese and fowl. Fish-dishes and sauces were adopted from the Greeks. With the fall of the Roman Republic and the beginning of the Empire, the extravagant tastes of the rich stimulated the imports of luxury-foods, such as oysters from Britain, and refrigerating cellars packed with snow and ice were constructed for their preservation.

In the Imperial period Roman doctors began to consider the diet of the sick. Realistic, though empirical, rules were given by Celsus, whose *De re medica* of about the time of Christ was translated from the Greek and, supplemented by the 'Faculties or Powers of Aliments' of Galen (A.D. 129?–99), became the basis of medieval dietetics [36].

The over-refined taste of the Romans in Imperial times demanded improved grinding and sifting of flour and the rejection of the nourishing bran. The common people, however, and notably the army, still ate the old-fashioned food (p 120). Pork-fat, beer, wine, and oil supplemented this diet. Meat was considered an extra, prompting Scipio Africanus (236–184 B.C.) to grumble about the spits which the soldiers would have to carry to roast their meat ration. Even during the late Empire a tactician like Vegetius (fourth century A.D.) considered meat to be an extra to the soldier's diet.

IV. PRODUCTION OF OLIVE-OIL

The ancient empires of the Near East and the Mediterranean depended on the olive for their main supply of oil (figure 90).[1] Indeed the word oil in the western languages can be traced through the Latin *oleum* and Greek *elaion* probably to the more ancient Semitic *ulu*, all meaning 'olive-oil'.

In Crete the olive was cultivated as early as 2500 B.C. (vol I, ch 14). The wealth of the kings of Crete was certainly partly based on the export of olive-oil to Egypt and the eastern Mediterranean littoral [37].

[1] They used small quantities of animal fats in addition to oil.

From the eastern Mediterranean the use of olive-oil in cooking slowly spread westwards. The cultivation of the olive reached Rome from Greece by way of south Italy after 580 B.C., when the vine also came to that city. By the classical period it was common in all coastal regions of the Mediterranean. Its use in ecclesiastical ritual aided its spread into northern Europe, but it never rivalled the

FIGURE 90—*Olive harvest. From a Greek black-figured vase of the sixth century B.C.*

animal-fat supply of western Europe, as the plant was not suited for northern latitudes and imports were too expensive. Rape-seed and poppy-seed were cultivated in Europe from Neolithic times for their oils (vol I, pp 358–9), which came to compete with animal fats during the thirteenth century.

There is no essential difference between the production of olive-oil in the Near East and in the classical world [38]. Figures on pottery vases show how the olives were first tested for ripeness and quality, by squeezing the juice out of a few fruits through a funnel into a little flask and inspecting the taste and smell of the separated oil.

Extraction of the oil was best carried out directly after picking, although sometimes the olives were stored on the floor of the press-house. The pulp had first to be separated from the kernel. As the skin of the olive is fairly tough, this was achieved by crushing in the way previously described (p 112).

The liquid extracted by pressing (pp 115 ff) was allowed to rest in vats until the water could be drawn off through taps at the bottom of the vats, leaving only the oil. Good separation of oil from aqueous liquid was essential, since the latter contains a bitter principle (amurca[1]) which would spoil the taste of the oil.

Second and third pressings, each of lower quality than the preceding, could be obtained after soaking the pulp in hot water. Usually only three qualities of oil, corresponding to the three pressings, were made. The many other qualities distinguished by ancient authors are related to the use of olive-oil not only for cooking but in unguents, cosmetics, and toilet preparations, which demanded an oil of high purity to serve as the carrier or body of perfumes and essential oils.

V. MEDIEVAL DIET

From the fall of the Roman Empire onwards we have good documentary evidence on the diet of the different classes, its variation according to region and period, and the gradual changes of taste.

The diet of the masses was mainly vegetarian and frugal, two meals being usual. There is a medieval saying that angels need to feed but once a day, mankind twice, and beasts thrice or more. The impression we have of colossal meals washed down with great quantities of wine, mead, and beer derives from occasions in the lives of royalty and aristocracy which chroniclers have recorded for special reasons. Such feastings were the exception, even for the rich (figures 91–94). The many feast-days of the calendar were counterbalanced by even more fast-days, when the morning and the evening meals were limited by strict rules.

Except that they feasted more frequently, many of the richer classes shared the diet of the artisans and townspeople, which consisted of some kind of soup in the morning and of porridge or soup, fish, and vegetables in the evening. Apart from bread and cakes, cereals in the form of pastry were still a luxury because of the cost of fats. Meals of fish—fresh, dried, or salted—were much more frequent than nowadays, for meat was still expensive and was taken hardly more than once a week. Water and milk were drunk, as were beer and wine by those who could afford them. The choice was largely regional; thus beer was more common in the northern districts of Germany, wine in the south.

Medieval cities took great pains to obtain regular and cheap food supplies for their artisans. The rise of agriculture and husbandry in Flanders was a leading factor in the independence of the Flemish towns and their successful resistance to the encroachment of royal power. Both river- and coast-fisheries were important. Herring, cod, mackerel, salmon, crab, lobster, eels, sardines, lampreys, oysters, mussels, and even seals, porpoises, and whales, were preserved by drying and salting or were consumed fresh. A technique for gutting herrings

[1] An anglicized Latin word meaning olive-dregs.

FIGURE 91—*Medieval kitchen. Figures 91–94 are from the fourteenth-century Luttrell Psalter.*

invented by William Beukelszoon of Biervliet (*c* 1375) improved methods of preserving sea-food.

The townsmen usually dined on vegetables and beans, dough-cakes, bread, and soup, with a reasonable proportion of fish and meat. In the thirteenth century they could generally add some bacon, beef, pork, chicken, and eggs to this diet, as well as beer or wine. Wages gradually rose during the Middle Ages; in the fourteenth and fifteenth centuries the working-classes were better off and could more often enjoy wheaten bread, mutton, pork, and goat's flesh. The diet remained essentially the same throughout this period. Alexander Neckam (1157–1217) says that cereals were consumed as leavened bread, unleavened cakes, and dumplings [39]. 'Sauces', he asserts, 'contain the true secret of culinary art.' 'A domestic fowl', for example, 'needs a strong garlic sauce, diluted with wine or vinegar.' 'Let fish that have been cleaned', he continues, 'be cooked in a mixture of wine and water; afterwards they should be taken with green "savoury" which

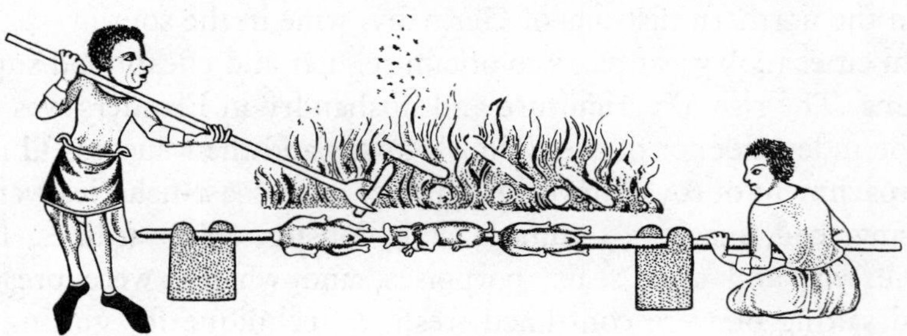

FIGURE 92—*Roasting on the spit.*

FIGURE 93—*Preparing the roast for the table.*

is made from sage, parsley, dittany, thyme, costus, garlic, and pepper; do not omit salt. One who takes this is especially exhilarated and restored by a raisin wine which is clear to the bottom of the cup.'

Peasants, as we should expect, took much simpler meals, which varied greatly from place to place and period to period [40]. Often the diet was very poor. The poor man's cereal, millet, remained the food in many regions. In the fifteenth century buckwheat was adopted from the Mongols. It is not a cereal but an annual plant, *Fagopyrum esculentum*, belonging to the Polygonaceae. The one-seeded fruits can be milled. Buckwheat appears first in Mecklenburg (1436) and soon becomes a staple crop in Normandy and Brittany. By 1500 this 'Saracen corn' was used as a cheap substitute for a cereal in poorer districts. The Bavarian peasants of the thirteenth century lived mainly on porridge of millet and oatmeal. Turnips (*Brassica rapa*) and rapes (*B. napus*) were eaten in poor districts of central France and southern Germany. The so-called dredge-corn was a mixed dish of vegetables and barley, oats, or rye. It was eaten green, and was essentially the same as farrago, a popular miscellaneous food of the classical world. Cress, radishes, carrots, and parsnips also formed part of the medieval diet.

FIGURE 94—*Medieval meal. Knives and spoons are used. Note the trestle table.*

A new type of food was made from carefully prepared and desiccated dough of cereal or rice flour. This, in the form of macaroni, vermicelli, and the like seems to have originated at the court of Naples towards the end of the Middle Ages; its invention contributed to the possibilities of preserving cereals.

Many medieval authors complain that the rich rob the peasants of their wine, wheat, oats, oxen, rams, and calves and leave them with their rye-bread only. However, we know that the peasant became better off as the Middle Ages

FIGURE 95—*Meal with knife and fork. The fork was an uncommon implement in medieval times. From an eleventh-century manuscript from Monte Cassino.*

advanced. The rise in wages and prices, particularly after the Black Death (1348–50), and the fixed rents, which transferred much of the former profits of land-ownership from the owner to the tenant, were important factors in this improvement.

Shortly before the Peasants' Revolt in 1381, the poet Gower complained that the peasants in the good old days 'were not wont to eat wheaten bread, their bread was either of corn or beans, their drink the spring. Cheese and milk were then a feast to them' [41]. The German humanist Johannes Boëmus records a similar diet in 1520: 'The German peasant feeds on brown bread, boiled beans or peas, drinking water or whey' [42] (plate 3 A). Other fifteenth-century sources tell of the gradual improvement of the peasants' diet. At Yule-tide they were often entitled to a fine loaf or two of white bread, a substantial helping of meat, and the right to sit drinking after dinner in the manorial hall.

The wages of tenants were often paid in kind—for example, three herrings and a loaf for a day's work, the age and condition of the herrings being carefully specified. In the days of 'Piers Plowman'[1] cheese, curds of cream, and the like

[1] 'The Vision of Piers Plowman' is a Middle English poem attributed to William Langland (second half of the fourteenth century).

enlivened a diet of oat-cakes and bread, often baked of 'peases and beans'; when times were hard, mussels, cold flesh, and cold fish were as precious as baked venison [43]. The continual struggle of the medieval peasant to raise fodder for his livestock made the supply of meat precarious. In the abbeys, the diet differed little from that of the peasant, though eggs, cheese, wheat-bread, and beer were more abundant.

Olive-oil, in northern climes, was consumed only by the rich. Rape-seed oil and colza (cole-seed) oil, together with beef-fat and lard, were commonly used by poorer folk. Butter, little valued in classical antiquity, was produced in ever greater quantities, but its price was too high for the common man and it did not reach his table until much later. The common sweetener was honey, and bee-keeping was an important medieval activity. In the absence of butter and sugar most medieval pastry was of a crumbly nature.

FIGURE 96—*Small mill for grinding or mincing. From a twelfth-century church at Vézelay, France.*

In the Middle Ages, and for centuries later, food was eaten with knives (of the hunting-knife type), spoons, and the fingers. Forks were less common (figure 95) and generally belonged to the kitchen or served to transport food from pot to plate. Wooden bowls served as plates, and were often shared between two people. Trenchers or roundels of wood on which food was placed or cut also served to take the boiled vegetables, which were piled on them. The meat was usually brought to table on the spit or in the pot (figures 97–98). Salt-cellars, often boat-shaped, were essential to the medieval dinner table, as were sauce-bowls. Medieval folk used much more salt than we do, and flavoured their food heavily with sauces.

The inventory of the kitchen underwent changes. Iron grills [44] and pots began to replace bronze vessels. Mills for spices, and flour-mills of different kinds, were introduced in many kitchens, and some of them were even mechanized to a certain extent (figure 96). Towards the end of the fifteenth century we find portable and wheeled baker's ovens. Earthenware and metal bell-shaped covers were used to bake food in embers. Wood, bronze, and iron spoons, ladles,

FIGURE 97—*Before the battle of Hastings Duke William of Normandy feasted on supplies levied from the country-side. Preparations for the meal include cooking on an open fire and roasting fowl on spits which a servant is carrying into the hall.* (Continued on p 129.)

hooks, and sieves were used. Wooden and pewter dishes were gradually ousted by glazed pottery, but glass was still uncommon in the kitchen [45]. Cooking-pots hung over the fire from iron chains, or stood over it tripod-wise; kitchen stoves begin to appear only in the fifteenth century.

As the Middle Ages advanced there was more concern about adulteration and its detection by analysis [46]. Strict local controls against adulteration had a strong influence on the standardization of weights and measures. Thus in England the Company or Guild of Grocers (incorporated 1345) not only had the 'custody of the Great Beam', that is of the scales, but were guardians of the public food and health through the office of 'garbling' [47]; they supervised the removal of impurities found in merchandise or themselves removed them. Controls were still very ineffective, and particular efforts were made to supervise butchers and fishmongers. In the control of such commodities as pepper, ginger, salt, and sugar weighing was the most important test.

We hear of many adulterants. Starch was mixed with whitening, and vinegar with weak sulphuric acid obtained by distilling green vitriol and diluting. The weight of the loaf of bread was carefully checked to detect whether bran had been added or a meal of acorns and beans used. Beer was very commonly adulterated, and vintners were sometimes found to add tragacanth, alum, tartar, or other inferior flavouring matter to their wines.

VI. VITICULTURE

(*a*) *The Vine in Greece.* There is a background of truth in the Greek tradition that Dionysus, god of wine, fled to Thrace from Lydia, country of 'froth-

FIGURE 98—*The inscriptions run: 'Here the meat was cooked, and here the servants served, here they had their breakfast, and here the bishop blesses the food and drink.' Duke William is seated at the table with his half-brother, Bishop Odo, and four knights. From the Bayeux Tapestry. Eleventh century.*

blowers', for the art of viticulture spread to Greece from the Near East[1] (figure 99). The Latin, Greek, and west Semitic words for wine are independent borrowings from Asia Minor, viticulture having reached Greece and Rome by different routes. Archaeology suggests that the vine was cultivated in the Ægean in the Early Bronze Age. It may have arrived there from Syria and Egypt through Crete, but it is more likely to have come by way of Asia Minor. Rome probably inherited viticulture from the Etruscans, who migrated from the eastern Mediterranean about 900 B.C.; later Greek colonists (from *c* 760 B.C.) also played a part in the development of vineyards in south Italy and Sicily.

The Greeks were great wine-lovers even from the days of Homer. The Eupatrids, the leading family in the state of Attica, were owners of vineyards. When the reforms of Solon, the law-giver of Athens, broke their power it became clear how deeply the wine trade had come to influence the economy of Attica: densely populated and with an average farm of only ten acres in extent, it could ill afford the use of land for crops other than corn. Thus it had to pay for imports of cereals by exports of wine, olive-oil, figs, and pottery. However, only in Attica among the Greek states did the vineyards encroach seriously on the corn-land. In other Greek states the area of corn-land remained practically constant, the vineyards expanding over forests and scrub-covered hillsides.

Greek colonization and the introduction of Greek wine proceeded hand in hand. Thus soon after the founding of Massilia (Marseilles), about 600 B.C., the Greek colonists started to export wine to the Celtic tribes of Gaul. Their wine-vessels (*amphorae*) and bronze flagons with beaked spouts, of Bronze and Early Iron Age, are found over a wide area, and show that wine penetrated along the

[1] In point of fact, Dionysus, who combined in himself two older gods, one from Thrace and one from Asia Minor, only later became god of the vine.

Rhône and Saône and through the Belfort gap to the upper Rhine. Along with the wine went not only the vessels connected with carrying, storing, and mixing it, but the rituals associated with drinking. The beautiful pottery or bronze

FIGURE 99—*The journey of Dionysus, god of wine. According to one version of the legend, he changed an attacking band of pirates into a shoal of dolphins, to continue his lonely but festive voyage with drinking-horn in hand, the vines happily surviving. From an Attic bowl of* c 500 B.C.

containers (figure 407) so deeply influenced barbarian art that it has been said that 'La Tène art may have largely owed its existence to Celtic thirst' [48].

By the end of the fifth century B.C. viticulture in Greece was becoming industrialized. Professional wine-merchants bought the produce and sold it overseas in sealed jars stamped with their own marks. There still exist the trade regulations for the Isle of Thasos in the north Ægean, forbidding the sale of the crop before its harvesting in May and June. Wine could be sold in sealed jars only, and Thasian ships were not allowed to carry foreign wine between certain ports of the Greek mainland. Greek wine was extremely popular in Egypt, where local viticulture was less successful, and was imported through Naucratis in the Delta.

It has been claimed that the expansion of Greek civilization in the wake of Alexander the Great's conquests, which we call Hellenism, was determined by the natural limits of viticulture. The Chinese general Chang Ch'ien, on his mission to the west, spent a year in Bactria in 128 B.C. and reported that grape-wine was made in Ferghana and Parthia. This was in fact the beginning of viticulture in China, for seeds were then sent home and planted by the Emperor [49].

As Alexander's successors dumped the wine of the great new domains on their natural markets—the many new Hellenistic cities throughout the Near East—they were slowly undermining the export trade in wines from Greece itself. Many foreign areas began to grow their own wines. This was the case in the important Black Sea markets, which however still imported the better wines from Greece. Again, the profitable markets on the Adriatic were cut off by the appropriately named Dionysius, tyrant of Sicily, who started to flood them with his native products. Large quantities were still shipped to the east and to Italy, but only wines of quality could hope to maintain the Greek trade.

With Roman domination in 146 B.C. the Greek wine-trade ebbed slowly away; several islands continued to export, but wine-growing for export disappeared from the mainland of Greece.

(*b*) *Techniques of the Greek Vine-Grower.* We have no early Greek documents on viticulture and can therefore only guess that it was developed from both the experience of the wine-growers and the theoretical interests of botanists. This assumption seems well founded if we turn to the first essays on vine-growing, namely those by Theophrastus of Eresus (*c* 371–*c* 287 B.C.), a pupil of Aristotle. His remarks on the vine and viticulture blend experience and theory and contain many first-hand observations [50].

Theophrastus believes that plants derive their vital spirit (*pneuma*) from the soil and draw it up through the pith, together with water. From this theory he deduces the correct way of striking cuttings from good vines, the conditions under which they should be planted, the porosity and moisture of the soil, and the care of the cuttings. Grafting he rejects, but he discusses the uses and methods of pruning. Despite the defects in his knowledge of plant physiology, his advice is generally good and often so much in accordance with modern views that we may reflect how little the practical experience of vine-growers has advanced during the last 2200 years. It shows that Greek genius raised viticulture to a very high level of achievement.

Greek vine-growers avoided propping, trellising, or festooning their vines, a practice common in Egyptian vineyards (where the pergola originated, vol I, figure 185), and later in Italy. They favoured letting the vines lie along the soil, except where summer moisture made a little propping necessary. As a result, much damage was done by mice and foxes. Low-growing vines saved money and

FIGURE 100—*Greek and Italian amphorae for storing wine. These jars were salvaged from a cargo ship of* c 200 B.C. *sunk off Marseilles.*

escaped the summer winds, but demanded careful hoeing two or three times a year to keep down weeds and to break up the soil. In summer all unnecessary leaves were pruned to avoid excessive transpiration. Sometimes green manure was used, or vegetables were planted between the rows of vines. Women often worked in the vineyards.

September was the vintage-month. The baskets of grapes were brought to the treading-floors, which were usually of cement but sometimes of acacia-wood. They were raised above the ground and gently inclined towards an outlet. The first must that was collected, particularly that squeezed from the grapes by their own weight, was especially valued, and remained wholesome for a fairly long time if well sealed. The rest of the juice was extracted by a press (p 113). A third-grade wine, drunk only by the very poor, was prepared by mixing or even cooking the lees with water and expressing this mixture.

Some of the must was consumed after treating it with vinegar, but the larger part was stored in cellars in huge pottery fermenting-vats. These *pithoi* (cf vol I, figure 252 A) were the common storage-vats for oil, wine, and the like in the Greek world. The wooden barrel was a northern invention as yet unknown. Often the *pithoi* were buried deep in the ground after being smeared with pitch or resin inside and out. This treatment reduced porosity and gave to the wine a peculiar and much appreciated flavour, still characteristic of the Greek *resinata*. The *pithoi* were so large (up to 10 ft high with a mouth 3 ft in diameter) that after the civil wars, when there was a severe housing shortage in Athens, many citizens slept in them [51]—including the philosopher Diogenes (412?–323 B.C.), who is fabled as having lived in an anachronistic tub.

During the six-month fermenting period, the liquid was constantly skimmed. In the following spring the wine was filtered into skins or pointed pottery amphorae (figure 100) for transport and sale. The handles and stoppers of the amphorae were stamped to indicate brand, origin, and vintage.

The main difficulty of the trade was the instability of the wines. Corks were then unknown as stoppers, and it was impracticable to prevent fermentation altogether during storage; hence wines had to be consumed within three, or at the most four, years. Homer recounts that Nestor drank a ten-year-old vintage, and Athenaeus (*c* A.D. 200), a wine-expert, mentions one 16 years old; but these must have been exceptional. A related problem that engaged experts was the acidity of their wines. They tried to correct it by treating with sea-water, turpentine, pitch, resin, chalk, gypsum, lime, and aromatic herbs. The Greek poets have much to say on the different types of wines, and they are discussed by many authors—above all by Athenaeus, who mentions no fewer than eighty-five varieties [52].

The wine-merchants were accustomed to test a sample by dipping a sponge into one of the large *pithoi* and from it estimating taste and bouquet. The taster would at times let a few drops fall on the back of his hand to test it, as with olive-oil [53]. Wine was sold over the counter in the wine-shop. It was never drunk neat but always mixed with water. At dinner-parties the toast-master indicated the appropriate dilution; certain wines were said to be so strong that they needed a twenty-fold dilution.

(*c*) *Roman Wine-Production*. Though the soil in many parts of Italy is eminently suitable for viticulture, and though there was indeed some wine-growing in early times, it was not until about 200 B.C. that the art became well established. It is said that in the second Punic War (218–201 B.C.) Hannibal and his African troops washed their horses with wine at Picenum, and cut down the vineyards of Campania, much to the chagrin of the Romans. After the war, the Roman policy of importing wheat favoured large increases in the areas under vine and olive.

The Roman playwright Plautus (254?–184 B.C.) thought only Greek wines worthy of mention. Times were changing, however. Soon works on viticulture from the Greek (Theophrastus) and the Carthaginian (Mago) were available in Latin translation, the latter since *c* 146 B.C. Choice vine-cuttings were imported, and Italian wines began to display their vintage and place of origin. 'The year when Opimius was consul' (121 B.C.) was famous for its vintage; a few amphorae of its wines survived until Imperial times, when they were the admiration of connoisseurs. During the first century B.C. Italian wines came into their own. Greek quality-wines, however, held their place, and were employed by Caesar and Lucullus to cajole the plebs of Rome. The yield per acre was about 880 gallons of high-grade wine, or up to double this amount of poorer quality. Generally speaking, 18 per cent per annum could be expected on the capital invested; consequently vintners became wealthy and substantial citizens. Over 660 million gallons were consumed yearly in the homeland, and more was exported [54].

When the Roman wine-growers displaced the Greek they failed to learn the lesson of history. Spain was conquered by the Romans in 133 B.C., and at that time its tribes were drinkers of mead. In a few generations, however, viticulture made such rapid progress there that Spanish wines became successful competitors of the Italian varieties not only in Gaul, Britain, and Germany but even in the east.[1] Enormous quantities of Spanish wines were imported by Italy; indeed, the sherds of their jars form the largest proportion of the 150-ft Monte Testaccio, near the Porta Ostiensis.

In the face of this competition, the Emperor Domitian (A.D. 81–96) tried to forbid viticulture in the provinces abroad, and even ordered vineyards in Gaul and Spain to be destroyed [55]. But in the Imperial period free wine had become so important in the policy of *panem et circenses* that large imports had to be tolerated for fear of mob risings. During the economic and political troubles of the third and fourth centuries the vineyards became neglected. The days when Italy could exchange her wine for Danubian slaves, cattle, and hides, and for Celtic silver, had gone for ever.

(*d*) *Techniques of Viticulture*. Cato, Varro, and Columella relate many details of wine-growing in the second and first centuries B.C. They describe the labour, traction-animals, and equipment needed to run a vineyard. Pliny devotes one

[1] The vine was introduced into Britain during the Roman period. Vine-stems have been found near a Roman villa at Boxmoor, Hertfordshire.

book of his 'Natural History' to the art, and is supplemented by Virgil and Strabo and others [56].

In contrast to Greek practice (p 132) the Romans propped or trellised their vines. Sometimes the vines were trained on trees, such as maples or figs. In close-set vineyards chestnut or willow poles with cross-pieces of cane were commonly used. The leaves of the trees were used for fodder in the dry seasons, and corn was sometimes sown between the vines. The maturity of the grape was judged by the colour of its pips. The first grapes harvested were held to give most juice, while the second lot gave the best wine and the last the sweetest. In some cases the choicest grapes were first harvested singly. It appears that the terms *spätlese* (late-picked) and *auslese* (selected), to be found to-day on the labels of Rhenish wines, have a venerable history.

The Romans carefully adjusted the conditions of fermentation to suit their grapes. In Campania the vats remained in the open air. Falernian, a very famous Campanian wine, with a high initial sugar-content, fermented quickly and completely[1] in the sun and air; the grapes of the Po valley would have turned sour under this treatment. The wines were then stored in special carefully sited buildings, for it was affirmed that proximity to dung-heaps and, less reasonably, to trees would adversely affect the flavour.

The fermenting-vats were drawn and the wine was transferred to amphorae (of 26-litre capacity) in May to July, when the dry north wind prevailed. To prevent any further fermentation during storage, the wine at this stage was filtered and afterwards heated. The effect of this last treatment would, of course, depend on the temperature which, if efficiently controlled, could first speed the fermentation to its end, then sterilize the wine in a manner equivalent to pasteurization. Some wines were simply exposed to the sun, or partially evaporated over a hot water-bath, though this lowered the content of alcohol.

Other treatments included sweetening with concentrated must (*defrutum*, p 139), and the addition of flavours such as resin, sea-water, herbs, and flowers, or of purifiers such as white of egg, lime, and gypsum. Wormwood (*Artemisia*) was imported from the Pontic and Alpine regions for the manufacture of vermouth. Essences—in the absence of distillation they could hardly be called liqueurs—were made from *defrutum*, herbs, and berries. Labels on the jars included such information as vintage, vineyard, date of bottling, purity, colour, type of grape, names of maker and cellarer, number of jar.

The clay stoppers[2] of amphorae were sealed with pitch, resin, or gypsum,

[1] Pliny claims that this was the only wine that would take fire when a flame was applied to it. Fermentation does not continue after the concentration of alcohol reaches a certain level.

[2] Cork was now occasionally used.

FIGURE 101—*Packing pottery wine-jars in straw. Operations take place on a barge whose tow-rope can be seen above the man on the right (cf figure 102). First century A.D.*

and the amphorae themselves were sometimes coated with oil to keep out the air. Amphorae protected during transport by plaited straw may be the ancestors of the Italian *fiasco* (Chianti bottle) (figure 101). Wines of the highest quality were occasionally sold in glass vessels.

The wooden cask, though known in ancient Egypt (vol I, figure 500), was to the Roman world a gift of the northern peoples. Its manufacture in Europe dates from the Late Bronze Age in Switzerland. Iron tools brought great proficiency in the art of cooperage. Good barrels, built up from dowelled staves and strengthened with metal hoops, penetrated southwards at the beginning of our era (p 100). Strabo (*c* A.D. 21), the first to mention these 'wooden *pithoi*', says that the Celts are fine coopers 'for their casks are larger than houses [!] and the excellent supply of pitch helps them to smear [the cracks of] these *pithoi*' [57]. The cask soon became common in Italy (figure 102, plate 3 B), and the Emperor Maximus is said to have bridged a river near Aquileia with them.

These technical advances gave the Roman wine much greater keeping-power

FIGURE 102—*Towing a cargo-boat on the Rhône, loaded with wooden casks. Second century A.D.*

than the Greek. Though a period of three to four years was the general rule, some wines were allowed to mature over 10 or 15.

(*e*) *Spread of the Vine into Europe.* Roman penetration transformed Gaul from a beer-drinking country into the traditional land of wine-lovers. The Romans had furthered the production of second-class wine in the province of Narbonensis, that is, Gaul south of Lyons. In the trade-centre of Lyons itself the powerful wine-merchants and their allies, the ship-captains, handled the import and export trade to northern France and Germany. The decline of Lyons dated from A.D. 197, when the Emperor Septimius Severus sacked it. Meanwhile local production of better wines by more scientific Roman methods had started all over Gaul. In the third century A.D. the vines of Beaune in Burgundy were said to be 'of a hoary age', while the wines of Bordeaux and of the Moselle are certainly several centuries older than the fourth century A.D. [58]. Caesar's legions brought viticulture to the Rhine (figure 103).

FIGURE 103—*Galley on the Rhine loaded with casks of wine. First century A.D.*

The fall of the Roman Empire and the incursions of German and Asiatic tribes, Muslim armies, and Viking hordes were a set-back for viticulture. Vines disappeared almost, but not quite, completely from southern England, northern France, Swabia, and eastern Bavaria, in all of which they had grown in Roman times. Medieval viticulture was but a slow rediscovery of the classical technique. In many places Christianity assisted the survival of the vineyards, for every parish tried to produce its ritual wine. The recovery was assisted by the fact that Spain, France, and part of Germany had been converted to wine-drinking before they were converted to Christianity. Wine also remained a major ingredient in pharmaceutical and medical recipes, for with oil it was the best organic solvent available at that time.

The barbarian invaders had never completely destroyed the vineyards, and soon, as with Charlemagne (768–814) and his successors, they became keenly interested in the wine that many Imperial Roman edicts had formerly withheld from them. During the ninth century viticulture began to flourish again in the regions of Worms and Speyer, spread along the Rhine, and reached Swabia, Franconia, and Thuringia (figure 104). By 1200 it had crossed the Elbe into the middle Oder district. During the Middle Ages wine-growing was practised to some extent in England, Flanders, and the Pyrenees. The wine of England must have been rough, and matched the poverty of diet of the Anglo-Saxons. In France there was a considerable medieval extension of vineyards in regions where the soil yielded less corn, particularly in the west. Religious houses held vineyards as distant as the Rhine, the Moselle, and Champagne, as well as in Paris.

FIGURE 104—*The vine-grower in Europe. From an album portraying the inmates of an almshouse in fifteenth-century Nuremberg.*

Both in France and Germany the flavour, thinness, and roughness of second-quality wines were often concealed by special additions of honey and of herbs or spices such as cinnamon, sage, and coriander. Grapes that did not ripen and remained green were used to make verjuice.[1] Classical methods were adopted in these regions. A full programme based on classical experience is found in the eleventh-century charter of the abbey of Muri in Canton Zürich, and in those of similar foundations. Willow-plantations usually accompanied the vineyards to supply withies and barrel-hoops.

The great medieval physician, scholar, and alchemist Arnald de Villanova (1235?–1311?), when shipwrecked on the coast of Africa, wrote in Latin a book 'On Wine', describing medical wines and their uses. In 1478 von Hirnkofen added to this book a Latin text on vine-growing and the production and preservation of wine, translated the whole into German, and had it printed at Esslingen [59]. It became a best-seller. Arnald's contemporary Pietro dei

[1] Verjuice, the juice of unripe fruit, was much in demand at this time for cooking.

Crescenzi (?1233–1320) compiled a more comprehensive book from observations in Italy and Provence. The Muslim historian and physician Ibn al-Khaṭīb (1313–74) was the first (1360) in the eastern world to extol wine openly and to justify its use—at least for Christians and Jews.

A side-line of viticulture is concerned with sweetening-agents. Sugar was little known and was an imported luxury occasionally given medicinally. The must pressed from grapes was locally a more copiously available substitute for honey. It could be kept for a year by the Romans, who, to prevent its fermentation, tightly sealed the concentrated liquid in jars and immersed them in cold water. It could be concentrated by evaporation, and, reduced to half its original volume, was known as *defrutum*. An appreciable proportion of the wine-harvest was thus treated, for, as well as being a sweetening-agent, the product was used to preserve olives and other food.

VII. CIDER, BEER, AND SPIRITS

(*a*) *Cider and Perry*. Other plants besides barley and vines yield decoctions suitable for fermentation. Early attempts, on apples and pears in particular, often failed because of too low a sugar-content. Though wild pears and apples grew in prehistoric Europe, no appreciable amount of alcoholic drinks was produced from them. Cultivated apples, like *Pyrus malus* L, were introduced from the east by way of the Mediterranean, and the Romans knew many kinds.

Serious cultivation of apples and pears hardly started before the fifth century A.D. The first cider and perry seem to have been poor men's drinks, brewed from wild fruit. St Guénolé of Brittany (414–504), wishing to chastise himself, lived on water and perry, while Ste Radegonde (519–87) and Ste Ségolène (fl 770) confined themselves to perry undiluted. Another second-grade brew, called *dépense*, was fermented from sliced apples, grapes, and other fruit. In regulations for his estates Charlemagne speaks of brewers who prepare fermented drinks from various types of fruit.

In the Basque country cider-making was old-established by about the twelfth century. It is claimed that better apples and pears came west with the crusaders, but several kinds appear likely to have come from Spain. Normandy became a centre of cider-making, and the industry spread thence to England in the thirteenth century. Cider was soon the particular favourite of the farmers and country people, whose cereals were often too valuable and too scarce to be used for brewing.

(*b*) *Beer*. Malting, the basic process in the making of beer, was practised in

Mesopotamia from the third millennium B.C. (vol I, pp 277–81), but as a popular drink beer got no nearer to the Hellenistic world than Crete [60]. The Greeks knew beer as a barbarian beverage, and consequently rarely drank it, while a passage by the Roman historian Tacitus reveals that beer was a curiosity to him: '[The Germans drank] a liquor drawn from barley or from wheat and, like the juice of the grape, fermented to a spirit' [61]. On drinking beer for the first time the Roman Emperor Julian the Apostate (361–3) was prompted to write a satirical poem on its tendency to cause flatulence [62]. This deep-rooted aversion to beer in Greece and Italy persisted until the nineteenth century and even later.

In its homeland the Near East, on the other hand, beer continued in popularity. In Hellenistic Egypt brewers were organized in local guilds and were in the service of the state. A tax on beer was levied. The trade also flourished in Babylonia, Syria, and Palestine. Zosimos, the famous alchemist of Panopolis in Egypt (*c* A.D. 300), is reported to have written a book on brewing [63].

Millet-beers were known in China several centuries before Christ, but it is uncertain whether the cultural exchange between the Near and Far East involved techniques of brewing. The art of making beer came to Tibet from China about A.D. 650 and included both barley- and millet-beers.

Medieval Brewing in Europe. In the west, the Celts knew the art of brewing before the Egyptian methods could have been transmitted to them by the Greeks and Romans. The classical authors report little on their methods, except that they used barley or wheat. 'The nations of the West have their own intoxicant made from grain soaked in water; there are many ways of making it in Gaul and Spain and under different names though the principle is the same. The Spanish provinces have taught us that these liquors will keep well' [64]. The German tribes adopted brewing in the first century B.C.

The Celts had also discovered the usefulness of barm or beer-yeast as leaven, for Pliny writes: 'When the corn of Gaul and Spain is steeped to make beer, the foam that forms on the surface is used for leaven, in consequence of which those peoples have a lighter kind of bread than others' [65].

In the classical world the young shoots of the hop were eaten as a vegetable, but the plant belongs primarily to the herb-garden and was used in medicine [66]. In 822 the abbot Adalhard of Corvey released the millers from their duty of grinding malt and hops: probably the first time hops and brewing are linked together in a document [67].

Not until the Middle Ages did beer come generally to have the taste that we now associate with it, for it was then that hops were first commonly used for flavour and preservation—although we know from Pliny that hops were an ingredient of Iberian beer. From time immemorial a diversity of herbs and aromatic substances had been added to alcoholic infusions to flavour and preserve them. Hops were introduced in competition with several other substances, among them heather, and a mixture called *Gruit* in medieval Germany, which contained bog-myrtle, marsh or wild rosemary, and yarrow. Hops contain a resin, lupulin (which yields on analysis two bitter compounds, humulone and lupulone), responsible for taste and preservative properties, and also tannin, which helps to clarify the beer.

Although hopped beer is said to have been made in Bavaria between 859 and 895 our earliest definite reference to it dates from a passage by the twelfth-century mystic Hildegard of Bingen [68]. By the thirteenth century mentions of it have become numerous.

The cultivation of the hop extended to north Germany, Bohemia, and Flanders, where it gained commercial importance in the fourteenth century some four hundred years after its introduction. In 1304 a letter from the bishop of Liége and Utrecht complains that hopped beers 'are now competing on the market some 30 or 40 years'. A century later French brewers at Dieppe were importing hops from Holland and England, so that the plant must have been introduced into England shortly before or after 1400. The word 'hoppe' is not known in English documents before the fifteenth century.[1] The use of hops in brewing was forbidden by Henry VIII, and the ban remained until the last part of his son Edward VI's reign (1547–53). Hopped beer can be said to have gained supremacy by 1500, when the cultivation of hops was mainly centred in western Germany.

(*c*) *Alcohol and Distillation.* The production of distilled alcoholic drinks was closely related to the development of distillation as a chemical operation. The distillation of alcohol requires, however, more refined apparatus than the forms available to the Alexandrian alchemists from the first century A.D. (ch 21). The distillate must be cooled, by air or water, and a proper procedure for successive distillations must be worked out. Quicklime was used in the early days as a dehydrating agent, together with other ingredients of more doubtful value such as salt, sulphur, and tartar [69].

The available texts show that alcohol was discovered by distillation about 1100, and evidence as to locality points to Italy, where the medical school of Salerno was the most important chemical centre. The common name for a fraction containing enough alcohol to burn was *aqua ardens*, very concentrated

[1] In the *Promptorium parvulorum*, *c* 1440, 'hoppe' is described as 'sede for beyre'.

alcohol often being denoted by the term *aqua vitae*, sometimes identified with the quintessence.

From the twelfth century onwards alcohol and its purification are mentioned frequently by the alchemists [70]. An unknown writer assuming the name of Ramón Lull or Raymond Lully (*c* 1235?–1315), by distilling alcohol thrice with quicklime, produced the liquid which we call absolute alcohol [71]. Arnald de Villanova (1235?–1311?) gives recipes for distilling wine with spices and herbs to make liqueurs [72]. By the fifteenth century alcohol was well known as a solvent for the essences of herbs and spices.

The production of these distilled liqueurs was originally in the hands of the apothecaries. Up to the fourteenth century apothecaries were closely connected with local governments or universities, but from then on they gradually acquired private shops and laboratories. An increasing number of wine-makers and inn-keepers now began to produce such drinks, although to the end of the sixteenth century the apothecaries and the monasteries—which also specialized in the production of medicines containing herbs and spices—were the main producers of both liqueurs and alcoholic extracts for perfumes, medicine, and so forth (figure 105). The herb-gardens of many monasteries were the birth-places of special liqueurs such as Chartreuse and Benedictine.

Physicians sought to combat the terrible plague of the Black Death (1348–50) and several later outbreaks by prescribing strong alcoholic drinks—which gave at least a feeling of warmth and confidence. The taste thus encouraged, gin (distilled with juniper berries) and liqueurs soon became serious competitors of beer and wine. The favourite distilled drink of the Italians was *rosoglio*, a very sweet liqueur prepared from raisins and, it is said, sundew (*ros solis*). From Italy the appreciation of liqueurs and the secrets of their manufacture were brought to Paris by apothecaries and distillers about 1332, and from France they spread throughout Europe. Originally the alcoholic distillate of wine and macerated leaves, herbs, and spices had been strongly sweetened, but there was later a tendency to reduce the sweetness and thus the viscosity.

In the wake of these liqueurs, brandy (Dutch, *brandewijn*, burnt wine, i.e. distilled from wine) came to France and Germany. By the end of the fourteenth century it was drunk all over Europe, and certainly from 1360 onwards strong measures were necessary to cope with the 'Schnaps fiend' (*Schnapsteufel*, as contemporary German documents have it) in both country and town. Increasingly stringent regulations against drunkenness and unruly behaviour, and heavy taxation of alcohol, acted as provocations to consumption rather than brakes on it. Such police regulations date back to the thirteenth century.

Early production of alcohol was from wine. In the fourteenth century, the production of alcohol from fermented cereals was first attempted. This was important to beer-producing countries, which had to import the wine from

FIGURE 105—*Botanical garden in 1500, with many kinds of distillation apparatus. Distillation of* aqua vitae. *The still-head is of a tall conical shape to effect air-cooling of the distillate. From Braunschweig's* Liber de arte distillandi, *1512.*

which they had hitherto made brandy. The distilleries were first attached to the breweries but soon began competing with them and became independent as soon as they learnt to grow the yeast needed for the fermentation of their barley. Though some Scandinavian countries tried to forbid the production of barley-alcohol they did not succeed in stopping growth of a taste economically favoured by the expense of wine. Alcohol was in medicinal use by physicians and apothecaries; this added to the pressure on the authorities, and their laws were largely disregarded. A town ordinance of Nuremberg (1496) reads: 'As many persons in this town have appreciably abused the drinking of *aqua vitae* the town council warns earnestly and with stress that henceforth on Sundays or other official holidays no spirit shall be kept in the houses, booths, shops or market and even the streets of this town for sale or consumption against cash payment.'

Perhaps it was wiser to warn people, as did a surgeon of the same city in a pamphlet of about 1493: 'As at present practically everyone becomes accustomed to drink *aqua vitae* . . . one should remember how far one can go and learn to drink of it as much as one should as a gentleman' [73].

REFERENCES

[1] JASNY, N. 'The Wheats of Classical Antiquity.' Johns Hopk. Univ. Stud. hist. polit. Sci., new series, Vol. 62, iii. Johns Hopkins Press, Baltimore. 1944.
[2] YEO, C. A. *Trans. Proc. Amer. philol. Ass.*, **77**, 221–44, 1946.
[3] PLINY *Nat. hist.*, XVIII, xx, 86–87, 89. (Loeb ed. Vol. 5, pp. 244 ff., 1950.)
[4] JASNY, N. *Wheat Stud. Stanf. Univ.*, **20**, iv, 157, 1944.
[5] PLINY *Nat. hist.*, XVIII, xxviii, 108. (Loeb ed. Vol. 5, pp. 256 ff., 1950.)
[6] *Idem Ibid.*, XVII, xxix, 115. (Loeb ed. Vol. 5, pp. 256, 262, 1950.)
[7] VICKERY, K. F. 'Food in Early Greece', p. 88. Ill. Stud. social Sci., Vol. 20, iii, 1936.
[8] PLINY *Nat. hist.*, XVIII, xxvi, 102–4. (Loeb ed. Vol. 5, pp. 252 ff., 1950.)
[9] *Idem Ibid.*, XVIII, xii, 68. (Loeb ed. Vol. 5, pp. 232 ff., 1950.)
[10] *Idem Ibid.*, XVIII, lxxii, 298. (Loeb ed. Vol. 5, p. 376, 1950.)
[11] MAURIZIO, A. 'Histoire de l'alimentation végétale depuis la préhistoire jusqu'à nos jours', French trans. by F. GIDON, p. 388. Payot, Paris. 1932.
[12] POLYBIUS I, 22. (Loeb ed. Vol. 1, p. 60, 1922.)
[13] BARTHELEMY, R. E. *Min. & Metall., N.Y.*, **19**, 244, 1938.
[14] ROSTOVTZEFF, M. I. 'The Social and Economic History of the Hellenistic World', Vol. 1, p. 176, and Pl. xxv. Clarendon Press, Oxford. 1941.
OPPENHEIM, M. VON. 'Tell Halaf', trans. by G. WHEELER, p. 206. Putnam, London. 1933.
[15] CHILDE, V. GORDON. *Antiquity*, **17**, 19, 1943.
[16] OVID *Fasti*, VI, 318, 470. (Loeb ed., pp. 342, 354, 1931.)
VARRO *De re rustica*, I, 55. (Loeb ed., p. 290, 1934.)
[17] CATO *De agri cultura*, XI, 4. (Loeb ed., p. 26, 1934.)
[18] VIRGIL *Moretum*, lines 19–29. (Loeb ed. Vol. 2, pp. 452 ff., 1918.)
[19] JACOBI, H. *Saalburgjb.*, **3**, 81 ff., 1914.
[20] MS. Lat. Munich 197, fols 18b, 42b.
[21] BLÜMNER, H. 'Technologie und Terminologie der Gewerbe und Künste bei Griechen und Römern', Vol. 1 (2nd rev. ed.), p. 337. Teubner, Leipzig. 1912.
[22] PLINY *Nat. hist.*, VII, lvi, 199. (Loeb ed. Vol. 2, p. 640, 1942.)
[23] DRACHMANN, A. G. *K. danske vidensk. Selsk. archaeol.-kunsthist. Medd.*, **1**, i, 7, 1932.
[24] ROBINSON, D. M. and GRAHAM, J. W. 'Excavations at Olynthus', Part VIII: 'The Hellenic House', p. 338. Johns Hopk. Univ. Stud. Archaeol., no. 25. Johns Hopkins Press, Baltimore. 1938.

[25] VICKERY, K. F. 'Food in Early Greece', pp. 58, 59. Ill. Stud. social Sci., Vol. 20, iii, 1936.
[26] EVANS, SIR ARTHUR (JOHN). *Annu. Brit. Sch. Athens*, **8**, 10, Session 1901–2.
BOSANQUET, R. C. *Ibid.*, **8**, 306, Session 1901–2.
DAWKINS, R. M. *Ibid.*, **11**, 276, Session 1904–5.
[27] PLINY *Nat. hist.*, XVIII, lxxiv, 317. (Loeb ed. Vol. 5, pp. 386 ff., 1950.)
[28] CATO *De agri cultura*, x–xiii; xviii–xix. (Loeb ed. pp. 26 ff., 32 ff., 1934.)
[29] HERO *Mechanica*, III, ii, 13–21. (Arabic ed. and French trans. by BERNARD CARRA DE VAUX, *J. asiat.*, neuvième série, **2**, 499, 1893.)
[30] PLINY *Nat. hist.*, XV, ii, 5–6. (Loeb ed. Vol. 4, p. 290, 1945.)
[31] DRACHMANN, A. G. *J. Hell. Stud.*, **56**, 72, 1936.
[32] HAMILTON, R. W. *Quart. Dept. Ant. Palest.*, **4**, 111, 1935.
[33] MAU, A. 'Pompeji in Leben und Kunst' (2nd ed.), Pl. IX and fig. 185. Engelmann, Leipzig. 1908.
[34] OVID *Fasti*, VI, 315. (Loeb ed., p. 342, 1931.)
[35] PLINY *Nat. hist.*, XVIII, xxix, 112–16. (Loeb ed. Vol. 5, pp. 260 ff., 1950.)
[36] RIDDELL, W. R. *Med. J. Rec.*, **134**, 247, 1931.
[37] HOOPS, W. *Forsch. Fortschr. dtsch. Wiss.*, **23**, 35, 1947.
Idem. 'Geschichte des Ölbaums', *S.B. heidelberg. Akad. Wiss.*, phil.-hist. Kl. 1942–3, no. 3, 1944.
[38] BLÜMNER, H. 'Technologie und Terminologie der Gewerbe und Künste bei Griechen und Römern', Vol. 1 (2nd rev. ed.), p. 332. Teubner, Leipzig. 1912.
DALMAN, G. 'Arbeit und Sitte in Palästina', Vol. 4: 'Brot, Öl und Wein', pp. 206, 221. Schr. des Deutschen Palästina-Inst., Vol. 7. Bertelsmann, Gütersloh. 1935.
[39] HOLMES, U. T. 'Daily Living in the Twelfth Century', pp. 87–92, 113–14. University of Wisconsin Press. Madison, Wisc. 1952.
[40] COULTON, G. G. 'The Medieval Village', pp. 113, 236, 278, 310. Cambridge University Press, London. 1925.
[41] GOWER, JOHN. "Mirour de l'omme", lines 26449 ff., in 'Complete Works', ed. by G. C. MACAULAY. Vol.: 'French Works', p. 293. Clarendon Press, Oxford. 1899.
[42] BOËMUS (BEHAM?), JOHANNES. *Omnium gentium mores, leges, et ritus*, III, 12 (fol. LVI). Grimm and Wirsung, Augsburg. 1520. The work was translated into English and supplemented from other sources by EDWARD ASTON. 'The Manners, Lawes and Customes of all Nations.' Eld, London. 1611.
[43] LANGLAND, WILLIAM. 'Piers Plowman', pp. 112–17. Dent, London. 1912.
CHADWICK, D. 'Social Life in the Days of Piers Plowman', p. 59. University Press, Cambridge. 1922.
[44] MAURIZIO, A. 'Histoire de l'alimentation végétale', French trans. by F. GIDON, pp. 426 ff. Payot, Paris. 1932.
[45] BAUDET, FLORENCE E. J. M. 'De Maaltijd en de Middeleeuwen.' Sijthoff, Leiden. 1904.
[46] FILBY, F. A. 'A History of Food Adulteration and Analysis.' Allen and Unwin, London. 1934.
[47] 'Calendar of Letter-Books . . . of the City of London', ed. by R. R. SHARPE. 'Letter-Book D', fol. 86 (p. 196). Corporation of the City of London, London. 1902.
[48] DE NAVARRO, J. M. *Antiquity*, **2**, 423, 1928.
[49] LAUFER, B. 'Sino-Iranica', pp. 220–45. Field Mus. Nat. Hist. Publ. 201. Chicago. 1919.
HUBER, E. *Umschau*, **31**, 530, 1927.

[50] THEOPHRASTUS *De causis plantarum*, III, xi–xvi. (Ed. F. WIMMER, with Latin trans., pp. 231–36. Firmin-Didot, Paris. 1931.)
SENN, G. *Gesnerus*, **1**, 77, 1944.
[51] ARISTOPHANES. 'The Knights', line 793. (Loeb ed. Vol. 1, p. 200, 1924.)
[52] ATHENAEUS *Deipnosophistae*, I, 26–34. (Loeb ed. Vol. 1, pp. 112–50, 1927.)
[53] IMMERWAHR, H. R. *Trans. Proc. Amer. philol. Ass.*, **79**, 184–90, 1948.
[54] REMARK, P. 'Der Weinbau im römischen Altertum', pp. 10 ff. Heimeran, Munich. 1927.
DALMASSO, L. 'La viticultura ai tempi dell' Impero romano', pp. 3–5. Quaderni dell' Impero. Le scienze e la tecnica ai tempi di Roma imperiale no. 7. Istituto di Studi Romani, Rome. 1941.
[55] SUETONIUS *Vitae XII Caesarum. Domitianus*, VII, 2. (Loeb ed. Vol. 2, p. 352, 1914.)
[56] CATO *De agri cultura*, x. (Loeb ed. pp. 25 ff., 1934.)
COLUMELLA *De re rustica*, v and vi.
PLINY *Nat. hist.*, XIV. (Loeb ed. Vol. 4, pp. 186–284, 1945.)
[57] STRABO V, C 218. (Loeb ed. Vol. 2, p. 232, 1923.)
[58] PLINY *Nat. hist.*, XIV, iv, 27. (Loeb ed. Vol. 4, p. 202, 1945.)
AUSONIUS *Epistolae*, no. 18. (Loeb ed. Vol. 2, pp. 24–25, 60, 1921.)
Idem Mosella, lines 21, 160. (Loeb ed. Vol. 1, pp. 226, 236, 1919.)
[59] VILLANOVA, ARNALDUS DE. *Tractatus de vinis*. (Trans. by H. E. SIGERIST, 'The Earliest Printed Book on Wine.' Schumann, New York. 1943.)
[60] EVANS, SIR ARTHUR (JOHN). 'Palace of Minos', Vol. 1, p. 415. Macmillan, London. 1921.
[61] TACITUS *Germania*, xxiii. (Loeb ed. p. 296, 1914.)
[62] *Anthologia Graeca*, IX, 368. (Loeb ed. Vol. 3, p. 201, 1917.)
[63] GRUNER, C. R. *Zosimi de zythorum confectione fragmentum*. Sulzbach. 1814.
[64] PLINY *Nat. hist.*, XIV, xxix, 149. (Loeb ed. Vol. 4, p. 284, 1945.)
DIODORUS V, xxvi, 2. (Loeb ed. Vol. 3, p. 166, 1939.)
STRABO III, C 155. (Loeb ed. Vol. 2, p. 74, 1923.)
[65] PLINY *Nat. hist.*, XVIII, xii, 68. (Loeb ed. Vol. 5, pp. 232, 234, 1950.)
[66] *Idem Ibid.*, XXI, xlix, 86 (*lupus salictarius*). (Loeb ed. Vol. 6, p. 222, 1951.)
MARTIAL V, lxxviii, 21 (*lupinus*). (Loeb ed. Vol. 1, p. 350, 1920.)
[67] ADALHARD, Abbot of Corvey. *Statuta abbatiae Sancti Petri Corbeiensis*, I, 7. (*Pat. Lat.*, Vol. 105, col. 542.)
[68] HILDEGARD of Bingen. *Physica*, III, 27. (*Pat. lat.*, Vol. 197, col. 1236.)
[69] FORBES, R. J. 'A Short History of the Art of Distillation.' Brill, Leiden. 1948.
[70] ALBERTUS MAGNUS. *De mirabilibus mundi*, bound with *De secretis mulierum libellus*, &c. Iodocus Iansson, Amsterdam. 1648.
[71] LULL, RAMON. *Testamentum novissimum* in *Libelli aliquot chemici*, pp. 1 ff. Petrus Perna, Basle. 1572.
[72] VILLANOVA, ARNALDUS DE. *Tractatus de vinis*. (Trans. by H. E. SIGERIST, 'The Earliest Printed Book on Wine.' Schumann, New York. 1943.)
[73] FOLZ, HANS. 'Wem der geprant Wein nutz sey oder schad, un wie er gerecht oder falschlich gemacht sey.' Ayrer and Pernecker, Bamberg. 1493.

5

LEATHER

JOHN W. WATERER

I. THE TREATMENT OF PELTS

THE use of skins dates from Palaeolithic times (figure 106), but when, where, or how man learned to transform stiff perishable pelts into flexible and nearly incorruptible leather is unknown. Probably the first step was to soften the dry, harsh skins with fat and brains. This is the basis of the process by which such leathers as chamois and buff were later made. Another early step was removal of hair. If putrefaction had begun the hair-bearing outer skin or epidermis would have begun to loosen from the main substance or dermis, and could then be scraped away before the dermis was damaged. It would be found that putrefaction could be promoted by damp heat, as in the method followed by Arctic peoples, who grease the skin, cover it with saliva, and then use it as a pillow. A comparable method is still used for sheepskins, which are sweated in warm, damp chambers.

Palaeolithic sites have yielded bone skin-scrapers showing clear signs of use. The curved shapes of some of these suggest that skins were scraped over a tree-trunk corresponding to the tanner's beam of later times. The form of these early bone implements appears to be perpetuated in the de-hairing and scraping knives of the ancient civilizations and of the present day (figure 107). Many kinds of stone scrapers have been found, from the most primitive types to the Aurignacean 'planes' (figure 108), but some may be 'slickers' used for working-in fat or brain-substance or, as with sealskins, for squeezing out surplus oil. Direct evidence of oil-tannage has been found in leather from Egyptian tombs.

Raw hide and skin putrefy very slowly when dry, so that in a suitable climate they may often be useful; however, only a few ancient Egyptian raw-hide articles have been found intact. Among them is a Dynasty XII (?) rectangular container, moulded into shape and provided with holes for hanging, and a four-string harp (1250 B.C.), the body of which is covered with raw-hide (figure 109).

The ease with which raw-hide, when wet and glutinous, can be moulded over hard cores or wet sand, led to its widespread use for all kinds of containers. In the pan-graves of Mostagedda of Tasian date, and perhaps as early as 4000 B.C.

FIGURE 106—*'Trousers' and 'skirts' of skin. From wall-paintings of the Upper Palaeolithic in the caves Els Secans near Mazaleon, Teruel, and Cogul near Lerida, Spain.*

(vol I, p liii), there were clay shapes, believed to be cores over which raw-hide pots were moulded; but Egypt has so far provided little evidence of the use of this process, which is, however, still employed in the Sudan, the Sahara, Abyssinia, India, and elsewhere in the east. Articles thus made are quite stiff and hard when dry (figure 121).

The fibrous dermis, or corium, is the central layer of a hide or skin[1] which the tanner must preserve by converting it into leather. It is composed of long, fine, collagen fibrils, of which twenty to fifty are grouped together to form fibres. These in turn are grouped into bundles bound by net-like or reticular tissue (plate 4 B). Thus the corium is a complex substance of great strength and flexibility. Above the corium is the epidermis, which, together with the hair and wool growing out of it, must be removed as a preliminary to tanning. Below the corium, subcutaneous tissue, fat or flesh, must also be removed. The discovery of what had to be retained and what discarded must have been spread over a long period. In ancient times hides and skins were probably converted into leather immediately after slaughter, but traffic in hides soon developed.

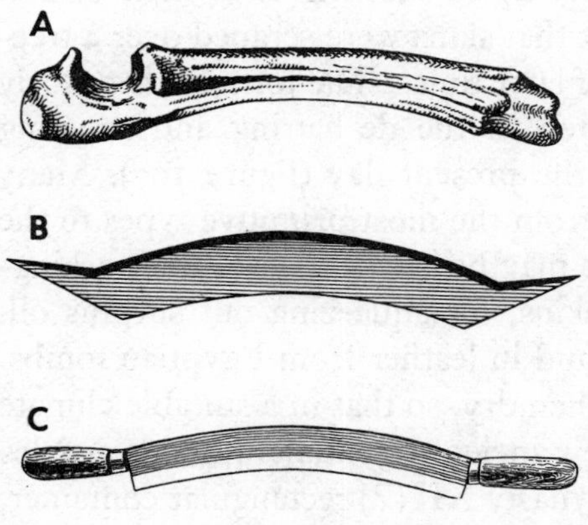

FIGURE 107—*Scrapers for removing hair from hide, (A) of bone, with sharpened edge, Palaeolithic; (B) of iron, found at Pompeii (before A.D. 79); (C) modern steel scraper.*

The transformation of putrescible skin into true leather requires thorough penetration by tanning-agents, of which the earliest were fats. This treatment would be achieved by working the materials into the skin, stretched on the ground, with a blunt tool or slicker, of bone or flint (figure 111). Homer describes how hide

[1] Trade usage confines the word 'hide' to the pelts of the larger animals such as horse ,buffalo, and cattle, and 'skin' to those of the smaller animals, such as calf, pig, goat, and sheep.

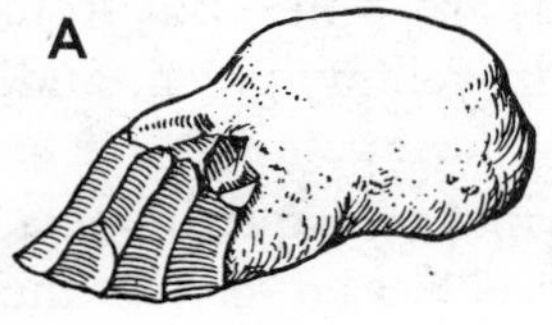

FIGURE 108—*'Plane' scrapers:* (A) *Aurignacean, from the Dordogne;* (B) *modern Eskimo tool with removable blade.*

is impregnated while held round the edge by several men, straining and pulling it in all directions [1]. The modern currier piles up hides, to both sides of which dubbin (cod-oil and tallow) has been liberally applied. When the pelts are thoroughly impregnated, he places them singly on a bench and removes wrinkles and surplus dubbin by working and smoothing them with a blunt-edged slicker of stone or glass.

Of methods of curing, the earliest was doubtless smoking, the action being similar to that which occurs in smoking bacon. Skins are still smoke-cured by Eskimo and North American Indians. The former first chew the skin to soften it, and the teeth of older women are often worn to the gums by this service. The latter make a more permanently soft leather by combining oil-tannage with smoke-curing.

Salt was early known as a preservative (vol I, p 256), and was doubtless used for hides and skins. Decay can be arrested or delayed by the ancient method of salting, or by sun-drying, or, today, by means of antiseptics.

Alum was also early used (pp 367 f and vol I, p 262); its application, known as tawing, yields a white, stiff leather which can be softened by working over a curved blunt edge, as is depicted in Egyptian tomb paintings (figure 111). The manufacture of tawed leather was widespread in antiquity and in the Middle Ages. From Egypt many objects of tawed leather even of predynastic date have

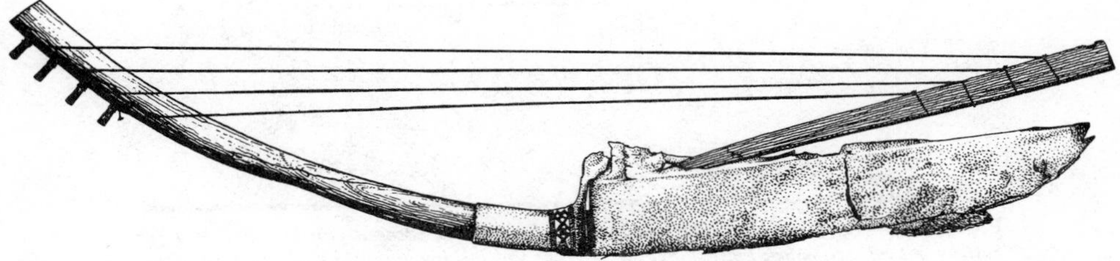

FIGURE 109—*Egyptian harp, the body covered with raw-hide.* c *1250 B.C. Scale* c *1/8.*

been recovered. The earliest Egyptian sandals (figure 112) are white, but from somewhat later periods come sandals and other objects in yellow, green, and, more rarely, blue leather, presumably alum-dressed.

There is clear evidence of the early use of tawed leather in Assyria, Babylonia, Phoenicia, and India. The Greeks used it for the uppers of footwear. The Romans called it *aluta* and applied it in many ways; Caesar mentions sails of *aluta* among the Veneti in Gaul [2]. Analysis of the sheepskin lining of a Neolithic dagger-sheath from Stade, near Hamburg, suggests tawing, and there are other indications that the process was known in that region before the use of metals (figure 119).

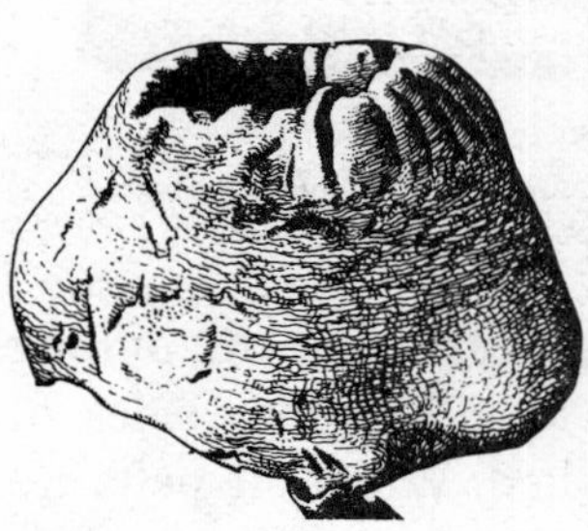

FIGURE 110—*Skin pot, from a pre-dynastic Nubian grave. Scale 1/10.*

In Spain in the eighth century A.D., some time after the Moorish conquest, there arose an important development, namely the manufacture of Cordovan leather ('cordwain'). The combination of imported and indigenous skills resulted in leather of unique character, with qualities that became famed all over Europe. The original leather of Cordova was made from the skin of the mouflon (*Ovis musimon*). This haired sheep, 'horned like a ram, and skinned like a stag', now survives only in Corsica and Sardinia. But there were several kinds of 'Spanish leather', involving different methods of preparation, among them tanning with sumac and tawing with alum. One type was silvered or gilded and embossed, but a greatly esteemed brilliant scarlet is believed to have been obtained by tawing with alum and then dyeing with kermes (p 366 and vol I, p 245).

The secret of making Spanish leather leaked out and its manufacture spread from its original home all over Europe. The cordwainers at first made many

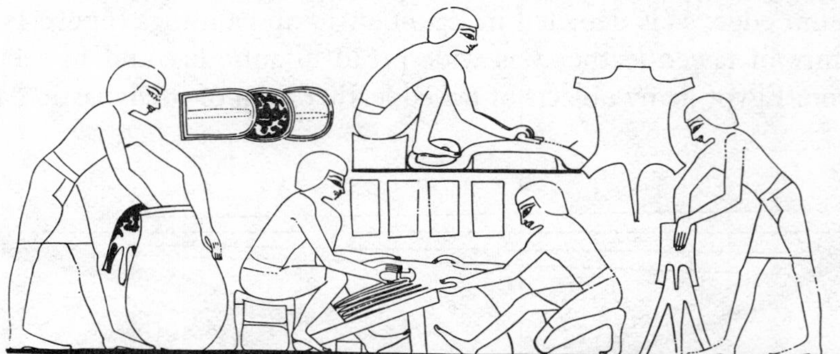

FIGURE 111—*Egyptian leather-dressers.* (Left to right) *Dressing a panther-skin in a jar cutting with a half-moon knife; 'slicking' a skin; and staking over a 'horse'; shields covered with skin can be seen near the jar. From the tomb of Rekhmire, Thebes,* c *1450* B.C.

kinds of wares, but in the early Middle Ages footwear played an important part in their trade. The manufacture of other articles became gradually less important, and the so-called cordwainers specialized as shoemakers. Goatskins imported from Norway and Denmark became those chiefly used for making Spanish leather. It was in use well into the eighteenth century, but by that time its character had long since changed. It was still made principally of goatskin, but was now tanned only with sumac (*Rhus coriaria* L) and finished with oil.

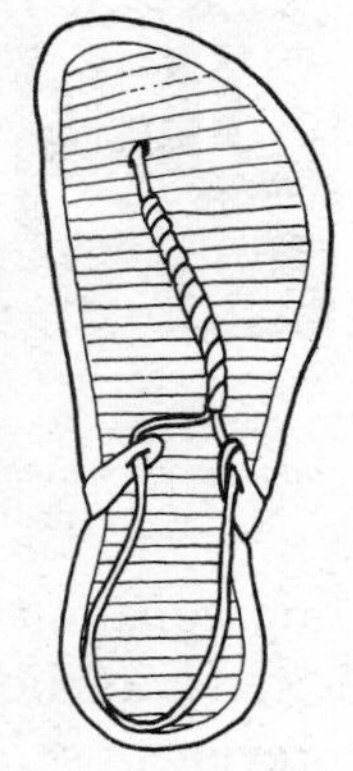

FIGURE 112—*Outline sketch of a sandal found at Balabish. Second millennium* B.C.

The most important process for preserving pelts employs the chemical properties of tannin. The origin of this process is still obscure. It may have resulted from the immersion of pelts in forest pools containing tannin, or from attempts to colour skins with vegetable matter. Oak-bark and oak-galls remained virtually the only sources of tannin in the west until sumac was introduced from the eastern Mediterranean region in the fifteenth century. Babylonian and Hittite inscriptions refer to groves of oaks, cultivated presumably for their tanning materials.

The ancient tanners of Egypt and the Middle East had used oak-galls. These are caused by an insect depositing its egg in the dwarf-oak (*Quercus infectoria*). The resulting spherical swellings or galls contain a higher percentage of tannin than oak-bark. Such 'green gall-nuts of Aleppo', mentioned by Pliny (d A.D. 79) [3] as best suited to the preparation of leather, provided an important export. In Egypt pods of acacia (*Acacia nilotica*) were also used as a source of tanning in the Roman period, as was the bitter rind of the pomegranate (*Punica granatum*), sometimes called leather-apple.[1] In addition, Pliny mentions sumac, 'the currier's plant' [4], which appears to have been known to the Babylonians.

Two or more different processes were sometimes combined. In some Egyptian inscriptions and paintings there is a hint of a combination of alum and oil processes for sandal leather. A Babylonian recipe also may perhaps be regarded as such a combined tannage: 'The skin of the kid thou shalt feed with milk of a yellow goat and with flour; thou shalt anoint [it] with pure oil. . . . Thou shalt dilute alum in pressed grape-juice, then fill the surface with gall-nuts [presumably in the form of a liquor] of the tree-cultivators of the Hittites' [5]. The 'feeding' of leather with lubricants such as oil, grease, egg-yolk, flour, and other substances is still used to produce special leathers. In Baghdad at the present day, leather is first alum-dressed and afterwards treated with gall-nuts and fat.

[1] Pliny may have confused the use of the rind for tanning with its use as a yellow dye.

The first stage in manufacture is to cleanse the pelts and to soften them by soaking. An early invention was some sort of 'stocking' machine which violently pummelled the pelts, often to their detriment (figure 113). Loosening of the epidermis by putrefaction would suggest that it be scraped off together with the hair. Sweating, to induce bacterial action and thus facilitate removal of the epidermis, was also an early discovery. When or how the action of lime in loosening hair was discovered is unknown, but it was certainly in use in the Middle Ages. Lime is also important in plumping the skins, which facilitates impregnation with the tanning materials. For liming, pelts are passed through a series of liquors made from slaked lime, the first having been mellowed by previous use.

FIGURE 113—*The stocks as used in the eighteenth century in chamoising leather. From Diderot's* Encyclopédie. *Scale c 1/100.*

The scraping-away of epidermis and hair was done over the beam (a section of tree-trunk) with a blunt-edged concave tool of a kind known in antiquity (figures 107, 114). The removal of the adhering flesh was similarly performed, but with a sharp concave blade. Cleansing ('scudding') of the hairy or grain side from remaining fragments of hair, sheaths, and other impurities, by forcing them out of the skin, is also carried out with a curved, blunt knife over the beam. This relatively late development produced an even finer and more perfect material.

FIGURE 114—*De-hairing of cowhides. Woodcut by Jost Amman. 1568.*

So far, all hides and skins followed much the same procedure. For certain purposes, however, such as clothing and some types of bags and purses, a particular degree of softness or stretchiness was required. This was attained by treating with a cold infusion of poultry- or pigeon-dung, or with a warm infusion of dog-dung. The resulting chemical and bacterial action removed lime and dissolved certain albuminous matter,

FIGURE 115—*Eighteenth-century curriers' workshop.* (Left, wall) *Using the* lunette *to soften or to level the hide;* (front left) *using the currier's knife to shave a skin;* (front right) *trampling it on a hurdle and beating with a* bigorne*;* (back, left to right) *stretching and flattening a wet skin with the slicker, softening and graining leather with the pommel.* (Below) *Currier's tools: knife (view and section);* lunette (*ditto*); bigorne*; slicker; pommels. From Diderot's* Encyclopédie.

leaving the skin flabby. The requisite degree of action was judged by feel. Scudding was carried out after these operations. Where quality and uniformity of colour were important, care would be taken to remove all traces of lime and other impurities, either by washing, or by drenching with the mildly acid liquor produced by fermenting bran.

The thickness of a tanned cattle-hide is from 4 to 6 mm. To produce thinner leather—as for covering coffers, for footwear uppers, or for bags—the flesh-side, after tanning, was shaved over the beam with the curious and ancient currier's knife (figure 115). The blade, of soft steel, is ground and set on both sides to a sharp edge; then with a special instrument the edge is turned over at a right-angle to the blade. Kept in order with a steel, this tool can be used to remove thin shavings from even soft hides. Today the splitting-machine divides a hide horizontally into two or more layers, each of which has a use, and the wastage of shavings is thus avoided.

By the eleventh century A.D. the early discoveries and practices in the manufacture of leather had been improved into well established techniques. From then to the nineteenth century there were no changes in the principles of leather-production, and but few in methods. The three basic processes remained: (*a*) the oil process or chamoising; (*b*) the mineral (alum) process or tawing; (*c*) the vegetable process or tanning. Whatever the method or combination of methods, the aim was to preserve the fibrous structure from which the principal characteristics of leather arise, and at the same time so to modify the proteins in the skin as to avoid decay in damp conditions.

In modern oil-tannage, as with vegetable and mineral processes, the fibres themselves are rendered insoluble and permanently non-adhesive, through chemical action. In medieval methods this effect was induced by oxidation-products of the oils or fats. The resulting leather is flexible but tough; its characteristic shade is known as buff because this typical oil-dressed leather was at first manufactured from the skin of the European buffalo or wild ox, the surface of which was suèded. Buff leather, serving the warrior at least as early as the thirteenth century, was still much used during the sixteenth to eighteenth centuries for protective garments. The modern process is now chiefly used for the conversion of sheepskin split into 'shammy' ('chamois') or wash-leather.

Chamoising is an important oil-oxidation process. Its history is obscure. Whatever the original connexion with chamois-skin—which can never have been very common or accessible—the term was early applied to the dressing of any leather in which oil predominated. The process was probably based on primitive softening with fatty materials. While some measure of oxidation may have produced a partial tannage, the primitive method probably resulted in little more than separation and dehydration of the fibres and the provision of a water-resistant film.

Tawing, now almost extinct, was important in the Middle Ages. Alum, when used without admixture, produced a stiff, 'empty', and imperfect leather which had to be softened by staking, or by stretching the skin and working it with either a spade-like tool or the circular 'lunette' (figure 115). The latter, provided with a sharpened and turned edge, as for the currier's knife (p 153), was also used for thinning a skin. It was more usual to add to the alum about half its weight of common salt; this mixture yields a superior leather, which still, however, lacks certain qualities. There accordingly grew up a practice of feeding this rather empty leather and rendering it water-resistant. In one method the skins were trodden in shallow tubs containing a pasty mixture of alum, salt, egg-yolk, flour, and oil. After hanging for several weeks, the skins were damped, softened by treading on a hurdle (figure 115), and finally staked.

In a grant of 1593 to one Edmund Darcy [6] to carry out the 'searching and sealing of leather'—a right claimed by the Leathersellers' Company—a list [7] indicates that in Britain alum and oil processes had become so closely associated that all the principal leathers produced by both processes, or by combining them, were regarded as tawed. The table of fees shows that white tawed leathers were made from the skins of sheep, goat, deer, calf, horse, and dog; oil-dressed leathers from buffalo, chamois, sheep, 'right Spanish skins' (perhaps mouflon), seal, deer, and calf. This list also proves that in the manufacture of cordwain, originally an alumed leather (p 150), with or without oil-dressing, alum had been wholly or partly replaced by oil. White-tawyers were also employed by the Skinners' Company to prepare fur-skins. The process, where the fur was to be preserved, was to stretch the skin on a frame and sponge the flesh-side several times with a solution of alum and salt.

The term tanning (p 151) has long been loosely used; strictly it applies only to the process which employs tannin or tannic acid. The weak tannic liquors, once obtained by macerating certain plant members in warm water, acted slowly but thoroughly. For cattle-hide leather there is still nothing better than oak-bark, of which supplies were plentiful in early days, particularly in Britain. It has most of the advantages of both the pyrogallol and the catechol groups of tanning-materials and few, if any, of their disadvantages.[1] Sumac (p 151), an excellent tanning agent, chiefly used for goatskins, was introduced into Britain during the fifteenth century.

Tanning was usually performed in pits, but later in some form of vat. As with liming (p 152), a series of liquors was employed, the first being old and mellow, the last fresh and strong. This method ensures full penetration and avoids a case-hardening effect. From time to time the hides would be hauled out, drained, and put back in a different liquor. In the last stages, they would be laid flat in a pit of liquor with crushed bark between them. This process was thorough but slow, and might take fifteen months or even longer. In more

[1] The latter group is particularly liable to produce a leather that disintegrates when attacked by atmospheric sulphuric acid.

modern processes the hides were suspended from poles, and the tannage was accelerated by giving them a rocking motion. In the process known as 'bottle-tanning' skins were trimmed and sewn up, then filled with a strong infusion of sumac and immersed in the same liquor. After a time the distended skins were removed, and, when all the liquid had oozed out, the tannage was complete and skins so tanned were virtually free from stretch. Bottle-tanning is still occasionally employed for morocco leather of the finest grade.

II. FINISHING

or some purposes little further treatment was necessary once the above processes had been completed, but from very early days it became customary to apply various finishing operations, to make the leather more adaptable or decorative. Oil-dressed leather is buff-coloured, alumed leather white, while vegetable-tanned leather varies from a light biscuit colour to reddish brown, according to the tanning-agents used. Among the earliest examples of decorated leather are fragments painted with blue and yellow chevrons, from Naqada and Ballas in Egypt (figure 124). Many examples of red-dyed leather have been found in Egypt dating from predynastic times onwards.

Examples of Egyptian finished leathers exist in clothing, braces, wrist-tags, mummy-labels, sandals, and bags. The dye used was either kermes (vol I, p 245) or archil, with alum as mordant. Black dyeing for leather was much used in early times and throughout the Middle Ages. For this, tannin with iron sulphate (copperas) provided one of the earliest principles—later logwood was used.[1] Vegetable dyes provided other colours, such as brown (brazil), blue (indigo), yellow (pomegranate), and green (source uncertain).

Additional methods were introduced for giving character to certain leathers. Thus it was found that the structure of goatskin caused it to develop the typical pebbly or granular surface of morocco leather when 'boarded', that is, folded hair side in and rubbed backwards and forwards with a curved board. The same treatment applied to calf merely produced a fine creasing on the flat surface, when it was known as willow calf. This process is used today for the softening of upholstery hide. Naps were raised on buff and chamois leather, arising originally from the scraping or 'frizing' of the surface to facilitate impregnation with oil. The agreeable scent of the 'Russian' leather, mentioned by Marco Polo (? 1254–1324?) [8], arose from the use of birch-tar oil as lubricant. It was diced with a wooden pummel, a practice adopted for most of the hide-leather for covering coffers and for upholstery in the sixteenth and seventeenth centuries.

[1] Other dyes are considered elsewhere (ch 10, and vol I, pp 245 ff).

In addition to those finishes the prime purpose of which was improvement of appearance, there were others directed at imparting specific properties. Sole-leather was hammered to compress and consolidate it. Harness-leather, which needs great flexibility and strength, was 'curried', that is, impregnated, after tanning and while still damp, with dubbin (p 148). As the water was driven out it was replaced by grease, providing a complete and lasting protection. Other leathers, chiefly calf, for the long riding-boots of the eighteenth century, were often waxed on the surface.

FIGURE 116—*Shagreen spectacle case, eighteenth century.*

'Shagreen' is a term the application of which has changed greatly. The word, it seems, is the Turkish *saghrī* and means the croup of a beast. The original shagreen was made in Persia from the hides of asses, horses, and camels, probably untanned. Seeds of a species of *Chenopodium* were trampled into the skin when it was moist and shaken out when it dried, thus leaving granular indentations. The material was then stained. In the seventeenth century and later, shagreen was made either of finely granulated shark-skin or of the skin of a ray-fish, whose pearl-like papillae were ground flat, leaving a lovely pattern (figure 116).

For the most part, tanning remained for millennia a widely distributed local industry. Thus in late sixteenth-century England, 'in most villages . . . there is some one dresser and worker of leather and . . . in most of the market towns, 3, 4 or 5 and many great towns 20, and in London and the suburbs . . . 200 or very near' [9]. Even in the early nineteenth century the wide diffusion of this essential industry persisted. At that time 131 towns and villages of Cornwall and Devon still possessed tanners. The methods described in early trade-books remained substantially those of the Middle Ages. Leather manufacture was one of the last industries to be affected by power-driven machinery, though leather played an important part in the design of various machines (figure 117).

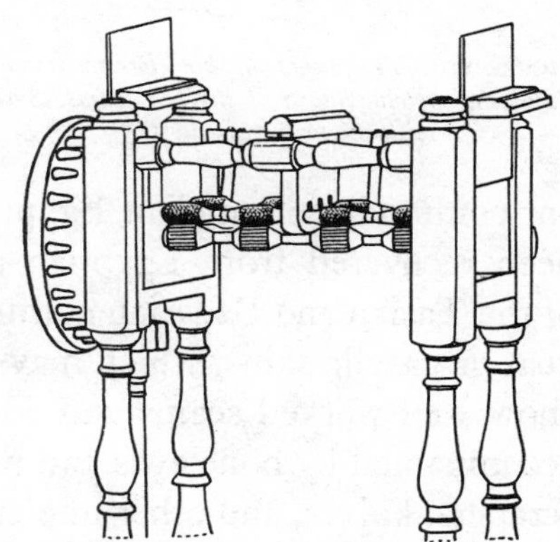

FIGURE 117—*Detail of Arkwright's spinning-frame, showing leather-covered rollers. 1769.*

III. UTILIZATION IN ANCIENT TIMES

Leather has been used for so many and such various purposes that, at one time or another, many of the processes employed in the industry have been used also in the working of other materials, such as wood, metals, pottery, and textiles. Some of these were almost certainly employed originally for leather. Thus stitching (figure 118) was an elaboration of the prehistoric method of joining pieces of skin by passing sinews or narrow thongs through a series of holes pierced with sharp-pointed flints. Primitive weaving was possibly carried out with thongs of leather as well as with vegetable fibres. Moulded raw-hide almost certainly preceded pottery (vol I, p 399). Leather can be as tough as hard wood, and can then be sawn or turned in a lathe; or it can be as soft as a fine weave and amenable to the best needlework. It has been used for armour and for feminine underwear; for the driving-belts of great machines and for the dainty purse; for the drinking-vessels of early times and for the indispensable washer in pumps.

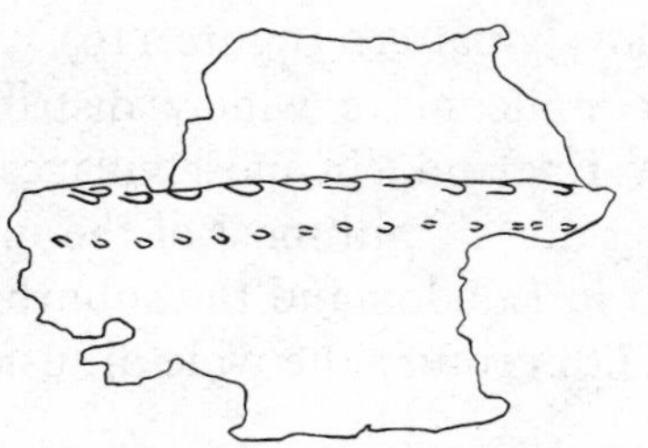

FIGURE 118—*Fragments of skin clothing found at Balabish, showing the well stitched seams. Second millennium* B.C. *Scale 2/5.*

The Mesolithic is the earliest period to which the use of leather can be safely attributed, though a carving in ivory of a figure clothed in skins, recovered from a Siberian site (vol I, figure 22), and certain Spanish cave-paintings of clothed figures are thought to be Upper Palaeolithic (figure 106). Leather was a very important commodity in ancient Egypt, and many leather articles of great age have been recovered from Egyptian tombs, cities, temples, and palaces. In graves of the Tasian and Badarian cultures (vol I, p liii) many bodies were wrapped in goat or gazelle skins, which may have been garments rather than shrouds; they show well worked seams and edges oversewn with thongs, and are sometimes accompanied by bone awls and needles with which the tailoring was done. Flint scrapers, knives, and other implements suitable for working leather are common in predynastic Egyptian sites.

Palaeolithic man probably heated water in skin bags by dropping in red-hot stones. In Norwegian Mesolithic dwelling-sites vestiges of vessels of leather have been found together with boiling-stones. These fragments perhaps represent the genesis of a process for working leather, by moulding, or 'blocking' as it is often called, by which was made the *cuir bouilli* of the Middle Ages, used for such things as armour, sheaths, bottles, flasks, and drinking-vessels. The stiffening of leather pots under the action of hot water would suggest a deliberate fabrication of suitable vessels.

A Neolithic flint dagger in a moulded and decorated leather sheath, with a strap for looping round a belt, comes from Stade, near Hamburg (figure 119). The sheath and strap are apparently of cattle-hide, well tanned, though by what method cannot be determined. A feature of special interest is a strip of sheepskin, perhaps alum-dressed, originally glued inside the sheath at each side to protect the edge of the dagger. The sheath was made by methods still in use. It was cut in one piece, seamed up the back with extremely fine leather thongs, and afterwards moulded into the desired shape.

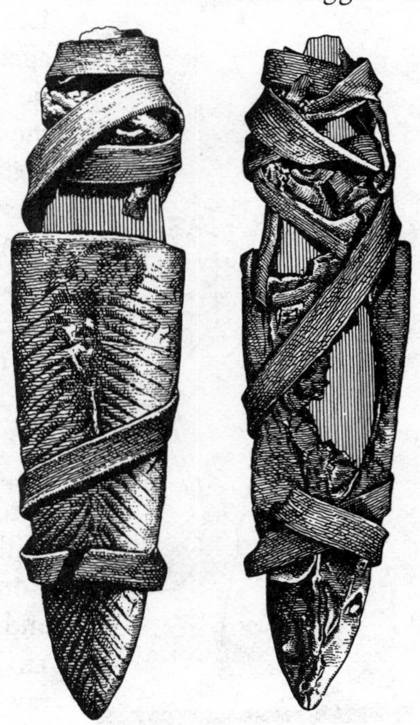

FIGURE 119—*Dagger-sheath of tanned cattle-hide from Wiepenkathen, Stade, near Hamburg. Believed to be Neolithic. Scale* c 3/5.

There is little evidence from Egypt of the early making of vessels of moulded leather or untanned skin. Several excavators have suggested that certain early pot-forms follow the shapes and styles of skin bags or pots (vol I, p 398). If this be so, the technique of making moulded leather pots or beakers may have disappeared long before identified predynastic cultures developed, for none has been discovered—although an object in the paintings of the Rekhmire tomb, *c* 1450 B.C., may perhaps be a leather bottle (figure 120). In an early dynastic Nubian grave a leather vessel was found near the head of an adult negroid male, in a position often occupied in such graves by a clay pot (figure 110). It seems to be a very primitive type of leather pot formed by drawing a circular piece of skin over a withy-ring, perhaps shaping the body into a spheroidal form with the aid of clay or wet sand. In pan-graves of about 4000 B.C. at Mostagedda were solid forms of the shape of clay pots, which were, perhaps, cores over which leather, wet raw-hide, or intestinal membrane was moulded and then hardened by drying (pp 147–8). This process is still employed by the Sudanese, Tuareg, and Abyssinians, and by tribes of northern India (figure 121).

From predynastic cultures (earlier than about 5000 B.C.) there has come evidence of leather clothing, sometimes fastened with a copper pin. The skin bag for liquid was used in Egypt throughout her history, in the form familiar all over the east, namely, a sewn-up goatskin filled and emptied by the neck (figure 122). Leather sandals are known from this predynastic age, as well as bags and cushions of painted leather, tied-up rolls of leather, and knotted thongs. Fragments of leather painted with chevron patterns are possibly the remains of the garments indicated on archaic statuettes (figure 124).

FIGURE 120—*Egyptian painting of a man carrying a leather bottle. From the tomb of Rekhmire at Thebes. c 1450 B.C.*

At Mostagedda tanned leather skirts of goat or gazelle skins have stitched seams neatly uniting small pieces, and are ornamented with blue and white beads or finely cut fringing. Corded and plaited leather girdles were found. Leather sandals are quite common, and some are of soft tawed leather unsuitable for any rough wear. The leather was often dyed red. Altogether, the evidence points to a considerable skill in leather-working at this early period. Objects from the same period and sites include leather ropes and cords, some used as girdles and anklets (figure 123); a fine polished stone axe-head lashed with raw-hide plaiting to a wooden handle; archer's 'bracers' to protect the arm from the lash of the bow-string; leather bags for grain and for toilet articles, one having a handle of vegetable fibres covered with fine plaited leather thongs; a dress of small pieces of red leather neatly sewn together; a plaited head-band ornamented with parallel slanting rows of small white beads; and leather buckets of a type still used with the ancient water-lifting shaduf (vol I, figures 344–7). Pieces of leather were joined by very narrow leather thongs or sinews. The methods employed were: a button-hole stitch which left small ridges at the joins; a narrow overlapped seam with a single running stitch; and a wide overlap of about $\frac{1}{2}$-in with two rows of running stitches (figure 118).

The techniques of these early periods can often only be inferred. While tools were still only occasionally of copper, the degree of skill in both the preparation

FIGURE 121—*Tuareg jewel-case of moulded camel-intestine, with reddish stain decoration. Scale 3/10.*

FIGURE 122—*Silenus with wine-skin. From a Greek wine-cup, fifth century* B.C.

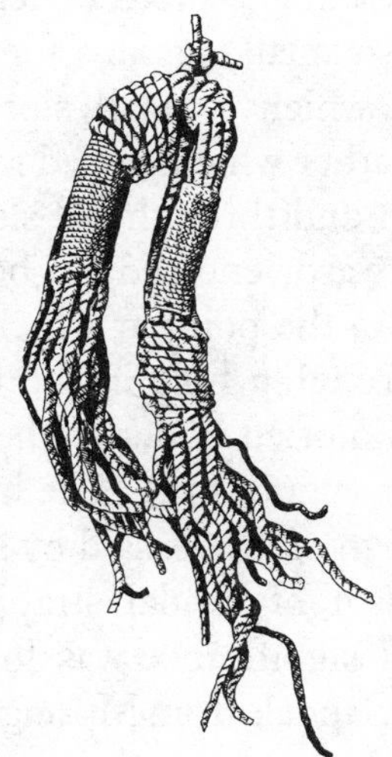

FIGURE 123—*Leather girdle from Balabish. Second millennium* B.C. *Scale 2/5.*

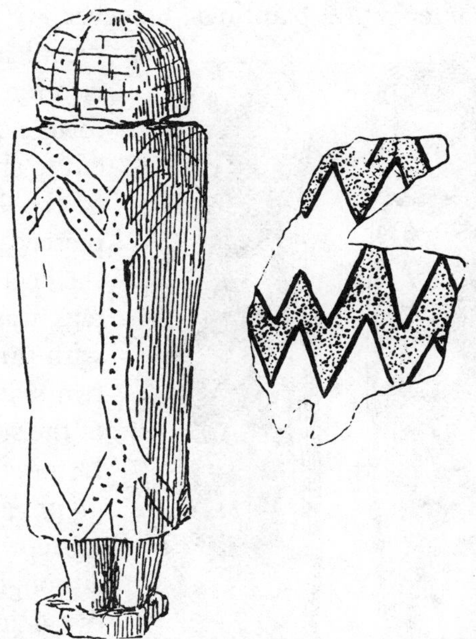

FIGURE 124—*Ivory statuette,* c $4\frac{1}{2}$ *in high, of a woman wearing a chevron-patterned cloak, from Hierakonpolis, and a similarly patterned fragment of leather from Naqada. Protodynastic.*

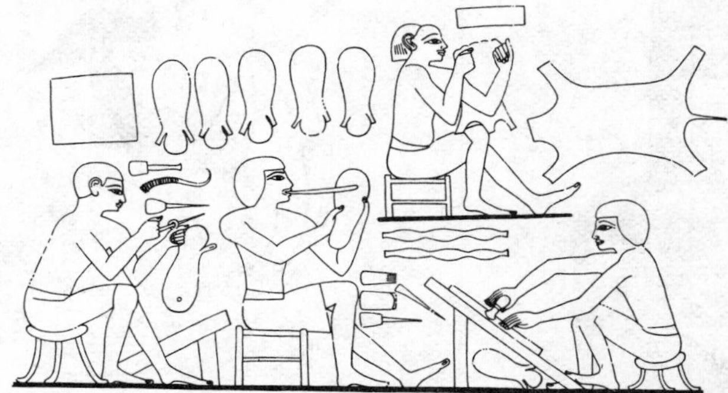

FIGURE 125—*Egyptian sandal-makers.* (Left to right) *Making holes in the side-lugs for the straps; pulling through the toe-strap, which is knotted under the sole; cutting out the leather with a half-moon knife. From the tomb of Rekhmire at Thebes,* c *1450 B.C.*

and the use of leather indicates a long development. The use of raw-hide for lashings, employed anciently for hafting flint axe-heads, was later used to secure joints of furniture and chariots. It was applied wet, and contracted as it dried. From later mural paintings, however, we can learn something more definite of technical processes. The making of sandals is often depicted (figure 125). In the earliest form a sandal was of a single piece of leather cut to shape, with two small tags just forward of the heel end, in which a small slot was pierced. A strip of leather which passed round the heel of the wearer and through the slotted tags was carried forward, encased or bound with thin leather from the point at which the two sides joined, through a hole in the front of the sole, and then knotted underneath, thus forming the toe-strap (figure 112). The hieroglyph ☥, *ankh*, the sign of life, is said by some to represent this form of sandal strapping, although its original significance was lost to dynastic Egyptians. Sandals from Mostagedda are made of alumed goatskin, the roughly finished flesh-side underfoot. Above, the skin retains the epidermis with some hair remaining, but the outer edge is very neatly bevelled,

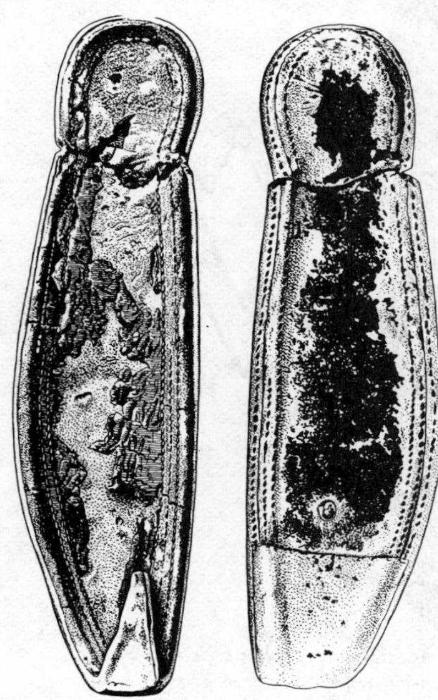

FIGURE 126—*Red-dyed and alumed goatskin sandals of Tutu, wife of the scribe Ani. From Thebes,* c *1300 B.C*

presumably with a flint knife. The pure white leather thus revealed provided a decorative border about $\frac{1}{4}$-in wide all round.

There were more elaborate types of sandals later. Those of the New Kingdom (from *c* 1580 B.C.) are more robust forms, with soles of two or more thicknesses of leather, and pointed toes. A fine pair of Dynasty XIX, from Thebes, with the turned-up point found in early Hittite sculpture, are of red alumed goatskin leather with brown inner soles of tanned hide and with two rows of neat thread stitching, the edges covered with a binding (figure 126).

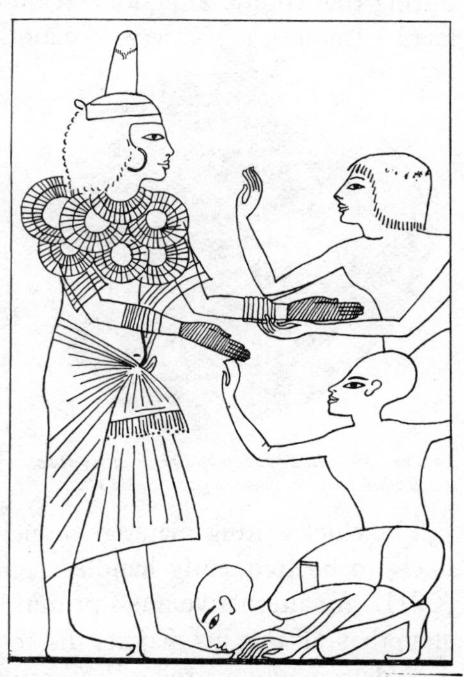

FIGURE 127—*Ay wearing red gloves. From a tomb at Tel-el-Amarna,* c *1370 B.C.*

Illustrations of gloves in ancient Egypt are rare. There is one in a Dynasty XVIII tomb at Tel-el-Amarna, showing gifts from the king to a high official whose duties included the management of horses. The gloves are red and therefore probably alumed and dyed with kermes. In another relief the official, Ay, is shown proudly wearing the royal gift (figure 127).

Details of the harness of the chariots which, in the New Kingdom, were important in war and ceremonial, can easily be examined. Bow-cases and quivers were of leather; one from Tutankhamen's tomb was of wood covered with green leather. In his chariot, as in other woodwork, raw-hide thongs were used to strengthen joints. In some cases there were leather or raw-hide stays running from the top of the body-frame to the pole. The felloes of the wheels consisted of two half-circles of bent straight-grain wood, mitred and glued together, and bound with raw-hide. Around the outer perimeter went a leather tire (plate 5 D).[1] The hub and several inches of the spokes were covered with raw-hide, which was also used as a bearing for the bronze axle. That this use of raw-hide was purely constructional is shown by the fact that in the more elaborate examples it was covered with dyed leather or thin sheet-gold beautifully embossed. The floor of the chariot was a mesh of raw-hide strips (vol I, pp 727 f). The frame-

[1] Traces of alumed leather tires were also found on chariots at Ur.

work of the body was partially filled in, sometimes with thin wood and sometimes with leather covered, in regal examples, with ornamented sheet-gold. The gold-refiners themselves used leather bellows.

Shields of panther-skin, with quivers of the same material, are illustrated during the Middle Kingdom. A mural of Dynasty XVIII depicts their manufacture (figure 111). One workman thrusts a panther-skin into a jar presumably containing a liquor, perhaps a solution of alum. The skin, still dry and stiff from the tawing-process, is shown having legs and tail trimmed off. A man using the Egyptian form of the half-moon knife then cuts the skin into the required shape. The skin is next stretched over forms of wood or wicker, sometimes edged with metal. Leather tents are referred to in the great Karnak inscription dealing with the abortive Libyan invasion of Dynasty XIX (*c* 1300 B.C.) [10]. The leather funeral-tent of Queen Isimkheb of Dynasty XXI (*c* 1000 B.C.) still exists.

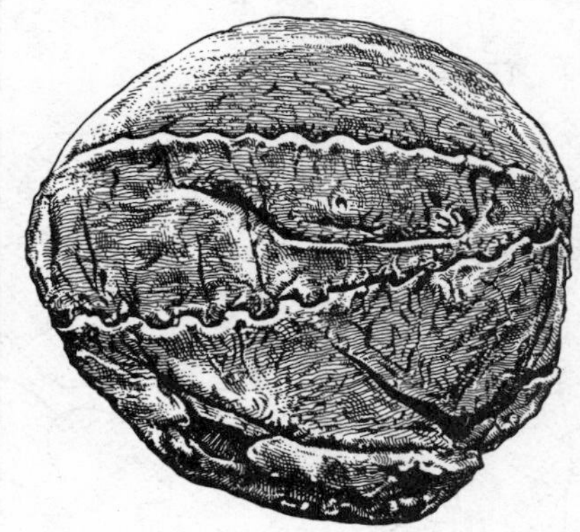

FIGURE 128—*An Egyptian ball made of six sewn leather sections. From Thebes. Scale 1/2.*

The chief writing-material of ancient Egypt was papyrus, but for documents likely to be frequently handled leather was used at least as early as Dynasty XVIII. An alternative was a primitive kind of parchment. The annals of Karnak state that 'all that his majesty did to this city, to that wretched foe was recorded each day . . . upon a roll of leather in the temple of Amen' [11]. Such rolls were used for the 'Book of the Dead' (plate 5 C). Assyrian sculptures also show leather rolls, and at Niya, in central Asia, official documents written on oblong pieces of sheepskin dating from about the fourth century A.D. have been found [12]. Coloured leather was used for ceilings and walls, and many examples remain of its use for stools, chairs, beds, flails, whips, dog-collars and leashes, trinket boxes, and sheaths for knives and daggers. There are play-balls of white and red-dyed alumed leather, consisting of six or more panels neatly sewn together (figure 128).

The Sumerians, like the Egyptians, were familiar with white alumed leather, which they used as tires for chariot wheels. The same leather formed the basis of an elaborate court head-dress of lapis lazuli. Men wore skirts of sheepskin, and hide leather was used for harness. Both Babylonia and Assyria were noted for ornamental footwear of brightly-coloured goatskin, often elaborately embroidered and garnished with jewels. Red goatskin, though no doubt made over a wide area, was specially associated with Babylon. The Assyrians used inflated skins as floats for rafts, and skin covered their large circular *quffa*s (vol I, figure 537). 'Red Babylonian leather', goatskin dyed with kermes as used for the laced top-boots of Roman emperors, was later valued in Greece. This was the original 'morocco' leather later tanned with sumac, which also appears to have been known to the Babylonians.

An interesting use of leather by the Phoenicians is recorded by Strabo (*c* 20 B.C.), who states that fresh water was collected from springs rising under the sea by dropping over the source a bell of lead to which was attached a leather pipe.

FIGURE 129—*Neolithic bowl of moulded leather, made in two halves laced together. From Schleswig. Scale* c *6/7.*

The ubiquity of flint leather-working tools demonstrates the continuing use of skins in Neolithic Europe. Leather beakers and a small bowl of that culture have been found in Schleswig. The bowl is of buff-coloured deerskin, perhaps vegetable-tanned, in halves laced together (figure 129). Around the outside are strands of twisted black horsehair, ornamenting it or serving to prevent spread. The edge was originally bound with leather. The moulding of the separate halves suggests a primitive technique incapable of moulding even so small an article in one piece. A primitive beaker from West Smithfield, London, thought to be Neolithic, is of tanned cowhide with some hair still adhering.

Skins were largely used in Neolithic and Bronze and Iron Age boats, both in the east and in Europe. These are discussed elsewhere (p 168 and vol I, pp 730–1). Another use for leather was the sewing together of planks with thongs, as in the Iron Age boat of Hjortspring, Schleswig (p 578 and figure 527). Roman writers state that British chieftains wore a cloak of blue leather. There is no evidence that leather was used in the construction of British chariots, as it was in New Kingdom Egypt (p 163), but it was required for the harness, and some shields are believed to have been made of wicker covered with hide. Bellows, used in smelting, were of leather. These, too, are discussed elsewhere (ch 2 and vol I, figures 382–3). Swords and daggers involved the use of leather for sheaths and scabbards, of which early examples have been found in many localities. Fragments of scabbards of two Bronze Age swords at Heeshugh, Schleswig-Holstein, were of hide dressed with inorganic salts.

Some of the most important leather relics of the Iron Age were found in the salt-mines near Hallstatt. They include a miner's kit-bag with tools; a leather bag for carrying salt, of conical shape with withy stiffeners to which it is secured with lacing (vol I, figure 376); and a fur cap with seams like those still used by furriers. At Dürrenberg, in the same locality, was a goatskin cap or hood of

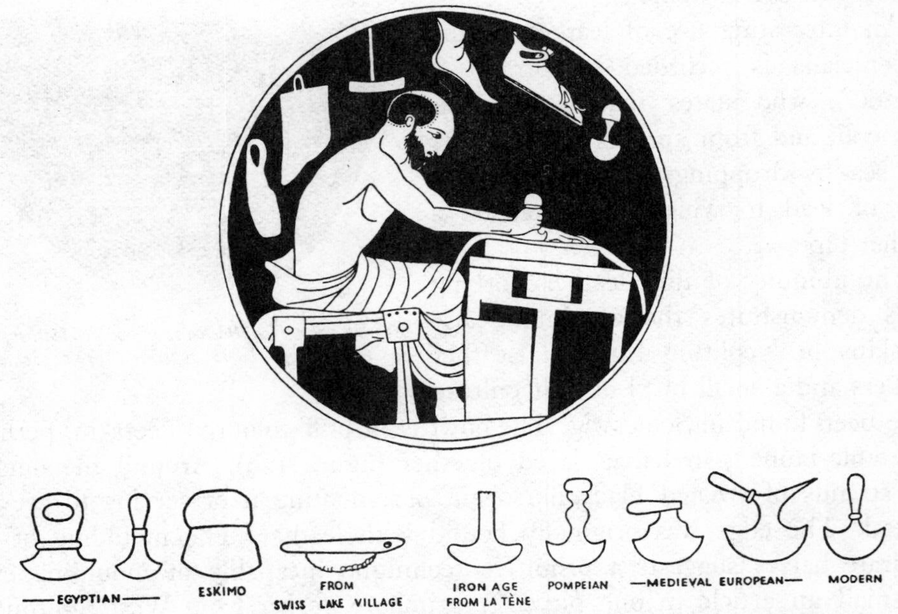

FIGURE 130—*Picture on a Greek bowl of a shoemaker cutting leather (restored). On the wall hang shoes, hammer, and half-moon knife. Sixth century* B.C. (Below) *Evolution of the half-moon knife.*

conical shape. Here also was a bag, cut from one piece of leather, laced up on one side and closing with a draw-strap. From La Tène, an Iron Age settlement at the eastern end of Lake Neuchâtel, came many leather-working tools, together with a leather bag containing the kit of a saddler and harness-maker, consisting of chisels, gouges, awls, punches, and half-moon knives.

The extension of the Roman Empire introduced a rationalized quantity-production into north-west Europe. Finds of leather in Roman settlements consist chiefly of sandal-soles and fragments of clothing. The pattern of life in Rome was followed as closely as possible in her colonial territories. Since a substantial volume of the requisite apparatus of this life, notably leather-ware, was imported from Italy, the Roman occupation left little impact on local leathercraft. We must therefore rely on Roman and to some extent on Greek sources for information.

In Homer's time leather was a well-known commodity. Odysseus wore a cap of strapped leather, lined with felt and 'armoured' with boars' teeth [13]. Contemporary statuettes show this conical headgear, while boars' teeth, pierced for sewing, were found at Mycenae. In Homer's Odyssey Laërtes wears gloves in his garden. The spear of Achilles strikes the shield of Aeneas 'near the first rim where the brass and o'erlaying hide were thinnest' [14], and the tug-of-war struggle for the body of Patroclus is compared to the labours of curriers stretching a hide to cause grease to penetrate (p 157). Odysseus finds Eumaeos making cow-hide shoes for the winter. Under the metal helmets, leather caps were probably worn to prevent chafing. Skin bags were commonly used to hold wine and oil.

FIGURE 131—*Greek shoemakers cutting a pattern from a customer's foot. Greek amphora, sixth century* B.C.

About the fifth century B.C. the Greeks adopted as armour a leather cuirass to which were wired small plates of bronze. The Romans later developed a similar protection plated with iron. On a kylix of the sixth century B.C. is depicted a shoemaker using a half-moon knife; another knife hangs on the wall with lasts and finished wares (figure 130). An amphora of about the same date shows a customer standing on leather, on the work-bench, while the shoemaker cuts a sole to the shape of her foot (figure 131). Elsewhere we find measuring-sticks in use, similar to those of the present day. The Etruscans were skilful dressers and workers of leather and produced luxurious footwear during their supremacy (seventh and sixth centuries B.C.). The styles resemble the Greek.

The chief types of shoe used by Greeks in later times were common to most countries of the Near East. They included sandals, shoes, boots, and buskins. The shoe with upturned toe and pompom shown on an Akkadian sculpture of about 2600 B.C. is still in use in Greece. A similar shoe, without pompom, is depicted in a Hittite sculpture from Carchemish. A form of footwear which must have descended from earliest prehistoric times was the *monodermon*, originally a mere bag made from a single piece of leather or skin, gathered round the ankle by a thong threaded through slots.

The details of Greek sandals can be learned from sculptures. They differed from Egyptian in having a series of straps by which they were firmly attached to the foot. There are many different designs. In classical times leather seems to have been worn by the well-to-do, the poorer classes using wooden soles. Uppers were usually of coloured leather, probably goatskin; the soles were of cattle-hide. The better type of sole was made of several layers. The *kothornos*, said to have been invented by Aeschylus to raise the stature of his gods and heroes on the stage, had a sole of many layers up to about 3 inches in height, with a shoe on the top. Fashionable women adopted it.

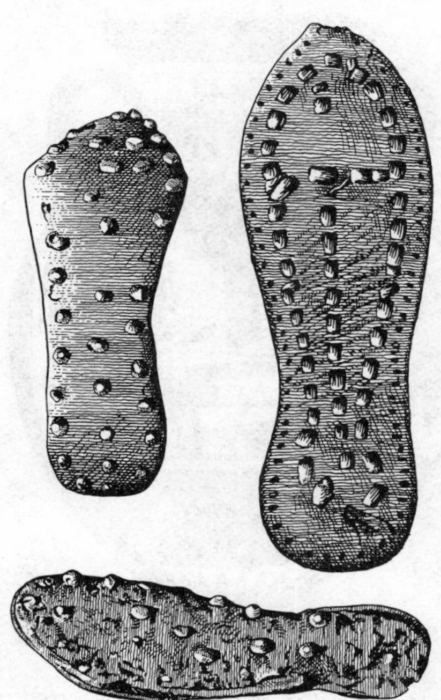

FIGURE 132—*Soles of* caligae, *one showing lacing, the other two iron nails. From Fleet Ditch, London.*

The Romans were great makers and users of leather. Their guilds included leather-dressers and makers of straps, harness, shields, wine- and oil-bags, and footwear. Roman footwear closely followed Greek models but developed an individual sturdiness for military use. It was produced in large quantities in organized factories, and a substantial export trade grew up. The *caliga* of the foot-soldier had a thick sole of two or more layers of vegetable-tanned hide sewn or laced together and studded with iron nails or spikes (figure 132). The upper consisted either of a system of straps or a kind of lace-up sock which left the toes bare. The cutting-out was done, as we learn from paintings and sculpture, with the half-moon knife (figures 130–1). Whereas ordinary footwear appears usually to have been black, that for the aristocracy was frequently brightly coloured—yellow, red, and green. Purple was reserved for imperial use. It is probable that the dyes were employed upon alumed leather (*aluta*) (p 150).

To what extent Roman colonists produced their own requirements in leather is uncertain, but a large percentage of finished goods was imported from metropolitan sources while hides were exported thither. Evidently some of the uses found in Britain and Gaul were strange to the invaders. Caesar notes, as though novel, the use by Britons of skins for clothing [15], their hide-covered boats, and the soft leather sails of the seafaring Veneti.

The age that succeeded the break-up of the Roman Empire is deficient in direct evidence of leathercraft. So far as we can judge, there was little change, either in preparation or use, from methods already established in earlier times. In Britain several examples have been found of a tool that forms an odd link between the highly developed Roman civilization and earlier prehistoric cultures. It is a bone scraper, improved by the addition of a lead core. One such instrument from Verulamium (St Albans) was made of the cannon-bone of an ox.

FIGURE 133—*The binding of the Stonyhurst Gospel, of reddish goatskin over carved lime boards. Seventh century* A.D.

As evidence that leathercraft continued unimpaired through these dark and troubled times we have the 'Lindisfarne Gospels' (*c* A.D. 700), the superb eighth century 'Book of Kells', the 'Benedictional' of St Æthelwold, and other such magnificent manuscripts, which illustrate the importance of vellum in the laborious and costly recording of knowledge. The 'Stonyhurst Gospel' (figure 133), recovered from the coffin of St Cuthbert, who died in 687, and the *Liber Winton* (twelfth century) (figure 147) exemplify the value of leather as a protective binding-material and as a medium for artistry. Leather satchels or budgets have been found at the ninth-century Breac Moedoc shrine and one or two other places. The Sutton Hoo ship burial (*c* A.D. 660) exhibits leather for shoes and bags and, following a much earlier tradition, for the covering, back and front, of great wooden shields (figure 134); while more than a century ago (1848) there were found at Buxton the remains of a Saxon drinking-cup of

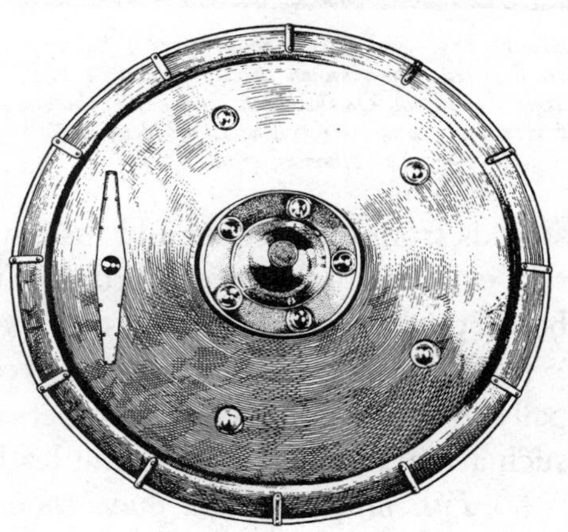

FIGURE 134—*A Viking shield of wood covered with hide (reconstruction). Diameter 35 in.*

moulded leather similar to a Neolithic bowl already mentioned (p 165), and having silver ornaments.

IV. METHODS OF WORKING

Early medieval references to leather indicate its importance in daily life. In Ælfric's 'Colloquy' (*c* 1000) the Saxon 'shoe-wright' says: 'I buy hides and skins and prepare them by my craft, and make of them boots of various kinds, ankle-leathers, shoes, leather breeches, bottles, bridle-thongs, flasks and budgets, leather neck-pieces, spur-leathers, halters, bags and pouches, and nobody would wish to go through the winter without my craft' [16]. After the Conquest and throughout the Middle Ages the pattern becomes progressively clearer, and it is possible to discern a wide variety not only in well developed techniques for working leather, but also in its application to common purposes. The chief techniques that appear to be firmly established at this period are:

FIGURE 135—*Shoemaker's shop: types of footwear are depicted. The man on the left 'closes' a shoe, using the stirrup. On the bench are a ball of thread, knives, awls, and bone polishers. Woodcut by Jost Amman, 1568.*

(*a*) *Sewing.* Thread of hemp or flax had replaced the earlier sinews or thongs of hide or skin (figure 135). Sewing-threads, stranded according to the thickness required, were twisted by rolling them on the thigh or knee—usually protected by a leather apron—and waxed. The wax, generally beeswax, both prevented the strands from unravelling and made them grip in the work, so that it would not slip even if stitches were broken. The fibrous nature of leather causes it to be very tenacious of stitches even when they are close to an edge.

(*b*) *Riveting* was used at least as early as the La Tène culture, at first principally to secure supporting members of wood or metal, but later to put together such articles as leather hose and leather buckets (figure 136).

(*c*) *Fastening to a rigid foundation.* Animal glues were used in ancient Egypt (vol I, p 695), and from very early times must have been employed for sticking leather to wooden boxes, coffers, and caskets: perhaps also to metal objects such as sword-hilts and scabbards. The early medieval heavy travelling-chests were covered with hides protected and kept in place by iron bands. From the sixteenth century these were replaced by brass nails, which resulted in a lighter construction. This technique was employed by the coffer-makers (figure 149), who also made furniture covered with leather and velvet (figure 231). Stretched leather nailed to frameworks was weight-saving, and was used for litters, coaches, and, later, sedan-chairs. It was also employed for mural panels, screens, and chairs.

FIGURE 136—*Moulded and riveted fire-bucket. Early nineteenth century.*

(*d*) *Moulding* was used in Britain probably in Neolithic and certainly in Saxon times, and both here and in central Europe it became of vital importance in the Middle Ages. The term *cuir bouilli* was known at least as early as the fourteenth century (Chaucer and Barbour). The basic principle rests on two properties of vegetable-tanned leather: (i) capacity to mould to a surprising degree, after thorough softening with water; in this state it can be coaxed, pressed, or beaten—in moulds or over cores of hardened clay, wood, or wet sand—into extraordinary shapes; (ii) tendency to remain permanently set on drying with moderate heat, the degree of which determines the rigidity. The process is still used for such things as scabbards, trunk-corners, cigar-cases, and a great variety of cups, packings, and washers for machinery. A quicker and harder setting is obtained if the moulded article be dipped, momentarily, into very hot water. This practice presumably gave rise to the term *cuir bouilli*. In some cases moulded objects were impregnated with wax, but leather drinking-vessels (black-jacks), bottles, and jugs were lined with resin or pitch (figures 138–9).

(*e*) *Laminating.* Another form of moulding for both raw hide and leather was to glue or paste successive thin layers over a core. With an adhesive, such as flour-paste, greater plasticity can be obtained than by wet-moulding, and laminating was later used for light articles such as sheaths, bottles with screw-tops, and spectacle-cases. Sometimes some of the laminations were of canvas or paper, but the leather always presented an unbroken surface, joints being often undetectable because the edge of the leather was shaved to wafer thickness and the actual join sometimes hidden by decoration.

(*f*) *Cutting and punching.* For cutting out strong, firm heavy leathers to a required shape such tools as knives, chisels, gouges, and punches were used. For bevelling edges, either the ancient half-moon knife (figure 130), or a flat rectangular one sharpened to a slant on one side, was used at an appropriate angle, bearing on a hard smooth base such as stone. All these tools were found in an Iron Age kit, with awls of various kinds for stitching (p 166).

(*g*) *Draping.* The softening of leather, particularly tawed sheepskin or goatskin, by the ancient method of 'staking' (figure 111), provided leather suitable for clothing, or soft enough for hangings, of a particular durable character.

FIGURE 137—*Casket covered in tooled leather. French, fifteenth century. Scale* c *1/5.*

(*h*) *Punching and slitting.* To render leather suitable for clothing by providing ventilation and conformation to the figure a multitude of small holes were made by slitting or punching. This could be done without fear of the material tearing because of the great strength of the fibrous structure. A similar procedure was employed in ancient Egypt and was still in use in the sixteenth century (plate 5 A and B).

(*i*) *Thongs and straps.* Leather provided not only the first large sheets of material but the first long strips, thongs, and straps. Egyptian paintings show hides being cut spirally to provide long, continuous strands for rope-making (vol I, figure 284). The use of leather strapping for harness was fundamental for the taming and use of the horse (vol I, pp 720 ff).

A characteristic of leather is that when cut it yields an edge that will not fray and can be attractively finished. In the best practice the sharp corners are removed with a special edge-tool; the edge is then coloured with dye or pigment mixed with glue-water, and afterwards rubbed up with wax. In many cases a neat crease-line, as a finishing touch near the edge, is added, with a heated iron of special design.

(*j*) *Dressing for particular purposes.* Leather articles were often valuable and had to be kept in proper condition. Wax-polishing was common. Water-budgets and coffers were oiled or greased to keep water in or out; harness was oiled to keep it supple; bottles, jugs, and black-jacks were re-pitched or resined to prevent liquor soaking into the leather, and alumed leather was re-softened by staking after it had got wet.

FIGURE 138—Cuir bouilli *chalice-box with incised decoration. From Swefling church, Suffolk. Thirteenth century.*

(*k*) *Covering pre-formed shapes.* Originally leather was used for covering such things as coffers and caskets because of the protection it provided, a practice still continued even though improved methods of joinery and finishing make leather no longer essential. It was a satisfactory alternative to the widespread practice of painting or covering with thin sheets of precious metals.

(*l*) *Decorating.* Colouring was employed from the earliest times. It was found that, as already mentioned, lines and patterns can be impressed into leather with hot metal tools which darken it at the point of contact. In the Middle Ages surface patterns of a most elaborate nature were made by modelling, punching, incising, carving, or by bruising over a relief in wood, or by a combination of these processes. The relatively plain surface of cattle-hide was sometimes patterned with parallel lines or dicing by using a fluted wooden pommel while the leather was still damp. The natural tendency of goatskin to crease into a granulated pattern was exploited by 'boarding' (p 156), that is, folding the leather hair-side inwards and rolling it backwards in one or more directions, according to the pattern required, with a curved, cork-covered board. The practice of impressing designs with beautifully cut metal stamps—blind-stamping—was probably first evolved for bookbinding (figure 147). Patterns produced by combining different stamps sometimes extended over the whole surface. Gold-tooling, probably evolved in Persia, reached England by way of Venice (plate 7 A), and was later applied to all kinds of leather objects, including boxes, caskets, hangings, and chair-leathers.

Another method of enriching the appearance of leather was to gild it all over.

FIGURE 139—*Bombard, black-jack, and bottle of moulded leather. Seventeenth century. Scale* c *1/5.*

Bookbindings, hangings, and panelling were thus treated. The method was described in the tenth century by Theophilus [17] (p 64), and was unchanged when, 600 years later, Pepys (1633–1703) delightedly watched the leather covering of his first coach made to glow like pure gold in the sunlight. The Royal coach and that of London's Lord Mayor still remind us of the practice. Real gold-leaf was sometimes used, but the more general procedure was to stick sheets of silver or tin foil to the surface with white-of-egg or shellac, and then apply successive coats of yellow varnish through which the metal glowed. When used for mural panels, screens, or table and bed 'carpets' the embossed design was usually painted, in part with opaque colours and in part with coloured varnish, with delightful effect. Embossed wall-papers were to a large extent a logical development of embossed and coloured leather (plate 6 A) and, in Britain, were originally made by the leather-gilders.

V. UTILIZATION FROM MEDIEVAL TIMES

Knowledge of dress before the modern period is largely derived from effigies and drawings. A few sixteenth-century leather jerkins still exist, some pierced or slit all over to provide stretch and ventilation; and a few flat collars, about 6 in wide, of black goatskin with pierced designs, have been found. From paintings of Pieter Brueghel the Elder (*c* 1520–69) we can form an impression of leather working-garments, described in a petition of 1595 as the chief wear of 'the poorest sort' [18]. Leather 'points' and laces were used as fastenings, for example, to join doublet and hose and component parts of plate armour. The leathers principally used for garments were goatskin, which was light, soft, and tough; sheepskin, which was light and soft but not so tough; chamoised leathers, including buckskin and doeskin; and buff, which was thick but light and tough, used largely for protective tunics and gauntlets (figures 140, 145). Buff leather was perhaps introduced early, but no record of its use exists before the thirteenth century, from which period examples occur in armour. Buff leather was reputed an adequate guard against sword and dagger, and its toughness would probably ward off a glancing blow. Long buff coats were still being worn at the end of the eighteenth century. Gay embroidered silk doublets for outdoor wear were lined with sheepskin.

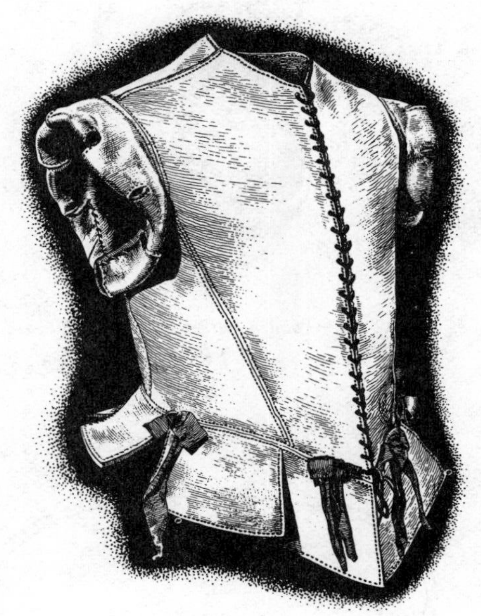

FIGURE 140—*Buff leather tunic with butt-jointed seams and silk points. Early seventeenth century.*

While boots and shoes have, at one time or another, been made of almost every imaginable material, leather has always been predominant. The ancestry of the boot is linked with a variety of methods of protecting ankles and legs—buskins, shin-guards, leggings, strappings—already found in Assyria and Babylon.

The uppers of most Roman sandals left portions of the feet exposed. They were therefore unsuited for north-western Europe, whence come examples in which a shoe-like upper is attached to a typical Roman sole—perhaps attempts by native shoemakers to provide something more suited to the climate. These had no permanent influence, for practically all medieval footwear, unlike the

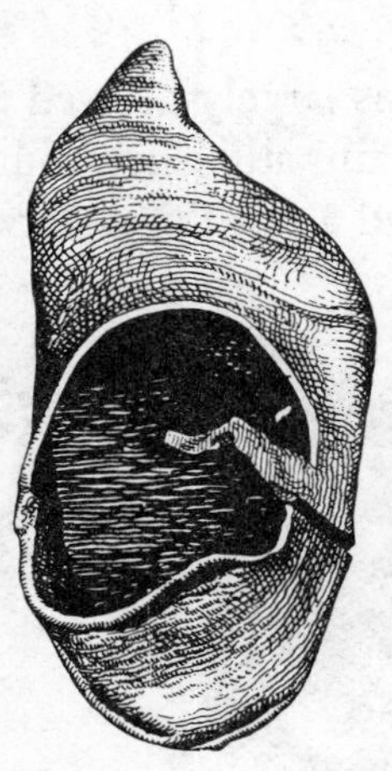

FIGURE 141—*Pointed leather shoe*, temp. *Edward III. From London.*

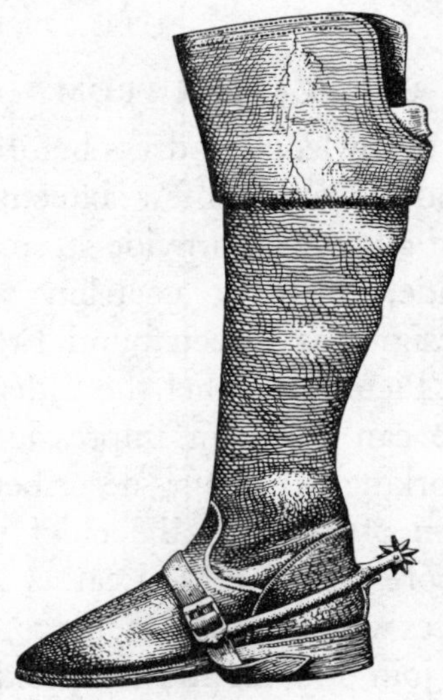

FIGURE 142—*English riding boot. Seventeenth century.*

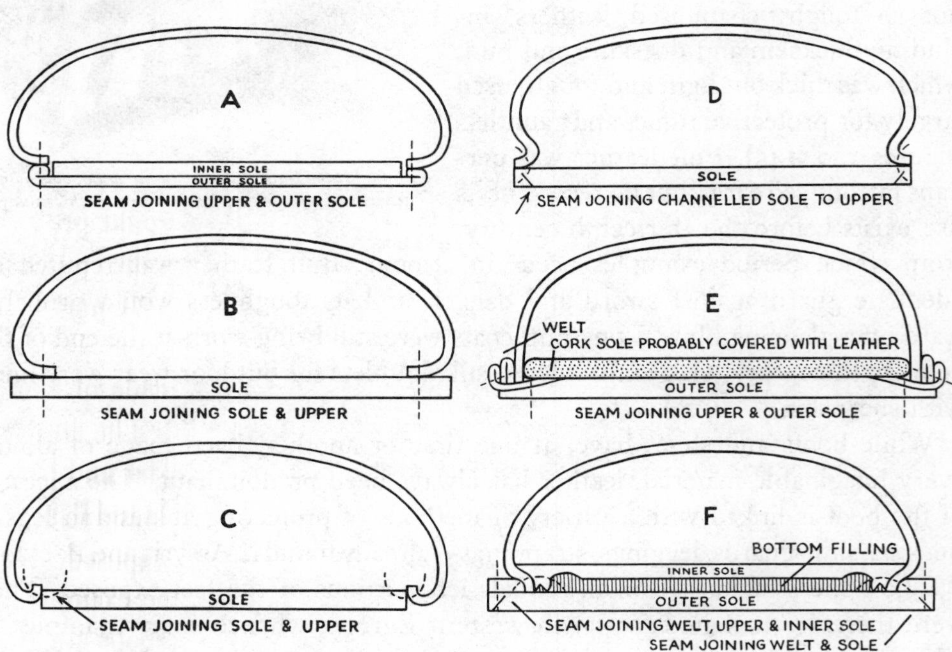

FIGURE 143—*Methods of joining upper and sole of a shoe.* (A–D) *Simple medieval methods;* (E) *a more elaborate one embodying a welt,* temp. *Edward III;* (F) *modern method of 'closing' a welted shoe.*

Roman, was soft and relatively insubstantial. Fashionable footwear was mostly of coloured cordwain (goatskin, p 150), and common wear of cattle-hide, usually black and about $\frac{1}{8}$-in thick. Yet even these seem to us little better than slippers, with uppers and soles of the same kind of leather seamed together (figure 141).

There appear to have been no significant changes in either materials or construction from Norman to Stuart times. Both boots and shoes were relatively light and simply made, despite numerous and fanciful changes in fashion. In the early seventeenth century the fashion arose of wearing long boots on all occasions. At first these were tight-fitting, of soft leather (cordwain and Russian calf), worn with short breeches. The heel probably appeared in the second half of the sixteenth century, and was the foundation of a more robust form of construction. As breeches descended to the knee, men's boots became shorter and fuller; they were made of Russian calf, cordwain, deerskin, and buff. The military campaigns of the seventeenth century caused development of a more robust form of footwear than was hitherto known (figure 142). From this period can be dated such common features of contemporary footwear as the welt, in the form known today; another form is found in some fifteenth-century shoes (figure 143).

FIGURE 144—*Gloves and mittens used in farming. From the Luttrell Psalter*, c *1338*.

Gloves, it is believed, were used by the early Germanic tribes. There are specific references to them in Celtic and Saxon writings, including the poem of 'Beowulf' (eighth century), although they do not figure in the list of leather goods given by Ælfric in the eleventh century (p 170). Liturgical gloves appear very early in the Christian era but were usually of linen or silk. The English psalters of Tickhill (*c* 1314) and Luttrell (*c* 1338) clearly indicate a widespread use of leather gloves and mittens by all classes (figure 144). The knight had his armoured gauntlets, the foundation of which was leather. Those of the Black Prince have hung over his tomb at Canterbury for over 600 years; their chamoised leather is still in an excellent state (figure 145). A hawking-glove of Henry VIII is of chamoised deerskin and is quaintly decorated. A number of gloves remain from the sixteenth and seventeenth centuries, usually of buckskin, kid, or cordwain, many with elaborately ornamented gauntlet cuffs. Noticeable features of the dress-gloves of these centuries are the shape of the thumb-pieces, which are cut straight at the bottom instead of in a deep curve as later, and the cutting of the fingers deep on the back but to normal length on the palm-side: this was presumably a device for creating the illusion of long tapering fingers.

FIGURE 145—*Left gauntlet of the Black Prince (d 1376). The fingers show decorative strips with zigzag edges that cover a round seam. Canterbury Cathedral.*

In Shakespeare's day the ancient half-moon knife was a regular tool of the glove-maker as well as of the saddler. Perhaps it was specially useful for cutting the heavy deerskin of which so many English gloves were made. The skill of the cutter, who worked in secret in medieval times, to a large extent determines all that follows. He must know precisely the degree and direction of stretch required, not only for a given type of glove but for every individual part of it. Three main types of seam are employed; the 'round-seam', which oversews the edges of two pieces of leather placed back to back, a kind of whipping that hides the cut edges; the 'prick-seam' for thick leather, which is joined edge-to-edge and sewn through and through in a manner that leaves the visible part of the stitch parallel to the edges; and the 'piqué-seam' used for overlapping seams. Considerable skill is necessary to ensure that tension is even, if the finished glove is to keep its shape.

Small Receptacles for Personal and Domestic Objects. From a remote period leather has been used for portable containers, to hold personal property. Even the savage had his leather or skin bag for tools, materials that served as currency, and so on, and Neolithic man had a leather sheath for his flint dagger. All the basic techniques described have been employed in the making of these receptacles, of which some examples are illustrated.

Stretched leather was used for seats of chairs and stools in Egypt (vol I, plate 25), but its use for upholstery is not definitely known to be older than the mid-seventeenth century A.D. Shaved cattle-hide, usually stained brown with a diced pattern, was then used. Contemporary covers for tables and terrestrial and celestial globes on stands were made from undyed sheepskin (basil), the design being produced by placing the leather, in panels, over a raised design carved in wood and rubbing the surface with a bone or wooden tool. Where the leather was crushed between tool and relief-design the surface was bruised, leaving the pattern in brown on the biscuit-coloured surface.

'Spanish' leather (p 150) was popularly used over all Europe in the sixteenth and seventeenth centuries for chair-seats and backs, hangings, screens, bed-covers, and many other purposes. Venetian leathers were usually painted flat, but

FIGURE 146—*Manufacture of gilt leather panels, as used for mural hangings.* (Left) *The surface of the prepared panels is being covered with gold foil, stuck on with white of egg.* (Right) *Shows the pattern embossed on the leather. From the* Encyclopédie méthodique, *1783.*

French, Flemish, Spanish, and Dutch leathers were embossed, using large wooden blocks cut in relief, after gilding but before painting or varnishing (figure 146, plate 6 A). Many leather-panelled seventeenth-century rooms survive; usually a number of panels of identical size were seamed together and the resultant sheet was fixed to the wall with leather-covered nails and a narrow leather border. Apart from the nailing all round the edges, the leather in effect was hung, and for a time supplanted tapestry because of its rich effect; it was easily cleaned and did not harbour insects. Leather hangings went out of fashion in France with the revival of tapestry-making under Colbert, and in England with the advent of wall-paper that simulated leather. Towards the end of the eighteenth century gilded and embossed leather was replaced, particularly for screens, by painted leather, much of it beautifully executed but leaving the character of the material obscured.

Decorative panels in high relief made by an adaptation of the *cuir bouilli* process (p 171) often display high technical skill (plate 6 B). In some cases they show a combination of repoussé and surface modelling, incising, and punching. A development of the embossing process, emanating perhaps from Spain, was the modelling in the round of leaves, flowers, and other motifs, which were then built up into an elaborate panel or frame. All manner of caskets and boxes were made for household use from the early Middle Ages onwards, in some cases of *cuir bouilli*, in others of leather-covered wood. They were often elaborately decorated by modelling, punching, or incising, and, from the sixteenth century, by gold-tooling, which followed the bookbinding technique developed in Venice. Bottles, drinking-vessels (black-jacks), and jugs (bombards) of leather were in common use by rich and poor well into the eighteenth century (figure 139).

Bookbinding. The manuscript roll was replaced by the codex early in the Christian era (p 188). It was found convenient to bind the book between boards, and to cover the unprotected back with leather which overlapped the boards and eventually covered them. Bindings were elaborately decorated, perhaps the most effective form being blind-stamping; this retains the character of the leather (figure 147). Originally, designs were built up with a variety of small tools. These offered great scope for artistry, but eventually designs were cut complete in metal plates, which made repetition possible and led, in time, to the dull mass-produced, plate-embossed book-covers of the nineteenth century. The leathers chiefly used were alumed deerskin and pigskin—the most lasting—or vegetable-tanned hide, calf, and goatskin. Most, if not all, early bindings were monastic, but commercial binderies were established at the seats of learning at least as early as the fifteenth century, though their work was not always of a high standard. Gold-tooling, which can be very beautiful when used with restraint and artistry, reached western Europe during the sixteenth century (plate 7 A and B). The gold is applied as gold-leaf impressed into the blind tooling by a hot iron. The superfluous gold leaf is wiped away with a cloth or sponge, from which the gold is ultimately recovered; the amount actually used is surprisingly small. A somewhat similar method is used for gilding the edges of books. These gilt edges are afterwards polished by rubbing with a smooth tool, still often of flint—perhaps the only surviving use of a flint tool in modern industry. Vellum, parchment, and an inferior quality of the latter called forel, were also used for bookbinding, vellum being sometimes used limp, without boards. There have been no fundamental changes in the principles of leather bookbinding from medieval times.

Musical Instruments. Music has uses for leather. Whether syrinx or bagpipe

FIGURE 147—*Blind-stamped binding of the* Liber Winton. *1148.*

came first is arguable; but it is not too fanciful perhaps to regard the 'windy organs' as a combination of the two. In quite early times organs employed alumed sheepskin for bushings and seatings for pipes, and for the gussets of bellows. The buff-stop of the harpsichord applied buff leather to the strings to produce a muffled tone. The shrill-toned cornet and the basset-horn were of wood covered with black shaved hide, often tooled, as was also the 'serpent', a popular feature of the amateur orchestras that in Britain led the singing in churches after the Puritans had silenced the organs. Each of these instruments of leather-covered wood had a distinctive tone that derived from the method of construction.

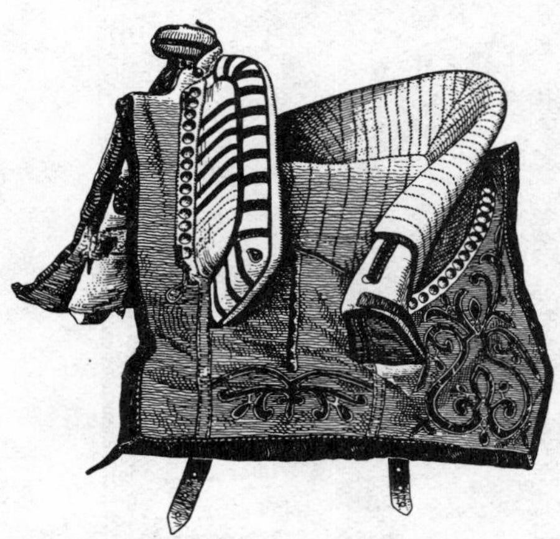

FIGURE 148—*Medieval riding-saddle of doeskin ornamented with appliqué leather and brass nails.*

Transport. The leather-using trades of the saddler, harness-maker, and coach-builder were essential to the community until quite recently, and still play an important role. Medieval saddles, remarkable for the support afforded to the rider by their high pommel and cantle (figure 148), were often of leather. The riding-saddle, as evolved from the end of the eighteenth century to the middle of the nineteenth, is often the high-water mark of elegance and craftsmanship, employing, as it does, all the basic techniques of leathercraft and its own special craftsmen.

The litter in its simplest form, as seen in ancient Egypt, was a couch attached to two poles carried by men at each end (vol I, figure 501). The provision of a canopy supported by four poles, and its gradual enclosure, at first by curtains, culminated in nailing sheets of shaved hide to the wooden framework. In this form, carried by two horses, it became common all over Europe by the seventeenth century. The earliest coach-body resembled the primitive litter. Corner posts supported a canopy, an apron was hung across the opening on each side, and cloth or leather was nailed to a light framework surrounding each end. The body was slung on leather braces, glass windows and hinged doors were fitted, and the vehicle was totally enclosed with leather.

The sedan-chair is said to have been introduced into England from Italy in 1634 [19]. Carried like a litter, it consisted of a single seat enclosed with leather, like a coach-body. It was often elaborately decorated with painting and brass ornaments. The earliest 'carriage', the so-called four-wheeled chariot, was in effect a sedan-chair on wheels. It appeared about 1744. Carriage-building then developed rapidly, and leather was used only for folding heads in such vehicles as the landau and barouche. For harness, leather was always essential, and in the eighteenth century its manufacture became a prosperous and highly skilled industry.

FIGURE 149—*Hide-covered coffer or travelling trunk, with brass-headed nails. 1672. Scale 1/12.*

Leather has, from the earliest times, been the chief protector of the traveller's belongings. In addition to the budget (plate 3 C) or bag, the luggage of the Middle Ages included the male, probably a kind of limp bale or holdall, the clothes-sack, and the gardeviance or food-carrier—all of leather kept supple with oil or grease, and of medium size to be carried on horse-back. The typical medieval travelling chest was the standard, often of vast size and weight, of wood covered with leather, protected with wide iron bands which in the lighter seventeenth-century coffer were replaced by brass nails helping to keep the diced hide in place (figure 149). Coffrets and forcers, used as strong boxes, were miniature coffers, often elaborately ornamented (figure 137). Other coffers were made of sheets of heavy hide, with designs modelled in relief, which were placed over pre-formed receptacles of wattle or wood and sewn together at the edges. For low-grade articles, deerskin or cowhide, rudely cured with the hair left on, was commonly used from the seventeenth century onwards.

FIGURE 150—*Leather lanthorn, and container with horn window. English, eighteenth century.*

In addition to various kinds of luggage, an infinite variety of leather bags and cases has been made from the earliest times. Those for personal use included ink-wells, pen-cases, bags and cases for books, hat-cases, and cases for basin and ewer, as well as flasks and bottles. Even the essential lanthorn, with its horn window (figure 150), was made of leather at least as early as 1517.

The use of water-skins was common all over the ancient world, but only in England is reference made to the water-budget (figure 151). No example of this contrivance survives, but it seems that it consisted of two bottle-shaped limp bags carried by their necks on a T-shaped pole. The Luttrell Psalter (*c* 1338) shows a large form of water-budget carried by a horse, as described by Lydgate (tailpiece). Water-budgets were maintained supple and waterproof by regular applications of grease.

FIGURE 151—*Norman stone-carving of Aquarius with water-budgets. From Hook Norton church font, Oxfordshire.*

Leather buckets, kept in large numbers in great houses, provided the chief defence against the menace of fire (figure 136). The hand-operated pump of the seventeenth century, fed by a suction-pipe, was improved when fitted with a leather hose to direct the stream of water; but its range was so small that the blaze often brought about its destruction. Leather hose-pipes with seams closed by copper rivets continued in use until a satisfactory textile hose-piping was produced. Firemen's helmets, introduced in the seventeenth century, were first made of leather.

REFERENCES

[1] HOMER Iliad, XVII, 389–95. (Loeb ed. Vol. 2, pp. 258 ff., 1925.)
[2] CAESAR *De bello gallico*, III, 13. (Loeb ed., p. 155, 1917.)
[3] PLINY *Nat. hist.*, XVI, ix, 26–27. (Loeb ed. Vol. 4, p. 404, 1945.)
[4] *Idem Ibid.*, XXIV, xi, 91.
[5] THUREAU-DANGIN, F. *Rev. d'Assyriol.*, **17**, 29, 1920.
[6] LANSDOWNE MS. 74, fol. 167.
[7] *Ibid.*, fol. 116.
[8] MARCO POLO. 'The Book of Marco Polo the Venetian Concerning the Kingdoms and Marvels of the East', trans. and ed. by SIR HENRY YULE, Vol. I, pp. 394 f. Murray, London. 1903.
[9] LANSDOWNE MS. 74, fol. 42.
[10] BREASTED, J. H. 'Ancient Records of Egypt, Historical Documents', Vol. 3, p. 251, § 589. University of Chicago Press, Chicago. 1906.
[11] *Idem. Ibid.*, Vol. 2, p. 164, § 392.
[12] STEIN, M. A. 'Ancient Khotan', Vol. 1, pp. 338, 340. Clarendon Press, Oxford. 1907.
[13] HOMER Iliad, X, 261–5. (Loeb ed. Vol. 1, p. 454, 1928.)
[14] *Idem Ibid.*, XX, 273–6. (Loeb ed. Vol. 1, p. 390, 1925.)
[15] CAESAR *De bello gallico*, V, 14. (Loeb ed., p. 253, 1917.)
[16] AELFRIC. 'Colloquy', ed. by G. N. GARMONSWAY, pp. 34 f. Methuen, London. 1939.
[17] THEOPHILUS PRESBYTER *Schedula diversarum artium*, I, 22 ff. (Ed. and trans. by A. ILG. Quell. Kunstgesch. Kunsttechn., Vol. 7, pp. 48 ff. Braumüller, Vienna. 1874.
[18] LANSDOWNE MS. 74, fol. 42.
[19] EVELYN, JOHN. 'Diary', ed. by W. BRAY; new ed. by H. B. WHEATLEY, Vol. I, p. 192 (8 February, 1645). Bickers, London. 1906.

BIBLIOGRAPHY

BAKER, O. 'Black Jacks and Leather Bottells.' Privately printed, Cheltenham. 1921.
BLACK, W. H. 'History and Antiquities of the Worshipful Company of Leathersellers of the City of London.' Privately printed, London. 1871.
BLÜMNER, H. 'Technologie und Terminologie der Gewerbe und Künste bei Griechen und Römern', Vol. 1 (2nd ed. rev.). Teubner, Leipzig. 1912.
BRAVO, G. A. "La lavorazione delle pelli e del cuoio dell'Egitto antico." *Boll. Staz. sper. Pelli Mat. conc.*, **11**, 75–94, 1933.
BRITISH MUSEUM. 'The Sutton Hoo Ship-Burial' (by R. L. S. BRUCE-MITFORD). Trustees of the British Museum, London. 1947.
BRUNTON, G. and CATON-THOMPSON, GERTRUDE. 'The Badarian Civilisation.' Egypt. Res. Acc. and Brit. Sch. Archaeol. Egypt, Publ. 46. London. 1928.
BRUNTON, G. and MORANT, G. M. 'Mostagedda and the Tasian Culture.' Brit. Mus. Exped. to Middle Egypt 1928–9. Quaritch, London. 1937.

BUDGE, SIR ERNEST (ALFRED WALLIS). 'The Dwellers of the Nile' (rewritten and enl.). Religious Tract Society, London. 1926.

CAPART, J. 'Primitive Art in Egypt', trans. by A. S. GRIFFITH. Grevel, London. 1905.

CARTER, H. MS. notes and sketches in possession of the Griffith Institute, Ashmolean Museum, Oxford.

CLARK, J. G. D. 'Prehistoric England.' Batsford, London. 1940.

COACHMAKERS AND COACH HARNESS-MAKERS, COMPANY OF. 'History of the Worshipful Company of Coachmakers and Coach Harness-Makers of London.' Chapel River Press, London. 1937.

DAVIES, NINA DE G. 'Paintings from the Tomb of Rekhmirē at Thebes.' Metropolitan Museum of Art, Egypt. Exped. Publ., Vol. 10. New York. 1935.

DAVIES, NORMAN DE G. 'The Rock Tombs of El Amarna VI.' Archaeol. Survey of Egypt, Memoir 18. Egypt Exploration Fund, London. 1908.

Idem. 'The Rock Tombs of Deir el Gebrâwi I.' *Ibid.*, Mem. 11, London. 1902.

Idem. 'The Tomb of Rekh-mi-rē at Thebes.' Metropolitan Museum of Art, Egypt. Exped. Publ., Vol. 11, New York. 1943.

FORRER, R. 'Archäologisches zur Geschichte des Schuhes aller Zeiten.' Bally Schuhmuseum, Schönenwerd. 1942.

GANSSER, A. "The Early History of Tanning." *Ciba Rev.*, no. 81, 2938–62, 1950.

GRASSMANN, W. (Ed.) 'Handbuch der Gerbereichemie und Lederfabrikation', Vol. I, i. Springer Verlag, Vienna. 1944.

HAWKES, JACQUETTA. 'Early Britain.' Collins, London. 1945.

HORNELL, J. 'British Coracles and Irish Curraghs.' Quaritch, London. 1938.

JÄFVERT, E. 'Skomod och skotillverkning fran medeltiden till våra dagar.' Nord. Mus. Handl., no. 10. Stockholm. 1938.

LAMBERT, J. J. 'Records of the Skinners of London (Edward I to James I).' Sir Joseph Clauston, London. 1933.

LUCAS, A. 'Ancient Egyptian Materials and Industries' (3rd ed. rev.). Arnold, London. 1948.

Idem. MS. notes in the possession of the Griffith Institute, Ashmolean Museum, Oxford.

MANDER, C. H. W. 'A Descriptive and Historical Account of the Guild of Cordwainers of the City of London.' The Company, London. 1931.

OAKLEY, K. P. 'Man the Tool-Maker' (3rd ed.). Trustees of the British Museum (Natural History), London. 1950.

PARTINGTON, J. R. 'Origins and Development of Applied Chemistry.' Longmans, Green and Co., London, 1935.

PETRIE, SIR (WILLIAM MATTHEW) FLINDERS. 'Diospolis Parva 1898–9.' Egypt. Explor. Fund, London. 1901.

Idem. 'Prehistoric Egypt.' Egypt. Res. Acc. and Brit. Sch. Archaeol. Egypt, Publ. 31. London. 1920.

ROSELLINI, N. F. I. B. 'I Monumenti dell'Egitto e della Nubia', Part II: 'Monumenti Civili', Vol. 2, pp. 355–64. Capurro, Pisa. 1834.

SALZMAN, L. F. 'English Industries of the Middle Ages.' Clarendon Press, Oxford. 1923.

SCHUCHHARDT, C. 'Alteuropa in seiner Kultur und Stilentwicklung.' Trübner, Strasbourg and Berlin. 1919.

SHERWELL, J. W. 'History of the Guild of Saddlers of the City of London' (rev. by K. S. LAURIE and A. F. G. EVERITT). Privately printed, London. 1937.

SHETELIG, H. and FALK, H. 'Scandinavian Archaeology', trans. by E. V. GORDON. Clarendon Press, Oxford. 1937.

STOKAR, W. VON. "Vorgeschichtliche Lederfunde und Lederverwendung." *Collegium, Haltingen*, no. 796, 433–37, 1936.

THOMPSON, R. CAMPBELL. "The Cuneiform Tablet from House D" in WOOLLEY, SIR (CHARLES) LEONARD *et al.* 'Carchemish II', pp. 135–42. Trustees of the British Museum, London. 1921.

THUREAU-DANGIN, F. "Notes Assyriologiques: XXIX. L'Alun et la noix de galle." *Rev. d'Assyriol.*, **17**, 27–30, 1920.

UNWIN, G. 'The Guilds and Companies of London.' Methuen, London. 1908.

VAN SETERS, W. H. "Shagreen on Old Microscopes." *J.R. micr. Soc.*, **71**, 433–39, 1951.

WAINWRIGHT, G. A. 'Balabish.' Egypt Explor. Soc., Mem. 37. Allen and Unwin, London. 1920.

WATERER, J. W. 'Leather: in Life, Art and Industry.' Faber and Faber, London. 1946.

Idem. 'Leather and Craftsmanship.' Faber and Faber, London. 1950.

WILCOX, RUTH T. 'The Mode in Footwear.' Scribner, New York. 1948.

Water-seller with horse carrying water-budgets. Luttrell Psalter, c *1338.*

A NOTE ON PARCHMENT

H. SAXL

As surfaces on which to write, different peoples at different times have used a great variety of substances. Among them have been palm-leaves, bark, wood, papyrus, clay, wax, lead, linen, and leather. In Europe parchment has been in growing favour for permanent record since early Roman Imperial times. By parchment is meant a specially prepared sheet-like material made from the dermis of various animals, notably calves, goats, lambs, and sheep (plate 4, A and B). Parchment differs from leather in that it is not tanned, is prepared by a special drying-process on a stretching-frame (figure 153, centre), and is usually thinned by shaving. It is also smoothed in some way and deprived of all or most of its grease. However, the high natural content of grease of sheep-parchments may cause a partial oil-tannage, producing a leather-like material suitable for binding and similar purposes. The finest parchment is made from calf-skins and is hence called vellum.

In Greek and Roman times sheets of parchment for use in writing were cut into rectangles; these were sewn together in long rolls, similar to those of papyrus used by Egyptian scribes and of supple leather by the Hebrews and other ancient peoples. The rectangles were inscribed in vertical columns and were added to as needed, the long strip being kept rolled up. For reading, a roll was unfolded by the right hand, the free end being held in the left, which rolled it again when the reading was completed (tailpiece). One or two columns of writing at a time were exposed for study. For languages written from right to left, such as Hebrew, the reverse process was of course followed. The rolls were kept in cylindrical cases, often represented in classical antiquity. For certain purposes, notably for legal documents, this method was maintained until the present century in many countries, including Britain, where the Master of the Rolls is still a very important legal officer.

FIGURE 152—*St Luke copying his gospel into a codex from a scroll written in Greek. On his desk are a double inkstand for red and black pigments, pens, pen-knives, a compass, and a semilunar knife. From a twelfth-century Byzantine manuscript.*

In the second century A.D. a new form of using parchment for literary purposes was invented. This device was called a *codex*, and was what we now describe as a book. The new scheme involved cutting parchment into rectangular sheets, each of which was then folded once into a *folio*, or twice into a *quarto*, or further into an *octavo*, and so on. These foldings were then collected and bound into a *volumen* or volume. The form of this word recalls the historic relation of the book to the scroll or *volvulus* (figure 152).

Parchment of a rude kind was used for making drum-heads in Asia Minor from very ancient times, perhaps as early as 1000 B.C. The word parchment, Latin *carta pergamena*, is said to derive from Pergamum, a town near Izmir (Smyrna) in Asia Minor. Pergamum was an important centre of Hellenistic culture, famous for its architecture and art and for the learning and industry of its citizens. There, from the third pre-Christian century onward, skins of animals were specially prepared as writing-material. Papyrus had been used in Egypt for millennia for literary purposes (vol I, pp 754, 757), and under the Ptolemaic sovereigns, from about 300 B.C. onwards, it was an article of export trade. Parchment, though more costly and difficult to prepare, had the advantage of being more durable, especially in the relatively damp Mediterranean climate.

As the preparation of parchment was perfected it gradually but slowly became the normal writing-material, its use spreading throughout the Roman Empire. It arrived in the north-west of Europe along with Christianity, being used first for liturgical and later for legal purposes. With increasing literacy it found other applications, such as in medical and commercial documents. From the twelfth century onward it was very slowly displaced by paper, which came to the west primarily from Islam by way of Spain, and ultimately from China (p 771). The adequate preparation and supply of paper were conditioning factors in the development in Europe of the art of printing in the mid-fifteenth century. Paper will be considered in a later volume.

The mode of preparation of parchment for writing has altered little in the last two thousand years. The modern invention of a machine that splits a thick pelt into two or

FIGURE 153—(Centre) *Stretching the parchment on a frame for drying. The workman shaves it with the half-moon knife which is of a special type, the blade being at right-angles to the handle (cf figure 130). From a twelfth-century German manuscript.* (Right) *Smoothing the finished sheet with pumice.* (Left) *Cutting the sheet with the help of ruler and set-square. Both from a thirteenth-century German manuscript.*

three thinner sheets has greatly reduced both the work of craftsmen, who used special knives for the purpose, and the wastage of valuable material. We have ancient accounts of the introduction, preparation, and use of parchment, notably from Pliny. The following may be taken to represent the normal series of operations adopted in the Middle Ages for preparing the fine article. They closely resemble the processes described by Theophilus.

The skins used were preferably those of calves and goats. The treatment here described is specially applicable to the skin of a week-old calf. Having been well washed, it was left undisturbed for 24 hours in clean water and then limed in a liquor containing about 30 per cent of freshly slaked lime, which would be of a paste-like consistency. It was left from eight to sixteen days in the lime-bath, depending on the temperature; the period would be about eight days at 18–20° C. The pelt was then freed from hair, which was pushed off by a blunt knife as with leather (figure 114). The pelt was next limed again for the same time in a fresh lime-mixture, washed, and then stretched on a frame.

The quality of the parchment depended on a careful control of the drying-process on the frame. It was dried at about 20° C, washed by pouring cold water over it, partly dried, and washed again, when a smooth glue-like surface was obtained. At the same time it was scraped and shaved thin with the traditional semilunar or circular knife. This was a highly skilled, delicate, and very characteristic operation; it is still practised. Whereas leather is composed of interwoven collagen fibres (p 148 and plate 4 B), in parchment the collagen is arranged in lamellae more or less parallel to the surface. This structure accounts for the function of the special semi-lunar knife used for scraping layers off the surface (figure 153, centre). After scraping, the skin was rubbed smooth or pounced[1] with powdered pumice or some similar material, allowed to dry in a stretched state, and finally cut into pieces of suitable size and shape (figure 153).

[1] From the French *poncer*, Latin *pumicare*.

BIBLIOGRAPHY

The subject is treated in full detail with an extensive bibliography in a M.Sc. Thesis, 1954, by Miss SAXL, prepared under the direction of Professor D. Burton and Dr. Reid at the Leather Industries Department, University of Leeds.

BIRT, T. 'Die Buchrolle in der Kunst.' Teubner, Leipzig. 1907.

DIDEROT, D. and D'ALEMBERT, J. LE R. (Eds). 'Encyclopédie ou Dictionnaire Raisonné des Sciences, des Arts et des Métiers', Vol. 11, pp. 929–31: "Parchemin." Neuchâtel. 1765.

KENYON, F. G. 'Books and Readers in Ancient Greece and Rome' (2nd ed.). Clarendon Press, Oxford. 1951.

MERRIFIELD, MARY PHILADELPHIA (Ed.). 'Original Treatises' (2 vols), with translations and notes. Murray, London. 1849.

SAXL, H. "Parchment." *Ciba Rev.*, 1957. (Forthcoming.)

THOMPSON, D. V. "Medieval Parchment Making." *Library*, 4th series, **16,** 113–17, 1935.

THOMPSON, SIR EDWARD (MAUNDE). 'An Introduction to Greek and Latin Palaeography.' Clarendon Press, Oxford. 1912.

A reader unrolling a scroll with one hand while he rolls it up with the other, exposing a space containing a column of writing. From a Roman tomb. c *A.D. 300.*

6

SPINNING AND WEAVING

R. PATTERSON

I. FIBRES AND THEIR PREPARATION

DURING classical antiquity the textile crafts developed from domestic occupations into an extensive and well organized industry. At the beginning of the period Egypt was already highly skilled in making fabrics of linen, India of cotton, and China of silk. The equipment was simple but the products were of very good quality. The Greeks, and particularly the Romans, carried the weaving of woollen cloths to remarkably high standards.

In the rest of Europe textiles of poor quality were made by primitive methods until the Romans introduced improved techniques. The fall of the western Empire brought a decline from this high standard, but Byzantium continued to foster the textile arts and, partly through contact with the Far East, became a most important trading and cultural centre. Persia owed her textile skills to her advantageous position between China to the east, India to the east and south, and Byzantium to the west. From the seventh Christian century onwards, the all-conquering Muslims spread the techniques of the cultivation of silk, linen, and cotton from Afghanistan in the east to Spain in the west (ch 22, figure 684).

The textile crafts were already slowly reviving in western Europe when in the eleventh century the Norman conquests and colonizing, from Britain to Constantinople, began to foster trade on an international basis. Succeeding centuries saw a great expansion of textile-manufacture in Europe. As specialization developed, state regulations followed on the formation of textile craft-guilds. Meanwhile new trade routes converged upon the great cloth-fairs, which were a most important feature in medieval life. The wealth of nations, and not least of England, became closely associated with control of raw materials of textiles, particularly wool, and of major centres of textile-manufacture. Thus specialized trades such as those of the weaver, fuller, dyer, and shearer developed from what had been a domestic occupation. The beginnings even of a factory system appeared in Italy quite early in the Middle Ages. While in Greek times the various processes were practised by women almost exclusively, they were now divided between the sexes, spinning alone remaining a feminine monopoly.

The fibres used in textiles fall into four broad categories: (*a*) animal coats, especially sheep's wool; (*b*) vegetable bast-fibres, the commonest being flax; (*c*) silk; and (*d*) vegetable seed-hairs such as cotton (vol I, pp 447–51, and plate 17 B).

(*a*) *Wool*

Wool was widely used by the Greeks and Romans, who appreciated the fineness, density, length, and colour of its fibres. They improved fleeces by selective breeding and imported high qualities from Miletus, Attica, Megaris, and Tarentum. Pliny (A.D. 23–79) classes wools as *molle* or *generosum* (soft wool), *hirsutum* (long coarse wool), and *colonicum* (rustic wool) [1]. Coarse wool was imported into Rome from Gaul. Another variety was termed *pelitum*, signifying that, during growth, the fleece was covered with skins to render it finer. The most valuable wool was white, then came the brown from Apulia (Canusian), and next a reddish variety from Asia Minor. Others were grey-brown and black. The hair of goats was used for coarse cloths, cloaks, rugs, and felt slippers, while that of beavers, camels, and rabbits had more limited application.

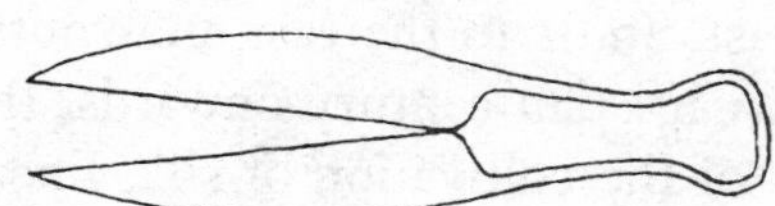

FIGURE 154—*Shearing sheep; from the Flemish Grimani Breviary*, c *1500;* (below) *a pair of Roman sheep-shears, identical in form with the medieval. From Italy, third century B.C.*

In the Middle Ages English wool was prized throughout Europe. It was grown largely on monastic estates, the Cistercians excelling among sheep-grazing pioneers. By the thirteenth century this raw material was exported direct to the great manufacturing cities of Flanders and Italy. Wool was as yet not classified by breeds of sheep but by district of origin. In 1454 some fifty-one grades of English wool are enumerated, the most expensive coming from Shropshire, Herefordshire, and the Cotswolds; medium from Lincolnshire, Hampshire, Kent, Essex, Surrey, and Sussex; and the cheapest from Cumberland, Westmorland, and Durham [2]. Towards the end of the fifteenth century continental wools improved. This was especially so in Spain, whence fine wools were exported to Flanders, then the great manufacturing centre. In the past, the improved quality of continental wool had been ascribed to the introduction of English rams, and their export from England was prohibited in 1338 and again in 1425.

Before raw wool can be spun into a thread it must be sorted into qualities, cleaned to remove dirt, burrs, and so forth, freed from natural salts and grease, and made into a homogeneous mass without lumps or tangles. According to Pliny wool was originally plucked from the sheep, and the Romans were probably the first to use shears. These were remarkably modern in design, with two triangular blades on a U-shaped spring. The fleece was taken off in one piece (figure 154). Little is known of Greek and Roman preparatory techniques, though Homer mentions carding (below) as an ordinary household occupation, and Pliny states that a comb was used for working silk and flax [3]. Combs were no doubt used to work the wool into an even fluffy state suitable for preparing the 'roving', a twisting preparatory to spinning.

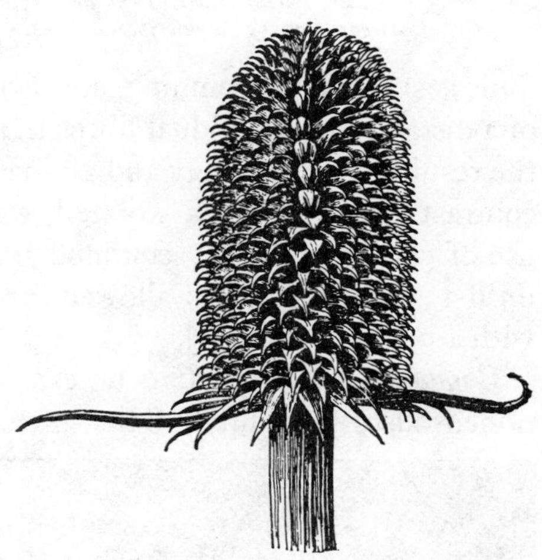

FIGURE 155—*Fuller's teazel*, Dipsacus fullonum. *Scale 1/2.*

In the Middle Ages there was careful preliminary treatment of wool. First, damaged portions were removed and the wool was sorted into three grades: fine, medium, and coarse [4]. It was then washed in lye to remove grease, dried in the sun, and spread on boards where soil and other particles were removed with forceps and small shears. In Flanders the wools were classed according to country, province, and even town of origin. The wool of the live sheep was kept separate from carcass wool, and that of young sheep from old, and it was an offence to mix the kinds [5]. The comparative inferiority of English cloth of this period, though England yielded the best wools, was partially due to careless sorting. When washed and dried, the wool was beaten with sticks to remove foreign matter and loosen the fibres, then oiled ready for carding or combing.

Carding is the loosening of tangled wool until it is uniform and the fibres are free. The word is derived from the Latin *carduus*, thistle, thistle-heads having been used for the purpose; more generally, heads of the teazel (*Dipsacus fullonum*) were employed (figure 155). Metal-covered 'cards' are recorded in France in the thirteenth century [6], and first illustrated in the early fourteenth (figure 167). They consist of a pair of wooden boards, about 9×6 in, each with a wooden handle at one edge. One surface of each board is covered with leather,

FIGURE 156—*Hand-combing wool, from an early fourteenth-century manuscript.*

through which protrude rows of inclined wire teeth. This type is still in use by hand-weavers.

A small quantity of wool is placed on one card and worked by pulling it apart gently with the teeth of the other. After a time the wool becomes evenly distributed among the teeth of the two cards; one card is then moved in the reverse direction and so collects the whole of the wool, which is released as a spongy roll. This is suitable for spinning, and is frequently seen attached loosely to the head of a distaff. The individual fibres in the roll are crossed in all directions, making the resultant yarn spongy and giving the final woollen cloth a soft texture, which contrasts with that of a worsted, where the fibres are parallel. In Flanders the use of carded wool was confined to the manufacture of coarse cloths and hats until 1377, when it was allowed for the weft of the finer cloths, in combination with a combed warp [6].

Combing, an alternative to carding, is as ancient as Homer and is mentioned also by Pliny. The ancient combs perhaps resembled the older Egyptian flax-combs. In the west wool-combing was probably practised in France from the ninth century, but there is no direct evidence of medieval European combing until the twelfth century, after which it becomes well known (figure 156).

FIGURE 157—*The bow as used to prepare wool for spinning. The workman holds in his right hand a tool for plucking the string. From a wall-painting in the Ducal Palace at Venice, showing the emblems of the wool merchants. 1517.*

A fifteenth-century Italian text [7] states that the wool was sprayed with water and soaked with oil before combing. The combs were heated to ease the passage of the teeth, and combing was continued until all the wool had been transferred to the free comb. The process was repeated, and finally the combed wool was drawn off in a mass containing only long, parallel fibres, leaving the short fibres still attached to the teeth. The thread spun from these long fibres had a firm, even structure and the worsted spun from it a hard clear surface. The short fibres were used for other purposes.

Bowing was a third medieval method of preparing wool for spinning. The bow had a long wooden frame on which was stretched a string of cord or gut. Its use is shown in the sixteenth-century emblem of the wool merchants of Venice (figure 157). The workman operates on wool laid out on a bench. He plucks the string of the bow with a piece of wood or bone and works the vibrating string amongst the wool. The rapid vibrations separate the tangles and give a similar result to carding. A bow of much the same kind is still used for this purpose in India and in parts of Europe. In 1409 the wool-workers of Constance complained that the bow, which should be reserved for wool, was being used for cotton, but the magistrates tolerated this new application [8].

(*b*) *Flax and Hemp*

Linen is woven from the fibres of flax. There are several wild species, but only the cultivated annual one, *Linum usitatissimum*, is of economic importance. Both Greeks and Romans imported their finest flax from Egypt, where its wide cultivation attained great perfection (vol I, index, *sub voce*). The cultivation of the plant was neglected by the Greeks, except for its seeds (linseed) which yielded an oil for cooking, and the classical writers on agriculture have little to say of flax except as a valued import. Virgil (70–19 B.C.) [9] and Pliny considered that it impoverished the soil. Pliny says that its cultivation had developed rapidly on the damp, fertile soils north of Italy [10]. However, Spain had early become famous for its flax, and Spanish linen tunics are reported to have been worn at the battle of Cannae (216 B.C.). According to Pliny the crop was common in Gaul and especially in the Low Countries. Along with Gallic and Germanic flax there came to Rome such new fashions in clothing as the linen shirt and hooded coat. The Low Latin word *camisia* (chemise) may be of Gaulish origin.

In the Middle Ages flax was cultivated throughout Europe, though the Egyptian fibre retained its supremacy until about 1300. It remained the most important vegetable fibre amongst western nations until the eighteenth century.

The use of flax was widespread in the Middle Ages. From illustrations of tools, and from the methods still persisting in remote regions of Europe, the technique of its treatment can be reconstructed. In Egypt the plant was pulled up rather than cut, and thus weed-fibres are occasionally found in Egyptian linens. The seed-capsules had to be removed, either by hand or with a comb. The separation of the fibres in the stalk from woody tissues was accomplished by the fermentative process known as retting. Pliny describes how, in retting, the stalks were immersed in water warmed by the sun, and held down by weights until the bark became loose [11]. They were then dried in the sun with frequent turning. This process, with modifications for retting by river, dew, or even snow, has continued to the present day.

The fibres were mechanically extracted from the retted and dried stalks, the first stage being to break up the woody tissues. In Egypt this was done by beating with mallets on wooden or stone blocks. The wooden blocks were sometimes

FIGURE 158—*Preparation of flax-stalks:* (left) *by beating with a wooden blade;* (centre) *in the flax-breaker. From a sixteenth-century woodcut showing the labours of the seasons.*

fitted with diagonal strips on the surface. A flax-breaker was invented in the fourteenth century, probably in Holland, and continued in use until recent years (figure 158). It consisted of two planks fixed with their edges uppermost, and a third plank or blade pivoted at one end to fall in the space between them. The stalks were laid across the fixed planks and the blade brought down with the handle to break the woody tissues. The handful of flax was turned and moved with one hand while the other hand operated the blade until the stalks were sufficiently broken.

In Coptic Egypt the broken stalks were combed with iron combs to remove the woody tissues. During the Middle Ages an intermediate process was introduced (figure 158, left), in which the broken stalks were laid over the edge of a bench and beaten with the edge of a flat wooden blade. Care was necessary to avoid injuring the fibres by striking on the bench. An alternative process imitated the threshing of corn with the flail (figure 65).

Next, the flax was repeatedly drawn through the teeth of a comb to split up and separate the bundles of fibres (figure 159). The comb used in Egypt has persisted with little change to recent years. Care was taken to split the fibres without breaking them. Waste fibres from the breaking-process were used in

Roman times for ropes and lamp-wicks, and later for sackcloth, while waste from combing was used for cheap coarse garments.

Hemp (vol I, p 373) is mentioned by Herodotus (*c* 480–*c* 425 B.C.) as used by the Thracians for garments that closely resembled linen ones in appearance [12]. There is, however, no evidence that the Greeks and Romans used hemp except for ropes and possibly sails. True hemp is obtained from *Cannabis sativa*, a native of the regions south-west of the Caspian Sea, but papyrus, esparto grass (*Stipa tenacissima* L), the bark of the lime-tree, and certain other plant-products have at various times been described as hemp.

FIGURE 159—*Combing flax. From a fourteenth-century illustrated manuscript of Pliny.*

(*c*) *Silk*

Silk is the filament of the cocoon of several species of moth of the family Bombicidae. The most important is *Bombyx mori* L, whose caterpillars, the silkworms, are best reared on leaves of white mulberry, *Morus alba*.

The use of silk for textiles originated in China at a very early date [13]. Under the Han Dynasty (202 B.C.–A.D. 220) silken textiles were exported and became coveted articles of trade in the west. Silk was probably unknown to ancient Greece and to Republican Rome. In the west it is first recorded by writers of the first century A.D. as an extremely expensive fabric imported from the Far East. The secrets of silk-production were sedulously guarded by the Chinese. Roman writers described silk as a fine down adhering to the leaves of certain trees or flowers, and as a very delicate species of wool or cotton. Even those who knew it to be made by an insect had no clear idea of its production. Silk fabrics were unravelled to obtain the yarn, which was then divided into finer yarns and rewoven, with a warp or weft of linen or wool, to make more fabrics. This silk-weaving industry began to flourish around the eastern Mediterranean, particularly in Egypt, Syria, and Palestine, in the second century, and in the fourth it developed in Constantinople and Persia. For many centuries both before and after silk-culture (sericulture) began in Europe, both raw silk and silk fabrics were brought to Europe across Asia along the 'Old Silk Road'. Many other things besides silk came and went by the same route (figure 684).

The production of raw silk in Europe was first attempted under the Byzantine Emperor Justinian in A.D. 530. Legend has it that eggs of the silkworms were conveyed to Constantinople, probably from Khotan, in a hollow stick. The Far East remained, however, the main source of raw silk until the twelfth century.

Silk-culture flourished in Spain in the tenth century and in Sicily and Italy in the twelfth. The weaving of fabrics of silk, or of silk and wool, was by then long established in the west. Italy became the chief region of the European silk industry, which was centred at Lucca. Towards the close of the Middle Ages silk-culture and silk-weaving began to develop in France, and on a smaller scale in England.

FIGURE 160—*Unwinding silk from the cocoons in a vat of boiling water. The filament is wound on the reel* (left). *From a Chinese woodcut of 1696, following a thirteenth-century illustration.*

A detailed and illustrated account of the processes of sericulture is given in a twelfth-century Chinese work [14]. The eggs of the silk-moth were dipped in a bath of salt-water at a carefully regulated temperature. They, and the young caterpillars that emerged from them, required constant attention during their long dormant and short feeding periods. Then came the 'awakening', and, after three days, the caterpillars fed voraciously and had to be supplied every half-hour with fresh leaves of white mulberry. The caterpillars for spinning were separated from the breeding-stock, and when fully mature were placed on trays of rice-straw. These were heated gently to stimulate production of the cocoons.

Caterpillars of poor colour or which had fed too rapidly were removed. In due course each healthy caterpillar enveloped itself in a cocoon of silk filaments formed, on contact with the air, from a viscid liquid secreted by special glands. Within two or three weeks the moth would hatch, rupturing the cocoon. During that time the cocoons had to be sorted and those that were tangled removed. One pound of caterpillars which survived the 'awakening' would in a normal year produce about 12 lb of silk.

The first process in the preparation of a silk yarn was reeling. The cocoons were plunged into boiling water to soften the gummy binding-material, and were stirred with rods to the end of which the silk filaments adhered. These filaments were wound on reels (figure 160), passing over a high bamboo frame to promote drying. Attached to the spindle of the reel was a crank which could be turned by the operator. A very similar arrangement was still used in sixteenth-century Florence (plate 8). The filaments of silk are too delicate to be wound singly, so a number of them—usually three to eight—were reeled together. The operator had to ensure that broken filaments, detected by observing that the cocoon had stopped rotating, were rejoined. To maintain a thread of even diameter demanded great skill, for the filaments become thinner towards the centre of the cocoon. One pound of silk was normally reeled by an operator in a day. The silk from perforated or tangled cocoons could not be reeled off and had to be spun like wool or cotton.

(*d*) *Cotton*

Cotton was imported into Europe from an early date (vol I, pp 373–4, 449). Indian cotton became familiar to the Greeks through the wars of Alexander the Great (d 323 B.C.). Theophrastus (*c* 371–*c* 287 B.C.) reports that the Indians planted the shrubs in rows [15], and his contemporary Aristobulus says that they separated the seeds from the capsules to obtain the fibres 'which are combed like wool' [16]. As cotton-fibres are too short to comb by hand, this probably refers to a kind of carding (p 193). In Roman days cotton-goods were made on a large scale in Malta, but those from India continued to be prized for whiteness and fineness. Cotton was also grown round the Persian Gulf and in Arabia, Syria, Asia Minor, and Egypt.

Of the many varieties of cotton-plant Pliny differentiates between the shrubby Egyptian species and the taller Indian one. The Greek lexicographer Pollux, a century later, confirms the distinction and describes the use of cotton for spinning weft-threads to be woven with a linen warp. This is the first reference to the mixed fabric which later, as 'fustian' (from Fostat, a suburb of Cairo), became a most important product of the European cotton industry.

In medieval records cotton is mentioned as growing also in Macedonia, Thessaly, on the shores of the Black Sea, and in China. In the eighth century it was introduced by the Moors into Spain, where its manufacture became of great importance. From Spain the industry had spread by the twelfth century into Italy and France, by the thirteenth to Flanders, by the fourteenth to Germany, and by the fifteenth even to England.[1] Of the many varieties of medieval cotton the best came from Epiphania (Hama) and Aleppo in Syria; second qualities were grown in Lesser Armenia and the Damascus region; and yet lower grades came from the Syrian coast and Cyprus [17]. Inferior to the cotton from the Levant was that from Europe, of which the better grades came from Apulia, the next from Calabria and Malta, and the poorest from Sicily. In the fifteenth century Syria and Cyprus largely supplied the south German industry, but Egyptian cotton was much prized for its fineness, long fibres, and durability.

[1] In England coarse woollen cloths were often called 'cottons'; this has caused mnch uncertainty in the interpretation of records.

FIGURE 161—*Girl spinning with a spindle and distaff. From a Greek wine-jug, fifth century B.C.*

Details of the processing of cotton are made clear by trade-regulations of the fifteenth century. Cotton from the bales, into which it had been compressed for transport, was spread out on a wicker-work or wire frame and beaten with rods to open it up and remove seeds and particles of sand and soil which passed through the mesh. A bow was often used to open up the fibres, as for wool (p 195). After being carded and spun into yarn it was boiled in lye of wood-ashes, sometimes with added lime. In Germany the waste fibres were felted by vigorous fulling (pp 214–16) and the product was used for rain-capes.

II. SPINNING

(a) *Wool and Vegetable Fibres*

Spinning is the process of drawing out and twisting fibres into a continuous thread. Except for silk, which is a naturally continuous filament, spinning is an essential preliminary to weaving. Its early history has been discussed in volume I (pp 424–5).

FIGURE 162—*Rolling wool-fibres on the leg preparatory to spinning. A foot-rest is provided. From an Attic cup, fifth century B.C.*

Spinning has almost invariably been done by women, and its tools are symbolically associated with them. In English an unmarried woman is a spinster and the female side of the family is the distaff side. In classical times, as earlier, the spindle was used for spinning (vol I, figures 268, 273, 276, 280, and 281). A simple mechanization of it, the spindle-wheel, was introduced in medieval Europe (p 202). At the end of the Middle Ages inventive genius yielded the more complex types incorporating the flyer (figures 168, 169) and the treadle—true spinning-wheels.

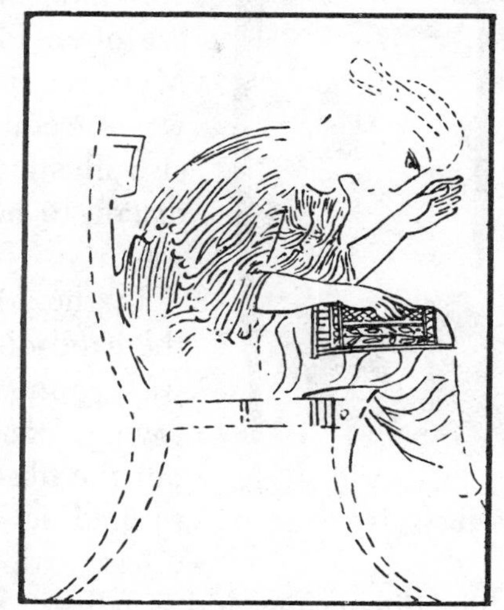

FIGURE 163—*Use of the* epinetron *in rolling wool, from a painting on a fifth-century* B.C. epinetron; (below) *a sixth-century* B.C. epinetron.

Hand-spinning among people of low culture involves only the simple twisting and drawing out of fibres with the fingers. The Greeks used a simple form of hand-spinning to produce a loose roving that was subsequently spun into yarn with a spindle (vol I, figure 281). The fibres were drawn out and rolled lightly on the leg (figure 162). The carded or teased fibre was held in the left hand and gradually drawn away as the rove was twisted between the palm of the right hand and the shin, with the foot supported on a special stand. A hollow tile, the *epinetron*, shaped to fit the bent knee, was sometimes used for this purpose (figure 163). Such tiles often have a roughened 'fish-scale' upper surface to facilitate the twisting of the rove. A basket of plaited reeds held the prepared rovings and spinning-equipment (figures 162, 163).

Spinning with the spindle arose from the need to wind the spun yarn around a stick. This, with a weight or whorl added as a fly-wheel, became the spindle. The instrument and its use changed little from ancient Greece to the fifteenth century, and very largely replaced other ways of spinning (figure 161).

The Roman method of spinning was very similar to the Greek. The figure is often seated and spins yarn from a hand-distaff. Details on a vase-painting from Florence indicate that the fibre is wound on the distaff as a previously prepared roving. Catullus (*c* 87–54 B.C.) describes the process and refers to a

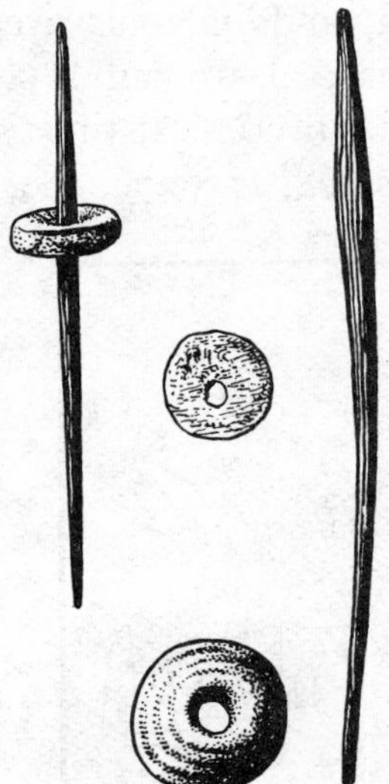

FIGURE 164—*Roman spindles and whorls, found in London. Scale* c 2/5.

slit or notch in the top of the spindle to attach the yarn, though most surviving spindles have smooth ends so that the yarn would need to be looped around (figure 164) [18].

Spindle-whorls of stone, clay, metal, bone, or wood are often found, occasionally in large numbers as at Mycenae. The short hand-distaff of Greece and Rome was introduced into Egypt, where it is still used, and is also illustrated in medieval Italy (figure 165). A longer distaff, to be inserted in a belt or held under the arm, was in use in the early Roman Empire.

In the Middle Ages this type (figure 166) replaced the hand-distaff for a new technique in which the yarn was spun direct from prepared fibre, instead of a roving; one hand rotated the spindle, the other drew out the fibres. The distaff often carried a loop or cage to hold out the fibres, and was occasionally fitted on to a base or low bench (figure 173).

The dimensions and weights of the spindle and whorl are determined by the strengths of the yarn desired and the fibres used. Stone whorls are usually attributed to flax, wooden to wool or cotton. Heavier whorls were for doubling or plying yarns to obtain greater regularity or strength. Spindle-spinning can produce extremely fine yet strong yarns.

FIGURE 165—*Spinning with a short distaff. From an Italian manuscript of 1023.*

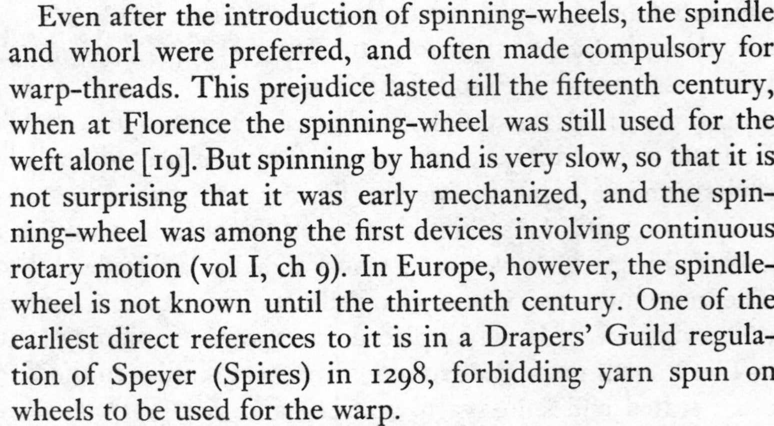

Even after the introduction of spinning-wheels, the spindle and whorl were preferred, and often made compulsory for warp-threads. This prejudice lasted till the fifteenth century, when at Florence the spinning-wheel was still used for the weft alone [19]. But spinning by hand is very slow, so that it is not surprising that it was early mechanized, and the spinning-wheel was among the first devices involving continuous rotary motion (vol I, ch 9). In Europe, however, the spindle-wheel is not known until the thirteenth century. One of the earliest direct references to it is in a Drapers' Guild regulation of Speyer (Spires) in 1298, forbidding yarn spun on wheels to be used for the warp.

The first known illustration of the spindle-wheel (early fourteenth century) shows it worked by hand. Another minia-

FIGURE 166—*Spinning with a long distaff supported by a belt and the crook of the arm, leaving both hands free. From a French fourteenth-century manuscript of Aristotle.*

ture, *c* 1340 (figure 167), shows a similar wheel with a pulley mounted on the horizontal spindle, and the driving cord crossed to produce an S-twist yarn (vol I, figure 268 B) with the normal clockwise rotation of the wheel.

In spinning with the wheel the right hand keeps it revolving while the left holds the prepared fibres. The unspun yarn extends from the fibres to the spindle, being held at an angle of about 45° to its axis. With each revolution of the spindle the last turn of yarn slips off the end of the spindle, imparting one twist. The fibres are drawn out as this twisting proceeds until the left arm is fully outstretched. The yarn is then brought at right angles to the spindle and the latter reversed for a few turns, to bring the yarn from the tip to the winding-position, before the spun yarn is wound on to the spindle. Thus the operation is intermittent and comprises two distinct actions—spinning and winding-on. The sequence is identical with that of the spindle and whorl. The spindle-wheel may be regarded as a mechanization of a spindle and whorl effected by mounting the spindle horizontally between bearings and grooving the whorl to make it a pulley. The spindle-wheel, once almost universal, is still occasionally used.

The origin of the flyer, which makes simultaneous spinning and winding possible, is unknown. When it first appears in an illustration of *c* 1480 it already incorporates a typical flyer mechanism too well designed to be other than the

FIGURE 167—*Spindle-wheel from the Luttrell Psalter,* c *1338. On the right a woman is carding wool.*

result of considerable evolution (figure 168). The flyer consists of two arms mounted on a hollow spindle with a driving-pulley at one end. Loosely rotating on this spindle, between the two arms of the flyer, is a spool with a separate driving-pulley. These two pulleys are driven from the driving-wheel by two cords, or rather a single cord passing round twice. Since the pulley on the spool is of less diameter than that on the spindle the spool will revolve faster than the spindle and flyer. The yarn passes through the hollow end of the spindle, out through a hole in the side, and then over one of a series of small hooks or hecks on the flyer, and is attached to the spool. When the flyer rotates it twists the thread once per revolution and, as the spool is turning faster than the flyer, the spun yarn is gradually wound on to it. The process is continuous, the flyer twisting while the spool winds on. The operator turns the wheel by hand and draws out the fibres at the correct rate to produce the yarn.

FIGURE 168—*Spinning-wheel with flyer illustrated in* Das Mittelalterliche Hausbuch, c *1480*.

Since the diameter of the spool or bobbin increases as yarn is wound on, it tends to wind more rapidly and so to produce yarn with fewer twists per unit length. To allow for this, the operator holds back the yarn so that it is not wound up before being sufficiently twisted. To prevent the yarn breaking, the spool must be free to slip, a condition that is secured by an adjustment to regulate the tension of the driving cords. This arrangement is that of a wheel with 'bobbin-lead'—that is, the bobbin travels faster than the flyer—but a similar result is obtained with 'flyer-lead', where the bobbin rotates more slowly than the flyer.

The small hooks permit regulation of the winding by reattaching the yarn periodically as the spinning proceeds. When the bobbin is full it is removed and replaced by unscrewing the nut on the carrier and removing the spindle-pulley.

Leonardo da Vinci shows a very similar flyer mechanism (figure 169) [20]. His device automatically distributes the yarn on the bobbin, thus eliminating periodic stops to move the thread from one hook to another. By a series of pegs and cage-wheels a lever is caused to oscillate slowly. The forked end of this lever is in contact with the flyer-spindle, so that the flyer automatically traverses the bobbin as the yarn is wound on. This mechanism is not known from any early spinning-wheel and was independently invented again in the eighteenth century. The flyer-wheel converted spinning and winding into a single continuous action, and also enabled the spinner to be seated. Thus a foot-treadle could be used to drive the wheel, and this is first encountered about the end of the fifteenth century.

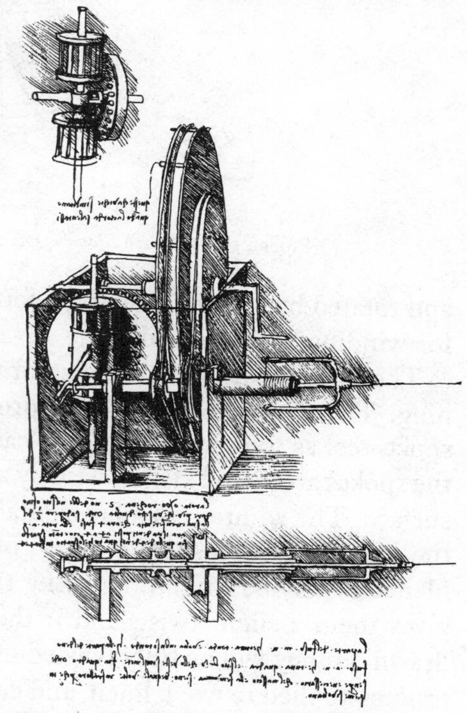

FIGURE 169—*Spinning-wheel sketched by Leonardo da Vinci. From the* Codice Atlantico.

The thirteenth-century Bolognese twisting-mills (p 206) are closely analogous to spinning-devices and suggest that a simple spindle-wheel must have been in use long before the earliest known records. They also throw some light on the origin of the flyer, for the system of rotating spindle and spool, with freely revolving S-shaped wings leading the yarn to the reel above, provides a possible intermediate stage between spindle-wheel and spinning-wheel. Used in reverse this system would certainly spin, with a bobbin-lead, if provided with hecks; otherwise an automatic traverse to distribute the yarn would be the only necessary addition. The yarn for the warp was frequently spun dextrally (Z-twist), while the weft was given a sinistral turn (S-twist) (vol I, p 424).

(*b*) *Silk*

After reeling, the threads of silk were twisted to prevent subsequent separation into individual filaments. The threads from the reel were wound upon spools, two or three being wound together for strength and to average out inequalities. The Chinese cage-spool (figure 177) could be mounted on a horizontal spindle

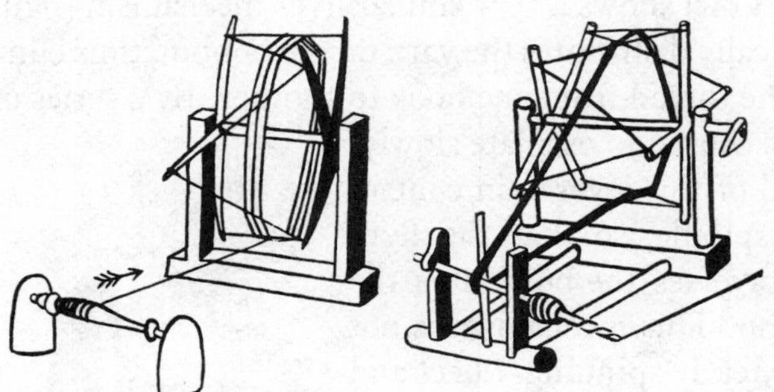

FIGURE 170—*Modern Siamese spoke-reel and spool-winder for winding silk thread.*

and rotated by turning one of the four arms. The early introduction of the wheel for winding was a considerable advance.

The final twisting or throwing of the silk threads is a process similar to spinning. The early spool-winder is probably represented by an adaptation of the spoke-reel as found both in the Near and the Far East (figure 170). The tips of the spokes are joined zigzag-wise by a continuous cord, which forms the winding-surface. The addition of a horizontally mounted spindle, with a driving-band to the reel, provides a spool-winder; on this spindle the untwisted bundles of silk filaments may be wound. Drawing the filament-bundles endwise off the spindle gives them a slight twist, and if the spindle is rotated while the filaments are drawn off they will be fully twisted into thread. This is a reversal of the spinning-process applied to wool, linen, and cotton, and is made possible by the continuity of the silk filament.

FIGURE 171—*Fifteenth-century sketch of a silk-throwing mill. From a Florentine manuscript, 1487.*

Whereas the spinning-process was intermittent with such a spindle-wheel, the twisting of silk is continuous. This accounts for the remarkably early invention of highly developed twisting-mills, such as those said to have been erected at Bologna about 1272 by an exile from Lucca. Certainly there were great technical improvements in the reeling and throwing of silk in Italy during the thirteenth century.

Strict secrecy surrounded these inventions, so that nothing is known of the history of the reeling-machinery. However, a Luccan document of the fourteenth century [21], and a crude Florentine sketch of the fifteenth (figure 171), show that the earliest twisting-mill was similar to one of the seventeenth (figure 172) [22]. This consists of two concentric wooden frameworks about 16 ft in diameter. The outer is fixed and bears two rows of vertical spindles with a row of horizontal reels above each. The inner rotates about a vertical axis, and its outer laths rub on the spindles and reel-mechanisms to rotate them. On each spindle is a rigidly fixed spool, and above this an S-shaped wire that rotates freely on a dome-shaped cap. The untwisted silk yarn passes from the spool and through eyelets on each end of the S-shaped wings to the reel above. When the spindle rotates, and with it the spool, the yarn is twisted, and as it is drawn up by the reel the wings automatically rotate and guide the yarn. The Luccan document indicates that the mill had two rows of twelve reels with ten spindles to each reel, 1000 spools, 240 spindles and a similar number of caps (*coronelle*), and glass sockets for the spindles. In 1385 a mill is described having four rows of spindles and reels. Another had five rows of 16 reels and six spindles per reel; each row thus had 96 spindles, and the mill a total of 480. The caps were of different weights: light ones for thin yarns and heavier ones for coarse yarns. A mill of 1331 had two rows of spindles with 120 in each.

FIGURE 172—*Water-driven silk-throwing mill illustrated by Zonca. 1607.*

These twisting-mills, driven by undershot water-wheels, were introduced by Luccans to Florence and Venice in the mid-fourteenth century. They were remarkable feats of mechanization, for they required only two or three operators to join the threads and replace the spools and reels, and thus to perform the work previously done by several hundred hand-throwsters. The design was little improved until the nineteenth century.

After twisting, the hanks of silk yarn were placed in small bags and boiled in soapy water, to remove natural gum which might interfere with the dyeing. They were then rinsed in clear water, and hung up to dry. The hanks, of a pearly colour, were next bleached in the fumes of burning sulphur (figure 331). The white yarn was then ready for dyeing and weaving.

III. REELING, WINDING, AND WARPING

After spinning come the necessary operations of removing the yarn from the spindle (reeling), filling the weft and warp spools (winding), and preparing the warp for attachment to the loom (warping). In silk-manufacture reeling and twisting replace true spinning.

FIGURE 173—*The cross-reel* (lower left); *spinning from a fixed distaff* (centre). *A Nuremberg symbolic woodcut of* c *1490.*

The spun yarn on a spindle cannot be slipped off, but has to be wound off by some means. In the Graeco-Roman world the yarn was probably wound by hand into a ball. A plaited, tapering basket of osiers or reeds, containing the balls of roving, the spindle and distaff, and probably the balls of yarn, is frequently portrayed, always associated with women (figure 162). Balls of yarn are also depicted in the Middle Ages (plate 9).

A simple instrument, used from the earliest Greek times, which facilitated the winding of yarn into a hank convenient for washing, sizing, or dyeing, is the stick-reel, which is a plain stick about 11 in long. The cross-reel, a development of the stick-reel, has two cross-pieces, one at each end, at right-angles to each other; these enable the yarn to be readily wound (figure 173). By counting the turns and knowing the length and weight of the hank, the grade or counts of the yarn can be calculated.

The rotary reel doubtless originated in the Far East, where reeling is essential to silk-manufacture. It entered Europe at about the same time as the spindle-wheel (figure 174). It came to be of standard dimensions, in which the circumference was a simple multiple of the unit of length. This enabled uniform hanks to be wound.

FIGURE 174—*Hank winding on a rotary wheel. A caricature from the fourteenth-century Ypres 'Book of Trades'.* c *1310.*

Whether the spun yarn was treated as a hank, or taken directly from the spindle for weaving, it was necessary to transfer the weft-yarn to the spool or shuttle used for weaving. The earlier types of loom and their working have already been discussed (vol I, pp 425–8). With the frameless and the vertical looms the spool was merely a short rod and the weft was wound on by hand. In weaving with the horizontal framed loom the spool was encased in a hollow shuttle to allow it to be thrown across between the warp-threads, but the spool was still wound by hand. With the aid of a pulley a more rapid spool-winder could be devised, and it is even possible that the spinning-wheel developed from such a device. The fourteenth-century spool-winder (figure 183) was no more than a spindle mounted horizontally and driven by a cord from a wheel. This type of winder was also used for winding cops (conical balls of thread) for the warp (figure 175). Often the spinning-wheel itself was used as a winder, as it sometimes is even at the present day.

FIGURE 175—*Nuns making cloth:* (right) *winding cops from a reel;* (centre) *warping a loom from 12 cops;* (left) *weaving on a horizontal loom with foot-treadles. From an Italian manuscript, 1421.*

Warping is the preparation of the warp for insertion in the loom. In a frameless loom, warping usually involves nothing more than winding the requisite number of threads around the beams of the loom. Generally, however, and invariably with the more advanced type of loom, the warp is prepared apart from the loom and then attached to the warp-beam. Before insertion in the loom the warp was dressed by treating with size, often prepared by boiling rabbit-skins in water, or with an adhesive made from the waste from corn-mills. By the ancient Egyptians warping-pegs were pushed into the ground or into a low wall, but there is no evidence of such pegs from Greece or Rome. Warp-weighted looms were often warped by a starting-border, as in recent Lapland examples. This border is woven separately with long weft-loops which become the warp when attached to the beam. There is no evidence of such practice in ancient Greece, and the rare examples of Greek cloth show no such features.

FIGURE 176—*Warping a loom with 12 threads. From the Ypres 'Book of Trades'.* c *1210.*

An illustration of about 1310 shows a woman preparing the warp on a simple frame with pegs (figure 176). The threads, passing through a hole-board held in the hand, are wound zigzag from peg to peg. Another drawing shows a similar warping technique, using twelve threads at a time (figure 175). The threads pass through rings to twelve spools mounted on a rack. Double pegs at the bottom of

FIGURE 177—*Winding threads for the warp from cage-spools in a rack. From a Chinese woodcut of 1696 following a thirteenth-century illustration.*

the frame enable the operator to separate the threads and cross them over alternately, preventing entanglement of the threads and simplifying the threading through the heddles and reed. This method of warping is still employed by hand-weavers.

The warping reel or creel naturally developed earlier in the east, and a thirteenth-century illustration from China shows a simple horizontal reel taking silk threads from cage-spools on a rack (figure 177). This type of reel was increased in diameter to accommodate greater lengths of warp and to minimize inequalities in winding.

IV. WEAVING

The essential of weaving is the interlacing of one series of filaments or threads, known as the warp, with another series known as the weft (vol I, pp 425–8). The warp-threads are stretched side by side and the weft-threads passed over and under them. Any apparatus for stretching the warp during weaving is a loom. Originally the loom was frameless, the weaver working upon a warp stretched

FIGURE 178—*Odysseus and Circe with a loom. Caricature from a Greek vase, fourth century* B.C.

from a tree or peg to his waist, or pegged to the ground between two bars or 'beams'. This very primitive type is still used by Near Eastern nomads.

The only loom known to the Greeks was vertical and warp-weighted. The upper beam was supported on two uprights, and the warp stretched by a series of weights (vol I, p 444). The cloth is woven downwards from the top, the construction suggesting that the beam could be rotated to wind up the cloth as woven, thus preserving a constant working height. Some Greek vases of the fifth and fourth centuries B.C. illustrate this type (figure 178). The early Roman loom was similar, for in the first century A.D. Seneca mentions the stretching of the warp by weights. This system was doubtless introduced into Egypt in the Roman period; it continued into Coptic times.

The later Roman loom was also vertical but the warp was stretched by passing round a second beam at the base. A loom in a fourth-century manuscript shows the two uprights with the warp thus stretched (figure 179). Across the warp lies a bar, and there is an indication that the cloth was woven upwards. The bar is possibly a heddle-rod, but perhaps there were two heddles—or more probably one heddle and a shed-stick as in the warp-weighted loom (compare vol I, figure 269 B).

FIGURE 179—*Late Roman vertical loom: some woven fabric is visible at the base. From a Virgil manuscript in the Vatican Library, fourth century* A.D.

The Greek shuttle was a short rod round which the weft-yarn was wound. The Roman shuttle was sometimes fashioned from a hollow bone, but it is possible that it was not used in cloth-weaving, and that the simple rod of the Greeks still served. The weft-threads were driven close together with a comb (figure 180), which could be used only on a loom weaving from below upwards. That the cloth was so woven in Egyptian two-beamed looms is indicated by Herodotus (*c* 484–425 B.C.) [23], while the Roman writer Festus (second century A.D.) regards standing at the loom as a traditional survival [24]. With the warp-weighted loom, which was necessarily worked from the top, the weaver had to stand.

FIGURE 180—*Roman weaving-comb from Kaar-el-Banaf, Egypt. Scale 1/7.*

The vertical two-beamed loom is illustrated in Egypt from the fifth century B.C., and survived in north Africa to recent times. In a modified form it is the standard tapestry-loom of today. With a device for rotating the beams as the work proceeds, a piece of cloth longer than the interval between the beams can be woven.

The next improvement in the loom was the horizontal frame, the origin of which is uncertain. In Europe it appears as a perfect instrument in the thirteenth century, but it may well be older in the east. It has a stout box-like framework stretching the warp horizontally between a warp-beam at the back and a breast-beam in front. This allows the weaver to work more conveniently, and also permits the introduction of a shedding-mechanism to lift and lower alternate warp-threads by treadles (figure 181).

FIGURE 181—*Horizontal loom with shedding-mechanism operated by treadles. From a thirteenth-century manuscript.*

In weaving, one treadle is depressed, the shuttle is thrown from one hand to the other through the shed so formed, and the fabric is beaten with the reed. Then the other treadle is depressed to open the counter shed, the shuttle again passed through, and the weft beaten home. As the weaving proceeds the warp is periodically paid out from the warp-beam by releasing the lever, and the woven cloth is wound on the cloth-beam.

With this loom only plain or tabby cloth can be woven. With two more heddles and treadles, various combinations of warp-threads can be lifted to produce simple patterns such as diamonds, twills, and herring-bones (figure 182). In this loom, which has no top frame, the shedding-harness is suspended from the ceiling, but most had a high frame from the top of which the reed was supported in a heavy batten that ensured close and regular beating (figure 175; plate 9). Figure 183 shows a four-shaft loom, with a ratchet-mechanism on the cloth-beam, which is wide enough to demand two weavers to throw the shuttle from one to the other. More heddles and treadles increased the range of woven patterns.

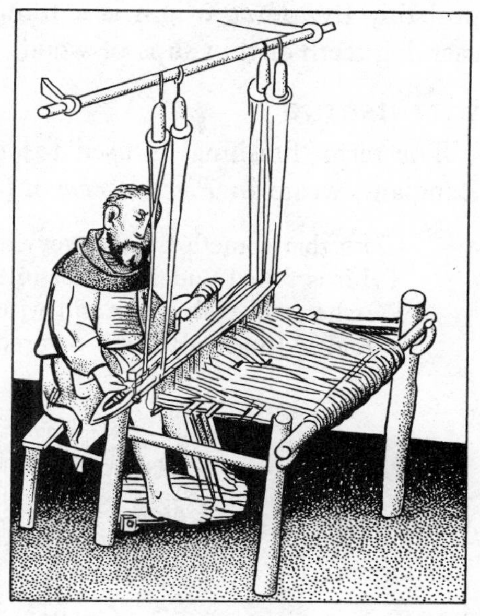

FIGURE 182—*Four-heddle loom of* c *1400, from the Mendel* Brüderbuch. *The illustration is of the fifteenth century.*

Details of the heddles are not clear, but other medieval illustrations show them as upper and lower rods with a series of looped cords between, the warp-threads passing alternately through these loops. The reed or beater is a comb, usually of reeds bound between two rods. An Egyptian example of

FIGURE 183—*Weavers at work, with a framed horizontal loom.* Below, *winding spools from a rotary reel on the right. From the Ypres 'Book of Trades'.* c *1310.*

probably Byzantine origin is a frame two feet long covered with leather and carrying teeth of thin slips of wood.

V. FINISHING

The term 'finishing' is used for the treatment of the cloth after weaving. Langland wrote in *Piers Plowman* (*c* 1362) [25]:

> Cloth that cometh fro the wevying is noght comly to were
> Tyl it is fulled under fote, or in fullying stokkes [fuller's stocks]
> Wasshen wel with water and with tasles [teazles] cracched [scratched],
> Y-touked [finished] and y-teynted [stretched], and under tailloures hande.

FIGURE 184—*Cloth finishing in a monastery:* (right) *removing imperfections from the fabric;* (left) *stretching cloth on a tenter-frame. From an Italian manuscript of 1421.*

First, fragments of extraneous matter must be removed from the cloth. This is carried out with the cloth hung over a rail or spread on a table (figure 184). The particles are picked out with forceps, knots in the yarn removed, and weaving errors repaired.

Woollen cloth has next to be fulled, in order to felt and thicken the cloth, to cause the fibres to adhere together, and to obliterate gaps in the weave. Fulling must be a very ancient process, for it is not dissimilar to felting, which probably preceded weaving. A milder washing and thickening of the cloth is called scouring. In an early representation of fulling, from Pompeii (first century A.D.), four earthenware bowls are in compartments separated by partitions; three workmen rinse the cloth, while a fourth stands in a large bowl trampling it. A Gallic fuller working in his vat is depicted on a Roman relief at Sens (figure 185).

Fulling-agents are described by Pliny and others [26]. Soap was not used by the Greeks or Romans, who employed various alkaline detergents including the lixivium of plant ashes, or of natron, and stale human urine. Fuller's earth—a natural finely divided hydrated aluminium silicate which takes up fats—was in general use. The juice of certain plants, notably *Gypsophila struthium* L and soapwort (*Saponaria officinalis* L), was also used (p 355). All these substances continued to be employed through the Middle Ages and until modern times. The cloth was trampled with the feet in some such substance until the wool-fibres had adequately felted (figure 186). It was periodically taken out and rearranged to ensure even action. When the fulling was completed the cloth was washed in a vat or in a stream and then beaten with sticks to increase the adhesion of the constituent fibres.

FIGURE 185—(Above) *Cloth shears;* (below) *fulling. From a Gallo-Roman tomb at Sens.*

In scouring in fifteenth-century Florence the fabric, fresh from the loom, was washed for two hours in hot water, worked with soap or stale urine and lye-water until it foamed, then rinsed in cold water and wrung out. There followed half-an-hour's treatment with fuller's earth in pits filled with boiling water. Finally the cloth was rinsed in running water again and hung up to dry. Flemish cloth was soaked for some days before fulling in a tub containing loamy earth. Stamping with the feet was then continued until gaps in the weave were obliterated. These Florentine and Flemish cloths were worsteds made either entirely, or at least as to warp, from combed wool—yet, contrary to usual practice both then and now, they received a fulling treatment normally associated with woollen cloths made from carded wool.

Linen cloth, according to Pliny, was beaten with clubs to improve the surface, and some form of fulling was usually applied to it during the Middle Ages to

FIGURE 186—*Fuller trampling cloth in the vat. From the painted window of the Clothiers' Guild, Semur-en-Auxois cathedral, Côte d'Or. c 1460.*

separate the stiff groups of fibres into smaller and softer ones. A fulling-trough for linen is recorded at St Gall in the thirteenth century.

Fulling-mills were known in the eleventh century and were widespread throughout Europe by the thirteenth, yet few details of their construction are known before the fifteenth. A mechanical device reproducing the action of trampling the cloth with the feet, the fulling-mill, consists of two heavy wooden mallets mechanically raised and allowed to fall upon the cloth in a trough or stock. Power was supplied by a water-wheel, such as had long been used for grinding corn (ch 17), but the application to fulling involved certain novel mechanical principles. A fulling-mill in Kent about 1438 was able to deal simultaneously with three half-length broadcloths 12 yds long [27].

Medieval fulling-stocks no doubt resembled the later mechanism depicted by Zonca in 1607 (figure 187). In this the shaft of the water-wheel bears tappets which raise alternately the heads of two wooden mallets. On disengagement the mallets fall into a wooden trough containing a fulling-agent. Periodically the cloth is removed and re-bundled to ensure even action (figure 553).

FIGURE 187—*Water-driven fulling-machine, as illustrated in 1607.*

After fulling and scouring, the wet cloth was stretched in the open air to dry on frames or tenters. These were constant in design from Roman times throughout the Middle Ages, and lines of post-holes at Silchester may have been for such frames [28]. Medieval tenters were typically two horizontal wooden rails supported on vertical posts (figure 184).

To give a woollen cloth a softer finish it was necessary to raise the fibre-ends to the surface. The process is merely an extreme form of brushing, and was known in classical antiquity. Pliny records that sometimes the skins of hedgehogs were used and sometimes thistles, generally of the genus *Carduus* (p 193). The process is seen in murals from Pompeii (figure 188), and persisted throughout the Middle Ages to this day. The teazel-heads (figure 155) were fastened in a light wooden frame to which a handle was fitted (figure 189). Sometimes the cloth was raised several times, harsh new teazels being followed by those softer and worn, so as to give a smoother finish. The teazels were specially cultivated in Europe and particularly in southern England.

After each raising, the nap of the cloth was cropped to remove the fibre-ends and give an even surface. The earliest cropping was probably with wool-shears. Cropping-shears have been discovered at Pompeii and at a Roman site in

FIGURE 188—*Fresco from Pompeii showing Cupids finishing cloth, fulling* (left), *and raising by brushing on a frame* (right). *Before* A.D. 79.

Essex (figure 191), and are illustrated in a Roman relief at Sens (figure 185). The method was to stretch the cloth over a bench and lay the shears across. In figure 190 the shearman inserts his left wrist through a stirrup-grip on the lower blade, opening and closing the blade with the fingers of the same hand. With his right hand he steadies the shears against his body. From a knowledge of later techniques it is evident that the shearman started cropping at one edge of the cloth and gradually passed to the other, with the blades opening and closing continuously. The bench was padded and the cloth attached to it by double-ended hooks or habbicks. When a section of cloth had been cropped it was unhooked and another section presented. Cropping and raising alternated, sometimes with the cloth wet and sometimes with it dry, until the desired effect was obtained. Cropping-shears continued in use, in an improved form, until the nineteenth century.

Linen cloth from the loom is grey-brown, and has to undergo bleaching and pressing. Regulations of the early sixteenth century from St Gall show that

FIGURE 189—*Raising cloth before cropping. From the Clothiers' window at Semur-en-Auxois.* c *1460.*

FIGURE 190—*Shearing (or cropping) cloth. From the Clothiers' window at Semur-en-Auxois.*

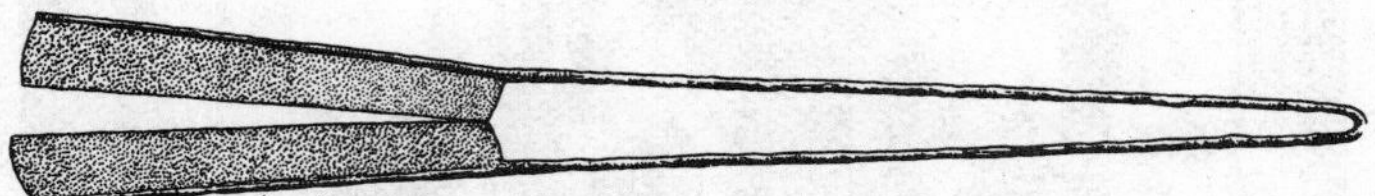

FIGURE 191—*Roman napping-shears from Chesterford, Essex.* c *4·5 ft long.*

the cloth was first boiled in a lye of ashes and then spread out on the drying-fields to bleach in the sun. These fields were necessarily extensive, and in the thirteenth century they were rented from the abbot of St Gall with permission to put up any necessary erections. This refers to the wooden posts on which the linen was supported. Thin linen cloths were exposed to the sun for eight weeks and thick linens up to sixteen weeks, according to the weather, being regularly sprinkled with water. The cloth was finally scoured in running water.

Cotton fabrics were similarly bleached by being laid on the grass in the sun and moistened by the dew. It was for its pure white colour that cotton was prized by the Greeks and Romans. Woollen cloth was bleached by fumes of burning sulphur. After fulling it was laid over a hemispherical cage beneath which the sulphur was burned (figure 192). Fine fuller's earth was often rubbed into the cloth as a final dressing to increase the whiteness.

Pressing gave the cloth a smooth, lustrous surface. A screw-press has been used for this purpose since Roman times, and Pliny indicates that it was invented in his day. This type of press is illustrated in a Pompeian fresco (figure 193). By rotating two upright screws with levers the top plate of the press is forced down

FIGURE 192—*Bleacher with cage and pot for burning sulphur. From Pompeii, before A.D. 79.*

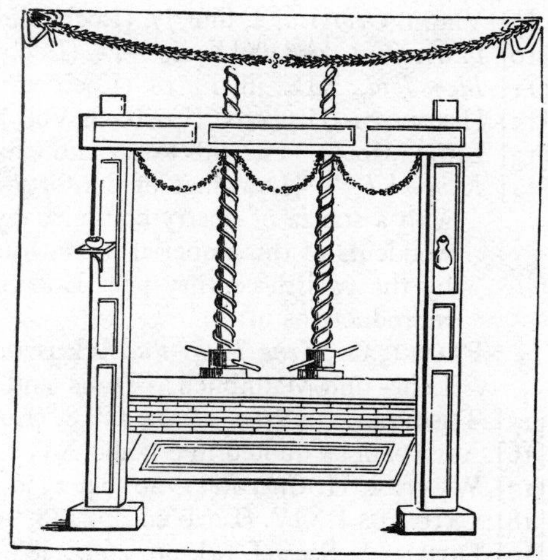

FIGURE 193—*Cloth-press operated by two screws. From a wall-painting in the fuller's workshop at Pompeii.*

on the base-plate. The cloth would be carefully folded backwards and forwards and placed between the plates. The left- and right-hand threads of the upright screws facilitated equal tightening. A second-century illustration shows that pegs in the cross-piece engaged with the threads cut in the uprights, so avoiding the need for a female screw or nut. Before pressing, it was customary to fill the mouth with water to sprinkle the cloth.

Linen cloth was smoothed by rubbing the surface with rods, as in Egypt, or by passing a wooden block over it under great pressure. Smooth stones were sometimes employed, and in Switzerland a heavy marble ball was made to slide on the stretched surface.

REFERENCES

[1] PLINY *Nat. hist.*, VIII, lxxii, 189—lxxiii, 191. (Loeb ed. Vol. 3, pp. 132 ff., 1940.)
[2] SALZMAN, L. F. 'English Trade in the Middle Ages', p. 307. Clarendon Press, Oxford. 1931.
[3] HOMER Odyssey, XVIII, 315 ff. (Loeb ed. Vol. 2, p. 218, 1919.)
PLINY *Nat. hist.*, VIII, lxxiii, 191. (Loeb ed. Vol. 3, p. 134, 1940.)
[4] DOREN, A. 'Studien aus der Florentiner Wirtschaftsgeschichte', Vol. 1: 'Die Florentiner Wollentuchindustrie', pp. 44 f., 486. Cotta, Stuttgart. 1901.
[5] GUTMANN, ANNY L. *Ciba Rev.*, no. 14, 485, 1938.
[6] WESCHER, H. *Ibid.*, no. 65, 2380, 1948.
[7] DOREN, A. See ref. [4], pp. 45 f.; 487.
[8] WESCHER, H. *Ciba Rev.*, no. 64, 2340, 1948.

[9] VIRGIL *Georgica*, I, line 77. (Loeb ed. Vol. 1, p. 86, 1916.)
[10] PLINY *Nat. hist.*, XIX, ii, 6–11. (Loeb ed. Vol. 5, pp. 424 ff., 1950.)
[11] *Idem Ibid.*, XIX, iii, 17–18. (Loeb ed. Vol. 5, p. 430, 1950.)
[12] HERODOTUS IV, lxxiv. (Loeb ed. Vol. 2, p. 272, 1921.)
[13] LIU, G. K. C. 'The Silkworm and Chinese Culture.' Osiris, Vol. 10, pp. 129–93, 1952.
[14] *Kêng-chih t'u* '[Drawings of the forty-six various] processes in tillage and weaving' each with a stanza of poetry compiled by the direction of Emperor KANG-HSI. 1697. The woodcuts of this imperial publication were based on illustrations, now lost, of poems by the twelfth-century poet LOU SHOU. For a complete set of the illustrations see the reproductions in:
FRANKE, O. '*Keng Tschi t'u.* Ackerbau und Seidengewinnung in China. Ein kaiserliches Lehr- und Mahnbuch.' [Trans. and comm.] Abh. Hamburg. Kolon. Inst. Vol. 2, 1913.
[15] THEOPHRASTUS *Hist. plant.*, IV, 8. (Loeb ed. 'Enquiry into Plants', Vol. 1, p. 316, 1916.)
[16] ARISTOBULUS quoted in STRABO XV, C 694. (Loeb ed. Vol. 7, p. 34, 1930.)
[17] WESCHER, H. *Ciba Rev.*, no. 64, 2336, 1948.
[18] CATULLUS LXIV. (Loeb ed., p. 118, 1912.)
[19] DOREN, A. See ref. [4], pp. 46 f.; 488.
[20] BECK, T. *Z. Ver. dtsch. Ing.*, **50**, 568, 1906.
VINCI, LEONARDO DA. *Codice Atlantico*, fol. 393 b. Facs. ed. by G. PIUMATI. R. R. Accademia dei Lincei, Milan. 1894–1904.
[21] BINI, T. 'Su i Lucchesi a Venezia.' Atti R. Accademia, Lucchesi, Vol. 15, p. 54, 1803.
[22] ZONCA, V. 'Novo Teatro di Macchine et Edificii', pp. 69–75. Pietro Bertelli, Padua. 1607.
[23] HERODOTUS II, 35. (Loeb ed. Vol. 1, p. 316, 1920.)
[24] FESTUS, SEXTUS POMPEIUS. *De verborum significatione* (ed. by W. M. LINDSAY), pp. 342, 364. Teubner, Leipzig. 1913.
[25] LANGLAND, W. 'The Vision Concerning Piers Ploughman', B Text, passus X, lines 444–8. Ed. with annot. by W. SKEAT, Vol. 1, p. 466. Oxford University Press, London. 1924.
[26] BECKMANN, J. 'A History of Inventions, Discoveries and Origins' (trans. by W. JOHNSTON. 4th ed. rev. and enl. by W. FRANCIS and J. W. GRIFFITH), Vol. 2, pp. 92–108. Bohn, London. 1846.
[27] GARDINER, DOROTHY. *Archaeologia Cantiana*, **43**, 202, 1931.
[28] THOMSON, J. 'The Book of Silchester', Vol. 2, p. 406. Simpkin, Marshall, Hamilton, Kent and Co., and Lloyd, London. 1924.

BIBLIOGRAPHY

ESPINAS, G. 'Recueil de documents relatifs à l'histoire de l'industrie drapière en Flandre' (3 vols). Commission Royale d'Histoire. Kiessling, Imbreghts, Brussels. 1906–20.
FELDHAUS, F. M. 'Die Technik der Antike und des Mittelalters.' Athenaion, Potsdam. 1931.
PODREIDER, FANNY. 'Storia dei tessuti d'arte in Italia. Secoli XII–XVIII.' Istituto Italiano d'Arte Grafiche, Bergamo. 1928.
ROTH, H. LING. 'Primitive Looms' (3rd ed.). Bankfield Museum, Halifax. 1950.
SALZMAN, L. F. 'English Industries of the Middle Ages' (new enl. ed.). Clarendon Press, Oxford. 1923.
SCOTT, E. K. "Early Cloth Fulling and its Machinery." *Trans. Newcomen Soc.* **12,** 31–52, 1933.
UCCELLI, A. 'Storia della tecnica dal medio evo ai nostri giorni', pp. 125–76. Hoepli, Milan. 1944.
YATES, J. '*Textrinum Antiquorum*: An Account of the Art of Weaving among the Ancients.' Taylor and Walton, London. 1843.

7
PART II
FURNITURE: POST-ROMAN

R. W. SYMONDS

I. THE 'GOTHIC' PERIOD

By the year 1000 Europe was settling down to what was to be the medieval order, and the barbarian realms of the north and west were developing some of the habits of civilized luxury. Their woodwork was still far coarser than that in the Mediterranean region, and especially in Italy, but they were imitating it in their own way. The inaptly so-called 'Gothic' period of furniture had begun.

The deterioration in the finer techniques of woodworking that in the west followed on the fall of the Roman Empire made little change in the ruder craft of the carpenter. His main tools—axe, adze, hammer, saw, and chisel—had been in use for millennia; his bag of tools hardly differed from that of an Egyptian carpenter (vol I, figures 487–9). In every township tree-trunks continued to be converted by axes, adzes, and pit-saws into beams, posts, and boards.

Throughout the Middle Ages, timber was in great demand for a variety of things—ships, building, wagons, the interior woodwork of houses and churches, and furniture. Timber-resources were unevenly distributed, so that while oak-framed houses were usual in some parts of Europe, fir- or pine-framed houses were normal in others. Only in districts where timber suitable for building was inaccessible was stone used. For good quality house-fitments and furniture wood could be transported from afar, but for ordinary houses and cheap furniture local woods were necessarily employed. Thus in England, where walnut was scarce, only the better quality furniture was made of it, but in France, Italy, and Spain, where walnut-trees grew abundantly, it was used even for common furniture.

Remains of furniture of before the fourteenth century are hardly to be found outside Italy, but, judging from pictures in illuminated manuscripts, the furniture of southern Europe, and especially of Italy, showed a high degree of technical achievement in the eleventh century and even earlier. Italian craftsmen were practising inlaying and veneering and using mortise-and-tenon and dove-tail joints long before the craftsmen of northern Europe. For example, the principle examples from Meir, Thebes, and the Tutankhamen tomb (1350 B.C.). Chests, if we are to judge from the well preserved specimens found at Abusir in Egypt, follow the pattern of Egyptian chests not only in design but in construction. In the gabled examples, for instance, one side of the gable pivoted upon side-pieces to form a sloping lid, exactly like boxes from Sedment of a millennium earlier (vol I, p 694), and the smaller caskets had detachable lids which were corded and sealed as in Egyptian examples. The hinges and automatic fastenings on some chests among the Tutankhamen furniture had no widespread congeners in the early classical world.

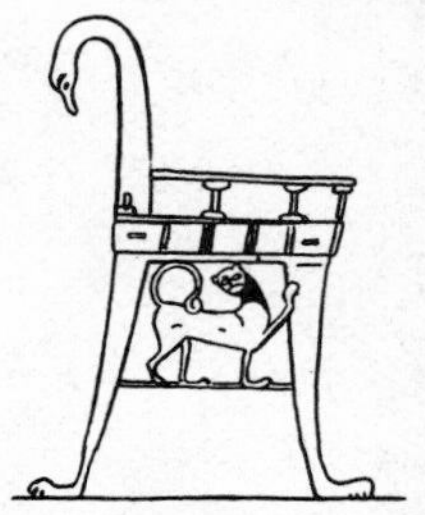

FIGURE 194—*Throne from a Greek amphora. Sixth century B.C. The legs with animal feet and the chair-back ending in birds' heads, as well as the rampant lion between the seat and stretchers, are in the tradition of the eastern Mediterranean. Note the tenons fixing the cross-rails to the legs.*

One article of Greek furniture, however, has some claim to be considered as an original product: this is the characteristic chair or *klismos*, which, though it introduces no new principles of construction, owes nothing in its design to Asiatic or Egyptian models. It consists of a deeply curved back-rail supported by two stiles, usually in a piece with the rear legs, and a central stile; the legs are curved outwards, and the seat appears to be woven from strips of hide or fabric (figure 195). The *klismos* is remarkable as being a notable attempt to ignore tradition in combining elegance with comfort. Its lines were not fixed by custom or ritual and could be varied to suit the taste of the joiner or his patron (figure 162).

The chief novelty introduced by the Greeks was the widespread use of turnery, particularly after the seventh century B.C. The art of turning on the lathe was probably developed in the wood-producing areas of the Middle East, whence it

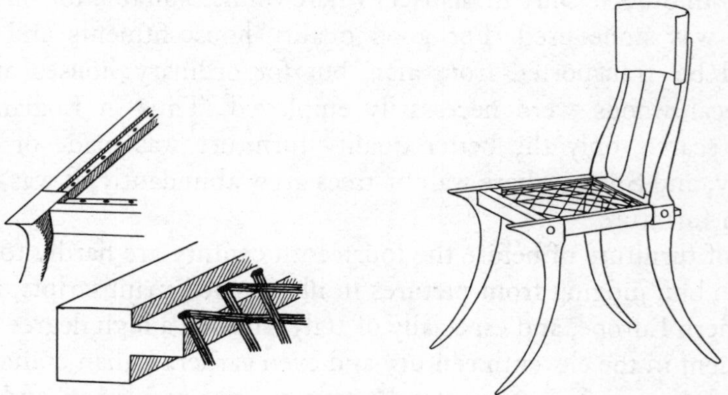

FIGURE 195—*Reconstruction of a Greek chair or* klismos *of the fourth century B.C., a design of purely Greek origin. The detail shows a suggested method of thonging the seat with hide or cord lashings.*

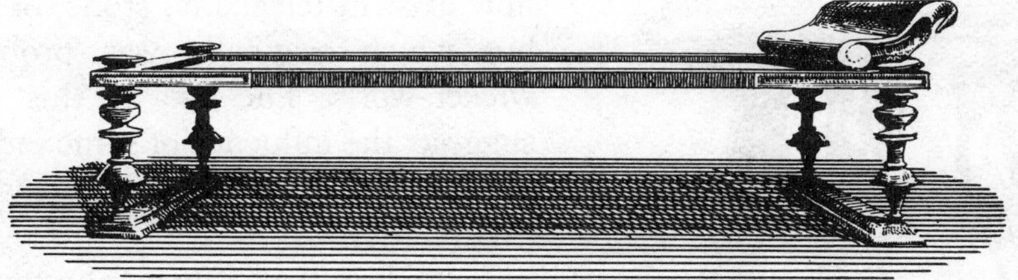

FIGURE 196—*Bronze couch from Boscoreale. First century* A.D. *The simple rectangular frame rests upon legs inspired by elaborately turned archetypes in wood: the plain headboard should be compared with a more developed example in plate* 10A. *A hide or cord mattress would originally have been woven diagonally across the frame.*

spread slowly to the Greek mainland (vol I, p 192). In the later Hellenistic period, turning was introduced into Egypt and other parts of the Roman world (vol I, p 688). Turnery encouraged the use of a lighter, more elegant form of construction (figure 196), which at times tended to the fantastic as the turner lost sight of the function of the various members in a desire to display his virtuosity. This form of construction may have been legitimate in ivory and metal but was inevitably flimsy in wood.

Etruscan furniture generally follows Greek models very closely, but there are two original Etruscan contributions to later design. The first is the use of bronze fittings and decoration (plate 10 A), and later of furniture made entirely of bronze; the second is the development of a tub-shaped chair or throne, of which models

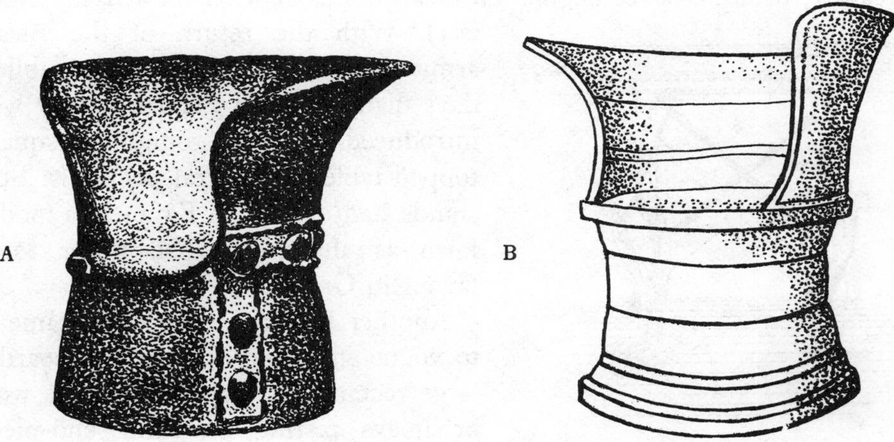

FIGURE 197—*A throne or arm-chair peculiar to the Etruscans and consisting of a rectangular or barrel-shaped base with curving sides and back apparently inspired by a woven original in wicker-work:* (A) *from a terracotta model;* (B) *shows the basis of a stone example, the Corsini chair at Rome, but without the relief decoration of the original. Fourth to third century* B.C. *Height* 81·5 *cm.*

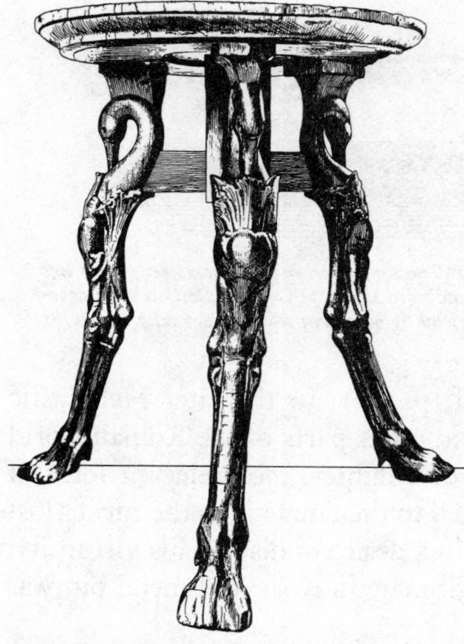

FIGURE 198—*Wooden table, partly restored, of Hellenistic design from Egypt, probably third to first century B.C. The antelope legs end in acanthus leaves from which emerge goose-heads to support the circular top.*

now exist in terracotta, stone, or bronze, but which originally was probably of wicker-work. The use of this material suggests the influence of some indigenous Italic elements and not Greek inspiration (figure 197).

It is clear that Rome inherited much Etruscan technique, particularly in the general use of bronze in furniture construction, but Roman furniture design is largely a development of the florid, later Hellenistic style which appealed to the Roman taste for luxury and opulence. Apart from certain technical refinements, the main contribution of Roman designers was in the development of substantial tables quite different in function and effect from the small and slight constructions that had served as stands hitherto (figure 198). Elegant tripods in metal (plate 10 B), and folding tables with inlaid stone or wooden tops on metal legs (figure 199), were in use in the homes of the wealthy, while stark rectangular tables appear as benches for artisans (figure 201). With the return of the Asiatic armies to Rome in the later Republican days many items of Persian luxury were introduced, such as round- or square-topped tables on central pedestals. Such stands had existed in Egypt in a modest form as altars, but had never found favour in Greece.

FIGURE 199—*Folding table from Pompeii. The legs are of bronze and enriched with devices showing satyrs holding rabbits; the cross-braces slide upon rings in guide-rails. The top of such a table would be of stone or wood, handsomely enriched. Before A.D. 79.*

Another Eastern novelty that came into vogue in Rome was the sideboard, a large rectangular top of expensive wood or inlays resting on solid end-pieces usually of marble elaborately carved (figure 200). Such furniture was used for the display of vessels in precious metal and other valuable objects. In some cases sideboards seem to have had partitions below, for the storing of additional plate. The large dining-table did not come into fashion until later, when the custom decayed of reclining on couches at meals. The cupboard also made its appearance in Roman times, as a large rectangular chest mounted on feet, with interior shelves and doors turning on hinges (figures 201, 210).

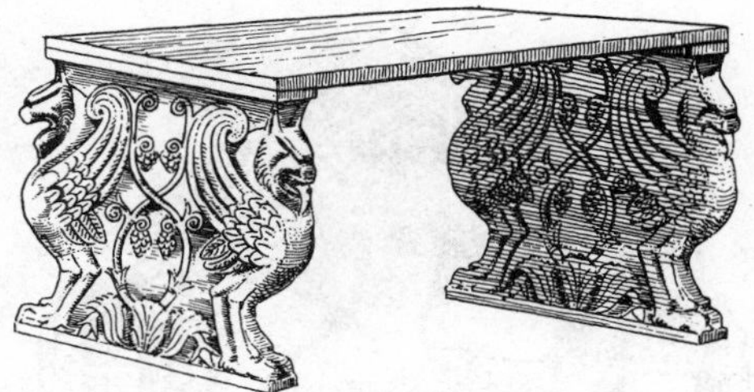

FIGURE 200—*Sideboard, consisting of two carved marble supports, from Pompeii; the top of expensive wood or stone has been restored in the illustration. Before* A.D. *79.*

With the steady impoverishment of the later Roman world, with unstable political and economic conditions, with the barbarian invasions of the fifth century, and with the spread of a more austere Christian outlook, the taste for elegance and luxury faded away. A general decline in artistic skill, in craftsmanship, and in design naturally followed a reduced patronage and a confusion of ideas. Classical ideals gradually gave place to an outlook that can be distinguished as medieval.

In Byzantium, however, the heritage of the classical world, particularly of the late Hellenistic age, was preserved—though infused with elements from the Asiatic hinterland. The craftsmen of Byzantium, especially in the imperial workshops, could continue to produce objects of refined classical taste, notably in metal-work and ivory. As far as furniture is concerned, however, very few actual examples have survived, and even these show later restorations (plate 11). The representations of furniture on the ivory consular diptychs suggest that Asiatic and classical decorative elements were applied to a structure which is basically late Imperial Roman in design (figure 202).

To the north of the Mediterranean civilization lay the alien world of the barbarian Teutons, of whose furniture we have not even vestiges. It may be deduced, however, from similar contacts between highly developed and primitive cultures in modern times, that the ruling caste would show a predilection for the objects exported by Rome, while the mass of the population followed a tradition

FIGURE 201—*Scene from a fresco from Herculaneum showing Cupids making shoes. A cupboard with shelves and hinged doors with folding leaves is shown at the right: to the left is a substantial plain work-bench or table. Before* A.D. *79.*

which has survived in the peasant crafts of modern Europe. These aspects are discussed in the sections that follow.

FIGURE 202—*Byzantine ivory panel showing a consular throne and footstool. The back of the throne is rectangular and surmounted by a pediment enclosing a scallop shell. Two pilasters act as the rear stiles. The elaborately carved frame is supported in front by two animal legs each ending in a lion mask holding a ring in the mouth. The arms are upheld by two caryatids. Sixth century A.D.*

II. MATERIALS

Unlike Egypt and Mesopotamia, the central Mediterranean region at first had adequate supplies of good quality home-grown timbers—maple, beech, oak, yew, fir, holly, lime, and the like. In addition, the Romans in Imperial times imported exotic woods, such as bird's-eye maple from central Europe and Germany, ebony from Corsica and Egypt, cedar and terebinth from Syria, and citron-wood from Africa. Climatic conditions in Europe, however, have been unfavourable to the survival of wooden objects, and we are dependent upon such authors as Pliny and Theophrastus for almost all our information on the timbers of classical antiquity [1].

It would seem that the special qualities of various woods were thoroughly appreciated. Ash, for example, is mentioned with beech as a 'moist' wood, particularly suitable for making resilient bed-frames, and that grown in Gaul for certain parts of carriages on account of its great pliancy. Beech was considered suitable for tables, chairs, beds, and caskets. Cedar was in great demand, that from Africa, Crete, and Syria being prized on account of its good weathering qualities, especially in ships and buildings. Fir and pine were easily worked, and were used for doors, panels, carriage-building, and roof-shingles: their quality of taking glue well is stressed. Several varieties of oak were used, holm-oak being common for axles and mortised members and whenever toughness was necessary. Elm was commonly employed for door-frames and for thresholds having sockets to take the top and bottom pins on which doors turned. Maple, particularly curly or bird's-eye maple, was used for high-quality beds and tables. Cypress, like cedar, was employed for making images of the gods, and for chests, doubtless on account of its tough, stringy texture, which is able to withstand rough usage. The buckthorn, holly, lime, and boxwood were regarded as best for turning, for which the sycamore also was doubtless used. The woods mentioned as particularly suitable for use in tools are the wild olive, box, elm, ash, and pine (p 85). Common woods, horn, and tortoise-shell were sometimes grained and painted to simulate more costly varieties of wood such as terebinth, citron, and maple.

Veneering, which in ancient Egypt had been used as much for economy as for appearance, appealed particularly to Imperial Roman taste. Citron-wood from Mauretania, for instance, was held in very high esteem; the mania for employing it reached such proportions that supplies became exhausted, and its value is said to have exceeded that of gold. A freedman of the Emperor Tiberius was credited with having a citron-wood table almost four feet in diameter, made from a single piece of wood [2]. This, however, was exceptional, for citron-wood was most commonly employed as a veneer. Wild and cultivated olive, yew, juniper, ebony, poplar, terebinth, box, maple, palm, holly, alder root, and holm-oak were veneered on fig-wood, willow, plane, elm, ash, cherry, or cork-wood.

Thin laminae of tortoise-shell were also used as veneers. Carvilius Pollio (*c* 80 B.C.), 'a man of lavish talent and skill in producing objects of luxury', is reported as the first maker of beds and cabinets thus decorated [3]. Horn and ivory, both natural and stained, besides coloured opaque glass and jewels, were used for inlaying and enrichment. Plating with gold and silver as well as bronze was common among the household possessions of the wealthy. The more extravagant even luxuriated in furniture entirely of precious metals. Bronze and wooden couches frequently had fittings inlaid with other metals; probably the most notable surviving specimen of this type is the bronze couch in the Palace of the Conservatori, Rome, with its ornate head and foot rests (plate 10 A).

In early Christian times economic stringency and a more severe taste resulted in less ornate and opulent furniture. Inlaying became more in the nature of overlaying, and such specimens as the so-called chairs of St Peter at Rome and of St Maximin at Ravenna gave full scope to the fashion for applied panels of carved ivory (plate 11). This was a craft at which Byzantium excelled, possibly because she inherited a long established eastern Mediterranean tradition of ivory-carving (vol I, ch 24).

To the north of the Roman Empire, and in uneasy contact with it for trade, tribute, and employment, was the barbarian world, stretching from Britain virtually to Siberia. This region, inhabited by various Teutonic and Slav peoples, had an entirely different climatic and physical environment. Above all it was a thickly forested area which, unlike the Mediterranean, did not yet suffer from dwindling supplies of timber, but on the contrary had ready access to unlimited quantities of it. On the free use of wood was founded the whole economy of the region.

Thus among these northern barbarians timber took the place of much of the stone, brick, pottery, glass, and metal employed in the cultures of the Mediterranean. For the most part the woods such as pine, beech, birch, alder, and willow were soft and could be worked with comparative ease, but, apart from a few specimens preserved by chance in bogs, we have little direct evidence on the subject, and literary sources are practically non-existent. We can, however, note the persistence of a tradition of wood-working that survived until the early years of this century in the cottage crafts of northern Europe.

Furniture was made of metal as well as of wood in the Mediterranean world, and this topic is considered later (p 235).

III. TOOLS

The tools employed by wood-workers in the classical world differed radically from those used previously in that they were of iron, not of copper or bronze. For this reason they were more rapid and effective in use, and a higher standard of accurate workmanship was more generally attainable, though the craftsmanship of earlier ages at its best could still only be equalled, not surpassed.

The hoard of Assyrian tools found by Petrie at Thebes in Egypt suggests that by the eighth century B.C. iron tools had reached a fully developed form (figure 203). It is, however, with Roman tools that we shall be chiefly concerned, since they, at the end of our period, show the greatest development and are the best represented. References in the literary collection known as the 'Greek Anthology',[1] suggest that the prototypes of these implements existed in Greece, for we read of one carpenter that his tools consisted of the rasp, plane, line and colour-box, hammer, ruler, bow-drill, heavy wood-axe, auger, gimlet, adze, and punch. Another workman speaks of his cubit-rule, rigid saw with curved handle, axe, plane, and gimlet [4]. It would be strange indeed if the Greek with his science and ingenuity had not improved traditional tools besides inventing new.

Nearly every iron tool, in fact, shows an improvement on the design of copper prototypes. The socketed axe- and adze-heads could be fitted into wooden hafts more firmly than tanged copper blades could be lashed with raw-hide on to their handles. More powerful and cleaner blows could be struck with such tools, speeding the work and lessening the need for subsequent rasping and filing.

[1] It was put together in the sixth century A.D. but contains elements of a much earlier date.

The chisels are also more massive and are fitted with tangs or sockets, allowing heavier blows on the wooden handles, which were frequently secured with a ferrule: at the same time a difference of function is clear between the bevelled

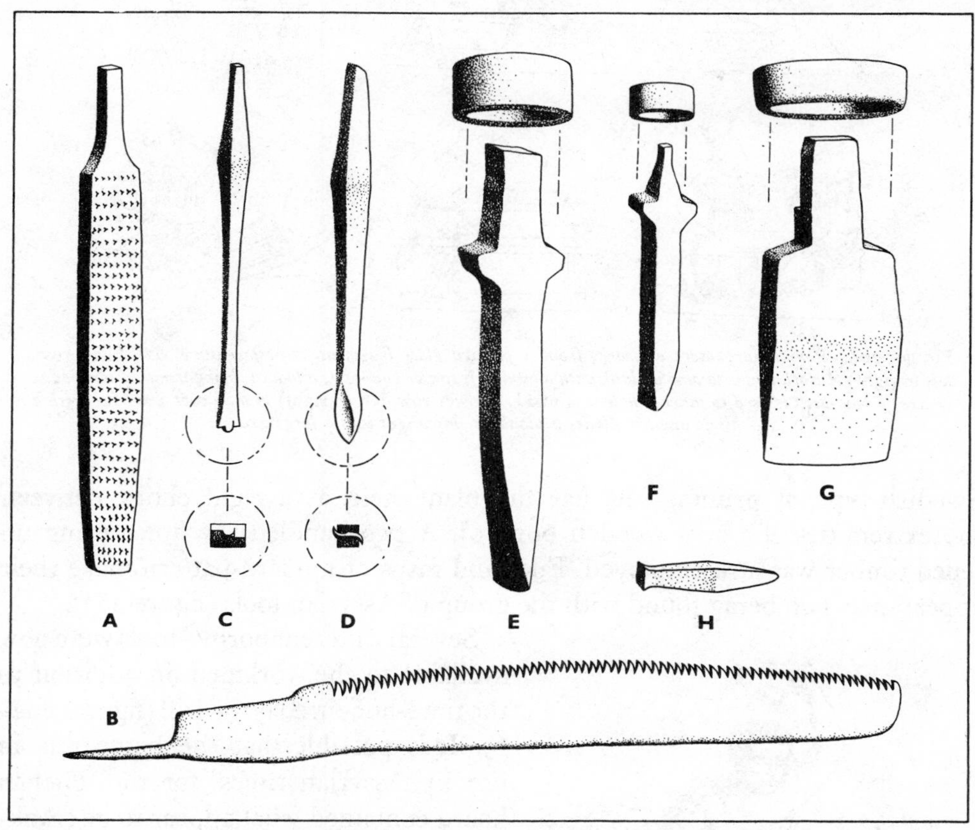

FIGURE 203—*Selection of Assyrian iron tools found at Thebes, Egypt, eighth century* B.C. (A) *Rasp with irregularly punched grating-surface, tapering and thinning at the end, with tang for insertion in a wooden handle;* (B) *saw-blade with teeth raked to cut on a pulling stroke;* (C) *a centre-bit for use in a brace or with a wooden handle like a modern gimlet; the two edges are formed as scrapers about a central pivot;* (D) *a spoon-bit for use as* (C), *the scoop lying on each side of the central point;* (E) *a heavy chisel with an equal bevel on opposite faces, and with a flange and ferrule to secure in a wooden handle;* (F) *similar but smaller chisel with bevel on one face only;* (G) *wide chisel with equal slope on opposite faces and accompanying ferrule to fit a massive wooden handle;* (H) *a small punch. Scale* c 1/3.

paring-chisel, the firmer-chisel, and the gouge [5]. The saw was greatly improved. Iron blades made possible raked teeth in place of the old irregular notching, and a pushing as well as a pulling movement could be used; examples exist in which the teeth were 'set' so as to form a channel for the continuous removal of wood-dust with each pull and thrust. A saw similar to the modern

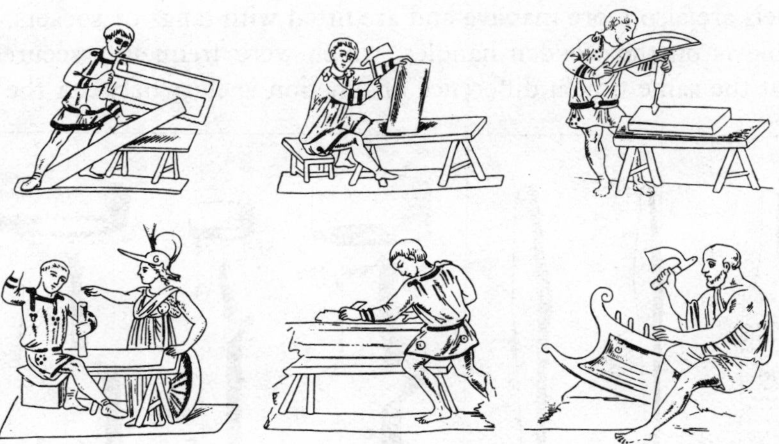

FIGURE 204—*Roman carpenters at work, from a painted glass fragment,* c *first century* A.D. (Top row, left to right) *A workman is sawing a plank with a kind of frame or bow-saw; a block is being trimmed with adze or axe; a bow-drill is used to make a hole in a plank.* (Lower row, left to right) *A man cuts a mortise with a chisel; another planes a plank; a shipwright uses a large axe.*

Swedish type of pruning-saw has the blade held as a rigid chord between the extremities of a bent wooden bow [6]. A two-handled saw for cutting up felled timber was also employed. Files and rasps of modern pattern make their appearance, two being found with the group of Assyrian tools (figure 203).

FIGURE 205—*Carpenter at work cutting mortises in a beam which is wedged in position on the bench by pegs; on the ground is a bow-drill. From a fresco at Pompeii. Before* A.D. *79.*

Several different boring-tools were now available to the workmen, in addition to the time-honoured bow-drill (figures 204–6). It is possible that the brace was in use by Assyrian times, for the Theban hoard contained what appear to be crank-pieces and centre-bits from a brace. The wooden parts had perished but the centre-bits remained [7]. Roman bits with square-section shanks are known, though braces have not so far been traced. The Assyrian bits were of a modern pattern, except that, instead of a cutter and scraper on either side of a central point, there were two scrapers; these would not make so clean a hole. Another Assyrian bit is a scoop-drill of

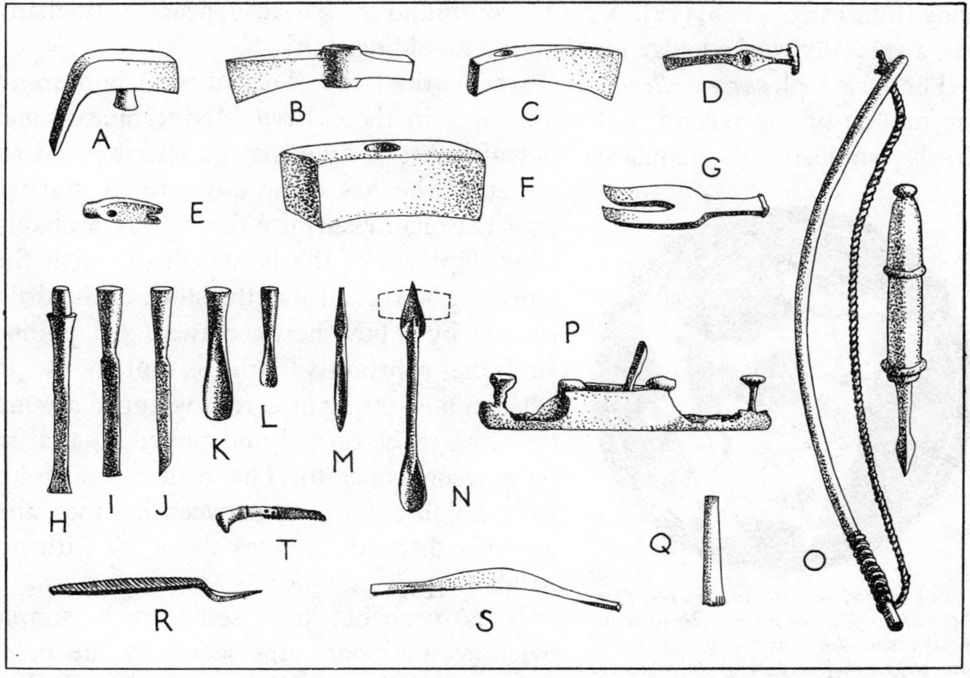

FIGURE 206—*Selection of Roman carpenter's tools, mostly of iron:* (A) *adze-hammer head;* (B *axe-hammer head;* (C) *axe head;* (D) *hammer head;* (E) *claw-hammer head;* (F) *wooden mallet head;* (G) *nail-lifter;* (H) *socketed chisel with deer-horn handle;* (I, J) *socketed chisels;* (K, L) *socketed gouges;* (M) *spoon-bit;* (N) *auger for use with a wooden handle;* (O) *bow-drill, doubtless the bit was interchangeable;* (P) *iron carcase of a plane;* (Q) *plane-iron for shooting mouldings;* (R) *rasp;* (S) *draw-knife;* (T) *double-edged hand-saw with deer-horn handle. Scale* c *1/10.*

S-shaped cross-section, suggesting that it was for use in a continuously rotating holder and not a reciprocating drill such as the bow-drill. Other scoop-drills like the modern quill-bit or spoon-bit come from Roman sites [8]. The bob-drill or pump-drill cannot be traced before Roman times, when a specimen is known from Hawara in Egypt [9]. It is worthy of remark that a primitive pump-drill was being used by the natives of Indonesia in recent times (vol I, figure 114). The shipwright's auger was in use by Dynasty XXVI, *c* 600 B.C., to judge from a specimen found at Tell Defenneh in Egypt [10].

Adzes, used for truing timber from earliest times, now had longer handles, allowing both arm and wrist to control the pressure and direction of each stroke (figure 207). The draw-knife is known from Roman times, though this essential tool of the wheelwright and the cooper was almost certainly in use much earlier [11]. The plane has also survived upon several Roman sites. Its cutter consisted of a single blade wedged at an angle in an iron-shod wooden body (figures 204, 206) [12]. A plane-iron found at Newstead, near Nottingham, had a specially worked edge for shooting mouldings [13].

The new tool *par excellence*, however, is the lathe. Turned work appears in the middle of the second millennium B.C. in the eastern Mediterranean, and slowly establishes its popularity thereafter. It is true that no specimen of an ancient lathe has come down to us, but we may reasonably surmise that it was probably a development of the bow-drill in which the object to be turned was the stock of the drill, rotated by a bow between two fixed points. In India, a primitive lathe is still in use in which a bow-string or cord is wrapped around the work to be turned and pulled to and fro by a second operator. The turner sits at his work holding the tools between his toes, and controls them and guides the work with his hands [14].

FIGURE 207—*Greek carpenter at work. From a dish of the fifth century B.C. He trims and smoothes a wooden beam with a long-handled adze gripped half-way so as to achieve a delicate, balanced stroke. The whole arm is employed with a regular circular sweep, the left hand controlling the application of the wood to the cutting stroke.*

It is probable that some such simple machine was soon superseded by the pole-lathe, which in essentials consisted of a bed with uprights between which the work, roughly hewn to a cylindrical shape, was revolved. A resilient pole, perhaps the branch of a living tree, projected over the bed, and a cord attached to its end was wrapped round the work and thence secured to a hinged board serving as a treadle. When the board was depressed the cord spun the work round in one direction; when the pressure of the foot was relaxed the bough sprang back, drawing the cord and treadle upwards and turning the work in the opposite direction. Motion was thus reciprocating, not continuously rotatory, and this allowed the wood cut by the simple hook-tools on the down-stroke to be cleared of shavings on the up-stroke, thus avoiding clogging and loss of control. It is probable that such tools were improvised in woods where timber was being cut, rural industries of this sort being carried on in Britain—in the Chilterns for instance—until very recently. The earliest representation of a pole-lathe occurs in the thirteenth century (pp 643 f), but this is almost certainly a medieval survival of a much earlier type [15].

Wood-turnery with its geometrical perfection appealed particularly to the taste of the Greek and Roman, and is the equivalent of wheel-thrown pottery.

The lathe for turning wood is mentioned by Aeschylus and Plato (*c* 400 B.C.) as well as by later writers such as Virgil. So much was turning considered an essential technique of wood-work that Plato refers to it specifically in mentioning the craft of carpentry. The use of the lathe spread from the Mediterranean to northern Europe, where a different tradition of wood-working prevailed. The old methods of turning wood, however, persisted until recent times, especially in peasant crafts concerned with such humble wood-ware as chair-legs, bowls, cups, and platters [16].

In classical times there was rather more specialization among wood-workers than there had been previously; and a distinction appears to be drawn between makers of couches, chests, and caskets. A chest-makers' quarter existed in Athens and there was a carpenters' quarter in Rome. The Greek word for carpenter, *tektōn*, is related to the idea of building, whereas the Latin, *lignarius*, means a worker in wood. *Carpentarius* was the name of the wood-worker who specialized in carriages (*carpenta*), so the present-day general term 'carpenter' originally meant a wainwright (p 540).

IV. TECHNIQUES

The techniques employed in the classical world of Greece and Rome appear to have been determined by the same economical use of wood that had been so large a factor in methods of wood-working in the eastern Mediterranean cultures. The classical desire for perfection, and the greater profusion of woods, made unnecessary such subterfuges as the replacement of flaws by patches or inserts, while the more general precision obtainable with sharp iron tools lessened the need for stopping and coating with gesso (vol I, p 685). Moreover, the better qualities of wood available in large sizes even for ordinary work obviated such devices as plywoods and the many ingenious joints that were used, for instance, by Egyptian wood-workers (vol I, p 692).

The seasoning of timber was now well understood. Pliny reports that some timber had its bark cut around while still standing so as to expose the wood and to allow the sap to flow out [17]. Other timber which tended to split naturally, especially in too drying a wind, was covered with manure, presumably so that the process of seasoning could be made more gradual [18].

Dowels, tenons, and tongues were still used for joining various members together. Thus in reliefs or sculptures showing Greek chairs the tenons and mortises uniting the seat to the legs are carefully depicted, as in examples of Egyptian chairs made a thousand years earlier. The dove-tail joint was known, under the name of the 'little axe-head', but is much less common than in ancient

Egypt, though the paucity of actual remains may give us an unbalanced picture. Metal nails were also used for joining different parts, and where the heads were left exposed they were given some embellishment. Glue and isinglass were used as adhesives, particularly for the veneers so common in Imperial times.

Wood was bent artificially, probably by steaming. Theophrastus states that most tough woods are easy to bend, the mulberry and wild fig in particular; the last two woods were used for making theatre-seats and the hooped foundations of wreaths or garlands [19]. The curved legs of the *klismos* (p 222), which was also the design for most theatre-seats, must have been obtained either by bending or by training branches to grow into the required curve.

The chief innovation in classical times, however, was the widespread turning of wood on the lathe. This technique appeared in the middle of the second millennium B.C., probably in the first place as a means of making elegant use of small pieces of timber; but it quickly established itself as an essential process appealing to classical ideas of symmetry. At first, and as long as considerations of utility and strength were kept in mind, such turning remained within reasonable bounds, but in the third and second centuries B.C. the love of the elegant and bizarre encouraged elaborate and even fantastic turning, which must often have been most impractical. Few examples in wood have survived, but metal furniture frequently shows in its design the influence of highly ornamental turning (figure 196).

Wood was appreciated for the quality and colour of its grain, the parts with wavy or knotted grain from near the roots and poles of pollarded trees being most in demand. Various grains were carefully distinguished, such as tiger-striped, panther-spotted, and a maculate kind described as resembling a swarm of bees. The expensive and scarce citron-wood was prized for a grain likened to wine mixed with honey; it is arranged in waves or small spirals. Special care was taken to finish wood so as to enhance its natural appearance. Even the painting of one wood to imitate the effect of another was but an expression of this same anxiety. In addition to the usual abrasives, the skin of the ray was used to get a fine finish. Polishes were made from oil of juniper or cedar mixed with beeswax.

In later times, increasing general poverty, political unrest, and a change of outlook resulted in much less ostentation and luxury, particularly in the west. The eastern part of the Empire with its centre at Byzantium still maintained its own opulent traditions, which were those of the later Hellenistic world modified by contact with the cultures of the Near East. Here furniture reveals a preference for the turned frame and for decoration of ivory and metal inlays.

Besides wood, metal was increasingly used for furniture, particularly in Imperial times. This may well have been a development of Etruscan techniques in the use and casting of bronze. A peculiar style was evolved for metal furniture, with its preference for light and elegant forms, frequently enriched with mouldings, figures, and other excrescences until the fantastic was reached. This in turn influenced the design of wood furniture, to judge from two table-legs from Roman Egypt which have been made so thin and attenuated as to be structurally unsound (figure 208): the companion leg was in fact found broken at its weakest point. Some of the metal furniture, however, is made to fold, and shows an early ingenuity in devising means whereby it may be opened from a collapsed state and yet be rigid in use (figure 199). Anticipating metal furniture there must have been a widespread use of metal furniture-fittings, which have, in fact, survived from the past in great quantities—legs, corner-pieces, locks, staples, hinges, and ornamental *appliqués* (figure 209).

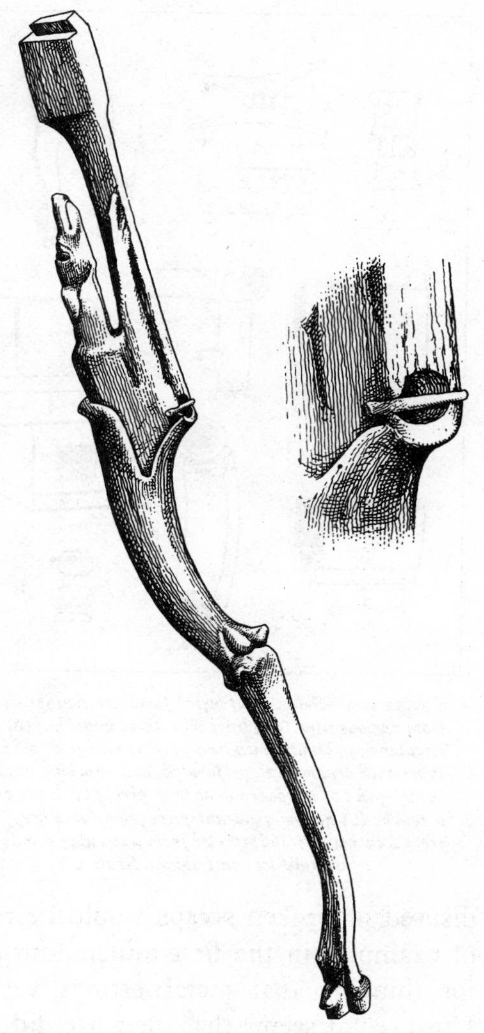

FIGURE 208—*Leg of a table or stand from Egypt; cf figure 198 and plate 10 B. The animal foot is surmounted by foliage from which emerges a coursing hound. The flimsy nature of the design suggests a bronze prototype. The detail shows the method of fitting and dowelling cross-braces.*

It is true that metal furniture existed as early as Assyrian times, for the sides of an elaborate bronze throne and footstool, dating from the reign of Ashur-nasir-pal II (*c* 850 B.C.) are preserved. Metal fittings such as silver feet and angle-pieces were also used on Egyptian caskets as early as the Middle Kingdom (*c* 1900 B.C.). The camp-bed of Tutankhamen (1350 B.C.) has massive copper hinged sockets with subsidiary hinges for folding legs. Since metal was a precious commodity,

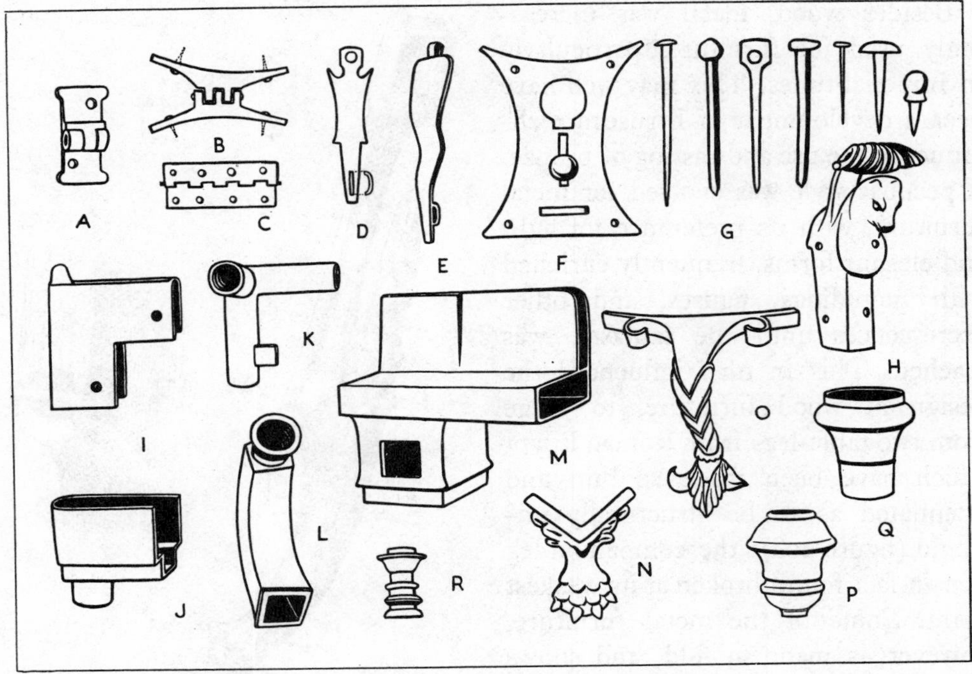

FIGURE 209—*Selection of metal furniture-fittings in bronze and iron, third century* B.C. *to first century* A.D., *from various sites.* (A) *Hinge with three knuckles;* (B) *hinge of five knuckles with curved leaves for attachment to circular box, bronze with iron pins;* (C) *hinge of nine knuckles;* (D, E) *hasps;* (F) *lock plate;* (G) *nails of various types with ornamental, perforated, and clinching heads;* (H) *handle of article of furniture, cast hollow;* (I) *upper corner-piece of a wooden door with pivot;* (J) *lower corner-piece with pivot of another door;* (K) *corner-piece of a couch;* (L) *hollow quadrant-piece from furniture;* (M) *corner-piece from a cabinet;* (N) *foot and corner-piece from a casket or stand;* (O) *leg from a circular stand;* (P, Q) *hollow caps from pieces of furniture;* (R) *ferrule for furniture leg, cast solid. Scale* c *1/4, except* (B), (F) c *1/2;* (D) c *1/3;* (K), (O) c *1/8.*

disused or broken scraps would be recovered for melting down; the absence of examples in the first millennium (B.C.) should not be accepted as evidence for thinking that metal fittings were not employed before Roman times. However, it seems that their use did not become general until the prosperity, resources, and improved techniques of Imperial Rome increased the supply of worked bronze. Fittings of a less decorative function were available in iron, though these have not survived in such quantities or in such good condition. Hasps, corner-pieces, bands, hinges, and ironmongery of a similar kind were made in all parts of the Empire, their design eventually becoming part of the stock-in-trade of the medieval blacksmith. The iron-bound wooden coffin of a Lombard chief (*c* seventh century A.D.) preserved at Innsbruck, is an example of the technique as adapted by the barbarian peoples of the north [20].

The processes of wood-working in the thickly forested region of northern

Europe were generally rather different. Here, as already mentioned, there were profuse supplies of suitable timber, together with methods of working it which, while they might employ the trade-tools of the Roman world, were really based upon a Neolithic tradition. This technique was inherited, though modified, by the medieval craftsman, and may be described in brief as axe-, adze-, and knife-work rather than the saw-, chisel-, and plane-work of the Mediterranean world: a carpenter's technique as distinct from a joiner's (p 241).

Unfortunately very few traces of such work have survived in the whole of this vast area, so that an estimate of its appearance has to be sought in the Viking ship-remains and in a wood-working tradition persisting almost to the present day in peasant crafts. From these we can see that a severe, framed, tenoned structure, depending for its appeal on proportion and clean jointing, was eschewed. The basis of construction was rather the pegged plank or pole, adzed to a faceted finish and then minutely chip-carved with the knife, supplemented for deeper carving with the chisel. On the whole, softer and easily worked woods were selected, and the final appearance was enhanced with bold colouring [21]. The clinker-built boats of Oseberg and Gokstad (figure 528), of *c* A.D. 800, and the mast-churches of Norway built like inverted ships, show these techniques at their most developed.

Until recent times, domestic ware in this northern region was also of wood carved from the solid. A Celtic box of the ninth century A.D. found in an Orkney bog is cut from a block of alder and decorated with chip-carving [22]. A chest at Carnarvon and another coffer in Wimborne Minster, both dating from Anglo-Saxon times, are hewn from tree-trunks, like dug-out canoes—the survival of a technique going back to Neolithic times [23].

Turning on the lathe was the only process from the classical world to invade the northern region on an appreciable scale. A coffin of an Alemannic chief of the sixth century A.D. is made of turned members. Anglo-Saxon churches, like that at Earls Barton, Northamptonshire, show a translation into stone of timber prototypes in which turned balusters are a prominent feature.

Lastly, some mention should be made of wicker-work, which seems to have been a widespread craft in the region, probably because ample supplies of osiers were ready to hand. This, too, was almost certainly the survival of a prehistoric craft that had not changed its techniques throughout the ages. A Gallo-Roman relief of the third century A.D. shows a basket-chair of surprisingly modern form [24]. In Italy wicker-work probably influenced the design of certain Etruscan seats or thrones, but it had ceased to be the furniture of an aristocratic class long before the end of the Republic.

During the period under review we have seen how the traditions of furniture design and technique evolved by the ancient cultures of the Mediterranean were absorbed and perfected by Rome, whose inclination was predominantly towards luxury and ostentation. Practically every technique rediscovered by the furniture-makers of modern Europe was already known to the Roman craftsman with his improved iron tools—veneering, inlaying, damascening, graining, turning, and the use of such materials as tortoise-shell, ivory, gold, silver, bronze, and semi-precious stones. This aristocratic taste for costly and splendid display passed away with its patrons, and was not revived until a later aristocratic age. The basic techniques of the Roman joiner, however, were not lost but were preserved by a new patron, the Church, even if Christian ideals were at most times inimical to opulence and worldly show. These techniques modified the more primitive methods of wood-working practised in northern Europe, to produce the medieval style.

REFERENCES

[1] PLINY *Nat. hist.*, esp. XIII–XVI. (Loeb ed. Vol. 4, pp. 98 ff., 1945.)
THEOPHRASTUS *De historia plantarum*, esp. III–V. (Loeb ed., 'Enquiry into Plants', Vol. 1, pp. 158 ff., 1916.)
[2] PLINY *Nat. hist.*, XIII, xxix, 91 ff. (Loeb ed. Vol. 4, pp. 152–4, 1945.)
[3] *Idem Ibid.*, IX, xiii, 39. (Loeb ed. Vol. 3, p. 190, 1940.)
[4] *Anthologia Graeca*, VI, 204 and 205. (Loeb ed. Vol. 1, p. 404, 1916.)
[5] PETRIE, SIR (WILLIAM MATTHEW) FLINDERS. 'Tools and Weapons', chap. 4. Egypt. Res. Acc. and Brit. Sch. Archaeol. Egypt, Publ. 30. London. 1917.
[6] *Idem. Ibid.*, Pl. LI, no. S 32.
[7] *Idem.* 'Six Temples at Thebes', p. 19 and Pl. XXI. Quaritch, London. 1897.
[8] CURLE, J. 'A Roman Frontier Post and its People', Pl. LIX, no. 12. Maclehose, Glasgow. 1911.
[9] PETRIE, SIR (WILLIAM MATTHEW) FLINDERS. See ref. [5], Pl. XLVIII, M 4.
[10] *Idem et al.* 'Tanis Pt. II: Nebesheh and Defenneh', p. 78, Pl. XXXVIII, 4. Egypt Exploration Fund, Mem. 4, 1888.
[11] PETRIE, SIR (WILLIAM MATTHEW) FLINDERS. See ref. [5], p. 39, § 107.
[12] *Idem. Ibid.*, p. 39, § 108.
[13] CURLE, J. See ref. [8], Pl. LIX, no. 2.
[14] HOLTZAPFFEL, J. J. 'Turning and Mechanical Manipulation', Vol. 4, pp. 1–25. Holtzapffel, London. 1879.
[15] SALZMAN, L. F. 'English Industries of the Middle Ages' (new enl. ed.), p. 172. Clarendon Press, Oxford. 1923.
[16] PEATE, I. C. 'Guide to the Collection Illustrating Welsh Folk Crafts and Industries' (2nd ed.), chap. 2. National Museum of Wales, Cardiff. 1945.
[17] PLINY *Nat. hist.*, XVI, lxxiv, 192. (Loeb ed. Vol. 4, p. 512, 1945.)
[18] *Idem Ibid.*, XVI, lxxxi, 222. (Loeb ed. Vol. 4, p. 532, 1945.)
[19] THEOPHRASTUS *De historia plantarum*, V, vi, 2; vii, 3–4. (Loeb ed. 'Enquiry into Plants', Vol. 1, p. 452, and pp. 456 ff., 1916.)

[20] BROWN, G. B. 'Arts and Crafts of our Teutonic Forefathers' (Rhind Lectures for 1909), p. 108; Pl. IV, fig. 15. Foulis, London and Edinburgh. 1910.
[21] SCHMITZ, H. 'The Encyclopaedia of Furniture', p. 8. Zwemmer, London. 1936.
[22] STEVENSON, R. B. K. *Proc. Soc. Antiq. Scotld*, **86,** 187, 1951–2.
[23] POLLEN, J. H. 'Ancient and Modern Furniture and Woodwork', Vol. 1 (rev. by T. A. LEHFELDT), p. 55 (Footnote). H.M. Stationery Office, London. 1908.
[24] COTCHETT, LUCRETIA E. 'Evolution of Furniture', Pl. 1. Batsford, London. 1938.

BIBLIOGRAPHY

BLÜMNER, H. 'Technologie und Terminologie der Gewerbe und Künste bei Griechen und Römern', Vol. 2, Zehnter und Elfter Abschnitt. Teubner, Leipzig. 1879.
PETRIE, SIR (WILLIAM MATTHEW) FLINDERS. 'Tools and Weapons.' Egypt. Res. Acc. and Brit. Sch. Archaeol. Egypt, Publ. 30, London. 1919.
POLLEN, J. H. 'Ancient and Modern Furniture and Woodwork', Vol. 1 (rev. by T. A. LEHFELDT). South Kensington Museum Art Handbook no. 3. H.M. Stationery Office, London. 1908.
RICHTER, GISELA M. A. 'Ancient Furniture.' Clarendon Press, Oxford. 1926.
SCHMITZ, H. (Ed.) 'The Encyclopaedia of Furniture.' Zwemmer, London. 1936.

7

PART I

FURNITURE: TO THE END OF THE ROMAN EMPIRE

CYRIL ALDRED

I. TYPES AND STYLES

In general, the furniture of the early classical world does not differ radically in its construction from that produced for centuries in the ancient civilizations of the Near East. Wood-working, in fact, may be considered as a craft having a common application and development throughout the eastern Mediterranean, with its forms and techniques determined as much by the general need to use timber economically as by function or taste. The articles of furniture—chairs, thrones, stools, beds, tables, stands, and chests—introduce few innovations. Even the cupboards of the Roman Empire can find an analogy in the wooden shrines and tabernacles of Egypt, except that the latter were not fitted with shelves. In Imperial Roman times, however, there was a considerable increase in the size and elaboration of furniture.

Only very scanty remains of wooden furniture have survived from the ancient world, and our knowledge is largely derived from representations upon reliefs and in vase-paintings, or from stone or metal versions of wooden prototypes. For the most part, the furniture of ancient Greece developed under Asiatic influences, almost certainly introduced from her colonies in Asia Minor. Early chairs and stools, for instance, have legs in animal form, with a distinction clearly made between the fore- and hind-legs; and stretchers under the seat are frequently carved as purely ornamental motifs (figure 194). Couches, too, appear with animal legs similar to those of Egyptian beds, the main difference being that the foot-board is replaced by a head-board.

Although furniture with rectangular members mortising into each other can be traced in the archaic period (*c* 700–500 B.C.) and although this style became more popular in later ages, its type had been long established in the Nile valley (vol I, p 692). Again, the folding stool (figure 163), consisting like the modern camp-stool of two sets of legs each pivoting medially on each other and having a flexible seat attached to two top stretchers, can be paralleled by earlier Egyptian of the drawer, with sides either dove-tailed, nailed, or pegged to the wooden back- and front, was not lost in Italy, which had inherited most fully the culture of the ancient world. The Vatican still possesses a cabinet with small drawers that is mentioned in an eighth-century inventory.

Dove-tailed drawers were known in ancient Egypt (vol I, figure 495). In Greece and Rome drawers were probably not uncommon, though metal handles of Roman date, believed to be drawer-handles, are the only material evidence. It is therefore likely that the drawer continued to be made in Italy without a break from Roman to Renaissance times. In northern Europe no evidence for the drawer is forthcoming from these early centuries, and it is only in the later Middle Ages that its use is suspected north of the Alps. In any event it came into Germany, France, and the Low Countries by the Rhine Valley, a route along which many southern devices reached northern Europe. Not until the fifteenth century was the drawer found in England, where it was called a till or a drawing-box (plate 13 A).

II. CRAFT-DIFFERENTIATION OF WOOD-WORKERS

Throughout the countries of northern Europe during the Middle Ages the chief house-builder was the carpenter (plate 30 A), the origin of whose title is discussed elsewhere (pp 233, 540). He constructed the frame, the floors, and the roof, and, after the plasterer had carried out the in-filling, he turned his attention to the internal woodwork. At the beginning of our period he also made furniture.

Like the frame of the house, early medieval furniture was heavy. Its construction was of stout posts and thick boards with simple joints held together by nailing and ironwork. Ambries (Latin *armaria*) and presses were formed of boards and the long dining-tables had tops of thick planks, often a single board, held secure upon loose trestles by their great weight. Chests were made of stout planks nailed or pegged together. Seating was mainly provided by benches fixed to the wall.

There were not many movable pieces of furniture: long trestle-tables, benches, forms, settles, stools, chairs, chests, ambries, and cupboards. The cup-board, from which comes our word cupboard, was at first literally a board on which drinking-cups were placed. A space enclosed by a door, which nowadays is called a cupboard, was in medieval times known as an ambry. Ambries were mainly for food-storage and either stood free or were built into recesses in the wall.

The massiveness and coarseness of carpenters' furniture was less noticeable than may be supposed, for the people of the early Middle Ages covered their tables with cloths and carpets, their bench-seats with cushions, their bench-backs with coverings known as bankers, and their chairs with chair-cloths. Where woodwork was exposed, as in the open timber roof, the doors, and the wooden linings of walls, it was painted in colours. Thus fabrics of bright hue and paint in primary colours added to domestic life a sense of cheerfulness and comfort.

The carpenter had need to work hand-in-hand with the blacksmith, for the external doors of houses were reinforced with ironwork to prevent them from being battered in, and chests for the storage of valuables had the heavy planks of the sides, bottom, and lid covered with ironwork. Moreover, the wrought-iron nails of the blacksmith were used by the carpenter for fastening together the boards of his chests, ambries, and wall-linings. An alternative was to use wooden pegs, but this, perhaps, was a refinement of a later century.[1]

The craft of carpentry is essentially one in which the disposition of the timber beams, posts, braces, collars, and joists gives strength to the construction as a whole. 'Carpentry proper', wrote Thomas Tredgold (1788–1829), 'is the art of constructing pieces of timber for the support of any considerable weight or pressure' [1]. Before furniture could be made with any degree of refinement, a new craft, that of joinery, had to evolve. In what century this first happened it would be hard to state, but it was certainly in Italy. As is indicated by the Vatican cabinet already mentioned, the joiner or his equivalent was known by the eighth century. In western Europe the joiner did not appear until the eleventh or twelfth century, and in England not until the thirteenth. By then the paramount importance of the joint (as opposed to the disposition of the timber) was realized by those carpenters who concentrated on the production of furniture, movable and immovable. 'Joinery', says Joseph Moxon (1627–1700), 'is an Art Manual, whereby several Pieces of Wood are so fitted and join'd together by Straightline, Squares, Miters or any Bevel, that they shall seem one intire Piece' [2] (plate 15 C).

At first the joiner's tools were the same as the carpenter's, but as the joiner's furniture became more refined, and of a smaller scale than the carpenter's had ever been, the design of his tools was modified accordingly (plate 12). The obvious difference was that the carpenter's tools, having to endure more vigorous usage, were heavier and stronger. For instance, the carpenter's axe was large, in order 'to hew great Stuff', and had a long handle, which was used with both hands, 'to square or bevil their Timbers' (plate 30 A; ch 11). The joiner, on the contrary,

[1] The fixing of hinges, locks, bolts, and other iron attachments to furniture, up to the late seventeenth century, was invariably done by nails and not by screws. Screws, though known in antiquity, were not used in carpentry until the sixteenth century, when they also took the place of wedges in the crafts of the locksmith and the clock- and watch-maker. The wood-worker of northern Europe (and probably of the south as well) did not make extensive use of screws until the seventeenth century.

wielded a light hatchet (figure 226 L). As it was used with one hand, it needed no more than a short handle. The adze was used by both carpenters and joiners for smoothing timber by taking thin chips from the surface (figure 225 B). A skilful joiner was so proficient with the adze that it was difficult to distinguish an adzed from a planed surface. Planes were certainly known in classical antiquity (figure 206 P) but cannot be recognized in the Middle Ages till the thirteenth century (figures 220, 356, 358).

For rough work, carpenters used chisels with stout handles to withstand the blows of the mallet. Joiners, when using a chisel, found that the pressure of the hand was usually sufficient. 'The joiners press the edge of the Blade into the Stuff, with the strength of their Shoulders, but the Carpenters with the force of the blows of the Mallet' [3] (plate 30 A, figure 350). The mortise-chisel of the joiner was struck with the mallet; so, frequently, were the chisels of the carver.

FIGURE 210. *Bookcase containing the books of the Bible, a holder for pens, and an ink-horn. The backs of the doors show a framed panelled construction. South Italian. Sixth century A.D.*

Of all tools, the hammer is the most essential and rudimentary (vol I, ch 6). The coat of arms of the London Blacksmiths' Company bears three hammers with the motto: *By Hammer and Hand, all Arts shall Stand.* In the carpenter's, joiner's, and carver's crafts hammer and mallet were basic tools, for joints in woodwork were held by iron nails or wooden pegs (figure 225 and plates 12, 30 A).

Joiners had other characteristic methods of construction apart from the mortise-and-tenon and dove-tail joints. The most important was the loose panel held in a frame; panelling not only saved wood but made furniture lighter (plate 13 A). Like the drawer, the panel is first found in the civilizations of the Mediterranean. It was known in ancient Egypt (vol I, figures 485, 490, 495–8) and in Rome, for an ivory throne of the eighth century with a panelled frame was found at Ravenna (plate 11). A picture in the famous *Codex Amiatinus* of the sixth to seventh century shows St Ezra seated before a bookcase, the backs of the open doors of which reveal a framed panelled construction (figure 210).

Centuries elapsed before the panelled construction reached the north. It does not seem to have appeared in Germany or the Low Countries till the thirteenth century. In England, wooden wall-linings were of boards which either overlapped or had a tongued and grooved joint to prevent cracks appearing through shrinkage between the boards. These boards were not replaced by panel construction until the fourteenth century.

The wood-lined wall was of special importance in the north as a means of insulating a room against cold and damp. Thus the making of panels, or ceiling as it was called, became a specialist activity. 'The Company of the Ioyners', says the chronicler John Stowe (1525–1603), 'called also *Ioyners* and *Seelers*, of ancient standing, and reputed to be a loving Society, were incorporated by Queene *Elizabeth*, in the thirteenth yeere of her Reigne' (1571) [4]. In other words, a subdivision of the joiner's craft was recognized. The verb to ceil—whence our ceiling—is a fifteenth-century term meaning to cover with a lining of woodwork, in fact to panel.

Panelling of the better kind was of so-called wainscot oak, a special quality of timber imported mainly from the Hansa region (figure 695). 'Of all oke growing in England', says William Harrison, in his 'Description of England' (before 1577), 'the park oke is the softest, and far more spalt [short-grained] and brickle [brittle] than the hedge oke. And of all in Essex, that growing in Bardfield parke is the finest for joiners craft: for oftentimes have I seene of their workes made of that oke so fine and faire, as most of the wainescot that is brought hither out of Danske [Denmark], for our wainescot is not made in England' [5].

Wainscot[1] was quarter-cut, that is, it was sawn into planks along radii from outside to centre. Although this method of conversion was extravagant, since the boards were not as wide as when the log was plain-sawn straight across, wood thus produced had little tendency to warp or twist and shrinkage was reduced to a minimum. Sawing the wood in this manner also revealed the silver grain of the rays. This feature, now much admired, was not appreciated in medieval times, for panelling and furniture of wainscot oak were usually covered with painted decoration. Wainscot was used in England for all house-fitments and furniture of the better sort, since little English-grown oak could compare with it in mildness of texture and straightness of grain. In time, the word wainscot became synonymous with panelling in general, without regard to the wood of which it was made.

Clapboard was another variety of oak. The quartered log was not sawn, but riven or split by a riving-iron in its weakest part, which was along the lines of the rays. The result was a board flecked with the silver grain. It also had little tendency either to warp or shrink and was quick to season. Clapboard was used

[1] Scot is LG *schot*, boarding; wain may be *waeg*, wall.

extensively by the coopers for their wine casks (p 136 and plates 3 B, 38 B), and was of the greatest value to the ceilers for their wainscot panels. The uneven surface was planed or adzed away on the front side, and each panel was formed of one unjointed board.

Panels were held loose in grooves cut in the framing, to avoid splitting when the wood shrank. In the fifteenth and sixteenth centuries, the demand for panelling was such that, in the workshops of many joiners, it was more or less mass-produced. In the last phase of this northern medieval or 'Gothic' period in furniture the well known linen-fold panel became the fashion in Germany, the Low Countries, France, and England. It has all the appearance of being carved, but the linen-folds were in fact produced mechanically by a moulding-plane, only the ends being finished by the carver's chisel (plate 15 D and figure 356).

III. THE CARVER

The carver worked either in wood or in stone but not in both, because he was trained either as a carpenter or as a mason and belonged either to the carpenters' or to the stonemasons' guild. The stone-carver was originally indistinguishable from the mason, as one would expect from the profusion of stone-carving in medieval cathedrals, just as wood-carvers were first and foremost carpenters or joiners.

It has been widely inferred that the wood-carver copied the stone-carver, but this is unlikely, for stone has a granular structure and its natural unit is the block, while wood is fibrous and its unit is a beam or plank. A slender length of stone would snap, and a wooden block, owing to the slow seasoning of the heart-wood, tends to split. The limitations and advantages of wood are determined by the grain. To work with the grain, the component parts have to be carved separately and then assembled. It was not possible to have a built-up construction of this kind before the rise of the joiners' craft. Thus the twelfth-century wood-carver had to ignore the limitations of the grain by carving from the solid block, thereby imparting to his work some of the character of stone-carving. It is true that fifteenth- and

FIGURE 211—*Carving in relief. Detail from an Elizabethan oak table-box (see plate 15 B).*

FIGURE 212—*Carving with ground sunk or recessed. From a frieze of an early seventeenth-century English press-cupboard.*

FIGURE 213—*Incised or scratch carving. Detail from an English cupboard of the seventeenth century.*

sixteenth-century panelling sometimes has the moulding with a stone-mason's joint, but this was only because it was easier to work the moulding round the corners than to mitre it diagonally.

When the joiners broke away from the carpenters to found their own craft-guilds, the carvers naturally followed them, since not only did a joined construction make the carver's work easier, but carving for the domestic house and its furniture was seldom on a large scale. Later it was agreed between the two crafts that 'all carved workes either raised or cutt through or sunck in with the ground taken out being wrought and cutt with carving Tooles without the use of Plaines' [6] should come within the province of joiners, not that of carpenters.

Apart from carving in the round or in relief (figure 211), there were several simpler kinds of carving that required more patience than skill. Thus there was sunk carving in which 'the grounde has been taken out'. To emphasize the design, the recessed background was either stippled or pricked with a pointed tool (figure 212). Stippling had also the effect of obscuring any unevenness on the surface. Another simple type was incised carving, done by a single line scratched on the surface with a gouge (figure 213). To relieve the dull effect, other forms of carving were often combined with it.

An important type of carving, much used throughout Europe in the Middle Ages and later, was gouge-carving. The wood was so scooped out by the gouge as to form a pattern which followed the half-circle of the edge of the tool, giving the effect of a range of flutes (figure 214). Sunk, gouge, and incised carving were

FIGURE 214—*Gouge-carving. Detail from an English chest of the early seventeenth century.*

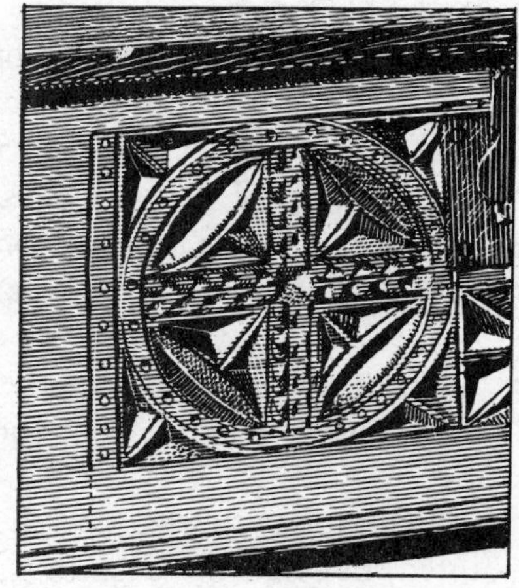

FIGURE 215—*Chip carving. From a small English box of the late sixteenth century* (*see plate 15 A*).

simple methods demanding more care than skill. The average joiner would be as efficient as the carver with these techniques.

An even more mechanical method of decoration was to stamp the wood with a steel punch, producing circles, stars, or rosettes. Stamping was used in combination with gouge, incised, and sunk carving (plate 13 D). It is found on fifteenth-century furniture and is probably of much earlier origin. Yet another type of carved decoration extensively used in Europe and England from very early times was chip-carving (figure 215). Unlike carving in relief, it was drawn not freehand but with set-square and compass. Chip-carved roundels formed a favourite decoration for the fronts of early chests.

IV. THE PAINTER

Medieval carving was not, as might be thought, left in the varnished wood, but was painted and often gilded as well. Medieval peoples, particularly those under the dull northern skies, loved bright colours and it seems never to have occurred to them to admire the natural grain of woods. Carved stone-work, to a lesser extent, was also painted and gilded. From the early thirteenth century onwards there are ample records of wall-linings and other interior wood-work copiously painted with images, histories, flowers, stars, and so on. Much of the wainscot panelling of the sixteenth and seventeenth centuries which has been preserved

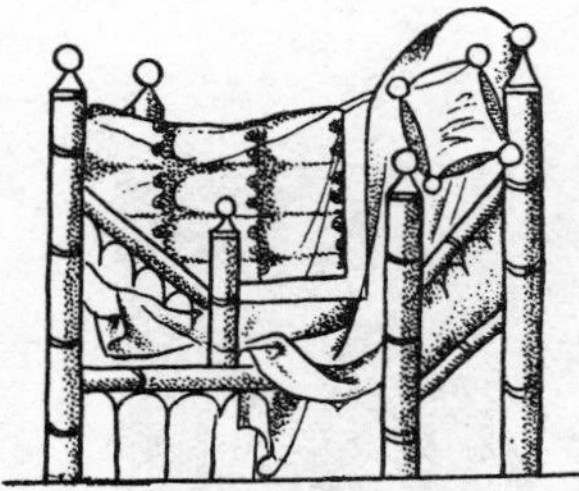

FIGURE 216—*Turned legs on a bed. From a fourteenth-century English astrological manuscript.*

with a waxed and polished surface was originally painted in colours, and decorative designs were often added to plain panels. The reason why little of this decoration has survived to our time is that when the paint became shabby it was not restored but stripped off. Furniture also was finished with coloured varnishes, and the flowers, masks, and human faces carved on it were realistically painted.

All painting and gilding had to be done by a member of the separate craft of painters. Thus a joiner was not allowed to paint the furniture he had made. To have done so would soon elicit a protest from the painters and an investigation by the city authorities.

It is difficult to say to what degree the furniture in the ordinary medieval home was painted, but that the custom was general is inferred from surviving pieces and by casual entries. Thus in the hall of one John Cadeby, mason of Beverley, Yorkshire, there stood, about 1530, 'j tabula mensalis coloreum nigri et glauci, cum rosis depictis' [one dining-table, coloured black and yellow, decorated with roses]. In the hall of Stevyn Bodyington, grocer, 1547, there was a 'Cobberd Joyned payntyd grene with a halpace [step]'. And in a north-country inventory of 1564 is mentioned 'a littel paynted ambry with ij doores'. But such items from inventories are not usual. What is far more suggestive is that although little medieval or sixteenth-century English furniture has survived, most of it bears signs of painting. What is now a plain oak chest was probably brightly decked a few hundred years ago. The wood-work of parish churches was profusely painted, and parishioners employed the painters to decorate the walls of their homes. The chest is the commonest article to be described in English inventories as painted; the colour, when mentioned, is usually red or green.

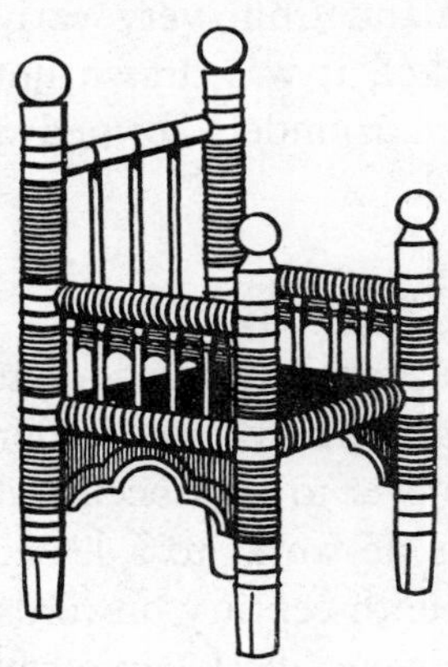

FIGURE 217—*Turned chair, as represented in a thirteenth-century relief from Chartres.*

The Royal Wardrobe accounts supply copious evidence of the painting and gilding of the furniture of Henry VIII and Elizabeth I, executed by the royal painter. Thus in 1582 he was paid for painting the wood-work of a walnut bedstead with 'good gold and silver, with beasts and flowers embossed in colour, well gilt, the ground silvered', and again 'for painting drawing and working of 347 flowers' on a satin tester and valances.

The best painted furniture had the wood treated with a thin gesso coating (p 363, and vol I, p 701) before the paint was applied, while other pieces were treated with a thin coat of priming instead of the gesso. Italian painted and gilded furniture sometimes had the wood carcass covered in a very fine stitched canvas as a foundation for the gesso, to prevent shrinkage in the wood and cracks in the gesso. With the full bloom of the Renaissance in England in Elizabeth I's reign, and under Italian influence, the northern medieval tradition of painted woodwork declined. In the citizen's parlour and bedchamber its place was taken by carving and inlaying in the natural wood, protected only by a thin varnish.

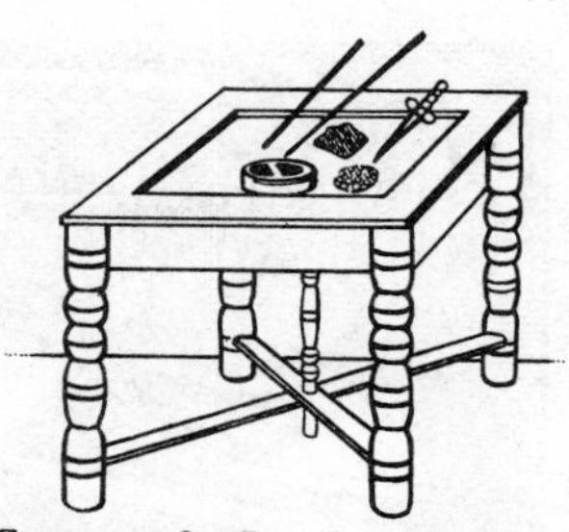

FIGURE 218—*Details of a turned table from a painting of St Luke in a tenth-century Greek manuscript.*

V. THE TURNER

By the sixteenth century the joiner's craft, with its pegged mortise-and-tenon joint and its framed panelling (plate 13 A), had reached a highly developed state. The joiner now became the complete furniture-maker. He produced all manner of ambries, presses, tables, court-cupboards (plates 13, 14, and 15),[1] benches, settles, chairs, stools, and beds, as well as house-fitments and panelling. But joinery was not the only craft that made furniture in Europe during the Middle Ages. There were several others, the chief being the turner's, which was by far the oldest. It had been known in the Near East from the eighth century B.C.

Turned furniture must have been made continuously in northern Europe from Roman times, and illuminated manuscripts from the seventh century A.D. show many turned chairs, stools, lecterns, and beds (figures 216–18). The seat of the turner's chair was supported by three or four heavy turned posts; the back rails held turned spindles, fixed by dowel joints. This design of a turned chair was familiar at least from Saxon times to the seventeenth century. During this period of a thousand years there was but little change in its design, though the earlier chairs were more massive and heavy than the later. Three-posted turned stools, judging by their appearance in medieval pictures, must have been common in Europe throughout the Middle Ages (figure 219).

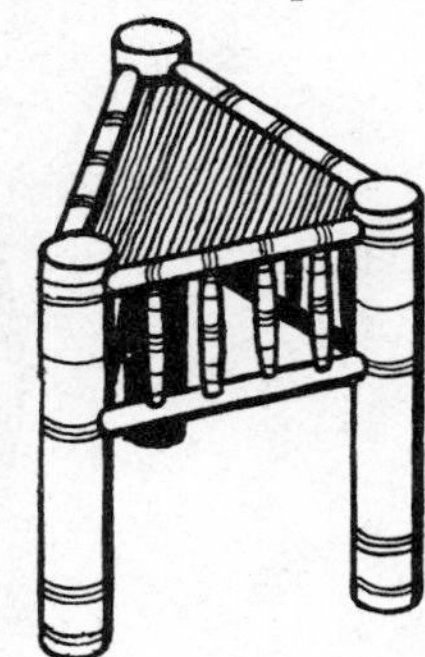

FIGURE 219—*Detail of turned three-legged stool. From a painting of c 1430.*

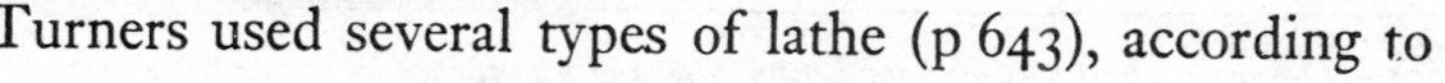

Turners used several types of lathe (p 643), according to

[1] Cupboards with open shelves.

FIGURE 220—*Carved panel depicting a joiner and a turner at work. The former is planing a board; the latter is turning a table-leg using a pole-lathe (string restored). Note the chisels and planes on wall-racks. The legs of lathe and bench are held by wedge-construction; cf figure 224. Low Countries,* c *1600.*

the size of the work to be done. For the turning of chair- and stool-legs a lathe known as the 'pole-and-tread' was used (figure 220). For heavy work there was a 'great wheel' lathe, turned by one or two handles according to the weight of the work (figure 221). A third type of lathe, the 'treddle wheel' (figure 222), was used for fine and light work.

We are now approaching a time that is, for furniture in north-west Europe, the end of the medieval or 'Gothic' period. The Gothic table, originally just a board laid on stout trestles (figure 94), was metamorphosed into one supported on a frame which at the corners had posts held by frieze-rails at the top and

FIGURE 221—*The great wheel was used for heavy work and was turned by one or sometimes two iron handles. From Moxon's 'Mechanick Exercises', 1703.*

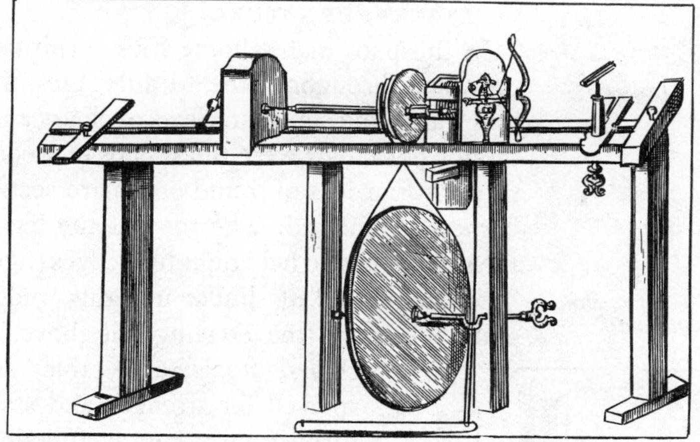

FIGURE 222—*The treadle-wheel was used for small work only. The bow was sometimes used as a spring, in oval- or rose-turning. From Moxon's 'Mechanick Exercises', 1703.*

stretcher-rails at the bottom (figure 229 and plate 14 A). Chairs, stools, and benches were becoming radically changed, for instead of being upheld by boards on edge, they were now designed with a frame formed of rails, legs, and stretchers. The posts or legs of this post-Gothic furniture cried out for some ornamentation. Turning, which could give endless variety of designs, was the obvious answer. By the sixteenth century the turners not only made their own chairs and stools, as they had been doing for centuries, but assisted the joiners in the decoration of the better chairs, stools, tables (plate 14 B–D), and bed-frames. Thus after three or four centuries of separation and competition the joiners and turners began to work together, though turners continued to produce turned furniture with dowelled joints for the cheap market.

FIGURE 223—*Example of stick-furniture. The stools are probably eighteenth-century, but are of traditional construction.*

VI. STICK-FURNITURE

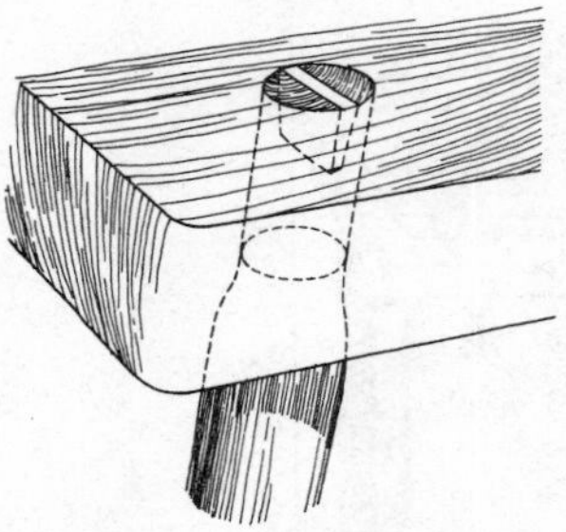

In the poor man's home stick furniture (plate 3 A) prevailed throughout the Middle Ages. Its construction was of the simplest. Seats of chairs and stools and the tops of tables were thick slabs of wood upheld on three or four legs of round or square section with the corners chamfered. The tops of the legs penetrated the seat and were held tight by wedges (figures 223–4). The backs were of similar uprights wedged into the seats below and the cresting-rail above. The legs of chairs, stools, and tables were always splayed for strength and stability. Such furniture was purely functional and the style therefore constant; but it was made throughout Europe, and local materials and tastes led to national or local versions. The great political economist Adam Smith (1723–90) had the makers of this furniture in mind when he wrote: 'A country carpenter deals in every sort of work that is made of wood: a country smith in every sort of work that is made of iron. The former is not only a carpenter, but a joiner, a cabinet maker, and even a carver in wood, as well as wheelwright, a plough wright, a cart and waggon maker' [7]. For stick-furniture in northern Europe, beech, ash, elm, and oak were the most common woods used.

FIGURE 224—*Detail of construction of stick-furniture.*

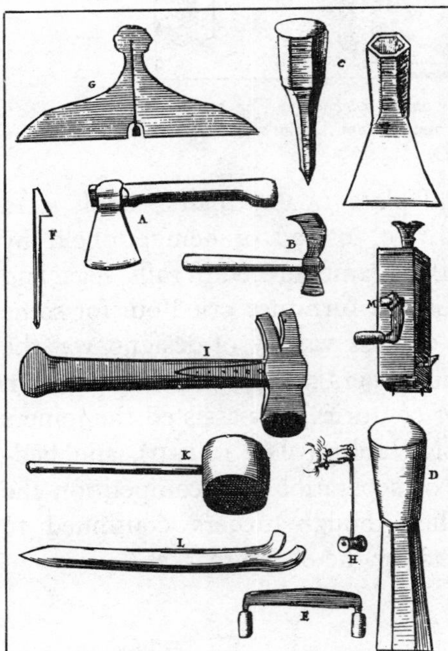

FIGURE 225—*A selection of tools used by the house-carpenter:* (A) *axe for chopping heavy timber;* (B) *adze with sharp edge on inner curve. This was used to remove chips of wood from surfaces such as floors or posts that were rather rough for planing;* (C) *chisels;* (D) *ripping chisel or wedge for splitting wood;* (E) *draw-knife for smoothing round articles such as rungs of ladders;* (F) *hook-pin to hold frame of floor or roof in position;* (G) *level and plumb-line: from 2 to 10 ft long to stretch over distances;* (H) *plumb-line;* (I) *claw-hammer;* (K) *commander or mallet for wooden piles;* (L) *crow (i.e. crow-bar) to raise heavy timber;* (M) *jack for moving and placing large pieces of timber. From Moxon's 'Mechanick Exercises', 1703* (see also *plate 12*).

The chief tool for making the legs of stick-furniture was the draw-knife (figure 225 E). Moxon writes that it 'is seldom used about House-building, but for the making of some sorts of Household-stuff; as the Legs of Crickets [low stools], the Rounds of Ladders, the Rails to lay Cheese or Bacon on, &c. When they use it, they set one end of their Work against their Breast, and the other end against their Work-bench, or some hollow angle that may keep it from slipping, and so pressing the Work a little hard with their Brest, against the Bench, to keep it steddy in its Position, they with the Handles of the *Draw-knife* in both their Hands, enter the edge into the Work, and draw Chips almost the length of their Work, and so smoothen it quickly' [8].

The chests associated with stick-furniture were also naturally of the simplest construction, being formed of six planks, one for each side, the lid, and the bottom. The back, front, and bottom boards were held together by nails, and the front was decorated by simple chip or incised carving—the kind of carving that in later times has been called peasant art. It was not done by a carver but by the maker of the chest. Many English elm or oak chests of this type have survived from the time of Elizabeth I and the following century.

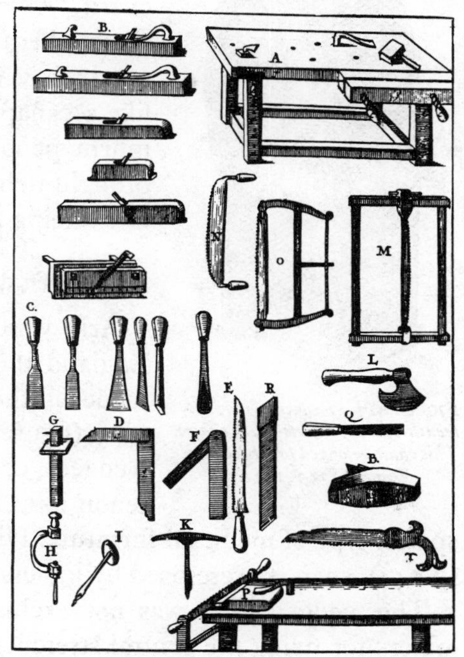

FIGURE 226—*A selection of joiner's tools:* (A) *work-bench with* (B) *a variety of planes;* (C) *various chisels;* (D) *square for measuring right-angles;* (F) *bevel with movable arm for measuring angles other than right-angles;* (G) *gauge for marking parallel lines;* (H) *brace for piercing holes;* (I) *gimlet for holes of various sizes;* (K) *auger for making large round holes;* (L) *hatchet;* (E, M, N, O, Q, T) *variety of saws;* (P) *whetting-block;* (R) *rule. From Moxon's 'Mechanick Exercises', 1703.*

VII. WICKER-WORK

Wicker or wand chairs and cradles were to be found in the ordinary man's home during the Middle Ages. They seem to have been the work of basket-makers, whose technique was that of plaiting osiers, that is young shoots of the willow. The coat of arms of the English fraternity bears a wicker cradle and wicker baskets. Beds, mats, bottle-casings, bird-cages, beehives, lobster-pots, and baskets were all part of this craft, which was common throughout Europe. Many of the Huguenot refugees to England in the years after the revocation of the Edict of Nantes in 1685 were members of this craft.

The frames of wicker chairs were entirely covered with wicker-work, which extended like a skirt to the ground. For comfort, wicker chairs had rounded backs, a design to which the pliant osier readily lent itself. From inventories we learn that such chairs were also made for children. The back was sometimes further prolonged upwards as a wicker hood (figure 227). Randle Holmes (1627–99), whose heraldic 'Academy of Armory' (1688) is a mine of such information, gives an account of a formidable variety of this hooded chair. 'A kind of these chairs [is] called Twiggen chaires because they are made of Owsiers, and Withen twigs: haveing round covers over the heads of them like a canopy. These are principally used by sick and infirm people, and such women as have bine lately brought to bed; from whence they are generally termed, Growneing [groaning] or Child-bed chaires' [9].

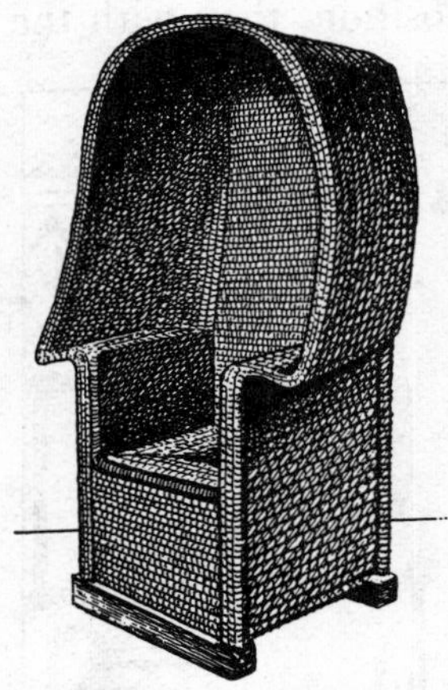

FIGURE 227—*Example of eighteenth-century wicker-work chair. This chair belonged to Dr Edward Jenner (1749–1823.)*

VIII. ARKWRIGHT AND COOPER

Arks were receptacles for meal or corn—in southern England they were called hutches or bins—with a lid made of three overlapping boards hinged on wooden pins (figure 228). The arkwright used a pegged and wedged construction instead of the mortise-and-tenon joint of the joiner. He thus provided his own special type of medieval furniture. His material consisted of riven or cleft boards (p 244), such as were used by joiners and ceilers for panels.

The wedged joint was not exclusive to the arkwright, for it was used by joiners for fixing the central stretcher to the two trestle supports of long tables, the ends of the stretcher coming through each of the supports (figure 229). Late 'Gothic' chairs, stools, and cradles with this wedged construction are extant.

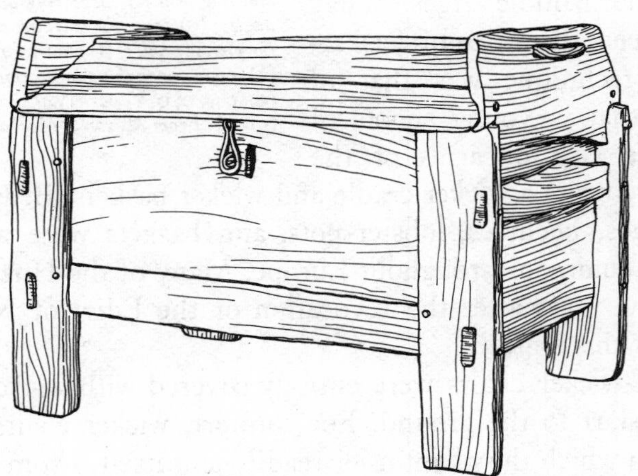

FIGURE 228—*Medieval meal-ark of riven wood with pegged joints. There are pin-hinges to the lid.*

Medieval illuminations figure a type of chair of the same construction as that of a wine-cask (figure 232). It had a rounded form and a high back and was probably made by coopers, who produced also bath-tubs and buckets (figure 230 and plate 3 B).

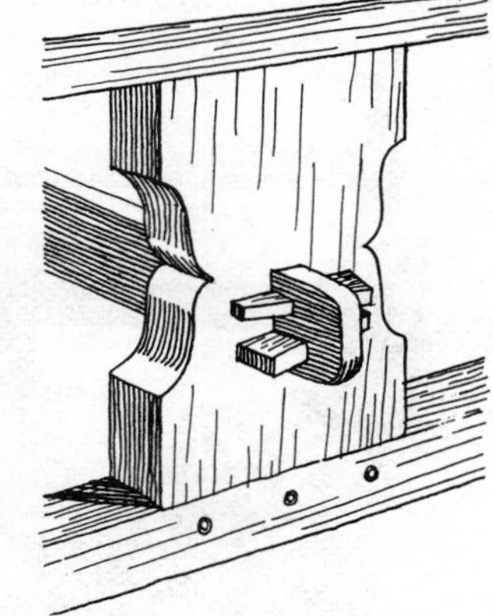

FIGURE 229—*The wedged joint used for fixing the central stretcher to the trestle-supports of a sixteenth-century table from the Low Countries (see plate 14 A).*

IX. THE COFFER-MAKER

The coffer-maker was another important craftsman in the Middle Ages, for he made clothsacks, barehides, standards, and coffers, all of which were essential for transport and travel. A clothsack was a large leather bag that was strapped to the pack-horse; a barehide, a leather cover for protecting goods transported by barge or cart; a standard, a large travelling-trunk much used in the Middle Ages. Standards were made of wood and covered with leather to keep them waterproof; they were banded with iron to protect them from the knocks and jolts received in transport and to make them more difficult to break open. Coffers, smaller and lighter, were mainly used for clothes, and were the medieval equivalent of the trunk. The leather element in all these is considered elsewhere (p 183). In constructing his coffers the coffer-maker fixed stretched leather on to the wooden carcass by close nailing with brass

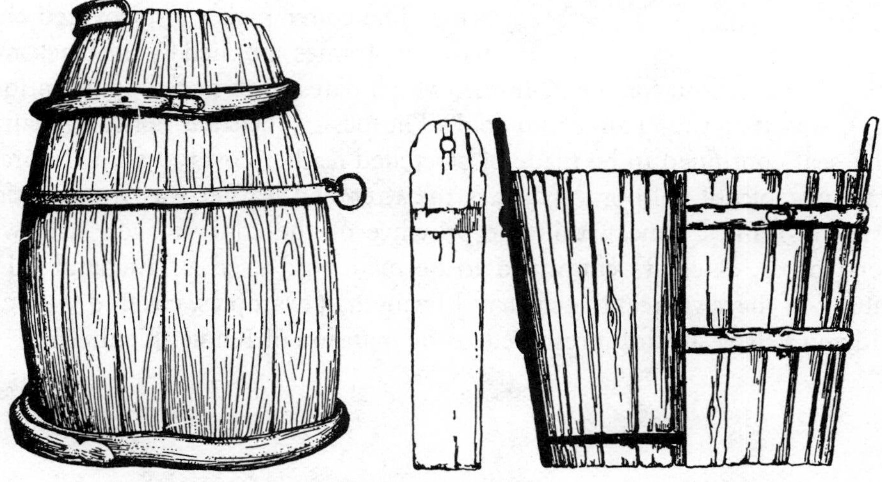

FIGURE 230—(Left) *Milk-churn from Lissue Rath, Co. Antrim, N. Ireland. Tenth century. Scale* c *1/8.* (Right) *Stave-built bucket from Ballinderry Crannog, Eire. Scale* c *1/8.*

dome-headed nails along the outer edge of the leather; at the corners were fixed protective iron or brass mounts. The coffer-maker further made that essential domestic utensil the close- or necessary-stool. This was in the form of a box with a lid, and utilitarian ones were covered with leather and garnished with nails. More costly examples were covered with velvet and other rich fabrics, and trimmed with ribbons and fringes. Sixteenth-century English coffer-makers also made coffers, desks, virginals, and jewel-boxes completely covered with rich fabrics and garnished with nails.

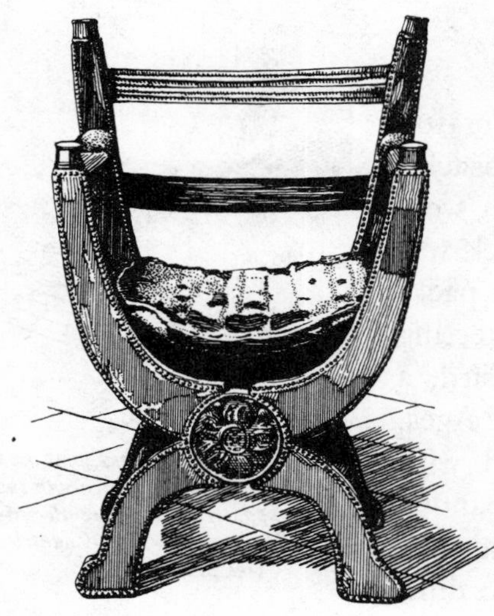

FIGURE 231—*The coffer-maker's X-chair in Winchester Cathedral, said to have been used at the marriage of Philip of Spain with Mary Tudor. The back and seat originally had a foundation of stretched leather covered with blue velvet. The chair-frame was also covered all over with velvet; the edges for protection were garnished with gilt nails.*

This technique would have had little bearing upon medieval furniture in general had not the coffer-maker also made a special type of chair, with an X-frame (figure 231). Many such chairs with kings or great churchmen seated on them are shown in medieval illuminations (figure 347). The frames are completely covered in rich material and no wood is exposed. A large cushion is upon the seat, and pommels decorate the uprights of the back and the ends of the arms. The coffer-maker also covered chairs with iron frames supplied by the blacksmith.

The original reason for the X-frame, which dates back to Egyptian antiquity (p 221), was that the chair could fold. The design became conventionalized, but the seat continued to be made of stretched leather covered with decorative material of cloth of gold or velvet. As the stretched leather was the functional material, X-frames came within the province of the coffer-maker, who was a leather-worker. X-chairs continued to be made and used in England till the beginning of the seventeenth century; Elizabethan examples, covered with cloth of gold and velvet, are still preserved at the mansion of Knole, Kent.

FIGURE 232—*Chair of coopered construction with staves held by bands. Flemish picture for the month of January. From a calendar, c 1500.*

REFERENCES

[1] TREDGOLD, T. 'Elementary Principles of Carpentry', p. 1. Printed for J. Taylor by Bartlett, Oxford. 1820.

[2] MOXON, JOSEPH. 'Mechanick Exercises' (3rd ed., to which is added "Mechanick Dyalling"), p. 122. Daniel Midwinter and Thomas Leigh, London. 1703.

[3] *Idem. Ibid.*, p. 63.

[4] STOW, JOHN. 'The Survey of London', p. 636. Printed by Elizabeth Purslow, London. 1633.

[5] HARRISON, WILLIAM. "An Historicall Description of the Iland of Britaine", II, xxii, in R. HOLINSHED (Ed.). 'Chronicles of England, Scotland and Ireland', Vol. 1, p. 357. Printed by Richard Taylor, London. 1807.

[6] JUPP, E. B. 'An Historicall Account of the Worshipfull Company of Carpenters of the City of London' (2nd ed. with suppl. by W. W. POCOCK), p. 297. Pickering and Chatto, London. 1887.

[7] SMITH, ADAM. 'An Inquiry into the Nature and Causes of the Wealth of Nations' (ed. with introduction, notes, marginal summary and enl. index by E. CANNAN. 6th ed.), Vol. 1, p. 19. Methuen, London. 1950.

[8] MOXON, JOSEPH. See ref. [2], p. 122.

[9] HOLME, RANDLE. 'The Academy of Armory or a Storehouse of Armory and Blazon', Vol. 2 (ed. by I. H. JEAYES), p. 14. Roxburghe Club, London. 1905.

BIBLIOGRAPHY

BOBART, H. H. 'Records of the Basket-makers' Company.' Dunn, Collin, London. 1911.

COOPER, W. D. (Ed.). 'Lists of Foreign Protestants and Aliens Resident in England 1618–1688.' Camden Society Publ. 82, London. 1862.

FALKE, O. VON and SCHMITZ, H. (Eds). 'Deutsche Möbel vom Mittelalter bis zum Anfang des neunzehnten Jahrhunderts', Vol. 1: 'Deutsche Möbel des Mittelalters und der Renaissance.' Hoffmann, Stuttgart. 1924.

FEULNER, A. 'Kunstgeschichte des Möbels seit dem Altertum.' Propyläen-Verlag, Berlin. 1927.

FURNIVALL, F. J. (Ed.). 'Harrison's Description of England in Shakespeare's Youth.' New Shakespeare Society, London. 1877.

HOLME, RANDLE. 'The Academy of Armory or a Storehouse of Armory and Blazon.' Printed for the Author, Chester. 1688. Vol. 2 (ed. by I. H. JEAYES). Roxburghe Club, London. 1905.

INNOCENT, C. F. 'The Development of English Building Construction.' University Press, Cambridge. 1916.

JUPP, E. B. 'An Historicall Account of the Worshipfull Company of Carpenters of the City of London' (2nd ed. with suppl. by W. W. POCOCK). Pickering and Chatto, London. 1887.

MOXON, JOSEPH. 'Mechanick Exercises' (3rd ed., to which is added "Mechanick Dyalling"). Daniel Midwinter and Thomas Leigh, London. 1703.

STOW, JOHN. 'The Survey of London.' Printed by Elizabeth Purslow, London. 1633.

TREDGOLD, T. 'Elementary Principles of Carpentry.' Printed for J. Taylor by Bartlett, Oxford. 1820.

Idem. Ibid. Appendix to 2nd ed. with additions by S. SMIRKE, J. SHAW, J. GLYNN *et al.* Weale, London. 1840.

WARBURG and COURTAULD INSTITUTES (Eds). 'England and the Mediterranean Tradition. Studies in Art History and Literature' by various authors. Oxford University Press, London. 1945.

8

PART II

CERAMICS: MEDIEVAL

E. M. JOPE

I. SOURCES OF INFORMATION

THE general principles of pottery-making, some properties of clay, and the early history and use of the potter's wheel and turn-table have been discussed in vol I (chs 9 and 15) and in the first part of the present chapter. Here we have to consider the post-Roman developments of European techniques, but we shall frequently need to refer to oriental influences, the general character of which is discussed in the last chapter of this volume (pp 755 f). It was not only techniques, however, that came from the orient; Byzantium and Islam considerably influenced also the decoration of pottery in Europe. On this latter topic we shall say little since it belongs to the history of art rather than of technology. For a like reason we shall pass over many indigenous European changes of style and taste in so far as they are not technological. We are concerned with new techniques or methods, whether indigenous or inspired by eastern practice.

Following the fall of the western Empire, the large-scale production of good wheel-made pottery ceased in many parts of the north-western provinces, and barbarian hand-made ceramic traditions once more came into their own. In some areas, however, such as the Rhineland, and parts of Gaul and Italy, the methods of the Roman ceramic industry were continuously developed through the Middle Ages. In Britain, outside Kent, the making of good pottery on a wheel lapsed from the fifth to the ninth century, when the technique was reintroduced from the Rhineland. Many new techniques and methods in the north and west resulted from influences coming directly or indirectly from the east from the eleventh century onwards. It is still difficult to separate the parts played by Byzantium, Spain, north Africa, the Levant, and Italy in disseminating this oriental influence through European ceramics.

Most of our evidence on medieval ceramic technology comes either from the detailed study of the pottery itself or from excavated kiln-sites, though a few potters' workshops have been found (figure 266). There have been few chemical and mineralogical analyses of medieval ceramic fabrics and glazes, and far too little determination of such matters as the temperatures or conditions of firing. In documentary sources there are incidental references to the activity of potters

FIGURE 233—*Corinthian vase thrown on the wheel. Seventh century B.C.*

various potteries supplied not only their own home markets but exported their wares extensively, especially to the west. At first this commerce was dominated by Corinth, whose geographical position on the Isthmus made her an appropriate link between east and west. Gradually, however, Athens became the chief ceramic centre in the Mediterranean—a position she maintained until *c* 400 B.C. By then other countries began to produce their own wares, though these were mostly imitations of the Attic. Athenian vases of the sixth and fifth centuries B.C. represent the acme of Greek pottery, both technically and artistically. On them, therefore, our technical analysis will centre.

FIGURE 234—*Moulded vase, east Greek. Seventh century B.C.*

Information regarding the technique of Athenian pottery is derived from several sources: statements of ancient writers, inscriptions, ancient representations of potters at work, and, above all, the vases themselves. Modern practice is also often helpful, for clay has not changed its nature or properties.

As is generally the case in technical matters, references to the craft by ancient writers give little definite information. They consist chiefly of general remarks on the nature of clay, the invention of the wheel, the hazards of firing, and the

FIGURE 235—*Athenian pottery-establishment. From a* hydria; *sixth century B.C.*

status of potters. Among the inscriptions, most important are the signatures of potters (plate 18 B) and vase-painters, which supply some artists' names and incidentally show pride in workmanship and public appreciation. The representations of potters at work on vases and terracotta plaques also provide valuable information. There are scenes of potters fashioning and refining vases on the wheel, attaching their handles, decorating them, carrying them away, and firing them in the kiln (figures 235–7, plates 16 A–C, 17 A). Furthermore, a few actual wheel-heads (figure 238), tools, and kilns have been discovered in excavations, supplementing and reinforcing the testimony of the pictures.

FIGURE 236—*Youth decorating a cup. From a Greek cup; fifth century B.C.*

From the vases themselves we learn the nature of the clay and of the pigments used, something of the methods of application, and the temperature to which the wares were fired. In some ways the work corresponded to modern practice. The chief difference lies in the shiny black 'glaze' that was universally used on this ware. It constitutes a refinement on that of the earlier pots. The nature of the Attic black 'glaze' has been recognized only lately and its discovery has thrown much light on this ancient craft [1]. The following is a summary of the technical processes in the production of Athenian vases, as suggested by the evidence at present available.

FIGURE 237—*Potter stoking the fire of his kiln. From a votive tablet. Sixth century* B.C.

Athenian pottery was made of a sedimentary or detrital clay which has acquired various ingredients during its movements on the Earth's surface; it is composed largely of the finer particles resulting from the decay of rocks. Having a high iron-content, it fires to a reddish colour under oxidizing conditions. The Athenian shapes—the wide-spreading cups and the jars with incurving shoulders—testify to great plasticity and toughness. The white residual clay, that is, clay that resided in its original pocket, is non-plastic and so was not used by the Athenians for forming vessels, but was employed only as a coating or slip on their white-ground vases (p 265). It may, however, also have been added as an ingredient to the red clay, as is done nowadays by potters in Athens.

After the clay had been washed and wedged, a ball of it was placed on the wheel-head, centred, and thrown to the desired shape. Smaller vases were thrown in one piece, larger ones in sections. The joins of the sections were generally at structural points, as between neck and body, or body and foot. On the inside the joins are often visible, but to conceal them on the outside thin coils of clay were added. For section-work in particular, proportions had to be checked with rule and callipers. By such means vases could be made to correspond to a carefully executed design.

FIGURE 238—*Terracotta wheel-head from Crete, underside. Minoan period.*

Potter's wheels represented on Greek vases are not worked with the foot but are propelled by a helper. The kick-wheel, however, is referred to by a biblical writer of the third to second century B.C. (the Apocryphal Ecclesiasticus xxxviii. 29–30). It may of course have been used earlier, for the mechanics involved are simple enough. The only wheel-heads of the period that have survived are of terracotta (figure 238). Nowadays plaster, metal, and wood are the materials preferred.

After the vase was thrown it was taken off the wheel with knife or string (figure 239) and left to dry until leather-hard. Then it was replaced on the wheel, recentred, and turned, that is, refined with metal cutting-tools (figure 240). In narrow-mouthed vessels the insides were not turned but left as thrown. Wide-mouthed vases and cups, however, where the inner surfaces showed, were carefully turned throughout. After removal of the marks of the turning—which was probably done as nowadays with scrapers and moist sponges—the handles were attached. These were all separately made by hand, not moulded.

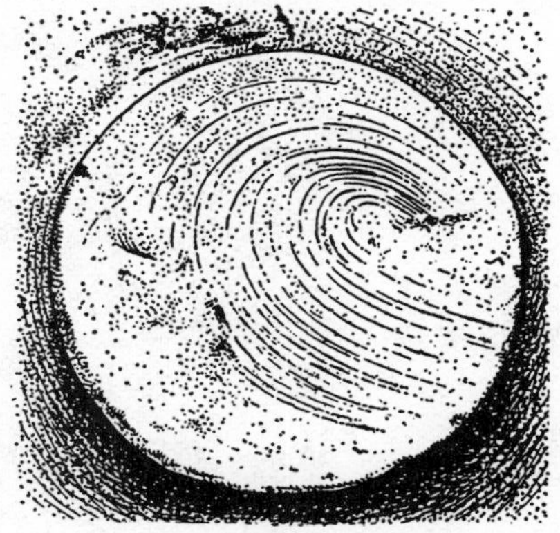

FIGURE 239—*Foot of a vase left as thrown, showing string-marks. Minoan period.*

There were relatively few standard shapes, the most popular being the amphorae or storing-vessels, the capacious jars for mixing wine with water, the three-handled water-jars, the narrow-necked oil-jugs and ointment-jars, the wide-mouthed cups, and the dainty toilet-boxes. Yet each vessel was a separate creation, differing slightly in curve and proportions from the others. It is only among the plain black vases and the plain household vessels that almost exact repetition can be found.

Decoration was applied before firing, while the vase was in leather-hard condition. In this respect Greek vases differ fundamentally from those of our time. Nowadays the practice is to apply the glaze to fired ware (biscuit) and refire, but Greek pottery is—with a few exceptions mentioned later—fired only once. The evidence for this is convincing. The incised lines in the drawing would never have had their characteristic smoothness and swing if made on fired clay, nor would the applied clay used for certain details have adhered to fired clay. The hazards in handling the unfired ware while decorating it were minimized by the toughness of the clay.

FIGURE 240—*Turned foot of a cup. Sixth century B.C.*

The black used for the decoration is not a glaze in the modern sense, since it contains insufficient alkali to render it fusible at a definite temperature. It is rather a liquid iron-containing clay, similar to that used for the pottery itself, but peptized, that is, with the heavier particles eliminated, and with enough

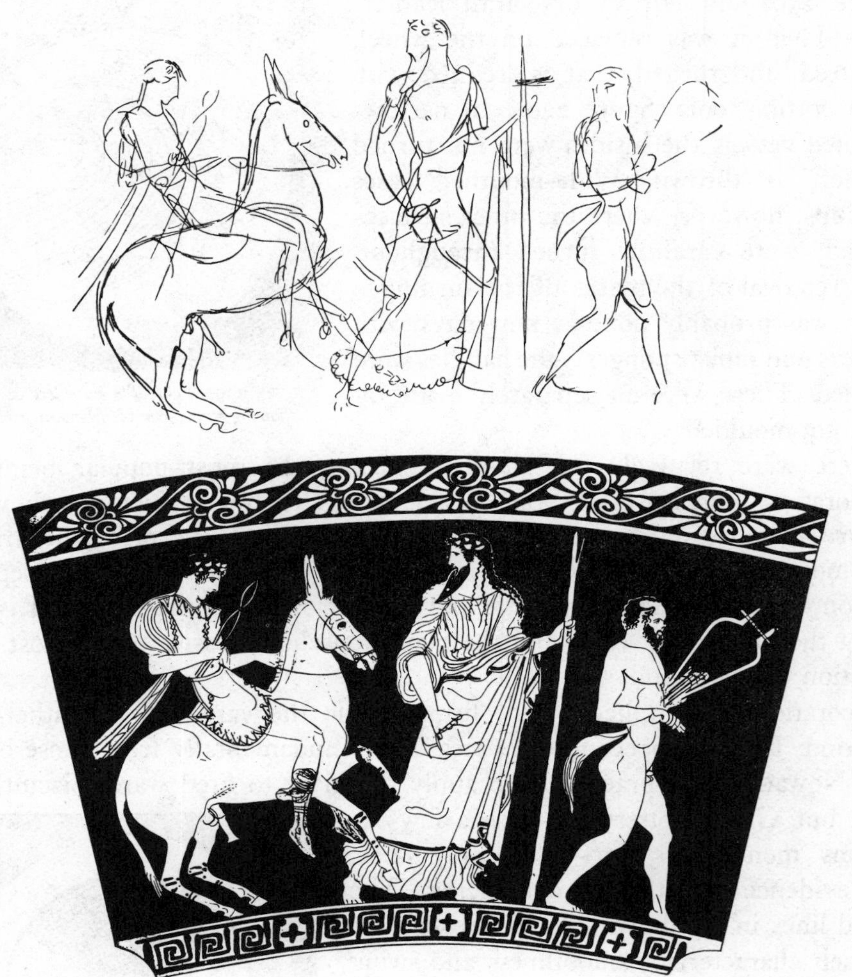

FIGURE 241—*The return of Hephaistos to Olympos. From an Athenian* krater. (Above), *preliminary sketch;* (below), *finished drawing. Fifth century* B.C.

alkali added to give it considerable gloss [1]. The resultant 'glaze' was remarkably smooth and thin, yet had sufficient body to make single lines stand out in relief.

With this medium the Athenian artist produced on his vases decorations that are unmatched in quality. Having the story-telling instinct, he chose for his designs not only the floral patterns and rows of animals of his predecessors, but scenes from mythology and daily life. As the contemporary panel and mural paintings have practically all disappeared, these vase-paintings have become a specially precious heritage.

The vase-painter's technique was as follows. First, the whole surface was generally covered with a thin protective wash (*Lasur*) made of diluted, peptized clay, which imparted a slightly glossy, reddish hue after firing. In the earlier, black-figured vases the design was painted in black against the red clay background, then the details were incised, and lastly the red and white accessory colours were added (plate 18 A, B). In the later, red-figured vases, in which the design is 'reserved' in red against the black (plate 18 C), the preliminary sketch for the design, made with a blunt tool, is often clearly visible (figure 241). The figures were painted in outline, first with a narrow line, then with a broader stripe (figure 242). The lines of detail within the red silhouettes were then drawn and the background was painted black. Finally, red and white accessory colours, or incisions, or applied clay for certain details, were added if needed. Sometimes a red ochre application was rubbed into the clay to intensify the colour.

FIGURE 242—*Unfinished red-figured cup. Fifth century B.C.*

Along with the black-figured and red-figured vases the Athenian potter made plain black ware, that is, vases painted entirely black, with only a few reserved areas at lip, handles, and foot. During the second half of the fifth and in the fourth century B.C. an all-black ware with incised and stamped decorations was popular. In addition, from the sixth century B.C. onward, there was a limited but continuous production of the so-called white-ground ware. In this, a slip of white clay [2] was applied on the central part of the vase, and on it the decoration was painted in solid black with details incised, or in outline in diluted glaze or matt colour (plate 17 B). First the whole decoration was in black but, as time went on, there came to be used, for draperies and other areas, tempera colours in red and yellow, and later also in blue, purple, green, mauve, and pink. These vases, in fact, resembled in miniature the larger panel and mural paintings that are lost to us.

After the decoration was completed, the vase was allowed to become bone dry before being placed in the kiln. The temperature at which Athenian vases were fired was rather lower than is usual today—it has been computed to have been about 900–960° C.

The process of firing was also different from that used commonly today, in that it was first oxidizing (that is, with air freely admitted), then reducing (with air limited), and lastly reoxidizing. During this last stage the clay that, together with the glaze, had turned black during the reducing fire, became red again (since it was sufficiently porous to re-admit the oxygen), whereas the denser glaze remained black [3]. The red spots that frequently occur on what are intended to be black areas (figure 243) can be explained as due (*a*) to accidental protection from reduction in the kiln, either through stacking or through contact with a jet of air, or (*b*) to the fact that the glaze was applied too thinly and so was porous enough to reabsorb oxygen during the third stage of the firing. The *Lasur*, as well as the red ochre application, and the red accessory colour (red ochre mixed with peptized clay) were, like the terracotta body, sufficiently porous to reabsorb oxygen, while the white accessory colour (= peptized white clay) contained practically no iron [2] and so was unaffected by the re-ducing fire [4].

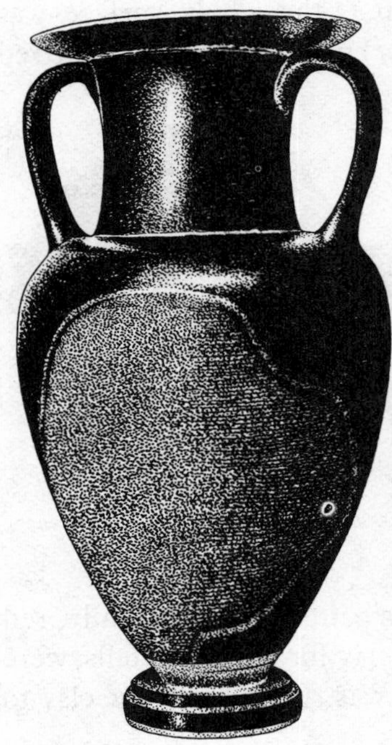

FIGURE 243—*Black amphora with large red spot. Fifth century B.C.*

Finally, an explanation has perhaps been reached of an old problem, namely, how the red and black glazes were made to contrast in precisely defined areas on certain vases—that is, intentionally, not accidentally (figure 244). How were the areas that were intended to come out red prevented from being reduced in the second stage of the firing, even when thickly applied? In these cases, it has been suggested that there was simply a second firing [5]. At first only those parts that were to come out black were painted, and fired in the regular way—at first oxidizing, then reducing. The firing was then interrupted and the parts that were to come out red were painted and fired in an oxidizing fire. That this is the correct explanation is suggested by the fact that in all the red-and-black vases examined, the red glaze, when in contact with the black, invariably goes over the black, showing that it was applied later [6]. That the red glaze is apt to peel more than the black also indicates that it was painted on clay already fired. The cumbrousness of the method would account for the relative scarcity of this red-and-black ware.

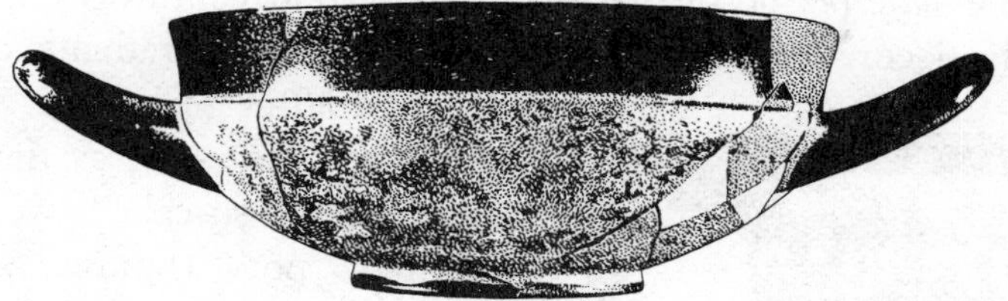

FIGURE 244—*Red bowl with black rim, handles, and foot. Fifth century* B.C.

To complete the picture of pottery produced by Athenians, two other techniques must be mentioned. Though work on the wheel was the favourite procedure, moulding was used for plastic vases in the forms of heads or animals or groups (plate 18 C). The moulds were of terracotta, each made in two parts. Clay was pressed into each part, and then the halves were joined with slip. The lip was often thrown separately. In a few instances several positives from the same mould have survived. The ancient technique of building (vol I, p 384) was sometimes employed for plain household pots and pitchers used for storing and cooking. Many of these vases have been found in the Athenian agora [7]. Simple incisions or painted bands were occasionally used for their decoration.

After the defeat of Athens in the Peloponnesian war (431–404 B.C.), her trade was crippled. Though she continued to produce excellent red-figured pottery and exported it here and there, she lost her primary position. As early as the second half of the fifth century Italy had begun to make her own wares; in the fourth she developed a flourishing industry. These Lucanian, Apulian, Campanian, and Paestan pots, though distinctive in style, are identical in technique with the Attic. They are thrown and turned on the wheel, their decoration is in red-figure, with accessory red and white, and the shapes correspond to the Attic. They form in fact a striking contrast to the native Messapian and other south Italian wares that never abandoned the time-honoured geometric ornaments on somewhat clumsy and fantastic shapes.

FIGURE 245—*West-slope ware. Third century* B.C.

II. HELLENISTIC GREEK POTTERY, *c* 300–*c* 100 B.C.

In the Hellenistic period of the third and second centuries B.C. there came a change in the decoration of Greek pottery. Relief-work gradually ousted painting. The latter technique was retained for a few wares such as the plain black vases and the so-called 'west-slope' and Gnathian pots (figures 245–6), thus named because many were found on the west slope of the Athenian acropolis, and at Gnathia in south Italy, respectively. The latter had ornaments painted in various colours on a black background, with occasional stamped or incised designs. Moreover, the Attic white-ground technique, with decorations in tempera colours on a white ground, continued to be practised both east and west. Thus the vases found at Hadra, near Alexandria, and the sumptuous wares from Canosa in Apulia and Centuripe in Sicily are in that technique (figure 247). Elsewhere, however, reliefs became the order of the day.

FIGURE 246—*Gnathian pot. Third century* B.C.

In Attica during the late fourth century B.C. there had been a short vogue for vases decorated with reliefs in *barbotine* or *appliqué* and painted with tempera colours; but for Hellenistic relief-decoration the chief inspiration must have come from embossed metal-ware. In some cases, for instance in the Calene bowls (called after their chief source, Cales, in Apulia), we can even determine that they were made from moulds taken from earlier Greek metal examples [8]. In others, new designs were created, often showing a mixture of styles. The reliefs were moulded either with the vase itself, or separately and attached to a wheel-thrown vase. The latter method was used in the so-called Pergamene ware of the second to first century B.C., the former in the so-called Megarian bowls of the third century B.C. The names are derived from the places where these vases were first found, Pergamon in Asia Minor and Megara

FIGURE 247—*Vase from Centuripe. Third century* B.C.

in continental Greece. Later, however, these wares were found in many other places.

The technique of the Megarian bowls is of great interest (figure 248). They were not pressed or poured in moulds, like the plastic vases, but present a combination of wheel-work and moulding not unlike that used today in the manufacture of platters and dishes. A mould in the form of a bowl with thick walls was first thrown on the wheel. When leather-hard, terracotta stamps were impressed on its inner side to form a connected design. After firing, the mould was firmly fitted on a jigger, that is, a wheel with a revolving head on which a mould could be fastened, and inside the mould the clay was spun. When the bowl was completed, the fingers were pressed into the hollows of the decoration in the mould, producing reliefs on the exterior of the bowl. The hollows within were then filled with clay and smoothed while the bowl was again revolving; that is why there are no depressions with finger marks visible on the interior as on the Etruscan ware of similar technique (p 270). After removal from the mould, finishing touches were given to the decoration. Finally, the surface was covered with glaze which became black in a reducing fire.

FIGURE 248—*Megarian bowl. Third century* B.C.

III. ETRUSCAN POTTERY, *c* 700–*c* 100 B.C.

The culture immediately preceding the arrival of the Etruscans in Italy is generally referred to as Villanovan, from a place near Bologna where extensive remains of that period have been found. In the seventh century B.C. the pottery of this early Italic culture, which succeeded the Terremare civilization of the Bronze Age, was still of primitive character. It is generally referred to by its Italian name *impasto*. Some of it was built by hand, but mostly it was produced by a combination of building and throwing (figure 249). Vases were first roughly shaped by hand, then finished on the wheel. They were of rather coarse clay

FIGURE 249—Impasto *ware. Two-handled cup found near Bologna. Eighth to seventh century* B.C.

FIGURE 250—Impasto *bowl on stand, from Latium. Seventh century* B.C.

containing iron oxide, which turned blackish brown or reddish in the firing. The variation in colour indicates variation in the extent of reduction. No glaze was used, but a sheen was obtained by polishing with wooden or stone tools, which produced a packed surface of levigated clay that crazed during firing. The decorations consist mostly of incised designs, sometimes reinforced with red ochre, and mostly geometric in character. Only occasionally is a motive from the animal world introduced. By the second quarter of the seventh century some magnificent vases were occasionally made in this ware (figure 250). Their decorations of lotus flowers and monsters show oriental influence.

The Etruscan *bucchero* ware was an outgrowth of the *impasto*. The chief change was one of refinement in technique. The clay was now well levigated and the colour no longer brownish but black throughout, being obtained by complete instead of partial reduction in the firing. The early examples are wheel-thrown, often with thin walls, and with decorations either incised or in relief (figure 251, p 271). Animal and plant motives are now popular, as is a fan-shaped design produced by a toothed stick [9]. Reliefs are mostly confined to the handles, or consist of narrow friezes produced by rolling cylinders along the surface of the vases while the clay was soft.

The later *bucchero* of the sixth century B.C. is of heavier make, with ornaments mostly in relief instead of incised (figure 252). Like the later Megarian and Arretine wares (pp 269, 272), it was regularly thrown in moulds on the wheel, as shown by the facts that wheel-marks are observable only on the insides of the vases, and that there are depressions with finger prints corresponding to the reliefs on the outsides [10]. As these *bucchero* vases were mostly not

FIGURE 252—*Etruscan* bucchero *jug thrown in moulds. Sixth century* B.C.

open bowls but vessels with upper parts incurving, more than one mould was required. The joins of foot and body, of neck and body, or of different parts of the body can often be detected. The foot was sometimes thrown, sometimes moulded. Occasionally, when there was no decoration in relief, the whole vase was thrown. At other times the reliefs were separately applied, especially when they were partly in the round, protruding above the rim. From a ceramic point of view these large black Etruscan pots are essentially satisfying, with colour, shape, and decoration harmonizing.

FIGURE 251—*Etruscan* bucchero *jug, wheel-thrown with incised decorations. Seventh century* B.C.

In addition to the black pots, there have been found many painted Etruscan vases imitating the Corinthian, Ionic, and Attic wares in technique (plate 18 D). This is not surprising, since pottery was extensively imported into Italy from these centres. The style is generally sufficiently distinctive to be easily recognizable. In the Etruscan red-figured vases the figures, instead of being reserved in the colour of the clay, are sometimes painted red (a technique also occasionally used in Attica). Vases moulded in the forms of heads and of whole figures were also produced. Though they imitate Greek examples in technique, here, too, the style is often refreshingly original.

In the Hellenistic period of the third and second centuries B.C., Etruscan pottery, like the Greek, shows the influence of metal-ware. The reliefs were as a rule separately moulded and applied while leather-hard, and the surface was sometimes silvered. The style is, as in the rest of Italy, Hellenistic Greek, except where moulds from earlier works were used.

Something of the age-long skill of the Etruscan potters must have been handed down to their descendants at Arezzo, once an Etruscan stronghold. There some of the finest vases of the Roman period have been found (p 272).

IV. ROMAN POTTERY OF THE REPUBLIC AND EMPIRE, *c* 100 B.C.–*c* A.D. 350

The art of the late Republic and of the Empire is a continuation of the Greek. The Romans, now masters of the Mediterranean, showed great appreciation of Greek art, and by their new demands gave it a fresh lease of life. This condition

FIGURE 253—*Arretine vase. First century* B.C.

is reflected in pottery also. The beautiful Arretine vases of the first century B.C. (figure 253) show the same combination of wheel-work and moulding that we observed in the Megarian bowls of Hellenistic times (p 269). Here, too, stamps (figure 254) were impressed into wheel-thrown moulds (figure 255) and the vases thrown inside the moulds. Moreover, the stamps themselves seem often to have been moulded from Greek metal-ware. This, at least, is suggested by their style and precision, and by the fact that two identical specimens exist, in New York and at Boston [11]. Details were retouched with wooden or metal tools (figure 256), and finally, as in the Megarian bowls, the whole surface of the vases was covered with glaze. The glaze, however, is no longer black but a brilliant coral red. Its constituents are the same as that of the black glaze of Corinthian, Chalcidian, Attic, and south Italian Greek vases, but it was fired under oxidizing conditions throughout and so shows no trace of black.

The same technique and colouring were continued for several centuries in the ware known as *terra sigillata*, the Roman pottery found all over the Roman Empire (figure 257). Black glaze was, however, retained for a few other wares—for instance, for that with painted inscriptions of convivial character, found in Germany, France, and Britain, and for the Castor vases with *barbotine* decoration. Another peculiarly Roman ware is that covered with white slip and decorated with veins in a marbled design. Lastly, some pots, both red and black, had reliefs applied instead of moulded, and many were left undecorated. As in Greek times, the common household pottery was often unglazed.

FIGURE 254—*Arretine stamp. First century* B.C.

The most important ware of the Roman period from the artistic point of view, apart from the Arretine, is that covered with blue-green glaze (figure 258). Though the shapes, the technique, and to some extent the decoration are similar to those of the Arretine, the glaze is fundamentally different. It is a true glaze in the modern sense, for it melts at a definite temperature. Blue (alkaline) and green,

FIGURE 255—*Arretine mould impressed with a sacrificial scene. First century* B.C.

yellow, and colourless lead glazes were effectively used. The glaze was generally thinly applied, by pouring, on the inside of the vases; more thickly, by dipping, on the outside. It has a matt surface, on account of the relatively low temperature at which the ware was fired, *c* 960–1030° C. Only where the glaze has pooled has it become transparent and shiny. As nowadays, the glazed vases were supported on stilts in the kiln, and the marks these left can often be observed. Stilts with glaze still adhering have also survived (figure 259). The favourite shape in this ware is the ring-handled cup, borrowed from metal-ware, and popular also in Arretine. Vases in the forms of heads and statuettes occur occasionally. The ware was produced widely through the Empire, though never apparently in quantity. It originated doubtless in the east. Blue alkaline glazes had had a long history in Egypt (ch 9) and a short one in sixth-century Ionia. Brilliantly coloured glazes have been favoured by potters ever since.

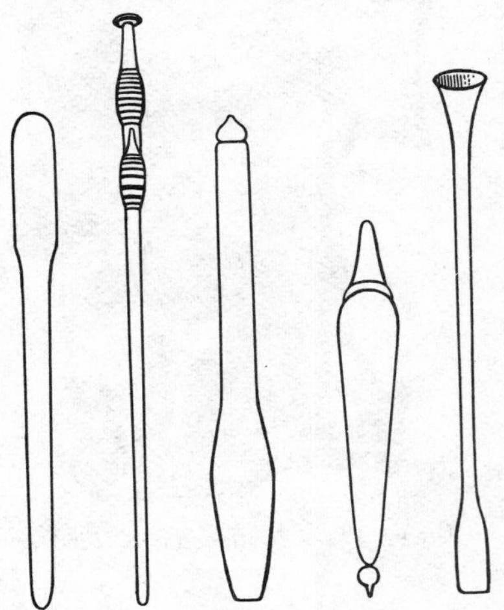

FIGURE 256—*Modelling-tools found at Arezzo. Roman period.*

V. STATUARY AND RELIEFS

Clay in the Mediterranean world was extensively used not only for pottery but for sculpture, large and small, in the round and in relief. The terracotta statuettes found on ancient sites are, next to pottery sherds, the most numerous relics. Statues and large reliefs in terracotta have come to light in considerable numbers in Cyprus, Etruria, and Greek southern Italy, where they were used in

FIGURE 257—*Bowl of* terra sigillata *ware. Roman period.*

the decoration of temples and as votive offerings; for before the opening of the Carrara quarries in the first century B.C. marble was scarce in Italy. Even in

FIGURE 258—*Cup with greenish glaze. Roman period.*

Asia Minor and Greece, however, where marble was ready to hand, clay was used for friezes, acroteria, and votive and cult statues.

The early Greek and Etruscan terracotta statues (plates 19 A, B, and 20 A, B) were not moulded but built in coils and wads of clay, with thick walls [12]. To prevent the distortion to which clay is liable during drying, and to diminish shrinkage, the clay was mixed with sand and grog (small particles of fired clay), which ensured porosity. The building was from the bottom up, starting with the feet and continuing to the head, without use of an armature. Probably the whole figure was first formed roughly and the modelling developed on this core. After the completion of the modelling a coating of finer clay was sometimes added, of sufficient thickness to conceal the foreign particles in the clay. While still leather-hard the surface was covered with a black glaze, consisting chiefly of peptized clay in diluted form (pp 263 f). Where this has shrunk in the fire, small cracks appear covering the surface. Over the glaze details were painted in red ochre, in liquid white clay, and in black glaze, in the same manner as were the contemporary Attic vases. The temperature of the firing was also similar to that of Greek vases, that is, about 960° Centigrade.

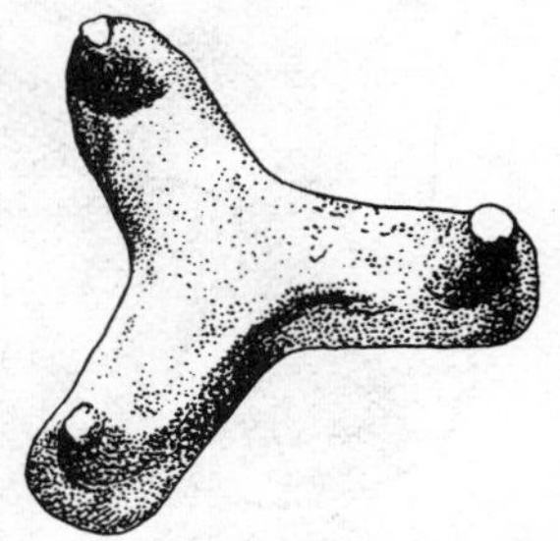

FIGURE 259—*Stilt. Roman period.*

In Greece proper have been found a number of examples that must have served as decorations of buildings and as votive offerings. Especially impressive are several figures from Olympia, including a Zeus carrying off young Ganymede (plate 19 A, B) and a head of Athena (plate 20 A), all datable in the first half of the fifth century B.C. A pedimental group representing a battle of Greeks and Amazons has come to light in Corinth. Figure 260 shows the reconstruction of a warrior's head found in Athens and put together from many pieces; on the helmet is painted a flying Pegasos in diluted glaze.

FIGURE 260—*Greek terracotta head of a warrior, under life size, from Athens.*

In Sicily and Italy have been found many terracotta figures in the round and in relief, both Greek and Etruscan. One of the most notable is a seated Zeus, 92 cm high, from Paestum. In the Etruscan terracotta sculptures the coating of fine clay was generally omitted. In the famous Apollo from Veii (cf plate 20B) the little black particles of sand in the clay body are visible on the surface. The revetments of temples often consisted of sculptural groups (cf plate 20 C). All were once gaily painted.

In Hellenistic and Roman times terracotta sculptures were often moulded instead of built. In the votive heads and busts from Etruscan and other sites (figure 261), the front and back were regularly made separately and then attached. This was sometimes done carelessly, so that the joins are clearly seen. Later retouching could change the appearance of examples made from the same mould. Thus the moulded Etruscan head of a young man has been changed into that of an old one by the addition of a beard, wrinkles, and furrows [13].

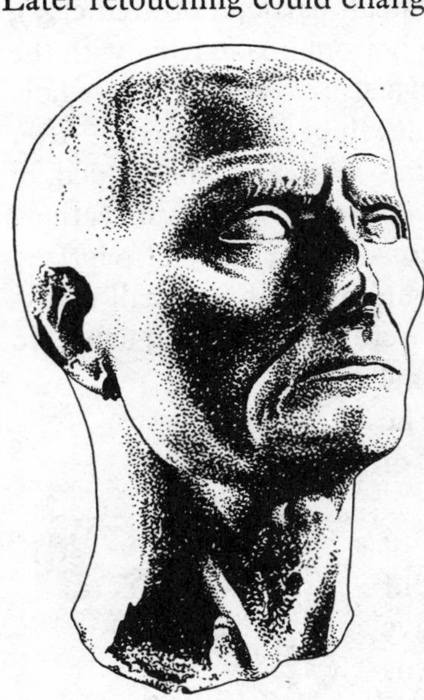

FIGURE 261—*Votive head of an old man. First century* B.C.

Most Greek, Etruscan, and Roman statuettes, from the seventh century B.C. on, are moulded, not modelled. Only occasionally was the older method of working the clay freehand retained. The use of moulds enabled the artist to diminish the weight considerably by making the figure hollow instead of solid, and to produce a number of copies rapidly. A transition stage between the two techniques is illustrated by some statuettes from Cyprus, where a wheel-made columnar body, modelled arms, and a moulded head were combined in one and the same figure (figure 262) [14].

The ancient technique of moulding statuettes

can be reconstructed in some detail, for, in addition to the statuettes themselves, many moulds have been found. These are mostly of terracotta, fired to considerable hardness; only occasionally, in Egypt, especially in the Hellenistic and Roman periods, were moulds of plaster used. In contrast to the statues, the terracotta statuettes were made of well levigated clay, since the vent-hole at the back sufficed in such small objects to avoid distortion in drying (plate 21 A, B). The clay was pressed into the mould first in a thin layer, so that it might penetrate into all recesses, then in slightly thicker applications. The shrinkage in drying made it easy to separate the figure from the mould. A characteristic of Greek terracotta statuettes is that they were generally made in several moulds. Individual moulds were used not only for the front and back, but for different parts of the torso, head, legs, arms, wings, and objects held. Thus great variety was obtained from a limited number of moulds. A flying Eros from Myrina was made from as many as fourteen moulds [15], but in more compact figures the number was of course less.

FIGURE 262—*Terracotta statuette from Cyprus.*

After the various parts of the statuette had emerged from their moulds, and while still leather-hard, they were joined together with slip. The finishing touches, on which so much of the ultimate fineness depended, were then given with wooden or metal tools. The bases, rectangular or round, were mostly made separately by hand. Finally, the whole figure, while still leather-hard, was covered with a thin layer of peptized white clay [16]. The temperature of the firing was fairly low, and the resultant terracotta is therefore relatively soft.

After firing came the addition of the colours. On the base provided by the all-over fired white, details were added in unfired tempera pigments—blue, yellow, different shades of red, mauve, and touches of gilding. The pigments have mostly flaked off, exposing the drab buff of the terracotta, but where they have survived they give an idea of the gay, harmonious effects obtained.

Terracotta statuettes have been found in many localities. Best known are those from Tanagra (figure 263) in Boeotia, and Myrina in Asia Minor; but numerous other sites, such as Athens, Corinth, north Africa, Italy, and Sicily

FIGURE 263—*Terracotta statuette from Tanagra. Fourth to third century B.C.*

have furnished innumerable examples, ranging in date from archaic to Roman times. In most Mediterranean countries clays are plentiful, so that those used were presumably local. One might have thought that it would be easy to determine the various sources by the appearance or analysis of the terracotta. This, however, has seldom proved possible. In statuettes, as in pottery, the clay was often mixed, and moreover the colour was influenced by the temperature of the firing. Among the Myrina statuettes, for example, nine different clays have been distinguished, ranging from red to light buff, all presumably from the same locality [17]. Much remains for further investigation in this field.

FIGURE 264—*Terracotta mould for a relief showing Herakles and a Maenad. Roman period.*

Terracotta reliefs play a large part in Greek, Etruscan, and Roman art. The majority were moulded (figure 264), and occasionally several examples from the same mould have been found. Sometimes a mould was taken from an existing relief, and the new positive was then naturally smaller than the original because of the successive shrinkages.

The large architectural reliefs, many partly in the round, which served as antefixes, acroteria, and friezes of temples were usually of coarse clay, like the statues (plate 20 C). The smaller votive reliefs, the so-called Melian (plate 21 C) and Locrian, the Tarentine of the fifth and fourth centuries B.C., and the Campana reliefs of the Roman period used in decoration of buildings (figure 265) [18] are made of the same fine clay as the statuettes. On their backs are sometimes found the marks of the strings used for removing the superfluous clay. Like the statues and statuettes all were once gaily coloured. Furthermore, moulded reliefs often decorate the little oil-lamps found in large numbers on Roman sites.

The purely utilitarian terracottas, such as roof-tiles, covering-slabs along the raking cornice of pediments, gutters, and so forth, were regularly made of fired coarse clay. Bricks in wall construction were either fired or merely sun-dried (p 407). On various Greek sites terracotta metopes and sarcophagi, often decorated with paintings, have been found.

With so many uses, clay remained one of the favourite materials throughout antiquity. Gradually, through increased use of metal and of stone, and especially after the invention of the blowing-tube for glass, probably in the first century B.C., the employment of terracotta somewhat diminished.

FIGURE 265—*Campana relief. Roman period.*

REFERENCES WITH NOTES

[1] C. F. BINNS (and FRASER, A. D. *Amer. J. Archaeol.*, **33**, 1, 1929) was the first to suggest that Athenian vases were fired successively under oxidizing, reducing, and reoxidizing conditions. This theory has been endorsed and amplified by T. SCHUMANN (*Forsch. Fortschr. dtsch. Wiss.*, **19**, 358, 1943). The latter's important new contribution was the discovery that, by peptizing the clay, i.e. eliminating the heavier particles (such as felspar, quartz, limestone), with the use of a protective colloid, and by adding a relatively small quantity of alkali, a substance could be produced equal in quality and appearance to the Attic black glaze. Microscopic and X-ray examinations by F. OBERLIES and N. KÖPPEN have substantiated Schumann's discoveries (*Ber. dtsch. Keram. Ges.*, **30**, 102, 1953).

[2] The yellowish hue of the slip in some of the examples must be due to the presence of iron as an impurity in the white clay (*cf* SCHUMANN, T. *Ber. dtsch. Keram. Ges.*, **23,** 423, 1942). In time the potters learned to eliminate the iron and thus produce a brilliant snow-white slip.

[3] The chemical formula involved in this process is: $3Fe_2O_3+CO = 2Fe_3O_4+CO_2$. In the oxidizing fire, the carbon of the fuel combines with two atoms of oxygen to form carbon dioxide, whereas in a reducing fire the carbon monoxide will extract oxygen from the red ferric oxide present in the clay and convert it into black magnetic oxide of iron. Both Miss Farnsworth and Mr Schumann think that this equation is preferable to that of $CO+Fe_2O_3 = CO_2+2FeO$, since ferrous oxide (FeO) is non-magnetic and unstable, whereas the Greek black glaze and black magnetic oxide of iron (Fe_3O_4) are magnetic and stable. (FARNSWORTH, MARIE, in SHEAR, T. L. *Hesperia*, **9,** 265, 1940. SCHUMANN, T. *Forsch. Fortschr. dtsch. Wiss.*, **19,** 358, 1943.)

[4] WEICKERT, C. *Archäol. Anz.*, **57,** cols. 523 ff., 1942.

[5] RICHTER, GISELA M. A. *Annu. Brit. Sch. Athens*, **46,** 146 ff., 1951.

[6] *Idem. Ibid.*, **46,** 149, 1951. *Idem.* 'Attic Black-figured Kylikes.' *Corpus Vasorum Antiquorum.* Metropolitan Museum of Art, New York, fasc. 2, p. 20. Harvard University Press, Cambridge, Mass. 1953. *Idem., Nederlands Kunsthistorisch Jaarboek*, **5** (Studies in honour of A. W. Byvanck), 127, 1954. Further investigation in this and related problems is now (1954) being carried on by chemists and archaeologists in America, England, and Germany.

[7] On this household ware, see BURR, DOROTHY. *Hesperia*, **2,** 597 ff., 1933. THOMPSON, H. A. *Ibid.*, **3,** 469–70, 1934. TALCOTT, LUCY. *Ibid.*, **4,** 493, 1935.

[8] RICHTER, GISELA M. A. *Amer. J. Archaeol.*, **45,** 383–9, figs. 23–26, 1941.

[9] A toothed wheel was suggested by former writers on this subject, but Miss Robinson showed me that a better result could be obtained with a toothed stick, for a wheel produces too regular a design. See RICHTER, GISELA M. A. *Studi Etruschi*, **10,** 64, fig. 1, 1936.

[10] *Idem. Ibid.*, **10,** Pls XXI, 2, XXII, 2. The depressions are clearly visible, for, being out of sight, they were not filled up as they were in the open Megarian and Arretine bowls.

[11] ALEXANDER, CHRISTINE. 'Arretine Relief Ware.' *Corpus Vasorum Antiquorum.* Metropolitan Museum of Art, New York, fasc. 1, p. 12, Pl. 1, 1. Harvard University Press, Cambridge, Mass. 1943. CASKEY, L. D. *Amer. J. Archaeol.*, **41,** 527, fig. 6, 1937.

[12] See Bibliography p. 283, esp. KUNZE, E. "Berichte über die Ausgrabungen in Olympia", and WEINBERG, S. S. "Terracotta Sculpture at Corinth".

[13] On such heads see KASCHNITZ-WEINBERG, G. *R.C. Pont. Accad. Archaeol.*, **3,** 325 ff., 1925, and especially Pl. XXI; also *Idem. Röm. Mitt.*, **41,** 133–211, 1926.

[14] MYRES, J. L. 'Handbook of the Cesnola Collection of Antiquities from Cyprus', p. 329. Metropolitan Museum of Art, New York. 1914.

[15] POTTIER, F. P. E. 'Les statuettes de terre cuite dans l'antiquité', p. 251. Hachette, Paris. 1890.

[16] In most accounts of the technique of Greek terracotta statuettes it is stated that the white was applied, like the tempera colours, after firing. But investigations kindly made for me in 1954 by Miss Maude Robinson in New York, as well as under the supervision of Mr R. A. Higgins, in the laboratory of the British Museum, have shown that the white is diluted clay (= slip), fired. *Cf* BIMSON, M. in HIGGINS, R. A. 'Catalogue of the Terracottas in the Department of Greek and Roman Antiquities in the British Museum', Vol. I, p. viii. Trustees of the British Museum, London. 1954.

[17] POTTIER, F. P. E. See ref. [15], p. 248.
[18] On the use of Campana reliefs for the decoration of temples, tombs, baths, and private houses, *cf* ROHDEN, H. VON. 'Architektonische römische Tonreliefs der Kaiserzeit', Pl. XLVII. Spemann, Leipzig. 1911. The term is derived from the name of the Marquis Giovanni Pietro Campana (1808–80), the famous collector.

BIBLIOGRAPHICAL NOTE

The wares mentioned in the text are described in the standard works. The following is a short list of books and articles dealing specifically with the techniques of Greek, Etruscan, and Roman ceramics. In my own writings I have had the invaluable help first of Mr Charles F. Binns and later of Miss Maude Robinson, both professional potters. In this article I have also profited from lively discussions of various problems with Miss Lucy Talcott of the Athenian Agora Staff.

BLÜMNER, H. 'Technologie und Terminologie der Gewerbe und Künste bei Griechen und Römern', Vol. 2, pp. 23–139. Teubner, Leipzig. 1879. Though now antiquated, it contains much useful information.

More recent are:

On Greek pottery:

RICHTER, GISELA M. A. 'The Craft of Athenian Pottery.' Metropolitan Museum of Art, Yale University Press, New Haven. 1923.
HUSSONG, L. 'Zur Technik der attischen Gefäßkeramik.' Dissertation, Heidelberg. 1928.
RICHTER, GISELA M. A. 'Attic Red-Figured Vases, a Survey', pp. 23 ff. Metropolitan Museum of Art, Yale University Press, New Haven. 1946.
BEAZLEY, SIR JOHN (DAVIDSON). "Potter and Painter in Ancient Athens." *Proc. Brit. Acad.*, **30,** 87–125, 1946.
LANE, A. 'Greek Pottery.' Faber and Faber, London. 1948.
BUSCHOR, E. 'Bilderwelt griechischer Töpfer.' Piper, Munich. 1954.

Specifically on the wheel:

RIETH, A. 'Die Entwicklung der Töpferscheibe.' Kabitzsch, Leipzig. 1939.

Specifically on the Attic glaze:

BINNS, C. F. and FRASER, A. D. "The Genesis of the Greek Black Glaze." *Amer. J. Archaeol.*, second series, **33,** 5 ff., 1929.
SCHUMANN, T. "Oberflächenverzierung in der antiken Töpferkunst." *Ber. dtsch. Keram. Ges.*, **23,** 408, 1942.
Idem. "Terra sigillata und die Schwarz-Rot-Malerei der Griechen." *Forsch. Fortschr. dtsch. Wiss.*, **19,** 356 ff., 1943.
WEICKERT, C. "Zur Technik der griechischen Vasenmalerei." *Archäol. Anz.*, **57,** cols. 512–28, 1942.
OBERLIES, F. and KÖPPEN, N. *Ber. dtsch. Keram. Ges.*, **30,** 102 ff., 1953.
RICHTER, GISELA M. A. "Accidental and Intentional Red Glaze on Athenian Vases." *Annu. Brit. Sch. Athens*, **46,** 143–50, 1951.
Idem. "Red-and-Black Glaze." *Nederlands Kunsthistorisch Jaarboek*, **5** (Studies in honour of A. W. Byvanck), 127–35, 1954.

Specifically on the 'Lasur' and accessory colours:

HUSSONG, L. 'Zur Technik der attischen Gefäßkeramik', p. 55. Dissertation, Heidelberg. 1928.

SCHUMANN, T. *Ber. dtsch. Keram. Ges.*, **23,** 422 f., 1942.

WEICKERT, C. *Archäol. Anz.*, **57,** col. 525, 1942.

On Hellenistic and Roman relief ware:

COURBY, F. 'Les vases grecs à reliefs.' Bibl. Éc. franç. Athènes Rome, fasc. 125. Boccard, Paris. 1922.

PAGENSTECHER, R. "Die Calenische Reliefkeramik." *Jb. dtsch. Archäol. Inst.* 8. Ergänzungsheft. 1909.

THOMPSON, H. A. "Two Centuries of Hellenistic Pottery." *Hesperia*, **3,** 311–480, 1934.

DRAGENDORFF, H. "Terra sigillata. Ein Beitrag zur Geschichte der griechischen und römischen Keramik." *Bonn. Jb.*, **96,** 18–155, 1895.

MUSEUM OF FINE ARTS, BOSTON. 'Catalogue of Arretine Pottery', by G. H. CHASE, pp. 8 ff. Houghton Mifflin Company, Boston. 1916.

OSWALD, F. and PRYCE, T. D. 'An Introduction to the Study of Terra Sigillata.' Longmans, Green and Co., London. 1920.

On West Slope ware:

WATZINGER, C. "Vasenfunde aus Athen." *Athen. Mitt.*, **26,** 67 ff., 1901.

THOMPSON, H. A. *Hesperia*, **3,** 438 ff., 1934.

On Centuripe ware:

LIBERTINI, G. 'Centuripe.' Guaitolini, Catania. 1926.

On the blue-green glazed ware:

ZAHN, R. "Glasierte Tongefässe im Antiquarium." *Amtl. Ber. preuss. Kunstsamml.*, **35,** cols. 277 ff., 309 ff., 1913–14.

TOLL, N. "The Green Glazed Pottery", with Technological Notes by F. R. MATSON in ROSTOVTZEFF, M. I. *et al.* (Eds). 'The Excavations at Dura-Europos, Final Report IV', Part 1, fasc. 1, pp. 1 ff., 81 ff. Yale University Press, New Haven. 1943.

ROBINSON, MAUDE. "Notes on the Glazes" in ALEXANDER, CHRISTINE. 'Green-glazed Ware.' *Bull. Metrop. Mus.*, new series, **3,** 133–6, 1945.

JONES, FRANCES F. "The Pottery" in GOLDMAN, HETTY (Ed.). 'Excavations at Gözlü Kule, Tarsus', Vol. 1, pp. 191 ff. Princeton University Press, Princeton. 1950.

On Roman pottery in general:

CHARLESTON, R. J. 'Roman Pottery.' Faber and Faber, London. 1955.

WALTERS, H. B. 'Catalogue of the Roman Pottery in the Departments of Antiquities, British Museum.' Trustees of the British Museum, London. 1908.

Idem. 'History of Ancient Pottery', Vol. 2, pp. 433 ff. Murray, London. 1905.

On Etruscan pottery:

On the black body: FORSDYKE, SIR (EDGAR) JOHN. "The Pottery called Minyan Ware." *J. Hell. Stud.*, **34,** 139 f., 1914.

BINNS, C. F. and FRASER, A. D. *Amer. J. Archaeol.*, second series, **33,** 9, Pl. II, no. 6, 1929.

On the decoration: RICHTER, GISELA M. A. "The Technique of Bucchero Ware." *Studi Etruschi* **10,** 61–65, 1936.

On terracotta sculpture:

KUNZE, E. "Zeus und Ganymedes, eine Terrakottagruppe aus Olympia". *Winckelmanns-programm der Archaeologischen Gesellschaft zu Berlin*, No. 100, Berlin, 1940.

Idem. 'Terrakottaplastik' in the "Berichte über die Ausgrabungen in Olympia". *Jb. dtsch. Archäol. Inst.*, **3**, 119–32, 1941; **5**, 103–27, 1956; **6**, 169–97, 1958.

WEINBERG, S. S. "Terracotta Sculpture at Corinth." *Hesperia*, **26**, 289 ff., 1957.

GIGLIOLI, G. Q. 'L'Arte etrusca.' Plates CXC ff. and *passim*. Fratelli Treves, Milan, 1935.

Statuettes: POTTIER, F. P. E. and REINACH, S. 'La Nécropole de Myrina.' Bibl. Éc. franç. Athènes Rome, second series, Vol. 8, i, pp. 125 ff. E. Thorin, Paris. 1887.

POTTIER, F. P. E. 'Les statuettes de terre cuite dans l'antiquité', pp. 247–62. Hachette, Paris. 1890.

BURR, DOROTHY. 'Terra-cottas from Myrina in the Museum of Fine Arts, Boston.' Dissertation, Bryn Mawr College. 1934.

Small reliefs: JACOBSTHAL, P. 'Die Melischen Reliefs', pp. 101 ff. Deutsches Archäologisches Institut, Berlin. 1931.

ZANCANI MONTUORO, PAOLA. "Note sui soggetti e sulla tecnica delle tabelle di Locri." *Atti Mem. Soc. Magna Graecia*, nuova serie, **1**, 1–36, 1954.

WUILLEUMIER, P. 'Tarente des origines à la conquète romaine', pp. 393 ff. Bibl. Éc. franç. Athènes Rome, fasc. 148. Boccard, Paris. 1939.

ROHDEN, H. VON. 'Architektonische römische Tonreliefs der Kaiserzeit.' Spemann, Leipzig. 1911.

Lamps: BRONEER, O. 'Terracotta Lamps.' Corinth. Results of Excavations conducted by the American School of Classical Studies at Athens. Vol. IV, part II. Harvard University Press, Cambridge, Mass. 1930.

8

PART I

CERAMICS

FROM *c* 700 B.C. TO THE FALL OF THE ROMAN EMPIRE

GISELA M. A. RICHTER

THE preparation of clay for use in pottery, the techniques of wheel-work, building, and moulding, and the construction of ancient kilns have already been discussed (vol I, chs 9 and 15). We now consider the ceramic achievement of the Greek and Roman worlds. It is convenient to consider under ceramics not only pottery proper but also all moulded and shaped works in baked clay except bricks, tiles, and building-materials. The convenient term 'ceramic' was introduced into English in the nineteenth century from the Greek *kerameikos*, the potters' quarter of Athens. The noun *kerameia* was the general term for the potter's craft.

I. GREEK POTTERY, *c* 700–*c* 300 B.C.

The seventh century B.C. was a period of general awakening in the Hellenic world. Following the upheavals of the preceding centuries and the founding of Greek colonies all over the Mediterranean, a new Greece was in formation. The buoyant spirit of the age is reflected in its ceramics. Distinctive wares were produced in many different centres. Chief among them were Asia Minor, the islands of Rhodes, Chios, and Cyprus, and the mainland cities of Sparta, Corinth, and Athens. Shapes and styles of decoration differ. The technique is essentially the same in all—that is, the vases are made of well levigated clay, buff or pink after firing, and are fashioned on the wheel, except when moulded in the form of heads and animals. Moreover, all have a brownish-black glaze with occasional touches of red and white (figures 233–4).

The seventh century in Greek art is generally called by archaeologists the period of oriental influence, because, after the geometric patterns prevalent in the preceding age, the decoration, in its stylized plant-forms and animal friezes, shows borrowings from Asia Minor, Egypt, and Mesopotamia. The technique, however, is directly evolved from the earlier Greek, as are the shapes also. The

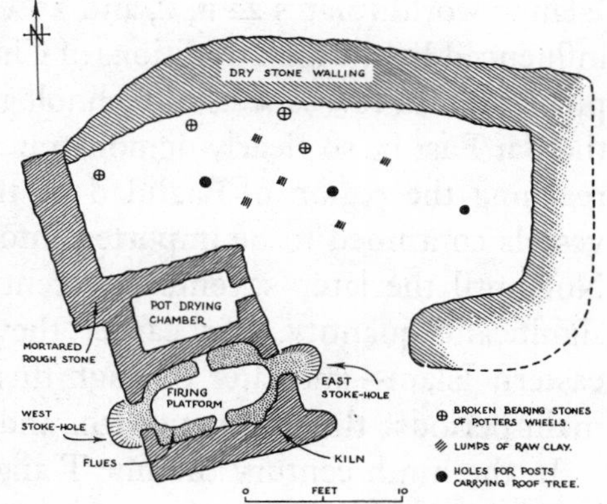

FIGURE 266—*Plan of a potter's workshop at Limpsfield, Surrey. Thirteenth century.*

and a few recipes for glazes. There are also some useful illustrations in manuscripts and early prints of potters at their wheels (figures 270–3).

From the end of the period comes an outstanding account by Cipriano Piccolpasso (1524–79) of 'all the secrets of the whole potter's art', as practised in north Italy about 1550. He lived in Castel Durante (now Urbania), a pottery-making centre, and his brother Fabio was a potter. This first-hand illustrated account is the earliest comprehensive treatise on the subject in the west (figures 267, 273, 274, 279, 280, 286, and 287), though there is a Persian treatise of 1301. Much of the information in it had hitherto been concealed as trade-secrets. Piccolpasso deals largely with the production of tin-glazed majolica, a fine-pottery tradition of Near Eastern origin, but much of his information concerning preparation of clay (figure 267), making wheels and kilns, and shaping and firing vessels, is applicable to fine ceramics in general.

II. ISLAMIC AND CHINESE INFLUENCES

The story of the development of finer ceramics in medieval Europe is largely that of the persistent influence of the orient. The direct inspiration was from the

FIGURE 267—*Sequence of settling-tanks for the levigation of clay. From Piccolpasso's manuscript on the potter's art.* c *1550.*

Islamic world (plates 22 B, C, and 23 A), the tastes and techniques of which were influenced by the high traditions of Chinese ceramics (see the concluding chapter, pp 767 ff). In few aspects of technology can this remote and indirect influence of the Far East be so clearly demonstrated. Chinese porcelain and stonewares were reaching the region of Baghdad by the ninth century, and these highly prized vessels continued to be imported into the Near East through the Middle Ages. Not until the later seventeenth century did Chinese wares reach Europe in significant quantity, but earlier they gave a vital stimulus to the potters of eastern Islam—and later through them indirectly to those of Europe—at three main periods: the ninth, twelfth, and fifteenth centuries.

In the ninth century the fine T'ang mottled stonewares and creamy porcelain created an appreciation of ceramic art in the Near East. At first Chinese colour, texture, and form were merely imitated, but new themes soon appeared. The Islamic achievement of whiteness by means of opaque tin-glaze (pp 301–2) opened the way for an entirely original flowering of ceramic painting. The Caliph's court at Baghdad was the focus of Islamic fine pottery production. By the eleventh century, however, Baghdad potters were migrating to Cairo, while Persia was rising to the forefront in ceramics. This tin-glazed pottery, and the painting styles developed thereon, greatly influenced the taste for fine pottery in Europe, notably in Spain and Italy, being thus the precursors of majolica and delft wares.

In the twelfth century Chinese whiteness was imitated in Persia by a white fabric, and not as in ninth-century Baghdad by an opaque surface glaze. Composed largely of powdered quartz and glass, with less than ten per cent of white clay, this Persian material lacked plasticity, and, though more resonant and durable than tin-glaze wares, had not their subtlety. This technique probably originated in Persia and spread to Egypt, but had no appreciable influence on European ceramics, though for centuries Persia showed the most consistently high achievement in ceramic art and techniques.

In the fifteenth century imported Chinese blue-and-white porcelain (itself partly influenced by earlier Islamic wares) once more revitalized the Islamic potter's art of the Near East. It provided, both through Islam and directly, the inspiration for much post-medieval European ceramic development. We thus owe the trends of European fine ceramics largely to both Islam and China.

The two most typical Chinese ceramic substances were porcelain and stoneware. *Porcelain* is formed from a fine white clay and china-stone. The latter is a felspathic mineral fusing to a glassy matrix at the high temperatures (1250°–1350° C) essential to the production of porcelain. The particular qualities of porcelain are whiteness, translucence, and metallic resonance. It was made in

China from about A.D. 700. Porcelain wares were reaching the Near East and notably Baghdad about a century later.

Stonewares of good resounding quality, though lacking the whiteness and translucence of porcelain, had been made in China from the fourth century A.D. They, too, were imitated in the Near East. Stoneware was made by firing fine plastic clay to a high temperature (*c* 1250° C). This implies great skill in packing and firing the kiln. The clay being partially vitrified, vessels formed from it were impervious to water even without glaze. Stonewares were developed in the Rhineland, quite independently of eastern influence, during Carolingian times (*c* 800 onwards). These early Rhenish stonewares were fired in very simple kilns (figure 281), and their pre-eminence may have been due more to particularly plastic clays than to abnormally high firing-temperatures. In the later Middle Ages the stonewares of other places, notably Siegburg (near Cologne), became renowned (figure 268).

FIGURE 268 — *Rhenish salt-glazed stoneware jug. From Siegburg, Rhineland. Fifteenth century. The spiral striations show the twisting of the fine plastic clay under the rotary process of throwing. Scale 2/5.*

In the fourteenth century Rhenish stonewares were further developed by adding a salt-glaze. Several pounds of common salt were thrown into the kiln at a late stage in the firing, and reacted with the minerals of the clay to form a glassy surface. The characteristic dark mottled effect of certain Rhenish stonewares was produced by reducing conditions in the kiln; blues and purples were also produced on some of these wares (Cologne). Such stonewares were produced in England from about 1684.

III. BYZANTINE INFLUENCE

It is usually held that the fine polychrome wares of medieval Italy were inspired by those of Moorish Spain. But during the twelfth and thirteenth centuries Venice and south Italy though no longer under the control of Byzantium were receiving some Byzantine and Levantine pottery. The influence of these imports in the development of Italian glazed wares from the early thirteenth century on, such as those of Apulia or Orvieto, has been underestimated (plate 22 E). The renewed use of lead-glazes in the eleventh century in Germany, the Low Countries, and England probably owes something to the continuous Byzantine tradition. However, we have little detailed knowledge of the development of Byzantine ceramics (figure 269). Byzantine kiln-sites have been explored

FIGURE 269—*Plaque of polychrome ceramic in green, brown, and yellow on a white ground, from Patleina, Bulgaria. Tenth century.*

at Corinth and in Bulgaria, but none are yet known at Constantinople.

The Mediterranean lands generally, through their maritime trade, were the scene of this infusion of eastern influences into western ceramic art. Byzantine ceramic influences did not penetrate up the Danube beyond Bulgaria. Although overland trade flourished between the Black Sea and the Baltic, ceramic art in the north was not sufficiently developed to be receptive of such oriental refinements as glazes and polychrome painting. Fine cream pottery, bearing, under a thin lead-glaze, painted brown and green birds and leaf-ornament of Mediterranean and ultimately eastern origin, was being produced in south-west France (Saintes) during the thirteenth century, and these wares were exported to Britain (figure 276), where they had some influence on local pottery. They even reached Scandinavia.

IV. SHAPING OF VESSELS

Hand-built pottery of various types was used by peasant communities in post-Roman Europe as well as in the northern trading towns of Viking age (sixth to ninth centuries) (figure 275 (1)). In remoter parts, such as the north of Ireland, the practice continued until recently. This pottery was built either from clay lumps, or long 'sausages' bent in rings or spirals, or on a table turned in jerks (vol I, figures 226, 242). A potter is seen in a thirteenth-century manuscript operating a wheel turned by a stick (figure 270)—an inefficient method, for the potter is much handicapped by having only one hand fully free for shaping the clay. The method was still in use in recent times (vol I, figure 233).

FIGURE 270—*A potter at a turn-table. From a French thirteenth-century manuscript. The right hand is used for turning the wheel with a stick and is thus not free for throwing. Contrast the kick-wheel (figures 271–3).*

Various forms of the kick-wheel are seen in illustrations in western sources from the thirteenth to sixteenth centuries (figures 271–4, plate 23 C). The great advantage of this device is that the vessel can be continuously rotated by the potter himself at a controllable rate. Much medieval pottery shows evidence of having been formed on such fairly rapidly rotating wheels. In the Rhineland, for example, where the factory tradition was continuous from Roman times, even cooking-pottery was throughout our period made on fast wheels. In England, however, no continuous tradition of wheel-throwing can be shown from Roman times, the potter's wheel being reintroduced from the Rhineland to East Anglia about the ninth century (figure 275 (2)), whence it spread throughout Britain by the thirteenth century.

FIGURE 271—*A potter's kick-wheel. From a fifteenth-century playing-card. The potter uses a notched tool for making ornamental grooves on the vessel.*

Piccolpasso (figures 273–4) records that the hard steel pivoting-point of the wheel ran on a flint stone or a very hard steel plate, and that the bearing towards the top of the axle was cased with oiled leather. Pivot-stones have been found at kiln-sites (figure 266).

Throwing on the wheel was practised for taller shapes of vessel of all sizes. Rims were usually strengthened by a thicker moulding formed in the throwing. Bringing the clay in from a bulbous body to make a narrow neck is the most difficult part of the throwing-process (figure 275 (8)). Vessels of waisted forms and shaped necks or pedestal bases were sometimes made in separate sections and joined after removal from the wheel, either by luting while still damp or even (according to Piccolpasso) after biscuit-firing, by a layer of glaze in the second firing. Shallow vessels, such as platters, were formed upside down by putting the ball of clay on a convex mould placed on the top of the turn-table or potter's wheel (figure 273).

Turning and finishing. Templets and other tools were sometimes used, largely to remove surplus clay, for finishing the pottery shape while still on the wheel (figures 271–2). This device was especially adapted for forming foot-rings and producing more elegant shapes by accentuating waists and hollow mouldings. Turning was also used, like burnishing, to give the surface a finish after the clay had somewhat hardened. Nevertheless, many regular and hard rim-profiles on medieval vessels, which at first sight look as though formed with tools, will be found to fit the fingers and were formed by finger-shaping on a fast wheel. As platters were formed upside down, foot-rings on them were easily formed by turning while still on the wheel. On other vessels, foot-rings were usually added after removal from the wheel.

The convex base is a particular characteristic of medieval cooking-pottery,

FIGURE 272—*Potter at a kick-wheel. He is seen finishing a pot with a 'rib' or smoothing-tool. Pots are drying on shelves protected by a roof. He is making deep pans with lids and sieve bottoms. They are used in the process of roasting pyrites and abstracting sulphurous or bituminous vapours from it. They are seen being heated in a furnace. From Agricola,* De re metallica, *1556.*

such as pans, bowls, and even pitchers (figure 275 (3)). This base was formed by pressing and working the inside of the base while the clay was still soft, either with the hand or with a pad, at the same time perhaps lifting the vessel slightly away from the wheel-head. The base was thus strengthened by compacting the clay, and the convex surface was less liable to crack on expansion or contraction with change of temperature, as is the modern round-bottomed flask of our laboratories. Vessels were otherwise usually removed from the wheel-head with a knife or with a string (like a cheese-wire, which leaves characteristic off-centre marks on the under surface (figure 239).

Handles were usually made from strips or rods of clay, sometimes of several strands twisted together. As in earlier (vol I, p 399) so in medieval times the influence of metal-work may be seen in the animal heads ornamenting the handles; indeed, in Italy such pottery flagons were denoted by the words *bronzo antico*. Handles were most frequently fixed by luting on, that is, by wetting the

FIGURE 273—*Italian potters at kick-wheels. From Piccolpasso, c 1550. The man to the left has shaped a platter on a convex mould (see p 289) placed on the wheel-head; he is now finishing the base.*

junction with a slurry and working the handle and body together with the fingers while the clay was still soft. The thumb and finger pressure-marks were often accentuated at the top and bottom of the handles as part of the plastic ornament style. At the lower junction these impressions are sometimes drawn down into hollow channels (figure 278). The handle-junctions were also strengthened by jabbing a pointed stick through from the inside, or, most elaborately of all, by forming plugs on the handle and fixing them through holes in the neck and body of the jug—a mortise-and-tenon joint.

Lips were formed by merely pinching the rim, though the large 'parrot-beak' spouts of French medieval jugs had to be added. Tubular spouts (figure 275 (4)) were added to the vessel either by piercing it from outside with a stick, or from inside with a finger, a strip of clay being then wrapped round stick or finger to make the spout. The joins were made by pressing the spout and body together. Fixing of both handles and spouts was no doubt frequently done while the vessel was still on the turn-table. Bung-holes were similarly made near the base of large vessels used for storing liquids.

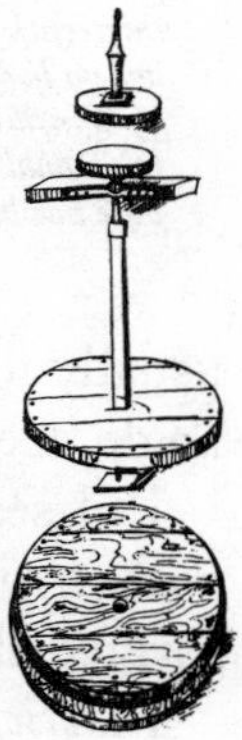

FIGURE 274—*From Piccolpasso's diagram of a potter's wheel.* (Top) *Pivoting point.* (Middle) *Note the casing for the bearing towards top of axle.* (Below) *Construction of the wooden wheel of two built-up disks with their joints set at right angles.*

Legs, usually three, short or long, were sometimes added to pottery vessels, either in imitation of metal styles, or to make convex bases stand properly. Applied frilled strips on the base-angle served the same purpose. In Germany, particularly on stonewares, the massive frilled base was developed, and was

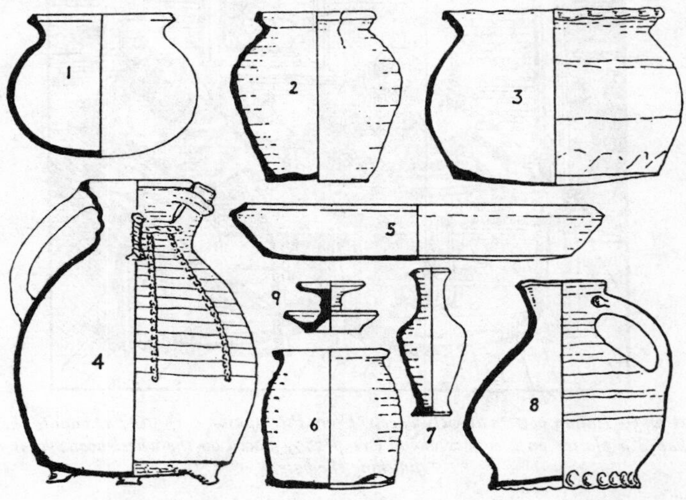

FIGURE 275—*Some typical European medieval pottery-shapes.* (1) *Globular cooking-pot, part of the hand-making ceramic tradition of northern and western Europe. Eighth to twelfth centuries.* (2) *Flat-based, narrow cooking-pot, wheel-thrown, perpetuating the Roman industrial tradition, a style spreading north and west from the Rhineland and its coastal areas. Eighth to eleventh centuries.* (3) *Baggy cooking-pot, with everted rim flange and convex base, on the outer edge of which knife-trimming marks may be seen. Britain, particularly south and midland England. Eleventh to fourteenth centuries.* (4) *Globular baggy pitcher with tubular spout, strap-handle, and sometimes small side-handles and three small feet; glazed. Southern and midland England. Twelfth and thirteenth centuries.* (5) *Wide shallow dish, used for serving meats or birds, or sometimes as a drinking-vessel; unglazed. Southern and midland England. Eleventh to thirteenth centuries.* (6) *Smaller tall cooking-pot; particularly north Britain. Twelfth to fourteenth centuries.* (7) *Elegantly shaped pottery bottle; glazed. England, thirteenth to fifteenth centuries.* (8) *Pitcher, with wide body and narrow neck. The narrowing of the neck from so wide a body is one of the more difficult operations in pottery-throwing* (*see also* 4)*; the base shape has been formed by pinching-in after removal from the wheel-head; glazed. Britain, later thirteenth to fourteenth century.* (9) *Double-shell lamp; the stem and some of the shape have been tool-turned on the wheel after throwing, and the base has been scooped out with a knife after removal from the wheel-head; glazed. English, thirteenth to fifteenth centuries. Scale 1/6.*

usually formed by pinching and moulding a thick ridge of clay left on the base-angle itself (figures 268, 271, 275 (8)).

Lid-seatings were made as hollow rim-mouldings, either with the fingers or with a simple tool. The lids were drawn out from a narrow pedestal, which formed the knob when inverted (figure 272).

Plastic ornament. Markings made on the soft clay during the throwing gave the potter's most obvious ornament—body-rilling, rim-moulding, and tooled grooves. Non-rotational plastic ornament could be applied while the formed vessel was still on the turn-table; it included application of vertical or horizontal pinched strips of clay (figure 278), stamping, rouletting, comb-marking, and scoring. Ornamentation of base-angles by nicking, thumb-pressing, and applica-

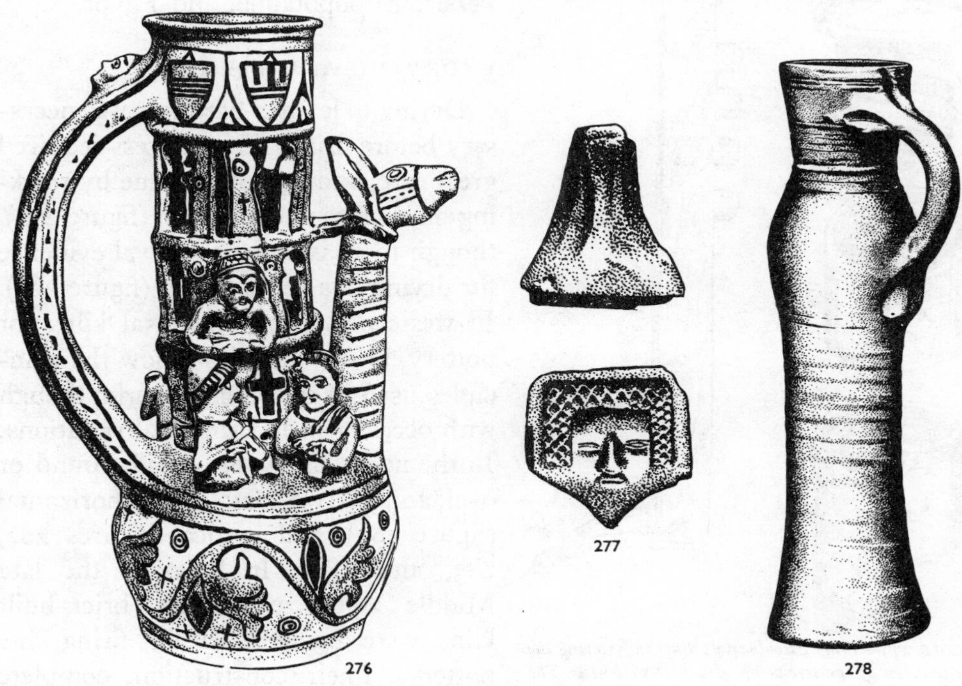

FIGURE 276—*Puzzle jug, polychrome painted in green and brown on cream ware with transparent glaze. Thirteenth century. It was found at Exeter, but made at Saintes, Charente, south-west France. Scale 1/5.*
FIGURE 277—*A mould for making a face-mask on a pitcher. From Lincoln. Probably fourteenth century. Scale 1/2.*
FIGURE 278—*Tall baluster jug from York. Late thirteenth to fourteenth century. Scale 1/5.*

tion of thumbed clay strips, legs, or feet, was, of course, carried out after removal from the turn-table. All these are common. Stamps or moulds of bone, wood, stone, or baked clay for impressed patterns, such as rosette-flowers or face-masks, have been found (figure 277). Plastic ornament, secondary to the rotation-process, was used in medieval Italy and France (figure 276), and became common in thirteenth-century England (figure 292), but declined during the later Middle Ages. Laborious plastic modelling was sometimes utilitarian; thus Piccolpasso describes the making of screw-stoppers on pottery bottles (figure 279, plate 22 D).

Vessels built partly or entirely in moulds were made by the Romans (p 272), and by medieval Islamic potters. Ornament was also carved on the leather-hard vessels, particularly in twelfth-century Persia, Mesopotamia, and Egypt.

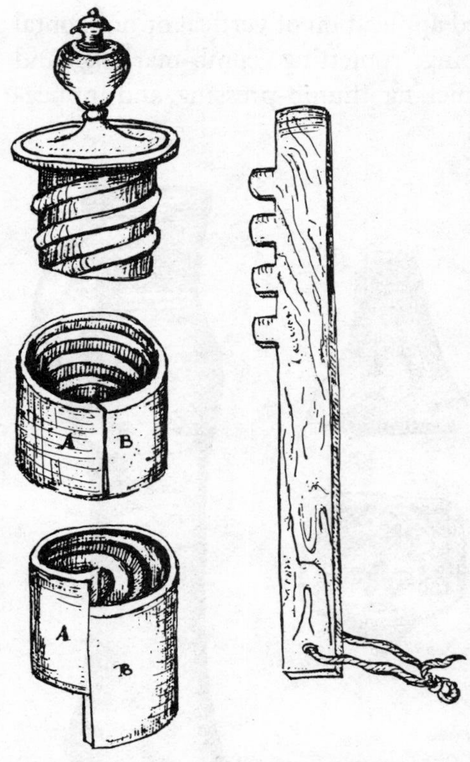

FIGURE 279—*From Piccolpasso's diagram showing the construction of the screw-top of a pottery bottle. The top is made as a cast from the neck of the vessel.* (Right) *Tool for making the thread (see plate 22 D).*

V. DRYING AND FIRING

Drying to leather-hardness was necessary before firing. This process required great care, and was often done by stacking on shelves under shelter (figure 272), though there is some medieval evidence for drying in a special oven (figure 266). In western Europe, medieval kilns for pottery, brick, and tiles follow the principles used in the Roman world, though with occasional experimental variations. In the north they were small, round or oval, and of two main types, horizontal (figure 281), and vertical (figures 282, 283, and 284). In Italy by the late Middle Ages rectangular brick-built kilns were being used for firing fine pottery. Their construction, complete with spy-holes, is described by Piccolpasso (figures 280, 287). On the other hand, much peasant pottery continued to be fired on open hearths or in clamps (vol I, figure 237).[1]

[1] G. M. Knocker has excavated the site of such a pottery-baking clamp of pagan Saxon age in Suffolk.

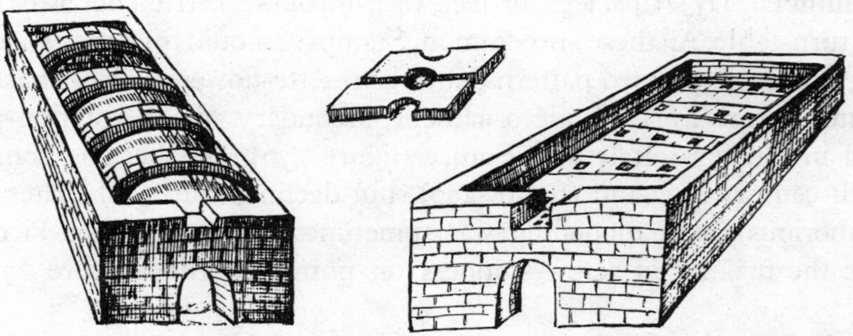

FIGURE 280—*From Piccolpasso's diagram showing two stages in the building of the potter's kiln seen in figure 287.*

In horizontal kilns the pots were stacked on the floor of the chamber, over which a temporary vault of turf, or of clay and wicker, was built. These simple kilns were used even for some hard-fired stonewares of the Rhineland (figure 281). Most vertical kilns had a round or oval chamber, and supported above the kiln-floor was a platform on which the pots were stacked. The simplest vertical kilns had a central pillar, with loose fire-bars laid across from the kiln-wall (figure 282), though in some Roman kilns (figure 291) the firing-platform was a permanent clay floor penetrated by small holes, as was used in medieval Corinth. Sometimes this floor was carried on arches arranged as in tile-kilns (vol I, figure 241 B). The central support was often a solid structure (figure 283) or a wall. Fire-arches permanently built in, springing from the kiln-wall, round in section, and made of a lime-clay mixture (figure 283), seem to be a post-Roman development, and were in use by late Saxon times, as at Thetford in Norfolk. The spaces between these arches were filled with loosely-joggled broken sherds, and through the interstices came the hot gases, which were deflected towards the centre of the platform by the domed roof. The vessels were stacked over the whole area of the platform and fire-arches. Some of the Pingsdorf (Rhineland) stonewares of the ninth to twelfth centuries were fired in such kilns.

Experiments in design, of the thirteenth century and later, presumably to

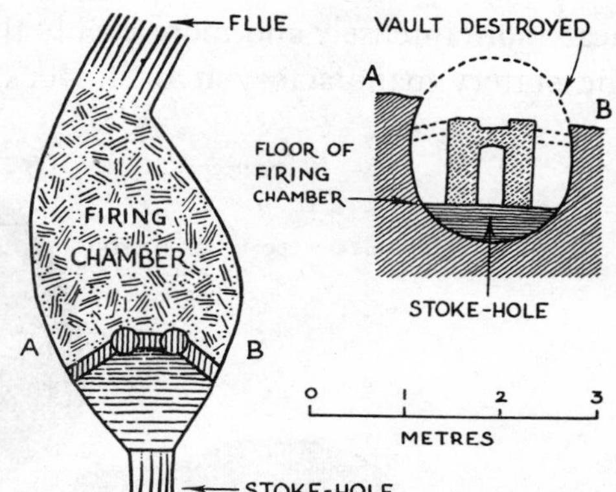

FIGURE 281—*Plan* (left) *and section* (right) *of a medieval pottery kiln of horizontal type. Galgenberg, near Siegburg, Rhineland. The hard Rhenish stonewares (see figure 268) were made in these simple kilns.*

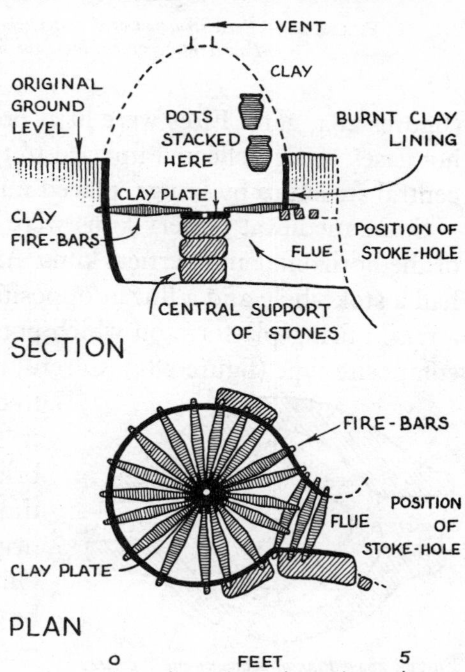

FIGURE 282—*Romano-British pottery kiln of vertical type. Weston Favell, near Northampton. The pottery was stacked on a platform made up of individual clay fire-bars resting on a clay plate supported on a central pillar of stones. First century A.D.*

heat more intensely and more evenly the whole of a larger kiln, may be seen at the pottery manufactory at Brill, Buckinghamshire, and at Limpsfield, Surrey

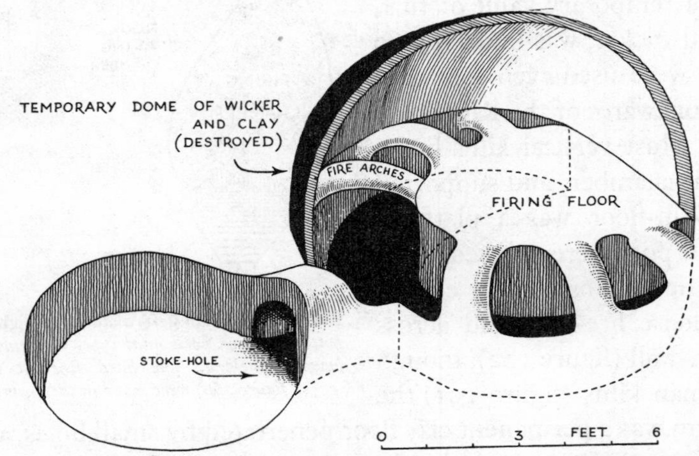

FIGURE 283—*Reconstruction of a potter's kiln of vertical type at Potterspury, Northamptonshire. Fourteenth century. Note the solid structure supporting the firing floor.*

(figures 284, 266). Fires were lit at both ends of double-ended vertical kilns, the hot gases being deflected upward to the firing-platform round either side of the central structure by bungs, placed midway along each flue.

Some medieval pottery kilns were apparently built to combine the functions of the horizontal and vertical kilns. Any kiln which on excavation proves to have had a stoke-hole and a flue at opposite ends, and in addition a structure to carry a raised firing-platform on which pots could be stacked, was presumably of this composite type (figure 285). Careful manipulation of the vents would be needed to control the firing in such kilns.

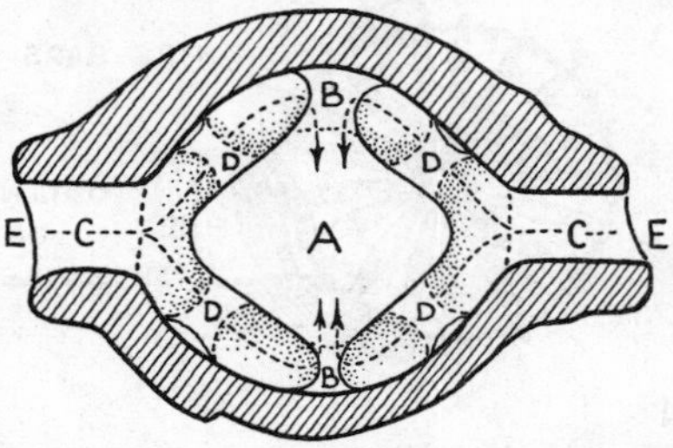

Tile- and brick-kilns were rectangular brick-built structures in Roman as in later times (figure 291). They had perforated firing-floors carried on strongly built arches. Only in the later Middle Ages can such kilns be shown to have been used for pottery-making. Ridge-, roof-, or paving-tiles and bricks were all made in such kilns.

FIGURE 284—*Plan of a double-ended pottery-kiln of vertical type at Brill, Buckinghamshire. Thirteenth century.* (A) *Firing-platform;* (B) *flue-bungs;* (C) *stoking-arches;* (D) *fire-arches;* (E) *stoke-holes. The arrows mark the path of the hot gases. The overall length of the kiln is 14 ft.*

Capacity and packing of medieval kilns. Most medieval kilns could take something of the order of 200 normal pots at a firing. Brick-

and tile-kilns were larger (p 305). The pots were stacked and spaced from one another by clay rings or placed in saggars (fine clay containers) (figure 286) to achieve more even heating, as they still are today.

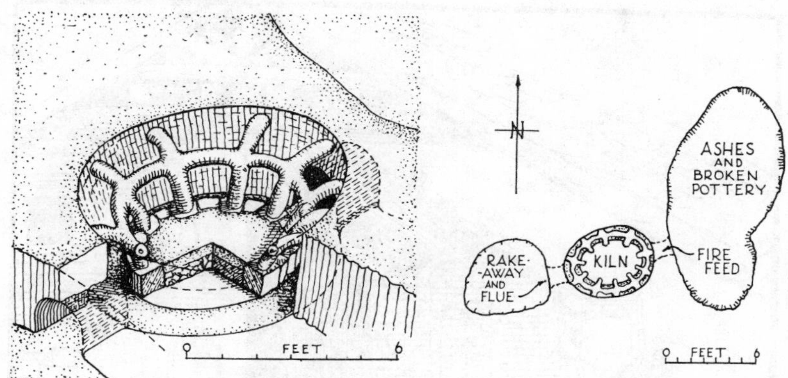

FIGURE 285—*Isometric reconstruction and site layout of a pottery kiln, apparently of combined vertical-horizontal type, at Cheam, Surrey. Fifteenth century.*

Firing-temperatures and their control. Open-hearth and clamp firing reached only about 700°–800° C, at which temperatures few even of the fine clays will fire to a hard, impermeable fabric. Temperatures of up to 1050°–1150° C were achieved for firing Roman *terra sigillata* (p 273), and even higher temperatures may have been reached in the horizontal kilns in which the early Rhineland stonewares were fired (figure 281). The types of kiln, and hence the temperatures they gave, were much the same in medieval western Europe as in Roman times, the fuel still being wood.

FIGURE 286—*Saggars for stacking plates* (middle) *and taller vessels* (right) *in the kiln. The stacking of plates on conical supports, a method of great antiquity, is seen on the left. From Piccolpasso,* c *1550.*

In the baking of ceramics in intermittently fired kilns only 5–10 per cent of the heat produced is actually taken up by the ceramics. Modern continuously fired kilns—one of which has been burning for 60 years—are said to utilize as much

FIGURE 287—*Piccolpasso's picture of a pottery kiln (cf figure 280). The master-potter controls the firing of the kiln by means of an hour-glass.* c *1550.*

as 30–40 per cent of the heat. Continuous firing was thus an important advance in making brick an economical building-material. Some pottery in the early Near East was fired to a temperature unnecessarily high for the production of a serviceable ware. Not until Islamic times (seventh century A.D. onwards) was it realized that this was a waste of the fuel, especially precious in those regions.

It is essential that the raising of kiln-temperature should be slow. Between 350° and 500° C organic matter burns out and water vaporizes off. The gases thus produced will damage the fabric if they are generated too rapidly. At 570°–600° C silica itself changes physically and expands. Above 600° C heating may be more rapid unless lead-glazed wares are in the kiln, when the glaze may break into holes, a defect sometimes seen on medieval pottery. Early potters were unable to inspect the interior of their kilns during firing. There is little evidence that medieval potters employed any aids to control firing, such as the modern Seger cones, though the Roman potters of Cologne seem to have used sherds to test their lead-glazes. Piccolpasso tells the potter to observe when the mortar on the fire-arches begins to drip. He gives the duration of firing, and shows the master-potter directing the operation with his hour-glass before him (figure 287), an early example of instrument-control of an industrial process.

Slow cooling—over 24 hours or more—is also essential. The contraction of silica at 573° C is a particular danger-point, and many medieval kiln-wasters, especially those of sandy fabrics, show cracks probably caused by this change. When the maximum temperature was reached, the fires were damped down, and the kiln was sealed and allowed to cool naturally, the dome being dismantled when cold for removal of the vessels. Thus much medieval pottery was cooled under reducing conditions, that is, out of contact with the air, though the extent depended upon the efficiency of the sealing.

Shrinkage of the clay over the whole process of drying and firing is usually about one-eighth to one-sixth of the linear dimensions. The production of fairly accurate measures, such as pint mugs, towards the end of our period implies that some potters had an excellent knowledge of their materials.

VI. GLAZES, SLIPS, AND PAINTED ORNAMENT

Vitreous glazes were primarily applied to pottery vessels to render them impervious to water, though the patchiness of much medieval glazing reduced it to a mere ornament; often, no doubt, glaze was applied by the potter from mere force of habit, without thought of its purpose. Glazes were commonly used on both the inside and the outside of all classes of pottery. Thus cressets, oil-lamps with the wick floating in the oil, which in England in the eleventh century were unglazed, were glazed inside the bowl by the thirteenth. In the east unglazed pottery, permeable to water, continued to be made with the deliberate purpose of keeping water cool by continuous evaporation from the outer surface. Such vessels sometimes had strainer-tops to keep flies out.

True vitreous glazes in Roman times were produced in two ways. The alkaline silicate glazes of the east (analogous in composition to glass) had no true medieval successors. Lead silicate glazes, on the other hand, were widely used in Roman times from Asia Minor to Britain, and form the basis of most medieval glazes (p 274). They adhere to ceramic surfaces better than alkaline silicate glazes, and soften at lower temperatures. Mixed with a glass they provide the vitreous matrix of 'tin' glazes made opaque with tin oxide, on Islamic, majolica, and delft wares. After the fall of the Roman Empire, however, lead-glazes were not made in the west until about the tenth century. They then appear again on pitchers of typical local shapes in western Germany, the Rhineland, the Low Countries, and England. The technique was presumably reintroduced into Germany from the Byzantine world (which included Adriatic Italy), for at Constantinople lead-glazes can be traced continuously from antiquity. In this transmission no more may have been involved than a knowledge of recipes, such as those of Theophilus or Heraclius (p 351), provided that they would actually work. From the twelfth century, lead-glazed pottery became common in western Europe, particularly in Britain and France.

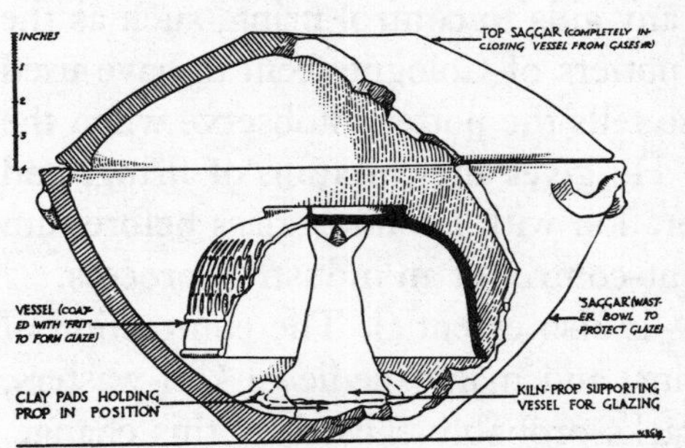

FIGURE 288—*Applying a lead-glaze to a vessel by firing it upside down on a pronged support inside another vessel (making a muffle-kiln). Romano-British. Holt, Denbighshire. Scale 1/6.*

In Roman times the lead-glaze mixture (a lead compound mixed with gum) was applied to the already fired vessel (biscuit), which was then refired to about 900°–1000° C inside another pot with a lid acting as a muffle-kiln. This method as practised in Roman Britain (figure 288) much resembled the procedure in the east.

Some medieval lead-glazed ware was probably produced by applying the glaze-mixture to the dried leather-hard vessels, so that fabric and glaze were fired at one operation. However, it is clear from kiln-excavations that much medieval glazed pottery was fired in two operations, as directed by Heraclius However, no separate glaze-firing kiln has yet been recognized, and Piccolpasso describes the packing and firing of a kiln for majolica with the finished painted vessels going in on top of those undergoing the first firing, so that both processes were done in the one kiln.

In a recipe of Heraclius for lead-glazing, a slip compounded of powdered unbaked and red-baked potter's earth, glue, and oil was applied to the already baked pot (perhaps for painted designs); it was then covered with boiled flour-water paste and immediately sprinkled with lead oxide, after which it was fired 'neither too strongly nor too slowly'. This gave a yellow glaze. For a green glaze, copper or brass filings were added to the lead while it was being oxidized. The green-speckled orange or yellow glazed jugs commonly found show that the medieval potters also sprinkled copper filings on to the glaze mixture on the pot before the glaze-firing.

Copper introduced into a lead-glaze was used in Roman times and earlier. It produces a green or, in the alkaline glazes of the Near East, a turquoise blue. A suspension of red cuprous oxide in a lead silicate was much used as red enamel on metal-work from the third century B.C. in the east and the west, but no red glazes were so produced in medieval Europe except for the Spanish ruby lustre, a technique described in a fifteenth-century recipe now at Munich, which was used brilliantly but fleetingly in Italy. A paler red was produced by iron rust (ferric oxide) remaining in suspension. Dispersed through the glaze it gives a yellow or orange; under reducing conditions, as ferrous oxide or the more fusible silicate, it gives a pale green. Iron in ferrous state is chiefly responsible for the subtle variety of greenish glazes.

The palette of the medieval potters in northern Europe was limited to the yellow and orange of ferric iron and the greens of ferrous iron and copper, the common olive green colours being due to copper with varying amounts of ferric iron. Opaque paint and applied clay stripes in colours contrasting with the body were used under a transparent glaze. The palette was enlarged through the influence of painted wares ultimately of Islamic inspiration, made in south Spain and the eastern Mediterranean from the tenth century, in Italy from the thirteenth, and in Flanders and England in the sixteenth. In these techniques the use of opaque ferric oxide for red was extended, and, mixed with antimony, gave more varied yellows. The range of colours was further widened from the same source with a deep blue produced by cobalt, purple by manganese, and yellow by antimony. A coarse mixture of antimony with copper produced a brilliant light green of attractive texture; a deep blue-green could also be produced. Cobalt blue, though known in the east for many centuries, was not used in Italy until early in the fifteenth century, when it was mentioned in a Latin recipe now at Munich. It was imported as the impure oxide, *zaffre*, from the Levant through Venice (*colore damaschino*, Damascus pigment). Since it was dutiable in many parts, there was considerable smuggling in it.

Though the technicians of the early empires had occasionally rendered glass and glazes opaque by incorporating tin oxide as a fine suspension in the lead silicate matrix, the device was first extensively used by Mesopotamian potters from the ninth century A.D., when they produced successful imitations of the

T'ang creamy porcelain then being imported from China. As well as providing a passable imitation of its texture, it had a great advantage in that paints applied to its unfired matt surface did not run when fired as they did when applied direct to the pottery surface under a more fusible and flowing lead-glaze[1] (plate 22 A). Imitation thus flowered into a most fertile new tradition of painted ornament, leading on the one hand to the finest achievements of the Near East, and particularly of Persia (figure 289), and on the other to the profusion of 'majolica' and 'delft' painted wares of Europe from the fourteenth century onwards, among which may be found the most remarkable medieval achievements of western polychrome ceramics (figure 276, plate 23 B).

FIGURE 289—*Polychrome painted bowl from Nishapur, Persia. Ninth century. Scale 1/2.*

The mixture for the white opaque glaze used in both eastern and western practice consisted of finely powdered potash-glass, oxides of lead and tin, and a little salt. The once-fired vessel was dipped in a suspension of this mixture and allowed to dry. On the resulting porous surface the colours were then painted with a brush. Brushwork on so absorbent a surface demanded a sureness of touch comparable to that needed for tempera painting. The whole was then given a second firing, set in the kiln above other pots receiving their first firing. Some finer tin-glazed wares received an overall film of lead-glaze as a sheen (*coperta*), which required yet a third firing.

The finest achievements of Islamic painted pottery are to be seen during the twelfth and thirteenth centuries in the elaborate 'seven colour' process. It is described in a treatise of 1301 written at Tabriz, then seat of the Mongol court, by one Abu'l-Qāsim of Kashan, member of a Persian family of potters. It is the ware now known as *minai* (figure 293). Paint went on both under and over more than one application of glaze, and ceramic quality is sometimes lost in complexity.

Slip is a semi-liquid fine clay mixture. It is used extensively in pot construction to join the various parts while still soft, and especially to fix spouts and

[1] The Samarkand potters solved this difficulty of painting under ordinary lead-glazes by mixing the pigments with a fine clay slip, or painting them on to a coating of such a slip.

handles to the body. Ornamentally it is used in two main ways. First, as a fine plastic cream applied to pottery at many periods and places, it gives a pattern in relief, in monochrome or in contrasting colours, usually white on a dark body. The barbotine (p 268) and trailed-slip of Roman times and the slipware of the seventeenth-century Staffordshire potteries are examples. Secondly, a thinner slip-mixture, painted on, or applied by dipping (French *engobe*, now anglicized), was widely used to give a fine light-coloured surface to ceramics made from commoner dark-firing clays. Both in the Near East and west, this practice was stimulated by the growing appreciation of Chinese porcelain from the ninth century onwards. Used thus and covered with a lead-glaze, it was however a poor imitation of porcelain textures, and engobe had its own qualities which were also appreciated. Painting in liquid slip, either white on dark, or brown or black on a pale body, was common practice among English medieval potters, as was also the application of modelled plastic ornament.

FIGURE 290—Sgraffiato *ornamented bowl from Salonika. Fourteenth century. Scale 1/3.*

One of the chief ornamental uses of slip was as a ground for *sgraffiato*[1] ('scratched') patterns, by cutting through the dipped-slip coating to reveal the dark body beneath, the whole being afterwards covered with a transparent lead-glaze (figure 290). Sometimes colour was painted into the scratched lines, but it was apt to run in the fusible lead-glaze (plate 22 A). This technique became one of the most prominent ceramic styles of Islam and Byzantium alike, and was also widely used during the Middle Ages in Spain, Italy, and occasionally in France and England. As for its origin, it is a simple idea, but some of the earliest examples are on ninth-century Persian imitations of mottled Chinese stonewares, on which this ornament is also found.

Lustre-painting, a metallic sheen, so characteristic of Islamic pottery and notably of the medieval wares of southern Spain, seems to have been originated by the glass-painters of Egypt during the seventh and eighth centuries A.D. It was taken up by the pottery-painters, the technique spreading east to Mesopotamia by about 850. At the end of the tenth century the best potters of Baghdad

[1] This word is first found in Piccolpasso's treatise (1556–9), which provides us with many potters' terms.

appear to have moved to the Fatimid court at Cairo, whence the technique was presumably brought to Spain by the thirteenth century.

The lustre effect was produced by applying silver or copper oxides, mixed with fine ochre as a medium, to the already glazed surface, and re-firing at a fairly low temperature (about 800° C) under reducing conditions, produced in a muffle-kiln or by simply piling brushwood on to the fire to produce a dense smoke. The metallic oxides were thus reduced to very fine particles of metal. The effect is variable; a thick film will look like solid copper or silver, while a thin film is iridescent and may be gold-coloured. In Mesopotamia during the ninth century the technique was used to give a metallic sheen to several colours, but later only greenish or brownish types were made. The fourteenth- and fifteenth-century Hispano-Moresque[1] wares of north Africa, Malaga, and Valencia (plate 23 A) were valued in Europe, and similar wares began to be produced in north Italy about 1500. Both copper and silver lustre techniques are included in a fifteenth-century manuscript of recipes now at Munich.

VII. BRICKS AND TILES

Utilitarian bricks and tiles. Flat bricks, hollow hypocaust-bricks, and half-round or S-section roof-tiles used by Roman builders (figures 374, 381, 386) ceased to be made over most of Europe after the fall of the western Empire. Only in north Italy and the Byzantine area is there evidence that brick-making continued. From these lands, or perhaps in imitation of Islamic brickwork in Spain, the use of brick spread again to southern France in the eleventh century and reached other parts of Europe, including eastern Britain, by the thirteenth (pp 384 ff). Bricks became the foundation of medieval building-style from the Low Countries to the Baltic area, the region most deficient in stone. Clay tiles were used in the Middle Ages for roofs, even beyond areas where building in brick was in vogue. The plasticity of clay was particularly useful for making Λ-sectioned ridge-tiles, used on roofs covered with stone slates or wood shingles.

The clay for tiles and bricks was allowed to weather through the winter, with occasional turning. It was then kneaded, often with bare feet. Bricks were shaped by pushing the clay into a wooden frame laid on a table covered with sand or chopped straw to prevent sticking, the surplus clay being removed from the top with a wooden stick. These bricks were then dried for a month or more in stacks up to ten bricks high, covered with straw, or roofed with canvas supported on sticks, or under open-sided wooden roofs. Both bricks and tiles were

[1] Strictly the term Hispano-Moresque includes the non-lustred painted wares of Valencia, Seville, and Granada; it is used also for the continuing tradition during the sixteenth century after the expulsion of the Moorish rulers.

fired in rectangular kilns (figures 291, 351), which were sometimes of unbaked brick and became baked during firings. At Hull, in the fifteenth century, 10 000 bricks were fired at a time. This would represent a kiln-capacity of 20′ × 20′ × 10′, which may have been divided between several kilns.

Bricks thus made were serviceable and cheap. The municipal brickyard at Hull in the fourteenth and fifteenth centuries turned out about 100 000 bricks a year. This was the result of about ten firings, each requiring about five days, with dried turf as fuel. The digging and preparing of the clay, the shaping of

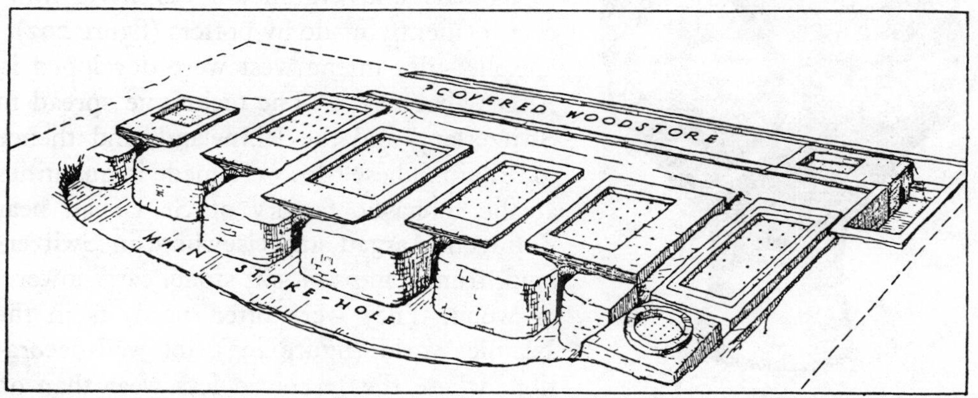

FIGURE 291—*Isometric diagram showing layout of the kilns at a Romano-British combined tile and pottery factory at Holt, Denbighshire. The circular kiln* (front, right) *is the pottery kiln. Second century A.D.*

the bricks, loading and unloading of the kilns and their firing were all done by three or four unskilled men.

Inlaid and printed paving-tiles. Inlaid tiles were developed in northern France about 1200, as a substitute for floor-designs composed of contrasting stone slabs. The practice spread to the Low Countries, and to England, where such tiles are seen at their finest. Some of the earliest tile-pavements were a mosaic of interlocking geometrically shaped plain tiles. At Clarendon Palace (1234–6) near Salisbury, and in the chapter house (1253–9) of Westminster Abbey, the pavements were of square tiles with designs in white pipe-clay set in channels cut or stamped in the soft clay matrix. These were then fired with a transparent lead-glaze over the surface. The technique of white inlay reached its peak at Chertsey Abbey (*c* 1270), Surrey, and Halesowen Priory (*c* 1290), near Birmingham, where scenes from romances are depicted with vigour and refinement.

Inlaid tiles continued in use for several centuries, but about 1350 the Chiltern tile-factories developed a cheaper method of printing the design from a wooden stamp coated with a pipeclay slip, closely comparable to the use of an inked woodcut block. This device brought the tile-maker very near to the wood-block printing which preceded that with movable type. Because of slipping or smudging the results were often inferior. This printing technique never entirely replaced the inlay method. The paving-tile industry seems to have been developed somewhat independently of the pottery industry. Ornamental roof-finials and chimney-cowls were, however, evidently made by potters (figure 292).

FIGURE 292—*Ornamental roof-finial. Dark red body, partly covered with green glaze. From Nottingham. Late thirteenth to fourteenth centuries. Scale 1/4.*

Relief-tiles in the west were developed in Alsace about 1150. The technique spread to Switzerland and the Rhineland, and thence to Britain. These tiles were made in quantities at the monastic factory of St Urban near Zofingen, Aargau, and elsewhere in Switzerland, from stamps of soft stone, earthenware, or wood. They were often used, as in the Islamic world (figure 293), for wall-decoration, where they received less wear than on floors. From the fifteenth century, houses in Germany, Switzerland, Holland, and Scandinavia were generally heated by large stoves of glazed bricks with relief-ornament, the deep hollows increasing the heat-radiating surface and giving something like black-body radiation. In the east the ceramic prayer-niches (*miḥrāb*) and decorative floor-tables to take drinks show another use of the medium. Large hollow cisterns of glazed earthenware (plate 37 B) to take wine were sometimes made, an English example bearing the badge of Henry VII.

Oriental traditions in ornamental tiles. Wall-surfaces were decorated with glazed tiles in the ancient empires of Egypt, Babylonia, and Persia. Under the Sassanians in Persia (221–641) painted stucco was used, and in early Islamic times glazed earthenware tiles began to be inserted among the painted patterns on the stucco. The techniques followed in general those developed by the potters, except that, on tiles, moulded or carved relief-decoration had more place. Tiles were made as hexagons and interlocking star-patterns, as well as squares. In the combination of relief-pattern and vivid painting up to seven colours—the *minai* technique (p 302)—may be seen some of the most brilliant ceramic achievements of Islam (figure 293), giving the characteristic rich effect to its buildings. The peak came in Persia in the fifteenth century, with such examples as Tamerlane's mausoleum at Samarkand (1404) and the Blue Mosque at Tabriz (1465).

Much of the cutting of panels on Islamic tiles was necessary to prevent the running-together of the adjacent areas of heavy glaze in contrasting colours. By the eleventh century Islamic tile-painters were separating their glazes by a line of purple pigment, mixed with a grease that disappeared on firing. This technique was used in Muslim Spain, and also made possible the most elaborate Persian tile-painting of the sixteenth century.

FIGURE 293—*Relief tile in* minai *painting technique Persia. Fourteenth century. Scale 2/7.*

The eastern techniques in ornamental tile-making had an effect in Europe comparable to that which they had on pottery. Spain developed a brilliant exotic tile-series, while Italian fifteenth-century tin-glazed tiles have painted arabesques as well as Renaissance motives, and provided the source of inspiration for this style of refined tile-painting in the Flemish, Dutch, French, and English series of the sixteenth and seventeenth centuries. Thus these painted tin-glazed European tiles have an ultimate oriental ancestry.

The end of the Middle Ages is not marked by any particularly significant change in ceramic techniques, and some of its pottery traditions have been practised until recent times with little change. On the other hand, technically advanced stonewares were already being produced in the Rhineland and Dutch Limburg even by the fourteenth century. They represent an indigenous tradition with roots extending backward for some five hundred years. Even salt-glazing was added to these wares by the fourteenth century. The Islamic device of tin-glazing for imitating the white translucence of the valued early Chinese porcelain wares was already being practised in Italy by the fourteenth century, and became widespread in western Europe during the sixteenth. It was not until the full secret of Chinese translucent porcelain was rediscovered at Meissen near Dresden in 1709 that a new era of European ceramic industry began. Coal-firing combined with improved kiln-design was first used at Dublin in 1752. These devices yielded more readily both the higher kiln-temperatures needed for porcelains, and a smoke-free atmosphere to preserve their whiteness. By this time Chinese wares were coming to Europe in bulk with the tea-trade and were freely imitated.

The effect of the Renaissance was less marked on ceramics than on most arts, being overshadowed by the oriental influences at work from the thirteenth century. Not until the eighteenth century was the potter's interest revived in the ceramics of ancient Greece and Italy, giving, for example, Wedgwood wares.

BIBLIOGRAPHY

Comparatively little has been written on the technology of medieval pottery, either European or oriental. The reports of such work as has been done are scattered through the periodical literature, often appended to the accounts of excavations. The more important, many of which contain further references, are listed below.

BRADFORD, J. S. P. "The Apulia Expedition: an Interim Report." *Antiquity*, **24**, 91–94, 1950.

BRITISH MUSEUM. 'A Guide to the Islamic Pottery of the Near East' by R. L. HOBSON. Department of Oriental Antiquities and Ethnography, British Museum, London. 1932.

BROOKS, F. W. "A Medieval Brickyard at Hull." *J. Brit. Archaeol. Assoc.*, third series, **4**, 151–74, 1939.

BROWN, G. B. 'The Arts in Early England', Vol. 4, pp. 488–508. Murray, London. 1915.

CHARLESTON, R. J. 'Roman Pottery.' Faber and Faber, London. 1955.

CHRISTIE, A. H. "Islamic Minor Arts and their Influence upon European Work" in 'The Legacy of Islam' (ed. by SIR THOMAS [WALKER] ARNOLD and A. GUILLAUME), chap. 4. Clarendon Press, Oxford. 1931.

DUNNING, G. C. "Polychrome Jugs Found in England and Scotland." *Archaeologia*, **83**, 126–38, 1933.

Idem. "Medieval Pottery" in KATHLEEN M. KENYON. 'The Jewry Wall Site, Leicester', pp. 222–48. Res. Rep. Soc. Antiq., no. 15. 1948.

Idem. "Medieval Pottery Kilns." *Archaeol. News Letter*, **1**, xi, 8, 1949.

Idem. "Trade Relations between England and the Continent in the Anglo-Saxon Period" in 'Dark Age Britain. Essays presented to E. T. Leeds' (ed. by D. B. HARDEN), chap. 13, pp. 218–33. Methuen, London. 1956.

ENGEL, F. VON. "Die mittelalterlichen Töpferöfen von Dümmer und Granzin." *Hammaburg*, **3**, 78–81, 1952.

FROTHINGHAM, ALICE, W. 'Lustreware of Spain.' Hispanic Notes and Monographs, Peninsular Series, Hispanic Society of America, New York. 1951.

GANZENMÜLLER, W. "Über die chemische Zusammensetzung mittelalterlicher Ziegelglasuren." *Angew. Chem.*, **50**, 260–3, 1937.

GARNER, F. H. 'English Delft Ware.' Faber and Faber, London. 1948.

GARNER, SIR HARRY (MASON). 'Early Chinese Blue and White.' Faber and Faber, London. 1955.

GRAY, B. 'Early Chinese Pottery and Porcelain.' Faber and Faber, London. 1953.

GRIMES, W. F. "Holt, Denbighshire: The Works-Depôt of the Twentieth Legion at Castle Lyons." *Cymmrodor*, **41**, 1–235, 1930.

HABERLY, L. 'Medieval Paving-tiles.' Blackwell, Oxford. 1937.

HONEY, W. B. 'European Ceramic Art' (2 vols). Faber and Faber, London. 1949–52.

JOPE, E. M. "A Later Medieval Pottery Kiln at Potterspury, Northants." *Archaeol. News Letter*, **2**, x, 156–7, 1950.

Idem. "Medieval Pottery Kilns at Brill, Buckinghamshire." *Rec. Bucks.*, **16**, i, 39–42, 1953–4.

Idem. "Medieval Pottery in Berkshire." *Berks. Archaeol. J.*, **50**, 49–76, 1947.

Idem. "Regional Character in West Country Pottery." *Trans. Bristol Archaeol. Soc.*, **53**, 61–76, 88–97, 1952.

JOPE, E. M. and HODGES, H. W. M. "Medieval Pottery and Kiln Evidence from Carlisle." *Trans. Cumberland Antiq. Soc.*, **53**, 1955 (forthcoming).

KOETSCHAU, K. 'Rheinisches Steinzeug.' Wolf, Munich. 1924.

LANE, A. "Medieval Finds at Al Mina in North Syria." *Archaeologia*, **87**, 19–78, 1938.

Idem. 'Early Islamic Pottery. Mesopotamia, Egypt and Persia.' Faber and Faber, London. 1947.

LEACH, B. 'A Potter's Book' (2nd ed.). Faber and Faber, London. 1945.

LEEDS, E. T. "A Saxon Village at Sutton Courtenay, Berkshire." *Archaeologia*, **92**, 79–93, 1947.

LONDON MUSEUM. 'Medieval Catalogue' by J. B. WARD PERKINS, pp. 229–53. London Museum Catalogues, no. 7. 1940.

LUNG, W. "Die Töpferöfen von Paffrath." *Bonn. Jb.*, **155**, 1955 (forthcoming).

MARSHALL, C. J. "A Medieval Pottery Kiln discovered at Cheam." *Surrey Archaeol. Coll.*, **35**, 79–97, 1926.

MATSON, F. R. in TOLL, N. 'The Green Glazed Pottery.' Excavations at Dura-Europos. Final Report 4 (ed. by M. I. ROSTOVTZEFF *et al.*), Part 1, fasc. 1, pp. 81–95. Yale University Press, New Haven; Oxford University Press, London. 1943.

MORGAN, C. H. 'Corinth: Results of Excavations conducted by the American School of Classical Studies at Athens', Vol. 11: 'The Byzantine Pottery.' Harvard University Press, Cambridge, Mass. 1942.

MYRES, J. N. L. "Some Anglo-Saxon Potters." *Antiquity*, **11**, 389–99, 1937.

PICCOLPASSO, CIPRIANO. 'The Three Books of the Potter's Art.' [Facs. ed. of the Italian manuscript] with trans. and introd. by B. RACKHAM and A. VAN DE PUT. Victoria and Albert Museum, London. 1934.

RACKHAM, B. 'Medieval English Pottery.' Faber and Faber, London. 1948.

Idem. 'Early Staffordshire Pottery.' Faber and Faber, London. 1952.

Idem. 'Italian Majolica.' Faber and Faber, London. 1952.

RICE, D. TALBOT. 'Byzantine Glazed Pottery.' Clarendon Press, Oxford. 1930.

Idem. "Persia and Byzantium" in 'The Legacy of Persia' (ed. by A. J. ARBERRY), chap. 2. Clarendon Press, Oxford. 1923.

Idem. 'Byzantine Art' (2nd rev. ed.). Pelican Books, no. A 287. Penguin Books, Harmondsworth. 1954.

RITTER, H., RUSKA, J., *et al.* 'Orientalische Steinbücher und persische Fayencetechnik.' Istanbuler Mitt., hft. 3. Istanbul. 1935.

SALZMAN, L. F. 'English Industries of the Middle Ages' (2nd enl. ed.), chap. 8. Clarendon Press, Oxford. 1923.

Idem. 'Building in England', chap. 8. Clarendon Press, Oxford. 1952.

SAVAGE, G. 'Porcelain.' Pelican Books, no. A 298. Penguin Books, Harmondsworth. 1954.

SELLING, D. 'Wikingerzeitliche und frühmittelalterliche Keramik in Schweden.' Petersson, Stockholm. 1955.

STEVENSON, R. B. K. "The Pottery" in 'The Great Palace of the Byzantine Emperors' (ed. by G. BRETT, W. J. MACAULAY, and R. B. K. STEVENSON), chap. 2, and p. 289. Oxford University Press, London. 1947.

Idem. 'Medieval Lead-glazed Pottery: Links between East and West.' *Cah. Archéol.* **7**, 89–94, 1954.

TREND, J. B. "Spain and Portugal" in 'The Legacy of Islam' (ed. by SIR THOMAS [WALKER] ARNOLD and A. GUILLAUME), chap. 1. Clarendon Press, Oxford. 1931.

VICTORIA AND ALBERT MUSEUM. 'Guide to the Collection of Tiles in the Victoria and Albert Museum' by A. LANE. H.M. Stationery Office, London. 1939.

A set of jugs in graded sizes produced at Cheam, Surrey. Fifteenth century. The largest jug is almost 12 in high.

9

GLASS AND GLAZES

D. B. HARDEN

I. THE ORIGIN OF GLASSES, GLAZES, AND FRITS

OUR subject must include both glazing and glass, though the former is additionally treated in ch 8. Glazing is the use of a vitreous substance as a surface-film to cover a core of another material. The three main core-materials used by the ancients were stone, including especially quartz and steatite (soapstone); powdered quartz, the resulting ware being known to Egyptologists as glazed-ware or faience (plate 24 A);[1] and clay. The vitreous glaze was usually applied in powdered form moistened with water, and fired *in situ*. When vitreous material is applied in similar fashion to metal it is called enamelling; this is dealt with elsewhere (pp 458–68).

Under the heading of glass must be considered objects or vessels manufactured from molten glass by moulding or blowing the material when hot and viscous, or by grinding it when cold and solid. The resulting object consists of glass alone, without any backing of other material, though for its manufacture a core or mould of some other substance may have been used and subsequently removed.

Glazing of small objects was practised as far back as *c* 4000 B.C., or even earlier, while the manufacture of free-standing glass objects is thought to have begun about 2500 B.C., both in Egypt and in Mesopotamia. The first glass vessels are about 1000 years later.

Glass has become one of the most important materials in the development of technology and experimental science, and is even now hardly replaceable by plastics, which should in any case not be regarded as substitutes for it. The gradual development of its effective manufacture from early Egyptian glass-making is a great achievement, in which the earlier advances, already effected by Roman times, were scarcely at all improved until the rise of modern science created a demand for specialized types of optical and resistant glass.

The Nature of Glass. Glass is a rigid but non-crystalline substance, usually transparent. It is made by fusing together alkalis (or their salts), lime, and sand

[1] The term 'faience' is properly applied to objects of earthenware when highly decorated with glazes or enamels, like the majolica ware from Faenza (whence the name), but its use for the glazed powdered quartz of antiquity is too firmly established to be discarded.

(or flint). It is a mixture of silicates of alkalis (soda or potash) and alkaline earths (in practice almost always lime, with some alumina and magnesia: table on p 313). If the lime originated from a dolomitic limestone, magnesium also became a constituent; baryta, too, is sometimes found in the alkaline earth portion. Lead silicate glasses were not significantly developed within the period covered by this chapter.

The fused mixture, when slowly cooled, gradually becomes more viscous until it congeals to a hard solid, remaining clear if made from clean materials. If cooled too fast the silicates are liable to crystallize, making the mixture opaque, brittle, and no longer a glass. Hence the process of slow cooling or annealing, often in a special oven, is a vital one in effective glass-production. Ideally, the material in the crucibles or pots needs also to be heated slowly, as well as cooled slowly. Glasses have no sharp, definite melting-point. On heating, soda-glass gradually softens to something like a fluid as the temperature rises to about 1000° C (that produced by an ordinary charcoal fire). Potash (*Wald-* or forest-) glasses, which are more readily fusible, were made particularly in the Middle Ages in the west. To the mixture of raw materials some scrap glass, known as cullet, was often added to assist fusion and to avoid waste.

Until the middle of the nineteenth century, glass-production involved two stages, first the fritting of the raw materials, and second the melting. The first was carried out in a separate furnace or in the lower-temperature part of the melting-furnace, the materials being raked from time to time to expose a fresh surface to the flames playing over them. This fritting-process brought about the initial stages of reaction, eliminated some of the gaseous products, and assisted the subsequent melting, which had to be achieved at temperatures lower than those used today. In Egyptian Dynasty XVIII times (*c* 1500–1300 B.C.) the temperature available did not reach 1100° C, which was insufficient to expel the bubbles of gas; hence much of this early glass is opalescent or opaque.

When glass is kept for a considerable time, or when it is rapidly cooled after strong heating, some of its ingredients may crystallize. This process is known as devitrification, since the glass then loses its transparency or translucency, and readily falls to pieces on shock or further heating.

Most glass made in the ancient world is predominantly a sodium calcium silicate mixture, colours being produced by small amounts of metallic oxides, as follows. *Blue*: paler blues are due to copper as cupric oxide (CuO), or even apparently sometimes to iron; deeper blues may also be due to copper (CuO), but cobalt (CoO) began to be used sometimes from Dynasty XVIII onwards in the Near East (table on p 313, *1*), though it has never so far been detected in glasses throughout our period in the west. *Green*: shades of bottle-green are due to

Table of Glass Analyses

	Egypt, Dynasty XVIII. 1400 B.C. Blue	Assyria, VIII–VI centuries B.C.		Cologne—Roman Period		Cologne, V century A.D. Greenish-yellow	Valsgärde, central Sweden c A.D. 700. Blue	Birka, Sweden		Glasshouse, Cordel, near Treves. IX–X century. A.D. Green	Arabic glass weight Egypt. VIII century A.D.	Derbyshire window glass XIV century A.D.
		Nearly colourless	Sealing-wax red[1]	I century A.D. Bluish	IV century A.D. Greenish-yellow			IX–X century A.D. Green	IX–X century A.D. Colourless			
	1	*2*	*3*	*4*	*5*	*6*	*7*	*8*	*9*	*10*	*11*	*12*
SiO_2	64·06	71·54	39·50	68·94	68·1	66·35	not determined	not determined	not determined	not determined	71·4	54·01
P_2O_5	0·21	0·11	..	..	..	..	not determined	..	..	..	..	..
Na_2O	15·47	12·7	9·71	18·8	18·61	21·47	16·3	16·8	22·1	8·1	16·98	1·70
K_2O	1·25	0·88	1·91	0·15	0·95	0·65	0·91	1·0	1·67	11·5	0·27	13·20
CaO	8·56	4·82	4·4	6·69	6·87	6·52	5·8	1·5	4·0	23·8	2·74	17·37
Al_2O_3	3·51	0·48	4·35 with Fe_2O_3	3·9	2·31	2·80	2·0	7·3	4·8	1·9	4·75	2·41
MgO	2·73	3·07	..	1·18	0·84	1·25	0·5	0·7	0·6	3·7	0·81	5·33
Fe_2O_3/FeO	1·71	0·91	see Al_2O_3	0·66	0·63	0·45	0·4	1·34	0·6	present	2·02	0·81
MnO	0·16	0·02	..	trace	1·52	1·03	0·05	0·03	0·017	1·26	0·3	1·03
CuO	0·005	..	..	..	..	..	0·91	..	0·028	0·076	0·10	..
Cu_2O	..	..	13·58	..	..	..	..	..	..	..	..	..
CoO	0·55	..	..	..	..	..	..	..	..	..	..	..
PbO	0·01	..	22·80	..	..	..	0·15	..	0·05	trace	..	..
Sb_4O_6	0·05	0·25	4·07	..	..	..	..	..	..	..	..	..
SnO_2	0·02	..	0·32	..	..	..	..	..	..	..	..	..
TiO_2	0·08	0·19	..	..	..	..	trace	trace	trace	trace	..	..
SO_3	0·16	0·99	..	..	0·60	..	..	..	..	..	0·12	..
Cl	1·17	..	..	..	..	..	..	..	..	..	..	..
Loss on ignition (= H_2O)	0·27	4·58	..	..	..	..	..	..	..	..	0·28	0·21

[1] The analysis of the Assyrian sealing-wax red is similar to that of red enamel used on metal work in various parts of Europe from the Iron Age onwards.

Sources of Analyses

[*1*] Geilmann, W. *et al.* *Glastech. Ber.*, **28**, 146–56, 1955.
[*2*] Turner, W. E. S. *J. Soc. Glass Tech.*, **38**, 445–56, 1954.
[*3*] *Idem.* (Analysis by Brit. Mus. Lab.), *Ibid.*
[*4, 5, 6*] Neumann, B. and Kotyga, G. *Z. angew. Chem.*, **38**, 776–80, 857–64, 1925.
[*7, 8, 9, 10*] Arbman, H. 'Schweden und das karolingische Reich', p. 252. Stockholm, 1937.
[*11*] Matson, F. R. *Glass Ind.*, **30**, 548, 1949.
[*12*] Heaton, N. *J. R. Soc. Arts*, **55**, 468–84, 1907.

ferrous (reduced) iron (FeO), but the deeper greens of Egyptian glass to copper (CuO). *Transparent amber* is due to ferric (oxidized) iron (Fe_2O_3); *amethyst* or *purple* to manganese (MnO_2). *Red opaque* (sealing-wax red) is due to a suspension of cuprous oxide (Cu_2O) in a vitreous matrix, and its high lead-content (table on p 313, *3*) reveals it as akin to the red enamel on Celtic, Roman, and medieval metal-work. *White-opaque*, found occasionally from Dynasty XVIII onwards, is due to a suspension of tin oxide (SnO_2), or sometimes to minute air-bubbles, and *yellow-opaque* usually to antimony compounds (Sb_4O_6). Egyptian *black glass* contains large amounts of iron, or of copper and manganese mixed.

Most sands contain some iron oxide, sufficient to give a dirty greenish or brownish glass. This can be counteracted by the addition of manganese dioxide (pyrolusite, the so-called glassmakers' soap) which oxidizes the iron and neutralizes the resultant yellowish colour with its own purplish tinge. It is doubtful, however, if this was ever deliberately added in antiquity, though particular sands that actually contained a little manganese may have been found by experience to give a colourless product. Many of the numerous excellent colourless glasses of antiquity, particularly Alexandrian crystal, were probably achieved by using carefully selected fine silver sands free from iron, as is suggested by at least one analysis.[1]

Evidence on Ancient Glass-making. We know all too little of how the ancients made glass, and of what kinds of glass they made at different periods and in different places.[2] Our insight into their practices is befogged by the darkness that surrounds so many of the more elaborate technical processes of antiquity. It will be well to start by considering the nature of the evidence.

First there are the objects themselves. For certain periods and countries these are plentiful—for example, the glass of Dynasty XVIII in Egypt, that of the sixth century B.C. onwards in the Mediterranean, and that of the Roman imperial age, which is found in profusion throughout the Empire. But for other periods and from other lands there are fewer pieces, and it is often uncertain whether they are native or imported. Thus the objects themselves give a dubious view of the extent, scope, and distribution of the industry.

Secondly, there are ancient texts and sources. These are mainly in Greek and Latin, but ancient Semitic literature (including the Old Testament) and texts in other languages also contain many interesting allusions, though usually disjointed and incidental. Latterly some further ancient evidence of a highly

[1] I am greatly obliged to Professor W. E. S. Turner and Mr. E. M. Jope, as well as to the editors, for much help here, and elsewhere in this article, on technical and chemical questions.

[2] The only major general account of ancient glass and glass-making is A. Kisa's *Das Glas im Altertume* (Leipzig, 1908). It contains much of permanent value, but needs revision in the light of more recent researches.

important kind has come to light, in the shape of formulae and chemical details in Mesopotamian cuneiform tablets of the second and first millennia B.C. These last sources make it surprising that hardly any references to glazing and glass-making have been found in the numerous Greek papyri from Egypt, of a period when glass-making was one of the major industries of that country.

Thirdly, there is evidence on ancient practices in medieval and early Renaissance writers, such as Theophilus and 'Heraclius' (p 351) at the end of the tenth century, and Agricola in the sixteenth, who describe the art as it was practised in their times. Their writings show that formulae and technical processes descended unchanged for centuries in this most conservative of industries; we can recognize, for example, that some of the recipes of Theophilus are the direct descendants of those in use in Mesopotamia nearly 2000 years earlier [1].

Glazing. The earliest known examples of vitrification are the glazed coatings on steatite beads from Badari in Upper Egypt. How and where the invention was made is unknown; it has been suggested that it took place when it was found that a malachite grinding-stone, heated in the presence of a little alkali, acquired a glazed surface. Once invented and applied to stone and silica, the alkaline glaze became of very general and frequent occurrence throughout all periods of Egyptian history [2].

Glazing on quartz, as a stone, seems to have died out in Egypt after *c* 2000 B.C., but glazing on steatite, for scarabs, kohl-pots (plate 24 B), and so on, continued throughout the dynastic period, while the Egyptian manufacture of faience did not cease until Byzantine times or later. In Egypt faience was used for all kinds of small objects and jewelry, as well as for vessels (plate 24 A), figurines, and amulets, for inlay patterns on large objects, such as coffins and furniture, and on walls of houses and temples. Several varieties of different colours and textures are distinguished, but the substance is essentially the same [3].

The normal colour of the glaze was a deep blue or green, but other colours were used at times, especially during Dynasty XVIII. A group of glazed objects from El-Amarna may contain eight or nine different main colours, not to mention various shades of each. The green glaze, the commonest of all, takes on different shades at different periods, even to the palest.

Lucas [3] gives some analyses of typical examples of the base material of Egyptian faience. It was normally about 95 per cent silica, with the remaining content divided more or less equally between alumina, oxide of iron, lime, and magnesia. He can cite only one analysis of the glaze itself, a piece of deep blue Roman glaze from the Fayum, which contained 75·5 per cent silica, 10·7 potash, 5·6 soda, 3·8 lime, 1·8 oxide of copper (for colouring), and small quantities of the oxides of aluminium, iron, manganese, and magnesium. This glaze is similar in composition to ancient glass, though its content of lime is lower and of silica higher.

Since there was glazing in Egypt, not only on stone but on a base of powdered quartz, by early predynastic times, and before Egypt is known to have experienced any Asiatic influences, it seems that it was an Egyptian, or at least an African, invention. In Mesopotamia it appears as early as the Jemdet Nasr period, *c* 3000 B.C., and we must thus assume either independent invention there or a very early spread from Egypt. But alkaline glazing, though well known from then on in Mesopotamia and neighbouring lands, and also in India and Crete in the early third millennium B.C., was neither so common there, nor used for so many different kinds of objects as in Egypt. In these lands it was common enough for beads and seal-stones, and to a certain extent for small amulets and figurines, but was not used so much for vessels or for inlay.

It may be that in Mesopotamia, and perhaps elsewhere in Asia, a different method of glazing with a lead glaze was developed fairly early and was preferred for vessels and larger objects. This lead glaze—unlike the Egyptian alkaline glaze—was used on a pottery base.[1] Examples of it have been found at Atchana in Turkey [4] in a level of the seventeenth century B.C., a date which happens to fit with that of the earliest recipe for it (below). Lead glaze is more vitreous and glossy in appearance, and usually thicker than the earlier alkaline glazes; it is inclined to flake off, adhering poorly to its pottery background (plate 24 C). If it did not hold on Mesopotamian pottery, still less would it do so on the cruder ware normal in Egypt after the Old Kingdom. For that reason, perhaps, as well as because of the great popularity of the rival alkaline method, lead-glazed pottery did not take root in Egypt before the Arab period, though a certain amount of lead-glazed faience is known there from Dynasty XXII onwards [5].

For the Mesopotamian lead glaze we possess an ancient recipe on a cuneiform tablet of the seventeenth century B.C. from near Tall ʿUmar (*Seleucia*) on the Tigris [6]. It may be thus rendered in modern notation:

	Parts
Glass . . .	243·0
Lead . . .	40·1
Copper . . .	58·1
Saltpetre . . .	3·1
Lime . . .	5·0

[1] Modern lead glaze consists of lead oxides and silica (flint), in the ratio of about 4 : 1, ground and mixed to a creamy consistency with water. Suitable pigments are added. It dries rapidly on the surface of an unglazed fired vessel of pottery, and is vitrified by a second firing.

The tablet also describes the preparation of the clay base, and it has been possible to make experimental glazed pottery objects from these recipes [7].

Other variant types of lead glaze, much thinner and more adhesive, were developed at some time in the late Hellenistic period in the Levant, where there is evidence of workshops at Tarsus and elsewhere [8]. The practice spread to other parts of the Mediterranean and Europe, notably St Rémy (Allier, France). Throughout Roman imperial times and into the Byzantine age these factories produced well characterized, though never very common, glazed pottery vessels (plate 24 G), figurines, and lamps—mainly yellow, brown, and green in surface-colour. Much of this glazed ware looks remarkably modern—a fact that, combined with its rarity, has often prevented it from being recognized as ancient by excavators [9], but it is surprising that so serviceable a ware was not in more common use.

Frit.[1] A substance variously termed by archaeologists frit, composition, or compost is frequently mistaken by them for glass or faience. It is not vitreous, though fashioned from the same, or similar, raw materials as glass, and objects made of it have the matt aspect of a partially fused powder. It has received little attention from historians of technology and should, therefore, be given some mention here, though it does not come within the strict scope of this chapter.

It first became common during Dynasty XVIII in Egypt, and contemporaneously in Mesopotamia, Syria, and Crete. It was used mainly for small objects such as amulets and seals, but occasionally, especially in Egypt, it appears that blue frit was added in a powdered form to a fine clay to produce blue or green pottery vessels (plate 24 F). These would remain blue only if fired below 685° C; above this temperature they would turn greenish owing to vitrification of the blue crystalline structure. In a powdered form, blue frit was used as a pigment. For vessels its use seems not to have outlasted the second millennium B.C., but for amulets, and especially for seals, it continued to be popular till much later. Compost scarabs, scaraboids, and conoids are among the commonest varieties of seal-amulet from Mediterranean sites of the middle of the first millennium B.C.

The ware is of one colour all through and usually has a smooth surface. Analysis of a blue specimen of Dynasty XIX yields silica 57·2, copper oxide 18·5, lime 13·8, soda 7·6 per cent, with small quantities of magnesium, iron, and aluminium oxides [10]. This crystalline mass, if heated too strongly, would fuse and become a greenish glass. It is, in fact, one stage farther from true transparent

[1] The term 'frit' is now used for the mixed raw materials in glass-making, for molten glass after its first fusion, and for ground-up glass used in the glazing of pottery.

FIGURE 294—*Sand-core bottle of Mesopotamian or Syrian fabric with marvered trails, from Ur. Latter part of second millennium* B.C. *Scale 2/3.*

or translucent glass than the so-called glass-paste of which many 'gems', amulets, ring-settings, plaques, and even vessels of Greek and Roman times were made. These, being vitrified but opaque, should be given the title of 'opaque glass' rather than 'paste', a misleading modern trade-term for vitreous imitations of precious stones which are clear and translucent.

We have little evidence of how or where the frit objects were made. Doubtless Egypt produced most, but Syria and Mesopotamia and other lands must have had factories, at least for small objects if not for vessels. As so often in antiquity, it is even possible that the frit was made in one place and exported for fashioning elsewhere. At any rate the blue variety, much used in Egypt as a powder for pigment, was exported in balls or ingots, notably to Minoan Crete and Mycenaean Greece. In the first century B.C. Vitruvius described the manufacture of blue frit balls for pigment from sand, copper, and soda [11]. Examples of Roman date are recorded from several sites in Britain, as well as from Italy and the Near East [12].

II. GLASS-WARE TO 750 B.C.

No glass vessels are known in Egypt earlier than *c* 1500 B.C., but some glass objects of other kinds are certainly older. Glass became a recognized product about the middle of the third millennium B.C. [13], though it seems that for the first thousand years or so of its existence glass objects were rare, and were treasured as imitations of precious stones. They perhaps cost as much as the stones they resembled.

FIGURE 295—*Sand-core unguent-flask in shape of a pomegranate, from Enkomi, Cyprus. Mycenaean period. Scale* c *3/4.*

About 1500 B.C. there was a sudden change. Glass vessels and other objects, as well as glass-inlay, now occurred with rapidly increasing frequency. By the mid-fourteenth century B.C. (Amarna period) Egyptian factories at El-Amarna, Lisht, Gurob, and elsewhere were producing in quantity glass of many kinds and colours. The evidence from Mesopotamia is less strong, but it is clear from finds at Ur (figure 294) and as far north-east as Nuzi [14] that by Kassite times (mid-second millennium B.C.) glass vessels were not uncommon there. From Syria, Palestine [15], and Cyprus (figure 295) [16] the evidence is also strong, but the soil of these countries is less kind to glass than that of Egypt and the examples are mostly very fragile and friable. No glass-factory of this period is known outside Egypt. Nevertheless, it is plain from variations in shape that not all the glass just mentioned was made in that country, and it is quite likely that there were factories in many Near Eastern lands. We do not know whether the manufacture of glass vessels originated in Egypt, or whether it was developed elsewhere and brought there.

FIGURE 296—*Greenish moulded figurine of a mother goddess, from Lachish, Israel. Sixteenth century B.C. Scale 4/3.*

Small glass objects of the second millennium B.C. were shaped in clay moulds, just as their faience counterparts were. Even figurines of some size were made in this way (figure 296). Some vessels, too, especially little open bowls of hemispherical and other simple shapes, were made by pressing viscous glass into a clay mould, or perhaps by spreading it like icing-sugar over an inverted mould.

Another process practised at this early period was the cold-cut technique, that is, grinding or carving a lump of glass as if it were a piece of stone. The designs on seals were certainly carved in this way—even if the blanks were moulded, which is doubtful. It may be that the blanks were carved and polished like their stone counterparts. Larger objects, such as a head-rest (figure 297) and a palette of Tutankhamen (*c* 1350 B.C.),[1] and perhaps some vessels, may also have been so made.

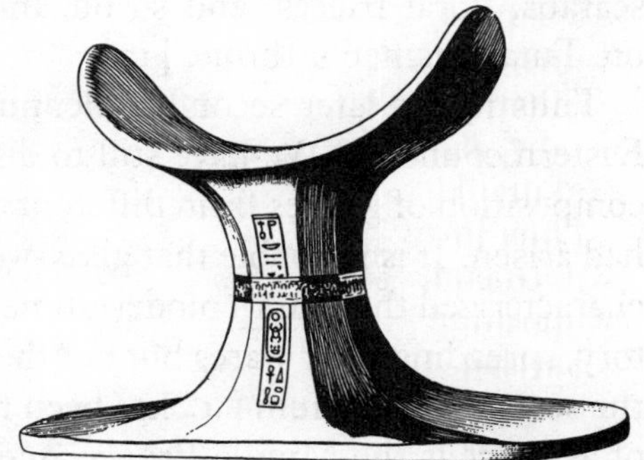

FIGURE 297—*Head-rest carved from a cold block of glass and inscribed with name and titles of Tutankhamen. From his tomb at Thebes.* c *1350 B.C. Scale 1/4.*

The vast majority of glass vessels, however, at this time and for the next

[1] Two head-rests of vitreous aspect were found in the tomb. One has the appearance of being faience, the other is turquoise-blue in colour and seems to be glass, as Professor W. E. S. Turner, who recently examined both carefully, kindly informs me.

thousand years or more, were fashioned round a core of sand tied up in a cloth bag, by dipping the core into a crucible of viscous glass-metal and afterwards modelling it by rolling on a flat stone bench or marver (French *marbre*). Decoration was normally added by winding trails or applying blobs or rings of other colours on the walls of the vessels, and then sometimes combing and marvering the added glass flush with the surface (figure 298), but occasionally (plate 24 D) matt painted decoration occurs.

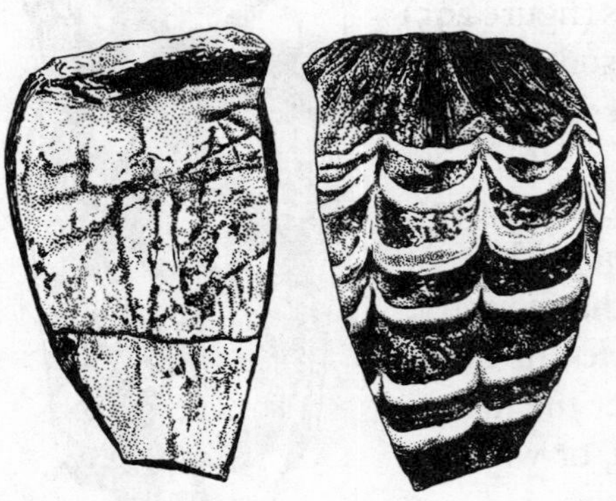

FIGURE 298—*Inner and outer views of fragment of sand-core bottle, showing marks of string-binding of the core or of the tool which scraped out the core after the vessel was moulded. From El-Amarna.* c *1360* B.C. *Scale just over 1/2.*

The ingredients for glass-making were quartz sand, calcium carbonate, and either natron (native sodium sesquicarbonate) or plant-ash, together with colouring matter. The calcium carbonate was in some cases contained in the sand and thus used unknowingly.

In Egypt at this period a common use of glass was for inlay work in imitation of lapis lazuli, red jasper, and other stones. The furnishings of Tutankhamen's tomb, notably the several coffins, are elaborately decorated with inlay work, much of which is of glass (vol I, frontispiece). The pieces were made in specially designed moulds and fitted together like mosaics, sometimes *cloisonné*, to form hieroglyphic texts, wings of scarabs, floral friezes, and so on. Inlaid figures were also made, such as those on Tutankhamen's throne [17].

Thus in the later second millennium B.C. glass was well known in all Near Eastern countries. We have still to discover, perhaps by a study of the chemical composition of glasses from different countries, in how many places glass-houses had arisen. It is probable that glass-workers already had the migratory habit that characterized them until modern times, and that they moved from factory to factory, spreading their wares but not their knowledge. As yet, however, no vessel of the second millennium B.C. has been found north or west of Greece. Glass beads of as early as the Middle Bronze Age are known from various parts of Europe, and faience beads of Mediterranean affinities are known from Bronze Age graves (middle of the second millennium B.C.) of the Wessex culture in Britain, but the earliest glass beads in these islands are from 'Bronze Age' burials in peripheral areas such as Scilly, Scotland, and Ireland. Most, if not all, of them are much later than 1000 B.C. [18].

After the heyday of the later second millennium B.C. a dark age fell upon the ancient world, and the story of glass, like that of so much else, becomes obscure. Few fragments of glass, other than beads and amulets, can be firmly dated between 1100 and 750 B.C., and we get the impression, perhaps wrongly, that manufacture of vessels and larger objects almost ceased.

III. FIRST RENAISSANCE OF GLASS, FROM ABOUT 750 B.C.

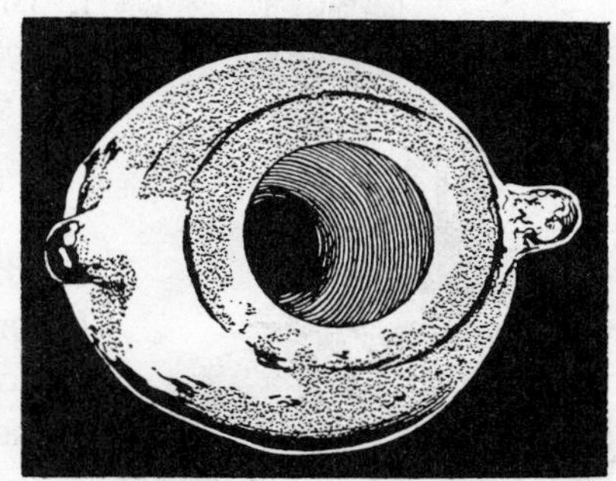

FIGURE 299—*View looking into a green carved alabastron, showing rotatory grinding-marks. The vase bears in cuneiform characters the name of Sargon II of Assyria, 722–705 B.C. Scale 2/3.*

During the eighth and, still more, the seventh century B.C. glass vessels began to be common again. The old techniques of cold-cutting, mould-pressing, and sand-core continued, but the shapes of the vessels, and perhaps also the centres of fabrication, changed. This was the era of sea-green cold-cut glasses such as the late eight-century alabastron of Sargon II (plate 26 C, D, and figure 299). We encounter, too, mould-pressed bowls of the seventh century, such as one from Fortezza in Crete, and the earliest of certain new types of sand-core vessels [19].

These sand-core vessels continue in an unbroken series, with but minor changes of shape and decoration, to the end of Hellenistic times. They are found from Mesopotamia to Spain and from the African coast to the transalpine lands. Some were perhaps exported in the fourth century B.C. as far as Cerne on the Atlantic coast of Africa [20]. Their main centre of fabrication was probably Phoenicia, and there is little doubt that their diffusion was due to traders from that country. As so little of this later sand-core glass has been found in Egypt, it seems unlikely that much, if any, of it was made there. Equally, the cold-cut glasses of sea-green or even at times almost colourless transparent glass were probably an Asiatic rather than an Egyptian product. Their manufacture continued after the seventh century B.C., and a fragmentary bowl with a pattern of cut petals on the base was found at Ephesus in a fourth-century B.C. context [21].

Before leaving this period we must consider a group of glasses quite outside the main line of development. In northern Italy and round the head of the Adriatic (Aquileia) glass decorated with trails of a second colour was being used by the sixth and fifth centuries B.C. for beads and brooch ornaments. A parallel group of unusual bowl-shaped vessels with vertical ribs and handles, of the same date, has been recovered from Hallstatt, St Lucia (Istria), and elsewhere [22]. By this time, too, glass beads and bangles, both in plain blue and green and with trailed and 'eye' decoration in a second colour, were being made in central and northern Europe and, a little later, perhaps even in Britain.

FIGURE 300—*Amber jug bearing the trade-stamp of the maker Ennion, who owned factories in Syria and Italy in the early first century A.D. Scale 2/5.*

IV. GLASS-BLOWING BEGINS, FIRST CENTURY B.C.

The sand-core technique, essentially limited to small vases with closed forms, lost popularity in the second or first century B.C. This coincided with a major turning-point in the history of glass. There arose new technical methods which combined mould-pressing and cold-cutting (tailpiece), and also introduced the 'cane' technique (p. 335) [23] for elaborate inlay designs (plate 24 E) and bowls with mosaic patterns (plate 28 A–E). The Egyptian industry, now firmly rooted in the Delta at or near Alexandria, came into its own again.

An even greater new development occurred towards the end of the first century B.C. This was the invention of glass-blowing, which doubtless began by blowing glass into a mould but soon developed into free-blowing. This invention almost certainly emanated from Syria, for mould-blowing seems never to have been prevalent in Egypt, if we may judge from the types of glass current on Romano-Egyptian sites.

The Roman imperial peace, which dawned with Octavian's victory at Actium in 31 B.C. and the annexation of Egypt in the following year, provided a most favourable political opportunity for the rapid spread of the new glass-techniques, accompanied by their craftsmen, from Syria and Egypt to Italy and other parts of the Roman world. Writers in the latest years of the Roman Republic have little to say of glass, while the earliest Imperial writers are very conscious of its importance. So rapid, indeed, was the spread into Italy of Alexandrian cane technique and mould-pressing, and of Syrian mould-blowing and free-blowing, that it is impossible, without much more research, to differentiate glasses made in Italy from those of the east—except for one specific instance, the products of Ennion's factories (figure 300) [24].

Pliny (d A.D. 79) says that glass was made in his day near Cumae in Campania with sand from the Volturnus, while Strabo, writing before A.D. 18, speaks of clear glass being made in Rome. If, as seems most probable, an independent glass industry had long flourished at the head of the Adriatic around Aquileia and what was to become Venice, it too must have had some part in these early Roman developments. It may be that Alexandrians, settled in Campania and in Rome itself, supplied the Roman market with costly mosaic and mould-pressed wares and cameo-pieces such as the famous Portland vase (plate 26 A), while the Syrians went farther north, thus obtaining the opportunity for an easy westward and northward advance across the Alps during the first century A.D.

FIGURE 301—*Painted bowl of dark green glass from Locarno, Switzerland. Mid-first century A.D. Scale 5/8.*

It has been necessary thus to hazard guesses because, though finds of glass in the east, and in Italy and the west, indicate a rapid spread of glass-making, we cannot yet fill in the details. We do know, however, that by the middle of the first century A.D. glass had become common in the Alpine area; thus no fewer than 350 vessels, including a facet-cut beaker and a painted bowl (figure 301), have been found in the Locarno cemeteries alone [25], and indeed throughout northern Italy glasses that can be ascribed with certainty to this period are no less numerous. The soil of Rome and central and southern Italy, and the ruins of Pompeii, have yielded numerous similar pieces, with even larger quantities of mosaic and of mould-pressed vessels of the most diverse colours and composition.

The story of glass-making during the Roman imperial period to A.D. 400 is simple [26]. Once the glass industry burst its Alpine bounds it moved rapidly up the Rhône and Saône valleys and down the Rhine until during the second century it was planted firmly round Cologne and Treves. It spread quickly northwestward to what is now Belgium and the neighbouring lands, where it settled in the forest-country on the modern frontier between Belgium and France in the valleys of the Meuse, Sambre, and Oise. Thence it came to Britain, where it existed at Colchester, Warrington, and Caistor-by-Norwich, and no doubt elsewhere. That factories also existed in Spain, not only in Roman times but earlier, seems certain from the very peculiar and individual types of glass often found there.

Meanwhile, other waves of workers must have moved from Syria towards Mesopotamia, Cyprus, Greece, and south Russia, for divergent shapes and

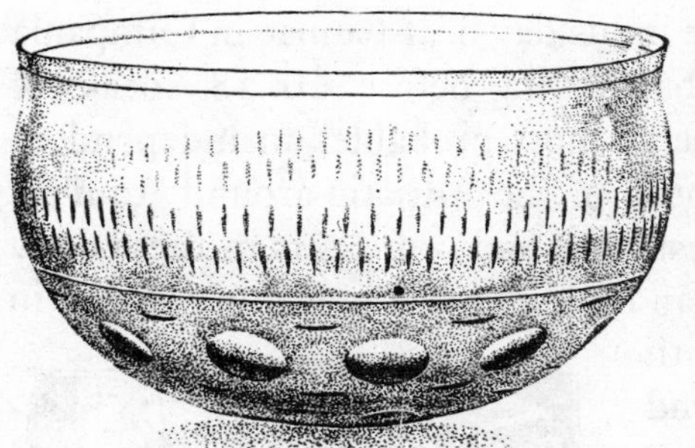

FIGURE 302—*Colourless bowl with facet cutting, probably Alexandrian. Himlingoie, Denmark. Second century A.D. Scale c 2/3.*

designs in those countries from the second century onwards are almost certain proof of local factories [27]. This is the greatest known spread of the glass industry in Roman times, though Roman glasses were undoubtedly dispersed by trade or looting to places far beyond the bounds of the Empire. Recent finds at Begram (ancient *Kapisa*) in Afghanistan, and others deep in the Sahara, as well as numerous examples in Scotland north of the Antonine Wall, in Scandinavia, and in northern Germany, prove that glass was popular with the barbarian neighbours of Rome and that they delighted specially in the better-class wares, including cut (figure 302) and painted glasses [28].

The manufacturing centres, once established, remained to some extent in touch with the parent factories in the Levant. It is clear that they all received constant reinforcements of Syrian and Alexandrian craftsmen, for in no other way can we explain the similarities of shape, pattern, and technical detail of manufacture that appear from time to time in eastern and western factories during the first four centuries of the Christian era. The similarities extend beyond mere art-fashions to the details of rim-forms and minutiae of shaping. The same likenesses are not found on ceramics of the period to any comparable degree, except for such international wares as Samian.

FIGURE 303—*Mould-blown green hexagonal jug, Syrian fabric. Fifth to sixth century A.D. Scale c 3/8.*

In Egypt and in Syria the glass industries ran to some extent in parallel, especially in the matter of shapes, but each specialized (though not exclusively) in one or more varieties of glass-ware. Thus at Alexandria in early imperial times emphasis on mould-pressed, ground, and polished wares led later to a development of cutting and engraving. Some of the finest cut and engraved bowls of colourless glass with figure-scenes (plate 26 B, figure 321) clearly came from Alexandria, however far away they may have been discovered. Alexandrian workers seem also to have specialized in painted glasses such as those recently found at Begram and the Daphne vase (plate 27 A) [28, 29].

There is also a well defined and easily recognizable plain table-ware, of bowls, cups, jugs, and other vessels, that was made in Egypt in the fourth century A.D. [30]. On the other hand, following on Ennion (p 322) and his contemporaries and their predilection for mould-blown glasses, the Syrian industry concentrated on mould-blowing (figure 303) and free-blowing, with forms becoming more and more elaborate and with trailing and tooling as the normal decorative processes (figure 304).

Two classes of glass-workers are frequently referred to in Roman literature. *Vitrearii* or glass-makers were clearly differentiated as a trade from the *diatretarii* or cutters and polishers [31]. If the distinction between Alexandrian and Syrian specialities is correct, we should expect the *diatretarii* to have flourished in Egypt but to have had only a small foothold in Syria. That they were flourishing in Italy is clear from Ulpian (A.D. 170?–228), who refers to their existence at Aquileia.

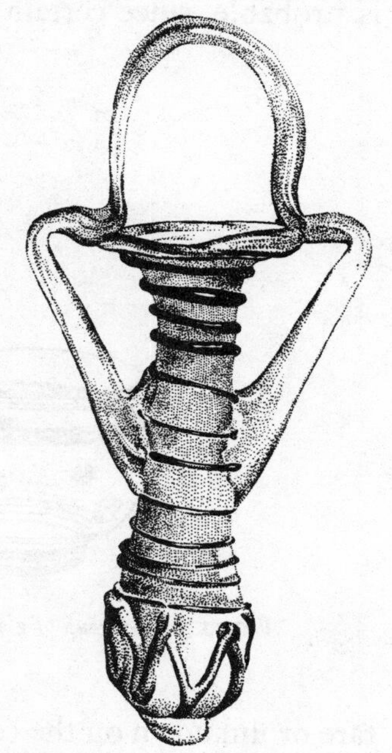

FIGURE 304—*Green unguent-flask with blue-green trails, from Syria. Fourth century* A.D. *Scale c 1/2.*

V. POST-ROMAN GLASS

We turn now to glass-making after the break-up of Roman power. In the west, omitting Spain, where the history of the industry is little known, the main centres of glass-making in the fourth century were in Italy itself, Belgium and its neighbourhood, and the Rhineland. The last especially had been a main centre of the trade for two or two-and-a-half centuries (figure 305), and both *vitrearii* and *diatretarii* were active there [32]. During Roman times the glass-industry was carried on within or just outside the towns, especially in the Rhineland and northern Gaul. By the ninth century it seems that glass-furnaces had been set up in the forests [33], but it is not clear when the change of location took place. The available chemical analyses of Rhineland glasses (table, p 313) seem to show that there was no interruption of the glassmakers' soda supplies in Merovingian times, and that they were not forced to make greater use of potash (plant-ash) from the forests than in later Roman times. In fact, the evidence of the analyses suggests no disruption of the industry in the Rhineland in the immediate post-Roman period, and that any shift to the forests may have been somewhat later. Not until the ninth to tenth centuries can the making of glasses relying largely upon potash for their alkali be demonstrated by existing analyses (table on p 313, no 10). The matter is worthy of much further study. All these post-Roman glasses do, however, differ from their Roman predecessors in being less varied in form and decoration, less clear and less free from colour, and, on the whole, less well made.

That glass was made in Britain, at least in Kent, during the pagan Saxon period is probable, since certain kinds of vessel common enough on Kentish sites are

FIGURE 305—*Group of glasses from a grave at Mayen, near Treves. Late fourth century A.D. Scale c 1/4.*

rare or unknown on the continent. In the north of England two abbots of Wearmouth, Benedict in 675 and Cuthbert in 758, sent requests to the continent, the former to Gaul for window-glass makers and the latter to Mainz for vessel-makers, 'because the art was unknown in this country', that is, in Northumbria. Benedict's appeal was granted, but the results cannot have been lasting there, in view of Cuthbert's later attempt [34]. Remains of a glass-furnace of late Saxon date have been discovered at Glastonbury by C. A. Ralegh Radford.

In medieval times in the west, *Wald* or forest glass continued to be made, and there is evidence in contemporary texts for the sites of factories near Treves and elsewhere [35]. In Britain, glass-house sites from the thirteenth century onwards have been revealed in Surrey and Sussex. These were founded by workers from Normandy, the first of whom was one Laurentius Vitrearius, who settled at Dyers Cross, near Pickhurst, about 1226 [36]. This industry, developed in the Weald and other places, held the field in England, both for windows and for vessels, until the mid-sixteenth century, when workers from Lorraine and Venice were brought in. Those from Venice originated English crystal glass.

We turn now to Italy. The Roman glass-industry, which was remarkably strong in the fourth century, making fine cut glasses with figure-scenes and even, perhaps, cut open-work cage-cups (figure 306) as well as gilt glasses and so forth, undoubtedly persisted. Seventh-century Lombard graves in Italy have yielded stemmed cups and other glasses quite unlike any from the Rhineland or Belgium. These apart, Italian glass of this period is unknown, and we must await the thirteenth century for frequent and firm references to Venetian products. From then on that industry grew rapidly, no doubt largely through an influx of workers and patterns from the east as a result of the crusades. Certain Venetian workers set up factories elsewhere in Italy, and beyond, and workers from Normandy opened rival factories near Genoa [37]. These two branches of the industry, the Venetian and the Norman, initiated the modern glass-industry throughout western and northern Europe in the fifteenth century.

FIGURE 306—*Open-work cage-cup with bronze handle, probably Italian fabric. Late fourth or fifth century A.D. Scale c 1/4.*

In post-Roman times in the east there was a definite change in tempo, shapes, and fabrics in all the glass-making centres, usually in connexion with the Arab conquest. Thus in Egypt by the seventh century the old Romano-Egyptian styles and forms gave place to new types—trailed or ribbed or cut scent-bottles with pointed bases, dark green bowls with lustre-painted (plate 25 D) or pincered (figure 307) decoration, and so on. In Syria and Mesopotamia, too, new varieties of vessel with trailed decoration (figure 308) or heavy facet-cutting, often in relief, arose in the sixth and following centuries, though the old blue-green blown ware of Romano-Syrian style continued at least into the seventh century, if not the eighth [38].

In the ninth and following centuries a remarkable renaissance of fine cutting on colourless crystal glass occurred in several Near Eastern countries. An equally high attainment of medieval eastern glass-ware came in the thirteenth and fourteenth

FIGURE 307—*Green bowl showing pincered decoration. Egyptian, ninth to tenth century A.D. Scale 1/2.*

centuries, with the prevalence of enamelling on colourless and coloured glasses (drinking-glasses, mosque-lamps, and other shapes), at Raqqa, Damascus, and other places in Syria (frontispiece). It was presumably from the glass-factories responsible for these masterpieces that workers moved westwards to revitalize the western industry at Venice and elsewhere.

It is often stated that Byzantium was a major glass-making centre in the fifth and following centuries. Numerous fine glasses have been ascribed to it, notably certain bowls with raised cut bosses in the Treasury of St Mark's at Venice. But these now seem more naturally to fall into place as eastern products, in parallel with glasses found at Kish and elsewhere [39], and we are left with no certain knowledge of what glass, if any, emanated from Byzantium. There is one Greek site, however, namely Corinth, where two glass-factory sites of the eleventh century A.D. have been examined [40]. It is believed that they were founded by Egyptian workers, for they show parallels with Egyptian styles of the period; and that, when they were destroyed at the Norman conquest of Corinth in 1147, the workers were carried off to the west, because some of the types of glassware are very like certain medieval western vessels. If so, this Corinth manufactory stands as an important half-way house in the renaissance of the western industry in medieval times.

FIGURE 308—*Light brown unguent-flask, with trails, from Syria. Sixth to eighth century A.D. Scale c 1/3.*

VI. MATERIALS AND IMPLEMENTS

The evidence for the materials used comes from two sources: (*a*) ancient recipes and other detailed information in technical and literary texts; and (*b*) the study and analysis of specimens of ancient glass.

Ancient sources are very disconnected and inadequate, apart from the Assyrian texts of the seventh century B.C. (p 315) [41]. Ancient writers indicate that the main ingredients were sand (of which Pliny lists that of Belus in Syria and of Volturnus in Campania as among the best), *nitrum* (natron) or soda, from Egypt, together with shells providing lime, and a substance called *magnes lapis*—perhaps magnesian limestone [42]; colouring agents are scarcely mentioned. The Assyrian texts give certain formulae, not always readily translatable, for making *uknu-merku* glass, thought to be moulded blue glass or *kyanos* of the Greeks, *dusu* glass, thought to be crystal, or clear glass, and *sirsu* and *zuku* glass [41]. The last variety is the basic ingredient of the glaze-formula from the seventeenth-century B.C. Mesopotamian tablet (p 316). Further study of these Assyrian texts, in the light of the early medieval recipes given by the *Compositiones Variae* [1], and by the texts of Theophilus and 'Heraclius' (p 351), which are all in the main line of descent of traditional shop-manuals, may well lead to greater elucidation of their difficulties. It is very remarkable that

FIGURE 309—*Three-storeyed glass furnace. From an Italian manuscript. 1023.*

medieval texts should be our basic literary sources for the composition of glass of 1500, and more, years earlier.

Our surest source for learning of the ingredients used in antiquity is to study the composition of existing pieces of early glass. Unfortunately analyses, both chemical and physical, are slow and costly. Few have so far been made [43–45, and references in table on p 313], and much more work must be done before definite conclusions can be reached.

The normal glass of Dynasty XVIII in Egypt was a soda-lime glass with an average percentage content of 55–65 silica, 15–22 soda, 1–3 potash, 3–10 lime, 3–5 magnesia, 1–3 alumina, and 1–3 iron oxides. This remained the general pattern of Egyptian glass into Arab times. Glass in other parts of the ancient world seems to have been basically of similar composition, though in general with more silica. It was from the start a soda-silica or potash-silica compound. These ingredients were probably used almost indiscriminately according to the

FIGURE 310—*Glassmakers and their furnace. The figures in the background are engaged in the production of rocchetta (see pp 356, 358). From an early fifteenth-century Bohemian manuscript.*

supply, and it seems useless to try to prove [45] that composition was changed by rule from time to time and that, for example, all Roman glass is of one kind and all early medieval glass of another. It may well be that in some countries, such as Egypt, soda was never replaced by potash, and that in others, when glass-workers set up furnaces in forested areas, they tended to produce a potash-glass; but ancient glassmakers at all times knew both these compositions and no other, except in the Far East [46]. The existence of true lead glass in antiquity

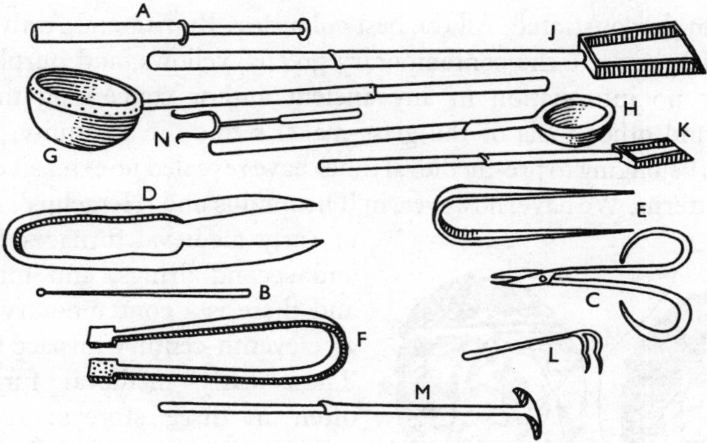

FIGURE 311—*Glass-making tools, from a work of 1699:* (A) *blowpipe;* (B) *pontil, or punty;* (C) *shears;* (D, E, F) *pliers or pucellas;* (G, H) *ladles, for lifting molten glass;* (J, K) *shovels for lifting vessels when hot;* (L) *hooked fork, for stirring glass in the pots;* (M) *rake, for the same, and for moving frit;* (N) *fork for carrying glasses to the annealing-oven.*

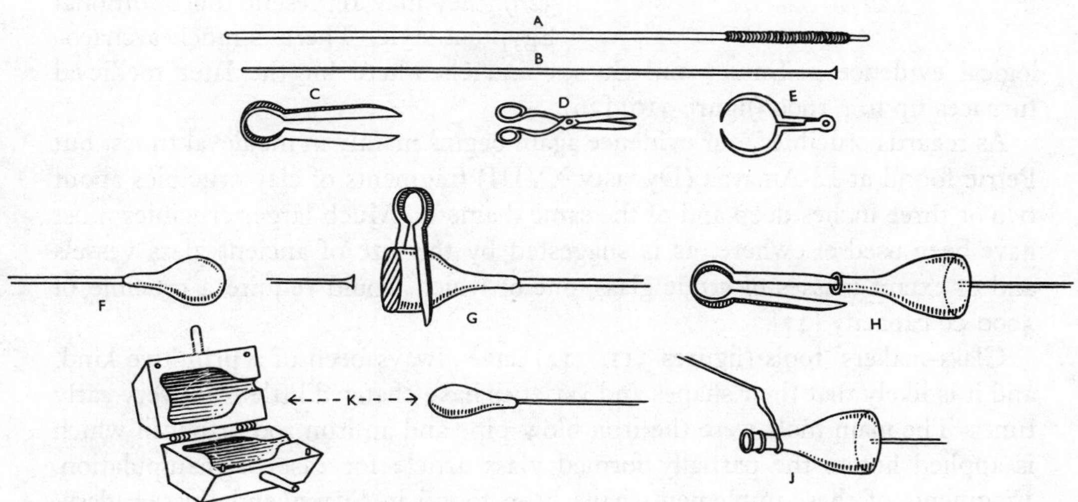

FIGURE 312—*Glass-making tools and processes, from a work of 1817:* (A) *blow-pipe, an iron tube about $3\frac{1}{2}$ ft long, covered at the mouth-piece end with yarn to protect the workman's hand;* (B) *pontil, or punty;* (C) *pliers or pucellas;* (D) *shears;* (E) *calipers;* (F) *a partially blown* paraison, *on the blow-pipe;* (G) *modelling a flask on the blowpipe with the pliers and preparing to snap off the blowpipe after attaching the pontil;* (H) *forming the rim of the flask with the pliers, while flask is held on pontil;* (J) *trailing drawn glass on the neck from one pontil while flask is held on another;* (K) *preparing to insert a* paraison *held on a blowpipe into an iron mould for forming a mould-blown bottle.*

has not been demonstrated. All the best colourless Roman and Arab glass seems to be a soda-glass like the contemporary greens, yellows, and purples.

There is no information in any ancient author concerning the furnaces, crucibles, and other tools of the glass-maker's trade in antiquity. Remains of glass-works belonging to pre-medieval times have revealed no extensive evidence of furnace-patterns. We have, however, in Theophilus and 'Heraclius', descriptions of early medieval furnaces for the first and second firings, and for annealing, and there is a contemporary drawing of an eleventh-century furnace (figure 309). These early medieval furnaces were often in three storeys: fuel-chamber, pot-chamber, and annealing-oven. The remains of a furnace at Corinth seem to conform to this pattern and, as these glass-works are thought to have been founded by Egyptian technicians (p 328), they may represent the traditional Egyptian style. There is much archaeological evidence in Surrey and Sussex and elsewhere for the later medieval furnaces up to *c* 1600 (figure 310) [36].

FIGURE 313—*Pottery mould (with plaster cast) for making mould-blown flasks. Arabic,* c A.D. 900. *Scale just under 1/2.*

As regards crucibles, our evidence again begins mainly in medieval times, but Petrie found at El-Amarna (Dynasty XVIII) fragments of clay crucibles about two or three inches deep and of the same diameter. Much larger crucibles must have been used elsewhere, as is suggested by the size of ancient glass vessels and of extant masses of crude glass, one of which would require a crucible of 5000 cc capacity [47].

Glass-makers' tools (figures 311, 312) have always been of a primitive kind, and it is likely that their shapes and varieties have changed little from very early times. The main tools were the iron blow-pipe and an iron rod or pontil which is applied hot to the partially formed glass article for ease of manipulation. Fragments of these implements have been found in Surrey and Sussex glass-houses [48], but none from any earlier date. Other tools were pincers, the reamer (a pointed instrument for tooling the vessel), and various shears or scissors for clipping rims.

It seems that moulds for glasses in ancient times were usually of clay. Examples are known for a square bottle, a corrugated flask, and a grape-flask [48] of Roman date from a Belgian glass-house site. Medieval Arab clay moulds also exist (figure 313). Later, iron or copper moulds were normal, as they are today [49].

VII. NATURE OF THE GLASS, AND WEATHERING

Glass in ancient times nearly always tended to be more bubbly, and more prone to carry impurities from the batch, than its modern counterpart. This is not surprising, in view of the methods of working and the great difficulty the workers must have had in controlling both the firing of the batch and the purity of the raw materials. However, some glasses, notably the best Alexandrian crystal of the second century A.D. and the best colourless Rhenish glass, reveal few bubbles except under a magnifying glass. Roman glass as a whole compares favourably in this respect with most other glass made before the nineteenth century. On the other hand, pre-Roman sand-core glass is often very bubbly, as is most Arab glass—even the best Syrian enamelled ware. Yet the Arab cut-ware of the ninth and following centuries (p 327) is often as good as the best Roman and modern crystal. As higher temperatures in the furnace help to drive out bubbles and to make glass of crystal transparency, this seems to indicate that the temperatures attained in Roman times were generally higher than before or after.

There are, however, some pre-Roman and Roman glasses, not only vessels, but objects such as medallions, bracelets, ring-settings, and amulets, which are quite opaque and reveal countless tiny bubbles under a magnifying lens. These opaque glasses, though often mistermed pastes, are no less true glass than their transparent or translucent counterparts. The opacity is due to the presence of the bubbles, which further firing under a higher temperature would drive out and so produce a translucent glass.

The main impurities in ancient glass are sandy and black particles, and a white scum which may be either 'stone', i.e. particles of clay from the crucible, or 'glass-gall' which forms on top of the frit after the first fusion. When blown-glass is bubbly or full of impurities it usually shows marked striations or 'blowing spirals', the bubbles or impurities lying in streaks caused by the rotation of the glass on the blowpipe.

Weathering is the term applied to any change for the worse on the surface or in the internal texture of glass, caused during the passage of time by contact with outside influences or by internal decomposition. It covers a wide range of phenomena (plate 28 F–I), and weathering is often important for determining date and place of origin [50]. The main varieties of weathering are:

(*a*) *Dulling of surface*, causing loss of clarity and transparency. This must be distinguished from dulling brought about by stains or scratches caused by usage.

(*b*) *Frosting* and *strain-cracking*. Various causes, including faulty annealing or cooling of the finished vessel, set up differential strains, which in time cause surface or internal cracks (plate 28 F). In its most advanced stages this amounts to devitrification, and ultimately the whole vessel disintegrates into tiny fragments. Sand-core glass found on Syrian and other Asiatic sites is particularly liable to such weathering, which partly explains why so few early glasses from Asiatic sites are preserved. Many Roman glasses, especially colourless pieces and window glass, are also prone to it. These develop a particular variety of strain-cracking which appears as linear gaps right through the glass.

(*c*) *Enamel-like weathering* is a surface change, manifesting itself in milky white, brown, or even black patches (plate 28 G–I) which spread until, in the worst cases, they cover the whole vessel, giving an appearance of an enamel coating. Decomposition of the surface then begins and may take two forms. First, the weathering may flake away in small round areas, leaving a pitted surface (plate 28 H) which is iridescent owing to the interference of light at the uneven surface. This form of weathering, though common on Roman glass from certain countries, such as Syria, is even more common on medieval glass both in the east and in the west. It occurs on Arab glass in Mesopotamia and elsewhere, and a particular form of it, with a blackish surface leading to deep pitting, is very prevalent on western forest-glass. The characteristic deep pitting found on stained medieval window-glass is another instance of this form of weathering. Secondly, the weathering may flake away in thin powdery layers (plate 28 I), again leaving an iridescent surface. Glasses from Syrian and Palestinian sites are markedly prone to such layer-flaking; through it they often develop the beautiful golden or purple rainbow tints much valued by collectors.

(*d*) *Iridescence* is the commonest and most prevalent form of weathering. It gives a variegated coloration with a rainbow-like effect and can be found, though often in the merest traces, on almost every piece of ancient glass. In its mildest form it is confined to filmy patches; in a more advanced state it begins to flake off, but when this happens it is usually combined with, and is a result of, variety (*c*). It is rare, except in incipient forms, on glass from Egypt. Roman glass from the west and medieval forest-glass are not usually seriously attacked by it, except as a concomitant of variety (*c*).

These weatherings are due to many causes: some external, such as exposure to water, sunlight, acids and other agents in the soil, and habitation-refuse; and some internal, such as chemical composition of the batch and faulty manufacture. External and internal factors interact to a considerable extent, so that, though we can recognize certain varieties of weathering as characteristic of certain countries and of glass-wares of certain dates, it is quite impossible to be sure what type or degree of weathering will appear on a piece of glass after preservation for a fixed time under seemingly fixed circumstances. Fragments of the same vessel discovered close to each other are frequently found to have weathered differently [51].

VIII. TECHNIQUES OF MANUFACTURE

(*a*) *Mould-pressing*. The probability is that the earliest glass-making technique was moulding. The ancients from very early days understood the use of moulds for objects in clay and metal. It is therefore natural that they should have adapted this process to the new material, glass. The production of glass in a crucible would show that, when cold, it took the shape of the crucible. Countless glass objects were made by pressing viscous glass into clay moulds from at least the middle of the third millennium B.C. How soon vessels began to be made by this process is not so certain. Some Egyptian open bowls of Dynasty XVIII were probably so made, though it is possible that they were modelled on a positive rather than pressed into a negative mould.

The main development of mould-pressing for making vessels was in the second half of the first millennium B.C. when an extremely active mould-pressing industry arose in Egypt (p 322). This, combined with the development of the cane process (below) and with a prolific use of grinding and polishing for finishing the vessels, brought the Alexandrian region into great prominence as a glass-making centre. From the first century B.C. or a little earlier a great quantity of fine ware, mainly open bowls and jars, with some closed shapes as well, was produced by Alexandrian workers both in Egypt, and, after 30 B.C., in Italy (plate 25 A, and tailpiece).

The polychrome glasses were built up of sections of multi-coloured and patterned rods or 'canes', fused together in a mould, and ground and polished after delivery. The monochrome glasses, which were made of fine emerald green, turquoise and cobalt blue, sealing-wax red, and other colours both opaque and translucent, were also mould-pressed and finished by rotary-polishing if plain, or by fire-polishing (that is, reheating to give a shiny surface) if ribbed. The Alexandrian industry of mould-pressing lasted into the second half of the first century A.D. From then on, mould-pressed glasses ceased to be common, except for blanks made for completion by *diatretarii* (p 336). After the first century A.D. the ancient industry, when it used moulds for finished glass-ware at all, seems to have preferred to use them for mould-blowing rather than for mould-pressing. This process is considered below (p 339).

(*b*) *Cold-cutting*. The production of blocks or 'blanks' of glass by moulding provided something resembling a lump of stone. This could be ground and polished by stone-cutting techniques that had been in use from at least the fourth millennium B.C. in Egypt and Mesopotamia. Gem- and seal-cutters had from early times cut such hard stones as rock-crystal and dolerite. They found no difficulty in working from a block of glass such things as a cylinder-seal or a ring-setting, and these are fairly common from the middle of the second millennium B.C. Petrie found cut and ground glass at El-Amarna [13]. Certain larger objects, such as Tutankhamen's head-rest (figure 297), were perhaps made by this technique from a moulded blank. Yet, with one doubtful exception [52], no vessel of the second millennium B.C. made by this process has been found.

The earliest datable cold-cut vessel is from Nimrud (p 321), inscribed with the name of Sargon II, *c* 720 B.C. (figure 299, plate 26 C, D). Thereafter a small but widespread use of this technique for vessel-making may be traced to Roman times and beyond, but it was more in demand as an adjunct to other processes, for finishing and decoration. We have noted its use for grinding and polishing Alexandrian and Italian mould-pressed ware (p 335). Its exponents became the *diatretarii* of the later Roman period, who worked in conjunction with the *vitrearii*, or blowers, in fabricating fine vessels with cut decoration. These men handed down their art to successors until Arab times, when cut ware during the ninth and following centuries became once more prevalent and popular. Gem-cutters, too, never ceased to use glass. Countless thousands of Greek, Roman, and later glass gems imitate precious and semi-precious stones.

(*c*) *Core-winding*. From the middle of the third millennium B.C. to late Roman times beads of glass were made by core-drawing, which involved either winding a viscous trail of glass on a wire or moulding viscous glass on a wire core, or perhaps dipping a wire in molten glass. The wire was withdrawn when the glass had cooled. From late Roman times onwards beads seem normally to have been made by cutting a blown tube into sections. This process was a very much quicker and easier one to manipulate, and it is not surprising that, once invented, it ousted the old wire-drawn method.

From core-winding of beads, workers in Egypt in Dynasty XVIII developed their sand-core technique for vessel-making, the principle being essentially the same. The method of manufacture is described above (pp 319 f). By its nature the process had to be confined to closed shapes and could not be used for bowls; yet the number of extant examples shows that it was by far the most popular process until the invention of blowing. Thereafter it rapidly died and has never been revived commercially. Blowing displaced it as far quicker and involving little or no limitation of shape or size.

(*d*) *Blowing.* Thus by *c* 1500 B.C. three basic glass-making techniques were in use. It was not for another 1500 years or so that a new process was developed, but that process—the invention of glass-blowing—was revolutionary (p 322). This, whether mould-blowing or free-blowing, is essentially one process, and was the last main technique to be invented. That there has been no further change in the basic techniques since Roman days is one of the many indications of the modernity of Roman industrial processes.

We cannot say for certain when or by what group of workers this process was invented. No examples of blown glass have been found which can be firmly dated on stratigraphic evidence to before the Christian era, and glass-blowing was probably a Syrian invention of the early first century A.D. Among the earliest examples are the mould-blown vessels of the Sidonian glass-houses of Ennion (figure 300), who also worked in Italy (p 322), and others of that time. We may guess that mould-blowing came before free-blowing, but there is no compelling reason why the first man to blow glass should not have worked without a mould. Once invented, the spread of blown glass was one of the most rapid movements of ancient techniques. By A.D. 50, probably only a few decades after its first introduction, blown glasses were in use throughout the Roman world.

The method of blowing a glass vessel need not be described in detail here, for neither the process nor the tools underwent significant modification for nineteen centuries (figures 311, 312). The only modern change has been the introduction of machinery of increasingly elaborate design for the mass-production of such objects as bottles and electric-light bulbs. Roman workers of the first century A.D. were as adept at blowing a finely shaped vessel as any of their successors; and those of the second to fourth centuries A.D. could produce as elaborate a piece of glass as any Venetian worker of the Renaissance. They knew and used all the tricks of the glass-blower's trade. They could, for example, make flashed glasses of two or more layers; they could spin broad, flat dishes up to two feet or even more in diameter; they could insert one small jug within a larger one, and they could draw out hollow lobes from the walls of a vessel by the drop-on process.

After the Roman decline in the west the blower's art was still maintained at a high pitch in the east. There, some of the early Arab glasses are as dexterously made as most Roman vessels, and it was from eastern workers that the Venetians presumably relearned their skilled methods.

IX. TECHNIQUE OF DECORATION

Decoration can be either an essential ingredient in the main process of manufacture (*a–c* below), or a secondary process performed when the glass is still warm (*d* and *e*) or after it has cooled (*f–h*), though in the last instance it may entail reheating the vessel.

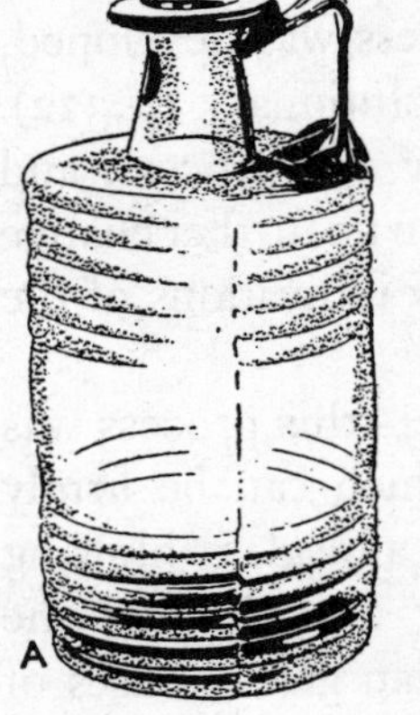

FIGURE 314—(A) *Mould-blown bluish-green jug made by Frontinus, in his northern Gallic factory.* c A.D. *300. Scale 1/3.* (B) *Bottom of the jug, showing the glass-maker's name FROTI, for FRONTINUS.*

(*a*) *Casing or Layering.* This was especially prevalent in the Alexandrian and Italian industry of the early Roman period. The method is extremely difficult: it entails gathering one or more layers on an inner *paraison*[1] so deftly that the walls coalesce at every point but do not sink into one another. At times no further decoration was attempted on these vessels: at others, notably in the famous 'cameo' glasses, a floral or figure pattern in relief was made by carving the outer layer or layers. Thus on the Portland vase (plate 26 A) we have a mythological scene in opaque white glass on a dark blue background.

(*b*) *Mosaic Glass.* Under this broad term we include a whole range of polychrome patterned pieces (plates 24 E, 28 A–E). Some were made by fusing together sections of canes (p 335) bearing floral or other patterns (plate 28 A), and are normally termed *millefiori* ware; others by mixing two or more colours of glass into a marbled (B) or mottled (C) pattern; others by fusing together sections of rods bound with spiral trails into a laced pattern (D); and yet others by fusing strips and squares of multi-coloured pieces into a design resembling that of a patchwork quilt (E). The last type often includes pieces of gilt glass. Though mosaic glass inlay-pieces were known in Mesopotamia and Egypt as early as the second millennium B.C., the main use of mosaic glass for vessels and inlays was centred in Alexandria and Italy from about 25 B.C. to A.D. 100 (pp 322, 335). Mosaic glasses were some of the costliest available, and were deemed worthy of a place in the emperor's own cabinet and on his own dining-table. The more elaborate mosaic glass had ceased to be manufactured by the second century A.D., but simple varieties continued to be made, and there was a revival of the art in early Arab times [38].

(*c*) *Moulded Patterns.* Pre-Roman and Roman mould-pressed glass rarely has any pattern, and such little as it has was usually made by cutting and grinding after delivery from the mould. This does not, of course, hold good for

[1] The glass-maker's term for the mass of glass collected on the blow-pipe.

figurines and objects of that kind, where the main pattern was produced in the mould, just as on terracotta or metal objects.

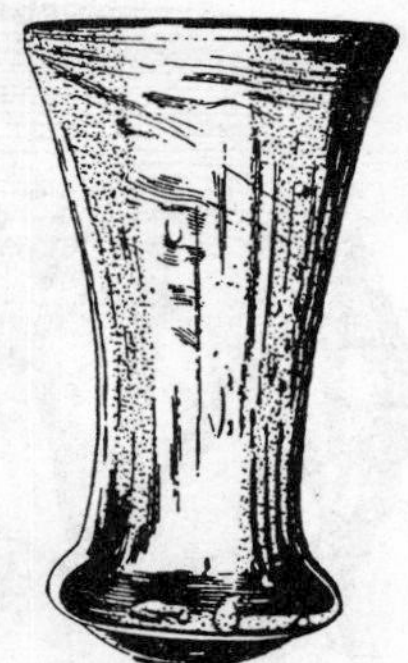

FIGURE 315—*Olive-green bell-beaker from Faversham, Kent, showing pre-moulded corrugations. Sixth century A.D. Scale 1/3.*

On the other hand, mould-blowing was essentially a process for applying a pattern to blown glasses. It could be done in two ways. In the first (plate 27 D, figure 314), the mould was given the shape of the final pattern (say a man's head, a barrel, or a series of knops) and the vessel was reckoned to be finished when delivered from the mould. In the second (plate 25 E, figure 315), the mould was of a cylindrical or other simple shape, with a corrugated or diaper pattern, and the vessel underwent a further blowing-process after delivery. By this means the mould-pattern became contracted or elongated in the narrow parts of the vessel, and expanded on the bulbous parts. Corrugations that were originally vertical could be given a writhen form by twirling the vessel during the second blowing. A variation on this method was to blow first into a patterned mould and then into a plain one; this caused the

FIGURE 316—*Two colourless bottles of 'snake thread' fabric:* (left) *from Syria,* (right) *from Cologne: both with colourless trails. Late second or early third century A.D. Scale 3/8 and 3/7 respectively.*

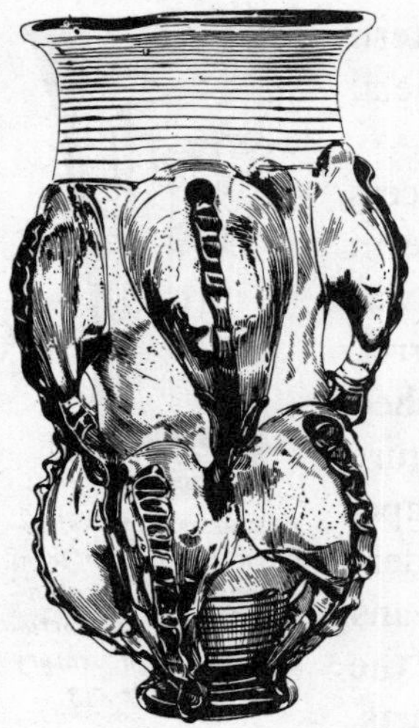

FIGURE 317—*Green claw beaker with blue trails from Castle Eden, Durham. Late fifth to sixth century* A.D. *Scale 1/3.*

pattern (usually ribbing) to appear raised on both the inside and outside walls of the vessels, and is called 'optic' blowing [53]. Mould-blown patterns of all these varieties occur at all periods of glass-making, and in all countries, from Roman times onwards.

(*d*) *Application of Trails and Blobs.* The earliest manifestation of this method is the use of combed trails and other applications of dots, circles, and so forth on sand-core ware of Dynasty XVIII. This continued to be the normal method of decorating sand-core down to Roman times and was soon adopted by the blowing-industry (plate 27 C, F). It was not used by the mould-pressed industry of Alexandria, but later Roman glass-blowers made great use of it. The Syrian industry and its second-century offshoot in Cologne (figure 316) often added the most elaborate and fantastic trails, especially on the so-called snake-thread ware of *c* A.D. 200. Trailing was also used quite considerably on Frankish and Merovingian glass in the west (figure 317), and on early Arab glasses (figure 308). The application of even, regular trails to a vessel is a very difficult process and calls for great dexterity in manipulating a long and slender thread of viscous glass while the vessel is rotated on the blow-pipe. Blobs were added by dropping on, and then reheating the glass so that the blob expanded and fused into the wall of the vessel. Trails and blobs were either left in relief or marvered flush. Trailing was also the normal way by which handles were added to a vessel, and quite a frequent way of attaching a base-ring.

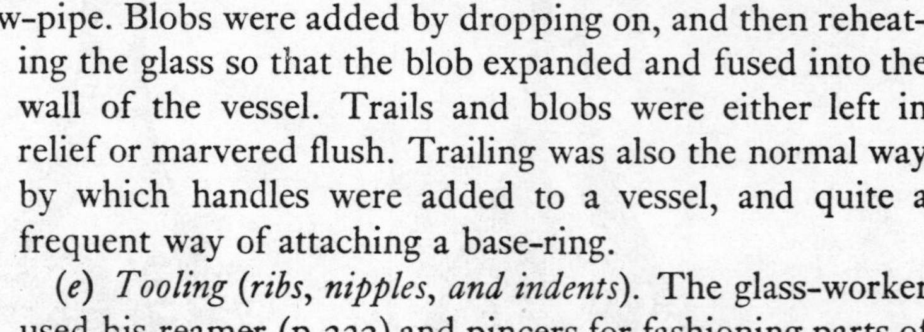

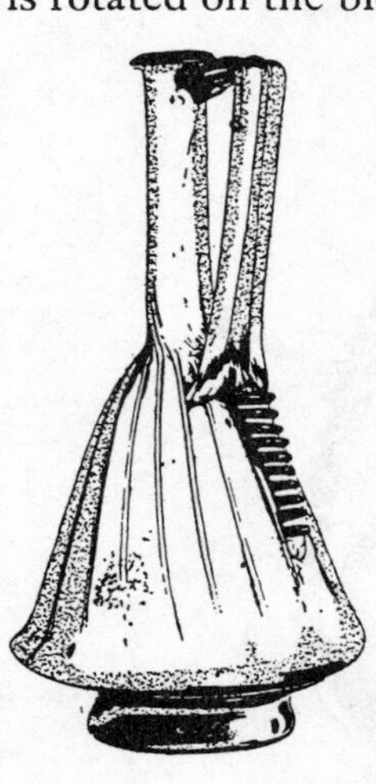

FIGURE 318—*Jug with ribbed decoration, western Roman. First century* A.D. *Scale 1/4.*

(*e*) *Tooling* (*ribs, nipples, and indents*). The glass-worker used his reamer (p 332) and pincers for fashioning parts of his vessel such as the rim, neck, and base. He also used the reamer and other tools for making vertical ribs (plate 27 C, figure 318) and indents, and the pincers for pinching out nipples and even short ribs (plate 27 E, figure 319). The methods employed are evident and need no description.

(*f*) *Cutting* is one of the commonest and at the same time most elaborate and successful methods of decorating glass. This was soon realized by ancient workers, who had had long practice in similarly decorating stone vessels and objects. It is likely that it was the gem-cutters and stone-cutters who first set themselves up as *diatretarii* for glass-work. Petrie [13] mentions cut patterns on glass of Dynasty XVIII at El-Amarna, but the earliest piece we know with a cut pattern on it worthy of the name is the Ephesus fragment of the fourth century B.C. (p 321) with petalled decoration [21]. Figured scenes do not begin before the 'cameo' glasses of the first century B.C. (plate 26 A), but from then on figured scenes—often of a most elaborate kind—are prevalent throughout Roman times. The methods of cutting these patterns were threefold: (i) facet- and other hollow-cutting (plate 26 B, figures 320–1), by means of a wheel, fed by water and emery; (ii) relief-cutting, also with a wheel; in its most elaborate form on the cage-cups (figure 306), this produced an almost free-standing design adhering to the base-cup only by means of small connecting bridges; (iii) graving (plate 26 E, figure 322) with a flint or other tool. The diamond was unknown in antiquity as a specific stone. Cutting of all these three varieties was used on post-Roman glass in the east, but not in the west until reintroduced in the later Middle Ages.

FIGURE 319—*Greenish bowl with pincered decoration. From Syria. Second to third century* A.D. *Scale* c *1/2.*

(*g*) *Painting, lustre, and enamelling.* The earliest example of painting on glass is the Thothmes III jug (plate 24 D) of *c* 1500 B.C. This was obviously a rarity, and painting on glass does not become a normal type of decoration until Roman times. We possess many Roman painted glasses. For the most part they are painted with enamel colours (plate 27 A, figure 301), which were fired on by reheating the glass after the design had been added. Other glasses have more fugitive colouring which does not seem to have been fired on, but has been protected by a matt background layer. The design was to be viewed through the glass, as for example on a series of lids found in Cyprus (plate 25 C), decorated with cupids and grapes and other subjects [27]. Painting on glass continued in post-Roman times in the east but, like cutting, not in the west. One particularly noteworthy eastern example of early Arab

FIGURE 320—*Colourless carved cup with facet decoration, Alexandrian or Italian work. First century* A.D. *Scale 1/3.*

times is the group of lustred bowls (plate 25 D) made in some quantities in Egypt and other Near Eastern countries from the eighth to the tenth centuries A.D. in imitation, no doubt, of lustred pottery (pp 303 f). The most important post-Roman group under this head are the Arab enamelled glasses (frontispiece). These glasses, made largely if not entirely in Syria from the late twelfth to the fourteenth century, are decorated with true enamel, applied as a powder and fired *in situ*, and not with an enamel paint such as we get on Roman glasses. The difference is mani-

FIGURE 321—*Colourless bowl with facet-cut scene of Hypermnestra, Lynceus, and Pothos. Alexandrian work. Second century A.D. Scale c 2/3.*

fested by the glossy sheen on the later glasses, which contrasts with the much more matt appearance of the Roman enamel-painted examples.

(*h*) *Gilding*. The use of gilded glass insets in mosaic glasses has already been mentioned (p 338). Of about the same date, or perhaps even earlier, is a group of glasses with an elaborate floral pattern in gold-foil, applied to the exterior of an inner shell, and framed, as it were, by another colourless bowl fitting tightly over the gilded portion; there are two examples (plate 25 B) from Canosa [54]. These cannot be true cased glasses (p 338), since the gilding is on the inner portion, and it is not clear how the fashioning was done; but they are, in general, prototypes of a variety of gilt sandwich glasses (plate 27 B) that became very common in the third and fourth centuries, with medallions, gilt and often coloured as well, sandwiched between colourless glass. On these the gilding is sometimes, though not always, added to the under side of the outer glass, and the colouring is used as backing for the gilding before the inner glass is applied. The two layers of glass were joined together by lightly warming their surfaces.

Gilding was also used without two layers of glass, by simply applying it to the surface of a vessel. This is, however, rarer. The gilding was perhaps applied in the form of powder or of tiny bits of gold leaf in suspension and fired at low temperature to bond it to the surface of the glass. Some glasses made thus, such as the fine Daphne jug (plate 27 A) [29], have retained their gilding remarkably well.

FIGURE 322—*Engraved flask with hunting-scene inscribed in Greek 'Drink and good health, Fortunatianus'. From Chiaramonte Gulfi, near Syracuse. Fourth century A.D. Scale 1/2.*

REFERENCES

[1] JOHNSON, R. P. 'The *Compositiones Variae* from Codex 490, Biblioteca Capitolare, Lucca, Italy. An introductory study.' Ill. Stud. Lang. Lit., Vol. 20, no. 3. University of Illinois Press, Urbana. 1939.

[2] LUCAS, A. *J. Egypt. Archaeol.*, **22**, 141–64, 1936.
BECK, H. C. *Ancient Egypt*, 69–83, December 1934.
Idem. Ibid., 19–37, June 1935.

[3] LUCAS, A. 'Ancient Egyptian Materials and Industries' (3rd ed.), pp. 184–92. Arnold, London. 1948.

[4] WOOLLEY, SIR (CHARLES) LEONARD 'Alalakh: an Account of the Excavations at Tell Atchana in the Hatay, 1937–49', pp. 299 f. Res. Rep. Soc. Antiq. Lond., no. 18. 1955.

[5] LUCAS, A. See ref. [3], pp. 190–2.

[6] GADD, C. J. and THOMPSON, R. CAMPBELL. *Iraq*, **3**, 87–96, 1936.

[7] MOORE, H. *Ibid.*, **10**, 26–33, 1948.

[8] GOLDMAN, HETTY (Ed.). 'Excavations at Gözlu Kule, Tarsus, I. The Hellenistic and Roman Periods.' Institute for Advanced Study, Princeton. 1950.

CHARLESTON, R. J. 'Roman Pottery', pp. 24 ff. Faber and Faber, London. 1955.

[9] JOPE, E. M. *Archaeol. News Letter*, **2**, 199–200, 1950.

[10] LUCAS, A. See ref. [3], pp. 392–4, 550.

[11] VITRUVIUS, *De architectura*, VII, xi. (Loeb ed. Vol. 2, pp. 122 ff., 1934.)

[12] JOPE, E. M. and HUSE, G. *Nature*, **146**, 26, 1940.

[13] NEWBERRY, P. E. *J. Egypt. Archaeol.*, **6**, 155–60, 1920.

BECK, H. C. *Ancient Egypt*, 7–21, June 1934.

LUCAS, A. See ref. [3], pp. 207–21.

FOSSING, P. 'Glass Vessels before Glass-blowing.' Munksgaard, Copenhagen. 1940.

PETRIE, SIR (WILLIAM MATTHEW) FLINDERS. 'Tell el Amarna', pp. 25–30. Methuen, London. 1894.

[14] STARR, R. F. S. 'Nuzi', Vol. 1, pp. 457 ff. Harvard-Radcliffe Fine Arts Series. Harvard University Press, Cambridge, Mass. 1937.

[15] FOSSING, P. See ref. [13], p. 26.

TUFNELL, OLGA *et al.* 'Lachish II', Pl. XXIV. Wellcome-Marston Archaeol. Res. Exped. to the Near East Publ. 2. Oxford University Press, London. 1940.

WOOLLEY, SIR (CHARLES) LEONARD. See ref. [4].

[16] SCHAEFFER, C. F. A. 'Enkomi-Alasia. Nouvelles missions en Chypre, 1946–50', pp. 210–14. Publ. Miss. archéol. franç. et Miss. Gouvt de Chypre à Enkomi, Vol. 1. Klincksieck, Paris. 1952.

[17] LUCAS, A. *Ann. Serv. Antiq. Égypte*, **39**, 227–35, 1939.

[18] BECK, H. C. and STONE, J. F. S. *Archaeologia*, **85**, 203–52, 1936.

O'NEIL, B. H. ST. J. and STONE, J. F. S. *Antiq. J.*, **32**, 21–34, 1952.

[19] FOSSING, P. See ref. [13], pp. 36, 42 ff.

[20] HARDEN, D. B. *Antiquity*, **22**, 147, 1948.

SCYLAX (Ps.-) *Periplus maris ad litora habitata Europae et Asiae et Libyae* in *Geographi graeci minores* (ed. by C. MÜLLER), Vol. 1, pp. 15 ff. and especially p. 94, para. 112. Didot, Paris. 1855.

[21] FOSSING, P. See ref. [13], p. 84.

[22] DEHN, W. *Germania*, **29**, 25–32, 1951.

[23] SMITH, R. W. *Bull. Metrop. Mus.*, **8**, 49–60 (esp. 51–52), 1949.

[24] HARDEN, D. B. *J. Rom. Stud.*, **25**, 164–86, 1935.

[25] SIMONETT, C. 'Tessiner Gräberfelder.' Birkhäuser, Basel. 1941.

[26] KISA, A. 'Das Glas im Altertume', pp. 163–255. Hiersemann, Leipzig. 1908.

MORIN-JEAN. 'La verrerie en Gaule sous l'empire romain.' Laurens, Paris. 1913.

[27] VESSBERG, O. *Opuscula Archaeologica*, Vol. 7, pp. 109–65. Skr. utg. Svonska Inst. i Rom. 1952.

DAVIDSON, GLADYS R. 'Corinth XII. The Minor Objects', pp. 76–122. American School of Classical Studies at Athens, Princeton. 1952.

BROCK, J. K. and YOUNG, G. M. *Annu. Brit. Sch. Athens*, **44**, 1–92, 1949.

[28] HACKIN, J. 'Recherches archéologiques à Begram.' Mem. Déleg. archéol. franç. Afghanistan no. 9. Éditions d'art et d'histoire, Paris. 1939.

Idem. 'Nouvelles recherches archéologiques à Begram.' *Ibid.*, no. 11. 1954.

CAPUTO, G. *Mon. ant.*, **41**, 201–442 (esp. 391–9), 1951.

CURLE, J. *Proc. Soc. Antiq. Scotld*, **66**, 277–397 (esp. 290–6), 1932.

EGGERS, H. J. 'Der römische Import im Freien Germanien.' Hamburg Mus. f. Völkerkunde u. Vorgeschichte, Publ. no. 1. Hamburg. 1952.

[29] MUSÉE DE MARIEMONT. 'Catalogue des verres antiques de la collection Ray Winfield Smith', p. 27, no. 123, Pls IV–V. Duculot, Gembloux. 1954.
[30] HARDEN, D. B. 'Roman Glass from Karanis.' University of Michigan Humanistic Series, Vol. 41. University of Michigan Press, Ann Arbor. 1936.
[31] THORPE, W. *Trans. Soc. Glass Tech.*, **22,** 5–37, 1938.
[32] FREMERSDORF, F. 'Römische Gläser aus Köln' (2nd ed.). Völker-Verlag, Cologne and Leipzig. 1939.
Idem. 'Figürlich geschliffene Gläser.' Röm.-Germ. Forsch., no. 19. De Gruyter, Berlin. 1951.
[33] RADEMACHER, F. *Bonn. Jb.*, **147,** 285–344, 1942.
HARDEN, D. B. "Glass vessels in Britain, A.D. 400–1000" in 'Dark-Age Britain. Essays presented to E. T. Leeds' (ed. by D. B. HARDEN), pp. 132–67. Methuen, London. 1956.
[34] TROWBRIDGE, MARY L. 'Philological Studies in Ancient Glass', pp. 113–14 (citing the original texts). Ill. Stud. Lang. Lit., Vol. 13, nos. 3, 4. University of Illinois Press, Urbana. 1930.
[35] STEINHAUSEN, J. *Trierer Z.*, **14,** 29–57, 1939.
ARBMAN, H. 'Schweden und das Karolingische Reich', pp. 26–86. K. Vitterhets Hist. Antik. Akad. Handl. 43. Stockholm. 1937.
[36] WINBOLT, S. E. 'Wealden Glass.' Combridges, Hove. 1933.
THORPE, W. A. 'English Glass' (2nd ed.), pp. 79–93. Black, London. 1949.
[37] VICTORIA AND ALBERT MUSEUM HANDBOOK. 'Glass', by W. B. HONEY, pp. 55–56. H.M. Stationery Office, London. 1946.
[38] LAMM, C. J. 'Das Glas von Samarra.' Reimer, Berlin. 1928.
Idem. 'Mittelalterliche Gläser und Steinschnittarbeiten aus dem nahen Osten' (2 vols). Reimer, Berlin. 1929–30.
Idem. 'Glass from Iran in the National Museum, Stockholm.' Fritzesbokh, Stockholm. 1935.
HONEY, W. B. See ref. [37], pp. 35–54.
[39] HARDEN, D. B. *Iraq*, **1,** 131–6, 1934.
[40] DAVIDSON, GLADYS R. See ref. [27], pp. 83–90.
Idem. Amer. J. Archaeol., **44,** 297–324, 1940.
[41] THOMPSON, R. CAMPBELL. 'On the Chemistry of the Ancient Assyrians.' Luzac, London. 1925.
Idem. 'Dictionary of Assyrian Chemistry and Geology.' Clarendon Press, Oxford. 1936.
[42] TROWBRIDGE, MARY L. See ref. [34], pp. 96–101.
[43] NEUMANN, E. and KOTYGA, G. *Z. angew. Chem.*, **38,** 776 ff., 1925.
PARODI, H. D. 'La verrerie en Égypte.' Thesis, Grenoble. Cairo. 1908.
LUCAS, A. See ref. [3], pp. 537–40.
LAL, B. B. *Ancient India*, **8,** 17–27. 1952.
[44] FARNSWORTH, MARIE and RITCHIE, P. D. *Tech. Stud. fine Arts*, **6,** 155–68, 1938.
[45] FAIDER-FEYTMANS, G. *Rev. belg. Archéol. Hist. Art*, **10,** 4, 1940.
SALIN, E. 'Rhin et Orient. Le haut Moyen-Age en Lorraine d'après le mobilier funéraire', pp. 46–50, 177–213. Geuthner, Paris. 1939.
[46] SELIGMAN, C. G. and BECK, H. C. *Bull. Mus. Far East. Antiq.*, **10,** 1–64, 1938.
[47] LUCAS, A. See ref. [3], p. 220.
[48] CHAMBON, R. 'L'Histoire de la verrerie en Belgique du IIme siècle à nos jours', p. 30 and frontispiece. Librairie Encyclopédique, Brussels. 1955.
[49] WINBOLT, S. E. See ref. [36], pp. 58, 79.

[50] FOWLER, J. *Archaeologia*, **46**, 65–162, 1881.
HARDEN, D. B. See ref. [30], pp. 8 ff.
[51] *Idem*. See ref. [30], p. 11.
[52] FOSSING, P. See ref. [13], p. 26. Vessel from Kakovatos, claimed to be moulded, but which is more likely to be cold-cut.
[53] LAMM, C. J. 'Glass from Iran', p. 10. See ref. [38].
Idem. 'Das Glas von Samarra', p. 40. Reimer, Berlin. 1928.
[54] DALTON, O. M. *Archaeol. J.* **58**, 225–53, Pl. v, 1901.

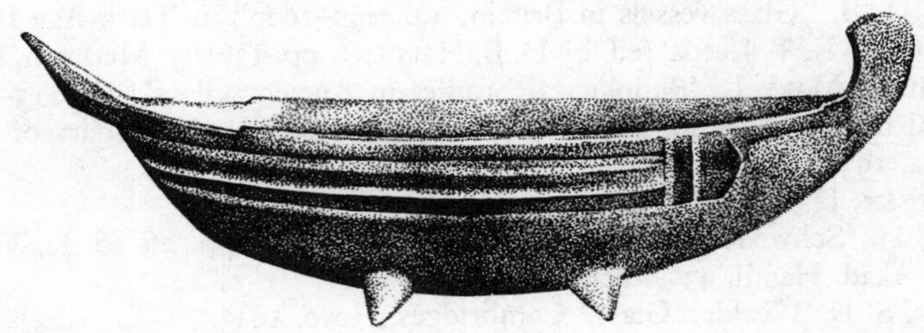

Dark green boat-shaped saucer, mould-pressed, ground, and polished (pp 322, 335). First century A.D., probably Alexandrian. Scale 2/5.

10

PRE-SCIENTIFIC INDUSTRIAL CHEMISTRY

F. SHERWOOD TAYLOR AND CHARLES SINGER

I. METHODS OF CHEMICAL INDUSTRY

NEITHER in antiquity nor in the Middle Ages was there any conception of the existence of an independent science, chemistry, and as little of an art, industrial chemistry. These things were perceived only dimly even in the sixteenth century. Nevertheless, throughout this long period artisans were often doing much the sort of thing that industrial chemists do now. They, or those who directed them, not seldom tried methods on a small and experimental scale before embarking on a larger industrial undertaking. Certain industries were then, as now, habitually concerned with converting one kind of chemical individual into another.

Among such industries the most important had to do with the manufacture of metals (ch 2); alkalis (p 354), soap, and other cleansing materials (p 355); acids (p 356); ceramics (ch 8) and glass (ch 9); pigments, dyestuffs, and mordants (p 359); combustibles, incendiaries, and explosives (pp 370, 374 ff); drugs, foodstuffs, and alcoholic liquors (ch 4). The pre-scientific chemical processes thus involved may be roughly divided into (*a*) those at high temperatures, and (*b*) those in solution and at ordinary or comparatively low temperatures.

High-temperature processes were involved in metallurgy, ceramics, glass, lime-burning, and the making of some pigments. The essential instrument was the furnace. Of these there were two main types: first, that with a combustion chamber, from which flame and hot gases passed into another chamber containing the matter to be heated; and secondly, that in which the matter to be heated was mixed with or surrounded by the fuel. The former type was used for glass and ceramics. Sometimes the two chambers were superposed vertically (figure 287) and sometimes they were arranged horizontally (figure 281). The normal fuel was wood. This yielded a long flame capable of passing into the second chamber. Coal was apparently not thought suitable for glass or ceramics, though it was used for certain other industrial purposes. The second type of furnace is typified by the smelting-furnaces of the metallurgists (figure 38), by lime-kilns (figure 323), and by the foundries of the period (figures 34, 35).

FIGURE 323—*Lime-kiln reconstructed from a painting by David Teniers the younger (1610–90). Chalk or limestone is brought up a ramp (indicated by dotted lines), heaped on the platform, and held in place there by a rough paling. The central fire is fuelled through the tunnel and the lime turned over from time to time.*

The temperatures attained depended on the fuel and on the draught. Coal was preferred to charcoal for most smith's work, but otherwise was little used until the sixteenth century. Draught was induced not by chimney-stacks but either by siting the furnace so that the prevailing winds entered the draught-hole, or by bellows (figure 36), or again by increasing the height of the furnace (figure 38 D). In the fifteenth century the development of mechanically operated bellows driven by water-power (p 73) enabled the temperature of furnaces to be so far raised as to permit of the casting of iron (p 74).

Crucibles were of various clays resistant to fire. We hear of white clays, and of mixtures of burnt and raw clay, being used, and of a hard stone which did not absorb liquid and withstood the fire. This may have been quartz, but it was probably used only for small crucibles.

Lower-temperature processes were employed for the large vessels of liquid used by dyers, brewers, soap-boilers, alum-makers, saltpetre-boilers, and others. Such cauldrons were usually of copper or brass set in brick, much like the more modern but now old-fashioned 'copper' (figure 324). Large vessels for which external heat was not required were commonly of coopered staves: such were casks, barrels, vats, and tuns (figure 328). Lead cisterns were also commonly employed.

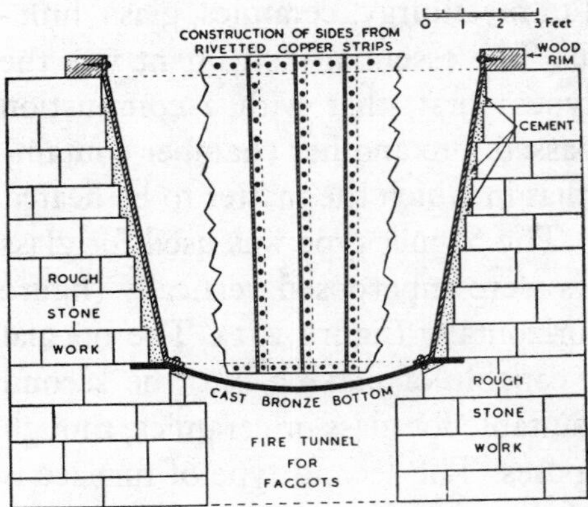

FIGURE 324—*The traditional alum cauldron reconstructed from various medieval specifications.*

Distillation in antiquity and the Middle Ages was carried out on a small scale (ch 21). Stills did not contain more than a few gallons and were often of a much lower capacity. They were of earthenware, copper, iron, or glass. Still-heads were of copper, pewter, or glass. Glass still-bodies were apparently a late medieval product. They had to be carefully protected from sudden changes of temperature, and were therefore usually coated with a lute of clay, horse-dung, blood, or other materials. As a rule, stills were embedded in sand or ashes to which heat was applied, or immersed in a hot water-bath (figures 660, 663). An air-bath of an empty earthenware pot around the still-body was also often used. Still-heads were sometimes water-cooled, but not by running water before about

FIGURE 325—*Preparation of woad, 1752. In the background is a plantation of woad plants. The leaves are being stripped, heaped, and shovelled into the mill* (in the foreground), *and there raked over and ground. The product is heaped* (left foreground) *and left to dry. It is then rolled into balls, which are further dried on the racks in the mid distance.*

1500. Pounding was done in mortars and by hand (figure 339), but stamps counterpoised by weights or springs (figure 345) were sometimes used in place of the ordinary pestle. Grinding-mills (figure 325) were used for tan-bark, woad, &c.

Industrial processes were on what we should regard as a small scale, not so much because of the difficulty of constructing large plant, but to keep all operations within the powers of human muscles. Despite the development of water-mills from early times (p 608), and of windmills from the fourteenth century, these were mainly applied to the grinding of corn and only rarely to industrial processes, such as the fulling of cloth (figures 187, 553).

There is an aspect of industrial chemistry, and indeed of industrial processes in general, to which special attention should be drawn. A method of preparation being once established, it may be continued to some extent for centuries or even millennia after it has been generally superseded. Thus, to take a single example, it was discovered in remote antiquity that certain dyes were faster and brighter when mordanted on fabrics (pp 366 ff). The usual mordant was alum, which is found native. Indigo, however, does not require this treatment nor does it gain thereby. Nevertheless, from ancient Egyptian times onwards fabrics to be dyed with indigo were often mordanted at great trouble and expense. The habit has continued in some backward areas almost to our time. Similarly, conservative modern leather-workers would recognize the substances, processes, and even tools used by a leather-worker of centuries before the Christian era (figures 107, 130). On the conservatism of even modern craftsmen a volume could be written. The reader will learn of many examples in the course of this work.

II. SOURCES OF TECHNICAL INFORMATION

The Greek and Roman worlds derived their chemical techniques mainly from the more ancient civilizations. The chief source of our information must be systematic chemical and physical examination of ancient materials. This is a slow and laborious process and, though much work has been done on it, we are still ignorant on many fundamental matters. Most of the older technical practices were handed on to the Middle Ages by tradition, but descriptions of some small part of them are to be found in written records. There exist a few early collections of recipes of a chemical character deriving from the Greek and Roman tradition.

By far the largest of these collections is that made by Pliny (d A.D. 79) in his *Historia naturalis*, an immense work divided into 37 books [1]. Especially in the last seven books there are scattered numerous recipes and descriptions of chemical processes. These include many in vogue in his day, and many others which were, even in his time, of purely historical interest. Pliny was neither scientific, nor critical, nor possessed of any practical knowledge; yet his work is incomparably our best single authority on ancient technology.

A certain amount of the traditional chemical technology of the Egyptian craftsmen had been accumulating in Greek-speaking Alexandria, perhaps from as early as the third century B.C. [2]. At least one work containing such material was available to Pliny. There survive also certain papyri containing 'formulae' written in Greek, probably at Alexandria and perhaps in the second or third century A.D. They contain much older material, and thus also represent the Egyptian technical tradition. With these must be classed several manuscripts written in Greek in the tenth or eleventh century but based on texts contemporary with the above papyri. They include many goldsmiths' and dyers' recipes. There is also the *Liber ignium* of the Byzantine Marcus Graecus (*c* 1250) [3].

Of Latin medieval craftsmen's handbooks the earliest is the text known as *Compositiones ad tingenda*, 'Recipes for colouring' [4]. Its relevant parts were collected at Alexandria about A.D. 600 and were translated into Latin some 200 years later. Despite its pre-Muslim origin it shows some Arabic influence. Certain of its terms for dyestuffs are Arabic or Persian, such as *luza* for weld, *lulax* for indigo, and *lazure*, whence our 'azure'. Another text, the *Mappae clavicula*, 'Key to painting' [5], was compiled about A.D. 800 in south Italy; many of the materials it mentions are oriental, from such sources as Alexandria, Ethiopia, Persia, and Cappadocia.

The *Schedula diversarum artium*, 'Notes on various arts', of Theophilus [6] was put together in Germany *c* 1110–40, probably from much older material. There is some evidence that its putative author was a Byzantine monk who migrated to central Europe. The *De coloribus et artibus Romanorum* of Heraclius [7] is a craftsman's handbook of about the same date as the preceding, or a little later. Finally, among the early medieval works there is a craftsman's manuscript written in Catalonia about 1130 [8], with many recipes for pigments, dyestuffs, inks, varnishes, and glues.

When we come to the end of the twelfth or beginning of the thirteenth century, we pass into a very different literary world, which is richly supplied with translations, versions, and adaptations of Arabic technological works. These continue to appear until the time of printing in the fifteenth century, and beyond. Many of them are alchemical or have a strong alchemical tinge. We are not here concerned with the history or nature of alchemy as such. So far as the west is concerned, it is a product of contacts with the Arabic-speaking world [9]. It is of interest, however, in the history of technology that the immense and unorganized number of alchemical experiments led to a development of apparatus from which that of scientific chemistry was at long last evolved (ch 21).

If there is a dearth of literary material on the crafts until the thirteenth century, from then on the number of technological documents is so overwhelming as to be an embarrassment to the historian. The great European libraries contain many thousands of manuscript collections of recipes dating from about 1200 to 1500, and Britain alone has several hundred [10]. The settled medieval way of life had matured and brought wealth and comfort. The numbers and literacy of craftsmen had increased, and many of their recipes were repeated for generation after generation. Formulae inherited from antiquity remained mixed in the collections with those of later introduction. It would be impossible to discuss the innumerable recipe-books of the thirteenth to the fifteenth centuries. Two must suffice, one dating from the beginning, the other from the end of this period.

The *Liber sacerdotum* is a highly typical product of the early Latin scholastic period [11]. It contains over 200 technical recipes. Many go back to Alexandrian practice in the early Christian centuries, and some are linked with yet more ancient Egyptian methods and provide evidence for the continuity of technical traditions. Its recipes were culled by an Arabic writer from works of many periods, and the substance of the collection was finally translated into Latin in the first part of the thirteenth century. The text is full of outlandish terms, many of badly transliterated Arabic and some of corrupt Greek that had passed through Arabic. It even contains a little dictionary of Arabic–Latin metallurgical and chemical terms. Its recipes divide naturally into four classes, the members of which are grouped together by the compiler himself:

(*a*) Directions for 'transmutation' of 'base' into 'precious' metals.

(*b*) A small number of recipes for soldering, somewhat analogous to those of the *Mappae clavicula*. One example may be thus rendered:

> A solder of copper. Take 3 parts of tin, one part of best lead. Melt the tin; add the lead; work it like silver; pound it if necessary; smear with it the work which you want to solder; add to it water of borax; melt the powder on a slow fire.

(*c*) A series of operations with the 'sulphurs' (p 736) of the metals (that is, marcasite, &c.), tutty (Arabic *tutya*, zinc oxide), vitriol, kohl (Arabic *kuḥl*, stibnite or galena), cinnabar (Arabic *zinjafr*), litharge, refuse from the extraction of gold and silver from their ores, 'vermilion' (pp 361, 366), minium (p 361), orpiment, pumice, &c. Here is a single example:

> How to prepare tutty. Tutty must be warmed at the fire until, through the fume of the fire, it takes on a golden-yellow colour. Then burn it and pound with best vinegar (the golden-yellow is very strong); let dry. The heating and pounding with vinegar must be repeated seven times.

(*d*) A number of obscure operations involving organic substances with which we are hardly here concerned.

From the first half of the fifteenth century there is a craftsman's handbook carefully put together by the skilled craftsman and painter, Cennino Cennini (*c* 1370–*c* 1440) [12]. It is a detailed account of the whole craft of painting and is dedicated to the memory of Cennino's master, Agnolo Gaddi (d 1396), to

Agnolo's father Taddeo, and to Taddeo's master, the great Giotto (1267?–1337). It presumably reports their methods, and contains hundreds of directions both for painters of pictures and for house-decorators, classes less sharply distinguished then than now. Some conception of its approach may be given by an extract:

The Way to Paint a Drapery in Fresco [i.e. on a freshly prepared surface which has still not dried out]. If you would paint a drapery, you should first draw it carefully with your *verdaccio* [crayon]; your drawing not showing too much, but only moderately. Then get three little dishes. Take one, and put into it whatever colour you choose, we will say red: take some *cinabrese* [a red pigment] and a little lime white; and let this be one colour, well diluted with water. Make one of the other two colours light, putting a great deal of lime white into it. Now take some out of the first dish, and some of this light, and make an intermediate colour; and you will have three of them. Now take some of the first one, that is, the dark one; and with a rather large and fairly pointed bristle-brush go over the folds of your figure in the darkest areas. Then take the intermediate colour; lay it in from one dark strip to the next one, and work them in together, and blend your folds into the accents of the darks. Then just using these intermediate colours, shape up the dark parts where the relief of the figure is to come, but always following out the shape of the nude. Then take the third, lightest colour, and just exactly as you have shaped up and laid in the course of the folds in the dark, so you do now in the relief, adjusting the folds with good draftsmanship and judgment. When you have laid in two or three times with each colour, never abandoning the sequence of the colours by yielding or invading the location of one colour for another, except where they come into conjunction, blend them and work them well in together. Then in a fourth dish take yet another colour, lighter than any of former three; and shape up the tops of the folds, and put on lights. Then take some pure white in yet another dish, and shape up definitely all the areas of relief. Then go over the dark parts again, and around some of the outlines, with straight *cinabrese*. You will now have your drapery, systematically executed, but you will learn far better by seeing it done than by reading [about it]. When you have finished your figure or scene, let it dry until the mortar and the colours have dried out well all over (figure 326).

FIGURE 326—*Delineating drapery for a lady's portrait. From a Flemish fifteenth-century manuscript. The painter works on a wooden panel. His assistant has mixed the ingredients of the paints in little dishes and is covering a second wooden panel with gesso for another painting in the manner suggested by Cennino Cennini.*

III. ALKALIS, CLEANSING-AGENTS, ACIDS

Soda and *potash* were in great demand as detergents and also for making glass, glazes, and soap. Their sources were native sodium carbonate, plant ashes, and calcined tartar.

Crude sodium carbonate or sesquicarbonate is found in quantities in various places in Egypt and notably in the famous Wadi Natrun which runs in the Western Desert parallel to the west limb of the Delta and about thirty miles from it. This native carbonate was known as *nitrun*. Of this word *natron* is a seventeenth-century European variant of the Arabic *naṭrūn*; hence *natrium* and our symbol for sodium. Important in remoter antiquity (vol I, pp 259–60, 268, 270) sodium carbonate was little employed in medieval Europe, though potash was in great demand.

FIGURE 327—*Scraping tartar from the inside of a wine cask. From a printed book, Strasbourg, c 1497.*

Plant ash or 'pot-ash' contains a workable proportion of potassium carbonate with a little sodium carbonate, unless it is derived from salt-marsh plants or seaweeds, in which case the proportion of the sodium salt is much larger. Wood-ashes were the normal material for making glass, lye, and soap in north Europe, and were even imported into England from the Baltic countries. For glass-making the ashes of beechwood were recommended by Theophilus, but those of the bracken fern were especially esteemed and in England are said to have been used exclusively. For cleansing purposes, the ashes were extracted with water, yielding lye. For soap-making, lime was added to the wood-ash and water allowed to percolate through the mixture; the potassium carbonate was thus converted into caustic potash ('sharpening the lye'). The caustic solution was then boiled, or in some recipes merely stirred, with oil or fat.

In the Mediterranean area there was in general use a supposedly superior product, namely the fused ashes of a low woody shrub, probably *Salsola soda* L. This, when imported from Syria and the Levant, was called polverine or *rocchetta*; when brought from Spain, *barilla*. It contained up to 20 per cent of sodium carbonate and was effective for all purposes for which wood-ash was used; moreover, it contained more alkali and had the additional advantage of yielding a hard soap with fats, rather than a soft.

Tartar or *argol* is the deposit in wine-casks (figure 327). It consists chiefly of potassium hydrogen tartrate. When calcined it yields fairly pure potassium carbonate. Its chief use seems to have been medical, though it had some application in mordanting. The dregs of wine were sometimes dried and carbonized, giving a product (*cineres clavelati*) containing a high proportion of potash.

Ammonia was available in quantity in the form of stale urine. This was much used for the purpose of cleansing from grease, for making alum (pp 215, 368), and for pigments (p 361).

Lime, formed by 'burning' chalk or limestone, was familiar. The kilns were simple, usually holes in a hill-side or low tower-shaped furnaces (figure 323), charcoal being used as fuel; coal, where available, was preferred in the Middle Ages. Lime was used in great quantity for plaster, mortar, and *attractum*, the favourite mixture of lime, sand, and stones for filling the interior of thick walls. Mixtures of lime and sand were the chief medieval cement, but gypsum cements were also used. In Italy and the Mediterranean islands, pozzolanic cements (p 407), which would set under water, were made from volcanic tuff, lime, and sand; they have been used from classical antiquity to the present time. *Pozzolana* derives its name from the port of Pozzuoli, near Naples (figure 473).

Cleansing-agents were used for personal hygiene and in the textile industry. The Greeks and Romans of the earlier centuries cleaned their persons by means of oil with or without mechanical agents such as bran, sand, ashes, juices of certain plants, and pumice. They were fond of hot, cold, and vapour baths. Various earths or clays were used for the hands. Soap, in our modern sense, seems to have been unknown to them; it may have been a Teutonic or Tatar invention. It became known about the beginning of the Christian era, and was at first regarded rather as a cosmetic or medical preparation (see below).

Clothes were cleansed in antiquity with fuller's earth (a hydrated aluminium silicate) and alkali. The former absorbs fat, the latter forms soluble compounds with it. The alkali usually employed was urine that had become ammoniacal, but sometimes lye from natron or wood-ash was used. The action of these substances was aided by mechanical beating in fulling-mills (figures 187, 553). These remained in use until the close of the Middle Ages, and for long after soap had become a familiar article of commerce.

Soap is of obscure origin. The Latin word *sapo* from which it comes is used by Pliny to describe a pomade invented by the Gauls. As this came to be extensively imported into Rome it is not safe to translate *sapo* by 'soap'. The first reference in which this rendering is fairly certain is of *c* A.D. 385 by the physician Priscian, who speaks of it as used for shampooing. By 800 its manufacture was a common domestic craft in Europe, and from then on we hear frequently of the calling of soap-boiler. From the twelfth century, at least, soap was produced in large quantities, and became a major export from several countries, notably Scandinavia. At first soap was made from animal fats and had a very unpleasant smell. This was overcome by the discovery that olive-oil could be used for the purpose, after which much of the trade passed to Spain and the Mediterranean region.

Hard soap was made around the Mediterranean from olive-oil and soda-ash (*rocchetta* or *barilla*). It was first produced by the Arabs and later made especially in Castille, Marseilles, and Venice, and exported to the northern countries. It was often perfumed and was regarded as an article of luxury.

Soft soap is typically a potassium compound. In the northern countries great quantities of it were made for textile-cleansing, at least as early as the twelfth century. The alkali used was caustic potash lye made by running water through layers of wood-ash and lime. The fat was tallow or whale-oil or other animal fat. The soap was of industrial application rather than a luxury article. Recipes of the sixteenth century indicate that some soap was made by simply stirring together hot lye with oil and letting them stand until they had thickened. Some was made by boiling fat with weak lye until saponified. The process of 'salting-out' is not recorded until the seventeenth century.[1]

Nitric acid. The notion of acids as a class was not developed until the sixteenth century, although certain acids had been familiar from antiquity. There is no Latin description of the preparation of nitric acid before the *Summa perfectionis* of Geber [13], though such a description is given by the Arabic Jābir. Nitric acid is also mentioned in treatises falsely ascribed to Ramón Lull (1235?–1315), but written after 1330. It was made by distilling nitre (p 370) with sulphates of aluminium, copper, or iron, previously partly dehydrated. A high temperature was required, and the distillation-apparatus had to be resistant to heat and acid vapours. The industrial use of the acid was in the separation or 'parting' of gold from silver, the silver dissolving and the gold remaining.

Aqua regia. Geber says, correctly, that the addition of sal ammoniac to nitric acid enables it to dissolve gold. This mixture, *aqua regia*, containing hydrochloric as well as nitric acid, could be used to separate silver from gold. The gold dissolved to a soluble chloride, while silver was attacked and precipitated as an insoluble chloride. The gold could readily be recovered by evaporation of the

[1] Oils and fats are essentially esters of glycerol with fatty acids. When an alkali such as soda is made to react with a fat, the sodium salt of the fatty acid is formed and glycerol liberated. In the earlier methods of soap-manufacture the glycerol remained in the product. It has, however, no detergent action and can be removed by adding salt to a soap solution. This throws the soap out of the solution, and leaves a proportion of glycerol in the brine. The operation may be repeated until the major part of the glycerol has been removed. Nowadays the glycerol is recovered, as a valuable by-product.

liquid and heating the residue, while the silver could be obtained by smelting the chloride with an alkali. Nitric acid was suitable for separating small quantities of gold from silver, and *aqua regia* for separating small quantities of silver from gold.

Sulphuric acid has a history that is far from clear. There is hazy evidence that it was made by Arabic-speaking alchemists as early as Jābir (p 736). However, it is not certainly known to have been isolated before *c* 1535. Thereafter it was made both by burning sulphur and condensing the traces of acid so formed, and by the distillation of green or blue vitriol (iron or copper sulphate). It did not become of industrial importance until the seventeenth century. Nevertheless, the acid had been employed in vapour form for many centuries, both in the east and in the west, in the cementation process for purifying gold by heating it with vitriol and salt.

Hydrochloric acid similarly had been produced, but was not isolated or recognized until the seventeenth century. Organic acids such as vinegar (*vin-aigre*) from wine turned sour, or merely the juice of sour grapes (verjuice), have been in general domestic use from an early date (vol I, pp 284–5).

IV. CERAMICS AND GLASS

The red and black ware of Greece and Rome was an indigenous Mediterranean invention (ch 8). Most later developments of the art in Europe were, however, prompted by the desire to imitate eastern products, notably those of the great age of ceramics in Persia and China, which corresponds roughly to our earlier Middle Ages (ch 8). The typical red and black ware had a polished surface without true glaze. It was made from a red-burning clay and was decorated with clay slip liquefied by addition of a little alkali and organic matter. Its colour depended on the conditions of firing—reducing or oxidizing. These could be controlled at will (p 266), though the chemical nature of the changes was, of course, not understood.

The secret of this ware was lost after the fall of the Roman Empire. The first important western ceramic development of the Middle Ages was the lustre-ware in Moorish Spain, the knowledge of which probably came from the Near East about the eleventh century (pp 303 f). Its glaze was based upon oxides of tin or of tin and lead, made by calcining the two metals separately or together. Cobalt, iron, and copper were used to give colours to the glazes, and a thin film of iridescent metal or metallic sulphide was imposed by coating the appropriate parts of the ware with a mixture of copper and silver sulphides, ochre, and vinegar, followed by firing in a reducing atmosphere. The whole process involves a very difficult and refined technique with a long history. In general, medieval ceramic art was deeply influenced from the Near East (ch 8).

Majolica, a ware characterized by an opaque white glaze containing tin, was developed in the fourteenth century and is probably named after the island of Majorca (Mallorca). The basic glaze was of potassium carbonate (calcined tartar, p 354) and pure white sand. For the white enamel, tin oxide was added. Green colours were given by copper compounds, yellows by antimony and lead, darker shades by red iron oxide (rust or *crocus Martis*), blues by *zaffre* (a cobalt mineral, see below), and violets by manganese dioxide.

Glass, though known far back in Egyptian times (p 311), owes its special development to the invention of the glass-blowing pipe, which ranks among the great technical achievements of history (figure 309). Blown glassware first appears about the beginning of the Christian era. Methods of glass-making described by Theophilus, *c* A.D. 950, probably differed little from these (ch 9).

Medieval glass was of two chief types, the potash-glass made with wood-ashes, and the soda-glass made with *rocchetta* (figure 310) or occasionally natron (p 354). These were fused with the whitest sand obtainable. Much medieval glass is coloured greenish or brownish owing to the presence of iron. At a later period this colour was removed by adding manganese dioxide, the 'glass-maker's soap'. In general three furnaces were employed, one for melting the sand and alkali, a second for working the glass, and a third for annealing. Glass plates, as for church windows, were made by blowing large bulbs, splitting them with the hot iron, and opening them out in the furnace as described by Theophilus. The work of the Florentine priest Antonio Neri on glass-making (1612) [14] shows the great care taken by the later glass-makers of Italy and the Low Countries to obtain a colourless product. Neri's alkali was carefully recrystallized and his silica was powdered white quartzite.

The production of coloured glass for jewelry had been well understood even in Egyptian antiquity. Coloured blown glass was used for church windows at least as early as the twelfth century. Blue glass was coloured with *zaffre* (? from the Arabic), a material made by roasting sulphide and arsenide ores containing cobalt, giving oxide of cobalt with various impurities. For green glass, copper was used, the metal being heated with sulphur and the resulting black sulphide roasted until it formed red cuprous oxide ('burnt copper'). Alternatively the metal was heated with copper sulphate, giving a basic copper compound. The dross formed when brass was melted or heated was used for the same purpose. Red glass also was made by the use of copper, but under reducing conditions. To obtain the finest reds, gold chloride, made from gold and *aqua regia*, was often used, as it is today. Yellows and browns were made by adding iron, usually as the red oxide or as various products of the action of acids on iron. A particularly fine yellow was given by metallic silver. The church glass of the twelfth and thirteenth centuries is witness of the early discovery of these techniques.[1]

V. PIGMENTS, VARNISHES, GROUNDS, AND BINDING-MATERIAL [15]

Colouring-matters, as used industrially, can be classified into three groups: pigments, dyes, and lakes. Pigments colour surfaces without penetrating much, if at all, below them. They are normally applied either as washes suspended in water, or as oil paints. Dyes and dyestuffs (substances producing dyes) differ from pigments in that when dissolved—normally in water or some aqueous medium—they penetrate the material involved. They are used especially on woven fabrics or skins. Lakes are pigments of a special type. They are derived from the interaction of dyestuffs and carriers, by which the colouring matter becomes insoluble. The process of rendering dyestuffs insoluble within fibres is known as mordanting, the principal mordant being a solution of alum. Many pigments and dyestuffs used in ancient and medieval times were native substances that required little treatment before use, except the removal of impurities and the requisite degree of grinding.

The ancient Egyptians painted their buildings and statues in stock colours, of which they had about fifteen. They made up their pigments with water and gum tragacanth or gum acacia. The surface of the painting was sometimes waxed, less often varnished. Their brushes were reeds soaked in water until the ends split into fine fibres. The Greeks, of whose painting we have hardly a trace, seem to have worked in much the same way. They also painted in wax, which was used particularly for boats. The tradition of painting with pigments in wax fixed by heat, 'encaustic', persisted into Byzantine and Roman Imperial times, when drying-oils and varnishes, as well as naphtha, were also used in the process. We have, however, woefully little information on the pigments used in classical times, and most of what we have to say on pigments refers to the Middle Ages, which doubtless inherited classical methods.

For *black pigment* the normal substance was carbon, either lamp-black or charcoal. Lamp-black was made by burning various oils, waxes, or resins and collecting the soot on a suitable surface. The finest charcoal was made by heating various plant-materials in earthen pots (figure 328).

Inks, among black pigments, have a special application. Two kinds were used from antiquity and through the Middle Ages. In one, the blackness was deter-

[1] The general history of glass is discussed in greater detail in ch 9.

mined by minute particles of carbon and is therefore permanent. In the other, the blackness was due to a gallate or tannate of iron which becomes black on exposure to air, owing to oxidation, but which is further oxidized into a brown in the course of time.

FIGURE 328—*Making fine charcoal for ink or other pigment. A supervisor watches workmen casting thin sticks into a tile-lined tun where they are burned; behind is a covered tun. German fifteenth-century manuscript.*

Lamp-black provides the carbon for the first of these types. Soot was collected from a cold surface on which was made to play a flame from an oil-lamp or a candle of beeswax or tallow. The fine-grained deposit was held in suspension with a watery gum or size. Another favourite form of carbon for inks was burnt vine-shoots, but its manufacture was not without difficulties. Burnt too much, the shoots were reduced to ashes and produced a correspondingly pale result; burnt too little, they retained certain tarry substances which yielded a brownish result. The carbonization of such material needs careful exclusion of air.

The other main type of ink was made by mixing crushed galls, usually of oak, containing much gallic acid and tannin, with one of numerous iron salts, usually ferrous sulphate. This was pale brown at first and was thus often coloured darker before use. The two types of ink were not infrequently mixed.

Brown pigments. That various argillaceous iron ores, when burnt, give good browns had been known since remote antiquity. The medieval painter usually made his browns by mixing other colours. In general, brown was less of a favourite in the Middle Ages than with the Renaissance artists and technicians.

Of *white pigments* the most important was white lead or ceruse (basic lead carbonate), a material first described by Theophrastus (d *c* 287 B.C.), the pupil and successor of Aristotle. The earliest process for making ceruse was to hang lead plates in earthen pots containing a little vinegar.[1] After a time the white deposit, a mixture of carbonate and acetate, which had accumulated on the lead was scraped off, finely ground, and boiled. No important improvement was made in this process until the sixteenth century when the reaction was hastened by keeping the pots warm in a bed of damp straw or dung. Bone-white, made by calcining bones and consisting mainly of calcium phosphate, was also used in

[1] In the presence of carbon dioxide some of the lead acetate first formed was converted into carbonate. Fermenting organic matter was always present in later processes.

the Middle Ages. Suspensions of chalk such as whitening, and milk of lime or lime-wash, were in common use for walls.

Of *red pigments* the commonest at all times was red ochre or ruddle, a clayey ironstone. It is very long-lasting and is still used. In classical antiquity red ochre from Sinope on the Black Sea had a special reputation which it retained through the Middle Ages, though much so-called *sinopia* had never seen Sinope. Red ochres from different sources vary greatly in colour. The red pigment *lac* is discussed below (p 362).

Red lead was known to the Greeks and Romans. It was made, as it sometimes is today, by roasting white lead exposed to air. It is not altogether permanent. Mercuric sulphide, generally called vermilion (p 366), either found native, as cinnabar (HgS), or made artificially, was the most brilliant of red pigments. In the Middle Ages it was made by grinding together mercury and sulphur, giving the black form of mercuric sulphide (*aethiops mineralis*); this was then sublimed, the sublimate forming as the red variety. Red lead is extremely stable if not exposed to strong light, which darkens it.

Red lead is now known as minium, but that word, both in classical times and in the Middle Ages, was used also for several other colouring substances of reddish or orange hue, including cinnabar and massicot (PbO). Many of these were used in decorating manuscripts, hence our word 'miniature'.

The *blue pigments* used by the Romans included azurite, a blue basic copper carbonate which continued to be used during the Middle Ages. Smalt (*smalto, esmaltus*), which is a dark blue cobalt glass, ground fine, was used from the end of the Middle Ages and was made by heating ores of cobalt with alkali and sand. The mixture on fusion separated into an upper layer of a blue cobalt silicate glass and a lower of metallic sulphides. The finest blue pigment of the Middle Ages was ultramarine. It was made by grinding lapis lazuli and mixing it with wax, oil, and resin to form a soft mass. This was kneaded for long periods in weak alkali (lye). The finely divided ultramarine found its way into the liquor, while the impurities adhered to the wax. The very costly ultramarine later settled out.

Certain other blues, though their lakes were used as pigments, are more naturally considered as dyes. Such were indigo, woad, and turnsole. The Romans used indigo intensively as a pigment, but not as a dye (p 365). The commonest blues, however, at least in the Middle Ages, were the artificial copper blues, for the making of which there are endless recipes. Most of these depend on the readiness of copper compounds and ammonia (stale urine), to react, yielding deep-blue solutions. If to these solutions lime be added fine blue compounds are formed, which may be obtained solid. These cuprammonium blues are not permanent.

Green pigments were mostly derived from copper. Malachite, another basic copper carbonate, was employed even in ancient times (vol I, p 293), and was well known to the Romans. The same is true of various copper frits (coloured vitreous compounds ground up). During the Middle Ages verdigris, a basic acetate of copper, was made by moistening copper with vinegar and exposing it to air, or by packing copper plates in fermenting grape-skins. The action of vinegar and salt on copper gave basic copper chloride, salt-green. Other green pigments were *terre verte* or *verdaccio*, a green earth, and those from the juices of various plant materials, such as buckthorn berries or iris flowers (p 366).

Yellow pigments. Yellow ochre (a clayey iron ore) has been used as a pigment in all ages. The Romans made much use of orpiment (arsenic trisulphide), which has a fine golden colour. The Greek word for orpiment is *arsenicon*, hence the name of the element. The medieval painters preferred ochre or organic pigments, such as saffron, weld, and Persian berries (*Rhamnus saxatilis*). Finely divided gold, made by grinding gold leaf with honey or salt, was much used. A good yellow pigment, used by painters from the later Middle Ages onward, was known as 'Naples yellow' (*giallolino di Napoli*). Its composition varied. Sometimes it was a mixture of oxides of lead and antimony.

Varnishes. Much painting, especially on wood, was protected by varnishing. For this purpose various resins were steeped or heated in oils that had long been exposed to the sun. Those mostly used for the purpose were linseed-oil, hempseed-oil, and walnut-oil. All tend to darken with time.

Layering of paints. Many of the results attained by the great painters were reached by adding one coloured layer to another. This process of layering can be examined by embedding fragments of a painting in a transparent medium and cutting sections of them. These may then be studied microscopically and submitted to chemical tests.

Lacquer—otherwise shellac—is a special type of varnish. The substance, like the word, is Indian. It was reaching Europe from the tenth century as 'gum-lac' and 'stick-lac'. It is the secretion of female coccid insects, notably *Tachardia lacca*, parasitic on certain trees. A widely used red pigment was precipitated from shellac by alum and became known as 'lac', whence our much more generally applied 'lake'. The similarity and overlap in sound and meaning of lacquer, shellac, lac, gum-lac, stick-lac, lake, lac-lake, and their variants cause much confusion. Ancient, like modern, lacquer-work was made by painting layer on layer of a solution of shellac on a wicker or light wood base, sometimes first covered by a very thin layer of fine clay.

Surfaces. Ancient and medieval painting of the major sort, both of the coarser variety as used in building and of the finer adapted for furniture, decoration, and representation, was almost always upon wood-panelling or walls of stone or brick. Both in classical and in medieval times people had a great love of colour but less appreciation of texture than is perhaps usual among ourselves. Thus neither the grain of wood nor the surface of stone attracted their admiration. Both were customarily concealed by paints where not exposed to weather.

The wood-panelling for finer work had to be jointed and smoothed with special care. To obtain large flat wooden surfaces it was necessary to joint very firmly. For this purpose an extremely strong and effective cement was formed of cheese crumbled into lime and a little water and allowed to dry slowly. The surface was carefully smoothed, and its permanency and continuity were often further ensured by gluing a sheet of linen over it. Over this again was spread a layer of gesso, a white paste of chalk, gypsum, or plaster with glue or gelatine as binding-material. By varying the components and rate of drying, the fineness and consistency of such a composition can be controlled. Gesso can be made of a hardness and grain suitable for moulding and carving. It was the common basis of most medieval, and of some ancient, painting (figure 326).

Canvas, that is hemp or linen cloth, now the common ground for oil-painting, was little used until the fifteenth century. It is unsuitable for outdoor work, and the earliest large-scale painting on canvas is said to be Botticelli's 'Birth of Venus', *c* 1480. Painting on cloth was common for objects not designed for durability, notably processional banners, widely used in the Middle Ages.

The main surface for decoration and portrayal both in Roman and in medieval times was the wall. Of Roman painted walls well known are those of Pompeii of the first century A.D., of the Villa Livia of some decades earlier, and at the farthest eastern limit of the Empire, those of Dura-Europos, on the Euphrates, of the third century A.D. Of medieval mural paintings there are innumerable examples in churches, abbeys, and cathedrals throughout Europe. The surface was usually dressed with a mortar of sand and lime. When this dries the lime forms crystals which both enclose the sand grains and penetrate into the wall, thus binding them to their background. No combination takes place between the lime and the sand, but the lime slowly absorbs carbon dioxide and becomes calcium carbonate. The mixture of carbonate and sand carries the painting, but this is very sensitive to moisture, especially that seeping through from the wall. Painted walls were therefore usually elaborately protected against damp.

Tempera. Pigments in the pure state were seldom used in mural painting. For that purpose most pigments require a binding-medium; this was usually mixed with the pigment. The binding-media hold the granules of the pigment together and in place, but they also have some effect on the impression created by the pigment itself; they 'temper' it. Such paintings are spoken of as tempera paintings and the process of mixing as tempering. Among the substances used for this process are gums, size, mortars, alumina, oils, glues, chalk, glair (whipped white of egg in water), egg-yolk, lime, honey, and even ear-wax. Under modern conditions it is possible, with even a minute fragment of a painting, to examine the various layers microscopically and to give a chemical and physical analysis of each.

FIGURE 329—*Dyers, from a mural painting at Pompeii c A.D. 70. Fabrics from the heated dye-vat are wrung out on trays from which the dye drips back. The surface of the trays is of some light impervious material, such as sheet bronze.*

VI. DYESTUFFS, MORDANTS [15]

The practice of dyeing goes back to remote antiquity (vol I, pp 245–50). Dyes in use in Graeco-Roman times and in the Middle Ages were almost entirely of vegetable or animal origin. They are most conveniently discussed under the colours that they produce in textiles. Most of them are not fast except with a mordant which, throughout the period under discussion, was normally ammonium alum, potash alum, or a mixture of the two. The dyer's craft was highly technical, many of his materials were imported and costly, and many of his processes remained trade-secrets (figures 329–31).

FIGURE 330—*Dyers preparing a figured fabric. From a French manuscript of the thirteenth century. The man on the right withdraws the dyed cloth from the washing vat, while he on the left displays the pattern which has been previously painted with alum solution.*

For *blacks* the only methods of dyeing were either to add extract of galls to iron sulphate (green vitriol or *atramentum*) or to superimpose several dark shades on the same material.

Of *blues* the most important dyestuff was indigotin,[1] derived either from the indigo plant (*Indigofera tinctoria* L) or from woad (*Isatis tinctoria* L). Indigotin is insoluble in water. It is obtained from its vegetable source by natural fermentation or bacterial reduction to the colour-

[1] To avoid confusion the modern term *indigotin* is here used for the chemical compound responsible for the blue colour, irrespective of whether it has been obtained from indigo or from woad plants.

FIGURE 331—*Bleaching and dyeing silk. From a manuscript 'Treatise on the Silk Art' written at Florence in 1458. On the left two men stir a cauldron of heated solution of sulphurous acid in which skeins of silk are bleached. The men wear wooden clogs and one holds his nose. The skeins, after a process of washing and aluming* (not shown), *are immersed in a heated bath of kermes solution* (centre), *allowed to drip dry on the two rods across the bath, and finally washed in cold water* (right) *to test for fastness.*

less and soluble compound indoxyl. Subsequent oxidation of the indoxyl to indigotin occurs on exposure to air. The fabric is steeped in the fermenting solution and hung out in the air, whereupon the deep blue appears. The genus *Indigofera* is tropical. The Indians fermented it in water, and the subsequent action of air on the solution precipitated insoluble indigo, which was exported. The Romans do not appear to have known how to bring indigo into solution, and employed it only as a pigment (p 361). But medieval dyers, in the mid-fifteenth century at latest, were reducing it to indigo-white with honey and lime, and were dyeing cloths with this solution.

Woad, on the other hand, was extensively grown throughout Europe. The young leaves were plucked and milled to a pulp (figure 325). This was formed into balls (figure 332) which, after a few weeks of drying, were powdered and allowed to ferment with water until they yielded a dark mass that was rammed into air-tight barrels and so sold. For use, the woad was fermented afresh in the heated woad-vat, and the cloth was dyed as described (figure 333). The indigotin content of the woad-plant is relatively small, and from the sixteenth century onward woad increasingly gave place to indigo. A traditional practice was to add indigo to the woad-vats, to be reduced in course of the woad-fermentation.

FIGURE 332—*Merchants selling balls of woad from sacks. Thirteenth-century French statuary.*

Of *yellows* the most important in antiquity and the Middle Ages was weld (*Reseda luteola* L). An extract gave a pure yellow when mordanted with alum. Other yellow dyes included safflower (*Carthamus tinctorius* L) and (young) fustic (*Rhus cotinus*).

Greens were normally produced on cloth by dyeing first with weld and then with woad. Some vegetable green dyes were used, but were not fast. A green colour could be attained by boiling fabrics with verdigris (p 362) and alum.

For *reds*, madder, obtained from the roots of one of the several species of *Rubia*, was the most commonly used dye. The genus is widely distributed in the Old World. Species were cultivated near Rome in classical antiquity, and its culture became a speciality of the Low Countries in the Middle Ages. Late in this period a very complex method involving madder and producing a vivid red was invented in the Near East. Goods dyed this famous 'Turkey red' were imported into Europe until the nineteenth century. The full secret of the process is still obscure. Mordanted with alum, madder dye gave a good red, somewhat inclined to purple. Another red dye was archil, made from Mediterranean maritime lichens (*Roccella tinctoria* and others) treated with stale urine and lime. All these gave rather unstable reds of a crimson or rose shade.

FIGURE 333—*Dyers stirring a heated woad-vat. In front lies a roll of dyed cloth. A youth displays another roll. From a fifteenth-century French manuscript.*

The finest and brightest reds, both in antiquity and in the Middle Ages, were from certain species of 'coccus' (hemipterous insects parasitic on various plants) mordanted with alum. The insects had several names. The ancients termed them *cocci*: to the Arabs they were known as *kermes*, whence our words 'crimson' and 'carmine'; medieval Latin writers described them as *vermiculi* (little worms), whence 'vermilion', and *grana*, whence 'grain' and 'dyed in the grain'.[1] In the sixteenth century a coccus (*Dactylopius*) parasitic on New World species of cactus (*Opuntia*) was discovered to yield a good red, and 'cochineal' from this source soon replaced all others. A true and permanent scarlet, however, was not obtained until the early seventeenth century, when Cornelius Drebbel (1572–1634) introduced tin salts as a mordant.

[1] How the red roses flush up in her cheeks,
And the pure snow with goodly vermeil stain
Like crimson dyed in grain! Spenser: *Epithalamion*, 226.

The above red dyes (except cochineal) were used from Roman times. In the Middle Ages there was added *brasil*; this is the wood of species of the tree *Caesalpinia*, brought originally from Ceylon and the East Indies but later also from Brazil, which was in fact named after it (*terra de brasil*).

Lac, *lacca*, much used by medieval painters, is discussed above (p 362).

Lakes—the word, though not the thing, is the same as *lac*—were made by precipitating soluble red dyes with alum. They were much favoured. Crimson lake was made by boiling scraps of fabric dyed with 'grain', a substance closely allied to cochineal (see above), with lye, and precipitating the colour with alum. *Brasil* and the madder plant yielded red extracts from which rose-coloured lakes could be precipitated. In antiquity, the Middle Ages, and later a great variety of lakes were in use. Most of them were very fugitive, the most permanent being a madder lake. Lakes, though obtained from dyes, are themselves pigments.

Purple occupies a special place among dyes. The famous Tyrian purple of the ancients is of a dark violet inclined to brown. The colouring matter is 6, 6′-dibromoindigo; it is derived from several species of whelk-like molluscs (*Murex* and *Purpura* spp), each of which secretes a few drops of a creamy liquid. This liquid in air and light changes through various shades to dark purple. The places principally associated with purple-dyeing in Greek, Roman, and Byzantine times were the ancient Phoenician towns of Tyre and Sidon. The secretion of the shell-fish was boiled and the cloth dyed in the liquor. The shade varied according to the proportions of the various species used. The final product could be topped with archil or kermes. True purple-dyeing was an extremely costly process.

The dyeing of purple in the Middle Ages was a secret of Byzantium and, after the fall of the city in 1453, the process became a lost art. There were various spurious purples prepared from early times and from various species of mollusc, yielding products similar to but not the same as the Tyrian or imperial purple. But a purple colour could, of course, always be produced by dyeing the same fabric first blue and then red. The medieval recipe books abound in substitutes for the real imperial purple.

Mordants—almost always aluminium sulphate—to cause dyestuffs to adhere to the textile fibres were in nearly universal use. Aluminium sulphate was normally applied as potassium or ammonium alum, but occasionally in its native form, then usually known as 'alum of Yemen'. Since any admixture of iron salts in the alum gives much darker shades the preparation of pure alum became important. The trade in it was large from the earliest classical times until the end of the nineteenth century.

Throughout antiquity and the Middle Ages any whitish astringent mineral substance that would act as a mordant was liable to be called alum, *alumen*, or *stypteria*. These terms therefore included not only aluminium sulphate, ammonium alum, and potassium alum, but some double salts of aluminium that are not alums in the modern chemical sense. The Greeks and Romans had access to potassium alum, $KAl(SO_4)_2.12H_2O$, which occurred native in certain volcanic regions, such as the Lipari Islands, Sicily, and Milos. They also classed as 'alum' the mineral alunogen (hydrated aluminium sulphate) and a number of minerals which contain aluminium sulphate together with compounds of iron, magnesium, or manganese. We have no record of attempts to prepare any pure material from any of these before the thirteenth century. At this period the Arab writers give us the first hint of methods of preparing an alum from alunogen or from native aluminium sulphate, that is 'alum of Yemen'. It was treated with stale urine containing ammonium carbonate and, when the liquid was concentrated, ammonium alum crystallized out. This was a much more convenient astringent or mordant than the native aluminium sulphate.

FIGURE 334—*A charcoal burner heaps logs. A similar heap to the left has been covered with turves and earth, leaving a few vent holes near the top, and then fired. Printed book; Venice, 1540.*

Though from at least 1450 alum was being prepared in Europe from the mineral alunite, and though the discovery of deposits of it is well documented, yet we have no account of the process involved until 1480. The mineral was roasted and weathered, and successive quantities of the product were boiled in the same water until the solution was strong enough to crystallize. On cooling, alum separated. The demand for alum in the Middle Ages followed the great increase in the textile and dyeing trades.

In the thirteenth and fourteenth centuries much of the alum came from the Greek islands and the Near East, and especially from Phocaea (Fogliari) near Izmir. After the Turks captured these regions, and more particularly after the fall of Byzantium in 1453, an alum-famine developed. It was opportune that vast deposits of alunite were then discovered at Tolfa within the papal territory in Italy. Attempts

FIGURE 335—*Nitre-gatherers removing efflorescence from stable-walls with a special type of brush. Late fifteenth century.*

were made to establish a papal alum monopoly. This was unsuccessful, but Tolfa was long the chief source of alum. The mines there are still active.

Alum is historically important as the first substance deliberately prepared in what the modern chemist would regard as a substantially pure state. This was not due to any chemical knowledge, but resulted from the dependence of alum for its effective use on its freedom from iron salts, the presence of which greatly detracts from its value as a mordant. Since these salts are coloured, since alum is one of the most easily crystallizable of substances, and since repeated crystallization eliminates the iron salts, the ancient method of getting rid of the colour in this way was in reality a process of chemical purification.

FIGURE 336—*A saltpetre works, Germany, 1580.* CCC *are beds of material gathered from sheep pens, mixed with decaying vegetable matter and rubble. The workman in the foreground is gathering the efflorescing nitre which is put into a wheelbarrow and leached out in vats in the hut* A. *The liquid is concentrated by boiling in the hut* B, *behind which logs are heaped as a fuel store.*

VII. COMBUSTIBLES

The principal fuels throughout the western world were wood, charcoal, and coal. Wood needs little comment, except that the depletion of forests consequent on its use has altered whole landscapes and initiated profound economic change. The burning of wood or charcoal produces ashes containing potassium carbonate, a main source of ancient and medieval alkali (p 354). Charcoal has been made in the same manner from remote antiquity to our own time, and was in very wide use until the eighteenth century. Logs are carefully stacked into a hemispherical heap (figure 334); this is covered with earth, or, better, turves; a hole is made in the centre and the heap ignited; and by opening or closing air-holes the mass can be evenly carbonized with the minimum wastage. The wood is in effect distilled, part of the resulting pyroligneous acid and wood-tar condensing and running to the bottom of the heap, where it was sometimes collected in drains. Wood-tar was used in considerable quantity for nautical purposes. Charcoal is an almost perfect fuel, yielding a high temperature with a minimum of ash and no smoke.

The use of coal for heating was known to the Greeks and Romans, but it was dug only from outcrops. In the Middle Ages coal, though not used domestically, was an important industrial fuel where available, as in England and the Saar. It was particularly used by dyers and brewers, who had to boil large quantities of

FIGURE 337—*Crystallizing saltpetre. Germany, 1580. The concentrated liquid is poured into the tall vat* A *to cool and is then emptied into* E. *Thence portions are run in, to be further concentrated over the furnace at* B. *This liquid is ladled out hot by the attendant* C *into a rectangular basket where some saltpetre is deposited in crystalline form. The rest of the liquid drips back and, ultimately, is crystallized in the open pans at* F *or in tubs* G *sunk in the ground.*

liquid, and by lime-burners (figure 323) and smiths who needed high temperatures. Coal was not used for smelting metals before the seventeenth century, nor is there any mention of coal-tar. Illuminants included fat, oil, resin, and beeswax. The tallow candle, rushlight, torch, and lamp were in use, though candles, sometimes burnt in lanterns, became especially popular in the Middle Ages. Petroleum, known over all the period, was not a customary fuel.

Incendiary and explosive mixtures are considered elsewhere (pp 374 ff). A very important element in their manufacture was saltpetre (potassium nitrate), and the discovery that this substance provided explosive and incendiary possibilities had far-reaching consequences. The earliest evidence for the manufacture of saltpetre is in Chinese records (p 377) before A.D. 1200, and Ibn al-Baiṭār of Malaga (d 1248) mentions it as 'Chinese snow'. Saltpetre was imported from India during the later Middle Ages, but was also extracted from European saltpetre deposits at least from about 1300 (p 379). These deposits consisted of earth from old, dry sheep-stalls and stables (figure 336), and incrustations on old walls and cellars (figure 335). Saltpetre was made by filling vats with alternate layers of saltpetre-earth, wood-ash, and lime, and trickling water through them. The resulting solution was concentrated by boiling, a little lye with alum being added. Common salt, contained as impurity, crystallized out from the boiling solution. The residual solution was run off, and on cooling yielded crystals of nitre (figure 337).

FIGURE 338—*Pharmacy from a thirteenth-century manuscript of an Arabic translation of Dioscorides.* Below: *an apothecary stirs drugs in a cauldron over a fire and fills a cup for the customer before him.* Above: *a store of jars and on either side a figure preparing their contents.*

Indian saltpetre was made in much the same fashion. The crude saltpetre, whether of European or Indian origin, still contained more or less common salt and other impurities, which caused it to become damp in a humid atmosphere. To fit it for use in gunpowder (p 380) it had to be recrystallized once or twice.

Sulphur. The method by which the Romans refined sulphur is not known. The Alexandrian alchemists distilled or sublimed it, but probably not on any large scale. In the sixteenth century the crude sulphur was distilled, the vapour passing into a large earthen pot, where it condensed to liquid sulphur which

FIGURE 339—*Pharmacy, from a thirteenth-century French manuscript.* (Left to right) *A seated figure weighs drugs over a round table on which are other drugs; a sitting figure pounds drugs in a mortar with two pestles; a container with drugs in small sacks; a standing figure stirs a mixture in a cauldron over a fire. Above is a series of waisted jars on a shelf.*

passed out through a spout near the base and then solidified (p 380). Sulphur was used as a medicament and for making incendiary mixtures. Sulphur matches are referred to in the early sixteenth century but may be older.

VIII. DRUGS, SUGAR, ALCOHOLIC LIQUORS, DISTILLATION

Drugs. The greatest part of the pharmacy of antiquity and the Middle Ages was mere grinding, solution, heating, or other physical treatment of plant substances (simples), but a few preparations involved chemical change (figure 339). Dioscorides (first century A.D.) (figure 338) explained in his *De materia medica* how to make several mineral remedies such as burnt copper, i.e. copper sulphide (p 358), and verdigris (p 362). He gives the first description of the extraction of mercury. His work was translated into Latin as early as the fourth century, and was extremely influential in the development of medicine. Even the current scientific pharmacopoeias contain many elements that can be traced back to it.

The drug-list *Liber servitoris* of the Spanish Muslim Khalaf ibn ʿAbbās al-Zahrāwī (Albucasis of the Latins, d 1013) was translated into Latin in the thirteenth century and also exerted much influence. It describes the preparation of

FIGURE 340—*Sugar-production. In the distance the canes in a sugar-plantation are cut down and loaded on an ass which brings them to the factory. There they are chopped into short pieces and shovelled into baskets which are emptied into the hopper of a water-driven mill. The product runs into a hand-press and the expressed juice is evaporated in cauldrons over fires. It is then ladled into cone-shaped moulds where it crystallizes. Finally the moulds are emptied and the saleable cones displayed. From Stradanus,* Nova Reperta, *Antwerp, c 1600.*

litharge, white lead, lead sulphide, copper sulphide, cadmia, the vitriols, crocus of iron, and other substances. A very complete work on the many drugs known in the Arabic-speaking world was that of Ibn al-Baiṭār (d 1248). Further progress with the preparation of mineral remedies is not apparent before the sixteenth century, when it came into fashion with Paracelsus (1493 ?–1541) and his school.

Carbohydrates and their products [16]. Sugar was the only foodstuff which involved any chemical preparation. Its existence was known in classical antiquity, but it was not much used in Europe until the period of Islamic influence of the twelfth century and after; the word sugar is itself from the Arabic. The Arabs cultivated the sugar-cane and introduced it into Palestine, Sicily, Spain, and the Greek islands. The juice was expressed and boiled until it set to a semi-solid mass. From this, the residual liquid was extracted by pressure. The date at which this crude sugar was first refined is doubtful, but is at least as early as the fifteenth century (figure 340). During the Middle Ages sugar was a luxury, the usual sweetening and preserving agent being honey.

Fermented liquors (ch 4) and *distillation* (ch 21) are considered elsewhere.

REFERENCES AND BIBLIOGRAPHY

There is no work on the general subject of chemical methods of the Classical and Middle Ages. Of the many sources here employed, the following may be mentioned:

[1] PLINY *Nat. hist.* For everything in this work related to chemical industry and technology the best reference book is K. C. BAILEY. 'The Elder Pliny's Chapters on Chemical Subjects' (2 parts). Arnold, London. 1929, 1932.

[2] The contents of Graeco-Egyptian technical papyri at Leyden and Stockholm have been translated and annotated by E. R. CALEY in *J. chem. Educ.* **3**, 1149–66, 1926; **4**, 979–1002, 1927. See also: P. E. M. BERTHELOT. 'Traités techniques (Traduction)' in 'Collection des anciens alchimistes grecs.' Troisième livraison, cinquième partie, pp. 307–75. Steinheil, Paris. 1888.

[3] MARCUS GRAECUS *Liber ignium ad comburendos hostes*, translated and analysed by P. E. M. BERTHELOT. 'Histoire des Sciences. La chimie au moyen âge', Vol. 1, pp. 89–135. Imprimerie Nationale, Paris. 1893.

[4] *Compositiones ad tingenda musiva.* Its history is discussed by R. P. JOHNSON. '*Compositiones Variae* from Codex 490, Bibliotheca Capitolare, Lucca' in Ill. Stud. lang. lit., Vol. 23, no. 3. University of Illinois Press, Urbana. 1939. The most recent edition with German commentary and translation is by H. HEDFORS. *Compositiones ad tingenda musiva.* Diss., Uppsala. 1932.

[5] "*Mappae Clavicula*, a Manuscript Treatise on the Preparation of Pigments and on the various Processes of the Decorative Arts practised during the Middle Ages" communicated by Sir Thomas Phillipps, ed. by A. WAY in *Archaeologia*, **32**, 183–244, 1847.

[6] The most recent and reliable edition and German translation and commentary of the major part of the treatise is by W. THEOBALD. 'Technik des Kunsthandwerks im zehnten Jahrhundert. Des THEOPHILUS PRESBYTER *Diversarum artium schedula.*' Verein deutscher Ingenieure, Berlin. 1933. For a complete edition and translation see A. ILG in Quell. Kunstgesch. Kunsttechn., Vol. 7. Braumüller, Vienna. 1874. The first English translation was made by R. HENDRIE. 'An Essay upon various Arts by Theophilus called also Rugerus.' Murray, London. 1847. Another and more satisfactory translation of the second book only is given in Appendix A of C. WINSTON. 'An Inquiry into the Difference of Style observable in Ancient Glass Paintings' (2nd ed.). Parker, Oxford. 1867.

[7] HERACLIUS *De coloribus et artibus Romanorum.* Edition with German translation and commentary by A. ILG. Quell. Kunstgesch. Kunsttechn., Vol. 4. Braumüller, Vienna. 1873. An older edition with English translation by Mrs. M. MERRIFIELD. 'Original treatises dating from the twelfth to the eighteenth Centuries on the Arts of Painting', Vol. 1, pp. 166–257. Murray, London. 1849.

[8] The Madrid Codex has been edited by J. M. BURNAM. 'Recipes from *Codex Matritensis* A. 16.' Palaeographic edition. University of Cincinnati Studies, second series, Vol. 8. University Press, Cincinnati. 1912.

[9] A short general survey of Latin alchemy with bibliography is given by F. SHERWOOD TAYLOR. 'The Alchemists. Founders of Modern Chemistry.' Schuman, New York. 1949. Reissued Heinemann, London. 1951.

[10] A preliminary attempt to list the later medieval texts on chemical technology has been made by D. V. THOMPSON. "Trial Index to some unpublished Sources for the History of Mediaeval Craftsmanship." *Speculum*, **10**, 410–31, 1953.

[11] The *Liber Sacerdotum* has been edited by P. E. M. BERTHELOT. 'Histoire des Sciences. La chimie au moyen âge', Vol. 1, pp. 179–228. Imprimerie Nationale, Paris. 1893. See also J. RUSKA. 'Studien zu den chemisch-technischen Rezepten des *Liber Sacerdotum*.' Quell. Gesch. Naturw., Vol. 5, pp. 83–125, 275–317. Springer, Berlin. 1936.

[12] Cennino's work has been edited and translated by D. V. THOMPSON. CENNINO D'ANDREA CENNINI. 'Il Libro dell'Arte' (2 vols). Yale University Press, New Haven. 1932–3. Quotation from Vol. 2, p. 49 (abbreviated).

[13] An English translation appeared in the seventeenth century, and was reprinted with an introduction by E. J. HOLMYARD. 'The Works of Geber, Englished by Richard Russell 1678.' Dent, London. 1928. A bibliography of Arabic alchemical works which have much on the acids is by O. TEMKIN. "Medicine and Graeco-Arabic Alchemy." *Bull. Hist. Med.*, **29**, 149–55, 1955.

[14] NERI, ANTONIO. 'L'Arte Vetraria distinta in libri sette.' Giunta, Florence. 1612. (2nd rev. ed. Rabbviati, Florence. 1661.)

[15] Several accessible books summarize the methods and materials of early painting, dyeing, and mordanting. We may mention:

DE WILD, A. M. 'The Scientific Examination of Pictures.' Trans. by L. C. JACKSON. Bell, London. 1929.

GETTENS, R. J. and STOUT, G. L. 'Painting Materials. A short Encyclopaedia.' With an introduction by E. W. FORBES. Van Nostrand, New York. 1942.

HILER, H. 'Notes on the Technique of Painting' (rev. ed.). Faber and Faber, London. 1948.

HURRY, J. B. 'The Woad Plant and its Dye.' Oxford University Press, London. 1930.

LAURIE, A. P. 'The Technique of the Great Painters.' Carroll and Nicholson, London. 1949.

SINGER, C. 'The Earliest Chemical Industry.' Folio Society, London. 1948.

THOMPSON, D. V. 'The Materials of Medieval Painting.' Allen and Unwin, London. 1936.

[16] On sugar see: DEERR, N. F. 'The History of Sugar.' Chapman and Hall, London. 1949–50. LIPPMANN, E. O. VON. 'Geschichte des Zuckers.' Hesse, Leipzig. 1890.

A NOTE ON MILITARY PYROTECHNICS

A. R. HALL

THE use of fire as a weapon of war, to destroy crops, huts, stockades, and, later, ships and siege-engines, is certainly very ancient. A burning brand in the hands of a savage is a dangerous weapon, but for more sophisticated pyrotechnical effects a knowledge of certain mineral substances is requisite. Of the minerals familiar in ancient and medieval times, those whose combustible or combustion properties were most readily applicable to military purposes were sulphur, saltpetre, bitumen, and the volatile petroleum constituents with low flash-points collectively known to the ancient writers as naphtha. All these materials were available in the ancient Near East, though it is doubtful whether refined saltpetre was prepared, and therefore its value as an aid to combustion would have been small. In the Roman period bitumen was largely replaced by resin, and by tar from the distillation of wood. Petroleum products, which could be procured in south-western

Asia from the natural seepages at Hit and elsewhere (vol I, pp 250–4), were of course known only by report in Africa and western Europe.

The use of these materials is discussed later, but first it is necessary to describe three tactical methods for exploiting fire as a means of destruction before the invention of gunpowder. The simplest was the fire-arrow or fire-lance. A development of this was the fire-pot, thrown by some form of projectile engine (p 700). Lastly, the more fiercely combustible ingredients of the fire-pot were, from the early centuries of the Byzantine era, projected against the enemy by means of tubes and perhaps rockets. Obviously very different levels of pyrotechnical skill were necessary for the command of these three methods.

For the first, it is necessary only to wrap an arrow or lance with tow, cloth, or dry vegetable material, to daub this with pitch, resin, or bitumen, and to set it alight, in order to prepare a flaming torch which may be hurled against wooden defences or enemy ships. Such a device may well have been known in the Near East before civilization began, and it has lasted throughout history. For the fire-pot to be effective, however, it is necessary that machines should have been constructed for throwing it a reasonable distance, and that a liquid filling should have been developed so that when the vessel fractured the burning contents would spread over a wide area. Such a combustible liquid might have been most easily utilized in the first place by simply pouring it upon the assailants of a fortification. It was furnished to the Near East by naphtha, which long remained the principal ingredient in the 'Greek fire', so mysterious to those unapprised of its simple secret.[1] No doubt the volatile petroleum was often commingled, either by accident or on purpose, with heavier oils, bitumen, and other substances; but, though we may trust the ancient authors who state that these artificial fires were very difficult to extinguish, it is impossible to believe that any of them ignited spontaneously on impact.

Fire-pots, probably containing naphtha, were used in war throughout the second half of the first millennium B.C. and perhaps even earlier. Literary references to the ingredients employed are somewhat late. Procopius (mid-sixth century A.D.) writes:

> The Persians were the inventors of this: having filled vessels with sulphur and that drug which the Medes call *naphtha* and the Greeks 'oil of Medea' and lighted them, they hurl them against the frame-work of battering-rams, and soon set them ablaze, for this fire consumes the objects which it touches, unless they are instantly withdrawn [1].

Similar recipes become more frequent in the Middle Ages, for the fire-pot had a long history before it was supplanted by the mortar-bomb in the sixteenth century.

The improved form of liquid combustible known as 'Greek fire' was apparently devised in the seventh century A.D. It was to the expert use of this weapon that the Byzantine Empire owed its survival through the Muslim attacks of 673–8 and later, and of western Europeans and Russians during the following centuries. Apart from its indubitable destructive power, Greek fire long aroused a superstitious fear in the adversaries of Byzantium, which the Emperors were careful to foster by preserving the nature

[1] It should be remembered that no liquid combustible was known in the west, where *aqua ardens* ('burning water', alcohol) was not prepared until *c* 1200–1300.

FIGURE 341—*Greek fire used in a naval battle. It is apparently being directed from a 'siphon'. From a tenth-century Byzantine manuscript.*

of the composition as a cherished secret. But there was in fact no single incendiary substance which may be called Greek fire, nor was it invariably used in the same manner.

Some of the methods employed by the Byzantines are well authenticated. At sea, they are said to have mingled finely ground quicklime with the petroleum mixture, so that spontaneous ignition would occur when the lime reacted with sea-water.[1] Addition of sulphur produced denser and more noxious fumes. That saltpetre was used in these fluid combustibles is unlikely.

Equally ingenious were the devices used to project the Greek fire against an enemy, such as the 'siphon' fitted to ships of war (figures 341–2). This was probably a bronze force-pump, from the nozzle of which a jet of burning liquid was hurled. According to the Emperor Leo VI (886–911), soldiers were equipped with tubes filled with combustible matter, carried behind their shields, which they lit and flung in the faces of the foe. Similarly the Byzantine princess Anna Comnena includes in her history a passage which some writers have taken to indicate the propulsion of such tubes as rockets—which is very doubtful [2].

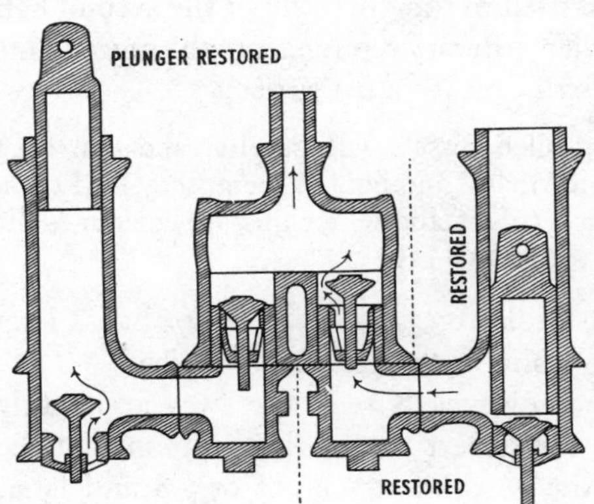

FIGURE 342—*Reconstruction of a late-Roman bronze force-pump such as could have been employed with the 'siphon'. From Bolsena, Italy.*

Recipes for pasty mixtures which could have been used for filling tubes of this kind occur at earlier dates, such as one given by Julius Africanus (*fl* A.D. 220–35): 'Grind unburnt sulphur, the salt extracted from dust, thunderstone, pyrites, in equal parts in a black mortar. Add the juice of the black sycamore and liquid bitumen [i.e. naphtha] to make a flowing paste' [3]. Some have been tempted to think that such terms as 'salt of dust', 'cooked salt', and so on indicate saltpetre, but there is no evidence to support

[1] This trick has often been ascribed to the Byzantines, but never tested. It seems unlikely that the reaction of quicklime and sea-water would ignite petroleum: more probably, fire-pots were used to set the floating spirit afire.

this opinion. Yet without reasonably pure potassium nitrate the manufacture of rockets (and, *a fortiori*, explosives) would have been impossible.

The only certain fact in this perplexing subject is that it is useless to base our ideas on exact identification of the names of 'salts' given by medieval authors with the substances clearly distinguished by modern chemists. The old authors made a thorough confusion of the nitrates and carbonates of sodium and potassium.

The Muslim states, by reason of their close contact with Byzantium, their control of the petroleum seepages in Iraq, and their study of Greek technological texts and methods, were not long ignorant of the methods of incendiary warfare, which they in turn opposed successfully to the Christian invaders of their territories. The chroniclers of the Crusades often mention pyrotechnics; thus Joinville (1224?–1317), describing the destruction by fire of wooden castles built by St Louis (Louis IX) to protect the crossing of a river (1249):

> The form of the Greek fire was that it came as big as a barrel of verjuice in front, emitting a tail of fire as long as a tall lance. It made such a noise in arriving that it seemed like thunder in the sky; it was like a dragon flying through the air. It cast such a brilliant light that we could see as clearly in the camp as though it were day, because of the great abundance of fire which shed a great light. Three times they hurled fire upon us that night, and four times they shot it from cross-bows [4].

There is good evidence that Muslim armies accumulated large stocks of fire-pots, which were fitted with slow-matches and flung by mechanical projectile-engines of the type of the mangonel and trebuchet (figure 583). But there is no solid evidence that they improved pyrotechnical recipes or ballistic methods.

The manufacture of fireworks was certainly known in Europe before the end of the thirteenth century, and the use of gunpowder in firearms—one of the most dramatic and far-reaching technical inventions of the Middle Ages—occurred in the early years of the fourteenth. If, as seems likely, knowledge of a saltpetre–sulphur–charcoal mixture as a pyrotechnical composition was diffused from the Far East, it must have come either through Byzantium or through the Islamic world. Admittedly Europeans had occasional direct contacts with the Mongol power in Poland, Russia, and Persia, but it was in western Europe that gunpowder first appeared, and the Mongols made little use of the pyrotechnical inventions of their Chinese subjects.

Though the route by which, it may plausibly be supposed, knowledge of black-powder compositions passed from China to Europe cannot be clearly established (as it can be for the linked inventions of paper-making and printing), it is certain that the appropriate formulae, and methods of applying explosive force, were known in China much earlier than in Europe [5]. One such preparation is recorded in the Wu Ching Tsung Yao (originally written in A.D. 1044):

> 1 chin 14 ounces of sulphur, 2½ chin of saltpetre, 5 ounces of charcoal, 2½ ounces of pitch, and the same quantity of dried varnish are powdered and mixed. Then 2 ounces of dry vegetable matter, 5 ounces of tung oil, and 2½ ounces of wax are stirred into a paste, and finally all the ingredients are stirred together. The mixture is then wrapped in five layers of paper which are tied with thread and covered with wax or pitch [6].

It will be observed that hydrocarbons form a considerable part of this substance, and that the recipe is more suitable for some kind of fiercely-burning firework than for an explosive. At this time the Chinese were well acquainted with fire-arrows and other incendiary devices, projected from mechanical engines much as were the liquid compositions of the Near East. The addition of potassium nitrate—long recognized, it is claimed, as a drug—was apparently fairly recent.

With such incendiary compositions in use, the accidental discovery of an explosive

FIGURE 343—*A bomb or grenade bursting near a Japanese bowman. From a Japanese woodcut of 1292.*

recipe would not be improbable. Bamboo tubes filled with gunpowder were soon brought into military service; these were the precursors of more effective grenades hurled by hand or mechanical engines (figure 343).

The invention of firearms with barrels of metal followed naturally from the compression of explosive compositions into hollow cylinders of bamboo or rolled paper along with stones, broken porcelain, and bullets. From this more devilish version of the 'Roman candle' it was not a long step to the filling of a tube with a series of charges of powder and projectiles, fired off successively by means of a row of touch-holes. This stage occurred in China about the middle of the thirteenth century. Soon afterwards, the first references to metal barrels are found, since it was now the object of the device to restrict the force of the powder to the expulsion of a projectile. Guns of a primitive kind had therefore been invented. The oldest surviving example comes from 1356 [7].

By this time guns were already becoming fairly commonplace in European war (p 726). Intercommunication between Europe and China cannot have had any effect at this moment, and in fact the European exploitation of artillery was very soon to outdistance the Chinese. Recipes incorporating saltpetre, sulphur, and charcoal occur in the west about 1300. Some sort of firework was apparently known to Roger Bacon (1214?–94), who, contrary to a common statement, certainly did not invent gunpowder. Almost at the same time, an Arabic-writing Syrian, Al-Ḥasan al-Rammāḥ (*fl c* 1280–90) wrote a military treatise in which he clearly indicated that saltpetre was the primary

FIGURE 344—*Manufacture of sulphur from pyrites by distillation. From Agricola, 1556.*

substance for pyrotechnic compositions, and described carefully how it was separated from other salts by solution and repeated crystallization [8]. Yet apparently he was ignorant of the propulsive force of gunpowder. Contemporary again is the 'Book of Fires for the Burning of Enemies' put together under the name of Marcus Graecus, which includes many pyrotechnical recipes of considerable antiquity and (in some manuscripts) directions for purifying saltpetre and making gunpowder, added *c* 1300 [9].

'Saltpetre', wrote Marcus, 'is a mineral found as an efflorescence on stones. This earth is dissolved in boiling water; then the liquor is poured off, filtered, and heated for a day and a night, when little plates of the solid clear salt are found at the bottom.' 'There are', he added, 'two ways of making a "fire flying in the air".' In the first, one part of rosin, as much sulphur and (six?) parts of saltpetre are ground thoroughly and dissolved in linseed-oil or (which is the better) laurel-oil. The mixture is inserted in a reed or hollow stick and lighted. It will then fly where you wish and set fire to anything. In the second recipe, 1 lb of sulphur, 2 of the charcoal made from the willow or the lime-tree, and 6 lb of saltpetre are very finely pulverized on marble. This composition can be used either to fill a rocket (like the former) or as an explosive [10].

There is still no hint in the 'Book of Fires' that incendiary compositions could be used to impel a projectile. It may be, therefore, that although knowledge of this new pyrotechnology was transmitted to Europe from the Far East, Europeans themselves re-invented (as tradition has long asserted) the application of gunpowder to cannon and other firearms.

FIGURE 345—*Grinding gunpowder in mortars, the pestles being suspended from springy poles. From a manuscript of the* Feuerwerkbuch, *before 1450.*

Of the ingredients required for gunpowder, charcoal had been known since ancient times, and no real improvements in the method of carbonization were effected before the end of the eighteenth century. The history of sulphur is obscure, though the substance was almost equally long familiar. According to Agricola (1556) it was obtained from an ore consisting of sulphur and earth (i.e. native sulphur, associated with volcanoes) or from pyrites—in any case by distillation. The native sulphur was heated in an earthen pot, the vapour allowed to liquefy in a second pot, and the liquid then run off to set solid in a third. Pyrites was heated in a pot with a hole in its base through which the liquid sulphur ran into another vessel containing water (figure 344) [11]. Sublimed flowers of sulphur was much used medicinally. For explosives, the ordinary sulphur of commerce could be improved by 'softening it in an earthenware vessel so that it is well melted. Take to one pound of sulphur half an ounce of quicksilver that is deadened with sulphur, and mix these thoroughly. And afterwards pour the sulphur into good brandy, thus it becomes the more dry, hot, and good' [12].

Saltpetre demanded the most careful treatment, set out at length and with many variants in the firework-books. Possibly the first source was indeed the efflorescence on damp walls (p 370). Later, it was leached out of the nitrogenous earth in stables, pigeon-lofts, pig-styes, and similar places (figures 336–7). One method of extracting the potassium nitrate from solutions containing many impurities was described as follows in the early fifteenth century:

> If you wish to purify saltpetre and make it good, freeing it from salt and alum, take two pounds of unslaked lime, one centner [cwt] each of verdigris and white vitriol, and two centners of salt, and make a lye of them with wine or vinegar. Let the lye stand three days, to clarify. Afterwards put the saltpetre in a copper, as much as you like, and add the lye till it is just covered.

Mix all together, boil down to half the volume, and pour off the liquor. You will find the alum and salt and all impurities at the bottom of the copper. Let the saltpetre-water (which was formerly the lye) grow cool, so that the saltpetre forms like frozen ice. When it has formed, pour off the lye and dry the saltpetre in the sun [13].

Another way of separating saltpetre from common salt was to wash the substance in cold water, the salt being more soluble at that temperature. The addition of stale urine was supposed to promote the crystallization of the saltpetre. By the sixteenth century it

FIGURE 346—*A large water-driven gunpowder-mill. Sixteen pestles or stamps are raised by tappets mounted on a shaft driven by the water-wheel. From an engraving of 1676.*

was well known that the whole process was improved by the use of potash in addition to the lime. Nitrous earth was laid in a vat with alternate layers consisting of a mixture of wood-ash and lime, and covered with water. After a time the solution was run off and concentrated by boiling it with more of the same mixture. Common salt—the chief impurity—was deposited first. Then the solution was run into another vessel to cool, whereupon the saltpetre crystallized out [14].

Apart from the correct preparation of the three ingredients, the manufacture of good gunpowder requires that they be thoroughly incorporated in due proportions. These are not the same for all purposes, cannon requiring a different powder from that best for hand-firearms. In the search for the strongest powder many different proportions were tried, as well as the addition to the mixture of traces of other substances, such as sal ammoniac, arsenic, and antimony, and the grinding of the powder with alcohol instead of water. In the early fifteenth century the sulphur and charcoal were commonly apportioned in the ratio of 2:1 by weight, and the saltpetre added in the proportion of 4, 5, or

6 by weight according to the strength desired. In modern times mixtures in the proportions sulphur 1, charcoal 1½, and saltpetre 7½, or thereabouts, have been used. The mixing was at first carried out with a pestle and mortar (figures 345–6), which was later replaced by an animal- or water-driven roller-mill. The mix was kept damp to reduce the risk of chance explosions, but these were far from uncommon. Until the middle of the sixteenth century the powder was used in the fine state as it came from the mills, but later it was granulated or corned by rubbing it through a sieve while still damp. Fine powder was still required for the priming, while the more freely burning corned powder permitted a considerable reduction in the size of the propulsive charge.

REFERENCES

[1] PROCOPIUS *Hist. bellorum*, VIII, xi, 35–38. (Loeb ed. Vol. 5, p. 160, 1928.)

[2] LEO VI *Tactica*, c. xix, 58 (p. 347). Elzevir, Leyden. 1613.
ANNA COMNENA, XIII, iii, 6. ('Alexiade', ed. and French trans. by B. LIEB, Vol. 3, p. 96. Les Belles Lettres, Paris. 1945. Engl. trans. by ELIZABETH A. S. DAWES, p. 329. Kegan Paul, Trench, Trubner, London. 1928.)

[3] MERCIER, M. 'Le Feu Grégeois', p. 36. Geuthner, Paris. 1952.

[4] JOINVILLE, JEAN DE. 'The History of Saint Louis', trans. by JOAN EVANS from the modernized French ed. by N. DE WAILLY, p. 61. Oxford University Press, London. 1938.

[5] DAVIS, T. L. *J. chem. Educ.*, **24**, 522, 1947.
GOODRICH, L. C. and FÊNG CHIA-SHÊNG, *Isis*, **36**, 114, 1946.
WANG LING. *Ibid.*, **37**, 160, 1947.

[6] *Idem. Ibid.*, **37**, 162, 1947.

[7] GOODRICH, L. C. and FÊNG CHIA-SHÊNG, *Ibid.*, **36**, 122–3, 1946.

[8] SARTON, G. 'Introduction to the History of Science', Vol. 2, ii, pp. 1039–40. Carnegie Instn Publ. no. 376. Washington. 1931.

[9] BERTHELOT, P. E. M. 'Histoire des Sciences. La chimie au moyen âge', Vol. 1, pp. 89–135. Imprimerie Nationale, Paris. 1893.

[10] *Idem. Ibid.*, Vol. 1, pp. 108–10.

[11] AGRICOLA *De re metallica*, trans. by H. C. HOOVER and LOU H. HOOVER, pp. 578–81. Mining Magazine, London. 1912.

[12] HASSENSTEIN, W. (Ed. and Trans.). 'Das Feuerwerkbuch von 1420', p. 60, col. 2. Deutsche Technik, Munich. 1941.

[13] *Idem. Ibid.*, pp. 53–54.

[14] AGRICOLA. See ref. [11], pp. 561–4.

BIBLIOGRAPHY

SARTON, G. 'Introduction to the History of Science', Vol. 1, p. 495; Vol. 2, ii, p. 1038. Carnegie Instn Publ. no. 376. Washington. 1927, 1931.

MERCIER, M. 'Le Feu Grégeois', pp. 151–8. Geuthner, Paris. 1952.

II

THE MEDIEVAL ARTISAN

R. H. G. THOMSON

I. INTERRELATION OF CRAFTS

It would be instructive, though it might also be very tedious, to trace the history of each of the common tools from early classical antiquity through the early and later Middle Ages into Renaissance times. Knowledge is at present quite inadequate for such presentation. There is, however, good reason to suppose that many of these tools remained little changed during this long period of history. In this chapter we shall content ourselves with a cross-section in time. With two exceptions, all the illustrations that follow are from documents of the twelfth to fifteenth centuries inclusive. They show artisans using the tools of their trade. Many of these tools had then not changed greatly for centuries or even millennia, and some are current to this day.

Systematically collected series of instructional pictures of artisans at work began to appear in the sixteenth century. Before that time a high proportion of the pictures were of biblical subjects, and the largest or most dramatic operations were naturally the most popular for illustration. Buildings under construction and forges in action are common themes. One of the best illustrations of carpenters at work shows Noah building the ark (figure 357); another shows sappers constructing fortifications in the front line of battle (figure 350).

In the building of a cathedral all types of artisan were involved. The stonemasons and builders had blacksmiths to construct, sharpen, or renew their tools. Carpenters put up the scaffolding and the timber work to support the arches. One or other of them used each of the common tools (for agricultural tools see ch 3). Other workers, such as plumbers (for some lead fittings see figure 404) and glaziers (figure 310), used only their own modifications of tools which will be described. Our comparative cross-section will thus attempt to set out how the work of the three types of artisan, the stone-mason, carpenter, and blacksmith, was carried out during the Middle Ages.

The reader should perhaps be reminded that during the period treated here, and indeed for centuries earlier, all skilled work was organized in guilds or crafts. Many of these were subdivided into special brotherhoods. Apprenticeship to them was universal. Thus labour conditions were excessively complex. Such matters are the business of the historian not of technology but of economics, and therefore cannot be described here.

II. THE STONE-MASON

Buildings in stone or brick were relatively rare in the Middle Ages; our mental picture of medieval buildings has naturally been formed largely from those that have survived. Even the houses of noblemen were usually built of timber, wattle, plaster, and thatch. But the predominance of timber for constructional purposes in no way diminished the scale of those buildings that were designed to be of stone, such as castles, churches, cathedrals, city walls, and a few other constructions.

FIGURE 347—*An early fifteenth-century picture showing Pharaoh setting tasks for the Jews. Here the scaffolding is of the simplest kind: a spiral walk rests on cross-beams let into the wall at one end and lashed at the other to uprights on the lower levels; higher up they are braced with struts. A worker carries up a hod of mortar. The rough-mason squares his stone with a scabbling-axe. From a French manuscript of biblical stories.*

The traditional quarrying-tools are described elsewhere (p 36). In the Middle Ages cutting-edges were of iron. These would have been unfamiliar to the ancient Egyptians (vol I, p 569), who performed all rough work by pounding with stone hammers and finished off with copper chisels. Today, by contrast, the quarryman's tools are universally of steel and often operated pneumatically.

Because of the difficulties and slowness of communications the cost of transport of building-materials was serious, and frequently much higher than the total paid to the quarrymen. Stone used at Eton College in the fifteenth century was worth about a shilling a load, but a further 6*s* 6*d* had to be added to this after its journey by road, sea, and river from Huddlestone in Yorkshire [1]. As a result, first, local stone was used where available, and secondly, as much preliminary mason's work as possible was performed at the quarry. In this way quarries became masons' nurseries, where mastery over less elaborate work was attained.

There was no exact medieval equivalent of the specialized architect, but design and unity were not the mere result of co-operation between the builders. In control of most large-scale operations was a master-mason, who had graduated from apprenticeship in his craft [2]. A very well known master-mason of the later fifteenth century was Henry de Yevele (d 1400), who worked mainly at Westminster Abbey and left in his will considerable properties and endowments to testify to his success. The master-mason or master-builder, as well as supervising, himself took a part in the work. Plate 30 B shows such a man who, perhaps on his way to check levels with a plumb-bob and square, has been called to attend some dignitary and discuss the progress of the building.

Directly under the master-mason worked a number of freemasons and rough-masons. The freemason was qualified to carve freestone, or fine-grained sandstone and limestone, which could be freely worked in all directions and undercut [3]. In his work there were many possibilities for individual artistry, though the templates for much of the moulded work were of fixed design and the property of the guilds. The name freemason was first used in the fourteenth century, and probably originally had no other connotation than that described. With the freemason worked the rough-mason who carried out the less skilled operations of shaping stones. The rough-mason did the straight moulded work, square ashlar, and any other stone work (figure 347), but the freemason cut the carvings and more intricate mouldings (plates 31 A, B). Both of these artisans had under them labourers to carry and heave.

FIGURE 348—*Cutting thin slabs of marble for facings with a frame-saw. From an eleventh-century Italian manuscript.*

In modern stone-masonry the block of stone is usually cut to size with a circular saw. Saws were occasionally used in the Middle Ages on the softer stones (figure 348), but otherwise the first shaping, as now, was with the scabbling-axe, a heavy type of sledge-hammer with an edge. Subsequent work was either with a lighter-bladed hammer for ashlar, or with a mallet and chisel for carving. Splayed out at the blade, the mason's chisel is known as a bolster (figure 25) and its edge may be toothed for planing surfaces. As our figures indicate, there was a substantial similarity in equipment between the medieval and the modern mason, and with the more ancient ancestor of both. Toothed bolsters and hammers are often displayed (plate 31 A). Some pictures show the hammer and chisel being used to square surfaces (figures 349–50), but otherwise there was a decided preference for bladed hammers of various shapes, both for squaring surfaces (figure 347), and for carving (plate 31 B).

Hammer, square, and plumb-bob were the symbols of the mason's trade. Very common was the right-angle plumb-bob for testing horizontal surfaces (plate 30 B). This can be traced back unchanged to Egyptian antiquity. Spirit-levels, which are instruments of great delicacy, were invented in the seventeenth century and the clumsy and less sensitive water-table of the Greeks and Romans would have been of smaller practical value. The simple vertical bob, as

FIGURE 349—*Scaffolding for a stone building. The poles are lashed together at the diagonals with tourniquets to tighten the binding. The master-mason greets a party of dignitaries led by Charlemagne. Stones are squared with mallet and chisel. Water for the mortar comes from a newly dug well (the village pump was adopted from a type only then being developed for mine-drainage, see figure 11). Note the stone-layer's trowel. From a French manuscript, 1460.*

seen today, was in general use. The square (figure 347, plate 31 A), was graduated along both its arms. Slopes were thus set off as ratios of the two sides of a right-angled triangle. Measurement in terms of angles has always been a sophisticated technique more suitable to the drawing-board than to the work-man's bench. Plumb-bob and square were naturally as familiar to the carpenter as to the mason. More peculiar to the mason were the compasses for marking the ever recurring arcs and circles of his work (plate 31 A).

The cement used in northern Europe was invariably a sand-and-lime mixture. Many pictures show this mortar being mixed and carried up in trough-shaped hods (e.g. figures 347, 349). The correct burning of lime was of some importance, and—until the nineteenth century—there was no successor to Vitruvius to lay down the best proportions in an aggregate.[1] As a result, much of the medieval and later stone-work is bonded with mortar in a dangerously powdery condition.

FIGURE 350—*The construction of a stone bridge and wooden fortifications under fire from the enemy. Trees are felled, squared by axe, and cut into planks with a pit-saw. The planks are cut to length, jointed, and pegged. Lying on the ground are a square, trimming-axe, pair of pincers, brace, and hand-saw. Masons square their stone with mallet and bolster, and the mortar is mixed and carried forward to the builders, who work behind screens to protect them from the enemy's fire. The scene shows the siege of a city by Charlemagne. From a French manuscript, 1460.*

After the end of the Roman occupation of Britain the manufacture of bricks and tiles had been discontinued [4]. Their successors, thatch and wooden shingles (figure 392), were fire-hazards, and for this reason the use of tiles as roofing material became compulsory in London in 1212. From that time the trade of tile-making expanded and by the fifteenth century its products were in

[1] The patent for Portland cement was taken out in 1824, and it was even then some decades before its optimum proportions were determined.

great demand. The tile-makers were unpopular, however, since their kilns, kept burning for several days at a time, expelled a quantity of smoke considered both disgusting and insalubrious in the unpolluted atmosphere of that time (figure 351). Bricks came into use as cheaper and easier to handle than stone. Many brickmakers and bricklayers were foreigners from Flanders, where the exhaustion of timber-supplies and absence of stone had early forced builders to use brick and tile (plate 31 B). The bricklayer's main tools were his trowel, a hammer to cut bricks, and a level. A bricklayer of today might take up a medieval trowel and fail to notice any change from his own (figures 347, 349, plate 31 B).

FIGURE 351—*Making bricks and tiles. The clay is mixed with water in the foreground, moulded by hand in the shed—a four-sided mould is probably used—and stacked for drying on the left. The kiln is on the right, and the finished bricks lie by the two gentlemen. In the coloured original flames can be seen coming from the lower door of the kiln, and the upper opening is coloured red. From a mid-fifteenth-century Flemish biblical manuscript.*

To raise heavier material into position use was made of the simple hoist. This normally consisted of a fixed pulley at the upper level, and at ground level a windlass turned by spokes (plate 30 B), or by the recently applied crank (plate 31 A) (p 652). An unlooped rope passed from windlass to weight. Little use was made of block and tackle, derrick, or treadmill, though these were known and employed in other fields (p 646, plate 38 B). Large block masonry, like that of the Romans and Egyptians, was seldom used as it vastly increased the size of the equipment needed to manipulate the stone (pp 658 ff).

Of particular interest is the nature and pattern of medieval scaffolding. It was neither as common nor as elaborate as that of today. Often the supporting beams of platforms were merely let into prepared holes in the wall already built, to be replaced or extended by successively higher additions (figure 347, plate 31 A). At other times a framework was constructed of lashed poles, similar to that used at any time before the age of tubular steel (figure 349, plate 30 B). An important example, with ladders of modern form between levels, is to be seen in plate 31 B.

III. THE CARPENTER

The construction and erection of scaffolding was the business of the carpenter (on the origin and meaning of the word carpenter see p 233), and on any stone building of importance the head carpenter was also a 'master' of his trade. He had in addition to make the frameworks that supported arches until their keystone was lowered into place, and all temporary wooden structures and huts. For the ordinary building he was in sole command (plate 30 A).

FIGURE 352—*Squaring a piece of timber with the axe. The young nobleman learning the craft was to become the Emperor Maximilian I. It may be judged from this picture that the standard of axe-trimming was high. The blade of such an axe was offset, so that one side—the far side in the figure—was perfectly plane. Note the two-handled saw, and the saw-blades that the two carpenters carry in their belts. On the ground is a cord, covered with powdered pigment, which is used for marking lines. This picture dates from the sixteenth century, but the tools differ in no essential from those used in the preceding two centuries. From Maximilian's autobiographical notes called* Weisskunig. *Vienna,* c *1514.*

Of his tools, the axe had the longest ancestry of all except the hammer. It was of relatively greater importance than even the saw before the latter became transformed into a circular power-driven instrument. Cutting a large log into planks with a frame-saw was arduous work. Riven beams, that is beams that had been split radially from the trunk with wedge and axe, were smaller and more wasteful, since each had to be squared from a wedge-shape. But they had an important advantage, for wood dries anisotropically in such a way that the only sections free from warp or stress are segments or planks cut radially. There is therefore no more reliable timber than a well seasoned plank of riven oak.

The squaring and trimming of planks were done with an axe. It is clear from surviving examples of his work that the medieval carpenter carried out this task with a skill quite beyond the powers of the carpenter or woodsman of today. The art appears to have died out in this country at least three generations ago [5]. The typical trimming-axe is shown in detail in figure 352. Earlier examples are illustrated in figures 353–4. The blade was hafted to a handle of oblong cross-section, preventing twist. Earlier, that is to say up to the end of the thirteenth century, a type with a curved splayed blade, rather more like a battle-axe, was favoured (figures 357–8). The felling-axe had not at this stage reached that present perfection of form which makes it an example of what is regarded as a completed evolution. The shape of the medieval axe-head was variable (figures 350, 355), and its shaft, though oval as in the modern axe, was uncurved.

FIGURE 353—*Ship-builders. Proportions are a little distorted, and while the clawed hammer and trimming axe are well represented we must suppose both the teeth of the saw and the nail to be properly very much smaller. From a late thirteenth-century French historical manuscript.*

The adze was no longer the all-purpose wood-shaping instrument of the ancient Egyptians (vol I, p 687, figures 134, 487), but for many operations was being replaced by the axe. It is probably represented for boat-building in figure 361. This was a field in which it always retained its usefulness, being especially fitted to cutting away in curves. A small adze may be seen in a carpenter's workshop of about 1500 (figure 356).

FIGURE 354—*Carpenters making a bridge. Squaring a beam with the axe, sawing, cutting a mortise-hole with mallet and chisel, boring a hole for a peg with an auger. A shell-auger is lying near the bridge. From a French manuscript, 1460.*

Next to the axe, the saw was the symbol of the carpenter's trade. The variety of its specialized forms was as great as it is today. The teeth of the hand-saw were raked away from the hand so that the cut was on the downward stroke, as today. To judge by its proportions—it was commonly sabre-shaped (figures 350, 354, plate 30 A)—it must have needed to be rather thick to maintain a proper rigidity. The absence of the pistol-grip should be noted. Two-handed saws (figure 352) in skilful hands are always in tension and therefore do not buckle.

Frame-saws are depicted on Roman frescoes, but without means of adjusting

Fig. 355

Fig. 356

Figure 355—*Woodcutters. The upper man lops off side branches with a felling-axe. The frame-saw is of modern design, tensioned with a Spanish windlass. From a Burgundian tapestry of the fifteenth century.*

Figure 356—*The carpenter and his family. Two braces are included among the collection of chisels, mallets, and pincers on his tool-rack. He is finishing a plank with his trying-plane. Before him lie a trimming-axe, a large chisel and small adze, a marking-tool, a mallet, two more chisels, a smoothing-plane, a square, and a pair of compasses. From a miniature by Jean Bourdichon.* c *1500.*

tension, which seems an early medieval device. Yet the carpenter on a twelfth-century Italian mosaic is working with a startlingly perfected instrument, whose blade may be twisted out of line with the frame so that a plank can be cut down its long axis (figure 357). The medieval frame-saw is also beautifully illustrated in a fifteenth-century tapestry (figure 355).

Large logs were sawn into planks with a pit-saw, as an alternative to being riven. The log was manœuvred over a large pit in which the lower of the two men worked. In our examples the log rests on trestles well above the ground, but this may have been due to the exigencies of the situations. The carpenters on the building site in plate 30 B may be a mile or two from their permanent establishment where most of their wood is sawn; those working under bow-fire in figure 350 might well prefer the security of a saw-pit if they had but the leisure to make one. The saw-cuts were marked top and bottom by stretching a string covered with chalk or other pigment to the line, then sharply lifting and releasing

FIGURE 357—*Noah building the ark. Two fine saws are shown: a frame-saw on the left, and a pit-saw. Below are two men trimming, on the left with an adze, and on the right with an axe. From a twelfth-century mosaic in the cathedral at Monreale.*

it (figure 352). The junior sawyer took the lower position and worked in great discomfort. Sawn planks could naturally be turned out to a more even thickness than those that had been riven, and therefore found a special use as floor-boards.

Chisels (figures 354, 356) and gouges, both tanged and socketed, had been developed in the Bronze Age (vol I, ch 22). The same forms were now used with the wood-turning lathe (pp 249 f, 643). The simple draw-knife, a forerunner of the spoke-shave, has a handle at either end of the blade and is drawn towards the body; it is the most effective instrument for roughly planing a convex surface such as a wheel-spoke or broom-handle. It was well known to the Romans, and must have been used in the Middle Ages, though pictures of it are hard to come by until the sixteenth century (figure 225 E). Another important bladed tool, the plane, has had a varied history. Examples of this highly developed wood-working tool have been found at Pompeii [6] and in Roman Britain, but it is supposed to have suffered an eclipse until about the thirteenth century (figure 358). A writer of the fourteenth century is confident that the plane was unknown in his youth and introduced in his time. The planing of a surface properly involves at least two sizes of plane: the small smoothing-plane, which is not able to remove large surface-curves, and the long trying-plane for the final levelling. A carpenter of *c* 1500 has fine examples of each of these (figure 356). At that time there may well also have been moulding-planes, which again were known to the

FIGURE 358—*Carpenters working on the ark. On the left is a very early medieval plane, in the centre a pit-saw, on the right an auger, and, below it, a trimming-axe with a very wide blade. From a thirteenth-century mosaic in the porch of the cathedral of St Mark, Venice.*

Romans [7] (p 232), and were very popular in the carpentry of the Renaissance (figure 220).

The Romans might also claim to have introduced all the drill-bits used up to the time of the invention of the spiral bit in the nineteenth century. This tool carries the wood-shavings out of the hole. The medieval spoon-shaped bits were round-ended or pointed (plate 30 A), and often screwed for a few turns (figure 359). The shell-bit was in the shape of a half-cylinder up to four inches long with sharpened sides (figure 354); a hole for it had first to be made with mallet and gouge. The centre-bit might be compared to a three-pronged fork with one or both of the side-prongs modified into a cutting-blade. For deep holes, shell-bits were preferred, even if screw-tips were available, for the latter tend to follow the grain and so throw a deep hole out of true.

Where medieval carpenters improved on all their predecessors was in their method of turning these bits. The crank—an invention of far-reaching importance (pp 652 f)—was first used, in the form of two right-angled bends, to turn wheels (figure 593) and windlasses (plate 31 A). But the idea was soon adopted by the tool-maker, probably in the fourteenth century, as four right-angled bends, for turning drills.[1] The well equipped carpenter in figure 356 had two braces in his tool-rack (see also figure 350, plate 30 A). For heavy work the

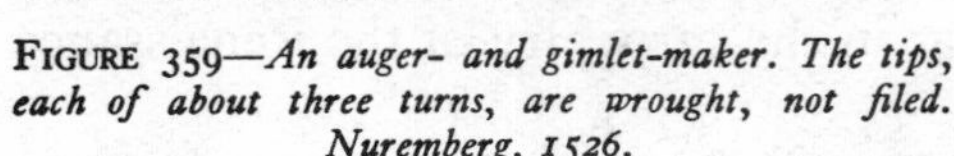

FIGURE 359—*An auger- and gimlet-maker. The tips, each of about three turns, are wrought, not filed. Nuremberg. 1526.*

FIGURE 360—*A file-maker. He cuts each line of the file with a blow from a sharp-edged hammer. His anvil is spiked into a large block of wood. Nuremberg. 1417.*

brace did not displace the T-handled auger (figures 350, 354), and there were always jobs that could most easily be performed with its small equivalent, the gimlet (figure 359).

Handsome claw-hammers are illustrated (figure 353). Such well balanced hammers were valuable, for the crude hand-made nails (figure 40) could not be driven in with the careless abandon of the amateur carpenter of today—a hole had first to be prepared with a bradawl. Both nails and wooden pegs can be seen in plate 30 A. The latter have always been considered more suitable for joinery and for large pieces of timber-work such as roof-trusses. Metal screws were a rarity in this period. Even in the early nineteenth century they were not made with points,

[1] A sort of half-brace made from a rod bent twice has been claimed for the Romans, and even for the Assyrians. See Petrie, 'Tools and Weapons', plate LXXVIII, M 3. Because of its similarity with the ancient eccentric-handled quern such an early date is not unlikely.

so that a hole had first to be prepared for the full length of the screw. The use of glue was prohibited by guild regulations. Such aqueous glues as were known (p 363) were unsuitable for anything that might come into contact with humid conditions.

Files were made by striking a series of closely spaced blows on a flat piece of iron with a sharp-edged hammer (figure 360). This operation must have required considerable skill. The file was then tempered and quenched. Leonardo da Vinci applied himself to the design of an automatic file-maker in which, after each blow of the hammer, the file was advanced a short distance by means of a screw.

FIGURE 361—*Ship-builders finishing off a hull with hammers and unhafted chisels, and boring with an auger. The man, upper right, takes from the tool-basket what is probably an adze. From a thirteenth-century high relief on the outside of the cathedral of St Mark, Venice.*

For sharpening tools various homogeneous stones, often metamorphic rocks, were used, but even the best-chosen of natural stones, such as slates, must have given comparatively inferior service. Nothing so hard or free from flaws as our synthetic Carborundum oil-stone was known until very recent times.

It should be realized how multifarious was the work of the medieval carpenter. Timber was used not only for the frameworks of houses (plate 30 A), and ship-building (figures 353, 361), but for very many small everyday objects, such as milk-buckets, wash-tubs (figure 230), and trunks, where we employ other materials. The wheelwright often prized curved pieces of wood passed over by the carpenter. Moreover water-mills, and from the fourteenth century wind-mills, involved large pieces of machinery made almost completely of wood (pp 608 ff, 623 ff). Great skill and experience went into the choice and erection of the huge posts which bore the whole weight of the early post-mills (figures 561–3). Europe, however, like the Near East before her, was faced with the ever increasing problem of shortage of timber.

FIGURE 362—*Blacksmiths. Two men beat out a large lump of red-hot iron on the anvil. Blast for the forge comes from two manually operated bellows working alternately. From an English manuscript of the fourteenth century.*

IV. THE BLACKSMITH

The third member of our triumvirate of artisans, the blacksmith, worked in close co-operation with mason and carpenter. His wrought iron, as well as being used in the manufacture and maintenance of all artisans' tools, including those of the farmer, and in the shoeing of horses (figures 41, 42), contributed to the beauty and utility of the carpenter's work and the stability and grace of the mason's.

The medieval blacksmith possessed a specialized kit of tools, which included hammers, tongs, chisels, and cutters. His forge was as compact, except in the means of raising blast, as those to be seen in rural districts today (figure 362). The bigger establishments were even acquiring heavy, water-driven hammers (p 73). Yet much skill and experience went into the welding and shaping of simple tools: the farmer's scythe (pp 95 f), for instance, whose body is of wrought iron for toughness, and whose blade is of steel. The art of the armourer and the locksmith were further specialized.

This short survey has perhaps revealed the basic differences between the medieval artisan's methods and those of today. Small changes and improvements of the common tools there certainly were. Nevertheless one factor, the universal application of mechanical power, has since dramatically changed their setting. And the beginning of that change had already made its appearance far earlier than is usually reckoned. Its herald was the windmill, which appeared well within the period treated in this volume and immensely increased the amount of power available.

REFERENCES

[1] KNOOP, D. and JONES, G. P. 'The Mediæval Mason', p. 51. University Press, Manchester. 1933.

[2] HARVEY, J. 'English Mediaeval Architects.' Batsford, London. 1955.

[3] KNOOP, D. and JONES, G. P. See ref. [1], p. 86.
ANDREWS, F. B. 'The Mediaeval Builder and his Methods.' *Bgham Archaeol. Soc. Trans. Proc. for 1922*, **48**, 63, 1925.

[4] SALZMAN, L. F. 'English Industries of the Middle Ages', p. 173. Clarendon Press, Oxford. 1923.

[5] ROSE, W. 'The Village Carpenter', p. 80. University Press, Cambridge. 1937.

[6] PETRIE, SIR (WILLIAM MATTHEW) FLINDERS. 'Tools and Weapons', Pl. XLIII, figure 39. Egypt. Res. Acc. and Brit. Sch. Archaeol. Egypt, Publ. 30. London. 1917.

[7] CURLE, J. 'A Roman Frontier Post and its People', figure 21. Maclehose, Glasgow. 1911.

12

BUILDING-CONSTRUCTION

MARTIN S. BRIGGS

I. THE GREEK PERIOD

THE subject discussed here is confined to Europe and especially to those areas most influenced by classical or Mediterranean culture. The date of this impact naturally varied from country to country.

Greek civilization did not cease abruptly at any given date, to be followed by Roman civilization. The two ran concurrently for several centuries before the Christian era. The so-called Hellenistic buildings of that period were Greek in character though erected under Roman auspices. Moreover, much of our knowledge of Greek methods of building is derived from the remarkable manual of building-construction of the Roman architect Vitruvius, written during the reign of Augustus (27 B.C.–A.D. 14). His name must therefore be mentioned in reference to many features of Greek building. He was an accurate observer, fully qualified to describe and criticize the examples of Greek architecture still extant in his day.

In the minds of most of us, Greek architecture is synonymous with great marble temples and theatres, but the earliest Greek buildings were constructed of mud-bricks and timber-framing. They must have differed little from such European buildings of the Neolithic and Early Bronze Ages already described (vol I, p 306). Nevertheless, many of the notable achievements in brick vaulting as well as in masonry construction, which the nations of the Near East had created in preceding periods (vol I, ch 17), were ignored by the capable architects of the Golden Age of Athens in the mid-fifth century B.C. The Athenian marble masterpieces were derived rather from timber prototypes than from the palaces of Babylonia or the tombs of Mycenae. In other words they are distinctively European rather than Asiatic in essential type.

Vitruvius says little of the timber-supply in Greece. It is certain, however, that the walls of the earliest form of temples were constructed entirely of mud bricks, usually resting on a stone plinth and having a timber roof covered with thatch. As the size, and especially the width, of such buildings increased, a row of wooden posts was erected down the middle of the longer axis. These posts carried transverse beams across the temple, and also struts or small posts which supported the ridge-beam.

In the next stage of development, timber posts were inserted in the outer mud-brick walls, one opposite each post of the central row. This formed a primitive type of timber-framing, an example of which was the temple of Artemis Orthia at Sparta (ninth or eighth century B.C.). When the temple at Samos was rebuilt, before 775 B.C., there was a central line of posts, with other posts at the gable-ends. In the temple of Hera at Olympia (*c* 640 B.C.)—the first temple surrounded by a colonnade—the entire weight of the roof was carried on a framework of timber posts, with a filling of clay blocks or mud-bricks, the whole being plastered over and painted. Subsequently these posts were replaced by stone columns, but the original stone plinth still survives. In Greece, as in many other countries, this ancient half-timbered type of construction has lingered on to this day.

As for Greek brick-work, it appears that mud-bricks or sun-dried bricks were always used in the early centuries, for public buildings as well as for private houses. One notable example was the half-timbered superstructure of the Older Propylon at Athens; others are the walls of most of the one-storey houses (fifth–fourth centuries B.C.) at Olynthus. Burnt bricks did not appear until the middle of the fourth century B.C., and then only occasionally. Vitruvius states that two sizes of bricks were used by the Greeks: they were either 'five or four palms every way' [1]. Public buildings were erected with the former, private buildings with the latter.

Greece was, however, a land of building-stone, chiefly limestone and marble. It has been called 'a marble peninsula', and all its important buildings in the classical period were of one or other of these materials. A notable variety of hard limestone, occurring round the plain of Argos, could easily be split into irregular blocks and was so used in the ancient polygonal masonry of Tiryns and Mycenae; but in classical times the material chiefly employed was *pōros*[1] limestone, found in the west and north of the Peloponnese (p 400). This stone had a rough surface and was full of cavities, thus providing an excellent key for plaster. Indeed, it was often plastered all over with fine stucco, inside and outside a building, and was then colour-washed. The stucco was made by burning limestone.

The marble most often used in Greek buildings, Pentelic marble, was and still is quarried at Mount Pentelicus, a few miles north of Athens. It consists

[1] *Pōros* stone is described by Theophrastus (*c* 371–*c* 287 B.C.) in his book 'On Stones' as a kind of marble suited to building. The verb *pōroein* means to make stony, to petrify. It is unrelated to Greek *poros*, a pore.

almost entirely of calcium carbonate, showing a brilliant white fracture when broken. This material was employed for the Parthenon, the 'Theseum' (figure 363), the Erechtheum, the Propylaea, the temple of Zeus Olympios, and most of the other principal public buildings of Athens. Its nature encouraged the adoption of miraculously fine joints, a very smooth finish, and all those

FIGURE 363—*Temple of Hephaistos, called the* 'Theseum' *on account of representations on its frieze of the life of Theseus. Athens. Fifth century* B.C.

amazing optical refinements which made the Parthenon one of the great masterpieces of architecture.

Having progressed from half-timbering, plastered over and colour-washed, to the much more durable limestone coated with stucco and painted, the Greeks now favoured marble, which gave them both durability and sparkling brightness in one medium. They never made much use of coloured marble in building as the Romans did in later days, and when they employed any other variety than Pentelic it was generally white Parian marble from the island of Paros; but friezes of black Eleusinian marble were introduced at the Erechtheum and the 'Theseum', with reliefs in white marble attached to them by means of iron clamps.

The Greeks showed no inclination to utilize the dome, arch, or vault, contenting themselves with developing trabeated architecture (Latin *trabs*, beam), that is, a system of posts and beams or lintels.[1] This policy was facilitated by the

[1] Lintels are beams spanning an opening.

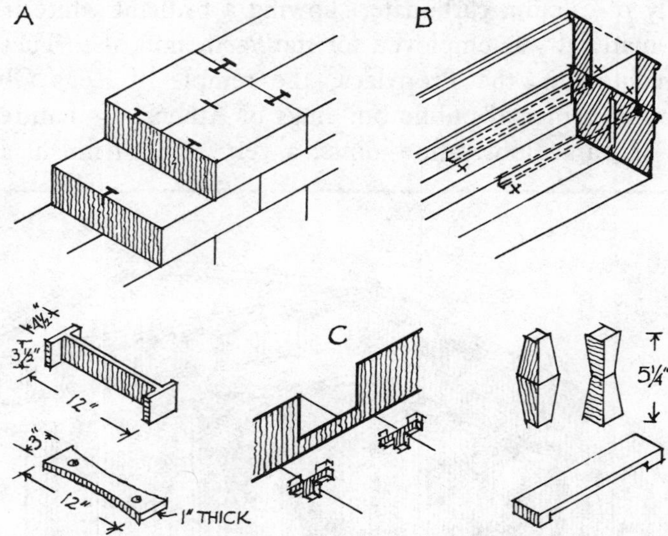

FIGURE 364—*Diagram showing use of metal in Greek masonry:* (A) *typical use of metal clamps;* (B) *structural iron beams (indicated by* X), *at the Propylaea, Athens;* (C) *various dowels and clamps in bronze or iron.*

use of *pōros* limestone and Pentelic marble, each of which could furnish beams up to 12 or 15 ft long, and because suitable timber for roofs was also available.

Much has been written on the excellence of Greek masonry, and certainly it was of a very high order, especially in the Golden Age of Pericles (d 429 B.C.).

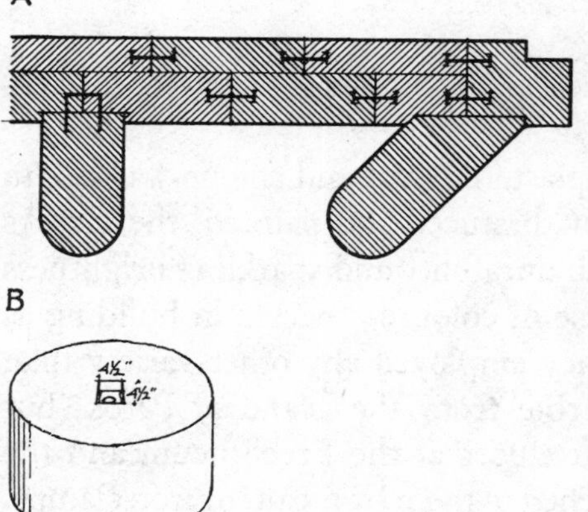

FIGURE 365—*Greek masonry details:* (A) *metal clamps, used in jointing masonry of the Temple of Apollo at Bassae. Fifth century* B.C.; (B) *drum of a column at the Parthenon, Athens, showing* empolia *(sockets). Fifth century* B.C.

Blocks of stone were laid with very fine joints and were clamped together with metal fastenings of various types (figures 364, 365 A). This fineness was obtained by grinding the blocks together until an extremely close contact was achieved. The pins, let into sockets (*empolia*) above and beneath each drum of the columns of the Parthenon, were not—as formerly supposed—to enable the two surfaces to be ground together by rotating, but were the cypress pins and sockets serving merely as dowels holding the drums together. The sockets were 4–6 in square and 3–4 in deep; the pins were about 2 inches in diameter (figure 365 B).

The beautiful finish of the echinus[1] of many Doric marble capitals of the Periclean period suggests that they were turned in a lathe, but there is no evidence for this. On the other hand, the 132 column-bases of the temple of Hera at Samos (probably *c* 575 B.C.) are of soft limestone which shows clear traces of working in a lathe, thus confirming a statement by Pliny.

As to the transport of the enormous blocks found in some of the Greek temples (figure 366), the rough bosses (*ancones*) of marble, 8–10 in square, left projecting some 6–8 in at four points of the circumference of each column-drum in certain of them, have been interpreted as hubs for attaching ropes to roll them from the quarry to the site. This was impossible for practical reasons, and they were surely intended only for lifting. For rolling over short distances devices such as that shown in figure 366 B were employed. Drums over 6 ft in diameter, at the Parthenon, must have been transported in carts drawn by 30 or 40 oxen. At Selinus in Sicily are far larger column-drums, 10 ft 8½ in across, and Doric capitals with a spread of 13 ft; in a neighbouring quarry are drums of 12 ft diameter, partially hewn out of the living rock. Agrigento (*Acragas*) built a temple with columns over 13 ft in diameter. This was, however, beyond the limits of single blocks, so each drum was built up in ashlar courses with radial joints.

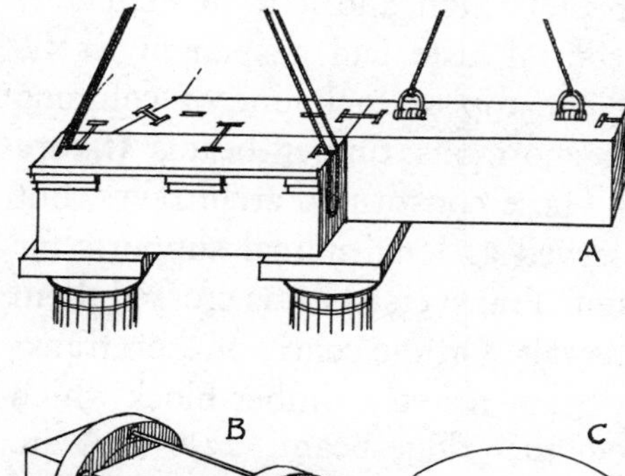

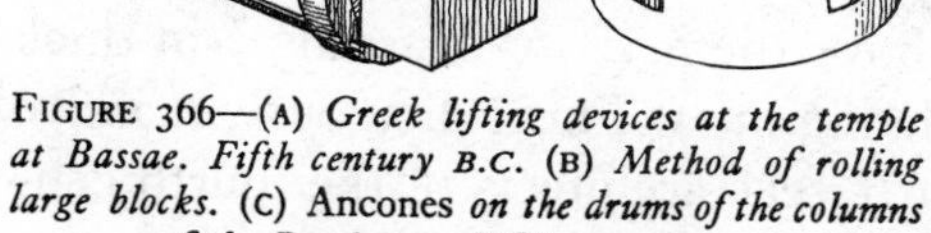

FIGURE 366—(A) *Greek lifting devices at the temple at Bassae. Fifth century* B.C. (B) *Method of rolling large blocks.* (C) Ancones *on the drums of the columns of the Parthenon. Fifth century* B.C.

Many attempts have been made to show in graphic form the origin of Greek trabeated architecture in timber prototypes copied in masonry (figures 367–8).

Apart from timber-framed buildings, there was much heavy structural carpentry in the roofs of most stone and marble temples and other public buildings, but not a trace of it now remains. Any attempt to picture it must be based on written records, or by inference from such details as surviving holes in masonry for the ends of joists, rafters, and beams. Some primitive Greek roofs were steep in pitch and covered with thatch, others flat, as in many eastern countries. Hipped roofs are illustrated in reliefs of *c* 570 B.C. The typical roof of classical times, was, however, of low pitch, about 30°, and covered with tiles of terracotta or marble.

[1] Moulding at the top of the capital, directly under the abacus.

The great wooden roof of the arsenal at the Piraeus, the port of Athens, built between 340 and 330 B.C., was completely destroyed in 86 B.C. A full contemporary specification has survived on a slab of marble. This building, about 434 ft long and 59 wide, had external walls of ashlar masonry. Internally, it was divided into three aisles by ranges of columns about 32 ft high and 3 ft in diameter. The central aisle had a span of 21 ft. Along the top of each line of columns rested enormous timber beams (figure 369). These constituted architraves, but also served as longitudinal supports for the roof. Transverse beams crossed them at intervals. On the centre of each transverse beam rested a timber block which carried the ridge-beam, $22\frac{1}{2} \times 17\frac{3}{4}$ in. Rafters 12×8 inches in cross-section, and spaced 16 in apart, were supported by the external walls, the longitudinal beams along the tops of the columns, and the ridge-piece. On the rafters were laid battens $6\frac{1}{2} \times 1\frac{5}{8}$ in, spaced $3\frac{1}{4}$ in apart, and, above them, close-boarding $\frac{7}{8}$-in thick on which terracotta roofing-tiles were bedded in mud. This ridiculously massive and costly form of construction proves that Greek architects of the fourth century were unacquainted with the principles of trussing.

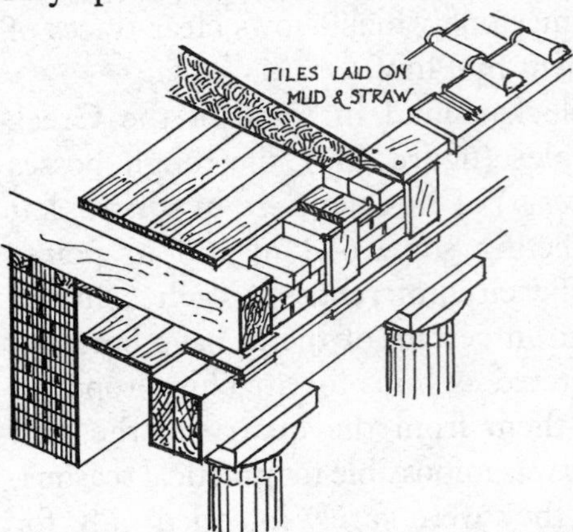

FIGURE 367—*Timber origins of Greek masonry construction.*

Over the roofs of the *cella* (sanctuary) of the Erechtheum (*c* 421 B.C.), however, where there was a clear span of over 32 ft, diagonal struts were used to brace the timber transverse girders carrying the wooden beams of the ceiling. At the arsenal of Pergamum there was a heavy timber-framed floor of beams, joists, and boarding.

The roofs of all important Greek buildings were covered with tiles of terracotta or marble. The former (figure 370) were of one of three types: Laconian or Spartan, concave pantiles with joints protected by convex cover-tiles; Corinthian, flat pantiles with raised rims and triangular or saddle-shaped cover tiles; and Ionian, closely resembling the Corinthian. There were also

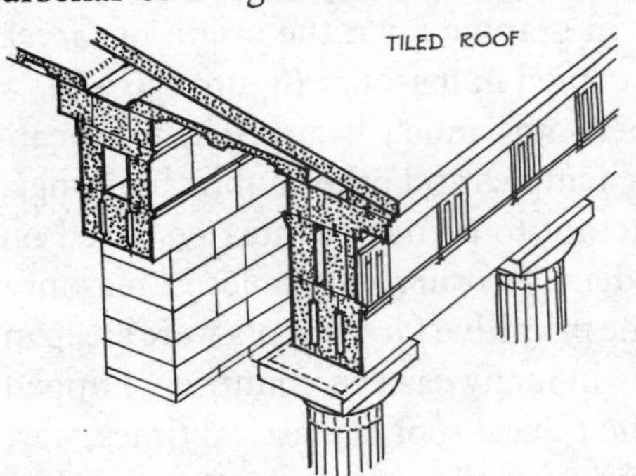

FIGURE 368—*Greek masonry construction, from the Parthenon at Athens. Fifth century B.C.*

special tiles for particular purposes, such as eaves and ridges; and, of course, local variations occurred. As early as *c* 530 B.C., when the temple of Apollo at Calydon was repaired, the tiles were supplied from Corinth, each numbered for its position on the roof. The use of marble tiles for roofing can also be traced back to the mid-sixth century B.C. at least.

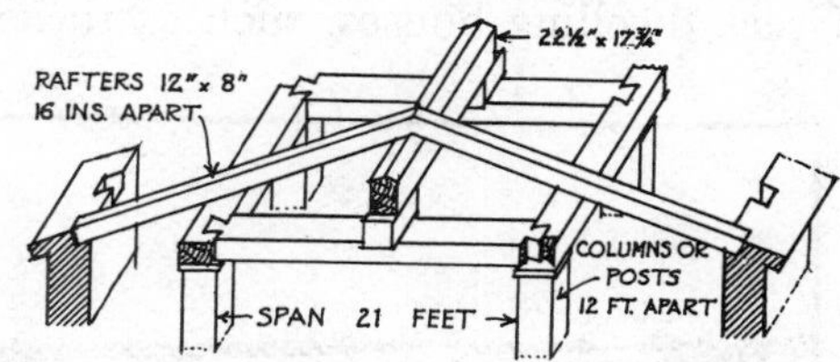

FIGURE 369—*Roof of the arsenal at the Piraeus Fourth century B.C.*

One of the most startling recent discoveries is the considerable structural use of iron in these great marble and stone temples, notably at the Parthenon. Here broad flat wrought iron beams were used as cantilevers to support the heaviest statues of the pediment, their ends being built into the masonry of its recessed face. Special provision was made so that deflexion would not cause them to bear directly on the marble cornice. At the adjoining Erechtheum, however, no such allowance for deflexion was made, and a marble beam did actually crack. In one temple of *c* 470 B.C. at Agrigento, iron beams of 5×12 inches in cross-section and 15 ft long were let into the under surfaces of the architraves and rested on the column-capitals. In another temple at the same place, iron cantilevers carry the topmost member of the cornice. In the Propylaea at Athens concealed iron beams 6 ft long transmitted loads from the heavy marble ceiling-beams on to the Ionic columns (figure 364 B). Here, too, due allowance was made for deflexion.

The Greeks, though conservative in most structural matters, seem to have had complete confidence in iron. They used it freely for lifting-tackle, such as lewis-irons and tongs, and also for many kinds of plugs and clamps (figure 364 A). Sometimes they used molten lead in their dovetail-clamps. Iron nails of various forms were employed to fix terracotta facings to timber or stone structures, and for other purposes. Bronze was used for plugs and dowels (figure 364 C). Doors often had bronze handles and enrichments. Some tombs of the Hellenistic period had marble doors as much as 10 ft 3 in high, with bronze enrichments. Limestone doors were occasionally used. There were window-openings in Greek buildings, but little is known of the windows themselves.

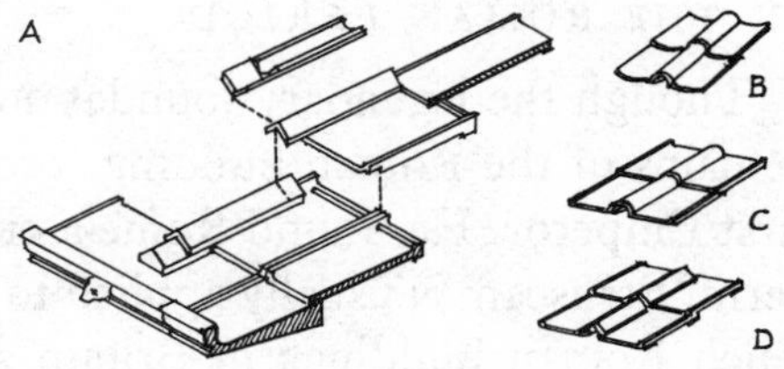

FIGURE 370—*Greek roofing-tiles:* (A) *from the temple at Rhamnus, fifth century B.C.;* (B) *Laconian;* (C) *Sicilian;* (D) *Corinthian.*

The Greeks never developed an elaborate system of central heating as did the Romans (pp 419–20). Their houses were warmed by portable stoves or braziers. Similarly, arrangements for water-supply and bathing were rudimentary compared with those of the Romans under the Empire (pp 418–19). The ordinary Greek dwelling-houses, such as those of Delos (second century B.C.), had tiled

FIGURE 371—*Roman aqueduct* (Pont du Gard) *near Nîmes. 19 B.C.*

roofs delivering rain-water through lead down-pipes; these were fixed on the columns surrounding the central courtyard, where a large rainwater-tank was sunk from which water was drawn through a well-head.

II. THE ROMAN PERIOD

Though the legendary foundation of Rome took place in 753 B.C., very little remains of the Roman buildings erected before Augustus (27 B.C.–A.D. 14), the first Emperor. He 'found Rome a city of brick and left it a city of marble'. The term 'Etruscan' is usually applied to buildings earlier in date than 150–100 B.C. Such Roman buildings in Britain as will be discussed were erected after the governorship of Agricola (A.D. 77–84) and before the evacuation of Britain by the Roman army (*c* A.D. 410).

Vitruvius (p 397) played a prominent part in the rebuilding of Rome, but, though a most reliable informant about the technique of building-construction in his own day, he can tell nothing of the greatest monuments of Rome itself or of its provincial cities, for they were erected after his time. Many facts about aqueducts and water-supply are given in the *De Aquis* (*c* A.D. 100) of Frontinus

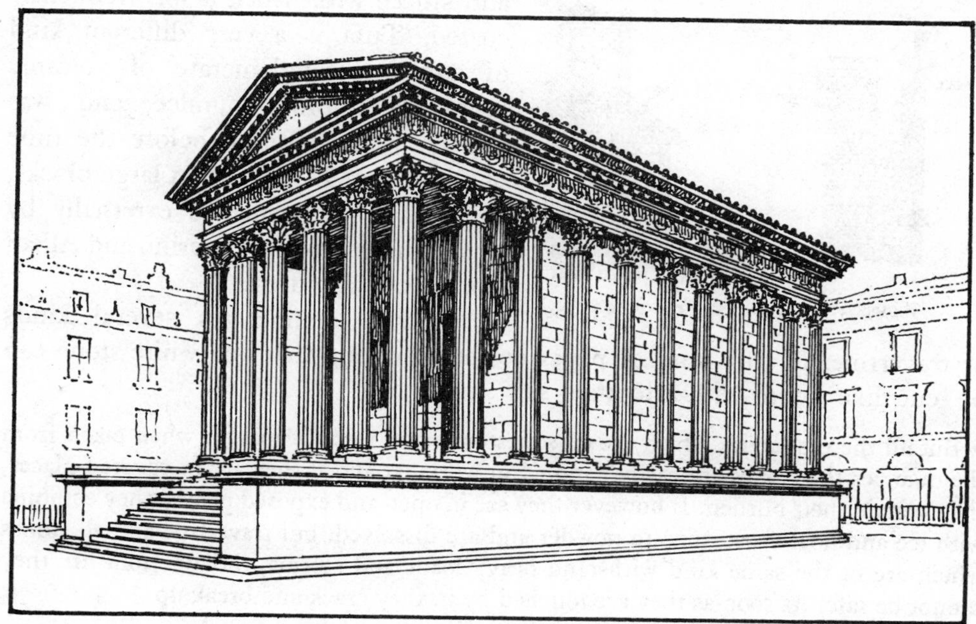

FIGURE 372—*Roman temple* (Maison Carrée) *at Nîmes. First century* A.D.

(p 419), water-commissioner of Rome. Much of our information on Roman building-construction must be deduced from the buildings themselves, mostly ruined. It may be assumed that all buildings mentioned here were erected between *c* 90 B.C. and *c* A.D. 410, unless otherwise stated. Greece became a Roman province in 146 B.C., and Greek building-methods exerted a powerful influence on Roman construction.

Generally speaking, the Romans derived their knowledge of arched construction from their Etruscan forbears, and developed the methods of vaulting and dome-building, in brick, concrete, and masonry, to a very high level under the Empire (figure 371). From the Greeks they borrowed post-and-lintel or trabeated construction, adapting it to their own purpose, especially for temples (figure 372) and public buildings, but often using it for decorative rather than for structural reasons.

Of building-stones (pp 27 ff), the most popular in Rome itself was travertine, *lapis tiburtinus*, so called because it was chiefly quarried near Tivoli (*Tibur*). It is a cream or brown limestone from deposits in the valleys of the Tiber and Aniene (*Anio*). It varies greatly in texture, being sometimes fine, sometimes coarse with numerous cavities, which afforded excellent binding for the plaster and stucco with which it was frequently coated. Tufa is a very different kind of stone, a conglomerate of volcanic origin, resembling pumice, and was largely used in Rome before the time of Augustus. It was laid in large blocks, and had to be protected externally by a covering of stucco. Peperino and Alban stone were also used.

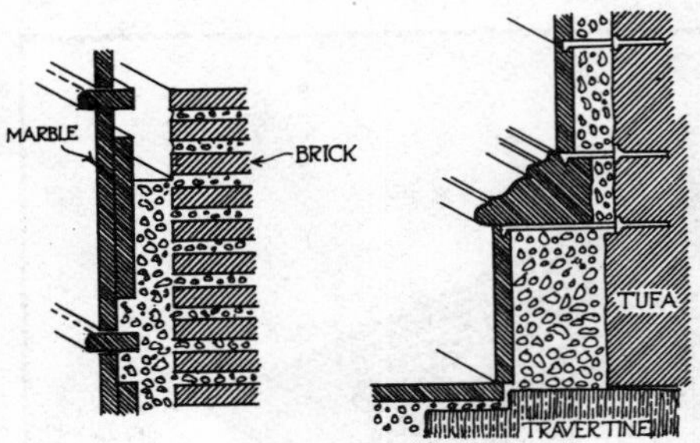

FIGURE 373—*Roman marble facings.*

Vitruvius enumerates several kinds of travertine, of tufa, and of peperino, commenting that one white stone can be 'cut, like wood, with a toothed saw'.

> But all these quarries which are of soft stone have this advantage: when taken from the quarries the stones are easily handled in working, and if they are in covered places, they sustain their burden. If however they are in open and exposed places, they combine with ice and hoar-frost, turn to powder and are dissolved, but travertine and all stones which are of the same kind withstand heavy loads and storms, though from fire they cannot be safe. As soon as they are touched by it, they crack and break up.

He proceeds to say that, if circumstances compel the use of some other varieties from quarries conveniently near Rome, certain precautions must be taken.

> Let the stone be got out two years before, in summer but not in winter, and let it lie in exposed places. Those stones, which in this time are damaged by weathering, are to be thrown into the foundations. Those which are not faulty are tested by Nature, and can endure when used in building above ground [2].

Marble was not introduced into Roman buildings till the first century B.C. and was rare before Augustus (figure 373). When Crassus built a house on the Palatine about 92 B.C. his use of a few small marble columns was ridiculed as Greek luxury. Soon after, the marble quarries at Carrara were opened. Other white marbles were imported from Mounts Hymettus and Pentelicus (pp 25, 398–9). The florid taste of the Romans led to the importation of many coloured foreign varieties. Granite from Egypt, Elba, and Naxos also came into use. Much red basalt, known as porphyry, was brought from Egypt for great monolithic columns and for the geometrical paving called *opus Alexandrinum*. The granite drums of Trajan's column (details of the frieze are seen in figure 467 and ch 20) are 6 ft 3 inches in diameter.

Among mineral products found near Rome, the most important was *pulvis puteolanus* (pozzolana), a volcanic earth occurring in thick strata in the Alban Hills and near Naples (p 410). This earth when mixed with lime forms a useful cement, which sets very strongly under water and also has fire-resisting qualities. It rendered possible the wonderful achievements of the Romans in building the concrete vaults and domes of the *thermae* (baths), for mortar made with it was as strong as the aggregate, whether that was of broken brick or of stone. Vitruvius gives a detailed description of the use of pozzolana [3].

He also writes that sand should be:

without the admixture of earth. If there are no sand-pits, then it must be sifted out from river-bed, gravel, or sea-shore.

Such sand is, however, less satisfactory, for it is hard to dry, sets slowly and uncertainly with cement, and is unsuitable for vaulting. Sea-sand, when plastered surfaces containing it are laid upon walls, is liable to effloresce or, as he expresses it, 'will discharge the salt of the sands' [4].

Of lime he observes that:

we must be careful to burn it out of white stone or lava; lime prepared from thick and harder stone will be useful for structural purposes; and that from porous material, for plastering. When slaked, let it be mixed with sand in the proportions: for pit-sand, three parts to one of lime; for river-sand or sea-sand, two to one of lime. . . . Also, with river-sand or sea-sand, if one part be added to two or three of crushed and sifted potsherds, the result will be a more effective mixture [5].

He describes the processes of burning and slaking lime, and what seem to him to be the principles involved. He points out that, after being burnt in the kiln, the lime has lost about one-third of its previous weight.

Bricks were largely used in Roman walling and vaulting. Up to the time of Augustus, Roman bricks seem to have been of sun-dried clay, but from then onwards kiln-burnt bricks came into general use. Of sun-dried bricks Vitruvius says:

They ought not to be made from sandy or chalky or gravelly soil, because they first become heavy, and then, when they are soaked by rain, disintegrate, and the straw does not stick to them because of their roughness. Bricks should be made of white clayey earth or red earth, or even rough gravel, for they will then be durable, light and easily put together. Bricks should be made either in the spring or in the autumn, so that they

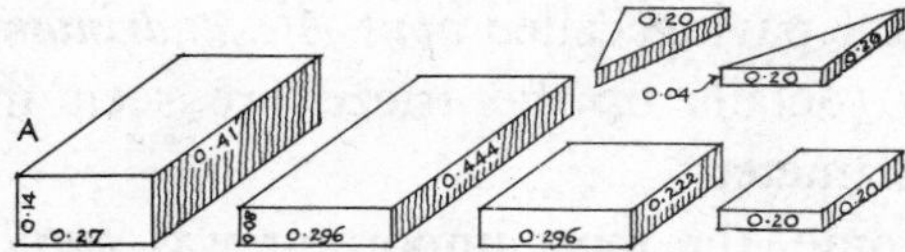

FIGURE 374—*Types of Roman bricks; (A) is Etruscan.*

will dry at the same time [throughout]. They will be far more durable if made two years ahead, for they cannot dry properly in less [6].

He adds that at Utica, near Tunis, the accepted age of bricks for wall-building is no less than five years.

Vitruvius describes three types of bricks, two of which are square and used by the Greeks (p 398), and a third oblong and used by the Romans: he gives its size as 1½ ft long and 1 ft wide, but square and triangular bricks were also largely used, the standard size being 2 Roman ft square, about 23 English inches. The normal thickness for all these, as well as the smaller bricks, was about 1½ in, but oblong 'Etruscan' bricks, *c* 16×10½ in, were as much as 5½ in thick.

The size of Roman bricks (figure 374) decreased progressively with the years, whereas the thickness of mortar joints increased, as the quality and strength of the mortar were improved. Yet, although Roman builders employed bricks so freely and so skilfully, they apparently never considered that brickwork as a

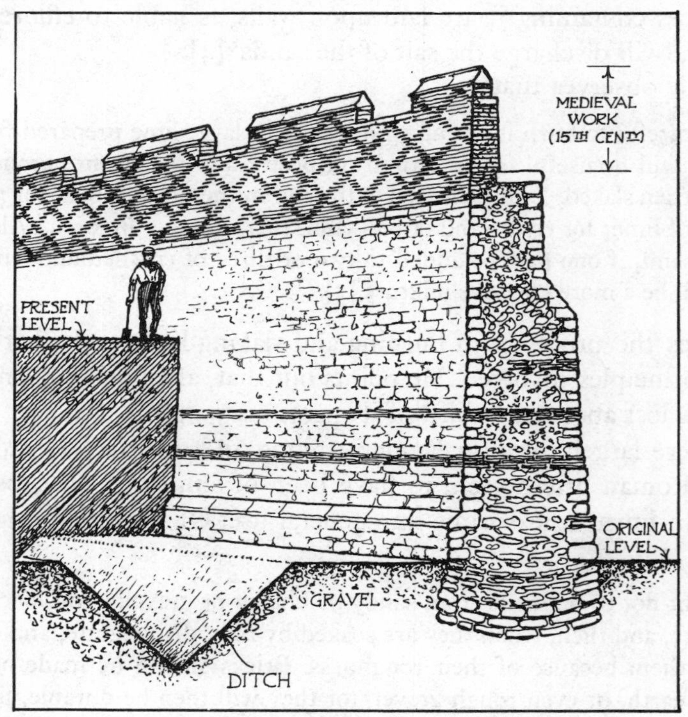

FIGURE 375—*The Roman Wall of London.*

facing-material might be pleasant to behold, and always covered it with plaster or marble veneer. Kiln-burnt bricks formed a useful coping to mud-brick walls, and they were often used as lacing-courses in rubble and flint walls, as may be seen in many British examples (figure 375). Apart from bricks of the size men-

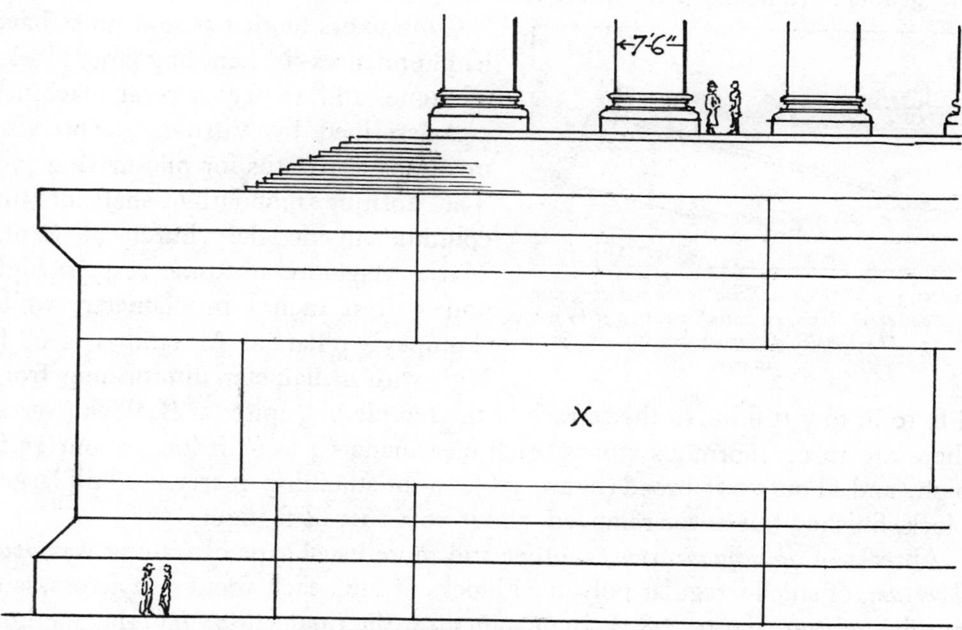

FIGURE 376—*Part of the terrace of the Roman temple of Jupiter at Baalbek in Syria (second century* A.D.*) showing the colossal masonry, including at 'X' one of the three great stones known as the* trilithon, *each of which measures 63 to 65 ft by* c *13 by* c *10.*

tioned above, the Romans in Britain and elsewhere also used small tiles about $7\frac{3}{4}$ in square for the columns of their hypocausts (figure 385), and large tiles about 23 in square for the floor of the rooms over the hypocausts. Except for these specific purposes, and for arches, the Romans in Britain seldom employed brickwork for solid walls, or even for facing walls.

Where stone was used, it was almost invariably laid in regular courses of squared blocks. For more finished work, dressing with a punch was often used, and at Corbridge, Northumberland, there is a rare example of the chisel-draughted margin, that is, a chiselled border round the face of each block. Random rubble is rarely found. In spite of its high reputation in Italy, Roman mortar in Britain was frequently poor. In some of the larger buildings, the interiors of stone walls were grouted and the face was flush-pointed. In Rome, homogeneous walls of squared stones, arranged in alternate courses of headers and stretchers, were known as *opus quadratum*; this was true masonry, derived from Etruscan practice. The blocks were jointed with iron clamps or dowels run with lead. In some famous surviving examples of this style of masonry the headers are nearly 2 ft square and the stretchers twice as long.

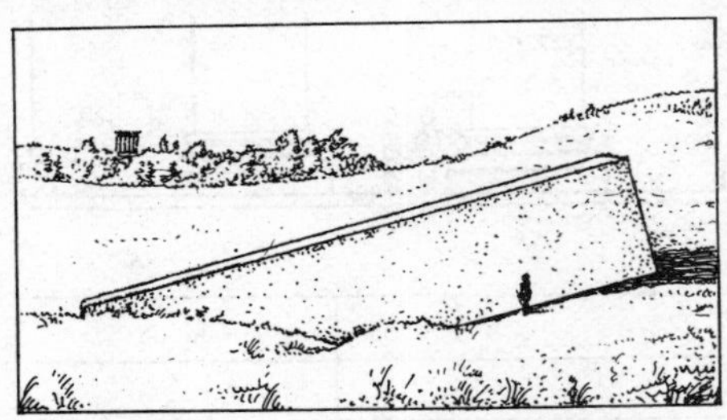

FIGURE 377—*The 'Great Stone' in the southern quarry at Baalbek in Syria, measuring about 70 ft by c 16 by 14. The temple of Jupiter can be seen in the distance.*

Contractors at that period must have had appliances for handling large blocks of stone, and in fact several machines are described by Vitruvius, who also mentions apparatus for pile-driving [7]. The antique monolithic shaft of the column outside the church of Santa Maria Maggiore at Rome is 47 ft high and 5 ft 9 inches in diameter, while 'Pompey's pillar' at Alexandria is 68 ft high with a diameter diminishing from 8 ft 10 in to 7 ft 6 in. In the terrace of the temple of Jupiter at Baalbek (Syria) there are three enormous stones, each measuring 63 to 65 ft long, about 13 ft high, and about 10 ft broad (figure 376). In an adjoining quarry is a still larger block, finished but never removed, about 70 × 16 × 14 ft (figure 377).

Abreast of *opus quadratum*, another and more usual type of walling was *opus incertum*, of small irregular polygonal blocks of tufa, each about 5 in across, laid in thick mortar (figure 378 A). In or soon after the first century B.C. this method yielded to *opus reticulatum*, in which the irregular blocks came to be roughly squared and laid in diagonal rows suggesting a net (Latin, *rete*) (figure 378 B). Special stone blocks were made for the quoins. Vitruvius states that '*reticulatum* looks better, but its constitution makes it liable to crack' [8]. Horizontal bonding-courses of flat bricks or tiles, at regular intervals, were soon introduced to strengthen it.

From the first century B.C. onwards, however, the commonest building material in Rome was concrete, not only for walling and foundations (figure 378 D) but for the vaults and domes which formed the finest achievement of later Roman architecture (figures 378 C, 379). Walls were cast between timber shuttering, which had the struts on the inside and was often tied together by timber battens while the concrete was poured in. The mixture consisted of pozzolana cement with an aggregate of broken peperino, tufa, or bricks, and was spread in alternating layers of large stones and smaller aggregate.

About the time of Augustus, wooden lintels were still in use over doorways and window-openings in small houses, but flat arches were replacing them in other types of building. Tiers of brick relieving-arches were introduced with great effect in the huge concrete walls of the Pantheon. Nero's enormous palace (A.D. 64 onwards) was a concrete structure faced with brickwork. Marble facings, as used in the mausoleum of Augustus (28 B.C.), later became a characteristic feature of the imperial palaces and of many public buildings; hitherto they had

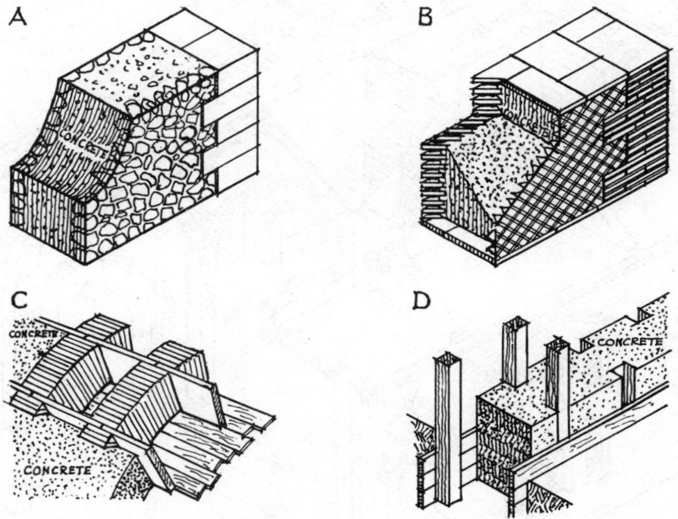

FIGURE 378—*Details of Roman brickwork, concrete, and masonry:* (A) opus incertum; (B) opus reticulatum (left)*;* (C) *brick ribs in concrete vaulting;* (D) *foundation-trenches for concrete walls.*

been confined to private houses. Figure 373 shows how they were attached to the walls.

Roman vaults, excluding domes, were invariably barrel-vaults, or intersecting barrel-vaults (figures 379, 396 A). A vault is simply a continuous arch. Even before Roman times, builders had learnt to construct vaults of wedge-shaped blocks so that the stability of the vault did not depend upon wide mortar joints. As in the building of arches, the erection of a vault involves temporary support during construction, usually by means of timber centring (that is, massive frames of the intended shape of the vault), which is costly.

As the Romans extended their vault-building ventures to large spaces, such as the great halls of the *thermae*, they were hindered by this difficulty of support, and also by the inherently defective form of the barrel-vault itself, which is not only dark but exerts an immense outward thrust upon its supporting walls and piers. Yet the vaulted roof over the hall of Diocletian's palace had the enormous span of 100 ft, and that over Constantine's basilica was 83 ft wide. These astounding feats of construction were made possible only by the addition of cross-vaults buttressing the walls and piers of the main vault against the outward

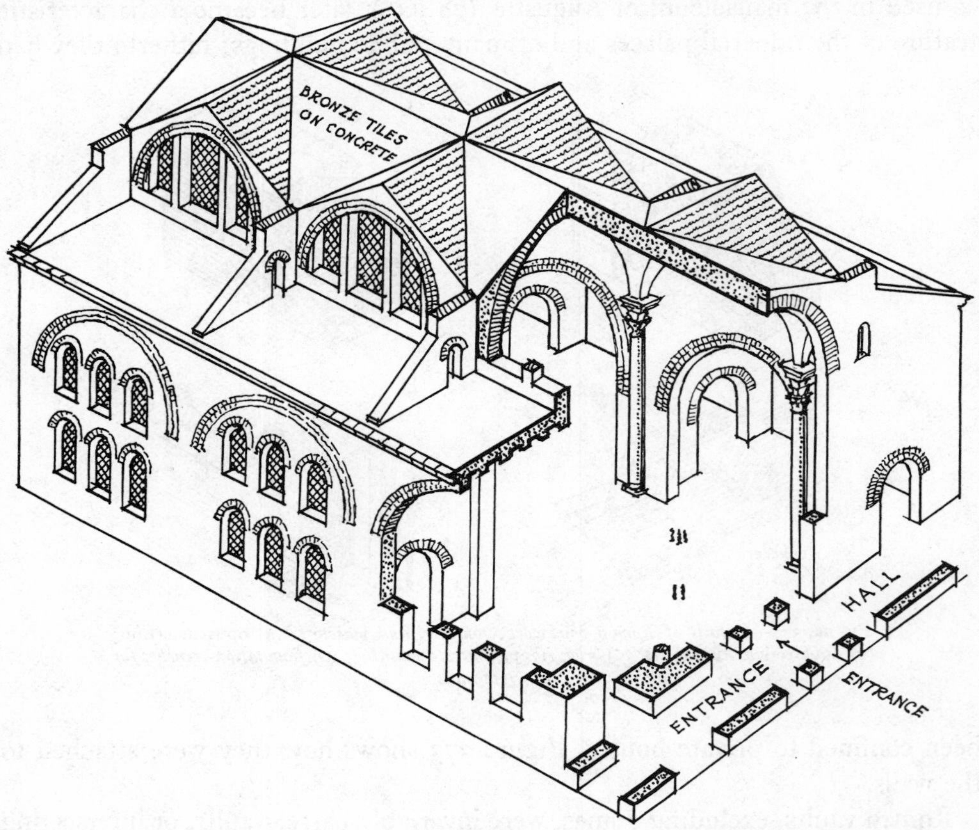

FIGURE 379—*Roman vaulted construction in concrete and brick. The* Basilica Nova *of Maxentius at Rome, c A.D. 313. Portion removed to show construction.*

thrust, as in figure 379, and by a lightening of the weight of the main vault itself. This latter object was achieved by the use of light pozzolana concrete introduced between brick arches or ribs (figure 378 C). Thus the amount of timber centring required to support the boarding for the concrete—in addition, of course, to that needed to support the arched brick ribs—was reduced to comparatively light timber frames or trusses spaced about 10 ft apart.

A dome is a vault that is circular in plan, and presents similar structural problems. It may be built up of bricks or stones, or be formed of concrete. The huge stone lid of Theodoric's mausoleum at Ravenna (figure 389) can hardly be called a dome. Domed structures date back to very early times; thus the treasury of Atreus at Mycenae, of about 1450 B.C. (vol I, figure 318), is constructed of stone in horizontal courses, each projecting slightly over the one beneath till the apex or crown is reached. The most famous Roman dome is that of the Pantheon, (A.D. 120–4), which has the immense internal diameter of 142 ft. It rests on a circular concrete wall, strengthened by a system of massive brick relieving-arches. The dome itself is hemispherical in form, but no one has yet discovered to what extent it is of brick and of concrete respectively, or precisely how it was constructed. Its upper part is 4 ft thick; the lower part is much thicker, but is lightened by rows of coffering.

Vitruvius discusses extensively the different varieties of timber and their uses in building [9]. He recommends felling during winter, when the leaves have withered and, he thinks, the roots having drawn sap from the soil recover their solidity. He gives advice about the conversion of the tree into planks, 'so that the sap may dry out by dripping'.

As regards the types of timber, Vitruvius states that fir is naturally rigid, and does not deflect much when used for flooring; however, it burns easily. The lower part of the trunk is generally quartered, the sapwood being rejected and used for inside work. Oak (*Quercus robur*) resists moisture well, even when buried in foundations. Winter oak (*Q. aesculus*), Turkey oak (*Q. cerris*), and beech are liable to decay. Poplar, willow, and lime are soft, white, porous, and suitable for carving. Alder has the remarkable quality of being 'imperishable underground', and will 'uphold immense weights of walling and preserve them without decaying. . . . All buildings at Ravenna, both public and private, have piles of alder under their foundations. . . .' Cypress and pine, containing an excess of moisture, tend to warp but last long without decay, and the oil in cypress resists dry-rot and worm. Larch contains a bitter sap which also resists dry-rot and worm, but it is poor fuel and so heavy that it will not float.

The Romans took steps to make timber fireproof, and were not alone in doing so: when they besieged the Piraeus, in 86 B.C., they were unable to burn down a wooden tower because the defenders had saturated it with alum.

Vitruvius describes many practical details of carpentry. Thus, referring to the so-called Etruscan order or architectural system, he observes that the beams above the columns must be 'so coupled with dowels and mortices that the coupling allows an interval of 2 in between the joists. For when they touch one another and do not admit a breathing-space and passage of air, they are heated and quickly decay' [10]. He explains the use of timber in the oldest

Roman temples, justly concluding that 'craftsmen imitated such arrangements when they built temples of stone and marble'. Dealing with wooden floors, he expresses a preference for winter oak and recommends the use of thin boards: 'for the weaker they are, the more easily are they kept in place by nails. Two nails should be driven into each joist at the edges of the board, so that the corners of the plank may not warp and rise up' [11].

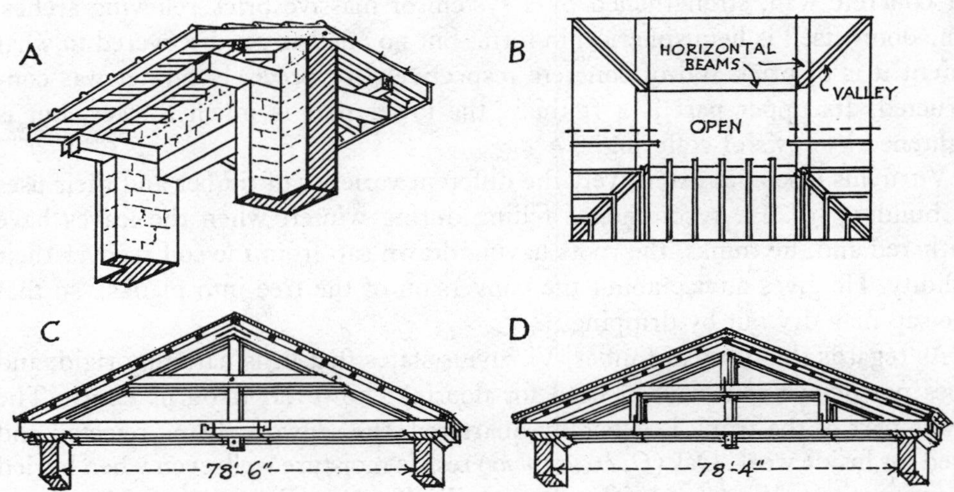

FIGURE 380—*Roman timber roofs:* (A) *simple span roof at Pozzuoli;* (B) *roof round the open court of a house, according to Vitruvius;* (C) *original roof of the basilica of St Peter at Rome. Fourth century A.D.;* (D) *original roof of the basilica of St Paul-without-the-Walls. Fifth century A.D.*

Timber-framed buildings were still quite popular in Rome during the Imperial period, as were also timber roofs and floors (figure 380); but since no examples of Roman carpentry have survived we depend on literary evidence for information as to its nature. Hence the description by Vitruvius of the roof of the basilica at Fano, which he himself designed and erected, is particularly interesting [12]. Its span was 60 Roman ft, somewhat less in English feet. He regarded it as an economical scheme, but the distances to be spanned were enormous. 'Above the columns are beams made of three 2-ft joists bolted together. . . . Cross-beams and struts support the ridge.' In some later and larger roofs, the important principle of trussing (figure 380 C, D) was introduced. There were outer roofs of timber over the great concrete vaults of the *thermae*, but in those buildings the rafters seem to have rested upon the extrados of the vaults.

In Britain, most of the Roman dwelling-houses at Silchester, and perhaps some also in London, were timber-framed and rested on a stone plinth; at Silchester there are traces of timber partitions. Rooms over hypocausts, however, had stone walls in which the heating-flues were built (figure 385).

The favourite Roman roofing-materials were terracotta tiles with joints protected by cover-tiles of semicircular section (figure 381 A); this fashion recalls that of the Greeks. The cover-tiles were tapered so that the lower end of each overlapped the upper end of the one below. The normal pitch of the roof was about one in three. In Britain there were often used similar tiles measuring about 22 × 16 in and 1 in thick; but slates were employed in Cumberland, and 'stone slates' in the Cotswolds, south-west England, and south Wales (figure 381 B, C). In shape and size they varied, a common form being an elongated hexagon about 11 in wide and 16–18 in long, held in position by a single nail on a boarded or wattled roof. The resulting effect was a lozenge pattern. Special slates were required at eaves and ridge.

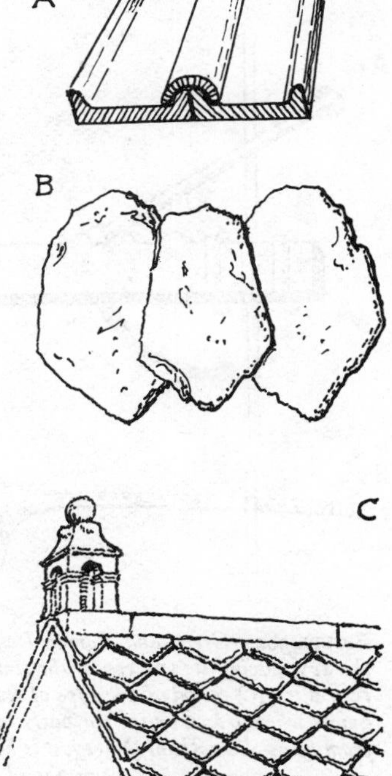

FIGURE 381—*Romano-British roofing materials:* (A) *tiles, London;* (B) *stone slates, Dorset;* (C) *stone slates and finial, Llantwit Major, Glamorgan.*

Doors were normally of wood, but very few examples have survived. One, found in the treasury at Chesters in Northumberland, was of oak studded with iron, and was so strongly constructed that it must have been an external door, but it fell to pieces almost as soon as discovered. In Roman Britain, as in Etruscan Italy, doors were usually hung on pivots let into stone lintels and thresholds, and often made in two leaves. Ordinary houses in Italy had wooden doors of softwoods such as cypress or deal, hardwoods being used for pivots and bolts. Door-fastenings consisted of bolts let into lintel and threshold, or a cross-bar let into holes in the door-posts. The doors of the baths at Pompeii had inclined door-posts, so that the doors closed automatically to exclude cold draughts and prevent the escape of warm air.

The primitive type of wooden lock used by the ancient Egyptians (vol I, figure 496 B) was also used by the Greeks, and even by the Romans in early days; but eventually locks of bronze and subsequently of iron came into common use in Roman houses. They were of two kinds, one of which could be operated by one hand; the other required two. A straight or curved metal key took the place of a clumsy wooden one, and a spring was introduced into the mechanism, which was rather complicated (figures 382 A, C). Not many hinges have survived from

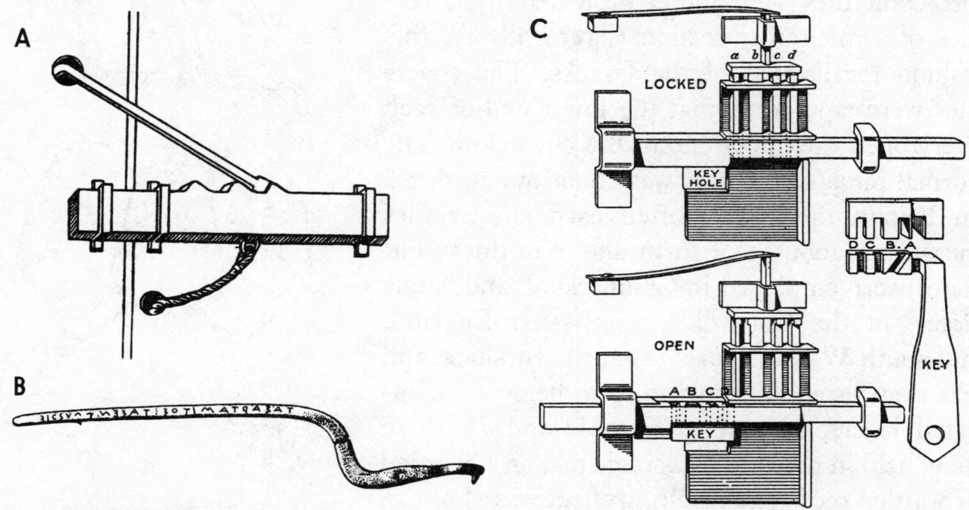

FIGURE 382—*Greek and Roman locks and keys:* (A) *diagrammatic drawing of a bolt with three bosses and the key moving it;* (B) *key of the temple of Artemis at Lusoi, Arcadia, according to the inscription; fifth century* B.C.; (C) *model of one type of Roman lock and key* (above) *shut,* (below) *open. The bolt slides in a pair of guides when drawn by the curiously cut key, whose projections fit slots cut through the bolt. To prevent the bolt from being slid without the key, it is held by four pins* (abcd) *pressed down by the spring into the slots, the pins being raised clear of the bolt when the proper key is inserted.*

Roman buildings in Britain, but several hook-and-eye fastenings have been discovered.

Glazed windows were in general use from the first century B.C., and panes of glass measuring as much as 2×1 ft have survived from the later periods. At Pompeii, even larger bronze window-frames have been found, which were glazed with sheets 21×28 in. One example, in a bath-house, was 40×28 in and $\frac{1}{2}$ in thick; it was frosted on one side, probably by rubbing with sand. Normally, however, window-frames were of wood. In Italy, for example at Ostia, the usual shape of a window resembled that of a door, with a height about double the width; but in England low windows under the eaves of the outer wall of the peristyle were favoured, separated by balusters. The fragments of Roman window-glass discovered in Britain are usually at least $\frac{1}{8}$-in thick and of a greenish-blue tint. The glass was probably cast as required, but blown glass in small roundels was also used.

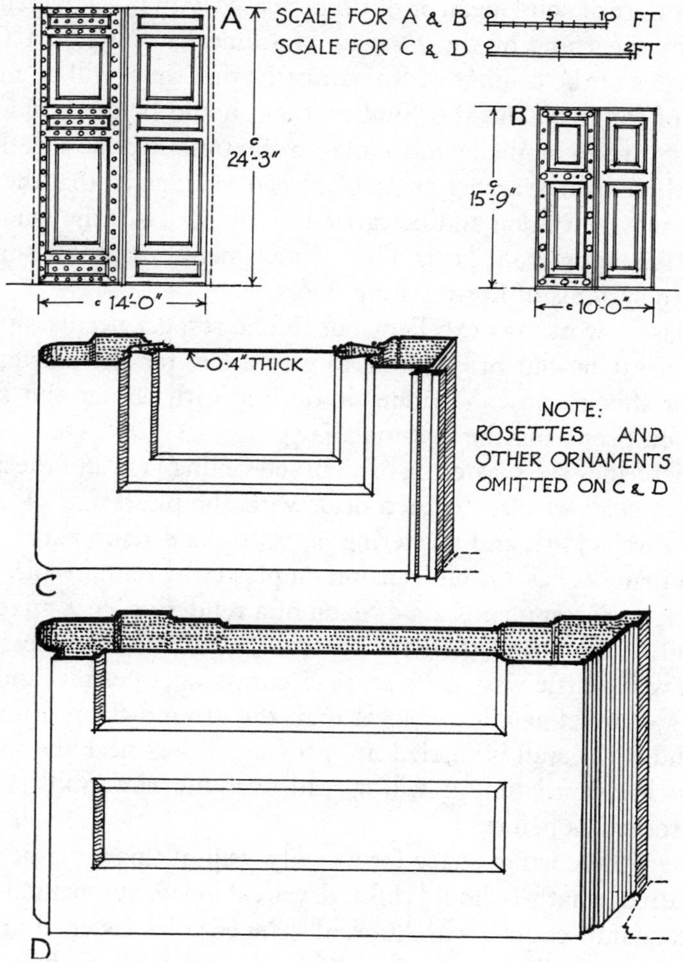

FIGURE 383—*Roman bronze doors:* (A) *and* (D) *from the Pantheon at Rome.* c *A.D. 125.* (B) *and* (C) *from the temple of Romulus (now the church of SS Cosma e Damiano) at Rome. Fourth century A.D.* (C) *and* (D) *are sections through the lower parts of the doors, showing the construction.*

Bronze and iron were employed for many purposes. The roof of the Pantheon was originally covered with gilt bronze tiles, but they were stripped off in the seventh century A.D. for use on the basilica of St Peter. A chapel of the church of St John Lateran contains four antique columns of the Corinthian order in gilt bronze. The magnificent bronze doors of the Pantheon are still in existence (figure 383). Dowels and clamps of bronze and iron were used in masonry (p 410). Classical writers mention roof-trusses of bronze, but it is uncertain whether they were of solid metal or of timber cased with it. T-shaped iron girders, discovered in the ruins of the *thermae* of Caracalla, suggest a framed floor. Vitruvius writes of the ceilings of Roman baths that they 'will be more convenient if made of concrete. But if of timber, they should be tiled underneath. Iron bars or arches are to be made and hung on the timber close together with iron hooks. And these bars or arches are to be placed so far apart that the tiles without raised edges may rest upon and be carried by them. Thus the whole vaulting is finished resting upon iron' [13]. This arrangement resembles some comparatively modern systems of fire-resisting floors.

Roman plaster-work was excellent but in one respect needlessly extravagant, for instead of raking-out brick joints to give a key for the plaster, plugs were driven in for this purpose. Vitruvius in dealing with plaster and stucco begins with advice on preparing the materials [14].

He next describes the plastering of a curved ceiling [15] and the application of the successive coats of plaster, then deals with the plastering of cornices [16]; plastering on walls [17]; and plastering on wattle-and-daub partitions [18].

Finally, Vitruvius has a useful section on plastering damp walls. For rooms at ground-level, he recommends a 3-ft dado of a rendering-coat mixed with burnt brick instead of sand, and then the stucco. 'But if the damp persists, build a second thin wall a little way inside it, as circumstances permit; and between the walls form a channel at a lower level than the ground-floor, with vents to the open air; and if the wall is carried up, provide outlets near the top. For unless the moisture has ventilation, it will spread over the new wall' [19]. Plastering may then proceed as before.

If, however, there is not space for a cavity-wall of this type, he recommends, as an alternative, what we should call a 'dry area' today, augmented by something like a vertical damp-course, with 'hooked [flanged] tiles fastened to the wall from top to bottom, carefully treating the inside of them with pitch so that they will reject moisture'. Vitruvius thus describes minutely a craft in which the Romans excelled. He also deals in some detail with fresco-painting, and with the preparation of the colours used therein.

Discussing the drainage of promenades and lawns, he seems to have in mind a herring-bone arrangement such as is still used today. 'In order that these walks may be always dry and free from mud, they should be dug out as deeply as possible, and drains laid in rectangular lines. In the sides of these drains are to be fixed earthenware pipes inclined to the drains. The trenches are then to be filled with charcoal; and, over this, sand is to be laid and levelled off' [20].

The chief authority on the marvellous system of aqueducts, cisterns, and pipes, which supplied the great city of Rome with water from as far as 60 miles away, is Frontinus (pp 405, 671 ff). In his book *De Aquis* he does not tell us much about the actual pipes; Vitruvius, however, devotes a page or so to the relative merits of earthenware and lead pipes (figure 384). He says that the latter should be in 10-ft lengths, and gives the weight of each according to its diameter. He remarks that earthenware pipes are cheaper; that they should have socketed ends, and that the joints between them should be made with quicklime mixed

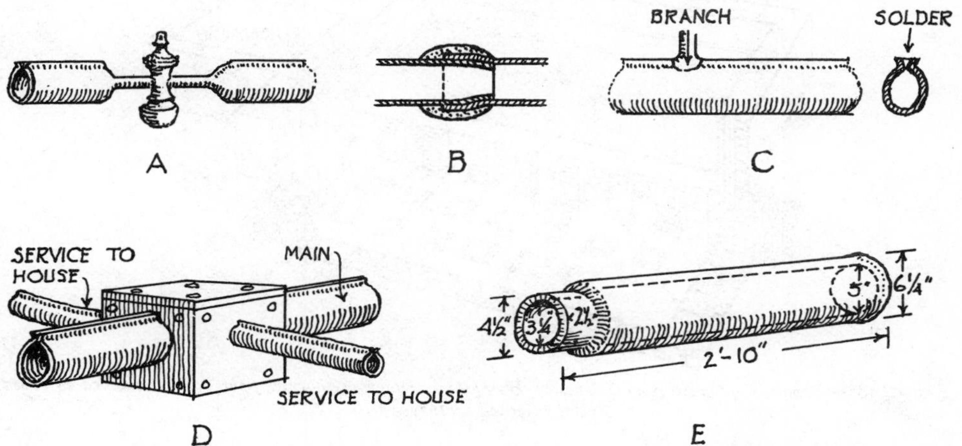

FIGURE 384—*Roman water-pipes:* (A) *stop-cock* (epistomium) *on a lead water-pipe;* (B) *and* (C) *joints in lead water-pipes;* (D) *junction-box in roadway;* (E) *earthen down-pipe for rainwater. From the Roman villa at Folkestone.*

with oil. He adds that these pipes can be quickly repaired and that water from them 'is much more wholesome than that which is conducted through lead pipes, because lead is found to be harmful to the human system. . . . This we can exemplify from plumbers, since in them the natural colour of the body is replaced by a deep pallor' [21].

For heating, the Romans often employed braziers, but the most efficient method of heating was by hypocausts. In the comparatively cold climate of Britain they were used extensively, as also in Gaul and Germany. Hypocausts were generally of either the channelled or the pillared type. In the former, the hot air from the furnace was carried through a main channel or trench into the centre of the room to be heated, and then conducted by other channels, radiating diagonally, to the four corners of the room, whence it was taken by trenches all round the bottom of the four walls, and finally upwards in flues formed in the walls. The channels were about 18 in deep and wide, with sides of rough masonry set in mortar or clay. They were covered with flagstones or large tiles, upon which was laid the concrete floor.

A pillared hypocaust (figure 385) was really a shallow basement some 2 ft or 2 ft 6 in deep, over which the floor of the room to be heated rested on short columns spaced 18 in apart or less. These were generally formed of bricks or

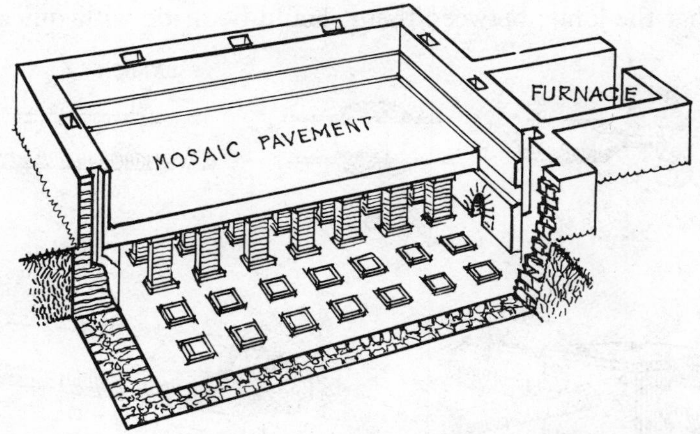

FIGURE 385—*Roman hypocaust from Silchester, Hampshire. Diagram showing floor partly removed to reveal* pilae. *Second century A.D.*

tiles about 7 in square, or occasionally circular. If the columns were made of stone, the sides were rendered with brick-dust mortar to prevent damage by heat. The floor beneath the hypocaust was usually of concrete and the floor above of flag-stones or of large tiles. The flues in the walls were of hollow box-tiles, oblong in section; their sides were about 1 in thick (figure 386). Usually they were from 12–17 in long, 6–8 in wide across the wider face, and $4\frac{1}{2}$–5 in wide across the narrower face. They were let into chases built into the walls, and their faces were invariably scored to provide a key for plaster or stucco, which completely hid them from the room.

III. ROMANESQUE PERIOD

The renaissance of Roman architecture was fully developed in Italy by 1450, but in England its effects were not in evidence until a century later. Thus the medieval period lasted at least a thousand years in Italy, where the Renaissance had its origin, and even longer in western Europe. For our purpose, it seems reasonable to subdivide medieval building into two main sections: Romanesque to about 1200, Gothic thereafter. Such division is both arbitrary and open to criticism, but the introduction of the pointed arch into European building, *c* 1200, forms a real landmark in architectural development. English Romanesque is divided into pre-Conquest Romanesque and post-Conquest Romanesque. These terms are preferable to the familiar but old-fashioned terms Saxon and Norman, which apply only to buildings in Britain. Romanesque, however, covers the whole of western Europe, and rightly implies that all this pre-Gothic architecture had its origin in the buildings of the Roman Empire.

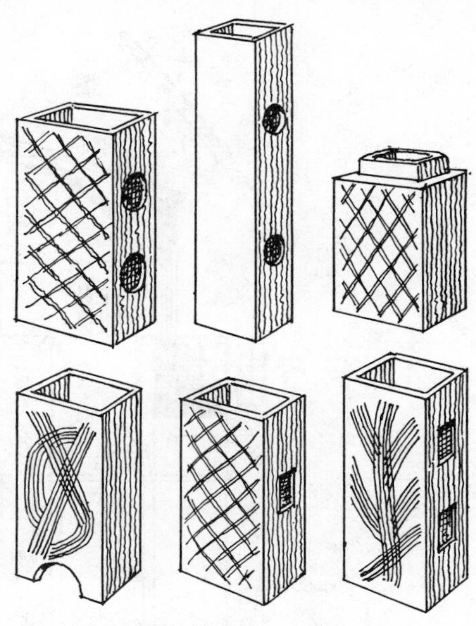

FIGURE 386—*Romano-British flue-tiles: box type.*

The earliest Christian churches are often called 'basilican', a term which has a somewhat complicated origin, and is derived from the Greek adjective *basilike*, royal. The Latin noun *basilica* means a royal building, not originally a religious building. Vitruvius uses it to describe a hall where legal and commercial business could be transacted under cover, normally adjoining the open forum which at first served for all purposes of public discussion. Gradually the word *basilica* came to signify the form of the building as well as its function. Indeed, Vitruvius prescribes rules for its planning [22]: its length was to be 1½ times its breadth, with two tiers of columns separating the nave or central area from the aisles, each of which was to be one-third of the width of the nave. He makes no mention of an apse at the end of the building, though later this came to be regarded as an essential feature of a basilican church. The covering of a Roman secular basilica might be a simple low-pitched roof of timber over the nave, with or without a flat ceiling, usually coffered, and with low-pitched lean-roofs over the aisles; or, in later examples, such as the huge and magnificent basilica of Constantine (or Maxentius) at Rome (A.D. 313) (figure 379), it might be vaulted throughout. The so-called basilican churches erected in Rome, Ravenna, and elsewhere from the fourth century onwards invariably consisted of a nave and aisles, generally with a narthex or vestibule at the entrance (west) end, and an apse at the other (east) end. Usually there was only one tier of columns on each side separating nave from aisles, and the wall above these columns, between the top of the aisle roofs and the underside of the nave roof, was pierced with windows, thus providing clerestory lighting for the interior.

The first stage after the fall of Rome up to *c* 900, during which period the chief buildings were basilican churches, is often classified as Early Christian.

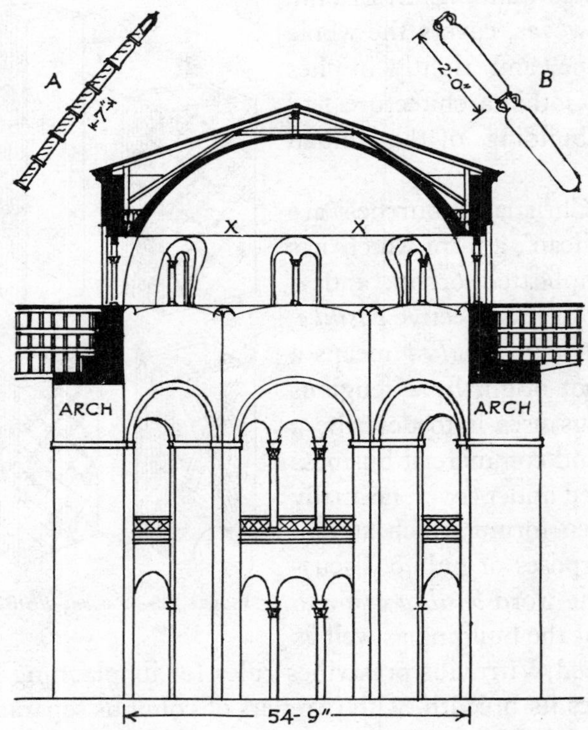

FIGURE 387—*Church of San Vitale, Ravenna (526–47): section of dome.* (A) *shows earthenware pots used in construction of dome above the line* XX; (B) *shows earthenware pots used in construction of dome below the line* XX. *The small pots* (A) *are laid horizontally in two layers; the large pots* (B) *are set vertically in six tiers.*

As for Lombard architecture, another common term, it merely means Romanesque architecture in Italy, with which Byzantine architecture is contemporary. The latter is chiefly confined to the lands of the eastern Mediterranean, extending into Italy at Venice and Ravenna.

The so-called Dark Ages—that is, the five or six centuries from the fall of Rome to about 1000—were indeed dark in the field of European architecture, except at Constantinople and Ravenna. In western Europe, including England and France, buildings of only the second rank were produced, far inferior in design and workmanship to the great monuments of Imperial Rome. In fact, the whole of western Europe now contains only two or three surviving churches erected between 450 and 1000, among them Charlemagne's cathedral at Aachen (Aix-la-Chapelle), built in 796–804; of secular monuments nothing whatever of importance remains.

The Byzantine churches of Constantinople (Byzantium), however, continued to follow Roman principles of dome-construction during the sixth century. The most famous examples are Santa Sophia (537) at Constantinople, and San Vitale (526–47) at Ravenna (figure 387). Other churches of the period were of simpler basilican type, with aisles and low-pitched timber roofs carried on arcades of semi-circular arches. Byzantine architects used brickwork on a large scale, notably at Santa Sophia, where not only the massive walls but the semi-domes, and even the great central dome itself, are all of brickwork. The dome has a flattish curve and is constructed with brick ribs, each 2 ft $4\frac{1}{2}$ in wide at the spring, the bricks measuring about $13 \times 6\frac{1}{2} \times 2$ in. The domes of St Mark's at Venice, afterwards concealed externally by Gothic leaden coverings of bulbous form, are also of brick; those of San Vitale at Ravenna are of earthenware pots about 2 ft long and $5\frac{1}{2}$ inches in diameter, used, in preference to brick, for the sake of lightness. Besides using brick for structural purposes, the Byzantines had no objection to it as a facing-material. They often treated it decoratively, panelling wall-surfaces with shallow brick arcading, variegating them with bands of bricks set diamond-wise, and even crowning them with brick cornices. They also used brick and stone in alternate courses to produce striped patterns in walling, and constructed arches with alternate voussoirs (wedge-shaped blocks) of brick and stone.

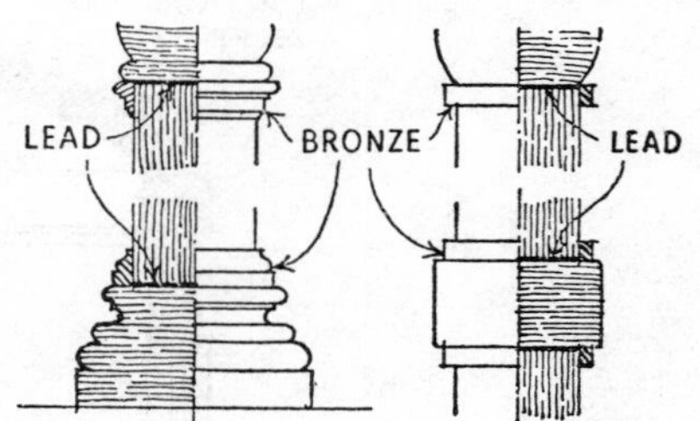

FIGURE 388—*Byzantine column-shafts, showing precautions taken against earthquakes.*

Byzantine masonry and concrete construction closely followed Roman practice. In some stone buildings no mortar was used, the joints being made very accurately. Marble columns were stripped from Roman temples to support vaulting or arches in Christian churches, an architectural use for which they were not designed. They were reinforced for this purpose with metal bands around the neck and base of the shaft, and, to lessen risk of fracture by earthquakes, a thin sheet of lead was inserted horizontally at the same points and held in position by the metal bands (figure 388). A cubical block (*dosseret*) of stone or marble, tapered downwards, was placed on the top of the capital to receive the vaulting springer from above. Very large blocks of stone were sometimes used in construction; thus the enormous 'lid' or flat dome of Theodoric's tomb at Ravenna (530) is about 35 ft in diameter (figure 389).

As a precaution against earthquakes, Byzantine walls were often faced with stone or brick and filled with rubble or concrete, brick lacing-courses or bonding-courses being used at frequent intervals. A further precaution was the insertion

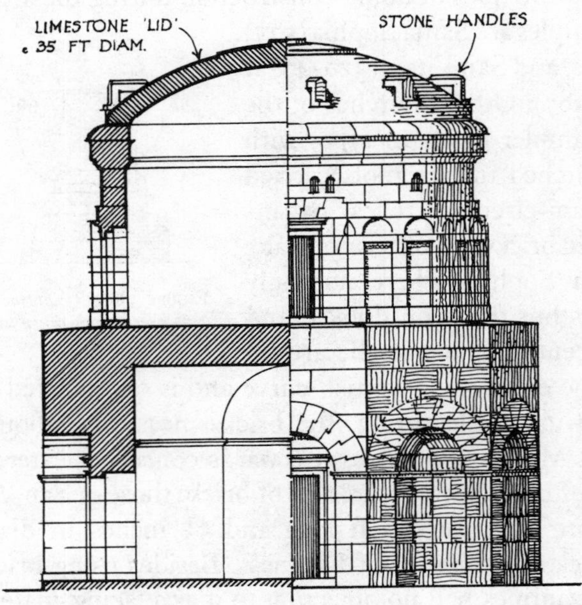

FIGURE 389—*The mausoleum of Theodoric at Ravenna* (c 530). *Section and elevation, showing the monolithic lid of Istrian limestone. Note the jointing of the arch-stones, joggled voussoirs, in the lower stage or storey. The 12 stone 'handles' are supposed to have been provided for hoisting the lid into position. The present ground-level is about 11 ft above the floor of the mausoleum.*

of bonding-timbers in walls. One example has a regular grillage of these timbers in layers about 6 ft apart, with longitudinal beams $5\frac{1}{2} \times 3\frac{1}{4}$ in, and transverse beams 7×4 in. For the same reason, unsightly tie-beams were inserted across the arches of arcades, level with the tops of the cubical blocks of stone, and also in vaults and domes. Iron ties were fixed across the arches of Santa Sophia and the vaults of St Mark's at Venice. Many domes were protected externally by low-pitched timber roofs, but some are exposed and are merely covered with cement or lead. In Egypt and Palestine, which formed part of the Byzantine empire to the seventh century, this type of building was practised; but after the Muslim conquests of those countries in 638–40 the brick construction of Mesopotamia was soon introduced there. The pointed arch, used previously in Byzantine

Syria, made its appearance centuries before the Crusaders transplanted it to Europe.

Building in England from about 410 to the Norman Conquest shows a sad falling away from Roman standards, as does contemporary work in France. The homes of the poorer people were shacks of timber or wattle-and-daub, or crude stone huts half sunk into the ground. Those of the upper classes were not much better. The highly civilized Roman villas, with their central heating and mosaic floors, were destroyed or allowed to fall into ruin. The Roman bricks were sometimes re-used for Christian churches, as in the fine tower of St Albans abbey and in the arcades of the remarkable basilican church at Brixworth in Northamptonshire (670); but, with these exceptions, the use of brickwork was completely abandoned in Britain.

FIGURE 390—*Greenstead church, Essex: angle of nave showing construction of timber walls (? early eleventh century). The brick plinth and oak sill are modern.*

There is a most interesting little 'Saxon' church at Greenstead in Essex, built of split logs, as doubtless were many others; but it is probably not earlier than 1015 (figure 390). All other surviving pre-Conquest churches in England are of somewhat rough masonry, even when executed by craftsmen brought specially from France or Italy. The narrow wooden doors of churches consisted of a single plank, hung in reveals and heavily armoured with iron strap-hinges and ornamental scroll-work, so fashioned as to spring from the main horizontal strap at a point behind the face of the door-jamb; additional resistance was thus offered to an enemy trying to tear off the ironwork.

Glass continued to be manufactured in Gaul after the Romans evacuated that province, and it was from France that craftsmen came in 675 to glaze the windows of the church at Monkwearmouth. As Bede relates, 'they not only did the work required, but taught the English how to do it for themselves' [23], and before the end of that century glass replaced the linen or perforated boards which had filled the windows of York minster. Yet in 758 the abbot of Jarrow had to send to the Rhineland for glaziers. Even in Italy, church windows were not glazed earlier than in England.

About the year 1000, building activity became very active again and soon spread all over western Europe. In England the first results were seen slightly before the Norman Conquest, in the new abbey church at Westminster erected by Edward the Confessor to replace an older church on the site. Edward seems to have borrowed his design from the abbey of Jumièges in Normandy (*c* 1040).

FIGURE 391—*St John's chapel, White Tower of Tower of London (at* A: *groin of vault).* c *1080.*

The latter and the abbey of Bernay (*c* 1020), both erected by the Italian abbot of a monastery in Burgundy, are the oldest important Romanesque buildings in France. Thus the civilizing influence of Roman architecture had only just begun to affect the Normans when they became rulers of Britain.

In all the crafts of building, Norman architecture soon showed a great advance upon native work. The examples now surviving consist almost entirely of monastic and parochial churches, cathedrals, and castles. There are a few stone houses here and there, for example at Lincoln and at Bury St Edmunds, but they are unimportant. As in Saxon times, most of the people, rich as well as poor, lived in wooden houses that have perished. There is no need to differentiate here between churches and castles, for the methods and materials of construction are the same, and brickwork had gone out of fashion. The principal building-craft throughout the 'Norman' period (*c* 1050–*c* 1200) remained masonry (figure 391).

Partly because fires often occurred in churches with wooden roofs, a stone-vaulted roof was desired for all important ones. As the early 'Norman' vaults were barrel-vaults (p 411) of great thickness, they were immensely heavy and exerted a powerful overturning thrust on the walls (figure 391). Even where a timber roof was used, it exerted a similar though lesser thrust. In the eleventh century the idea of strengthening the wall with a buttress opposite each roof-truss had not been fully exploited, and so the stability of the wall was dependent solely upon its thickness and mass. At Winchester cathedral, although there are rudimentary buttresses, the nave-wall between them is 7 ft thick. The walls of some castles are far more massive, being 17–21 ft thick at Dover, and 15 ft at the White Tower of the Tower of London.

Stone was obtained with great difficulty, for roads were few and very roughly made, and not well suited to wheeled transport-vehicles. The tendency was therefore to use blocks as small as possible. Dressed stone (ashlar) was sometimes employed for external facing, even when the interior of the wall was of rubble. More often, however, rubble walls were plastered both inside and out. Much of this plaster was later removed, especially in Victorian times, with the mistaken idea of exposing 'good honest masonry'. The rubble core has led to much trouble for restorers, and has often had to be reinforced by the modern and very difficult process of grouting with liquid cement-mortar.

A notable grouting operation was carried out at Winchester cathedral early in the present century, after serious settlements due in part to inadequate foundations: the first cathedral had been erected on marshy ground, with water only 10 ft below ground-level. When the second stage of building began in 1079, short oak piles—some salvaged from the first cathedral (*c* 980)—were driven into this ground, and foundations were laid on them. In the third stage of building, late in the twelfth century, the new builders had to advance still farther into the bog:

> Coming, like their predecessors, to water 10 feet below the surface . . . they were at a *non plus*; and, as the best that they could do, they cut down a wood of great beech-trees, laid them flat, and on them raised that building which is one of the gems of English art. The trouble which followed, and which nearly brought this part of the building to ruin, began as soon as it was built. The tree-trunks did not decay, but were pressed down into the soft ground, the vaults became disturbed and pushed the walls out, and

the whole building split off from the Norman part west of it, and slid eastwards, leaving gaping cracks at the point of separation and in several other places [24].

In the twentieth century divers had to be employed to underpin the walls and to lay an enormously thick bed of cement-concrete as a foundation.

Underpinning was similarly necessary at Peterborough cathedral in 1845, and at Salisbury from 1859 onwards. At Peterborough there is solid limestone only a few feet below the floor, but the original foundations were not carried down to it. Many other cases of bad Romanesque architectural planning could be cited, but Romanesque foundations in England were often excellent. Bonding-timbers, carefully framed together and embedded in the masonry, were used in Norman foundations of churches at York and Lewes, and in several castles.

Much of the stone used in the south of England was brought by sea from Caen in Normandy—for example, to Chichester and Canterbury cathedrals. But normally stone was either quarried locally, or floated to the site by river and sea. The best English stone came from quarries in Northamptonshire, south Yorkshire, and Somerset. In East Anglia, especially in the chalk districts, flint was much used for the filling of rubble walls and for facings. The flints were laid in courses. Rubble walls in small churches were generally about 3 ft thick, except in the very rare cases where vaulting had to be carried. They were normally plastered and lime-whited inside and out, and even the picturesque re-used Roman brickwork of St Albans abbey was thus concealed. To the end of the eleventh century the joints of Norman masonry were thick but thereafter soon became closer, so that Thomas Rudborne could write of the buildings erected by Roger, Bishop of Sarum (1107–42): 'the stone courses were so correctly laid that the joint persuades the eye that the whole wall is a single block' [25].

Many small churches, most dwelling-houses, and even many of the earlier Norman castles were timber-framed, but apparently none have survived, and there are very few timber roofs of the period still extant. One of the oldest is that of the hall of the bishop's palace at Hereford, of the twelfth century. It has supporting posts with wooden arches from post to post and across the hall. The roof of the south transept of Winchester cathedral is also Norman.

Romanesque church roofs may be roughly divided into those which protect vaulting, and those where there is no vaulting. Some of the Roman basilicas (p 421) and baths had their great vaults exposed to the weather, protected only by plaster or a sheathing of metal; in northern Europe, however, the sterner climate made a more durable covering essential. Where the timber rafters of a low-pitched roof rested directly on the outer surface of a light brick or concrete or masonry vault (figure 402 A), the weight of the roof distorted the vault. For this reason the timber roof was raised until the underside of its tie-beam, which prevented the feet of the rafters from spreading and so overturning the walls, cleared the vault (figure 397). However, it was not until the Gothic period that the construction of timber roofs, as well as of stone vaulting, began to develop rapidly.

FIGURE 392—*Builder hammering shingles to a roof. From an Anglo-Saxon twelfth-century manuscript.*

The covering of Romanesque roofs in western Europe, including England, generally consisted of half-round tiles where a flattish pitch was used; but in some parts of France, where a steeper pitch was favoured (up to 45°), slates may have been employed. Thatch, largely used during this period, was often the cause of the fires that destroyed so many churches. Roofs of thatch were forbidden in London in 1212. The safer, incombustible lead came very early into use for roofing in England and France. Bede relates of the church at Lindisfarne that in 688 the bishop 'took off the thatch and covered it, both walls and roof, with lead' [26]. York minster is said to have been roofed with lead about 669, as was the choir of Canterbury cathedral (1093–1130) and, somewhat later, the aisles. The roof of the famous church of St Martin at Tours was covered with tin. Manuscripts of the Norman period illustrate the use of shingles on roofs (figure 392).

Windows in Norman buildings were larger than those of Saxon times, and invariably had semicircular heads. Even church windows were not always glazed in the twelfth century. The oldest surviving stained glass in England dates from *c* 1170–80, and the finest examples are fragments at York minster and Canterbury cathedral. Surviving Norman doors are fairly numerous, as at Castor (with its lock and key), Sempringham (*c* 1133), and York chapter-house. The two latter are made of deal. Like the Saxon examples (p 425), such doors were at first formed of a single plank, very thick and heavy. Then they came to be made of two or more vertical boards, somewhat thinner, joined together by horizontal battens to which the long ornamental iron strap-hinges were attached. The usual method of fastening Romanesque doors in England was probably a stock-lock, that is, a wood-encased lock attached to the face of the door, not let into the edge (figure 393).

Besides its ornamental use in doors, hinges, and latches, iron was employed for many other purposes in churches—though not, apparently, in the actual construction of buildings. The sanctuaries and chapels of cathedrals and large churches were often enclosed by graceful grilles of wrought iron. Much more massive grilles, armoured with spikes, were used to protect the windows of church treasuries. In these grilles, the vertical bars passed through eyes welded

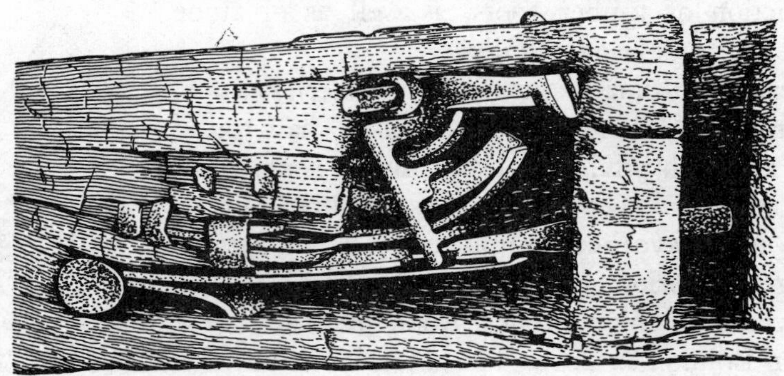

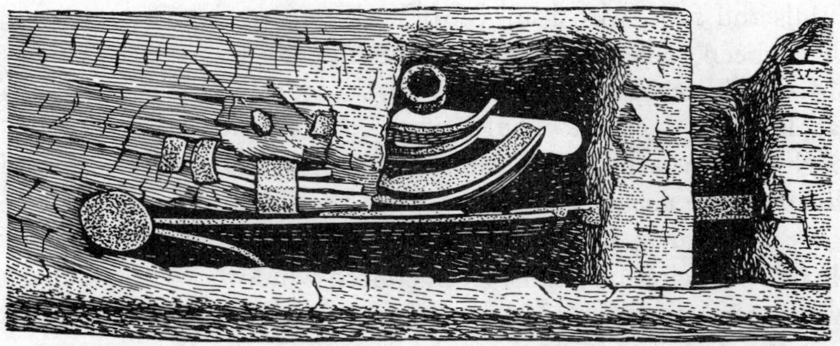

FIGURE 393—*Norman lock made of a wooden beam—only the centre of which is shown in the two views—to be bolted outside a door. The key is inserted from the far side, as shown in upper figure, and must be of a shape to pass between the curved iron leaves. First a projection on the key depresses the horizontal leaf-spring which is fixed on the left; this frees the bolt, which is then caused to slide to the left into the unlocked position by the same projection on the key. In the locked position* (lower figure) *the horizontal leaf-spring prevents the bolt from being withdrawn.*

in the horizontal bars. The doors of the treasuries were also protected by heavy ironwork, lined with iron plates, studded with nails, and furnished with strong bolts and locks.

IV. GOTHIC PERIOD

The dates of 1200 for the beginning, and 1540 for the end, of the Gothic period of architecture are only very broadly valid. The first manifestations of

FIGURE 394—*Salisbury cathedral, sectional view. 1237–58.*

Gothic in France are usually traced to the abbey church of St Denis near Paris, begun in 1140. This church was soon followed by the magnificent cathedrals of Paris, Chartres, Laon, Bourges, Beauvais, Rheims, and Amiens. In England the change from Norman to Gothic started with the rebuilding of Canterbury cathedral by a French architect in 1174. The change consisted primarily in the introduction of the pointed arch and its application to vaulted roofs of stone (figures 396 E, 397). Until about 1200 the process was very gradual, pointed arches being used abreast of round-headed windows and doorways, so that some

FIGURE 395—*King's College chapel, Cambridge. Sectional view. 1446–1515.*

scholars label the period 1150–1200 as 'Transitional'. For ordinary purposes, however, we may speak of buildings before 1200 as 'Norman' or 'Romanesque' and those of the next three and a half centuries as 'Gothic'; the rule applies well in England.

All through those centuries there was a progressive lightening of the structure, from the massive walls and heavy stone barrel-vaults of early Norman times to the amazing skill of late-Gothic construction. Ultimately a thin vault is supported on slender piers augmented by boldly projecting buttresses, the intervening wall becoming a mere screen of stone panelling pierced by huge windows. The process may be illustrated by a series of examples. The chapel of St John (*c* 1080) in the White Tower of the Tower of London is a tiny but typical specimen of Romanesque barrel-vaulting with vaults a foot or more thick (figure 391). The nave of Salisbury cathedral (1237–58, figure 394) is characteristic of Early English Gothic soon after the introduction of the pointed arch; and the chapel of King's College, Cambridge (1446–1515, figure 395) represents the climax of English Gothic skeleton-construction, comparable with Henry VII's chapel at Westminster (1500–12) and St George's chapel at Windsor (1473–*c* 1537). The method of lightening the stone roof, from 1080 to 1540, by the introduction of stone ribs, suggests the principle of an umbrella, with a thin 'web' or covering of stone not more than 6 in thick filling the segmental spaces between the stone ribs; all the downward thrusts arising from the weight of the roof were concentrated on to a carefully disposed arrangement of columns and buttresses, instead of on to thick stone walls.

A comparison of the three sectional views shows this progressive reduction in thickness from the massive Romanesque walls of St John's chapel (where buttresses are not used), through the bold buttresses and thinner walls of Salisbury, to the slender piers and immense pinnacled buttresses of King's College chapel, with walls virtually superseded by huge glass windows. This development was made possible only by the use of an elaborate system of stone ribs, and by the introduction of the pointed arch. Romanesque barrel-vaults were thick and heavy, exerting a powerful downward thrust and an overturning tendency. As in the Roman basilicas (figure 379), this difficulty was met by using cross-vaults, intersecting the main vault. A small-scale example may be seen in the lower aisles of St John's chapel (figure 391). The sharp line of intersection is called a groin. The next step was to build arched stone ribs across each bay of vaulting—that is, across the main vault from each pair of supporting columns of the arcade—and to rest the web of the barrel-vault on these transverse ribs. Next, diagonal ribs were constructed along the line of the groins, in the case of intersecting vaults. This method avoided much of the heavy timber centring required to support a barrel-vault during erection and was theoretically more soundly based, as well as being more economical of stone and timber (figure 396 B).

At first, these ribbed groined vaults were confined to narrow aisles, but, as the builders gained confidence, they sought to construct fireproof stone vaults over the higher and wider naves of their more important churches. Hitherto only semicircular arches had been used, but, as a glance will show (figure 396 C), if the transverse and wall arches are semicircular and of equal span, then the diagonal arches—being obviously longer—must also be higher, unless the semicircle is to be distorted into an ellipse. This did not matter much in the aisles; but when the wide and lofty vault over the nave had to be erected, it became necessary, for structural reasons and to enable it to be related to the vaulting-

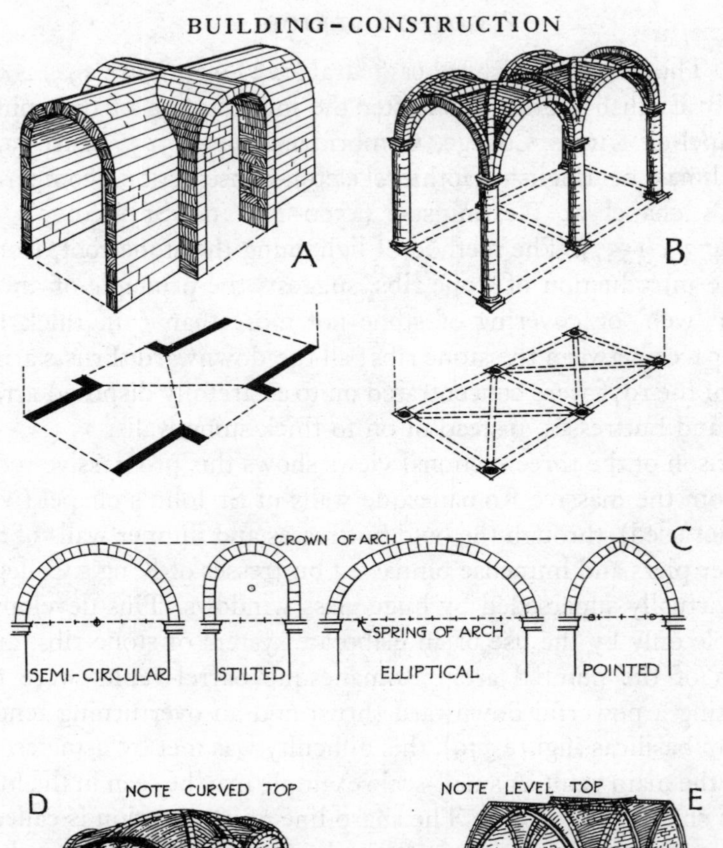

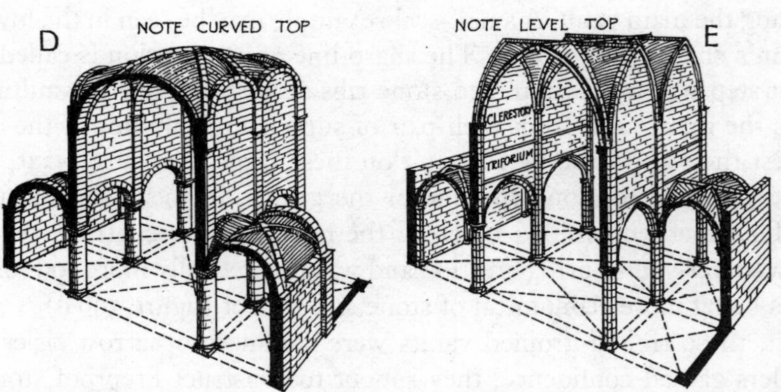

FIGURE 396—*Diagrams of vaulting:* (A) *intersecting barrel-vaults;* (B) *Romanesque ribbed vault,* (below) *perspective plans of* (A) *and* (B); (C) *vaulting arches;* (D) *Romanesque sexpartite vault;* (E) *Gothic vault (early period).*

system of the aisles, to have two arches in the nave-arcade and two bays of aisle-vaulting for each square bay of the nave-vault. The arch of the nave-vault would therefore be twice as wide as the arches of the nave-arcade, but much less wide than the diagonal rib. Hence, if all the arches were made semicircular and sprang from the same point at the base, their tops would be of three different heights and the effect would be unsatisfactory. These objections were met, once more, by distorting the arches (figure 396 D). For the nave, a true semicircular arch was used, for the longer diagonal rib a segmental arch of similar height but of depressed curve, and for the narrow wall-arches a 'stilted' arch of the same height. Each bay of a vault is thus sexpartite.

These problems of vaulting explain the great importance of the introduction of the pointed arch at the beginning of the Gothic period, for the width of a pointed arch can be varied a considerable amount while its height remains constant (figure 396 E). Thus it solved all the builders' chief difficulties: the wall-arches could be narrow and sharply pointed, the transverse rib across the nave could be twice as wide and more bluntly pointed, and the diagonal rib—perhaps half as long again—could be a true semicircle. All distortion of arches could be avoided, and the use of carefully designed stone ribs in combination with thin stone webs enabled the reduced weight of the roof to be transmitted on to columns, piers, and buttresses, while the wall could be pierced for windows to any extent.

FIGURE 397—*Section of nave of Amiens cathedral, showing a flying buttress.* 1230–40.

One particular type of buttress, the flying buttress invented by French architects in the late twelfth century, was much used in the larger French cathedrals, but is seldom found in England, the chief example being Westminster Abbey, a building erected under French influence and direction. The purpose of a flying buttress is to transmit the enormous thrust from the high stone vault of a nave or choir across the aisles and so down to the ground. Where no aisles exist, as at the Sainte Chapelle in Paris or King's College chapel, Cambridge (figure 395), the thrust is resisted by huge buttresses against the main walls; but where there are aisles, they must be bridged somehow. Massive arches across the aisles of Durham cathedral (late twelfth century) perform this function, but are hidden in the triforium. In the rebuilt choir of Canterbury cathedral (1175–8), where the architect was a Frenchman, they appear above the roof of the aisle but are almost invisible from below, so do not constitute a prominent feature in the design. The typical French Gothic example illustrated here (figure 397) is from the nave of Amiens cathedral (1230–40), where the vault is 140 ft high. Although, for the sake of clearness, only a single buttress is shown in the diagram, there is one corresponding to each pier of the nave-arcade, two piers being shown. It will be noted that the wall between each pair of buttresses and piers is largely composed of glass, masonry being reduced to a minimum. This result could have been achieved only by the use of flying buttresses.

It is commonly believed that Gothic masonry was invariably sound. Nevertheless failures frequently occurred even when the outer appearance of the stonework was honest enough. Several great towers collapsed. These and other accidents were sometimes due to the risks taken by the architects, but generally because foundations were inadequate or walls badly constructed. Thus at Chichester, 'before the spire fell, the mortar of the interior ran out like water when a stone was withdrawn' [27].

Contemporary records prove that occasionally great care was taken to carry foundations very deep. Footings were usually formed of a layer of rough stone, or sometimes of chalk or gravel. At Eton College in 1453 hard Yorkshire stone footings were laid in 'good and myghty morter'. Ramming of the subsoil is sometimes mentioned. Elm piles were used under buildings at the Tower of London, Hadleigh castle, and the Bulwark at Sandwich; beech piles at Winchester; and alder piles at York. Iron-shod piles were employed under London Bridge and Rochester bridge. Primitive pile-driving machinery is specifically mentioned in documents of the fourteenth to sixteenth centuries.

As in Romanesque times, walls might consist entirely of rubble, generally plastered on both sides, or of rubble with an ashlar facing on one or both sides; they were seldom composed entirely of ashlar. The rectangular blocks of ashlar were tooled on one face only. In thin walls of solid ashlar (for example, in parapets and in choir-screens), through-stones were provided, running right through the wall. Dressed stone quoins, window-tracery, and doorways were used in churches, especially in the chalk districts of East Anglia, where the main walling was generally of flint. In many East Anglian churches (as at Southwold) alternating squares of limestone and flint are used to give a chequered appearance (figure 401, left). In some of the northern counties of England, where hard mill-stone-grit and sandstone were the normal building-materials, the finish of the masonry was very rough and the carved ornament very crude; this was even more so in old Cornish and Breton churches where intractable granite was used. Moulded string-courses and dripstones around the arched tops of windows and doorways prevented rain-water from running down the faces of walls. Throated copings, unless the wall was sheltered by projecting eaves, protected the top

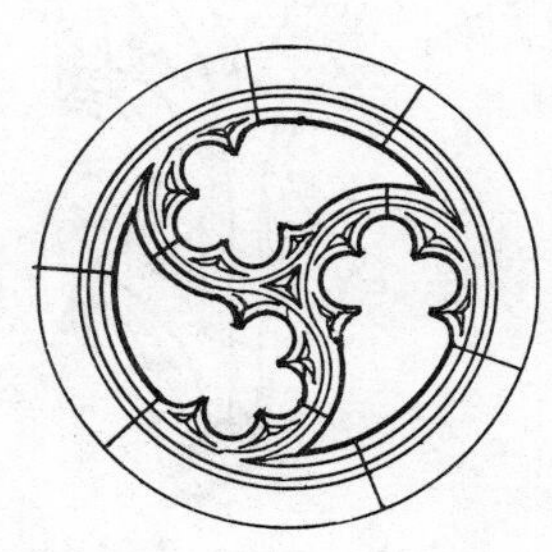

FIGURE 398—*English Gothic tracery, showing jointing of masonry.* (Left) *Easby abbey refectory, Yorkshire (thirteenth century);* (right) *Carlisle cathedral (late fourteenth century).*

of the wall and served a similar purpose. When wall-building was carried on during the winter months, or was temporarily discontinued during the winter, the tops of walls were protected from frost by a covering of straw, reeds, heather, or even thatch.

The stereotomy (stone-cutting) of Gothic masonry is a very highly skilled craft demanding knowledge of practical geometry. This was notable in Gothic traceried windows (figures 398–9). One of the largest examples is the east window of Carlisle cathedral (late fourteenth century), which is 51 ft high and 26 wide. The traceried part alone, above the spring of the arch, consists of 86 blocks of stone, some of them 4 or 5 ft long, and all elaborately moulded. They are so ingeniously jointed and arranged that any single stone can be cut out for repair or replacement without endangering the stability of the whole wonderful fabric.

Ironwork was used to reinforce these masterpieces of masonry. The various bars acted as a grille to deter robbers and to strengthen the glazing as well as the stonework. The whole system of bars consisted of a main horizontal stay-bar or transom set at the level of the springing of the arch and extending from jamb to jamb through the mullions; smaller horizontal bars to strengthen the glazing between the mullions; and small vertical bars, sometimes made to pass through holes or eyelets in the standards. External ironwork might be tinned, painted, or coated with black pitch. It is startling to learn that the slender central pillar of the chapter-house at Westminster Abbey had an umbrella-like arrangement of iron tie-rods radiating to the vaulting-ribs, and that iron ties were largely employed elsewhere in the Abbey. Iron clamps were used for joining stones

FIGURE 399—*French Gothic tracery:* (left) *portion of a window from Amiens cathedral (thirteenth century) showing elaboration of mouldings;* (right) *upper part of a window from the church of St Gervais, Paris (fifteenth to sixteenth century) showing jointing.*

together, especially in exposed features such as pinnacles, turrets, and machicolations, as well as for the stone hoods of fireplaces.

The revival of brick-building in England, after an interval of more than 800 years, began in the thirteenth century. A huge quantity of Flemish bricks was purchased at Ypres in 1278 for the Tower of London. It is possible that the tiles (*tegulae*) being made at Hull in 1303 were bricks rather than roof-tiles; but, at any rate, 'mural tiles', that is, bricks, were being made in England for use at Ely in 1335. Most of the bricks used in England till then, however, were imported from Flanders to Hull and other eastern ports. The large quantity used at Tattershall castle in Lincolnshire, from 1434 onwards, were manufactured in that county. Some of the bricks made at Beverley near Hull were chamfered for jambs and arches. Many of the first experts in English brick-making, as well as in brick-building, were Flemish or Dutch by birth. Fuel for the kilns was normally wood. Generally speaking the use of brick, up to the end of the Gothic period, was confined to the eastern counties, from Hull to Dover, and thus to those localities where most building craftsmen from Flanders had settled (figures 400, 351). The sizes of the bricks varied greatly, but the thickness was nearly always less than the standard thickness today; it ranged from 2–2½ in. The average length was 9 in, and breadth 4½ in, but the former varied from 8–10½ in, while the breadth might be as much as 5¼ in.

FIGURE 400—*The Guildhall at Blakeney, Norfolk (fourteenth century).*

Early in the sixteenth century, moulded brickwork came into use for brick windows with mullions and even simple tracery. There are several examples in Essex, notably at East Barsham (figure 401, right) and at Chignal Smealey church, where the mullions, jambs, and tracery are all moulded. The ordinary bricks used in walling at this church measure 9⅝ × 4½ × 2⅛ in, and vary much in colour. In early Tudor windows of this type, where transoms occurred, they were entirely of self-supporting brickwork, but in later Tudor work they were often reinforced by an iron bar, run either through or beneath them. At St Osyth, near Clacton in Essex, the fine parish church (early sixteenth century) has arcades and piers of brickwork. Another interesting detail of late Gothic brickwork is the occasional use of brick handrails built into the walls of spiral staircases, as in the Red Chapel (1482) at King's Lynn.

In continental Europe, the chief western centres of brick-building in the Middle Ages were north Italy, north Germany, Flanders (plate 31 B), Holland, and parts of Spain and south-west France. The Italian bricks were mostly red,

FIGURE 401—*English Gothic technique in stone, flint, and brick:* (left) *part of clerestory of Saxmundham church, Suffolk: shaded portions flint, the rest stone;* (right) *part of a brick window at East Barsham Manor House, Essex. Sixteenth century.*

built with rather wide mortar joints, and with dressings of finer quality. The sizes of bricks were, on the whole, larger than ours, but varied greatly, from $9\frac{1}{2}$ to $12\frac{1}{2}$ inches in length and from $2\frac{1}{4}$ to $3\frac{1}{4}$ inches in thickness, the normal width being 5 in; but in Venice bricks as small as 7×2 in have been noted. Moulded bricks and moulded terracotta blocks of every sort and size were used, and brick tracery is found occasionally. In north Germany many great churches were built entirely of brickwork, including crockets and tracery. The cathedral of Albi in France is another example of the kind.

Gothic carpentry may be classified as timber-framing in which the wooden roof formed an integral part of the timber frame (plate 30 A), and as timber roofs over stone or brick buildings, sometimes vaulted. Timber-framed buildings have been erected from Neolithic times onwards in nearly all parts of Europe, their construction improving through the ages in design and craftsmanship. Very few examples older than *c* 1500 remain in England, owing to frequent fires and to rebuilding; but as the principles and technique of timber-framing persisted right up to the seventeenth century in the wooded districts of England, reference will be made to them later in this work.

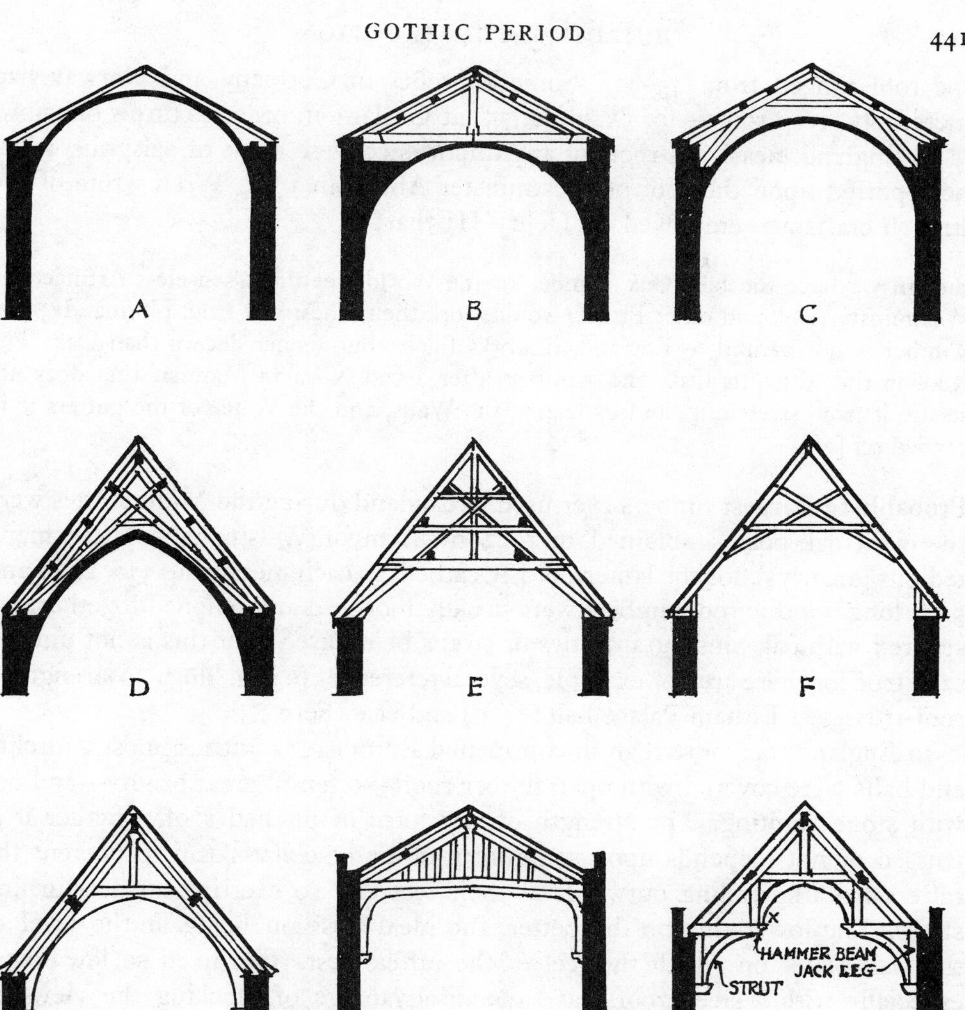

FIGURE 402—*Romanesque and Gothic timber roofs:* (A) *early Romanesque (Italian): rafters resting on extrados of vault;* (B) *Romanesque (French): tie-beam introduced to relieve vault;* (C) *Romanesque (French): collar-beam used instead of tie-beam;* (D) *Gothic (French): diagonal braces used instead of tie-beam;* (E) *early Gothic roof of Peterborough cathedral with wooden ceiling;* (F) *Gothic (English): trussed rafter roof with collar-beam;* (G) *Gothic (English): curved-brace roof with collar-beam;* (H) *late Gothic (English): so-called tie-beam roof. Note low pitch;* (I) *late Gothic (English): hammer-beam roof.*

Protective roofs over stone-vaulted churches were made self-supporting during the Gothic period, and the bad old Romanesque practice of resting the rafters upon the outside of the vault was abandoned (figure 402). Tie-beams were now generally used, just above the crown of the vault, and the pitch of the roof ranged from 45–60°. Some churches in Germany and Norway had incredibly steep roofs—for example, 75° at Goslar—in order to throw off snow.

In England, nearly all roofs of any importance were made of oak; but, when he reported upon the roof of Westminster Abbey in 1713, Wren wrote of the French craftsmen employed by Henry III that:

> though we have the best Oak Timber in the World, yet these senseless Artificers in Westminster Hall and other Places, would work their Chestnuts from Normandy; that Timber is not natural to England, it works finely, but sooner decays than Oak. The Roof in the Abbey is Oak, and wrought after a bad Norman Manner, that does not secure it from stretching, and damaging the Walls, and the Water of the gutters is ill carried off [28].

Probably the largest timbers ever used in England during the Middle Ages were the eight oak beams obtained in 1322 by Alan of Walsingham, 'after many tedious journeys', for the lantern of Ely cathedral, each measuring 33×21 in and 50 ft long. Gothic roof-timbers were usually mortised and tenoned together and secured with oak pins, no iron ties or straps being used; but this is not universally true for there are, for example, several references to iron 'dogs' to strengthen roof-trusses at Eltham Palace hall (1479) and elsewhere.

In England, far more than in continental European countries, most churches and halls were covered with open timber roofs—often of great beauty—and not with stone vaulting. The strength of any form of pitched roof, whether it is trussed or not, depends upon some form of tie or collar-beam to prevent the rafters from spreading outwards at their feet and so exerting an overturning stress. The lower this tie the better, the ideal position being at the level of the wall-plates on which the feet of the rafters rest. Ties fixed so low down, especially with a steep roof, have the disadvantage of blocking the views of the east and west windows. Various devices were adopted to avoid this drawback.

The hammer-beam truss, peculiar to England (figure 402, 1), was invented as a palliative, being somewhat more effective than a truss with its tie halfway up the slope. This device is more ingenious than scientific and has been defined as:

> an imperfect tie-beam of which the middle part has been removed. . . . The principal framed into it tends to drive it outwards, but this is resisted by the strut below, which presses firmly against the jack-leg, and so brings the thrust lower down the wall, where superior weight comes into play to steady it. From the point 'X', thus fortified, an upright post is framed into the principal, and from it spring curved struts to the collar, which is often hung up by a king-post to the ridge [29].

The timbers used in all types of Gothic roofs, trussed or otherwise, were very massive according to modern standards. Common rafters were sometimes tapered: for example, from 8×7 in at the bottom to 7×5 in at the top. Boarded or panelled wooden ceilings were often applied to roofs having trussed rafters or curved braces, thus producing the so-called wagon-roof much favoured in the west of England during the fifteenth century.

Panelling was first introduced into England in the thirteenth century, but only sparingly and not for doors, which continued to be constructed of battens and 'ledges' (figure 403). During the next century, doors came to be made in two thicknesses: continuous vertical battens externally, with the joints covered by thin moulded strips, and horizontal battens internally. Wrought iron nails with approximately square heads were arranged in vertical rows, and were driven through the vertical strips, the strap-hinges, and both layers of battens. The points were then clenched on the inner side and left exposed.

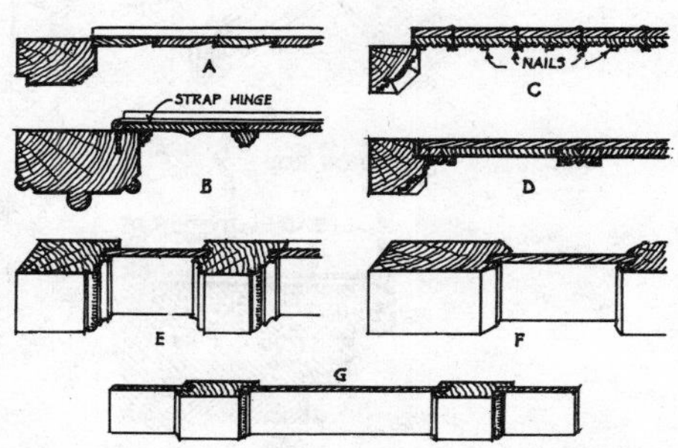

FIGURE 403—*Late English Gothic doors and panelling:* (A–D) *Doors from various sources;* (E–F) *late Gothic panelling*—(E) *moulded,* (F) *chamfered framing;* (G) *panelling* c *1530, showing thinner framing and larger panels.*

The first known record of panelling is of the thirteenth century (p 243). It was of Norway pine, specially imported for a chamber at Windsor Castle, and was coloured. Panelling began as half-timbered partitioning, with sturdy posts and rails about 4×3 in, the former fixed from 18 to 24 in apart, with a rail from 3 to 5 ft above floor-level according to the height of the room. The framing was tenoned and pinned, sometimes splayed but more often moulded. The panels were boards let into the framing (figure 403, E–G). This type of framing was progressively reduced in thickness, until by 1540 it was not more than an inch thick. The usual material was oak.

In the remaining crafts, advance was on decorative rather than structural lines. Ornamental ironwork was used for hinges, as in the Romanesque period. Plaster-work was of little importance during the Gothic period.

Glass appeared in the windows of most churches and the larger dwelling-houses. Chiddingfold in Surrey, where glass-manufacture began in 1230, was the chief centre of the industry. The two varieties made in England were *brode-glas* (broadglass) and Normandy glass, corresponding respectively to modern sheet and crown glass. At first the glass was thick, uneven, and somewhat green in colour (p. 326). When coloured, it was 'pot-metal', that is, the colour was mixed into the molten glass. Hardly any surface-painting was done. The use of the diamond was unknown, the glass being laboriously cut by passing the point

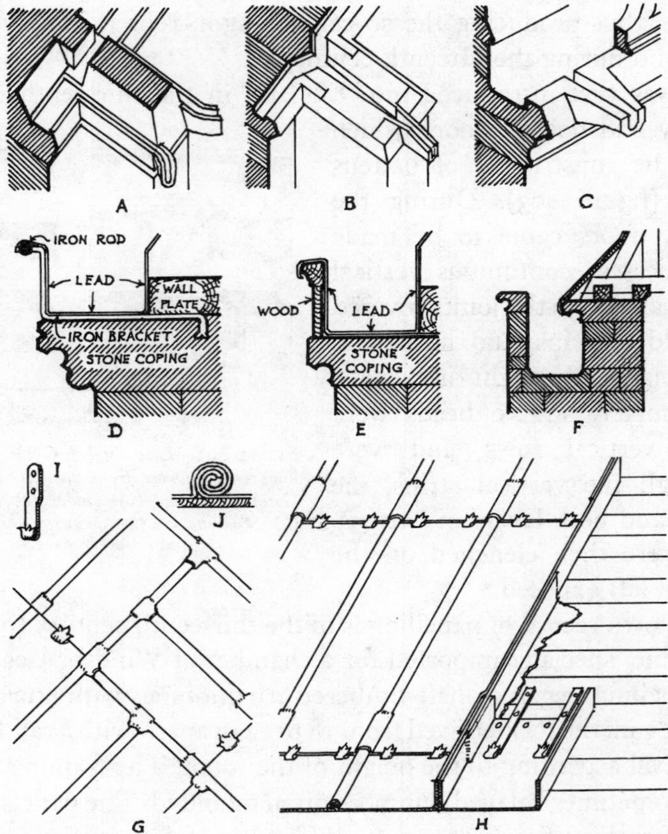

FIGURE 404—*Gothic roof-drainage and roof-covering:* (A), (B), (C) *stone gargoyles (French);* (D), (E), (F) *parapet gutters (French);* (H), (I) *diagonal leadwork on a spire (French);* (G), (J) *leadwork of Chartres cathedral roof (French).*

of a red-hot iron over the surface. In the fourteenth century, the art of glass-painting made great strides, especially in London and at York. Light saddle-bars were fixed inside the glass, between the stout iron framing-bars mentioned (pp 437 f). To these saddle-bars the glazing was attached by means of strips of lead, soldered to the leading of the glass and twisted round the bars. The use of copper wire for this purpose is modern.

External lead-work was almost entirely confined to roof-coverings for churches and other large buildings (figure 404, G, J). Water was often discharged direct to the ground by stone gargoyles (figure 404, A–G), thus avoiding the need for lead down-pipes; but lead gutters, 2 ft wide, laid on boarding, are mentioned from 1300 onwards; other lead gutters were carried on iron stirrups or brackets (figure 404, D–F). Lead down-pipes were, however, occasionally used: thus at Westminster Palace in 1532 there were 'square pipes of leade, garnysshid with the kinges armes and badges'.

Medieval lead was always cast, not milled. It weighed as much as 12 or 13 lb per sq ft. The casting was done on a bed of smooth sand, specially prepared for each sheet. The lead contained much silver and arsenic, hence the beautiful patina that appears on it with age. At Chartres cathedral in the thirteenth century the sheets were about $\frac{1}{6}$-in thick, and not more than 2 ft wide and 3–4 ft long. They were fastened to boarding at the top with large-headed iron nails. The side of each sheet was rolled up with the edge of the adjoining sheet so as to produce a roll of about $1\frac{1}{2}$-in diameter (figure 404, G, J). The bottom edge was secured by two iron clips to prevent the wind from blowing it up, and the tails of the clips were nailed to the boarding. All this work was in good order six centuries after it was first laid. Several lead-covered spires of the Middle Ages still survive in England, as at Godalming and Chesterfield, both of which have the lead rolls fixed diagonally. The most famous of all, that at Old St Paul's cathedral, London, is said to have been 260 ft high; it was destroyed by fire in 1561.

FIGURE 405—*Rustic building and wattle. From an Italian manuscript of 1083.*

Consideration of special types of building is beyond our concern, which is solely with constructional methods. Hence domestic buildings have been mentioned only when their construction differed from that of major public buildings. Nearly to the end of the Middle Ages, most of the dwellings of artisans and peasantry, whether in town or country, were so primitive as to consist either of wattle-and-daub, or of timber-framing (figure 405), or of rough stone. Their roofs were generally thatched.

The better-class Greek houses in the Golden Age (fifth–fourth centuries B.C.) were usually of two storeys and either built round a courtyard into which all the rooms faced, and with blank external walls or, as at Olynthus, in short terraces. At Delos there was primitive drainage, with latrines. Middle-class Roman houses, for the most part, were simple, and built round a courtyard; but, in Rome itself, the high cost of land led to the erection of apartment-houses or 'flats', three to five storeys high, with separate staircases to each 'flat'; and similar houses existed at Ostia, the port of Rome (plate 29 A). At Pompeii were houses displaying a higher degree of luxury and refinement than prevailed anywhere in the classical period (plate 29 B). These, too, were built round a courtyard or atrium. The numerous Roman 'villas' in Britain were really large farmhouses of complicated plan, heated by hot air and provided with mosaic floors in the principal rooms. The villa at Bignor in Sussex had 65 rooms, that at Folkestone had over 50, and there is another fine example at Chedworth near Cirencester. The earliest medieval stone houses in England are of the Norman period, as for example the so-called 'Jews' Houses' at Lincoln, and Moyses Hall at Bury St Edmunds. Such houses are, however, very few; and nearly all surviving medieval examples in stone or brick date from the fourteenth century onwards. Hardly any timber houses older than *c* 1400 have survived the accidents of fire or the ravages of weather and decay.

It may be said that, during the whole period of the Middle Ages, building was unscientific, in the sense that no theoretical rules of mechanics and statics were applied to problems of construction. Empirical methods of trial and error held the field, and collapses, especially of towers, occasionally resulted. On the other hand, medieval architects and builders did display a wide knowledge of applied geometry; but no technical treatise like that of Vitruvius has survived to inform us of the range of craft-lore available.

REFERENCES

[1] VITRUVIUS *De architectura*, II, iii, 3. (Loeb ed. Vol. 1, p. 92, 1931.)
[2] *Idem Ibid.*, II, vii, 1, 5. (Loeb ed. Vol. 1, pp. 106, 110, 1931.)
[3] *Idem Ibid.*, II, vi. (Loeb ed. Vol. 1, pp. 100 ff., 1931.)
[4] *Idem Ibid.*, II, iv. (Loeb ed. Vol. 1, pp. 94 ff., 1931.)
[5] *Idem Ibid.*, II, v. (Loeb ed. Vol. 1, pp. 96 ff., 1931.)
[6] *Idem Ibid.*, II, iii, 1–4. (Loeb ed. Vol. 1, pp. 88 ff., 1931.)
[7] *Idem Ibid.*, X, ii. (Loeb ed. Vol. 2, pp. 278 ff., 1934.)
Idem Ibid., III, iv, 2. (Loeb ed. Vol. 1, p. 180, 1931.)
[8] *Idem Ibid.*, II, viii, 1. (Loeb ed. Vol. 1, p. 110, 1931.)
[9] *Idem Ibid.*, II, ix. (Loeb ed. Vol. 1, pp. 130 ff., 1931.)
[10] *Idem Ibid.*, IV, vii, 4. (Loeb ed. Vol. 1, p. 240, 1931.)
[11] *Idem Ibid.*, VII, i, 2. (Loeb ed. Vol. 2, p. 80, 1934.)
[12] *Idem Ibid.*, V, i, 6–9. (Loeb ed. Vol. 1, pp. 258 ff., 1931.)
[13] *Idem Ibid.*, V, x, 3. (Loeb ed. Vol. 1, p. 304, 1931.)

[14] *Idem Ibid.*, VII, ii, 1–2. (Loeb. ed. Vol. 2, p. 86, 1934.)
[15] *Idem Ibid.*, VII, iii, 1–2. (Loeb ed. Vol. 2, p. 88, 1934.)
[16] *Idem Ibid.* VII, iii, 3. (Loeb ed. Vol. 2, p. 90, 1934.)
[17] *Idem Ibid.*, VII, iii, 5–7. (Loeb ed. Vol. 2, pp. 90 ff., 1934.)
[18] *Idem Ibid.*, VII, iii, 11. (Loeb ed. Vol. 2, p. 96, 1934.)
[19] *Idem Ibid.*, VII, iv, 1–2. (Loeb ed. Vol. 2, pp. 96 ff., 1934.)
[20] *Idem Ibid.*, V, ix, 7. (Loeb ed. Vol. 1, p. 300, 1931.)
[21] *Idem Ibid.*, VIII, vi, 4, 8, 10–11. (Loeb ed. Vol. 2, pp. 182, 186, 188, 1934.)
[22] *Idem Ibid.*, V, i, 4–5. (Loeb ed. Vol. 1, p. 256, 1931.)
[23] BEDE, The Venerable. *Historia abbatum* in 'Complete Works', Latin and English ed. by J. A. GILES, Vol. 4, p. 366. Whittaker, London. 1843.
[24] JACKSON, SIR THOMAS (GRAHAM). Letter in *The Times*, 7th October, 1907.
[25] RUDBORNE, THOMAS. *Historia major Wintonensis* in WHARTON, HENRY (Ed.). *Anglia sacra*, Part I, p. 275. Richard Chiswel, London. 1691.
[26] BEDE, The Venerable. *Historia ecclesiastica gentis anglorum*, III, xxv, in 'Complete Works', Latin and English ed. by J. A. GILES, Vol. 2, p. 362. Whittaker, London. 1843.
[27] JACKSON, SIR THOMAS (GRAHAM). 'Gothic Architecture', Vol. 1, p. 259. University Press, Cambridge. 1915.
[28] WREN, SIR CHRISTOPHER. "Memorial on Westminster Abbey", in WREN, C. *Parentalia*, pp. 295–303. T. Osborn and R. Dodsley, London. 1750.
[29] JACKSON, SIR THOMAS (GRAHAM). See ref. [27], Vol. 2, p. 125.

BIBLIOGRAPHY

General:

BRIGGS, M. S. 'A Short History of the Building Crafts.' Clarendon Press, Oxford. 1925.
CHOISY, F. A. 'Histoire de l'architecture' (2 vols). Gauthier-Villars, Paris. 1899.
STRAUB, H. 'A History of Civil Engineering' (trans. by E. ROCKWELL). Leonard Hill, London. 1952.

Greece and Rome:

CHOISY, F. A. 'L'art de bâtir chez les Romains.' Ducher, Paris. 1873.
COZZO, G. 'Ingegneria romana.' Casa Editrice Selecta, Rome. 1928.
DINSMOOR, W. B. 'The Architecture of Ancient Greece' (3rd ed. rev.) based on the first part of 'The Architecture of Greece and Rome' by W. J. ANDERSON and R. P. SPIERS. Batsford, London. 1950.
DURM, J. 'Die Baukunst der Griechen' (3rd ed.) in E. SCHMITT, *et al.* (Eds). 'Handbuch der Architektur', Part II, Vol. 1. Gebhardt [Kröner], Leipzig. 1910.
Idem. 'Die Baukunst der Etrusker. Die Baukunst der Römer' (2nd ed.). *Ibid.* Part II, Vol. 2. Kröner, Stuttgart. 1905.
MIDDLETON, J. H. 'The Remains of Ancient Rome' (2 vols). Black, London. 1892.
RIVOIRA, G. T. 'Roman Architecture' (trans. by G. McN. RUSHFORTH). Clarendon Press, Oxford. 1925.
ROBERTSON, D. S. 'A Handbook of Greek and Roman Architecture' (2nd ed.). University Press, Cambridge. 1943.
VITRUVIUS *De architectura*. (Loeb ed. 2 vols. Heinemann, London. 1931–4.)

Byzantine and Lombard:

CHOISY, F. A. 'L'art de bâtir chez les Byzantins.' Librairie de la Société Anonyme de publications périodiques, Paris. 1883.

DIEHL, C. 'Manuel d'art byzantin' (2nd ed. rev. and enl., 2 vols). Picard, Paris. 1925–6.

RIVOIRA, G. T. 'Lombardic Architecture' (trans. by G. McN. RUSHFORTH. 2 vols). Re-edited by EDITH E. RIVOIRA with additional notes by various authors. Clarendon Press, Oxford. 1933.

Medieval:

ADDY, S. O. 'The Evolution of the English House' (rev. and enl. from author's notes by J. SUMMERSON). Allen and Unwin, London. 1933.

ANDREWS, F. B. 'The Mediaeval Builder and his Methods.' University Press, Oxford. 1925.

BOND, F. 'Gothic Architecture in England.' Batsford, London. 1905.

CLAPHAM, SIR ALFRED (WILLIAM). 'English Romanesque Architecture' (2 vols). Clarendon Press, Oxford. 1930–4.

CORROYER, E. 'L'architecture gothique.' Ancienne Maison Quantin, Librairies-Imprimeries Réunies, Paris. 1891.

CROSSLEY, F. H. 'Timber Building in England from Early Times to the end of the Seventeenth Century.' Batsford, London 1951.

ESSENWEIN, A. VON and HASAK, M. 'Die romanische und die gotische Baukunst' in DURM, J., SCHMITT, E., *et al.* (Eds). 'Handbuch der Architektur', Part II, Vol. 4. Bergsträsser, Darmstadt and Kröner, Stuttgart. 1892–1903.

INNOCENT, C. F. 'The Development of English Building Construction.' University Press, Cambridge. 1916.

KNOOP, D. and JONES, G. P. 'The Mediæval Mason.' University Press, Manchester. 1933.

LLOYD, N. 'A History of English Brickwork' (new and abr. ed.). Montgomery, London. 1935.

SALZMAN, L. F. 'Building in England down to 1540.' Clarendon Press, Oxford. 1952.

VIOLLET-LE-DUC, E. E. 'Dictionnaire raisonné de l'architecture française' (10 vols). A. Morel, Paris. 1854–69. Especially "Construction" in Vol. 4, pp. 1–279, trans. by G. M. HUSS. 'Rational Building.' Macmillan, New York. 1895.

13

FINE METAL-WORK

HERBERT MARYON

I. GREEK AND ROMAN FINE METAL-WORK

A CHARACTERISTIC quality of early Greek and Etruscan jewelry is the extreme delicacy of the craftsmanship. The goldsmith's craft has never yielded better granular work, more delicate filigree, or finer plaited chain-work (vol I, pp 657–8). And the fine judgement displayed in the association of well proportioned beads and links between coloured stones in necklaces has been equalled only in the best works of the Renaissance. In contrast with the delicate craftsmanship of the Greek goldsmith, that of his Roman successor sometimes presents broader surfaces of bright gold—perhaps a medallion of some emperor or ruler—with wide settings and strong links. We are impressed by the sheer breadth of the gold-work displayed and its massive quality.

Greek and Roman mirrors were of polished bronze, often circular, with a handle of delicate ornament, or perhaps a figure. Sometimes the back of the mirror has a pattern of concentric circles turned on a lathe. Again, it may have some mythological scene worked with the tracer.

For simplicity of form and sheer beauty of line the silver and bronze cups from Greek lands have few peers (figure 406). These works are usually small but have the same qualities of good proportions and fine lines as may be observed in the large bronze *hydriae* and craters, such as that from Vix, France, whose neck is decorated with finely modelled figures (figures 407–8). Ten circular silver bowls, each 9 inches in diameter, among the treasure found in the Sutton Hoo ship-burial (*c* A.D. 655), were made on a lathe by spinning: a technique seldom found earlier than the fourth century B.C., though from then onwards lathe-work is regularly met with. It would seem that the craftsman took a piece of wood, perhaps 10 or 11 inches in diameter, and trued up its surface on a lathe. Then in the face of the disk he turned a recess of the exact size and shape of the bowl to be made. He annealed a sheet of silver and nailed it firmly over the recess. He would need a firm rest, with a steadying-pin on it, and a long and strong spinning- or burnishing-tool, with a handle about 2 ft long and a brightly polished, hard-steel tip. He would oil the face of the silver, and while the lathe was rotating, and using the tool as a lever and the pin on the rest as its fulcrum, he would press or burnish the silver into the recess. With long stroking movements he would work it gradually into the shape desired. On the Sutton Hoo bowls the marks of the spinning-tools are still visible. Finally, the craftsman would cut the finished bowl free from its nailed border.

Greek and Roman metal-work was made over a period of some 2000 years from the days of Knossos and Mycenae, through the great centuries of Greek

FIGURE 406—*Hellenistic bronze cup. Height c 5 in.*

and Roman supremacy, to the beginning of the fifth century A.D., when the settled life of the Roman peoples ended. But the men who won the succession to that civilization, in Gaul, in Britain, and—beyond the old frontiers—in Germany and Scandinavia, had learnt and retained the traditional knowledge of the crafts inherited by the Romans. They in their turn produced great works bearing the mark of their own native cultures.

II. INLAYING

Cloisonné enamels from a Mycenaean tomb at Kouklia, Cyprus (figure 420), anticipated by more than 2000 years that splendid flowering of the enameller's craft which gave glory to Byzantine art in the early Middle Ages. But that art had other precursors. Celtic mirrors (figure 423) and horse-furniture enriched by coloured enamels had succeeded the delicate enamels of Greece; the sturdy works of Rome were followed by the fine *champlevé* ornament of Saxon hanging bowls.[1]

Cloisonné jewel-enriched metal-work preceded true enamelling by many centuries. When Greeks of the fifth century B.C. and later visited the court of Persia they were deeply impressed by the splendour of its jewelled metal-work,

[1] In *cloisonné* work the ornament (jewels or glass) is gripped, cemented, or fused into a network of cells formed of metal strip or wire. The cloisons may be soldered to a solid metal background, or open to the light, when the work is said to be *à jour*. In *champlevé* the hollows to receive the decoration are formed in the metal background by chasing- or engraving-tools. Occasionally they are provided for in the cast.

as were the Goths who raided Asia Minor in the third century B.C. They, too, adopted the Persian and Sarmatian methods of decoration and carried them in their migrations throughout Europe. By the mid-third century A.D. Gothic rulers in the Danubian region were using golden vessels that imitated the rich, jewel-encrusted ornamentation of the Persian court. Sometimes, for example in the treasure of Petrossa (figure 409), the coloured stones were fastened *à jour* in their settings, or sunk deeply, as in a buckle of gold set with red stones and incised in Pehlevi characters with the name of Artaxerxes.

FIGURE 407—*Bronze crater found at Vix, near Châtillon-sur-Seine.* c 575 B.C. *Height over 5 ft.*

Among other pieces of the same kind are the sword of Childeric (d A.D. 481) (figure 410), and the engraved crystal cup of Chosroes II (589–628) (figure 411), of gold encrusted *à jour* with red and green jewels set in *cloisons*. The golden book-cover in the cathedral at Monza, given by the Lombard queen Theodelinde before her death in 628, is decorated with *cloisonné* borders of garnets and *cabochon* stones and is of as fine workmanship as the sword.

Similar treasures have survived from the Jutish-Visigothic kingdoms, such as the crowns of Fuente de Guarrazar (figure 412). Comparable works in England include the Kingston brooch (figure 413), and the gold purse-mount from Sutton Hoo (figure 414): the greatest jewel of its age. The cross of St Cuthbert

FIGURE 408—*Detail of the frieze from the Vix crater. Scale 1/3.*

(figure 415) provides an almost unique example of northern English craftsmanship of similar date.

Probably the most ancient surviving example of such inlaid jewel-work by Byzantine craftsmen is a reliquary containing a fragment of the 'True Cross', brought from Constantinople in the year 569 to the convent of St Croix, Poitiers,

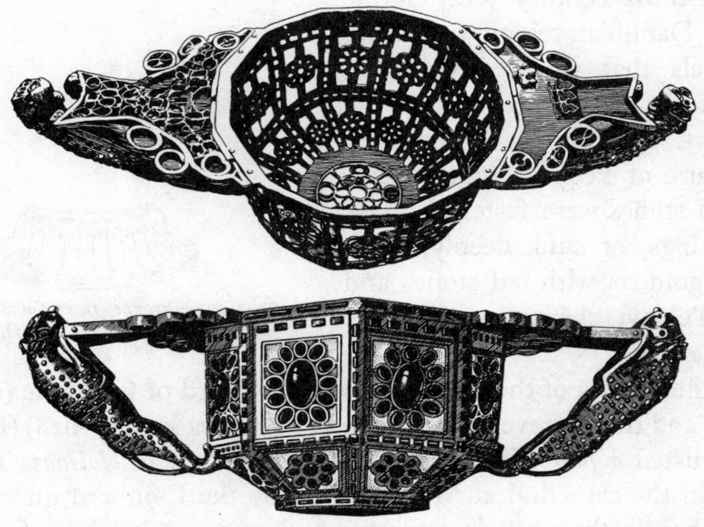

FIGURE 409—(Above) *Twelve-sided gold basket, and* (below) *eight-sided bowl, restored. From the Gothic treasure of Petrossa, Rumania. Third century* A.D. *Diameters:* c *6·9 in,* c *7·3 in.*

as a gift from the Emperor Justin II (565–78) to Queen Radegund, abbess of the convent (figure 690). It is mounted in the central leaf of a triptych and consists of a panel of *cloisonné* enamel, some 2¼ in square. In its centre the sacred relic is preserved within a cross-shaped recess, with a border of square green stones set in *cloisons*. A similar framework of green stones surrounds the panel. The enamel has a background of dark blue with floral scrolls in gold, leaves of turquoise blue, and flowers with red pistils. The style of both the jewelled border and the enamelled panel suggests an eastern origin. A contemporary work is the gold cross given by the same Justin II to Rome. The employment of a well designed inscription as an integral part of the design, in this and other works of the same period, is worthy of note.

The custom of inlaying iron buckles and like objects with ornamental patterns of silver was extensively practised in France during Merovingian times (431–751). Simple linear patterns were formed by hammering silver wire into lines indented by a chasing-tool, while broad spaces were covered with even layers of silver or gold. To gain the required effect, parallel grooves were cut deeply with a sharp tracer into the plate to be decorated. A length of silver wire was laid in each groove and flattened by hammering. Some of the silver entered the groove, but much of it spread over the surface. With closely spaced grooves the layer of silver or gold could be made smooth, continuous, and able to withstand rough handling. Cheap inlaid work is executed in the east today by roughening the surface of the metal with matting-tools, and then hammering or burnishing upon it one or more layers of gold leaf or foil.

FIGURE 410—*Reconstruction of the gold hilt of the sword of Childeric. The pommel and guard are set with garnets. Merovingian,* c *481. Scale 5/9.*

From the thirteenth century much fine inlaid work was produced in Persia, Arabia, and India. The design was generally worked upon the surface of a vessel with chasing-tools. The recess for each piece of inlay was then excavated with sharp chisels, the central portion of the recess often being allowed to remain at its original level. The metal to be inlaid was cut to shape and the craftsman used a thin, flat-ended punch, typically triangular in section, to drive the edges of each piece into the hollow carved to receive it. The punch was inclined to trespass a little upon the surface of the neighbouring metal, which was thus bent over to grip the inlay. The work, if carefully done, may survive centuries of wear and cleaning. Some brass and bronze vessels from Arabia and Persia inlaid in gold or silver with figures, heraldry, ornament, and inscriptions show craftsmanship of the highest order. The *baptistère* of St Louis in the Louvre is one of the most beautiful examples of such work (figure 416, plate 32 B).

Another method of assuring the stability of an inlay is to fill the recesses with silver solder, or to solder inlaid panels in position. Again, niello fused into a recess is stable. To roughen the ground in preparing sunken work the 'rocking' or *tremolo* treatment with a flat gouging-chisel may be employed. The tool is pushed forward and at the same time rocked from side to side. The cut is a characteristic zigzag. This pattern is found on Greek fibulae or pins of the eighth century B.C.; it proves the use of steel tools at that early age, since a bronze tool would not have served.

The great church of St Sophia, Istanbul (Constantinople), since 1453 a mosque and now a museum, has the oldest and most beautiful of a series of bronze cathedral doors of the early Middle Ages. They are of the ninth century and have well designed frames and borders in cast relief, with inlaid monograms in silver. There are other fine examples of bronze doors, in Germany at Mainz,

FIGURE 411—*Cup of Chosroes II, king of Persia; made of crystal, gold, and jewels. From the treasure of St Denis, Paris.* c *600. Scale 1/3.*

Hildesheim, and Cologne; and in Italy at Amalfi, Monte Cassino, Rome, Salerno, and Venice. Some are decorated with figures and ornament in relief, others with niello or inlays of silver. The doors of the church of San Michele at Monte Sant' Angelo near Foggia, given by the family of Pantaleone, have twenty-four panels of figures inlaid with niello and silver. The incised lines on the figures are filled with mastic, coloured black, blue, red, and green.

In the Sutton Hoo hoard are a number of jewels in gold *cloisonné* of unique construction. The patterns, often of interlaced dragons, are formed from shaped pieces of garnet, shown against a gold background. Apparently a little strip of gold was bent round each fragment of garnet after it had been ground to shape, to make a setting. These *cloisons*, arranged face downwards on a thin sheet of gold, were soldered in place. That part of the gold plate lying within the *cloison* was next cut away. Seen from the front at this stage the design would be represented by a gold background pierced with holes of the shapes required for the garnet dragon pattern. From the back, the *cloisons* would be seen closely surrounding each opening. Another plate of gold was then soldered at the back as a foundation for the jewel. Before the garnets were inserted, a piece of gold foil with a diaper of pyramids in relief was laid at the bottom of each cell to impart brilliance to the stones. In the shoulder-clasps filigree contrasts with the garnet-work (figure 417).

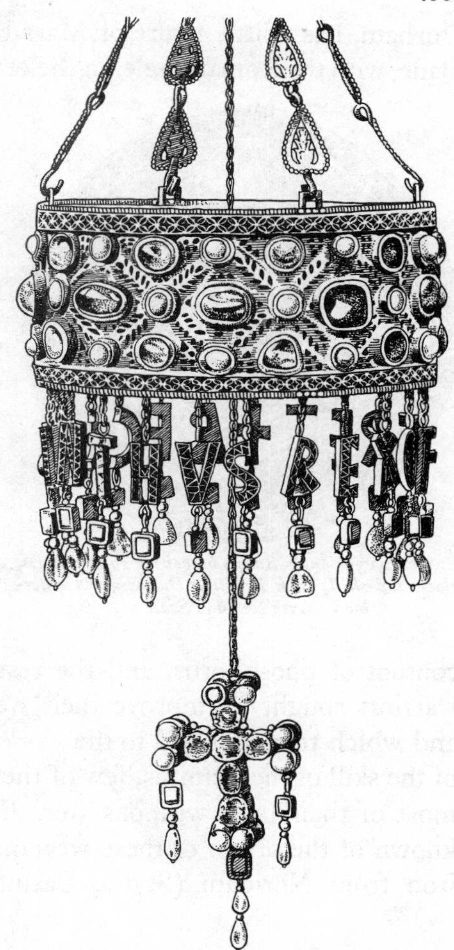

FIGURE 412—*One of the votive crowns from Fuente de Guarrazar near Toledo, of gold and stones cut* en cabochon *with jewelled pendants hanging from the Visigothic letters of the king's name and dedication:* RECCESVINTHUS REX OFFERET. C. *660. Scale 1/4.*

The crown in the Imperial Treasury at Vienna (figure 418), formerly known as the crown of Charlemagne, was probably first placed on the head of Otto the Great (in 962) and modified for the coronation of Conrad II (in 1027). It is of gold with eight round-topped panels joined by hinges, while an arch above, linking the front and back panels, is surmounted by a cross. Four of the panels bear large *cloisonné* enamel plaques, with figures of Christ between cherubim, kings, and prophets. With the exception of these plaques, the whole of the crown is covered with fine polished but uncut sapphires, emeralds, rubies, and pearls, all *à jour* in high settings, making together a work of great splendour.

III. ORNAMENTATION OF WEAPONS

From the beginning of the European Iron Age and typically at Hallstatt (*c* 900 B.C.) bronze weapons were occasionally inlaid with iron. Numerous later bronze weapons inlaid with gold or silver are known. Some Roman iron swords are inlaid with brass. One of the second century A.D., found at South Shields,

Durham, has a little figure of Mars between two standards on one side of the blade, with the Roman eagle on the other. It is also decorated by pattern-welding (pp 457, 458).

FIGURE 413—*Anglo-Saxon brooch of gold, garnets, glass, and shell, from Kingston Down near Canterbury. Sixth century. Scale 2/3.*

The production of steel suitable for weapons long remained a major metallurgical problem. An iron sword was too soft for efficiency, and had to be case-hardened and tempered to transform it into a useful steel weapon. Thus in the Icelandic Sagas a hero found that 'the fair-wrought sword bit not whenas it smote armour, and oft he must straighten it under his foot' [1]. This common experience indicates that soft iron weapons were less efficient than those of bronze. The bog ores from which the Scandinavian weapons were forged can be smelted at a low temperature but have a high content of phosphorus, and the result is a very soft iron. Hence the northern warriors sought to improve their weapons, to which they often gave names, and which they ascribed to the workmanship of gods and giants. Despite boasts of the skill of their smiths, few of them could, in fact, make laminated steel, and most of their finer weapons were brought from afar. Very little seems to be known of the origin of these weapons. Latin authors praise the high quality of iron from Noricum (Styria, Carinthia, and neighbouring districts). Swords

FIGURE 414—*Purse-mount from the Sutton Hoo ship-burial, of gold, ivory, garnets, and* millefiori *glass. Seventh century. Scale c 1/2.*

came also from the Romanized part of Germany on the middle and lower Rhine, where from the sixth century under the Frankish Empire the manufacture and export of sword-blades provided a staple industry.

FIGURE 415—*St Cuthbert's cross, of gold, garnets, and shell, buried with him in 687 at Durham. Scale 9/10.*

To produce a good weapon the smith tested iron from every available source, forging and welding, heating, quenching, and tempering, until at last he found a serviceable material. Then, having forged an efficient weapon he went a step farther by decorating the handle, scabbard, and baldric, and even the blade itself. The blade was decorated with strips of steel, twisted or plain, welded on to each side. The Sagas speak of snake-patterns running right down the blade. Beowulf's sword had an iron blade, 'patterned by twigs of venom, hardened with the blood of battle' [2]. A king of the Vandals sent to Theodoric the Great (489–526), then lord of Italy, sword-blades 'more precious than gold', seemingly covered with patterns of small snakes, whose steel gleamed with various colours [3].

Ninety steel swords found at Nydam, south Jutland, were decorated with twisted or coiled patterns wrought in the fabric of the weapon itself. Some show a central band composed of four twisted rods, or bundles of flattened wires, running parallel to each other the full length of the blade. Each alternate rod is twisted to the right, the others to the left. Another sword has but two bundles or rods down the centre (figure 419, left).

FIGURE 416—*Detail from the* baptistère *of St Louis showing gold and silver inlays, and recesses prepared for the missing pieces. The ground is of* brass. *c 1290–1310. Scale c 1/2.*

The smith would have no great difficulty in making steel strips about $\frac{1}{100}$-in thick by beating out a slender rod of the approximate length. With shears he could cut off the narrow strips required, each about $\frac{1}{8}$-in wide. Or, instead of dealing with so many loose pieces, he may have preferred to fold longitudinally into several layers a strip about $\frac{3}{8}$-in wide. This he would grip with two clamps, fixed perhaps an inch apart. After he had given the required twist to the bundle, in the space between the two clamps, he would move his clamps to the next position. After each bundle of rods had been twisted it was sometimes welded into one piece, and split longitudinally. A number of strips thus prepared were arranged down the centre on each side of the sword-blade, and welded there. Figure 419 (right) illustrates a modern attempt to make a 'pattern-welded' blade. The photograph was taken before welding.

The pattern-welding of these swords is an extremely difficult operation. It represents a forerunner of 'damask' steel.

FIGURE 417—*One of the pair of shoulder-clasps from the Sutton Hoo ship-burial, of gold decorated with garnets,* millefiori *glass, and filigree. Seventh century. Scale 3/5.*

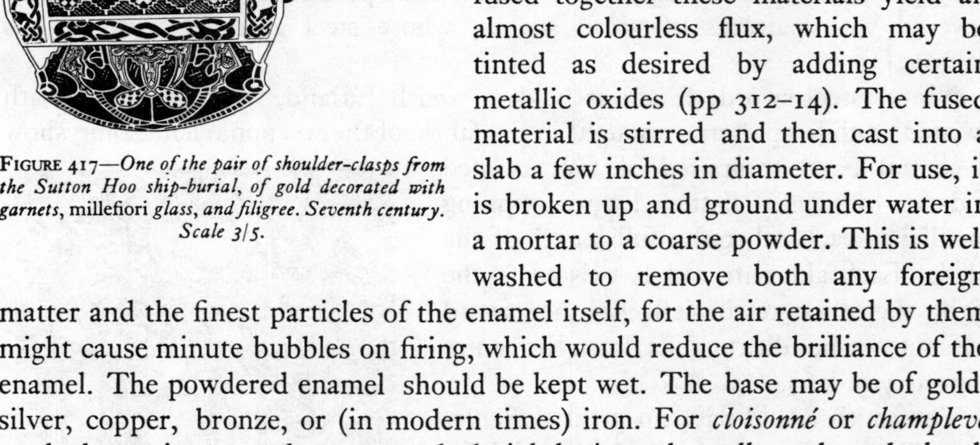

IV. ENAMELLING

Enamelling is the fusing on to a metallic base of a vitreous material. Generally this is a soft glass compounded of flint or sand (*c* 50 per cent), red lead (*c* 35 per cent), and soda or potash (*c* 15 per cent). When fused together these materials yield an almost colourless flux, which may be tinted as desired by adding certain metallic oxides (pp 312–14). The fused material is stirred and then cast into a slab a few inches in diameter. For use, it is broken up and ground under water in a mortar to a coarse powder. This is well washed to remove both any foreign matter and the finest particles of the enamel itself, for the air retained by them might cause minute bubbles on firing, which would reduce the brilliance of the enamel. The powdered enamel should be kept wet. The base may be of gold, silver, copper, bronze, or (in modern times) iron. For *cloisonné* or *champlevé* work the various powders are packed tightly into the cells and made level, thoroughly dried, and heated in the furnace until fused. The fired enamel has a highly polished surface and a very firm grip on the metal. Any hollows in its surface may be filled with more powder and the work refired. It may then be ground flat, and repolished by further firing or by friction.

Early enamels were fired in an open earthenware pot, laid on its side in a charcoal furnace, or, as recommended by Theophilus (p 351), covered with a perforated iron bowl [4]. Charcoal would be heaped over this, and the fire blown till the enamel fused. After cooling, imperfections could be cut out with a sharp tool, fresh powder added, and the work refired.

Though the Sumerians set coloured precious stones in gold or silver, and the Egyptians cemented fragments of glass and pottery in their jewelry, the earliest

FIGURE 418—*A crown, formerly known as the crown of Charlemagne, of gold enriched with enamels and jewels. Tenth century.*

examples of true enamel—fused in its *cloisons*—are late Mycenaean (*c* 1200 B.C.). These are six gold rings, decorated with roundels of enamel (figure 420), from a tomb at Old Paphos (Kouklia) in Cyprus. Each roundel is surrounded by a gold *cloison* and fitted in place within a band-setting. The area within the *cloison* was first covered with broken fragments of glass, of which the largest were about ⅛-in across. The disk was then fired. Upon its glazed surface a floral pattern of *cloisons* was arranged and stuck, probably with gum. Similar fragments of glass were inserted into the cavities between them. That the glass was not powdered is clear, for areas of different colour meet in a true line without penetration of one colour into the other. The whole disk was next fired and its surface ground and polished. It was then set within its band-setting in the finger-ring.

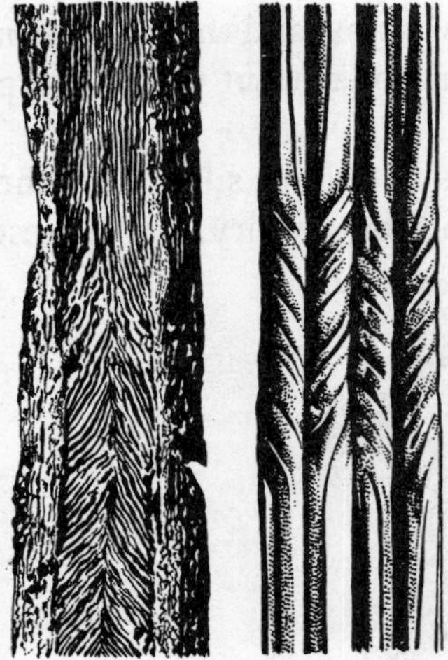

FIGURE 419—(Left) *A section of a pattern-welded steel sword from Nydam Moss, near Copenhagen. Scale* c *1/2.* (Right) *An experiment in pattern-welding; condition before welding.*

Few other examples of enamelling are known before the Greek enamels of about the sixth or fifth century B.C. A fine later example of such work is the golden sceptre with a Corinthian capital from Tarentum (figure 421). The shaft is covered with a diamond pattern of wires, and at each angle of the network a little ring of beaded wire is soldered. Each ring holds a touch of blue enamel, with quite magical effect. In all the earliest enamels yet discovered the enamelled portion of the work is small, and the material seems to have been used to simulate an inlay of coloured stones, such as that on the third coffin of Tutankhamen (vol I, frontispiece).

In the west, under the influence of Celtic craftsmen, the decorative use of small pieces of coral gave way to the employment of spots of enamel on horse-trappings, mirrors, and brooches. The famous bronze shield from the Thames at Battersea (figure 422), though exemplifying the Celtic craftsman's interest in pure line-ornament, has been mistakenly described as enamelled. Its coloured disks of red glass were cemented, not fused, into their sockets. A red enamel obtaining its colour from cuprous oxide was the earliest employed, but other colours, including blue, yellow, and green, were added during the first two centuries of our era, probably under Roman influence. The beautiful mirror from Birdlip, Gloucestershire (figure 423), again demonstrates the Celtic craftsman's interest in line, and the little touches of red enamel on its handle make a pleasant contrast with the golden bronze.

FIGURE 420—*Mycenaean gold ring set with* cloisonné *enamels. Kouklia, Cyprus, 1200 B.C. Original size.*

Many Roman fibulae and ornaments found in western continental Europe and in Britain are decorated with *champlevé* enamel. The Rudge cup is an important example. It bears an inscription containing place-names of stations on the Roman Wall, and, in relief, what may well be a contemporary picture of the wall itself. Heavy bronze armlets from Scotland, decorated with red and yellow enamels, carry on the *champlevé* tradition. A new and very beautiful flowering of the art is seen in the Anglo-Saxon hanging bowls, of which there are fine examples from the Sutton Hoo ship-burial (figure 424).

Characteristic differences in the structure of the metallic foundation and in its relationship to the vitreous covering enable us to group all known enamels under a few headings:

(i) *Cloisonné*. Each mass of powdered glass is placed in a separate cell, formed from narrow strips or wires of metal, to which, and to the background if any, the enamel holds firmly when fired. If there is no background, and the enamel is translucent, the work is known as *plique-à-jour*.[1] A fourteenth-century covered cup from Burgundy, now at South Kensington, is a beautiful example of this technique.

(ii) *Champlevé*. The enamel fills cells in the metal base-plate which have been sunk with chasing-tools, carved, stamped, or cast in the body of the work.

(iii) *Bassetaille* enamels, a subdivision of (ii). The design is chased or carved in low relief below the original level of the background. It is then covered with a smooth layer of translucent enamels. The modelled design beneath the surface remains visible through the enamel, and variations in the depth of its relief are reflected by gradations in the apparent strength of colour in the overlying enamel.

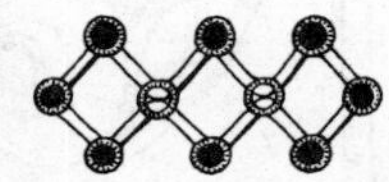

FIGURE 421—*End of a gold sceptre from Tarentum, Italy. The finial is a quince made of green glass.* (Below) *Detail of the enamelled gold network which covers the sceptre. Third century* B.C. *Length of sceptre 20·2 in.*

Bassetaille enamel is best illustrated by two well known cups (figure 425, plate 33) of the early fourteenth century. One is decorated with scenes from the life of St Agnes. The modelled surface to be covered with enamel was formed by chasing-tools on the front of a gold plate resting on a yielding bed, probably of pitch. The whole scene was driven down below the level of the background and was treated as a work in low relief and carefully finished. Here and there the surface was cut with a chisel or bevelled with a gouging-chisel to produce smooth-surfaced facets. The recesses were then filled with enamels of different colours, laid down side by side. When fired and polished the enamel occupied the recesses completely, the top of it being level with that of the ground alongside. The colours are of great splendour upon fine gold. A dotted diaper pattern enriches the otherwise

[1] 'Open plait'.

plain gold background. The other example of *bassetaille* enamel is the King's Lynn cup, described below (p 463).

FIGURE 422—*Celtic bronze shield, ornamented with red glass, found in the Thames at Battersea. First century A.D. Scale c 1/5.*

(iv) *Encrusted* enamels. A further subdivision of the second group. These may be employed to cover the surface of a figure beaten from sheet metal by repoussé and chasing, or to enrich an irregular surface, as on the mounts of a cup or the shoulders of a finger-ring, or to decorate a cast metal object.

(v) *Painted* enamels. These form the third main group. They were produced in great numbers from the later part of the fifteenth century onwards, and so fall beyond the scope of the present volume. Here, their name is a sufficient description.

From time to time other methods than those described above have been tried. The crown with enamelled figures of the Emperor Constantine Monomachos (1042–54), for example, has a plain gold background decorated with *cloisonné* enamel. The craftsman sank the area to be occupied by the figure about $\frac{1}{16}$-in below the surface of the background, driving it downwards with chasing-tools. In this recess he arranged the *cloisons* to complete the figure, and then filled them with enamel. A rare modification of this method was employed in the eleventh century on the portable altar in the treasury of St Foy at Conques, France. The material here is copper. The metal is in two layers, fastened together. The upper layer is pierced *à jour* to the outline of the device required. In the recess thus formed *cloisons* are fixed upon the lower plate, and the cells are filled with enamel, which is then fired and polished. All exposed portions of the copper are gilt.

Aesthetically, the *champlevé* technique lends itself well to sound design, for the metal forms left when the recesses for enamel are excavated need not be narrow or of even width, as in *cloisonné* work. They may extend laterally as desired. The reserved ground, often gilded, takes a more important share in the design, and the enamelled area becomes a decoration of the metal rather than

FIGURE 423—*Celtic mirror, of enamelled bronze, from Birdlip, Gloucestershire. Scale* c *3/10.*

an enrichment applied to it. In the Valence casket (figure 426), of gilded copper and enamel, there is a fine balance between the enamelled areas and the gilded background. Another example, with *bassetaille* enamels, is the King's Lynn cup (plate 33), which is the earliest piece of corporation plate in England (*c* 1325) and perhaps the most beautiful cup in existence. It is of silver gilt with enamels of dark blue, green, and purple. The occasionally wide reserved spaces of metal between the enamelled portions of each panel bind together the whole design in a manner impossible with *cloisonné*, and strengthen an impression of unity.

A characteristic of the Limoges enamelled caskets and reliquaries of the thirteenth and later centuries is the craftsman's custom of riveting to an otherwise flat-surfaced figure in *champlevé* enamel a head worked in high relief by repoussé. The effect produced is different from that of a sculptor's relief, where the change of planes is more gradual, but it enriches with contrasting relief the otherwise flat and rather bare surfaces of these works. Very rarely the head itself is enamelled, as in a figure of St Michael on an eleventh-century book-cover in the library of St Mark's, Venice. Here the figure and head are in repoussé gold, and both are encrusted with enamel.

FIGURE 424—*Foot of a hanging bowl from Sutton Hoo, with enamel and* millefiori *decoration. Seventh century. Original size.*

Yet another variety of enamelling technique developed later. On small glass medallions was engraved a pattern of narrow channels and cavities. These were lined with gold foil, the edges coming to the surface to form very delicate *cloisons*, into which enamel was filled and fired before the final polishing. An ornament of very delicate quality resulted.

A somewhat similar technique was employed in Ireland. On the Ardagh chalice (plate 32 A) are bosses of coloured glass into the surface of which are fitted disks of silver or other material, recessed to form patterns. The hollows cut in these insertions are filled with enamels of contrasting colour.

Twisted wires have been employed to enrich the border of an enamelled space where a plain *cloison* might appear too weak. Again, since the under-surface of an enamel is generally hidden by its setting, its colour there is of little importance. However, a few enamels, possibly intended for use in pendants, have both sides decorated.

Perhaps the most important early Byzantine enamel is the little gold reliquary cross found beneath the altar of the *Sancta Sanctorum* in Rome (figure 427). The sides bear an inscription in red enamel. The front is of five plates of *cloisonné* enamel representing the Nativity, the Journey to Bethlehem, the Adoration of the Magi, the Presentation in the Temple, and the Baptism. The colours are red, orange-red, dull red, lilac, blue, yellow, white, and black. The background is translucent enamel. The rude but expressive action of the figures, the pictorial style, and the character of the art are all eastern, and the cross was probably made at Antioch or in Armenia in the sixth or seventh century. It is a pioneer example of the *cloisonné* figure-work exported from the Christian east

FIGURE 425—*The Royal Gold Cup, decorated with* bassetaille *enamel. 1350. The two collars in the stem are later additions. Scale* c *3/5.*

to Rome when western craftsmen were decorating their work with inlaid stones and *champlevé* enamel patterns of ornament.

The Alfred Jewel (figure 692) found at Newton Park, Somerset, in 1693, is of gold, with a *cloisonné* enamel bust of a man. The enamel is covered by a plate of crystal. Round the border of the frame runs an inscription: AELFRED MEC HEHT GEWYRCAN, 'Alfred ordered me to be made'. Below is a boar's head covered with granulation-work. The under side is covered by a plate chased with floral ornament. The boar's head ends in a socket, which probably held a short rod of ivory or gold to be used as a pointer in reading a manuscript. The work shows strong Byzantine influence.

FIGURE 426—*The Valence casket, of gilt copper and* champlevé *enamel. 1290–6. Height 3·75 in, width 7 in.*

Equally Byzantine in character is the silver-gilt Beresford-Hope cross (figure 691). The figures of Christ and the Virgin and the busts of saints are in *cloisonné* enamel on gold. The drawing and workmanship are rather crude and probably of the eighth or ninth century. Another very important example of Byzantine enamelling is a reliquary formerly in the Oppenheim Collection and now in the Metropolitan Museum, New York.

Much of the Byzantine craftsmen's work in gold and silver was associated with coloured stones and enamel. Generally it is small in size, perhaps because the easily portable pieces had a better chance of survival. Occasional larger works were made, such as altar-fronts or reredoses, but these were still small in detail. Thus the altar-back at St Mark's, Venice, is of many small panels of *cloisonné* enamel in gold, set in a very elaborately decorated frame. The work varies in date from the tenth to perhaps the fifteenth century; its general effect is of innumerable small parts, glowing with gold and colour. The generally available furnaces could not cope with works on a large scale. In the Kircheriano Museum in Rome, however, is a gilded copper enamelled panel some 2 ft high, bearing a standing figure of Christ, perhaps of the twelfth century. The technique is unusual, for the greater part is in *champlevé* work, the remainder *cloisonné*. The cruciferous nimbus is jewelled, and edged with pearls. Below the feet is a border of intersecting circles, each containing a leaf in green and red enamel. The eleventh-century chalice of agate, with *cloisonné* enamelled panels and pearl borders and pendants (figure 693), in the treasury of St Mark, Venice, is a characteristic example of the work of the period. In this collection are many such chalices brought from Constantinople after its sack in 1204. The *cloisonné* enamel roundel (figure 428) is the base of a tenth-century paten in onyx from the same collection. It may be noted that in this example the cells were not completely filled with enamel.

FIGURE 427—*Cross-reliquary from the* Sancta Sanctorum, *St Peter's, Rome. Gold and enamel. Byzantine work of the sixth to seventh century. Scale c 3/10.*

The great European centres of enamelling in the Middle Ages were Limoges, in central France, and the valleys of the Rhine and the Meuse. Enamels were made, too, in northern France, Spain, England, and Italy. At first the art was confined to monastic workshops, but when growing demand led to the production of larger works in a cheaper material than gold, namely gilded copper, it was developed on a commercial scale in lay factories. Enamelling remained, however, closely related to church ritual; crosses, croziers, shrines, reliquaries, gospel-covers, and censers were produced in great numbers. Among secular

FIGURE 428—*Roundel from the base of a Byzantine paten.* Cloisonné *enamel. From St Mark's, Venice.*

objects with enamel were candlesticks, marriage caskets, and heraldic furniture.

Later the interest of the enamellers turned from *champlevé* in copper to *basse-taille* in silver. Vast numbers of little plaques were worked by chasing from the front and, when the modelling of the figures and draperies had been sufficiently indicated, the whole plaques, or selected portions of them, were covered with enamel. Blue and green enamels generally behave well on silver, as also do reds if the firing is skilful. Changes in technique were not limited to any one centre, for the craftsmen, like their brother stone-carvers and bell-founders, travelled from place to place with their tools.

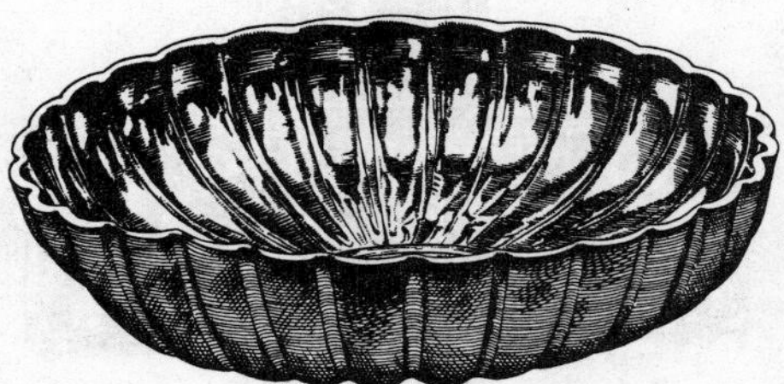

FIGURE 429—*Roman silver bowl, from Chaourse, Aisne, France. Diameter 7·2 in.*

V. REPOUSSÉ WORK

Repoussé is a general term to describe ornamental work produced by modelling sheet-metal with hammer and punches. The surfaces should be bent to and fro to catch the light at different angles. Some parts of the ornament may be lower in relief than the rest, and fade into the background. None of it should appear detached from the background. It should be obviously of one piece with it, but with surfaces tilted about to allow the play of light—an ornamentation of the metal itself rather than a decoration applied to it.

The tilting of the surfaces may be of the simplest kind, as in the fine bowl from Chaourse (France) (figure 429); or a flat sheet of metal may be enriched by skilful working of figures or floral ornament in relief, as on some of the French or Italian plaques of later times. Again, the subject may involve almost pure line-work, as on the remarkable Celtic 'Battersea shield' (figure 422). The term repoussé may be applied also to a subject such as the figure of St Foy at Conques, where a sheet of gold has been hammered, punched, and chased to form the head and figure of the saint, and afterwards nailed on to a wooden core. More elaborate works were sometimes undertaken. For example, the great altar-frontal in the church of St Ambrose at Milan has embossed figure panels and borders of *cloisonné* enamel. It was completed before 835. Another important work of the same class is the golden altar-front given by the Emperor Henry II to the cathedral at Basle about 1019, with figures in high relief in an architectural setting with stamped borders.

The aim of the craftsman in all such works is the production of an interesting pattern in light and shade, with pieces of rich pattern contrasted by larger, simpler lines and masses. He may even enhance the effect of light and shade by chasing what would otherwise be a smooth surface into one made up of many small waves, as on the King's Lynn cup (plate 33). The effect of this chasing is that as one's glance passes down, say, the stem of one of the trees, the light seems to ripple over the surface in a most charming manner.

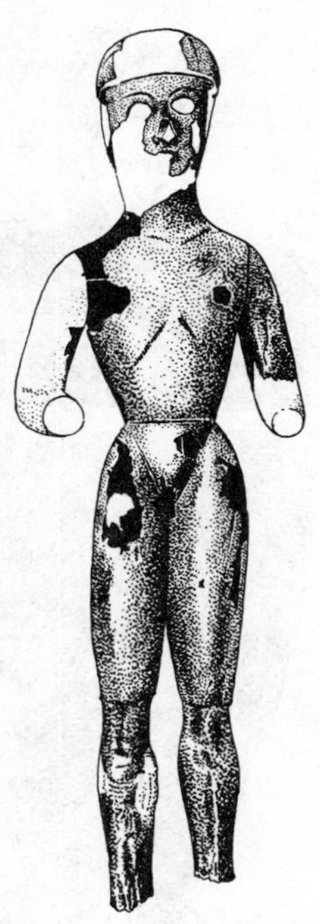

FIGURE 430 — *Human figure formed by* sphyrelaton *work, from the 'Geometric' temple at Dreros, Crete.* c 750 B.C. *Scale* c *1/10.*

For repoussé work the shape of the hammer is less important than its weight, though for fine work a light hammer with a flat face about an inch in diameter is generally used. The craftsman may have several hundred tools, though some 20 or 30 will serve for most of his work. Of these tools there are three essential groups: for tracing, for bossing, and for chasing. The tracers are blunt chisels varying a good deal in width of face. They are used to outline a form or to produce a narrow groove. Bossing-tools are of many shapes and sizes, but they all have rounded working-faces. They are used to boss up from the back any form which is to rise above the background. Chasing-tools are of many shapes and sizes, but in general their working-face is smooth and flat, though its edges may be rounded off a little.

In addition to the three main types of punches mentioned above a repoussé worker may have sets of number and letter punches, matt tools designed to produce a variety of textures, ring- and star-punches, and so on. While bossing up or chasing the metal it is usual to support it upon a rough wooden tray, about 1 in deep, filled with pitch; or on a hemispherical iron bowl of about 9 inches in diameter and ½-in thick, filled with pitch and weighing about 20 lb. The bowl stands in a ring of leather or coiled rope, and may be turned about or tilted to any angle as the work proceeds, always remaining steady. The weight of the bowl is important, for if it is much lighter it will be likely to joggle about when work is in progress.

From Hellenic lands, apart from small cast objects, the earliest metal figures are of sheet copper or bronze nailed to wooden supports. The later *sphyrelaton* work was of metal plates hammered and chased to shape, and held together by nails or rivets. Very few such works have survived, but three of them (*c* 750 B.C.), figures of a man and two women, were found in the shrine of the 'Geometric' temple at Dreros, Crete (figure 430). The two female figures have simple tube-like skirts, riveted up the front, with little modelling about the torso, and the arms straight down, with riveted seams. The principal parts of the work were shaped with the hammer, as was medieval armour. The man's arms are free from the body and bent at the elbow. The heads of the three figures have undergone very little shaping, though the straight locks of hair of the women may have been worked on pitch with chasing-tools. All the hands and feet are missing. Examples of this technique in gold are known from Etruscan sites.

FIGURE 431—*The silver vase from Chertomlyk, south Russia. Greek work,* c 400 B.C. *Scale* c 1/7.

Many figures of life-size or greater were made in this way in Greece and her colonies. The two most famous temple figures of antiquity, that of Athena in the Parthenon at Athens, and that of Zeus in the temple at Olympia, by the greatest artist of Greece, Phidias, were of ivory and gold. Their draperies were beaten out of gold sheets and removable from the framework. The largest statue of this kind, the Colossus of Rhodes (erected between 292 and 280 B.C.), representing the standing figure of the Sun god, Helios, was nude, with drapery hanging from its left arm to the ground. It was over 120 ft high and was placed at least 200 ft behind the quays. The modelled bronze plates were supported within by three columns of stone strengthened by heavy iron bars. After 50 or 60 years the figure was overthrown by an earthquake. Buckling at the knees, the upper part of the figure folded down, with its head and shoulders resting on the ground. In this collapsed state the figure remained a wonder of the world for nearly a millennium; ultimately it was broken up by the Saracens in 653.

The Chertomlyk vase (figure 431), found with many fine gold ornaments in a large barrow near the Dnieper in south Russia, stands nearly 28 in high. It is the finest surviving example of the work of the Ionian Greek silversmiths at a time when their rich rulers were living in close contact with the neighbouring Scythians, and before the degeneration of late Hellenistic art had begun. With the vase was a magnificent bow-case (figure 432) noteworthy for the excellent

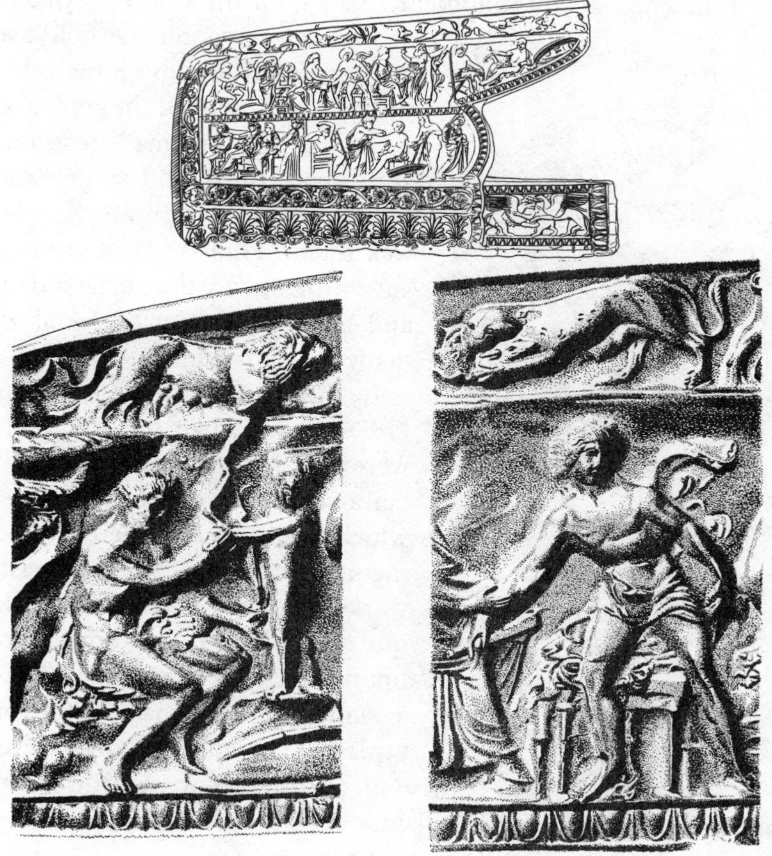

FIGURE 432—(Above) *A golden bowcase from Chertomlyk. Greek work; second century B.C. Scale* c 1/10. (Below) *Details of the repoussé work.*

primary modelling of the figures. They are in repoussé from a sheet of gold about 16 in long. The free execution shows a true understanding of three-dimensional relief. To carry this unfinished work to the level of the Siris bronzes (figure 433) might have been within the power of the craftsman who executed it, but for some reason the final chasing was never begun and much of the design is only roughly indicated.

In the Dublin Museum is a gold hair-ring of the Early Iron Age. Upon its surface the smith wished to simulate the effect of fine gold granulation, an operation for which he lacked the necessary skill in soldering (vol I, pp 657–8). On the underside of a very thin sheet of gold he drew with compasses about 50 concentric circles, in a band 10 mm wide, each circle being but 0·2 mm larger than the next. The circles consisted of innumerable dots, for the smith pressed one point of his compasses on the gold sheet, producing a small bump on the other side of the metal. He moved the compass forward about 0·2 mm and made a second bump, and then a third, and so on; and moved from circle to circle until the whole space was filled. Thousands of minute bumps gave a bloom to the surface of the gold, and from that sheet of metal the smith made his hair-ring.

FIGURE 433—*A very fine example of Greek repoussé work on the shoulder-piece of the bronze armour found near the river Siris, south Italy. Fourth century B.C. Height 7 in.*

By the beginning of the fifth century B.C. Greek smiths, like the Assyrians earlier, were making bronze armour—helmets, cuirasses, greaves, and other pieces—which would be hammered directly from ingots of bronze, perhaps cast by the worker himself. The smith would not require a pattern of wood or other material upon which to shape his work. He was a skilled hammerman, who could beat a plate of metal into any desired shape, and also adorn it. A little later some of the armour beaten from sheet metal was decorated with devices in repoussé, of a quality that has never been surpassed (figure 433).

Some of the silver dishes and bowls made during the last pre-Christian centuries in Persian lands show a curious development of repoussé. The ornament, generally of human and animal figures, was beaten in relief from sheet silver and the background cut away. The separate figures were then arranged on the surface of a thick silver bowl or dish. The edges of the figures were hammered or filed to fit neatly against the work. A line was next scratched round each, and a recess cut within that line so that when the figure was placed in position its edges would be sunk beneath the surface. Then the surrounding metal was driven inwards and downwards with chasing-tools, gripping the edges of the repoussé inlay. The result showed figures in high relief on a smooth background.

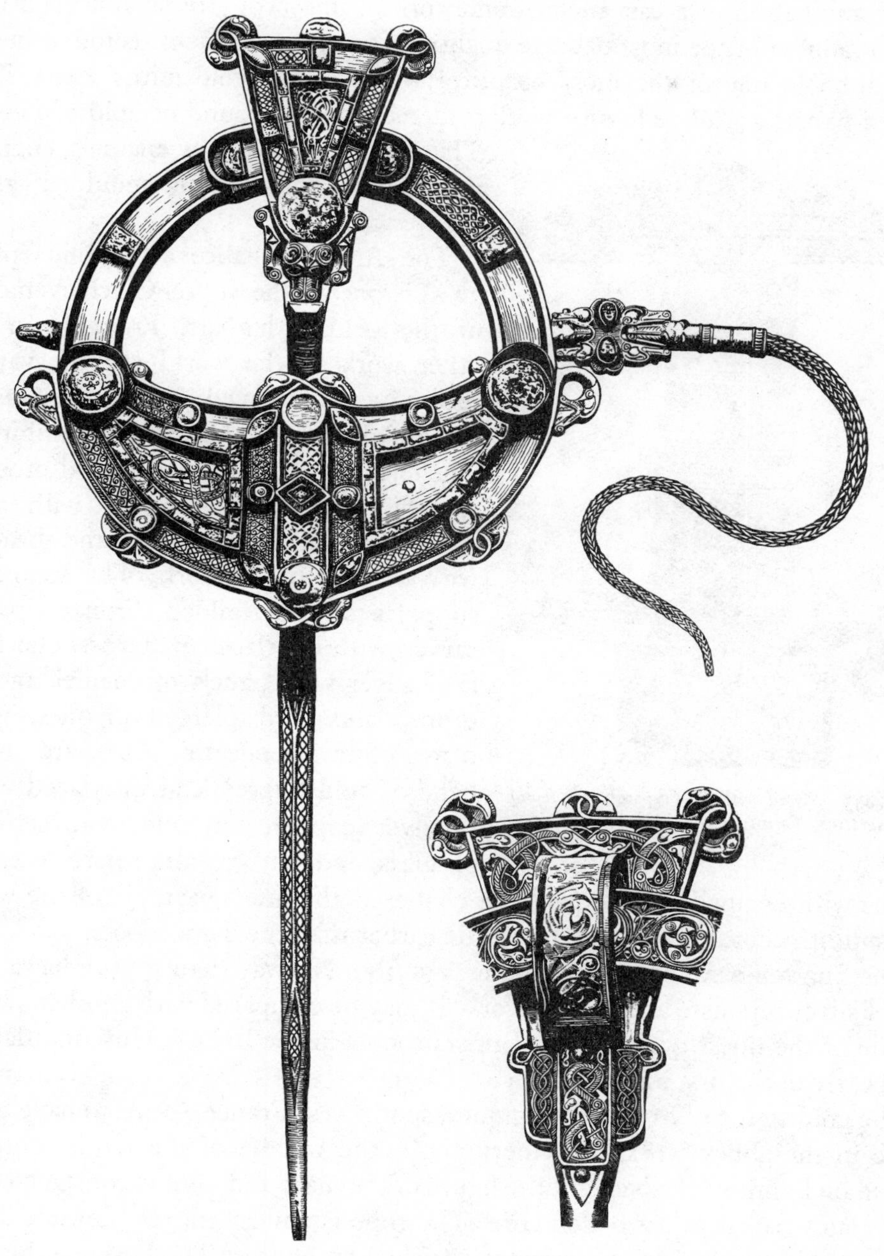

FIGURE 434—*The Tara brooch. From Bettystown near Drogheda, Co. Meath. Irish work of the ninth century.* (Inset) *Detail of the work. Scale* c 3/4.

A combination of a cast metal framework of fine proportions and good line with ornament of the most delicate quality, in varied techniques, is found in the Tara brooch, one of the most beautiful works of the goldsmiths' craft. The filigree panels are of rich wire-work soldered upon a ground of gold repoussé. The work is enriched by enamels, engraving, niello, amber, and moulded glass (figure 434).

The Ardagh chalice at Dublin (plate 32 A) is one of the supreme achievements of the goldsmith's art. Probably in no other work in the world are the varied techniques employed in metal-working more skilfully and beautifully combined. The chalice is of silver and gilded bronze, enamel, crystal, and amber, with rich decorations of gold filigree and beaded wire over repoussé work. The stem and supports are of gilded bronze, richly carved with interlaced patterns. The foot is of silver with panels of enamel, and of bronze and gold pierced plaques, with plates of mica beneath. There are also a band of gold filigree, and interlaced work in silver, copper, and gold, with settings of glass and amber, and pierced silver bosses with enamelled grounds. The chalice is the finest example of the work of the ninth century, and may be a little earlier than the Tara brooch.

FIGURE 435—*Sword, with silver repoussé hilt, found at Snartemo, Norway.* c 600 A.D. *Scale* c 1/2.

The Snartemo sword, of the Viking Age from Norway (figure 435), has a fine hilt of silver repoussé and carved work. It may be compared with another sword handle of the finest quality from Luristan, now in the British Museum, dating from early in the first millenium B.C.

The gold statue of St Foy at Conques, south-west France, seems to have been made in the abbey workshops there about the middle of the tenth century. Seated on a throne of silver-gilt the figure of the martyred saint is constructed of gold plates nailed to a wooden core. The robe is almost entirely covered with innumerable jewels, in band-settings enriched by filigree. The face is of beaten gold repoussé with beaded wire decoration. The eyes are of blue and white enamel, and the hair is of gold repoussé. The crown is of articulated plates with arches and fleurs-de-lis above. Both are covered with jewels in band-settings enriched by beaded wire, and there are a few blue, white, and green *cloisonné* enamels in gold. The feet and cushion are of leather. The throne is of gilded silver with four large balls of rock crystal and jewels in filigree-decorated borders. Its sides and back are enriched by a pierced all-over pattern of equal-armed crosses: a common decorative pattern of the time. The shrine for the iron bell of St Patrick at Dublin (figure 436) has a foundation of bronze plates decorated with panels of gold filigree, elaborately pierced gilt bronze work, and stones *en cabochon*. Its crown is enriched by fine beast ornament in silver and enamel. The back has a diaper of pierced crosses surrounded by an inscription.

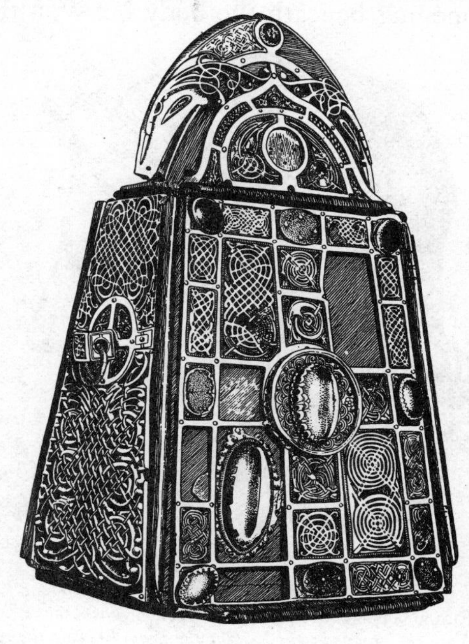

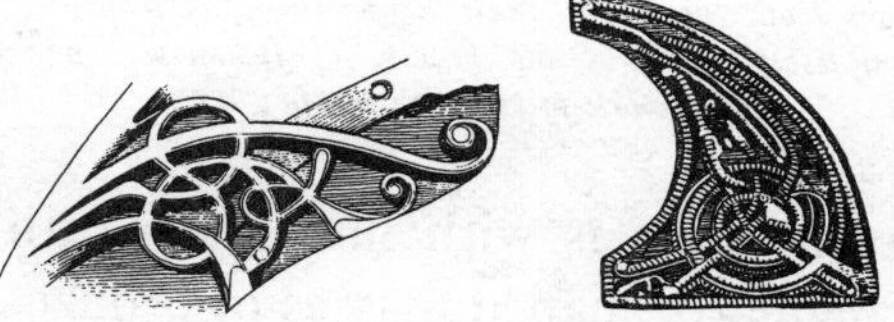

FIGURE 436—*Shrine for St Patrick's bell. Irish work.* c *1100. Height 11 in.* (Below) *Details of pierced bronze and filigree work.*

VI. METAL-CASTING

In early times the moulds for casting in metal were of stone or baked clay (vol I, pp 625 ff). Casting in sand with linked moulding-boxes, so general in modern foundries, was not known until the fourteenth or fifteenth century A.D., though cores of tightly compressed sand were used from Egyptian times onward. The first description of casting in sand is in Biringuccio's *Pirotechnia* (1540), and there is no evidence for its much earlier use [5]. The ancient smith could very easily prepare his moulds from local clay, without carrying moulding-boxes on his migrations. In the late Middle Ages, however, when large numbers of domestic and other objects were demanded on a commercial scale and a great expansion of iron-working stimulated the invention of new techniques, sand-casting developed naturally from the cumulative metallurgical experience of the ages.

From the earliest times the craftsmen in Asia Minor and Egypt had had experience in the preparation of moulds, large and small, and in the handling of masses of molten metal; thus the Greeks had a considerable reservoir of technical knowledge upon which they could draw. In the eighth to sixth centuries B.C. the 'geometric' bronze horses illustrate their methods. Seen from the front, the opening beneath the body between the legs of the horse almost always has the form of a wedge with a rounded top. The artist's method was to take such a wedge-shaped piece of dried clay and upon it to model the horse in wax, its head projecting above, its legs and tail down the sides and end. The ornamental base-plate on which the animal usually stands was made in wax at the same time, although examples are known in which, probably after a failure of the original casting, a new base was subsequently cast on. A pour and runners, also in wax, were fixed to the model and the mould was built up in the ordinary way.

FIGURE 437—*Life-size head of Apollo in cast bronze. Note the groove at the back of the head dividing the locks from the smooth hair above them; it was incompletely filled in the process of fusion-welding. Greek work of the fifth century B.C.*

An extremely interesting example of casting of the fifth century B.C. is a bronze head of Apollo at Chatsworth (figure 437). Life-sized figures were not then cast in one piece (figure 28); the parts were made separately and joined subsequently by riveting or, as with the Chatsworth Apollo, by fusion-welding. A number of irregularly shaped masses of bronze, part of the welding-material, may be seen inside the base of the neck where it broke from the body. It seems that the sculptor formed a core of well worked clay round an iron armature. To link up with the mould one end of this bar passed through the top of the head when modelled, and a square hole is still visible, though closed with a plug of bronze. A cord passing through the core from top to bottom provided ventilation. The sculptor then covered the core with a layer of wax, $\frac{3}{8}$-in or more in thickness. Upon this he modelled the features and formed the general shape of the head and neck. The eyes he left hollow, for stone or other insertions.

A striking feature of the work is the great mass of curls on either side and at the back of the head, hiding the ears. In parts the locks of hair are deeply undercut and provided the sculptor with an interesting problem. Should he cast them in one with the head, or make them separately and then attach them? He decided on the latter, made each curl from a rod of wax, and fitted it against the head. Leaving the detailed modelling of the hair to be executed with V-pointed chisels in the bronze itself, he cut each mass of three or four wax curls from the head and cast them in bronze, finishing them before replacing them on the head.

The sculptor considered the head, now shorn of its ringlets, and found that if he made his dividing-line at the widest part, just behind the ears, he might form a two-piece mould, easily separable without damage to his modelling. To separate temporarily the front and back halves of his mould, instead of employing the usual clay band, he might press into the wax along the line to be followed a number of slips of thin sheet metal, each about 2 in square. These would form a wall running over the top of the head and down each side to the base of the neck. After forming the pour and the vents in wax, he would make the mould of well beaten clay, mixed with powdered pottery or brick with some dung as binding. He would paint this slip over all the exposed surface of the head, and up both sides of the metal dividing-wall. Rougher material would be used as the mould grew thicker. When it was about 2 in thick, the mould would be allowed to dry, and then opened with a chisel driven in along the line of the wall. After removing the wax the sculptor remedied surface-defects in the mould, and before closing it fixed some chaplets—fragments of metal hammered into and projecting from the core or the mould to prevent them from touching, for contact of the surfaces would result in a hole in the casting. Chaplets are found in many spears and sword-chapes of the Early Iron Age and later.

The head being cast and its surface chased, the sculptor attached the curls by fusion-welding. He fitted each group of curls in position on the head, allowing a space or channel about $\frac{1}{4}$-inch in width between the top of the ringlets and the mass of hair to which the group was to be joined by new metal. He then formed a small mould over the spot, allowing a sufficiently large funnel and pour, and an escape-hole beyond the end of the little channel. His intention was to pour some 10 lb of molten bronze into and through the little mould. Most of it would pass along the channel and escape through the vent, but in passing it would heat up and melt the surface of the head locally, filling the little channel and fusing together the completed curls and the surface of the head from which they sprang (figure 437). Unfortunately, the metal heated the head insufficiently and the joint was a poor one, which in time failed. At the neck the thickness of the bronze is about $\frac{7}{16}$-in. This excessive thickness indicates early work. The head contains part of the core and much corroded matter, so it is difficult to measure its general thickness. At the back, where a hole has developed, it is only $\frac{3}{32}$-in.

When a casting has been thoroughly chased, the remains of runners and vents, and traces of the seams where the sections of a piece-mould came together, are removed. Thus it is generally impossible to prove whether such a casting was made by the *cire-perdue* process or in a piece-mould. This is the case with the bull and acrobat Minoan bronze shown in figure 438. Though repeatedly described as a characteristic example of *cire-perdue* casting this is probably erroneous. The group would be modelled in wax over an armature of wire. After the pour, runners, and vents had been fixed to the model a mould of three or four pieces would be constructed round it. A mould of this type would not only afford all the openings necessary for the removal of the modelling-wax by hand, but would provide the founder with a reliable means of inspecting the surface and finishing the preparation of the mould before pouring.

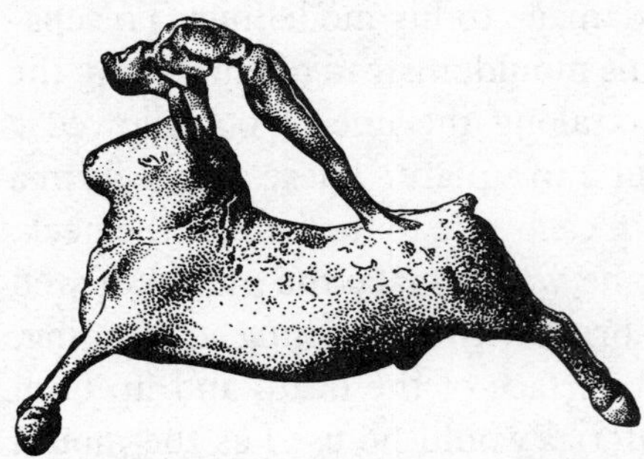

FIGURE 438—*Minoan cast bronze group of a bull and acrobat. Scale* c *1/3.*

From the fourth century B.C. the Greek founders went to work in another manner. They modelled a head, perhaps in clay, and formed the outer mould in the usual manner. They then lined the mould with thin slabs of wax, about $\frac{3}{16}$-in thick, pressing the wax fairly closely into all the forms. To ensure the free flow of the bronze they sometimes added small bridges of wax, say from the hair to the eyebrows and across the lips inside the head, before the core was inserted. A good example of such work is the bronze head of Aphrodite from Satala, Armenia, in the British Museum. In addition to the bridges a number of finger-marks are visible inside the head, proving that the core could not have been made first. After casting, the original modelled masses of hair were cut into curls with the chisel.

Henceforward this method of lining a mould with wax before preparing the core was generally adopted in Greek and Roman bronzes. With small variations it was followed throughout the Middle Ages and later. In a Roman bronze the thickness of metal employed for a life-size statue is normally about $\frac{3}{16}$-in, but is increased at points liable to strain. In the British Museum is a fine silver situla of Augustan age, found at Vienne, France, beautifully decorated in low relief in a casting a bare $\frac{1}{16}$-in thick. By contrast, two early royal effigies in Westminster Abbey, one of Henry III (1216–72), and the other of his daughter-in-law, Eleanor of Castile, wife of Edward I (1272–1307), have a thickness of about 4 in. These, the earliest royal bronze effigies cast in England, were the work of a goldsmith of London, who evidently called upon the bell-founders—craftsmen accustomed to handling great weights of molten metal—to do work beyond the powers of the bronze-founders of the time.

During the 'Wanderings of the Peoples'—third to eleventh centuries—very beautiful works in cast gold and bronze were produced in Hungary, Scandinavia, and Teutonic lands. Many of the castings are decorated with ornament of traced circles or with interlaced beast ornament characteristic of the art of northern Europe. Many moulds for such work have been found, usually of clay fired to terra-cotta hardness. Moulds of Viking age were found in the Orkneys, in the debris from a foundry, among them sword-moulds, which show how a sword-smith went about his work. First he obtained a wooden pattern for the bronze sword. On a flat board he formed a strip of well beaten clay, smooth on top, nearly an inch thick, and as long and wide as required. He dusted its surface with soot. He greased the wooden pattern, laid it in position, and pressed it to about half its thickness into the clay. He made a neat parting-line between the pattern and the clay on either side. Then with a rounded pebble or the end of a stick he made a number of hemispherical depressions, as register-marks, on each side. With soot or some other powder that would prevent the adhesion of clay to clay he dusted the exposed surfaces of the clay and the pattern. Then he pressed down on the pattern and clay strip a second strip like the first; this would take an impression

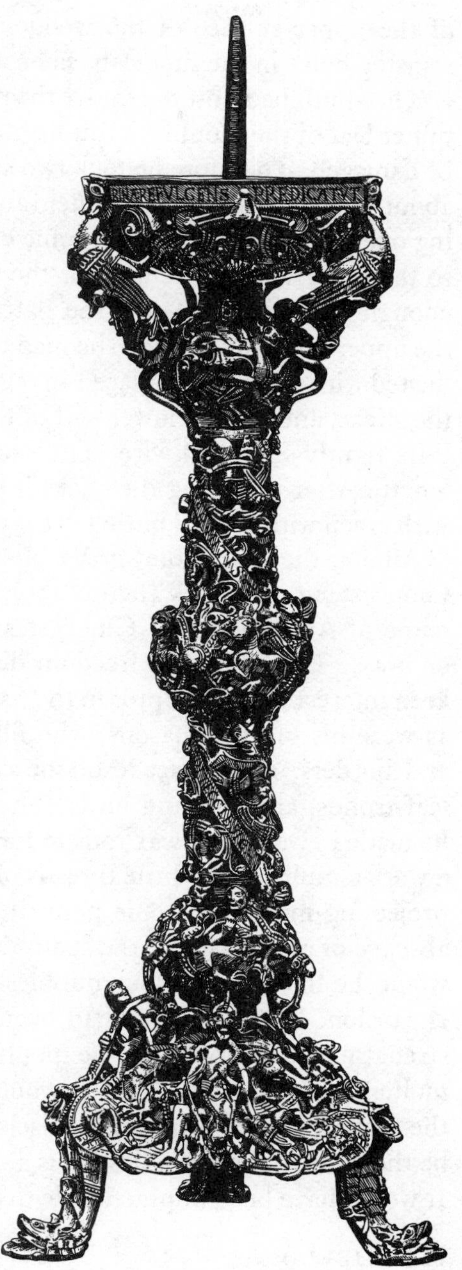

FIGURE 439—*Gilt bronze candlestick presented to the church of St Peter, Gloucester, by Peter, Abbot of Gloucester (1104–13). Scale c 1/3.*

of the upper surface of the wooden pattern, and some of it would enter the register-holes in the first slab, so ensuring correct registration afterwards.

The smith had now to remove the pattern from the mould. If he tried to lift the upper leaf of the mould—a thinnish piece of clay, more than 2 ft long—it might be damaged. Therefore he took two sticks, each rather longer than the sword and about $\frac{1}{4}$-in thick. He greased them and laid them on the upper strip of clay, leaving one end projecting. When some clay from the slab was worked over the sticks so that they were embedded in the upper half of the mould, this became stiff enough to be lifted off, and the pattern could be removed. The parting-line of the upper leaf would have to be made neat, then the two halves of the mould were dusted with soot and oil, placed in registration, and trimmed. The smith loosened the sticks, and cut the lower half of the mould free from the board by means of a thin, tightly stretched wire. The mould would shrink perhaps half an inch in length. After removing the sticks it would be baked hard in the fire, tied firmly with green withies, and buried in the casting-pit in front of the furnace to be filled.

Among the most noteworthy pieces of moulding of the Middle Ages is the Gloucester candlestick (figure 439), which bears an inscription containing the name of Abbot Peter of Gloucester (1104–13). It is of cast and gilded bronze or brass. The delightful freedom displayed in the modelling demonstrates the keen interest of the sculptor in the teeming world around him. He is as observant as were his brother masons, who filled the capitals and spandrils, the pilasters and borders, of the façade of the cathedral with images of men and demons, leaf-fronds, animals, and birds. The model for the complete candlestick would be made in wax, with wax rods to form the pour, gates, and vents. In so intricate a work a multitude of little threads of wax would be added to link every upward-projecting mass with some point in the modelling above it, thus ensuring the absence of air-locks. For the sculptor knew that air imprisoned by molten metal would be likely to produce bubbles in the cast, or worse trouble by its violent expansion. When the wax had been melted out of the clay mould, which was strengthened by iron rods, the mould would be heated to redness and filled with molten metal. After cooling, it would be broken open and the whole surface of the casting worked over with chisels and chasing-tools. The last process would be the gilding. This candlestick is a characteristic example of *cire-perdue* casting. It would have been impracticable to do the work by piece-moulding.

VII. NIELLO

Niello is a mixture of metallic sulphides, generally fused in a crucible. Its preparation and use are well described by Theophilus [6]. The niello is ground in a mortar, and washed to remove the finest particles, for the same reason as with enamel (p 458). Recesses in the metal-work, which is usually of gold or silver, are filled with the powder, which fuses like an enamel, forming a black infilling. Its surface may then be ground and polished, or it may be left as fired.

Niello has been found in Egypt, and at Mycenae in graves of about the sixteenth century B.C. Among the Mycenaean treasures were some bronze daggers bearing scenes in which the inlays are of gold, silver, and copper with a background of niello. On other daggers are ornaments in relief or in niello.

From a tomb at Dendra (southern Greece) came a gold-lined silver bowl, ornamented with inlaid bulls' heads in niello and gold. A still finer example of Mycenaean inlaid work is the bowl found at Enkomi, Cyprus (vol I, figure 421). When discovered in 1949 this bowl was completely covered by a corroded layer from which projected fragile ornaments in gold. An X-ray examination showed that the bowl was most elaborately decorated with a series of bulls' heads between floral ornaments, and an arcaded band of ornament below. It had a single wish-bone handle. The bowl was of silver with ornaments of gold, silver, and niello. The patterns of hair on the bulls' foreheads were incised with the tracer, and recesses for the gold and niello were prepared with chasing-tools. In some of the patterns the whole of the recess was filled with powdered niello into which the gold ornaments were inserted. When fired, the niello held the gold firmly in position. Again, a row of gold disks was required to decorate a narrow band of niello. They were first stuck in position on the floor of the recess, probably with gum, and the niello filled in round them. When the niello was fired it gripped the gold. There are indications that this bowl also was once lined with gold. The whole work is a most impressive illustration of the enterprise and skill of that early smith who has left us so beautiful an example of inspired craftsmanship.

Niello was regularly employed for the decoration of Roman silver-ware, and the traditional use of the material was not lost to craftsmen of later ages. For example, the golden ring of Æthelwulf (d 858), father of Alfred the Great, is decorated with confronted peacocks against a niello ground.

VIII. WIRE AND RIVETS

Decorative gold wire found in Sumeria and Egypt was probably cut from a sheet of suitable thickness, and hammered or rolled to the size required. Thick wire was also made by coiling lengths of thin sheet gold into a spiral roll or spill. Probably the earliest evidence for the use of a draw-plate may be found on a gold rhyton (drinking-vessel) of the fifth or sixth century B.C. from Persia. Round the mouth of this rhyton in a spiral coil are forty-four rows of twisted wires set closely side by side, forming a band of ornament $\frac{5}{8}$-in deep, soldered in position. More than 136 ft of wire were employed, no more than 0·072 in thick, and so uniform throughout that it could have been made only with the aid of some form of draw-plate. The rhyton is in the Metropolitan Museum, New York.

FIGURE 440—*Silversmith's workshop, showing wire-drawing and other operations. From an engraving on copper, 1576.*

A modern draw-plate is a stout plate of iron or steel through which a series of successively smaller holes has been pierced. A rod of metal pointed at one end is pulled through a number of the holes, gradually decreasing in size, so that it is extended and forced to conform to the diameter and shape of the last hole through which it has passed (figure 440).

It is a matter for surprise that with so many possible methods of forming interesting patterns by the twisting or plaiting of wires little advantage has been taken of them. Curiously enough, in this field, the sword-makers are far ahead of the goldsmiths. In the hand-grips of many of the finely mounted swords of the fifteenth to eighteenth centuries most beautiful combinations of twisted wires may be found.

In goldsmiths' work nothing finer in wire is known than the patterns on some of the buckles from Taplow, Buckinghamshire (figure 441). The strong colour and the smooth texture of the garnets and their settings are opposed to the delicate ripple of the gold-work. Near the ends of the buckle a row of large grains, surrounded by rings of small ones, is framed by a line of beaded wire. On the body of the buckle there are great bosses, and a panel of deep and unexpected repoussé lies beneath a pattern of beaded wire interlacement. The principal borders are formed from thirteen rows of beaded or twisted wires of various patterns. One must look far to find the equal of this beautiful work.

No finer instance of riveting has survived from the ancient world than a bronze trumpet of the Late Celtic period in Ireland; it is 8 ft 5 in long and made from sheet bronze bent round to form a tube. The abutting edges are riveted to a strip of bronze about ½-in wide. There are no fewer than 638 rivets along the seam. On another interesting piece of Irish riveted work the smith wished to make the heads of his rivets decorative in themselves. On a bronze cauldron hundreds of rivets rise perhaps ⅜-in above the surface of the metal, each with a sharp conical head.

FIGURE 441—*Gold buckle of Kentish type set with garnets, from a grave at Taplow. Early sixth century. Scale c 3/4.*

Some of the spiral wire armlets of the Late Bronze Age in northern Europe are formed of great lengths of wire measuring up to ¼-inch in diameter. Such wire could not have been pulled through a draw-plate without mechanical assistance. However, examination will generally show that the material was cast. A flat slab of fairly dry clay would be prepared. Then a long spiral or grid-shaped groove of suitable section would be cut or scraped into its surface, as many feet long as might be required. Suitable additional grooves for the pour and runners, designed to feed the molten metal to every part of the spiral or grid, would be provided. A second dry slab, similar to the first in size but with a smooth surface, would form the second leaf of the mould. The two would be fastened firmly together, fired, and buried in the casting-pit ready for the molten metal. The cast wire would be trimmed, bent, hammered, chased, or ground to its final shape.

IX. PATINATION

Love of colour is characteristic of oriental art, and the ancient craftsmen were not always content merely to juxtapose several differently coloured materials. On occasion they sought for even more delicate gradations by varying the colour of gold itself. It is not unusual to find work carried out in a combination of almost pure gold, of a fine yellow colour, and the gold-silver alloy called electrum, whose colour varied with the proportion of silver in it. The craftsmen who worked for Tutankhamen (*c* 1350 B.C.) even made use of an artificially rose-coloured gold prepared with iron salts. Many of the early alchemical recipes are concerned with the tinting of metals.

Newly finished bronze has the bright, reddish-golden, colour of a new penny. Exposure to the atmosphere soon forms a patina which may be of almost any hue, from a golden yellow, through browns and greys, to a deep black. The colour will depend upon the composition of the bronze and the chemical action to which it is exposed. The Greek sculptors and those of the Near and Far East were not content to rely upon a natural or an artificial patina to supply all the colour they desired in their bronzes. They might gild or silver the surfaces. Or a Greek might rely upon inlays to produce the contrasts he desired. The eyes of a statue were very often inlaid with stone or glass. The eyes of the gold hawk's-head from Hierakonpolis (vol I, figure 432) were made of obsidian. The charioteer from Delphi has lips of silver and a meandering pattern of inlays round his head. The Artemisium (north-west Euboea) Zeus has eyebrows of silver and nipples of copper. Again, a Chinese artist might inlay in a bronze spear-head a diaper pattern of bronze insertions differing but slightly in composition from the spear-head itself. After a patina had formed the inlaid portions would provide a delicate contrast in colour to the body of the work, sometimes with very beautiful effect.

REFERENCES

[1] MORRIS, W. and MAGNUSSON, E. 'The Story of the Ere-Dwellers' (trans. from Icelandic), p. 120. Saga Library (ed. by W. MORRIS and E. MAGNUSSON), Vol. 2. Quaritch, London. 1892.

[2] BEOWULF XXI, 1455 ff. (Ed. M. HEYNE. 10th ed. Schöningh, Paderborn. 1910.) For a modern Engl. trans. see E. MORGAN. 'Beowulf. A verse translation.' Hand and Flower Press, Ardington. 1952.

[3] CASSIODORUS *Variae*, V, 1, in *Pat. lat.*, Vol. 69, cols. 643–5. For a 'condensed' Engl. trans. see T. HODGKIN. 'The Letters of Cassiodorus', pp. 264–5. Frowde, London. 1886.

[4] THEOPHILUS PRESBYTER *Diversarum artium schedula*. Ed. and German trans. by W. THEOBALD, III, 53. Verein Deutscher Ingenieure, Berlin. 1933. Ed. and Engl. trans. by R. HENDRIE, III, 54. Murray, London. 1847.

[5] BIRINGUCCIO, VANOCCIO. *De la pirotechnia*, VIII, 3 and 4. Venturino Roffinello, Venice. 1540. Engl. trans. by C. S. SMITH and MARTHA T. GNUDI, pp. 327 ff. American Institute of Mining and Metallurgy, New York. 1943.

[6] THEOPHILUS. See ref. [4], III, 27 (ed. by W. THEOBALD); III, 28 (ed. by R. HENDRIE).

NOTE ON STAMPING OF COINS AND OTHER OBJECTS

PHILIP GRIERSON

AN important subsidiary form of metal-work was die-cutting, necessary for the making of coins, an activity that forms the one true example of the mass-production of a manufactured object in the centuries before the Industrial Revolution. The principle of coin-making is that a pellet of metal is placed between two engraved dies and has their designs impressed upon its two faces by one or more sharp blows. In the ancient world, where the designs of coins were in high relief, die-cutting was done mainly by means of graving-tools as used for engraving gems and seals. In the Middle Ages, when the designs were simpler and in low relief, the use of graving-tools was largely replaced by that of punches.

Only with the Renaissance was something corresponding to the naturalistic art of antiquity restored, but with the difference that the relief was always lower; the influence on the die-cutter was that of the maker of cast medallions, not that of the worker in precious stones. The method of casting has itself been rarely used in western coinage except by counterfeiters, whose moulds have survived in great numbers from the later centuries of the Roman Empire. Its use in legitimate minting was limited to the large bronze coins of the Roman Republic, which would have been difficult to strike by hand, owing to their size, and to some of the more barbarous series, like those of the early Celts.

The beginnings of western coinage go back to the seventh century B.C., when merchants in the Greek cities of western Asia Minor began to keep their stocks of precious metals in the form of small ingots, which were more convenient for trade than nuggets or gold dust. These ingots were not standardized in weight or in purity, though some bore stamped marks by which a merchant might recognize his own. The metal was generally electrum (p 484).

The kings of Lydia in Asia Minor, whose territory included gold-mines, were the first to stamp electrum ingots of a definite fineness and scale of weights with their own badge—a lion's head. This innovation, involving the use of fractions of some fixed unit of weight, was both convenient and important. In Greece there were no gold-ores, but silver, which was extensively mined (pp 1 ff; 43 ff), was minted instead. Tradition has it that Pheidon (seventh century B.C.), king of Argos, who was master of Aegina, began in this island the coining of silver pieces of definite weight marked with the badge of a turtle. He also laid down that these coins should be accepted as the equivalent of so many iron rods or spits, which up till then had been the usual currency of the state. The drachma and obol, as the basic denominations of Greek coinage were called, preserved in their names the memory of the iron currency, for the spits had been called *obeloi*, and *dragma* meant a handful, conventionally taken as the equivalent of six spits.

The earliest, unstandardized, ingots were slightly flattened on a rough anvil by punch blows on the back, the punches being of iron or hardened bronze. When the convenience of having an easily recognizable device on the face of the coin became apparent, this was cut in reverse into a block of bronze or iron, let into a recess in the anvil, and the pellet of metal was forced into the mould by punch-blows from above (figure 442). The earliest devices were probably made by the use of chasing-tools and a good hammer. Anyone who will set a stout piece of brass or even mild steel on an anvil, and try to produce a few depressions with chasing-tools, will need little convincing that the employment of graving-tools (which remove some of the metal) would have been unnecessary. Later, the Greek artist employed scrapers to soften or deepen details, and for the finest work every device known to him—drill, graver, and punches of innumerable shapes—was called into use (figures 443–5). There is evidence that sometimes a positive puncheon or

FIGURE 442—*Electrum stater of Phocaea, seventh century B.C. The design is the canting badge of the city* (phoke, *a seal*). *The marks on the other face are those of the heads of the punches used to force the pellet of metal into the die. Scale* c *2/1.*

hub was employed in die-making. On this pattern-piece of hardened metal the whole or part of the coin-design would be carved in relief, not in intaglio. From it a large number of dies could be made, all of them exactly alike—a matter of importance when a large coinage was required.

The earliest coins had a device on the obverse only, the reverse bearing no more than the mark of the head of the punch which had forced the pellet of metal into its mould. By the middle of the sixth century B.C., the use of a second device for the reverse began to come into fashion. It was cut on the head of the upper punch, which was held in the hand and struck by a hammer. This upper die is known as a trussel, as opposed to the lower pile or anvil-die. The earliest trussels were probably no more than thick iron bars long enough to be held in place by the hand or tongs, but in Roman Imperial times they often assumed the form of small conical or barrel-shaped objects (figure 447). These were set in iron blocks which took the force of the blow and were apparently fitted with some mechanical device to ensure that the designs on the two faces of the coin bore a regular relation to each other.

The two dies did not have the same length of life, for the upper dies wore out much more quickly than the lower ones. Many of the defects of early coinage arose, indeed,

FIGURE 443—*Athenian tetradrachm, fourth century* B.C. *Retention of an archaic style is common in commercially important coinages. The cut in the edge was made by a user of the coin who wished to assure himself that it was silver throughout. Scale 2/1.*

FIGURE 444—*Tetradrachm of King Eucratides of Bactria* (c 180–c 160 B.C.). *Portraiture was introduced by the successors of Alexander the Great. The reverse shows the Dioscuri, holding lances and palms. Scale 5/3.*

FIGURE 445—*Stater of the Bellovaci* (*northern France*), *early first century* B.C. *Copied from a coin of Philip of Macedon. Blundering and exaggeration of individual features, as with the hair on the head of Apollo here shown, are typical features of barbarian coinages. Scale 9/5.*

from difficulties either in making or in handling the upper dies. The die was prone to crack and difficult to repair. If it were not held exactly over the pellet of metal and the lower die, part of the coin would fail to register an impression and remain unstamped. If it were not held exactly upright, the side towards which it was leaning would receive most of the force of the hammer blow. The edges would consequently tend to give way

FIGURE 446—*A moneyer's equipment in Roman imperial times; in this relief the trussel appears to be of square section. From a funerary monument. First century* A.D.

under the blows, and the head of the die would become convex, which explains the slightly dished or watch-glass form of many ancient coins. Some of these defects were avoided by the use of the smaller inner dies of Roman Imperial times.

At first the blanks or disks which were to become coins were cast, perhaps in a slab of stone or baked clay with a dozen or twenty little hollows in its face, deep enough to make the blanks a little thicker than the coins, into each of which the molten metal was poured. For easy removal of the blanks, the little recesses in the slab might be made narrower below than above. On occasion a channel would be cut in the mould connecting one hollow with another, and a second (flat) slab provided to complete a two-leaved mould. Then all the hollows could be filled at one pouring, and the blanks separated with a chisel. Little effort seems to have been made in early mints to produce coins that were truly circular, and the provision of a milled edge as a protection against clipping seems hardly to have been imagined.[1]

In Ptolemaic Egypt, the circumference of the blanks was sometimes rounded a little on a lathe. On one or both sides of such a coin may be seen a conical depression where the tail-stock of the lathe pressed against the blank, holding it in place against the chuck. The turning-tool was not held rigidly in a tool-rest, but lightly, in the hand. The cut was intended merely to trim off the worst irregularities; sometimes a file was used instead.

Since, as already mentioned, the design of ancient coins was in very high relief (figure 448), great force was required to register the impressions properly. Care would have to be taken that the dies did not slip between

FIGURE 447—*Upper (trussel) and lower (anvil) die. German, sixteenth century. Scale c 1/4.*

[1] The milled edge on modern coins is formed by serrations or lettering inside the steel collar within which they are stamped.

FIGURE 448—*Sestertius of Vespasian, struck to commemorate the capture of Jerusalem in* A.D. *70. Vigorous naturalistic portraiture and designs of a propagandist nature are characteristic of Roman coinage. Scale* c *3/2.*

FIGURE 449—*Penny of Charlemagne (768–814) of the mint of Melle* (Metullo). *The monogram is that of the king. Scale 2/1.*

FIGURE 450—Carlino d'oro *of Charles of Anjou, King of Naples (1266–85). Scale 2/1.*

the blows, if more than one were necessary, and that they did not become hot enough to spoil their temper.

The ancient traditions of die-cutting and minting largely died out in western Europe during the centuries that followed the downfall of the Roman Empire, and new methods had to be evolved to deal with the lighter and thinner coinage of the later Middle Ages.

Medieval dies have survived in greater numbers than have ancient ones, and we thus know more about them (figure 447). The lower die was short and spiked at the bottom, to allow it to be let into a block of wood that would support it and hold it steady; it was thickened in the middle to prevent it from penetrating too far (figure 453). The upper die consisted of a bar of iron, about an inch in diameter and long enough to allow it to be held in place by hand. The upper end of surviving trussel-dies is usually splayed out and bent over by the repeated blows of the hammer falling upon it. The design of medieval coins was in low relief, and thus much less force was required in the striking than had been necessary earlier; contemporary illustrations show that it could be done by a single man in a standing or seated position, holding the upper die in his left hand and striking it with a hammer in his right (figure 451).

FIGURE 451—*A moneyer of the twelfth century. From the capital of a pillar at Saint-Georges de Bocherville, Normandy.*

The tradition that the dies should be mainly engraved died out in Carolingian times (eighth and ninth centuries), which saw a great simplification in design and the appearance of larger and more prominent lettering (figure 449). Both device and letters could be formed by a limited range of punches capable of producing straight lines, large and small crescents and curves, annulets, pellets, triangles, and so forth. The practice of committing the making of dies to local mints with largely untrained workers accentuated the tendency towards simplification. The lettering on the coins of eleventh- and twelfth-century France was often produced by the use of little more than a small selection of wedge-shaped punches, which yielded patterns that bore only the vaguest resemblance to the letters they were intended to represent (figure 452). Improvement did not come about until the thirteenth century, when the commercial cities of Italy began to take pride in the appearance of their money, and feudal princes conceived the idea of using gold coinage as an instrument of heraldic display. But the use of punches to produce details persisted; curved punches would serve to represent the curls on a facing head, and there might be more elaborate puncheons to produce such objects as a crown or a fleur-de-lis. It was not until the Italian Renaissance in the fifteenth century that the engraving of the dies after the fashion of antiquity again became common. The process of making the die has often left traces on the coins; there may be a central pellet representing the

FIGURE 452—*The lettering* (LUDOVICUS REX) *on a French coin of the twelfth century, produced by very few punches.*

FIGURE 453—*Interior of a German mint. From an early sixteenth-century drawing in Emperor Maximilian's autobiographical manuscript.*

depression on the die made by the point of the dividers used by the engraver when he was marking out the circumference of his pattern, and the fine lines that he incised on the die to ensure the symmetry of his design are sometimes visible.

The design of a coin was in principle decided by the issuing authority, but its directions, if any, were often of a general character, and the details were left to the artistic taste and skill of the engraver. Where issues were deliberately differentiated from each other, as in the later Middle Ages, certain details of punctuation, mint-marks, and private marks would have to be carefully specified and copied. Occasionally we have evidence of rulers taking an active interest in the design of their coins. Charles of Anjou, king of Naples (1266–85), rejected the first pattern of his beautiful *carlino d'oro* (figure 450) on the ground that the lettering was uncomfortably crowded and the obverse and reverse badly aligned with one another.

FIGURE 454—*Coin struck on one face only (bracteate) of Henry the Lion (1142–95), Duke of Saxony. More elaborate designs were possible on these large thin coins than on the bifacial pennies, and many of them are masterpieces of German Romanesque art. Scale* c 2/1.

Since medieval coins were very thin, especially in comparison with those of antiquity, the

FIGURE 455—*Corinthian die. Sixth century. Scale c 4/5.*

blanks were not cast but were cut with shears from sheets of metal hammered to a more or less uniform thickness. Generally the blanks were trimmed to a circular shape (figure 453), but in some places, to save time in cutting and the cost and trouble of remelting the discarded pieces of metal, they were left more or less square by the cutter, and a subsequent operation rounded them roughly either by inturning the corners or by extending the middle section of each side outwards by further blows from a hammer. The excessive thinness of the coins sometimes meant that the impression of each die came through the blank, spoiling the device on the other side. In many parts of Germany and the neighbouring regions of northern and eastern Europe the attempt to strike both faces of a coin was consequently abandoned (twelfth century), and there exist large series of thin coins struck on one face only (figure 454). The technique of these coins shows many analogies to that of contemporary jewellers' work, and their large area made possible much more elaborate designs than the restricted size of the more usual bifacial pennies permitted.

In antiquity and in the Middle Ages seals were incised by methods similar to those used for dies, and Greek and Roman craftsmen, like those of earlier times, used dies and punches carved in relief or intaglio for making jewelry as well as coins (vol I, p 648). Figure 455 shows a bronze bar, carved in intaglio on all sides with various devices, including a horse and rosettes. The northern European peoples of the fifth to tenth centuries A.D. were accustomed to carve in relief on a block of bronze such devices as warriors, interlaced dragons, and decorative borders, and to employ the stamp in conjunction with a die of lead for repetitive work. Impressions on foil, as on the gilded silver mounts of the horns from Sutton Hoo, were produced in a similar manner by squeezing or hammering. The tinned bronze plates that cover the Sutton Hoo helmet were stamped in the same way. Theophilus, in the tenth century (p 351), describes the manner of striking ornaments in low relief from gold foil.

BIBLIOGRAPHY

CASSON, S. "The Technique of Greek Coin Dies" in 'Transactions of the International Numismatic Congress, . . . London . . . , 1936' (ed. by J. ALLAN, H. MATTINGLY and E. S. G. ROBINSON), pp. 40–52. Quaritch, London. 1938.

FOX, SHIRLEY. "Die Making in the Twelfth Century." *Brit. numism. J.*, **6**, 191–6, 1909.

HILL, G. F. "Ancient Methods of Coining." *Numism. Chron.*, fifth series, **2**, 1–42, 1922.

MARÇAIS, G. "Un coin monétaire almoravide du Musée Stephane Gsell." *Ann. Inst. Étud. orient.*, **2**, 180–88, 1936.

MILNE, J. G. "Two Notes on Greek Dies." *Numism. Chron.*, fifth series, **2**, 43–48, 1922.

VERMEULE, C. C. 'Some Notes on Ancient Dies and Coining Methods.' Spink, London. 1954.

14

ROADS AND LAND TRAVEL

WITH A SECTION (VI) ON HARBOURS, DOCKS, AND LIGHTHOUSES

R. G. GOODCHILD (III, IV, AND VI) AND
R. J. FORBES (I, II, V, VII, AND VIII)

I. THE ANCIENT EMPIRES

MAJOR traffic on land was slower to develop than that on the water. This was partly because the earliest urban centres were valley cultures, well suited for river-transport. Even in Europe the earliest trade-routes were the big rivers. The ancient Egyptian foreign trade with the Syrian coast, Somaliland, and Arabia was mainly coastal. The trade with Crete necessarily involved some open sea.

But trade is related onlv partially to natural facility of transport. More powerful factors may sometimes both demand and create long-distance communication and road-construction. A religion with its holy wars, its missionary urges, and its pilgrimages may promote travel. Again, the centralized organizations of the ancient empires were based on rapid information by letter and messenger. The introduction of the horse-drawn war-chariot about 1800 B.C. (vol I, p 725) demanded a better track. A messenger-service needed stations and rest-houses at regular intervals. The needs of the merchant, following the extension of imperial power, also required a stable form of communication, and in considering the question of routes the economic factor is always important, though not always primary or overwhelming. Some idea of its character can be gathered from the fact that, in classical antiquity, every 100 miles of land-transport roughly doubled the price of heavy or bulky goods.

From the point of view of traffic the Mediterranean world was greatly privileged for travel by sea. Its climate, situation, and indented coastline were ideal for the evolution of long-distance communication. Divided into three zones by the peninsulas of Italy and Greece, it has four important gateways to regions beyond: Iberia, the Black Sea, Egypt, and lastly the Asiatic region of Syria and Anatolia, leading to Mesopotamia, Persia, India, and China.

In antiquity the centre of this world, which was united only by the Roman

Empire, changed from Alexandria in the east to Rome in the west; then back to the east again, first at Constantinople, and later, with the advent of Islam, to Baghdad. But wherever its focus, nature had shaped it for sea-traffic. Even the land-loving Romans, who were unwillingly drawn into naval warfare with Carthage, came to dominate their Empire by becoming perforce masters of the Mediterranean. Hence land-traffic and its conspicuous result, road-building, could gain ground only gradually by a series of strongly centralized political powers, first from the Persian Empire, then from the Hellenistic monarchies, and at the end from the Roman Empire.

In Roman times western Europe had hardly more than 5–12 inhabitants per sq km, as against 18–27 in the Mediterranean world. Therefore the network of Roman roads, dictated primarily by strategic considerations, was never as dense in the west as in the south and east, where there were over 50 million Roman subjects.

Early trackways and the most ancient roads have been considered in vol I, ch 26. The Egyptians built causeways and roads from the quarries to the Nile for the transport of building-materials. In Crete a track-system with guard-houses across the island from Knossos to Phaestos and a Minoan port on the Gulf of Mesara may be of *c* 2500 B.C.; it shows much masonry-work and grading [1]. Most of the main streets and market-squares of Egyptian and Mesopotamian towns were paved with flagstones. Some of these pavements were properly constructed with a brick foundation, the slabs being set in a mortar of bitumen. Specially designed joints prevented the bitumen from sweating out in summer heat. Some of these paved streets are processional roads connecting shrines in the city with temples outside it (vol I, figure 160 B). On them carts carrying statues of the gods were drawn at religious festivals. The oldest known processional road is at the capital of the Hittites, Hattusas; it is of *c* 1200 B.C. There is a series of processional roads in Assyria at Ashur of *c* 700 B.C., at Babylon of *c* 700–600 B.C., and at Uruk of *c* 300 B.C. Some of these roads have ruts, the gauge of which coincides with those of prehistoric roads in Malta and Greece (pp 498 f). The cart drawn by two oxen is responsible for this standardization (p 499).

Made roads covering long distances are later than those of the processional type. Even the much used great coastal road from Egypt to Gaza, Syria, and Mesopotamia was little more than a track, impassable for wheeled traffic. But the Assyrian attempt at world domination meant a mechanization of the army with a siege-train. Tiglath-Pileser I (*c* 1115–1102 B.C.), who extended his rule to include the old Hittite empire in the north-west, had an engineer corps that laid pontoon-bridges and levelled tracks for carts and siege-engines [2]. He says of one campaign in the mountains of Elam to the south: 'I took my chariots and my warriors, and over the steep mountains and through their wearisome paths I hewed a way with pickaxes of bronze, and I made passable a road for my chariot and my troops' [3]. More often, however, his army met 'a difficult country, where my chariots could not pass', and the army had to go on foot.

The Assyrian monarchs did not attempt to make a network of highways, though in establishing communications with distant parts of the empire they regularized communication by fire-signals, an old system long in use in Israel (Judges xx. 38; Jer. vi. 1) [4] and Mesopotamia [5]. Maps of towns and town-quarters, and lists of towns and countries with their distances in time, were at the disposal of 'men of letters of the king'. Some of these lists took a form similar to that of the later Roman itineraries. Such was a catalogue of towns from Ashur to the Persian Gulf, giving their distances (*c* 800 B.C.) [6]. There were ferries for the large rivers, and some permanent bridges such as that at Babylon, over which classical authors waxed enthusiastic [7].

As regards rate of travel, Hammurabi (*c* 1750 B.C.) could write to an official at Larsa (200 km away), 'Day and night you shall travel, so that you may arrive at Babylon within two days' [8]. Esarhaddon (681–668 B.C.) states that it would be his policy 'to open roads throughout the land so that traffic can be resumed with all neighbouring countries'. This meant little more than levelling tracks and policing them. It did not raise the speed of travel above that in Egypt, where the news of the rise of the Nile at Elephantine travelled to Memphis with a speed of about 11 km per hour (*c* 2000 B.C.). At about the same period the horse was introduced in the ancient Near East from the north (vol I, p 721). At first it served only to draw war-chariots. From the days of Sargon II (722–705 B.C.) horses were occasionally ridden, and his army had cavalry (710 B.C.), the earliest traces of which go back to Ashur-nasir-pal II (883–859 B.C.). In Egypt it was still a matter of surprise *c* 750 B.C. that a prince rode on his charger and 'asked not for his chariot'. The later Assyrian kings sent mounted messengers to foreign countries [9]. The art of horse-riding came from Persia, and for the first few centuries was confined to the nobility. The horse did not affect the development of road-building till much later.

After Cyrus (550–530 B.C.) had founded the Persian Empire the horse came fully into its own. The Persian kings, in their desire to amalgamate the peoples they ruled, sought an improved messenger-service for the whole empire. Cyrus ascertained how far a horse could travel in one day and then had a series of posting-stations built, one day's ride apart, with relays of horses and grooms and a superintendent at each station. The distance between the stations was some 25 km. The perfection of this messenger-service is ascribed to Darius I (521–485 B.C.). Echoes of it can be found in the Old Testament, which mentions letters sent by 'posts on horse-back, and riders on mules, camels, and young dromedaries' (Esther viii. 10).

The great Persian track-system led to the palaces of the Persian kings at Susa, Persepolis, and Ecbatana. Its main track, which Herodotus calls the royal road (figure 456), led from Sardis, the most western provincial capital, and the harbour at Ephesus to Susa by way of Issus, Laodicea, the Cilician Gates, Tarsus,

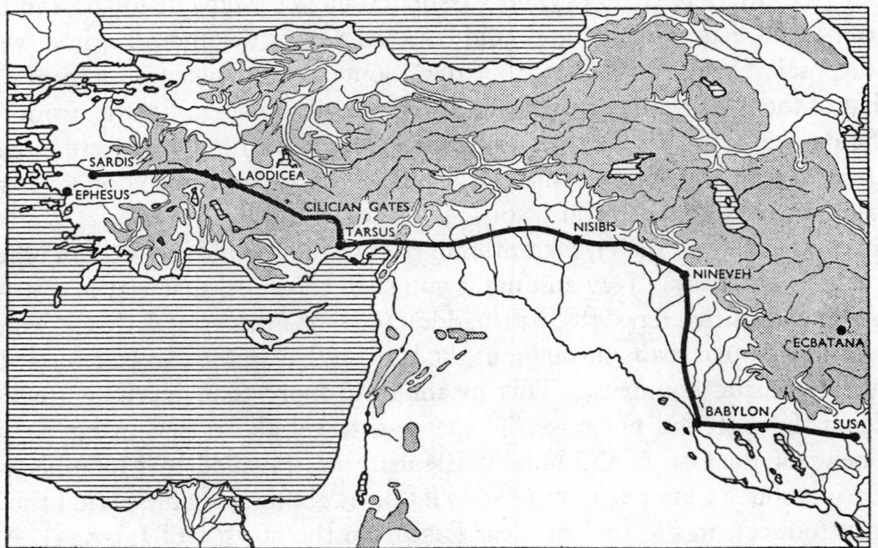

FIGURE 456—*Map showing the course of the Persian royal road from Sardis to Susa.*

Zeugma, Nisibis, and Nineveh [10]. This stretch of 2600 km was covered by messengers in nine days, ten times as fast as an army. The distance between Babylon and Susa was covered in $1\frac{1}{2}$ days, that is, at an average of 150 km a day. This speed of specially equipped messengers was not exceeded until after Napoleon's day.

Nevertheless this was no attempt at large-scale road-building. Stretches of the royal road had steep slopes, though it was practicable for carriages, but Xenophon, in his story of Cyrus's expedition and the retreat of the 10000 Greeks (401 B.C.), often mentions chariots stuck in the mud [11]. In fact the tracks were kept in repair by the simplest means. Cyrus the Younger (424?–401 B.C.) is reported as addressing his officers [12]:

You, gentlemen, who command the roadmakers, have the lists of soldiers I have disqualified from serving as javelin-men, bowmen, or slingers. You will make the old javelin-men march with axes for felling timber, the bowmen with mattocks and the slingers with shovels. They will advance by squads in front of the waggons so that if there be roadmaking to be done, you may get to work at once.

Such road-building simply means levelling a track, and that was its meaning until well into Roman times in the east.

The Persians also used the dromedary and the camel for riding and packs [13]. In Late Neolithic times the camel had been domesticated in central Asia and the dromedary in Arabia. The Egyptians by their contacts with the desert-dwellers knew the dromedary from the predynastic period, as did the Jews from the time of the patriarchs. Camel-breeding was the economic basis of the ancient southern Arabian kingdoms. The true camel, though introduced into Arabia about 1100 B.C., was never a success there, and the dromedary became the most important means of transport in the desert.

The rise of camel-transport became possible in Syria when the two horns of the Fertile Crescent (vol I, map 1 B) were held by powers willing to take the short cut through the desert for trade. The Arabs were by then sufficiently advanced in camel-breeding to have a stock of both riding- and pack-animals for trading agricultural products and salt, on which Palmyra was to grow rich. It was important that the routes should have a perennial water-supply, that they should involve no toll-barriers, and that the chronic threat of desert robbers should be controlled. These conditions were first fulfilled in Persian times (sixth century B.C. onwards).

The keys to this trans-desert traffic between Egypt and Syria on the one hand and Mesopotamia on the other hand are the two great desert emporia Palmyra and Petra, which arose under differing political conditions. Palmyra flourished in the Persian period but declined in Hellenistic times. The direct route from Syria to southern Mesopotamia was deserted until well into the second century B.C. Then Palmyra grew again in importance, especially at the end of the first century A.D., when Romans and Parthians came to an understanding on the importance of desert trade. The Romans developed a series of highways throughout Palestine, Syria, and Transjordania, and marked them with milestones (figure 459). These roads at first remained unpaved but well trodden and clear, and were supplied with wells or cisterns at 50-km intervals [14]. Palmyrene trade reached its zenith when Petra declined and Armenia became a battlefield between Romans and Parthians. Palmyra fell in A.D. 273. There was a revival in the eighth century; it lasted some 500 years, but dropped off sharply after the sack of Baghdad (1258) and ended with the fall of Constantinople (1453).

Petra first became important in the Persian period. Its prosperity rose with the decline of the Ptolemies and the Seleucids (second century B.C.). This tended to divert all the trade of south Arabia to Damascus and Syria instead of to Egypt. Pompey's conquest of Syria (66 B.C.) awakened Petra's interest in the Palmyra road, which continued until the reign of Trajan (A.D. 98–117). Petra declined as Basra began to take over her role as the 'desert port' of Arabia Felix.

II. MYCENAEANS AND GREEKS

The Mycenaean civilization of Greece, heir to the older Minoan traditions of Crete, had constructed some 'cyclopean' roads in the region of Argolis and Mycenae. These roads have a pavement of polygonal slabs, well drained foundations, and several bridges [15]. They may represent the *hamaxitos*, a road suitable for carriage traffic mentioned by Homer [16], for we find hardly any well built road in Greece before Roman times. This is clear from many passages in the 'Description of Greece' by Pausanias (second century A.D.). There many roads are said 'to grow steeper and more difficult to man on foot', or to be 'a mere footpath', 'easier for men than for mules', and 'impassable for carriages by reason of its narrowness' [17]. Only in a few cases are they 'suitable for carriages' or 'excellent carriage-roads'. Of the very important route from Megara to Corinth, he says: '[it] was first, they say, made passable for foot-passengers by [the mythical] Sciron, when he was war-minister of Megara; but the Emperor Hadrian (A.D. 117–38) made it so wide and convenient that even chariots could meet on it' [18].

In fact most Greek roads were hardly more than foot-paths or bridle-paths, generally unsuitable for pack-animals as steep slopes were taken by steps cut into the rocks. Carriages existed, but their use was considered effeminate and even forbidden in certain cases. A kind of covered wagon was used almost exclusively by women. Carts (*hamaxa*, p 539) were used in certain localities for the transport of grain, marble, and the like, if the roads permitted (figure 491). In a few cases special roads were built for such purposes from the mines and quarries to the coast. The best known is that to the quarries of the famous Pentelic marble in Attica (p 25).

Road-signs were formed by piles of stones. It was and is the custom for a traveller to add a stone to such a pile after saying a prayer. In historic times square stone pillars bearing a bust of Hermes, god of travellers, were often erected at such spots. There was a general sprinkling of rest-houses and inns along these tracks, for 'roads without inns are no better than life without holidays' [19].

Road-building was seldom considered as a task of the state, and indeed was sometimes held in disrepute. Yet there was one marked exception—that of the sacred roads. Such roads as those from Athens to Eleusis, from Sparta to Amyklae, and from Elis to Olympia were used for pilgrimages and religious festivals. Others led to the famous oracle of Delphi. The ruts of the sacred roads were carefully hewn, polished, and levelled, so as to form a perfectly smooth and easy track for the cart-wheels.

FIGURE 457—*Greek rut-road, showing a siding branching from the main track.*

In Greece and in the lands colonized by Hellenic immigrants, wheeled vehicles were used more extensively than elsewhere, and there was progress in the technique of road-construction. The most notable feature of the Greek roads is the depth and multiplicity of wheel-ruts. The deliberately made rut is, indeed, characteristic of the central and eastern Mediterranean area and appears first and most prominently in Malta, where it is associated with a period of megalithic building in a society still in a Neolithic stage (2000–1500 B.C.; see vol I, figure 514). Yet whereas the Maltese wheel-ruts wander somewhat erratically across bare rocky hill-sides, those in the Greek lands form part of a true road-system.

There can be no doubt that the Greek wheel-ruts were the result of an intentional engineering practice. This is shown by the consistent depth (7–15 cm), width (20–22 cm), and gauge (138–44 cm) of the ruts and by the fact that in some cases there exist paved roads in which the ruts occupy the centres of specially laid stone setts. Such rut-roads are akin to the 'stone railways' frequently used in British quarries and collieries before the Industrial Revolution. Indeed, their likeness to railways extends even to the provision of sidings to facilitate the passing of traffic on single-track routes (figure 457).

These features may be seen in Greece itself, in Sicily (Syracuse), and in Cyrenaica. The wheel-ruts are usually 140–150 cm apart, centre to centre. Over uneven country these rut-roads normally run in shallow cuttings, but traces of constructed embankments are few, perhaps owing to erosion of the surface. Equally rare are long stretches of artificial paving. The routes usually follow easy gradients at the expense of directness, and cross water-courses at places where bridges are not necessary.

III. EVOLUTION OF ROMAN ROADS

The art of road-construction in antiquity reached its highest level under the Romans. The source of their technical knowledge has been much debated. It was perhaps in the Etruscan civilization, though it is doubtful whether all the ancient paved roads claimed as Etruscan are really so. The Greek roads of Magna Graecia (the Greek colonies in southern Italy) may also have had their influence on Roman technical development. In the absence of any precise information as to the roads in pre-Roman southern Italy, however, these suggestions are merely conjectural.

The first of the great Roman highways intended to secure communications with newly acquired territory was the Appian Way or *Via Appia* (pp 29, 504), begun in 312 B.C. by the censor Appius Claudius. It was 162 miles long, its destination was the military centre of Capua (figure 458), and its surface was first gravelled, paving being added only at a later date (*c* 295 B.C.). The equally famous *Via Flaminia* (p 504), which linked Rome with the Po valley, came into continuous and coherent existence during the censorship of Gaius Flaminius in 220 B.C. Thereafter road-construction continued almost without interruption in Italy and Sicily; it was extended into Dalmatia in 145 B.C., into Asia Minor in 130 B.C., and into southern Gaul ten years later. The very existence of the Roman Empire depended on its road system.

Under the Roman emperors of the first and second centuries A.D. the great network of roads took full shape, stretching from Hadrian's Wall in north Britain to the edge of the Sahara, and from Morocco to the Euphrates (figure 459). After about A.D. 200 the pace of road-building fell rapidly. By the fourth century the Roman state was finding it difficult even to maintain existing roads. Imperial laws of the late fourth and early fifth centuries stress the obligations of landowners to provide for the repair of roads and bridges adjoining their estates.

Very few sections have been cut across Roman roads in the Mediterranean area, where they were most fully engineered and carried most traffic. Those dug in Britain, France, and Germany have only rarely revealed a highway in its pristine condition. Centuries of traffic and intermittent repair, followed by centuries of neglect and erosion, can greatly change the character of a road-bed. The general archaeological findings are, however, reasonably consistent with the description given by Statius (*c* A.D. 90) of the construction of the *Via Domitiana* in Campania. 'The first task', wrote Statius, 'is to begin the furrows (*sulci*) and to open out the track, and then with deep digging to hollow out the soil [between them]. Next, they fill in the hollow trench with other materials, and prepare a lap (*gremium*) on which the road-surface may be laid, lest the ground should give way or the spiteful earth provide an unreliable bed for the rammed blocks.

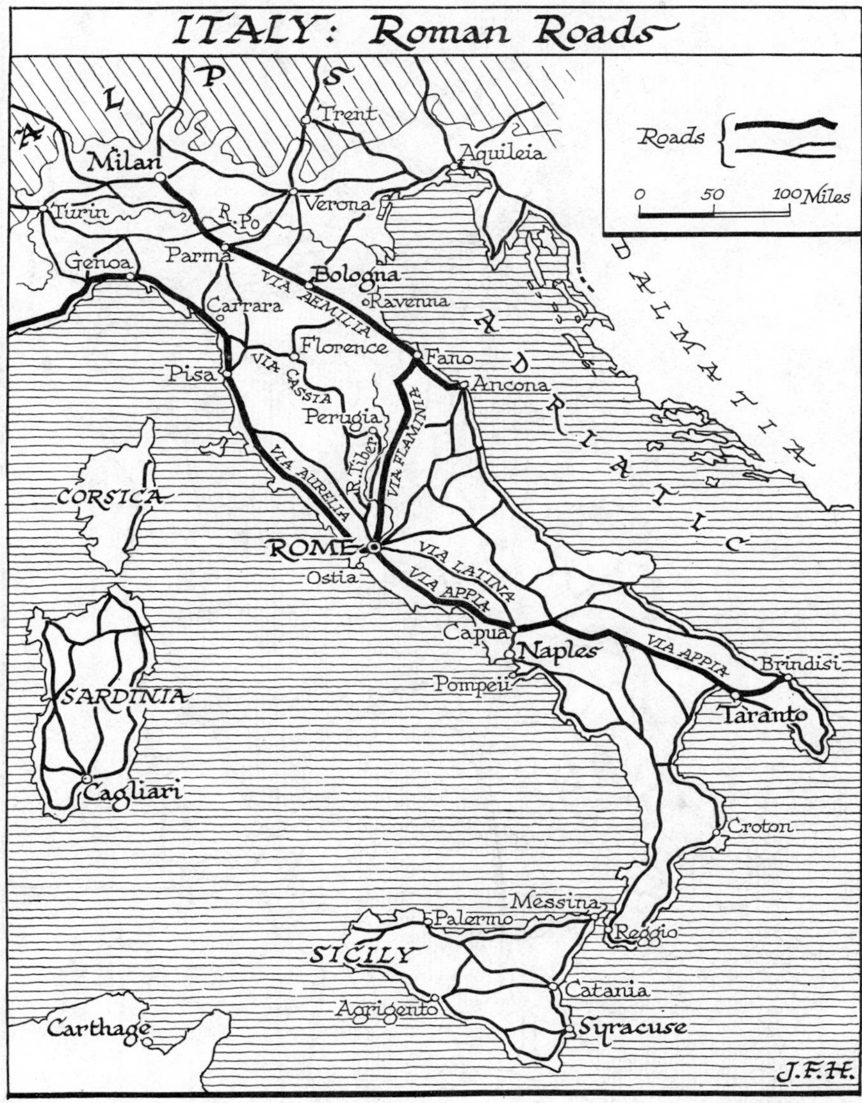

FIGURE 458.

Then, with close-set curb-stones (*umbones*), on both sides, and with many cramps, they bind the road together' [20] (figure 460).

The furrows first dug were probably parallel marking-out ditches, serving

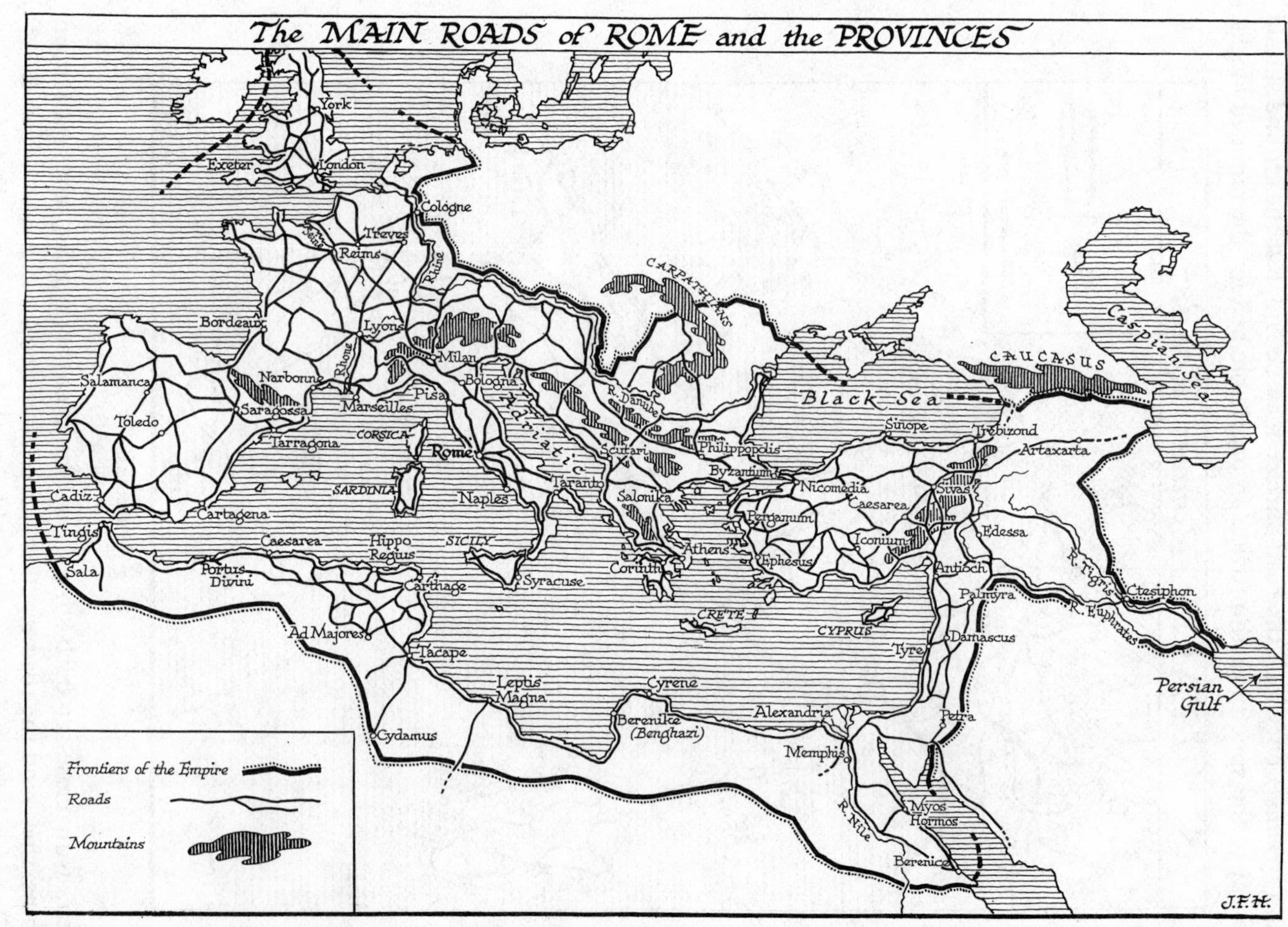

FIGURE 459—Note: *Some of the roads shown in north Africa and west Asia were merely cleared tracks or routes.*

also to drain the strip of land on which the engineers were working. They have been identified on several Roman roads in Britain as shallow trenches up to 25 metres apart, the actual road-bed, 6–8 m broad, occupying the centre of the strip bounded by these parallel ditches. On paved roads, where there was no gravelled surface to be washed away by winter rains, the drainage-ditches were most often immediately beside the road-embankment.

Since Statius describes the excavation of a deep trench to receive the materials of the road-bed, we may conclude that the road in question was not intended to

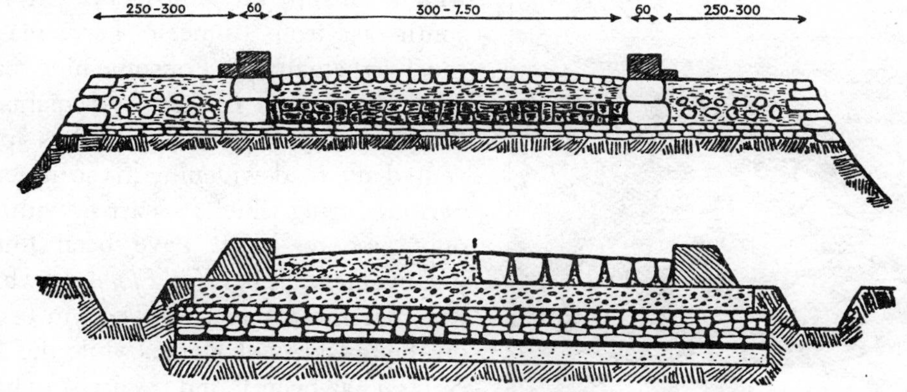

FIGURE 460—(Above) *Diagrammatic cross-section of a principal highway in Italy. The layers, from the top down, were: cobbles or slabs set in mortar; concrete or broken stone, sand, and trass; stone-blocks in mortar; flat stones.* (Below) *Diagrammatic cross-section of normal Roman highway. The layers, from the top down, were: stone setts or gravel concrete; concrete with crushed stone and mortar; slabs and blocks in cement mortar; mortar layer on top of sand course.*

be raised to any marked extent above the surrounding soil. The high embankments encountered among the Roman roads of Britain result, normally, from the laying of the road-material on the original ground surface, without preliminary deep trenching. Whether this provincial practice was dictated simply by speed of construction or had other motives (for example, uncertainty as to the reliability of the subsoil), is uncertain. It had the disadvantage of forming a road-embankment particularly exposed to erosion, and at the same time of necessitating ramps at points where the highway was met or crossed by minor tracks. In Italy, where the convenience of local landowners was probably given greater consideration, the Roman roads were rarely greatly raised above the surface of the ground, except in marshy terrain. Hence the necessity for the trenching described by Statius.

The materials used for the actual road-bed and its surface varied greatly from region to region, according to the character and availability of local stone. The one firm distinction that can be made is between the paved road (*via lapide strata*), and the sanded or gravelled road (*via glarea strata*) (figure 460). A third type, the unmetalled trace (*via terrena*), was also officially recognized, and many major highways in north Africa and Syria were no more than this, the hard surface of the desert requiring no metalling to make it suitable for wheeled traffic.

The paved roads are best represented by the famous consular roads in the vicinity of Rome itself (figure 461). Of these the best known is the Appian Way, which runs in an almost straight line south-east from Rome to Terracina on the Tyrrhenian Sea. For some nine miles from Rome it is flanked by remains of Roman tombs. These doubtless prevented any road-widening in the imperial period, during which its narrow width—only 3·60 m—must have been found inconvenient. The *Via Flaminia* which runs northwards was 4·20–5·20 m broad on the outskirts of Rome, while the *Via Salaria* has been found in places to have a width of about 6 m. These dimensions refer only to the paved surface of fitted polygonal blocks of hard volcanic stone, set sometimes in gravel mortar. On either side of the roadway were slightly raised sidewalks, unpaved and normally about 60 cm broad.

FIGURE 461—Via Praenestina, *near seventeenth ancient milestone looking west towards Rome. This was one of the consular roads paved with large blocks of hard volcanic stone.*

Solid paving 5–6 m broad, of large rectangular blocks resting directly on rocky subsoil, has been found in Syria on certain sectors of the Roman road from Antioch to Chalcis. Such massive paving was probably exceptional. Elsewhere in Syria polygonal paving-stones have been found, but the majority of the desert tracks were completely unpaved.

In the European provinces of the Empire, fully paved roads were mainly in the immediate vicinity of the towns, where traffic was more intense and dust or mud undesirable. The same applies to the raised side-walks. Elsewhere the roads were normally gravelled, and the numerous excavated sections give a fair idea of the method of construction.

The width of these gravelled roads was normally about 6 m, and their surfaces were cambered, the rain-water running down into the ditches that flanked each side of the road. Sometimes a thick gravel layer of uniform character constituted the whole road-bed, but very often the lowest layer was of larger stones rammed vertically. Revetment walls, separating the road-embankment from its ditch, are only occasionally found. In Britain, where investigation of Roman roads has been active, several minor roads of half the normal width (that is, 3 instead of 6 m) and very lightly metalled have been found. They often seem to end on emerging from damp soils to dry hilly country, and we may infer that their metalling was confined to those areas where it was essential.

A notable characteristic of Roman roads is their straight courses. The preference for straight lines seems to have developed in the Romans at an early date. Plutarch (*c* A.D. 46–*c* 120) writes that Caius Gracchus (d 121 B.C.) was particularly active in constructing roads,

> which he was careful to make beautiful and pleasant as well as convenient. They were drawn by his directions through the fields, exactly in a straight line, partly paved with hewn stone, and partly laid with solid masses of gravel. When he met with any valleys or deep water-course crossing the line, he either caused them to be filled with rubbish, or bridges to be built over them, so well levelled that, all being of an equal height on both sides, the work presented one uniform and beautiful appearance [21].

Yet this desire for straight alignments could rarely be exercised in Italy itself, except in the plains of Lombardy, where the *Via Aemilia* served as an axis for the extensive 'centuriation' or parcelling-up of the land distributed among colonists. Of the numerous Roman highways that converge on Rome, only the *Via Appia* maintains a true alignment in its final approach to the city. Straight alignments were therefore subject always to some modifications necessitated by geographical features. The Romans were never so rigid in their methods as to insist on straight roads at the expense of unnecessary bridges and cuttings.

It is in Britain that the straightness of Roman road alignments is most striking. Thus it has long been observed that Stane Street, the Roman road from London to Chichester, leaves the Thames on an alignment that would take it direct to its destination (figure 462). After 13 miles on this course, however, it makes rather grudging concessions to the terrain and swings well to the east of the direct line, though still maintaining long alignments. In more hilly regions the Roman road-engineers could not even attempt to follow straight lines, and were often content to use the most convenient water-shed routes, which minimized the costs of bridging. Reluctance to follow valleys was due not so much to a desire to dominate the countryside as to avoid marshy ground.

How the initial surveys of a proposed route were executed is unrecorded, though the Roman engineers had a fair armoury of instruments. They showed a masterly grasp of terrain, which is all the more remarkable in view of

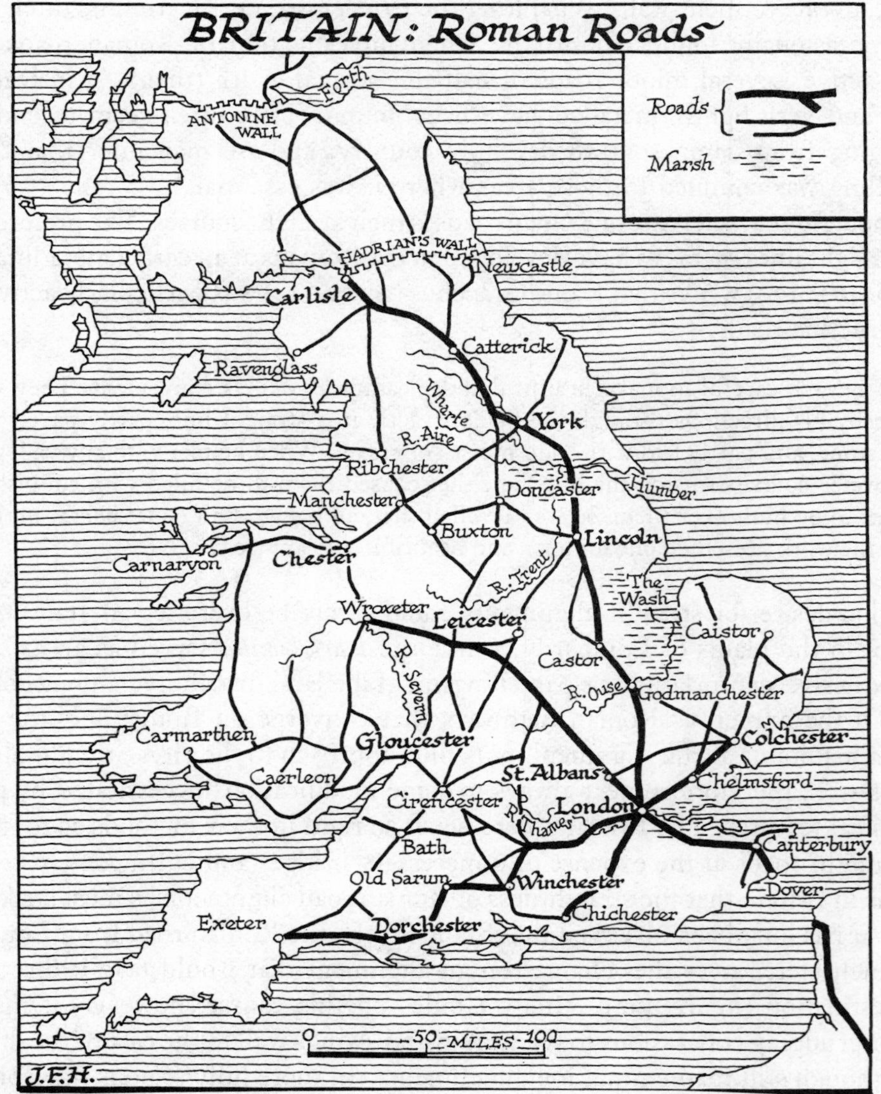

FIGURE 462.—*Based upon the Ordnance Survey Map with the sanction of the Controller of H.M. Stationery Office. Crown Copyright reserved.*

their having no accurate maps. In the case of the London–Chichester road, for example, the choice of alignments could have been determined only by accurate surveys and the use of intermediate sighting-marks, the erection and adjustment of which in heavily wooded country could have been no easy task. Nor were the Roman engineers so set on exhibiting their own technical competence as to ignore such pre-existing routes as suited their purpose. Though attempts in Britain to prove that the straight Roman roads followed equally straight prehistoric tracks have failed, in the eastern Mediterranean many Roman roads are demonstrably on the lines of Greek predecessors. Often the Roman engineer made no attempt to straighten irregularities in the ancient routes, but was content to construct bridges and erect milestones.

FIGURE 463—*Roman milestone found near Llanfairfechan, Carnarvonshire. According to the inscription it was set up during Hadrian's reign* (A.D. *117–138*). *Height 5 ft 5 in.*

Erection of milestones and provision of posting-stations at appropriate distances were essential parts of road-construction (figure 463). Plutarch's life of Caius Gracchus (153–121 B.C.) claims that he 'caused all the roads to be divided into miles. He erected pillars of stone to signify the distance from one place to another' [22], but the *Via Appia* seems to have been marked in this manner a century earlier. The first road outside Italy to be so marked was the *Via Domitia*, constructed in southern Gaul about 120 B.C. Under the early emperors, every new road was punctiliously marked with large inscribed columns at intervals of 1000 Roman paces (1620 yd = 1480 m), recording the distance from the road-head or from the nearest city. Only in Gaul was a different unit of measurement sometimes used—the Gaulish 'league' (2430 yd = 2220 m).

The distances marked on the milestones were normally calculated from the limits of the city to which they refer: in the case of Rome itself from the gates in the Republican city-wall. Yet in the Forum of Rome there stood a metropolitan milestone, the *miliarium aureum*, which gave the distances to all the main cities in the Roman world. An inscription at Autun shows that provincial cities were sometimes similarly equipped with detailed information as to routes. It was from these landmarks that accurate surveys of the Roman roads were compiled—the written itineraries or the schematic road-maps.

The milestones were not intended solely for the convenience of the traveller. They also served, like the mile- or kilometre-posts of modern roads and railways, to assist the organization of road-maintenance. Perhaps for this reason some of the principal highways had milestones giving long-distance mileages rather than local distances to the nearest centre. Later, when the upkeep of the roads was relegated to local authorities, these long-distance mileages were often replaced by local indications.

Since the primary purpose of the Roman roads was to facilitate communications between the seats of government, central and provincial, the maintenance of an efficient and rapid posting-system was vital to road-organization. This system, the *cursus publicus*, was restricted to the use of government officials. The private citizen had to make his own arrangements. The horses which carried the fast couriers bearing official dispatches, or which drew the two- or four-wheeled passenger vehicles, were changed in post-stations along the roadside at intervals of from 6 to 16 Roman miles. The rest-houses (*mansiones*) were at intervals of 20–30 Roman miles, and were situated either in convenient cities or on the roadside. In the more distant provinces the rest-houses were ditched and stockaded. The needs of the private traveller were provided for by inns.

IV. ROMAN BRIDGES, CUTTINGS, AND TUNNELS

The artificial works with which the Romans provided their roads rank among the major monuments of their civilization. Bridges in particular show the highest technical competence, but, as with road-beds, it is often hard to date them, for in their present form they may have been built long after the route that they serve. Thus, though Augustus provided the *Via Flaminia* with most of its stone bridges, yet several of the bridges now on it are of later, though still ancient, date. For these reasons it is not easy to establish a precise chronology for such public works. They can be discussed in general terms only.

Roman bridges vary in type and dimensions from the simple timber constructions that spanned British brooks to such monumental stone bridges as those of Alcantara in Spain and the *Ponte d'Augusto* at Narni (figures 465–6). The earliest bridges, including those on the great consular roads, were probably of timber, or had timber superstructures resting on stone piers, but they were mainly rebuilt with stone arches under the Empire. Timber or half-timber construction remained in use in the more outlying provinces where the influence of military engineering was most strongly felt. Trajan's bridge across the Danube, depicted on his column at Rome, is one of the most impressive structures of this type (figure 467). The Roman bridge across the Thames at London was evidently of the same character. Most surviving Roman bridges are of large stone blocks with inner cores of concrete (figure 468); but on the *Via Traiana* at least, from Benevento to Brindisi, the bridges were of brick-faced concrete. This form of construction was used for arches and vaults increasingly from the second century (p 411, figure 379). The bridge-arches were invariably semicircular, usually varying in span from 5 to 20 m. A 27-m span at Alcantara (figure 465) and one of 32 m at the *Ponte d'Augusto* (figure 466) are the largest recorded.

FIGURE 464—*A reconstruction of the Roman gate of London at Newgate as it existed at the end of the fourth century. It had been first built in the later part of the first century but was reconstructed and towers were added about* A.D. *300. Excavations have revealed that the wall was backed by an earthen bank behind which ran a road.*

The piers supporting the arches usually had cut-waters upstream (plate 34 A). They were often pierced, above the springing of the main arches, by small flood-arches. Where a single arch sprang from the extreme edge of a river, and its approach-causeways were high and lengthy, we sometimes find small passage-ways in the abutments of the bridge to allow cattle and pedestrians to pass up and down the valley. In a few rare cases the arches are askew to the road-way spanning them, but in such cases normal horizontal coursing was maintained.

The approaches to the bridges were usually earthen embankments. These have sometimes been almost completely washed away, so that the bridges appear

FIGURE 465—*Roman bridge at Alcantara near Lisbon.*

isolated. Sometimes the embankments were revetted in stone or brick. The *Via Traiana* provides several examples of long viaducts not unlike those of modern railways. That crossing the Cervaro near Foggia in the spur of Italy is 300 m long; even longer examples are known.

Cuttings and tunnels also occur on some of the major Roman roads, but normally only where the route could not be diverted to avoid natural obstacles. One of the largest road-cuttings near Rome is on the *Via Cassia*. Where it crosses the rim of the extinct crater of Barcciano—now a lake—it is about 1500 m long, 20 m deep, and 6 m broad. Shorter but even more spectacular is the cutting, 36 m in depth, on one side of the *Via Appia* near Terracina. Here the imperial engineers inscribed on the rock-face, at intervals of 10 Roman ft (approximately 3 m), the depth of the cutting. Another equally remarkable cutting is in the Val d'Aosta near the Great St Bernard pass, where a 200-m length of vertically cut rock-face includes not only a bridge with a span of 4 m but a milestone, numbered XXXVI, representing the number of Roman miles from Aosta. Both are cut in solid rock.

The best-known tunnel on the road-system of the Roman Empire is that in the Furlo pass of the *Via Flaminia*, which the Romans called *Petra Pertusa*, 'the pierced rock'. It was cut in A.D. 78 by order of the emperor Vespasian, and is

FIGURE 466—*The* Ponte d'Augusto *at Narni, about 50 miles north of Rome.*

40 m long and 5 m high and broad. A short tunnel at Bons, on the road from Grenoble to Vienne, is notable for having on its floor well marked wheel-ruts 1·44 m apart.

Many examples of deliberately fashioned wheel-ruts on rock-cut Roman roads have been found in the west, like those on the Greek roads in the east (p 499). Examples occur in France and in Jugoslavia, on roads constructed in Roman times. The distances between the ruts seem to be less consistent than in the Greek roads, the gauges ranging from 1·10 to 1·65 m. The regional distribution of such artificial ruts, which are not found in the paved highways near Rome or on the gravelled roads of Britain, corresponds to the differences of gauge. Such ruts would certainly have prevented heavily laden carts from slipping on the rock-surfaces of dangerous mountain-passes.

V. ROMAN CONTROL OF ROADS AND TRAFFIC

Many Roman roads were built by the army. The legions and their train, with large quantities of artillery like *ballistae* and *catapultae* (pp 711–13), needed a good surface. In peace-time the army and its engineering corps built roads as well as bridges, camps, and forts (figure 464). This was particularly true in

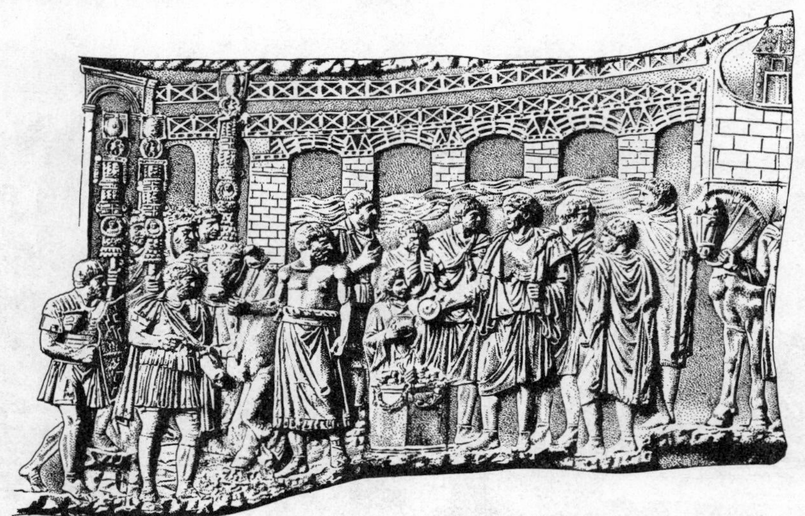

FIGURE 467—*Bridge built over the Danube near Orsova by Trajan* (c A.D. 99). *It was designed by Apollodorus and consisted of twenty enormous stone pillars with a wooden superstructure. The length was 1070 m. Relief on Trajan's column showing the bridge.* c A.D. 110.

Africa and the Danube (figure 652) and Rhine provinces. During the reign of Augustus there were several mutinies arising from such enforced road-building. Gladiators, slaves, and captives were also put to this kind of work.

From about 200 B.C. the provincial authorities began gradually to take over road-building, charging contractors with the actual work but retaining supervision. Such contracts became popular during the Empire and were the source of many frauds.

In early Roman times the task of building roads was entrusted to individual censors, who, like Appius Claudius and Flaminius, often attached their name to a particular road. Later emperors, generals, and rich private persons often did the same, prompted by rewards such as medals, statues, memorial stones, or even the right to erect a triumphal arch. Then special officers, *curatores viarum*, were installed as temporary functionaries, senators often being entrusted with the supervision of traffic. As legislation on public and private roads became more complicated Augustus created (22 B.C.) permanent boards, *curationes viae*, directly controlled by the emperor for particular roads, districts, or towns. Nero created the lower rank of sub-curator. Many local magnates served on these boards, particularly during the later Empire when the building of roads became less im-

FIGURE 468—*The* Ponte Amato *on* Via Praenestina *near Palestrina, about thirty miles east of Rome.*

portant than their repair. The provincial roads were administered by the provincial governor and his nominees, but during the later Empire all these functions were taken by state officials.

Financing the roads was an old problem. The earliest roads were often paid for from the booty taken in foreign conquests. An agrarian law of 111 B.C. aimed to make all those living along the road contribute towards its building and upkeep. Usually, building and repairs were paid by the treasury from direct taxes. In certain cases emperors or private citizens footed the bill. Thus Augustus paid for most of the *Via Flaminia*, but left part to the successful general C. Calvisius Sabinus, who built the emperor's triumphal arch at Rimini. Private individuals or governors have left us word of such acts on milestones. Certain towns like Nuceria were granted tolls to raise sufficient funds for road-building. The same forms of financing applied to secondary roads.

The *cursus publicus*, the state postal- and messenger-service, was the backbone of passenger-transport [23]. By it the emperors could rule from Rome by letter alone. Its origin lies in the third century B.C. Requisitions for travelling-facilities were even then possible for senators. Later, Caesar gave written permits to travellers and transports. The provincial governors had their messengers, and the tax-farming companies their own couriers.

This combined postal-, passenger-, and goods-service had strict regulations. Along the main highways there were stops where horses could be changed, as well as larger stations and hostels (*mansiones*). On rivers there were ferries, and the sea routes were covered by *dromones* or *cursoriae*, fast cutters. Passports rigidly prescribed the type and amount of transport one was allowed, as well as the scale of hospitality in the *mansiones*. For travel on the secondary roads or branch roads a special permit was needed.

The heavy ox-wagons, *clabularia*, each carried a maximum of 1500 Roman pounds (492 kg) of army goods, parchment and papyri, products from the imperial estates, and so on. The number of oxen or horses for these slow vehicles was eight in summer and ten in winter. Lower officials were allowed only this form of transport, which was also used by soldiers, if travelling far, and for all sick people. Express goods and precious metals were transported by several other types of carriages: such were (*a*) the *raeda*, a four-wheeled cart with mules (eight in summer, ten in winter) and a maximum load of 1000 Roman lb (330 kg); (*b*) the *carrus*, a four-wheeled cart carrying at most 600 lb (198 kg); (*c*) the *vereda*, originally used by women only, which was drawn by four mules and could transport two or three persons and carried a maximum load of 300 lb (99 kg); and (*d*) the two-wheeled *birota*, with three mules carrying 200 lb (66 kg) and one or two passengers (pp 540–5). The express service included riding-horses and pack-horses which carried a maximum load of 100 lb (33 kg). The animals served only between certain *mansiones*, and had to be exchanged at certain determined spots. Their feed was to be exacted from the local population. The *mansiones* also served as central offices for all taxes paid in kind. The express service could be used only by high officials, their families, and the imperial messengers (*tabellarii*) [24].

The average speed of the imperial post was some 50 Roman miles or 75 km a day (five miles an hour). Special dispatch bearers sometimes attained double that pace, but 240 km in 24 hours, allowing for halts, is the maximum ever reported [25]. Cicero did 24 miles a day in 51 B.C., as a provincial governor.

The Romans tried to reduce the time of land-transport and to increase its capacity, in order to overcome the difficulty of feeding the great cities. The solution adopted was a control of most of the land- and sea-transport by compulsory service. Stables were built by forced labour, and one-quarter of the horses needed for the *mansiones* were requisitioned yearly from the population living near by.

The great road-system broke down through official abuse and the fundamental mistake of attempting to provide for its maintenance by compulsory exactions. There was an increasing and often fraudulent demand for free transport of military and state officials. The municipal authorities, already loaded with work, could not exercise a proper control. By the time of Constantine (312–37) the imperial fiscal administration had already destroyed piecemeal all the chief sources of revenue, and, when private enterprise became completely absorbed by the state, traffic by land and sea gradually dwindled to a fraction of its former volume. Drastic reforms by Julian, Valentinian I, and Theodosius helped only temporarily.

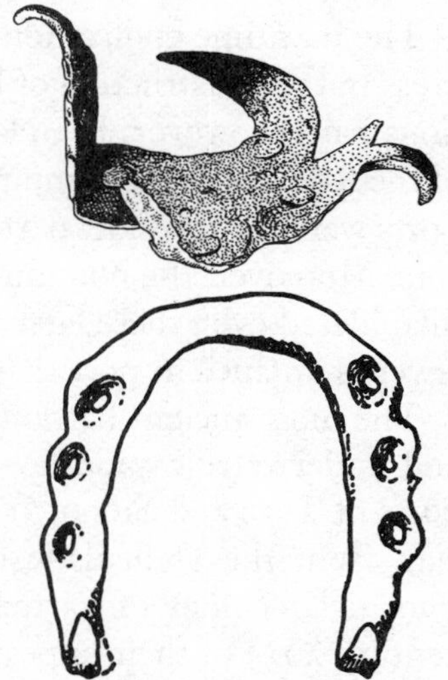

FIGURE 469—(Above) *Iron horse-sandal from Roman settlement at Hallstatt, Austria.* (Below) *Iron horse-shoe from pre-Roman level,* Camulodunum (*Colchester*).

Part of the troubles of ancient land-transport sprang from the inefficient use of draught-animals. This was first of all due to the ancient form of harness, which was not a shoulder-collar with breast-strap (pp 557 f), and which throttled a horse or mule that pulled too hard (figure 486). Each animal could pull a maximum of 62 kg—only a quarter of the modern figure. Moreover, the only way of yoking more than one animal to a cart consisted in augmenting the number of yokes. The Romans did not know how to arrange them in tandem, and thus not only complicated their manipulation by multiplying the number of reins but also considerably reduced the total energy-output (figure 506).

Horses were not shod in antiquity (p 561). Though the history of shoeing animals is still far from settled it seems that the ancients temporarily attached metal, leather, or straw shoes to the feet of horses, mules, or camels when the ground was hard or slippery (figure 469, above). Permanent iron horse-shoes, attached with nails (figure 469, below), seem to have been adopted from the nomads by German tribes of prehistoric Europe about the second century B.C. In the northern Roman provinces a few horse-shoes were in use before they became common about the eighth century A.D. Then a horse-shoe with a better grip on the hoof was generally used, from which type the different local European varieties began to evolve. In the ninth century the Emperor Leo mentions horse-shoes in his 'Tactics', together with the rein and the stirrup; these were unknown to the Romans, for whom horse-riding was thus more difficult. Lack of rein and stirrup hindered the evolution of cavalry as a heavy arm [26].

VI. HARBOURS, DOCKS, AND LIGHTHOUSES

The maritime civilization of the Mediterranean involved much technical progress in the construction of harbours and docks. Remains of the earliest installations have, however, been largely obliterated by the larger-scale harbour-works of the Romans, whose empire embraced all the Mediterranean shores, and whose ports were for the most part situated on the same sites as those of their predecessors. Moreover, the pounding of winter seas during many centuries of disuse has dilapidated even these later remains, and it is only when sand or alluvial mud has covered an ancient port that we can study its arrangements in detail.

The most ancient harbour-works are to be sought in Phoenicia. Recent aerial and underwater explorations have added considerably to our knowledge of the ports of Tyre and Sidon. It is still difficult, however, to disentangle the Phoenician from the Hellenistic or Roman constructions. For the harbours of the Phoenician colony of Carthage we have Appian's fairly detailed account (second century A.D.) of their appearance to the invading Romans:

> The harbours communicated with each other and had a common entrance from the sea seventy feet wide, which could be closed with iron chains. The first port was for merchant vessels and here were collected all kinds of ship's tackle. Within the second port was an island, and great quays were set at intervals round both harbour and island. These embankments were full of shipyards with a capacity for 220 vessels and also magazines for their tackle and furniture. Two Ionic columns stood in front of each dock. These gave the appearance of a continuous portico to both the harbour and the island. On the island was built the admiral's house . . . which rose to a considerable height. Thus the admiral could observe what was going on at sea, whilst those approaching by water could not get any clear view of what took place within. . . . There were gates by which merchant ships could pass from the first port to the city without traversing the dockyards [27].

Explicit as this is, it does not completely accord with the traces of harbours still visible at Carthage. There are, indeed, remains of a circular basin some 300 m in diameter with an island in the centre; but excavations have so far failed to reveal any clear vestiges of the admiral's house or of the adjacent docks, although this may be due to the extensive robbery of stone from the site in the course of the ages. One may also doubt whether the harbour-installations of Carthage would have remained unaltered throughout the Roman Empire, when there was no longer any threat of enemy attack from the sea, and when the merchant-vessels in common use were considerably larger than those used in the days of war between the Romans and Carthaginians (pp 573 ff).

Ancient Athens was, like Carthage, provided with complicated harbour-installations. They were at Piraeus, and of them something is known from both literary and archaeological sources. The main harbour, that of Cantharus, was

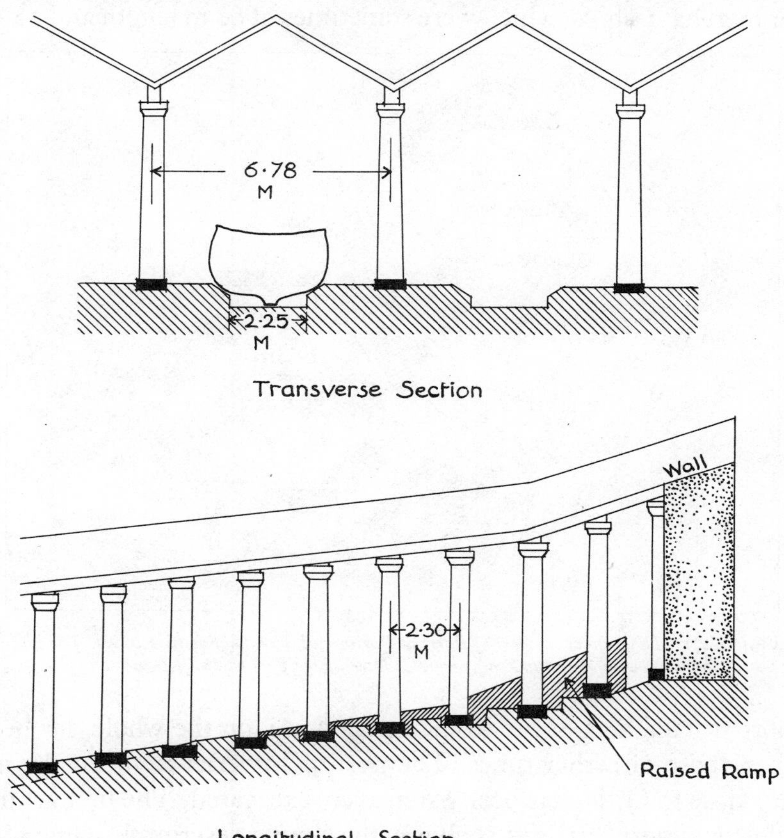

FIGURE 470—*Diagrammatic section of ship-sheds at Oeniadae in Acarnania, west coast of central Greece.*

on the west side of the promontory of Akte, and was protected by jetties forming an entrance of only 50 m width. Near this entrance lay the naval port with its dockyards, while farther in was the large *emporion* wharf where merchant-vessels berthed. The smaller harbours of Zea and Munychia, both on the east side of the promontory, were exclusively naval, and were provided with slip-ways up which the vessels could be drawn. These slip-ways, which enabled Piraeus to harbour no fewer than 372 vessels, are known from their remains to have been long, interconnected structures, with roofs supported on colonnades (figure 470). They contained raised ramps sloping down to the water, with central ridges for the keels of the ships. Vessels of up to 5 m beam and 30 m length could be accommodated in each. It must be stressed, however, that these slip-ways were for warships; it is very doubtful whether there was covered accommodation for the larger merchant-ships, which were sometimes of 60 m length and 12 m beam.

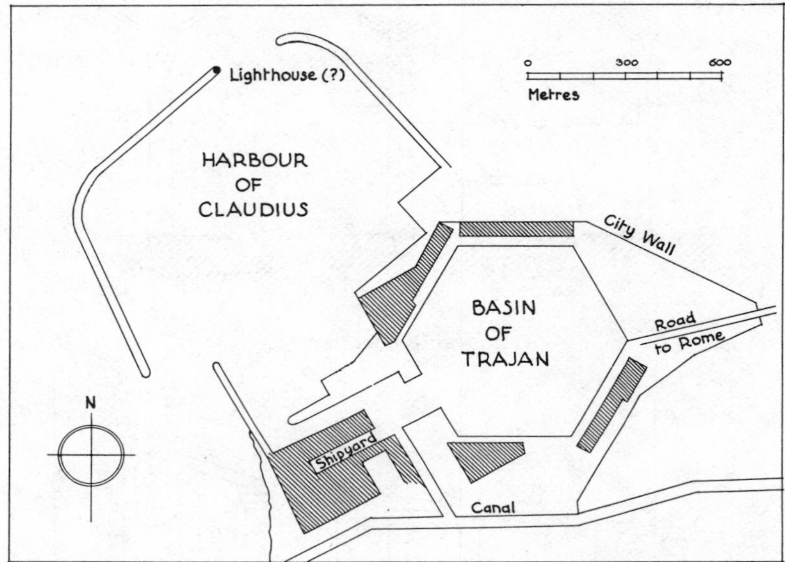

FIGURE 471—*Diagram of the port of Rome at Portus built by Claudius in* A.D. *46. The inner harbour with quays and shipyard was later added by Trajan* (A.D. *98–117*).

The port-installations of the Roman period are, on the whole, far better preserved than those of earlier times. The port of Imperial Rome at the mouth of the Tiber, close to Ostia, has been extensively excavated. The first artificial harbour at Portus (figure 471), for such was the simple descriptive name of the port of Rome, was completed by the Emperor Claudius in A.D. 46, after the old harbour of Ostia had begun to be silted by river-borne deposits. Two great moles were thrown out into the sea to enclose an area 1000 metres square. One of these moles was constructed on top of the hulk of an outsize vessel that Caligula had used to bring an Egyptian obelisk to Rome. On this mole was erected a lofty lighthouse, which became one of the most prominent landmarks of the Tyrrhenian coast. The site of the Claudian harbour is today marked only by mounds and concrete foundations, and its details cannot easily be reconstructed.

The port of Rome was further improved by Trajan, who constructed an interior harbour of hexagonal form, 700 m broad, with spacious warehouses and other buildings. The quays were built up in concrete and brick, and along their sides were projecting stones pierced with horizontal holes for mooring-ropes and bollards. The depth of the basin was 4–5 m. It was paved with large blocks of stone to serve as a hard bed for the dredging rendered necessary by the influx of mud and sand. A smaller basin, 250 by 40 m, near the entrance of the harbour, may have served as a dry dock, but there are no clear signs of how its narrowed entrance could be closed and drained, nor are there now any vestiges of the

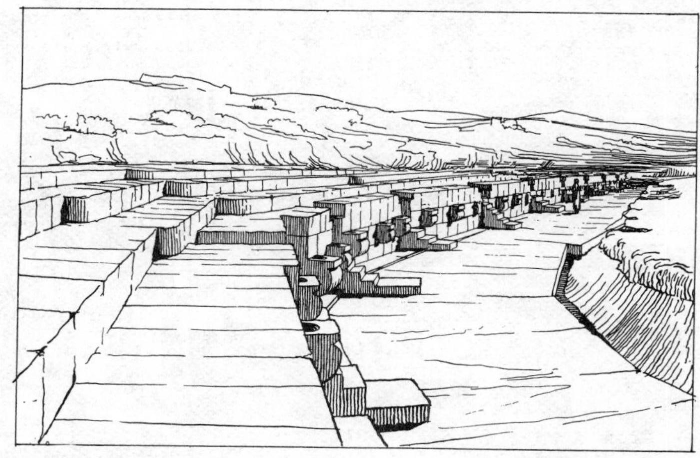

FIGURE 472—*Reconstruction of quays at* Leptis Magna, *north Africa. Built A.D. 193–211.*

cranes and tackle for unloading heavy materials from ships. The cargoes of corn in sacks and of oil or wine in jars were doubtless discharged manually, but the massive columns and blocks of marble imported for use in the capital must have demanded different methods.

Perhaps the best-preserved of all the ancient Mediterranean harbours is that of *Leptis Magna* (Lebda), on the African coast between Carthage and Alexandria (figure 472). Here the port in its present form was constructed in the reign of Septimius Severus (A.D. 193–211), a native of the city. Erosion of the sea has been confined to the outer edges of the moles, and the interior of the harbour is in remarkably good condition. It measures 360 by 300 m, and is enclosed by two moles with a passage 80 m broad between them. The western mole projects 100 m out to sea, and had a lighthouse at its point. The whole perimeter of the basin, little short of a kilometre in extent, was equipped with quays of large limestone blocks. Behind the quays were great warehouses with splendid colonnades. In places a flat platform only just above water-level kept vessels clear of the actual wharfside, which was equipped with projecting stones pierced to support vertical wooden bollards. Access from the lower platform to the quay was by staircases arranged at regular intervals.

Very similar remains, of a wharf equipped with projecting sockets for vertical wooden bollards, and backed by porticos and warehouses, have been found at Aquileia, at the head of the Adriatic. Here, however, the port was a river one,

FIGURE 473—*Roman harbour, probably Pozzuoli, from a fresco found at Gragnano near Naples. First century A.D.*

barely 50 m across, and there are no signs of the large basins encountered in the seaports. It is interesting to note that, as at *Leptis Magna*, the vessels did not moor immediately against the vertical quayside, but were kept clear of it by a low-level platform.

The harbours that we have described are technologically the most informative, but there are numerous other Mediterranean sites where the outlines of Roman harbours can still be traced. Anzio, Civitavecchia, Pozzuoli, and Terracina all have remains of great breakwaters, showing what vast labour was expended on port-construction. Such works require considerable maintenance, for winter storms can wreak much damage, while encroachment by sand is an ever-present threat. Some of the greatest ports went out of use early in medieval times, owing to failure to keep them in working order; medieval shipping tended to use the harbours that could be kept open with a minimum of maintenance. Partly through these causes, partly as a result of political events, Rome, Carthage, and Alexandria all lost their maritime importance. Alexandria alone has regained it, and that in relatively recent years.

Lastly we have a few representations of ancient Roman harbours and docks.

FIGURE 474—*Reconstruction of pharos and harbour-installation at Alexandria. Designed by Sostratus about 280 B.C.*

The best of them is a fresco from near Naples (figure 473). It is of the first century A.D. Even if the scene that it represents is imaginary, it is probably reminiscent of the neighbouring harbour of Pozzuoli. Pozzuoli was a main port of Rome, and Nero considered but abandoned the project of making a canal from it to the metropolis.

In the earlier days of Mediterranean navigation mariners were guided, if at all, by fires lit on hill-tops near the shore. As trade-routes became more clearly defined and ports were developed, lighthouses became a necessity. Their function in antiquity was not, however, to give warning of reefs or promontories, but to mark the entrance to the ports. They were therefore harbour-lights rather than lighthouses in the modern sense.

The great pharos of Alexandria, designed by Sostratus of Cnidos about 280 B.C., in the reign of Ptolemy Philadelphus, is the earliest true lighthouse of which we have any record (figure 474). It ranked as one of the wonders of the ancient world and was at least 85 m high. It rested on a great square base which occupied nearly half its height. Above this base was an octagonal storey, and above that a cylindrical one of smaller dimensions supporting the lantern. The latter was a fire kindled from resinous woods, with large mirrors of polished metal to concentrate the light; it is said to have been visible for 300 stades (about 35 miles). This lighthouse was destroyed in the fourteenth century, as the result of an earthquake, so that we are dependent for our knowledge of its arrangements on a few classical references and the somewhat garbled descriptions of Arab historians.

FIGURE 475—*'The Old Man': the pharos at Boulogne. From a drawing based on a wall-painting (now destroyed) of* c *1544, at Cowdray Manor, Sussex.*

This Alexandrian pharos gave its name and its basic form to all the lighthouses of antiquity. From coins and other representations it can be seen that a stepped structure was normally used for these lofty signal-towers (figure 524). The largest in Italy was at the entrance to Portus (p 518). Although nothing of it has survived above ground, and its height is purely conjectural, it appears to have been of four storeys, the lower three being square in plan, and the uppermost, supporting the lantern, cylindrical. The Roman reliefs and mosaics which attest this form show merely a fire blazing at the summit, and there is no evidence that reflectors were used as at Alexandria.

Of the other numerous lighthouses erected by the Romans we have but little detailed information. One of the most impressive to survive into the Middle Ages was that at Boulogne, which collapsed in the middle of the seventeenth century. Contemporary drawings show that it was a twelve-storey structure, nearly 60 m high (if we may believe our sources), each storey being of octagonal plan and of successively reduced dimensions (figure 475). A passage in Suetonius records that the Emperor Caligula (A.D. 37–41) celebrated his abortive invasion of Britain by erecting a 'very high tower from which, as from a pharos, a beacon might shine forth to regulate the course of ships', and it is reasonable to suppose that the Boulogne lighthouse owed its origin to the whims of this unbalanced ruler [28].

On the opposite side of the Channel, Dover boasted two Roman lighthouses, of which the westernmost one is now reduced to a shapeless mass of concrete embodied in the Drop Redoubt. The eastern pharos of Dover has been more fortunate, and still survives, adjoining the church of St Mary-in-Castro. It is octagonal externally, each side being 4·5 m in length, and has an interior shaft 4·3 m square. The surviving height is 13 m, but it has been reasonably conjectured that it must originally have been nearly twice that height (figure 476). The square-sectioned central shaft evidently held a timber structure incorporating a staircase. Although its facing has been much damaged, it is clear that it presented externally a series of stepped storeys, as at Boulogne.

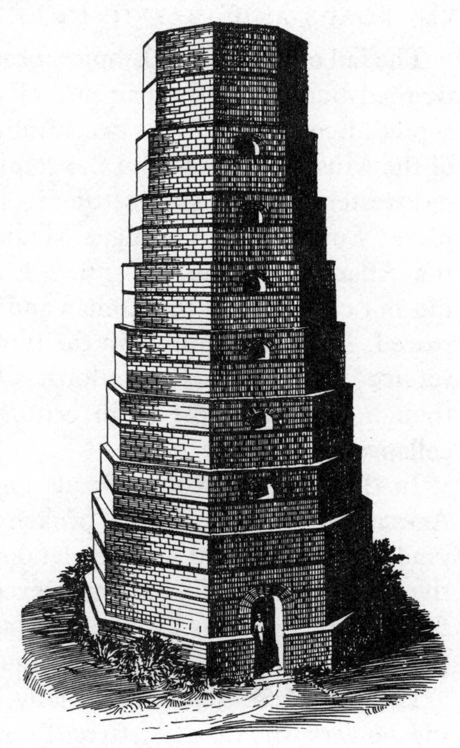

FIGURE 476—*Reconstruction of the Roman lighthouse at Dover.*

Our technical information regarding the manner in which the fires were kept kindled in these Roman lighthouses is very scanty. That wood was burnt primarily is beyond doubt, but combustible liquids must surely have been added to reduce the expenditure on fuel and to ensure a steady light during coastal gales.

The need for lighthouses of course continued into the Middle Ages, and long after the major Roman port-installations had become partly or wholly derelict some of the ancient beacons probably continued in use. When new ones were required they were perhaps initially timber structures supporting a brazier. Masonry lighthouse-towers were, however, certainly constructed soon after the first millennium A.D. (figure 530). Genoa was provided with a lighthouse in 1139, and Leghorn in 1304. That at Leghorn still survives; it is a stone-built tower 51 m high, consisting internally of seven drums of diminishing diameters and externally of two storeys. The earliest known lighthouse to be constructed on a rock fully exposed to the waves is that of Cordouan at the mouth of the Gironde, initially built in 1584. It was, however, only in the closing years of the seventeenth century that lighthouse-construction began in earnest, and on new and original lines that were to lead to the modern types of structure.

VII. ROADS AND TRAFFIC IN THE MIDDLE AGES

The fall of the Roman Empire meant the eclipse of central power. All systematic road-building and repair and all postal services centred on Rome were disrupted. Economic life was decentralized for several centuries. Only after repulse of the Muslim armies from the south and the Vikings from the west and north did western Europe start setting its house in order. A change was slowly taking place; Europe was no longer Mediterranean but started to look westward to the Atlantic. In the Byzantine Empire and that of Islam the traffic-lanes did not change, and the Roman and Hellenistic tracks were maintained and repaired. Ibn Khurdādhbih in the ninth century could still describe a messenger-service with 930 relay-stations. Only slowly did this system disintegrate, though as early as the tenth century the last ancient bridge over the Tigris collapsed.

In the west the Roman roads long survived, but by the end of the Middle Ages many were impassable, broken up by farmers in need of building-stones, or gradually encroached upon by landowners. For several centuries after the fall of the Empire they still remained excellent lanes for a much diminished traffic. Several are mentioned in the early sagas. They remained the routes for the dispersion of ideas and manuscripts and many types of goods.

Even in these 'dark ages' many towns, monasteries, and officials had their messengers who carried letters. In numerous cases there were regular services of such men, who had their passports and identification papers and in many ways resembled their Roman predecessors. These messengers travelled mostly on foot, but by the end of the thirteenth century horses were commonly used.

Our data on medieval traffic, travel, and road-topography are relatively scarce [29], but though in theory road-maintenance was an obligation of the landowners the obligation was little regarded. Hence compulsory labour was used for roads and bridges. This was indifferently successful, for only in cities was there sufficient control. The tax raised for the repair of roads and bridges helped in several cases to keep the traffic-system in fairly satisfactory repair. However, medieval trade was increasingly made to pay innumerable other tolls and imposts on roads, rivers, sea, and markets. In England such tolls were not as numerous as in France or Germany, though they tended to increase all over medieval Europe.

In Germany some of the great Hanseatic routes were comparatively free, but in the rest of the country the tolls gradually succeeded in clogging internal trade.

Though landowners frequently protested against being fined for not repairing roads, they did on occasion petition for turnpikes carrying tolls. Moreover, on many occasions money was given by municipalities and private individuals for

FIGURE 477—*Bavai–Tournai road from a fifteenth-century Flemish manuscript illustration, showing the use of the four-legged stool and the paviour's habit of working forwards.*

the specific purpose of repairing roads and bridges. This was a recognized pious duty. Thus the *Congrégation hospitalière des frères pontifes* built many bridges in twelfth-century France and Italy. In 1315 pious charity paid for the repair of roads near Durham. Monasteries, fraternities, and guilds were very active, but by the thirteenth century the crown often took the initiative. The *strata publica* was part of the legislation of thirteenth-century France, and in English common law 'the King's highway' meant not only military roads but all highways leading to towns and markets. In 1353 Edward III ordered the repaving of the road to Westminster, which had been paved in 1314.

Many of these roads were the old Roman roads, but in more cases deviations or new tracks were made with the purpose of evading tolls and taxes. Medieval traffic made many demands on the roads. There were the great pilgrimages to Rome, Compostella, and other famous shrines. To Compostella alone four big routes led from France. New markets and fairs demanded different traffic-lines. In many cases two or more new routes sprang up beside the old Roman road, as for example between Paris and Lyons.

In the eleventh century merchandise went in wheelbarrows or on porters' backs over the Alps from the Rhine to the Po valley. During the next century trade chose the route through the Rhône valley and then to Genoa, though it was thrice as long. In the thirteenth century the Alpine passes were repaved and improved and trade went back to its old routes. Sometimes the Roman road had become a mud-track and the new route was preferred. The concept of the ideal road was changing with the dominant means of transport. The Roman road was basically an ideal marching-road; vehicles played a less important part on it for reasons already discussed (p 515). Now, gradually, a highway of cobbles or broken stone on a loose foundation of sand, which could expand or contract with heat and cold and which was easily repaired, became the dominant type.

FIGURE 478—*German paviour using the one-legged stool. From the Mendel Album at Nuremberg.* c *1456.*

The medieval state built *chemins ferrés* with blocks cemented with mortar (a mixture of sand, lime, and river-mud) on a sand foundation. There were also the gravel or broken stone roads on a basc of sand or earth, akin to the ancient rammed *lithostrota.* Finally there were the cobble-roads, such as that between Verberie and Senlis (1322) which the town of Ghent paid for in order to promote its Paris trade, or the fifteenth-century Bavai–Tournai road illustrated in the *Chroniques de Hainaut* of Jacques de Guise (figure 477). Incidentally the French and Flemish paviours seem to have sat on four-legged stools and to have moved forwards as they laid the cobbles. The German paviours had the well known one-legged stool and moved backwards as they paved (figure 478).

The early Middle Ages saw the *coliers*, the medieval equivalent of coolies, carrying or wheeling their loads along the road. Much of the goods was transported in two-wheeled carts (*brouettes*), and more still in the four-wheeled cart or *charrette* (figure 497), a flat-boarded peasant cart which carried about three times the load of the *brouette*. The main alternative was not the pack-horse but the barge or other vessel. Goods travelled by land at about 22–35 km a day in flat country, because better horse-collars enabled faster horses to be used and thus longer journeys could be made. Thus, even with worse roads, road-transport was not over-expensive. The transport of wool from the Cotswolds to Calais raised its price by some 40 per cent. In the thirteenth century the price of wool rose in general about 1·5 per cent, and that of grain by 15 per cent, for every 80 km of transport. This was not unreasonable, though it compares unfavourably with sea-traffic, where Gascon wine was brought to Hull or Ireland at the cost of only 10 per cent of the lading-price. In fact, the transport service was sufficiently efficient to show a lower ratio of trading costs to total costs during the Middle Ages than at present. It suffered most from political instability.

Messages and travellers went by foot or on horseback. Travelling by cart was still mainly avoided. Journeys in flat country of about 50 km, in mountains of some 40 km, a day were quite common—about the same as, or not much more than, in antiquity. Though the lodging-houses of the *cursus publicus* had disappeared, private houses, inns, and *xenodochia* (pilgrim guest-houses established by religious communities) served the travellers. It has been calculated that most of the highways had to carry an average of 1000 tons a year, but some carried a good deal more; thus the salt-roads near Munich show salt-transport of 7000 tons on three roads in 1370.

VIII. CITY STREETS AND SANITATION

The central authority essential for the development and maintenance of a road-system was present in cities from the first. In the ancient cities of the Fertile Crescent and the Indus valley care was often given to the paving of streets, drainage, and lighting. Regard for traffic and hygienic conditions arose early.

The Indus valley had exemplary right-angled city-plans; it is clear that authority must have prevented the rise of the tortuous alleys characteristic of many later cities. The houses had a pronounced batter or slope and never encroached on the streets as they do in the bazaars of the modern east. All houses had latrines and bathrooms disposing their waste into street-drains (vol I, figure 296). Houses had rubbish-chutes, at the foot of which were sometimes bins at street-level; rubbish-bins stood at convenient places in the streets. The waste water entered the drains from tightly closed brick-lined pits, which had outlets to the drains about three-quarters of the way up. They seem the ancestors of our septic tanks and grit-chambers.

Each street or lane had one or two drainage-channels about 18–24 inches below the street level and 9×12 inches or 18×24 inches in area, covered with stone slabs or otherwise roofed. The streets usually ran from east to west or south to north. Some of the important streets were 15–33 ft wide, but the average was 9–12 ft. Most were unpaved, though some herring-bone brick pavements were found with the joints grouted with gypsum-mortar or bitumen. In later levels (800 B.C.) are found good floors and pavements made of pounded mixes of clay and potsherds. In still later levels there was much less care for detail.

In Mauryan India (323–160 B.C.) conditions were less good, though the ancient texts propagate ideal city-plans and wide streets [30]. Excavations at Bhita and Taxila proved that these towns had 10–12 ft streets with a 1–2 ft ditch in the middle. The basements of the houses could be entered from the inside only, as already reported by classical authors, and 'the town is divided by narrow irregular streets just like Athens' [31]. We find little of the slab-pavement or crushed stone (gravel) surface, 'which the King should have repaired every year by men who have been sued against or imprisoned', or of the 'drains on both sides [of the road] for the passage of water' [32].

In ancient Mesopotamia also there was, from the earliest times, a tendency to build houses in square or rectangular blocks. The streets and lanes were mostly unpaved, but many had sewers connected by laterals with the water-flushed latrines in the houses, as for example at Eshnunna (vol I, figures 295, 297). Most streets were rather narrow and much of the refuse was cast into them. Though floors of bricks in bituminous mastic were known as early as 2000 B.C., streets with stone-slab pavements were rare. Such efforts were mainly confined to the processional roads (p 494). During the Assyrian and Neo-Babylonian Empires stone-slab pavements became more common, and more attention was given to the lay-out and appearance of the city. Similar conditions were found in other countries of the ancient Near East and even in Egypt, though few entire towns have been excavated there.

The drainage and sewerage of the Minoan palace of Knossos (1900–1700 B.C.) were excellent (vol I, figure 349), but we know little of the settlements of the people. Conditions were far worse in classical Greece than at Knossos. Greek cities generally evolved without systematic town-planning, which was introduced by Hippodamos of Miletus in the middle of the fifth century B.C., according to Aristotle [33]. Such town-planning came into general effect in Hellenistic times. Spacious street-planning belongs to the same period. The main street of Alexandria was 30 m wide and the side-streets were 6–7 m. In the older towns they were seldom more than 4–6 m. Some processional roads like those in Miletus were about 20 m wide. The main road from Athens to its port at Piraeus was 14–15 m wide, but that from the agora to the acropolis was only 5 m. In general, city-streets were narrow, dirty, and unpaved. Thucydides writes of the 'muddy village streets' of Sparta [34], and the agora of Elis was unpaved and used for the training of horses. The Theban army got lost and was defeated in the darkness and the mud of the streets of Plataea (431 B.C.).

In some cases these Greek streets were paved with slabs or with chips rammed into the subsoil and sometimes grouted with lime mortar. This latter method produced a sort of 'macadam' surface, which required good drainage. According to Strabo drainage was unknown before the Roman rule [35]. The older metalled streets must have suffered not only from lack of drainage but from the refuse flung into the street with the cry *exito* ('stand clear', 'gardyloo'). Aristophanes complains about the muddy conditions of the Athenian streets. The authorities did something in the matter and the *astunomoi* or building-controllers in Athens had the duty of supervising the cleaning of drains and the sweeping of streets, which were executed by a corps of scavengers. An edict of 320 B.C. formally forbade throwing refuse into the streets. At the same time erection of balconies and encroachments on the street became punishable.

Lanes and streets were drained by mere ditches along the sides of the houses, and by slab-covered channels running irregularly through and across the streets. Pavements at the sides were non-existent except in the larger Hellenistic towns. Into most cities the entrance of carriages and carts was forbidden. The ancients preferred their streets to be narrow because they considered they were thus better protected from winds that brought humidity and diseases [36]. These streets were rarely lighted. Their maintenance was the duty of a board of officials whose services were paid for by the city treasury or by rich private individuals.

In the older Italian towns paved streets are mentioned from 400 B.C. onwards. In Rome itself no example has been found to antedate the Gallic invasion (390 B.C.). When Rome was rebuilt after the fire the square-block plan was adopted from the earlier Iron Age. The simple basic lay-out was a cross of two main streets. The new colonies and cities generally adhered to similar plans, which were rigidly adopted for all army camps and forts. The forum of Rome was first paved with rectangular slabs in the fourth century B.C. [37]. About 295 B.C. a foot-path from the temple of Mars to the Capuan Gate was paved with squared stone (*peperino*). It was made into a road paved with *silex* (see p 28) in 189 B.C. [38].

Streets in older Roman towns were narrow by our standards. They were usually not wider than 16 ft, except for the main streets, which were 23 ft and over. Augustus limited the height of buildings to 70 ft, a maximum reduced to 60 ft by Trajan. Thus building was limited to five storeys, and light could still enter the streets. Wheeled traffic was forbidden during the day except on such public occasions as religious festivals or games. The transport of building-materials (except for public buildings) and the like had to take place between dusk and dawn. Caesar had to repeat this order in 47 B.C., as did Claudius, Marcus Aurelius, and other emperors. Refuse and rubble had to be transported at night. Driving was considered ill-mannered: one walked, rode a mule, or was carried in a chair. Side-pavements (plate 34 B) blocked the house-entrances from carts driving into the inner courts of the houses. In Rome also, narrow streets were considered hygienic.

FIGURE 479—*Street in Pompeii, paved with slabs, showing wheel-ruts and stepping-stones.*

By 174 B.C. paved streets had become common in Rome, basalt slabs being generally used. Roughly half the width was taken up by the side-paths. About the same time streets in provincial towns such as Alatri or Puteoli (Pozzuoli) were also paved. We know much about the history of paved streets in certain fully excavated towns like Pompeii. There, till 200 B.C., streets with a 'macadam' surface were rather wide. During the next century adoption of the peristyle house (plate 29 B) led to the formation of narrow arcaded streets. These were typical of many cities of the Imperial period (Timgad, Lambaesis, Dugga). Slab pavement was now in regular use. Finally, from 50 B.C. onwards, the houses encroached still more on the streets, and house-owners possessing opposite blocks often closed the road to wheeled traffic by chains or stepping-stones. The latter allowed foot-passengers to cross the streets dry-shod (figure 479).

After Caesar's law of 47 B.C. the towns became more active in paving their streets. The law now forbade the throwing of refuse into the streets, which were regularly cleaned with water. Augustus fixed the width of streets of the first grade at 40 ft, of the next at 20 ft, side-streets at 12 ft, and lanes at 8 ft. This is in line with the dimensions found at Pompeii. The side-paths were held by curb-stones, 11–17 in wide, interrupted at regular intervals by drains leading to closed sewers.

In Rome, sewers for the surface-draining of streets were adopted from the

Etruscans. The *cloaca maxima* was 14 ft high and 11 ft across; it was built about 500 B.C. to drain the area of the Forum, and still functions. The three little streams that formed the natural drains of Rome were gradually converted into sewers. This system was extended about 300 B.C., and because of it Pliny proudly speaks of the 'hanging city of Rome' [39]. The foot- and shower-baths and the wash-stands in many Roman houses gave immediately into the sewers, as did the latrines invariably present. There were also public lavatories in many Roman cities—one for 45 persons in Puteoli and one for 28 persons in Timgad. Rome itself about A.D. 315 had 144 water-flushed public latrines.

Antioch appears to have been the first city in the world to have its streets lit. This was done about A.D. 450, tarred torches being used. Lanterns were well known in antiquity, and cylindrical and square prismatic forms with a conical top were quite common. They were fitted with thin plates of horn or mica, and later of glass. These lanterns were of metal, wood, or pottery [40]. Some were lit with candles, some with oil-lamps, of which there were many forms. For the wealthy a lantern-carrying slave preceded his master on his evening walks. Statuettes of such slaves with their lanterns are common. Portable lanterns were made in quantity at Capua from the first century B.C.

Before the days of street-lighting at Antioch the main streets of Roman towns must have been fairly well illuminated by the lamps in the shops and over the house-entrances. In the *Via dell' Abbondanza* at Pompeii there were at least 285 lamps on counters or above doors on a stretch of some 500 m [41]. 'Second Street' in the same town had 396 lamps in 132 shops over a distance of 576 m, and the 700-m 'High Street to Stabiae' had some 500 lamps. There were also lamps on the corners of streets under the statues of gods, or at the domestic shrines against which the Christian Tertullian protested. They were finally forbidden by Theodosius the Great (379–95). Many temples and often graves were brightly illuminated at night. Altogether traffic may not have been so hampered as is generally supposed. By the sixth century A.D. street-lighting had become common in the east.

In the Middle Ages conditions were certainly worse, though our information on this point is unfortunately scanty. Most towns of western Europe were new creations. They started without planning and with few such amenities as paving or lighting. In German towns paving did not start till the twelfth century [42]. It progressed very slowly and did not become common practice till much later (Lübeck 1310, Berne and Frankfort-on-Main 1339, Nuremberg 1368, Regensburg 1400). The smaller towns followed suit. Earlier streets were sometimes paved with 'macadam'. Later slabs, cobbles, or bricks came into use (plate 38 A, B).

Refuse was thrown into the streets, and, though the inhabitants had the duty of scavenging, the law was often ignored. When princes visited the town or processions were held there was a general clean-up. In some towns, however, the law was regularly enforced (Göttingen 1330, Mechlin 1348) until scavenging and cleaning became public services towards the end of the Middle Ages (Nuremberg 1490) (figure 480). Pigs wandered along the streets. Drainage was seldom adequate, though little streams were sometimes turned into the streets to clean them. Lights were scarce except on bridges. At night many streets were barred with chains.

FIGURE 480—*A German scavenger. From the Mendel Album at Nuremberg.* c *1456.*

In the days of Alexander Neckam (1157–1217) the streets of Paris were in disrepair and full of mud. Little remained of the Roman pavement and, indeed, doggerel verse derived the Latin name for Paris, *Lutetia*, from *lutum* (mud).[1] The sewer was an open channel in the middle of the road into which refuse was thrown, and into which the latrines often gave. Shoes had to be heavy, high, and thickly soled. As the streets were not lit one had to walk with a linkboy. There was some light from domestic lamps after dark. Horses but not carts were allowed in the town [43].

In England conditions were hardly better. In Cambridge the king commanded the mayor and bailiffs to repair the pavement of streets and lanes, and to compel everyone to pave the street before his tenement. The repairs were paid by levies on many classes of goods. Northampton was granted such a levy by Edward I (1284) and other towns were given the same powers (Nantwich 1277, Chester 1279, Cambridge 1289, Liverpool 1329). This pavage, as the levy was called, was based on the weight of goods and type of vehicle. Sometimes public-spirited citizens left money for this purpose, and thirty-six roads in or near London were thus paved between 1358 and 1509.

In London each alderman appointed 'four reputable men' to repair and clean the streets (1280). They were the ancestors of the later 'scawageours' or scaven-

[1] Instead of from the Celtic *louktaih*, the place of marshes.

gers (1364) controlled by a 'surveyor of streets' (1390). The actual paving, first done by the citizens, became an expert job superintended in the fourteenth century by a 'stone master'. Southampton had a 'town paviour' in 1482. The Paviors of London were organized as a City company in 1479. They earned respectable wages, fixed by the London ordinance in 1301, and several of them rose to some dignity and affluence [44].

These paviours used cobbles and sand for the markets and squares, and gravel and sand for the lanes. Pot-holes were often simply filled up with faggots of broom and chips. In the work of paving, the old surface was rarely cleared away and this made the whole street-level rise. Pavements were compacted with hand-rams (figures 477–8). The greatest wear of the streets came from iron-shod cart-wheels. When iron tires were forbidden the wheels were protected with nails; this made things worse, and the use of all but flat nails was forbidden (figures 506, 497). Wheelbarrows, drags, and sledges completed the range of vehicles [45].

There was much encroaching on the street by stalls and by overhanging balconies. Hence a mounted official with a standard lance across his saddle would now and then ride through medieval cities to see that the regulation width was being observed, and to fine offenders. Heavy traffic was ordered to follow certain streets and use certain gates, but these by-laws were often infringed. Traffic-jams at the toll-gates were frequent, as were accidents and law-suits about rapid or careless driving. Citizens were repeatedly ordered to clean the frontage of their house once a week (usually on Saturday) and cart the rubbish to the dunghills outside the city walls or dispose of it in rivers. Nevertheless it took many centuries to restore conditions to the level of the better conducted cities of antiquity.

REFERENCES

[1] EVANS, SIR ARTHUR (JOHN). 'The Palace of Minos', Vol. 2, Part I, pp. 60–92 and map. Macmillan, London. 1928.

[2] HUNGER, J. 'Heerwesen und Kriegführung der Assyrer auf der Höhe ihrer Macht.' Der alte Orient, Vol. 12, no. 4. Hinrichsen, Leipzig. 1911.

[3] LUCKENBILL, D. D. 'Ancient Records of Assyria and Babylonia', Vol. 1, p. 75, no. 222. University of Chicago Press, Chicago. 1926.

[4] TORCZYNER, H. *et al.* 'Lachish I. The Lachish Letters', p. 79. Oxford University Press, London. 1938.

[5] DOSSIN, G. *Rev. d'Assyriol.*, **35**, 174, 1938.

[6] JENSEN, P. *Z. Assyriol.*, **15**, 238, 1901.
LANDSBERGER, B. *Ibid.*, **35**, 215–16, 1924.

[7] HERODOTUS I, 186. (Loeb ed. Vol. 1, pp. 230 ff., 1920.)

[8] UNGNAD, A. 'Babylonische Briefe', no. 15. Hinrichsen, Leipzig. 1914.

[9] REISNER, G. 'Tempelurkunden aus Telloh', no. 129, IV, 9. Mitt. aus den orientalischen Sammlungen der Kgl. Museen no. 16. Berlin. 1901.

WIESNER, J. 'Fahren und Reiten in Alteuropa und im alten Orient.' Der alte Orient, Vol. 38, nos. 2–4. Hinrichsen, Leipzig. 1939.

[10] CALDER, W. M. *Class. Rev.*, **39**, 7–11, 1925.

[11] XENOPHON *Anabasis*, I, v, 7. (Loeb ed., *Hellenica* and *Anabasis*, Vol. 3, p. 288, 1921.)

[12] *Idem Cyropaedia*, VI, ii, 36. (Loeb ed. Vol. 2, p. 168, 1914.)

[13] FORBES, R. J. "The Coming of the Camel" in 'Studies in Ancient Technology', Vol. 2. Brill, Leiden. 1955.

[14] STRABO XVI, C 748. (Loeb ed. Vol. 7, p. 234, 1930.)

[15] STEFFEN. 'Karten von Mykenai', Textband, p. 8. Publ. of Deutsches Archäologisches Institut. Reimer, Berlin. 1884.

[16] HOMER Iliad, XXII, 146. (Loeb ed. Vol. 1, p. 352, 1919.)

[17] PAUSANIAS Description of Greece, X, v, 5. (Loeb ed. Vol. 4, p. 392, 1935.)

Idem Ibid., II, xv, 2. (Loeb ed. Vol. 1, pp. 322 ff., 1918.)

Idem Ibid., X, xxxii, 2. (Loeb ed. Vol. 4, p. 554, 1935.)

Idem Ibid., II, xi, 3. (Loeb ed. Vol. 1, p. 304, 1918.)

[18] *Idem Ibid.*, X, xxxii, 8. (Loeb ed. Vol. 4, p. 558, 1935.)

Idem Ibid., VIII, liv, 5. (Loeb ed. Vol. 4, p. 166, 1935.)

Idem Ibid., I, xliv, 6. (Loeb ed. Vol. 1, p. 240, 1918.)

[19] ARISTOPHANES The Frogs, lines 110–12. (Loeb ed. Vol. 2, p. 306, 1924.)

[20] STATIUS *Silvae*, IV, iii, 40–48. (Loeb ed. Vol. 1, p. 220, 1928.)

[21] PLUTARCH *Vitae parallelae. Caius Gracchus*, vii, 1. (Loeb ed. Vol. 10, pp. 212 ff., 1921.)

[22] *Idem Ibid.*, vii, 2. (Loeb ed. *ibid.*).

[23] KNAPP, C. *Class. Philol.*, **2**, 1–24; 281–304, 1907.

Idem. Class. Wkly, **28**, 177, 1935.

WELLS, B. W. *Class. J.*, **19**, 7; 67, 1923–4.

[24] DESJARDINS, E. "Les Tabelarii, courriers porteurs de dépêches chez les Romains" in Bibl. Éc. haut. Étud., Fasc. 35, pp. 51 ff. Paris. 1878.

[25] RAMSAY, A. W. *J. Rom. Stud.*, **15**, 60, 1925.

[26] JACOBI, H. *Germania*, **6**, 88–93, 1922.

[27] APPIAN Roman History, VIII, xiv, 96. (Loeb ed. Vol. 1, p. 566, 1912.)

[28] SUETONIUS *Vitae XII Caesarum. Caligula*, xlvi. (Loeb ed. Vol. 1, pp. 474 ff., 1914.)

[29] FEUCHTINGER, M. E. 'Der Verkehr im Wandel der Zeiten seit dem Jahre 1000.' Verein Deutscher Ingenieure, Berlin. 1935.

JUSSERAND, J. J. 'English Wayfaring Life in the Middle Ages.' Benn, London. 1950.

[30] DUTT, B. B. 'Town Planning in Ancient India.' Thacker, Spink, Calcutta. 1925.

[31] PHILOSTRATUS *De vita Apollonii Tyanensis*, II, xxiii. (Loeb ed. Vol. 1, p. 180, 1912.)

[32] *Sūkranītisāra*, ed. by B. K. SARKAR: 'The Sukraniti', p. 35, nos. 516–39. Sudhindranātha Vasu, Allahabad. 1914.

[33] ARISTOTLE *Politica*, II, viii (1267^{b}), trans. by B. JOWETT in 'Works of Aristotle', ed. by W. D. ROSS, Vol. 10, pp. 22 ff. Clarendon Press, Oxford. 1913.

[34] THUCYDIDES I, x, 2. (Loeb ed. Vol. 1, p. 18, 1919.)

[35] STRABO XIV, C 646. (Loeb ed. Vol. 6, p. 246, 1929.)

[36] NIELSEN, H. A. *Arch. Hyg., Berl.*, **43**, 85, 1902. HUGILL, W. M. *Class. Wkly*, **26**, 162, 1932. GARRISON, F. H. *Bull. N.Y. Acad. Med.*, second series, **5**, 887, 1929.

[37] VAN DEMAN, ESTHER B. *J. Rom. Stud.*, **12**, 4, 1922.

[38] LIVY X, xxiii, 13. (Loeb ed. Vol. 4, p. 446, 1926.) *Idem*. XXXVIII, xxviii, 3. (Loeb ed. Vol. 11, p. 94, 1936.) MOMMSEN, T. *Hermes*, **12**, 486, 1877.

[39] PLINY *Nat. hist.*, XXXVI, xxiv, 104.

[40] LOESCHKE, S. *Bonn. Jb.*, **118**, 370, 1909.
ROBINS, F. W. 'The Story of the Lamp and the Candle.' Oxford University Press, London. 1939.

[41] SPANO, G. Atti Accad. Archeol. Lett. Arti Napoli, new series, Vol. 7, Pt. II, pp. 1–128, 1920.
LAMER, H. *Philol. Wschr.*, **47**, 147, 1927.

[42] HEIL, B. 'Die deutschen Städte und Bürger im Mittelalter', pp. 104–7. Teubner, Leipzig. 1921.

[43] BOUTTEVILLE, M. R .*Rev. sci., Paris*, **71**, 609, 1933.
HOLMES, U. T. 'Daily Living in the Twelfth Century.' University of Wisconsin Press, Madison. 1952.

[44] THORNDIKE, L. *Speculum*, **3**, 192, 1928. SABINE, E. L. *Ibid.*, **9**, 303, 1934. *Idem. Ibid.*, **12**, 19, 1937.
SALUSBURY, G. T. 'Street Life in Medieval England.' Pen-in-hand Publ. Co., Oxford. 1948.

[45] BAUDRY DE SAUNIER, L. *et al.* 'Histoire de la locomotion terrestre', Vol. 2, p. 70. L'Illustration, Paris. 1936.

BIBLIOGRAPHY

General:

DAREMBERG, C. and SAGLIO, E. (Eds). "Pharus" in 'Dictionnaire des antiquités grecques et romaines', Vol. 4, Part I, pp. 427–32. Hachette, Paris. 1904–7.

FORBES, R. J. 'Notes on the History of Ancient Roads and their Construction.' Allard Pierson Stichting: Univ. Amsterdam. Archaeol.-Hist. Bijdr. no. 3. Amsterdam. 1934.

WYCHERLY, R. E. 'How the Greeks Built Cities.' Macmillan, London. 1949.

Africa:

GOODCHILD, R. G. 'Roman Roads and Milestones in Tripolitania.' British Military Administration, Department of Antiquities, Tripoli. 1947.

SALAMA, P. 'Les voies romaines de l'Afrique du Nord.' Imprimerie Officielle du Gouvernement Général de L'Algérie, Algiers. 1951.

Alexandria:

THIERSCH, H. P. 'Pharos: Antike, Islam und Occident. Ein Beitrag zur Architekturgeschichte.' Teubner, Berlin. 1909.

The Balkans:

BALLIF, P. 'Römische Straßen in Bosnien und der Hercegovina', Part I. Bosnisch-Hercegovinisches Landesmuseum, Vienna. 1893.

France and Germany:

DÉCHELETTE, J. (Ed.) 'Manuel d'archéologie préhistorique, celtique et gallo-romaine', Vol. 6, Part II, Section i: GRENIER, A. 'Les routes.' Picard, Paris. 1934.

Great Britain:

CODRINGTON, T. C. E. 'Roman Roads in Britain.' Society for Promoting Christian Knowledge, London. 1918.

MARGARY, I. D. 'Roman Ways in the Weald.' Phoenix, London. 1948.

Idem. 'Roman Roads in Britain', Vol. 1. Phoenix, London. 1955.

WHEELER, SIR (ROBERT ERIC) MORTIMER. "The Roman Lighthouses at Dover." *Archaeol. J.*, **86**, 29–46, 1929.

Italy:

ASHBY, T. 'The Roman Campagna in Classical Times.' Benn, London. 1927.

See also articles on individual roads which have appeared in the *Journal of Roman Studies* and the Papers of the British School at Rome.

Maps:

KIEPERT, H. *Formae orbis antiqui.* Reimer, Berlin. 1894.

ORDNANCE SURVEY. 'Map of Roman Britain' (3rd ed.). Ordnance Survey Office, Southampton. 1956.

Tabula Imperii Romani. International Map of the Roman Empire. Sheets published by various authorities in various countries in association with the Central Bureau of the Carte Internationale du Monde au millionième. 1930–53.

Bridges:

HORWITZ, H. T. "Über urtümliche Seil-, Ketten- und Seilbahn- Brücken." *Technikgeschichte*, **23**, 94–98, 1934.

JOHNSON, F. M. and GILES, C. W. SCOTT (Eds). 'British Bridges, an Illustrated and Historical Technical Record.' Public Works, Roads and Transport Congress, London. 1933.

THOMAS, W. N. 'The Development of Bridges.' Parker and Gregg, London. 1920.

WATSON, S. R. and WATSON, W. J. 'Bridges in History and Legend.' Jansen, Cleveland, Ohio. 1937.

A plan of the harbour of Calais drawn in 1546 by Thomas Pettyt. It shows the protective moles and a tower which may have acted as a lighthouse or merely to guard the entrance to the harbour. The presence of a post-mill near the sea is interesting, as it seems to indicate that corn was imported and ground in situ *before being carried inland.*

15

VEHICLES AND HARNESS

E. M. JOPE

I. GENERAL SURVEY

THROUGHOUT much of the two millennia considered here, wheeled vehicles were in regular use for transporting goods and people. This was so not only in the metropolitan areas of the expanding Roman Empire, but in many remote and rugged lands where the use of pack-animals and sledges still persisted. Thus western civilization came to rely on wheeled vehicles for transport services, and carts and wagons became a commonplace of medieval life. In medieval England their use was widespread, both on farms and for bulk transport, and they were used even in Anglo-Saxon times. Throughout the Middle Ages road-transport was often preferred to that by river, which was always erratic owing to flood and drought. Much of England's chief export, wool, was carried overland by cart from the west to the eastern and southern ports (p 527).

Medieval roads were better than has usually been assumed. It is probable that seventeenth-century complaints about the condition of roads sprang as much from their increased use during the previous century, owing to the rise in population, as from increased numbers of the vehicles themselves. Throughout the Middle Ages haulage-firms operated with some competition over much of Europe. To capture the carrying-trade between the great Italian towns and transalpine Europe, the Swiss about 1338 began constructing a road for carts through the Septimer pass, probably the first cart-route through the Alps.

Transport-vehicles rose in importance, but the use of chariots declined. Though the war-chariot formed a significant part of the fighting-arm of the Greeks of the heroic age, by classical times it was obsolete. It survived among the Romans only for ceremonial processions and for races, the effective mobile offensive arm being cavalry. The sole significant use of the chariot as a fighting-vehicle in Roman times was by the western Celts, with whom it was popular; in fact, many Latin terms for wheeled vehicles are of Celtic derivation. Both economic conditions and the spread of Christianity discouraged even the ceremonial use of chariots in the later Empire. In post-Roman Europe the only representative of the chariot tradition was the light private travelling-car of the wealthy. In the armies of Rome and of the medieval states the real function of wheeled vehicles was in the vitally important baggage-train.

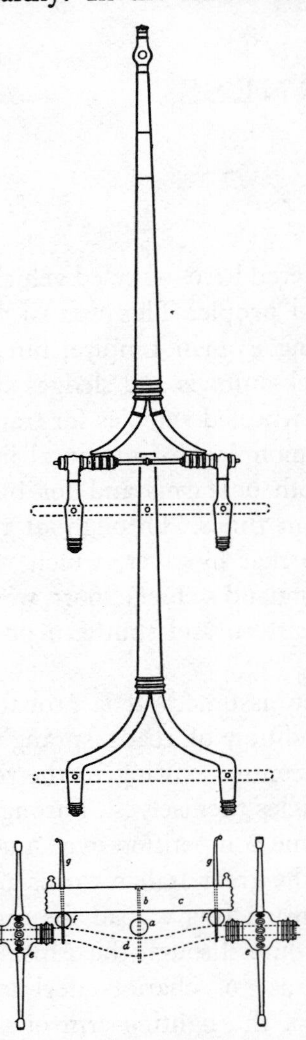

FIGURE 481—*Plan and front-elevation of the Dejbjerg wagon, probably a funeral vehicle, made in a Celtic workshop in south Germany or north-east France. Note the swivelling front pair of wheels. First century* B.C.

Many forms of vehicle used during the Middle Ages were based upon very old traditions of construction. The most important technical contribution to transport was undoubtedly the change in horse-traction harness. The breast- and girth-bands, yoke, and pole, by which until post-Roman times many horses had been made to pull—in great discomfort, as though they were oxen—were changed to the padded collar and shafts or traces. This improvement was achieved as part of a complex interaction between east and west. Though its beginnings can be traced in the Roman world, and contemporaneously in China, the full development to modern type was not complete until about the twelfth century A.D. (pp 552 ff).

The most significant advances in the effective use of horses by man, both for traction and for riding, seem to have penetrated to the west from the east. The horse is a native of steppe-land, and the new devices were developed perhaps by the Sarmatian, Parthian, Sassanid, and other inheritors of the steppe traditions. They first appear in the west in late Roman or immediate post-Roman times, perhaps stimulated by the strong eastern elements in the Roman armies, and associated with the rising importance of cavalry and artillery. This cannot be proved, however, and some improvements, notably the stirrup, may well have reached Europe through continued barbarian contacts between east and west during the succeeding centuries. Horse-riding in the west was transformed by the padded saddle and the stirrup, also contributions of the steppe-lands. The shoeing of horses was probably a Roman contribution.

A difficulty in discussing the history of wheeled vehicles is the confusion of the names for them and for their parts in most European languages, ancient or modern. Of the profusion of classical and medieval terms, there seems hardly one which can be associated exclusively with any particular type of vehicle. A typical example of this confusion concerns the meaning and origin of the common Greek word for a cart, *hamaxa*, said to be related to the *axon* or axle and to

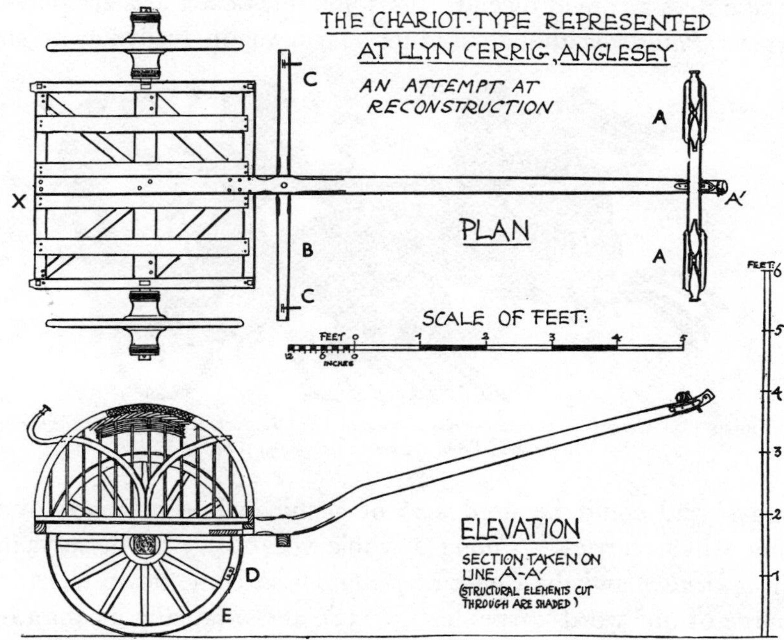

FIGURE 482—*Reconstruction of a Celtic chariot as used in Britain, northern France, and southern Germany:* (X–X′) *pole;* (A) *terrets;* (B) *swingletree;* (C) *trace-loops;* (D) *joint of felloe;* (E) *iron tire.*

imply that the vehicle had only one axle. Yet one dictionary after another maintains that the word refers to a four-wheeled vehicle. Though the word *hamaxa* is a general term in Greek, and though it evidently usually referred to a two-wheeled vehicle, the epithet *tetrakuklos* (four-wheeled) was sometimes added to it. The matter is of some importance with reference to the date of introduction of four-wheeled carts for other than ceremonial purposes. Such carts could hardly travel along the Greek rut-roads (p 499) unless provided with one swivelling axle, and not very easily even then. The question is complicated by the fact that there are no Greek terms which in themselves imply the number of wheels possessed by a vehicle, though in Latin the Theodosian Code (A.D. 438) has the term *birota* (two-wheeled) for a light vehicle able to carry 145 lb. Other

Latin terms such as *biga*, *triga*, *quadriga* (from *quadri-jugae*, four yoked together) referred originally to the number of the draught-team, not—as has sometimes been erroneously supposed—to the number of wheels.

In Latin literature there are many different terms for vehicles, most of them of Celtic origin. Thus *carruca*, originally a plough, came to mean a four-wheeled travelling-vehicle carrying both people and luggage, or even signified nothing more specific than a conveyance. It was sometimes used as a sleeping-place on long journeys. *Plaustrum*, though used for a farm wagon, really meant 'something

FIGURE 483—*Greek chariot with warrior dismounting to fight. Note the traces from the horses to the axle. From a Greek vase of* c 500 B.C.

which sways' and could be used also of a ship, while *vehiculum* was merely 'something which carries'. Among a whole vocabulary of these vague words the Latin *vehiculum* and the Celtic *carpentum* have come down to us; the latter is the source of our word 'carpenter' (p 233), and originally meant a light two-wheeled car but later, in post-Augustan times, even an agricultural wagon.

All these Celtic terms that passed into Latin are an unwitting tribute to the skill of the Celtic vehicle-builders, so amply seen in the Dejbjerg wagons (plate 37A, figure 481) and the Anglesey chariot-parts (figure 482). The fluctuating meaning of terms may be seen by considering the maximum load set out for each vehicle named in the Theodosian Code. Expressed in modern weights, a *birota* could carry 145 lb, a *vereda* 218 lb, a *carrus* 436 lb, a *vehiculum* or *carpentum* 726 lb, and an *angaria* 1089 lb. This last seems to have been the maximum load carried by a single vehicle at the time. Special arrangements were needed for moving large blocks of stone or tree-trunks (p 545).[1]

In medieval records many of these terms remained in use, often with even less precise meanings. To them were added both new terms and corruptions or

[1] It is often stated that heavy tree-trunks were transported by man-handling, but the evidence seems slender for a generalization; the stock illustration could be a battering-ram.

confusions of old. *Caretta*, whence our cart, originally meant a two-wheeled vehicle. The *caretta longa* was a four-wheeled wagon in the thirteenth century, anglicized by William Langland in 'Piers Plowman' (*c* 1380) as 'long carte'. Germanic languages had their own *wagen*, from which the English wagon and wain have taken their forms.[1] The English usage of the word chariot itself originates in a confusion between the French words *chariot* and *charrette*, the former a four-wheeled vehicle, and the latter a light two-wheeled cart.

FIGURE 484—*Model of a Roman chariot, found in the Tiber. Imperial period.*

II. VEHICLES IN CLASSICAL TIMES

Chariots. The chariot as a fighting-vehicle was a vital arm in Mycenaean Greece from the sixteenth century B.C. It was doubtless derived ultimately from Mitanni and the Hittite cities between the Black Sea and the Caspian. It had a long history in the Greek world, extending back a thousand years before classical times. One type, seen in Egypt in the fifteenth century B.C. (vol I, figure 525),[2] persisted almost unchanged through all this time, but by the sixth century B.C. the chariot had largely gone out of use for military purposes.

Most classical Greek representations set out to portray the chariots of the bygone heroic age. These carried a crew of two, charioteer and warrior, and were used to enhance the warrior's mobility; he may be seen dismounting to fight in figure 483. Except for occasional

FIGURE 485—*Chariot on an Attic geometric vase. Eighth century B.C.*

[1] As in vol I, 'cart' is here used for two-wheeled vehicles, and 'wagon' for those with four wheels, though general English usage is looser.

[2] The woods, among them ash and birch, of which this Egyptian chariot is made, betray a northern origin, since they do not grow south of the region between the Black Sea and the Caspian. It might have been imported ready-made from Mitanni, as were ten wooden chariots in the late fifteenth century, but an Egyptian inscription of 1435 B.C. records the making of chariots from woods imported thence. It would seem that both this Egyptian type and the similar Mycenaean one had a common origin in the Mitanni–Hittite region (Euphrates–Armenia area).

spear-throwing, fighting was rarely carried on from the chariot itself. Archery from the chariot was practised in hunting, but in warfare the archer was unprotected, there being no space for a shield-bearer (vol I, figures 526–7).

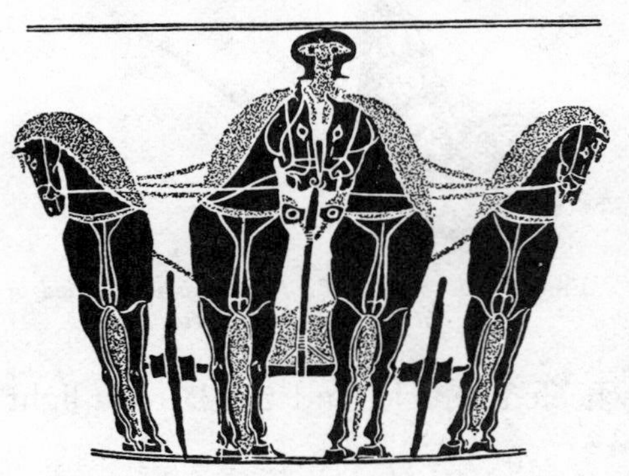

FIGURE 486—*Front view of a chariot. The strap passing between the legs to the girth-band keeps the breast-band well down on the neck. From a Greek vase of* c 500 B.C.

The chariot of the Greek heroic age was a two-wheeled car. Mounted on its axle was a light chassis, on which were erected frames of bent rods with their spaces filled in with sheet leather, woven thongs, or wicker-work to form breast- or side-screens (figures 484–5). The floor was probably of woven leather thongs, as in surviving Egyptian chariots. Chariots described as decorated with metal plates must have been of heavier construction. The draught-pole, with its characteristic curved shape, was attached to the chassis, and was sometimes supported at the yoke-end by a leather thong running back to the chariot breast-screen (figure 485; vol I, figures 525–6).

The wheels of these light chariots rotated on a fixed axle and had four, six, seven, or eight spokes, with hubs held in place by linch-pins. The draught-team was harnessed to the yoke, normally of box-wood, itself thrusting against a pin in the draught-pole and secured in place with a leather binding (vol I, figure 525). Mycenaean chariots were drawn by two horses, though three- and four-horse teams are sometimes found in the later heroic age (figure 486). Racing-chariots were similar to these vehicles, but doubtless lighter, being designed to carry the charioteer only.

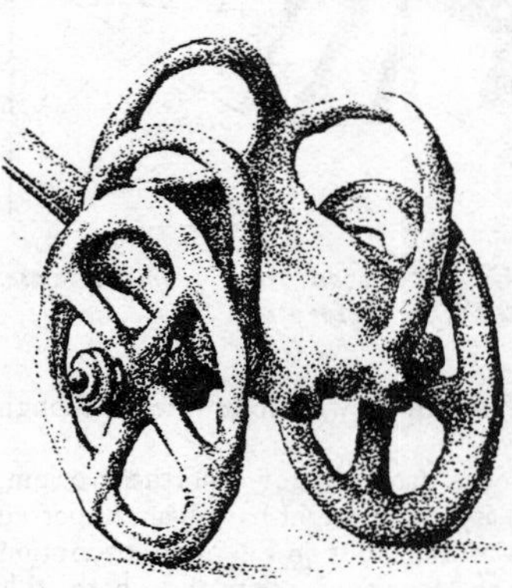

FIGURE 487—*Model of a chariot as used at the Olympian games.*

Etruscan chariots seem to have been modelled on those of Greece (figures 487, 488) and may have provided models for Roman racing- and processional-chariots, but the very skilled Celtic craftsmen beyond the Alps undoubtedly made a considerable contribution to the design. From the fifth century B.C. to the third A.D. the chariot was used for purposes of war among the Celtic peoples of the upper Danube to middle Rhine and Marne region. They must have taken over the custom during a formative period of their material culture (fifth century

B.C.) from the Greeks, probably indirectly through the Etruscans. The Celtic chariot is well known from the graves of chieftains and from other archaeological contexts, and may be illustrated by a reconstruction (figure 482). It was largely of wood and leather or wicker, as in Greece. No chariot axles survive, though many complete wheels have been recovered. They were held on the axle with iron linch-pins, often coated with bronze and with ornamental heads. The draught-pole extended in one piece to the back of the chassis. The shaped wooden yoke thrust against a stout square-sectioned iron peg inserted in the draught-pole near the tip, where it had a strengthening iron sheath. The reins were guided over the front of the chariot through a bronze rein-ring, and over the yokes through ornamented cast bronze rings or terrets. The surviving wooden yokes, however, show no vertical holes for these terrets, but only lateral ones, through which the reins themselves may have passed. Such yokes were probably for carts, and may have been used for oxen as well as for horses. Traces, when used, were attached by decorative loops to the swingletree. Horns with ornamented metal caps were provided for the warrior to grip.

FIGURE 488—*Etruscan chariot drawn by three horses. From a bas-relief. Sixth century* B.C.

As with the early Greek chariot, the Celtic chariot carried a crew of two. Caesar in 54 B.C. praises the skilled control of the British charioteers, and describes how the warrior could run out along the pole to the yoke, presumably either to hurl a javelin or for bravado, and jump back into the chariot. He makes it clear that the real tactical use of the Celtic chariot was, as with early Greeks, to provide a mobile arm, and that once the warriors reached the Roman cavalry they jumped down and fought on foot [1]. These chariots were technical masterpieces, composed of numerous individual components wrought at the hands of joiners, iron- and bronze-smiths, and enamellers. It reflects the scale as well as the quality of this achievement that Caesar implies that the British chieftain Cassivellaunus had at least 2000 chariots [2]. Thothmes III had recorded with satisfaction the capture of 924 chariots at the battle of Megiddo (1479 B.C.).

FIGURE 489—*Bas-relief of a boy's chariot with shafts. From a sarcophagus from Treves. Third century A.D.*

The processional- and racing-chariots of the Romans were often designed to carry merely a charioteer and sometimes had but one horse. As a two-wheeled vehicle needed some method of steadying, a pair of shafts then had to be added in place of a single pole. Shafts, already known in China, are seen for the first time in the western world in sculptures of the third century A.D. from Treves, one showing a boy's one-horse sporting-chariot (figure 489). They did not, however, automatically bring an improved system of harnessing.

In the late Empire the chariot tradition became merged with the general design of light two-wheeled ceremonial cars and personal conveyances such as may be seen carved on an Irish eighth-century cross-base (figure 490).

Carts and wagons. In classical Greece farm-carts were in common use where the terrain allowed. In the country-side the farm-cart served not only agricultural purposes but all ceremonies—weddings, festive processions (figure 491), and funerals. Similar vehicles served as rural coster-carts (figure 491, below). They always appear as having two wheels and drawn by two animals, usually mules, yoked on either side of a draught-pole. Their wheels are almost always of the characteristic cross-bar type (figure 491). The animals were harnessed to them exactly as to chariots, with a boxwood yoke lashed with leather thongs to a metal pin in the draught-pole. Traces were used only if more than two animals were pulling. The superstructure of these carts varied from the flimsiest wicker frame to carefully carpentered plank construction. It was sometimes separate from the chassis and strapped to it as needed, so that one chassis could

FIGURE 490—*Scene on the base of a cross at Ahenny, Co. Tipperary, Eire. Note the horizontal harness strap (cf figure 507). Eighth century A.D.*

be used for several purposes. Four-wheeled vehicles, though well enough known among the Scythians (figure 492), were rarely used in Greece itself, where the terrain was unsuitable.

There is little evidence as to how stone was transported from the inland quarries for the great buildings of Greece. Paved roads existed from some quarries (p 25), but only one example of an artificially grooved quarry-road is known, namely that from the marble quarries of Agrilesa to Kamaresa. Though Homer refers to the haulage of timbers from the tops of the high mountains by carts (*hamaxa*) [3], they were not necessarily large.

FIGURE 491—*Greek country-carts of the fourth century* B.C. (Above) *Greek wedding procession, from a Cabeiric vase.* (Middle) *A Greek cart carrying four people.* (Below) *A Greek coster-cart carrying wine-jars. Note the cross-bar wheel.*

In the Roman Empire, embracing the most varied terrains and peoples, it is not surprising to find evidence of a greater variety of transport-vehicles than in the Greek world. Bulk-transport was organized by large firms, and passenger carriages could cover as much as 100 miles a day. Four-wheeled vehicles were common (figures 493–4), while two-wheeled carts were still much used on farms. Light two-wheeled cars were available for rapid personal journeys, while four-wheeled vehicles, covered with a cloth or leather awning, were used by people and for baggage on long journeys, and even for sleeping in (figure 494). Some had a jointed, fully carpentered roof, almost like a coach (figure 506). Water or wine was carried in large barrels on both two- and four-wheeled vehicles. While spoked wheels were usual, solid wheels cut from planks were still in use on farms (figure 503). There is no evidence that Roman carriage-builders achieved the skill of the Celtic craftsmen as seen in the Dejbjerg wagons, with a pivoted fore-axle, giving a swivelling pair of wheels in the front (figure 481, plate 37 A).

FIGURE 492—*Clay model of a four-wheeled wagon from Kerch* (*Crimea*). *Third century* B.C.

III. VEHICLES IN MEDIEVAL EUROPE

Little is known of the structure of wheeled vehicles between the fall of the Roman Empire and the twelfth century. The carts and wagons of Europe in the eleventh and twelfth centuries seem to show little technical advance on those of the Roman world, and the vehicles used during the fifth to eleventh centuries must have been in the same tradition. On farm-carts, bodies of light staves, often filled in with wicker, were usual (figures 495, 499). Baggage-wagons had similar though stronger bodies, sometimes of planks or with panelled sides. Travelling-carts or wagons often had a fabric or leather cover over a wooden frame, and sometimes side-screens of woven leather strips (figure 496). Wheel-barrows, of simple but effective structure, seem to have been a medieval innovation, and a most useful one (figure 497).

FIGURE 493—*Relief of a four-wheeled baggage wagon from Trajan's column. Note the small collar-pads hanging from the yoke.* c *A.D. 110.*

In later medieval illustrations there appear carts consisting of a long frame, like a broad ladder, carried on a pair of large wheels at the middle (figure 498). By the fourteenth century the loading area of a cart or wagon was sometimes extended by projecting hurdles at the back (figure 499). A development of this principle, by adding surboards over the wheels, was much more recent and is peculiar to the British Isles, where it is very familiar.

FIGURE 494—*Roman relief of a four-wheeled travelling-wagon, from Klagenfurt, Austria. Imperial period.*

As paired draught for smaller carts gave way to single-horse draught, or a team in tandem, less obstructive on roads, shafts took the place of the draught-pole, one or the other being necessary to stabilize a two-wheeled vehicle. Shafts also came into use on larger carts and four-wheeled wagons, the draught team

FIGURE 495—*Late Saxon two-wheeled cart with spoked wheels and composite felloes. Calendar picture from an eleventh-century manuscript.*

FIGURE 496—*Travelling-carriage about 1317. From a French manuscript of the legend of St Denis.*

being arranged tandemwise, with the front animals pulling on traces (figures 497, 499). Rarely is there any sign that the shafts could swing vertically, as did the draught-pole on the Dejbjerg wagons (figure 481, plate 37 A). Usually the shafts were merely rigid prolongations of the two main long beams of the chassis, and their attachment to the rear horse by a body-strap must have caused it discomfort on uneven roads (figure 499).

Travelling-carriages with the body suspended on straps did not appear until the sixteenth century (figure 501), and were the ancestors of the sprung coach of later times. The term 'coach' is from the Magyar Kocsi, the name of a place in Hungary between Raab and Buda, and is found in some form in most European languages. Vehicles called Kocsi carts were in use in Hungary from about 1450, but it is not clear what were the features that caused the name to become so

FIGURE 497—*Baggage-wagon and wheelbarrow. From a French manuscript, 1460.*

widely diffused. In a German picture of about 1550 a Hungarian coach has no cover and is not suspended on springs.

The making of the front axle in one piece with the shafts, but swivelling by attachment to the chassis through a pivot to facilitate turning—a bogie[1] or turning-train as we now call it—came into prominence at the end of the Middle Ages. The device can be traced on war-engines in the fifteenth century (figure 502), and may be inferred when the front wheels are shown smaller than the back. The front turning-train is a feature of most travelling-vehicles, from the sixteenth century onwards, to which we now give the name coach; they typically had sprung suspension and well constructed bodies. A swivelling front axle was no new idea, for it was used by the Celtic craftsmen of about the first century B.C. as seen on the Dejbjerg wagons (figure 481). No history of the idea can yet, however, be traced between then and the fifteenth century.

FIGURE 498—*Cart with large wheels. From a painting by Brueghel. 1564.*

The final development in wheeled transport, the use of rails, an idea most pregnant for modern civilization, was also foreshadowed in antiquity, namely by the artificial ruts of the Maltese, Minoan, and Greek roads (figure 457; vol I, figure 514). In the sixteenth century solid-wheeled trucks are shown running on wooden rails at mining-centres (tailpiece).

IV. WHEELS

The radially spoked wheel was a great achievement of the joiner's skill. It was evolved in the period covered in vol I (p 212), but it is doubtful whether its full potentialities for sustaining heavy loads were realized before the later Middle

[1] Bogie is a north English dialect word for a turning-train, introduced into general usage through the northern railway workshops in the nineteenth century.

FIGURE 499—*Two-wheeled cart with hurdles added to increase the loading-area. From the Luttrell psalter*, c *1338.*

Ages. In classical Greece two types of spoked wheel were in use, and no doubt the three-part solid wheel was also sometimes used. Chariot wheels had four, six, or eight spokes. It is probable that the felloes were normally of several parts,

FIGURE 500—*Two-wheeled cart with studded wheels. From the Luttrell psalter, c 1338.*

though those of the (Egyptian) Thebes chariot, for instance, are of one piece of ash bent full circle (vol I, figure 525), as were those of Celtic chariots. There seem to have been attempts to strengthen the attachments of the spokes, often lathe-turned, at the outer ends (figures 482, 504 A). There is some little evidence that metal tires were occasionally used in the Greek world, and certain figures of chariot wheels could be interpreted as having four-part felloes with an added metal tire.

Greek carts usually appear with the cross-bar type of wheel (figure 491), a Late Bronze Age example of which has been found in Italy (vol I, figure 135). This type of wheel could still be seen in China and England in recent times (vol I, figure 42). Cross-bar wheels had the felloes in two parts, divided by the cross-bar, which extended right through to the outer edge of the wheel. The hub was formed by strengthening the cross-bar at the central hole with a plate on the outside, and perhaps one on the inside, which involves simpler joinery than a radially spoked wheel. The directions of Hesiod (*c* 700 B.C.) for cutting the timber necessary for making cart-wheels may refer to the cross-bar wheel, usual on Greek carts.

FIGURE 501—*German coach, showing leather-strap suspension. From a printed book of 1568.*

He directs the cutting of an *apsis* of three spans (about 2 ft 3 in) [4], which probably refers to the diameter (2 ft 8 in long in the example found in Italy). Then follows the cutting of 'bent timbers'—most likely the felloes, the rim of the Italian

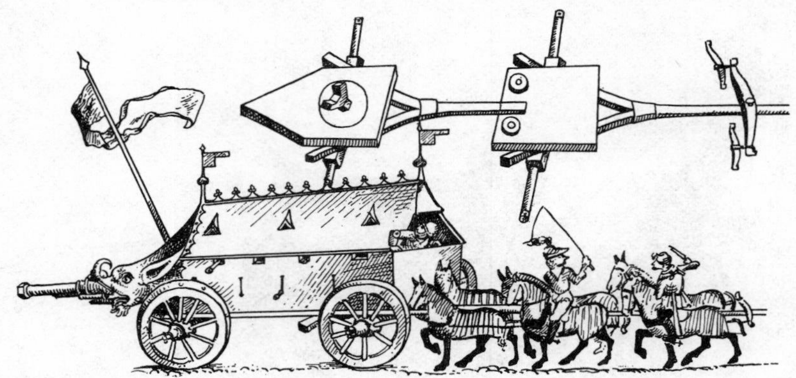

FIGURE 502—*A turning-train used in manœuvring a war-engine. From a German manuscript,* c *1480.*

cross-bar wheel being of two felloes. On Greek carts wheels seem to have been either rigidly fixed to a rotating axle,[1] or rotating freely on a fixed axle.

We know little in detail of the construction of wheels in Roman times. Solid wheels built up of planks were still much used on farm-carts (figure 503). Radially spoked wheels were usual, however, and since Latin has so many Celtic terms for wheeled vehicles and their parts we may assume a debt to Celtic craftsmen for much of the joiner's technique in making wheels. Wheels evidently made by Celtic wheelwrights have been found in the Roman forts of north Britain (figure 504 A).

Celtic wheels, of which many have been preserved, had ten to fourteen

[1] The part of the axle projecting out from the wheel is sometimes shown in Greek sources as rectangular in section, but usually looks rather slender to transmit the rotation of the wheel to the axle, with its inevitable friction on the bearings. When the vehicle was cornering the strain on the wheel-axle junctions would have been enormous. It is difficult to deduce the correct interpretation of this representation of the axle-end.

FIGURE 503—*A farm cart with solid wheels built up from planks. Walking backwards in front of the draught animals is the 'caller'* (*see p 92*). c A.D. *300.*

spokes. These were lathe-turned and of hornbeam, oak, or willow, mortised into the hub, and mortised or dowelled into the felloe. The hubs were turned, the wood being elm, as in modern cart-wheels, or oak. They were quite long, usually 12–15 in, and bound at either end with cast bronze collars. Wheels of the earlier Iron Age (Hallstatt) had beaten iron collars. Some wheels had the hubs lined with a bronze ring at each end, as bearings—probably a Roman improvement (figure 504A). The hubs of the Dejbjerg wagons had the most sophisticated refinement of all, roller-bearings created by channels inside the hub with rods of wood turning between hub and axle (figure 504C).

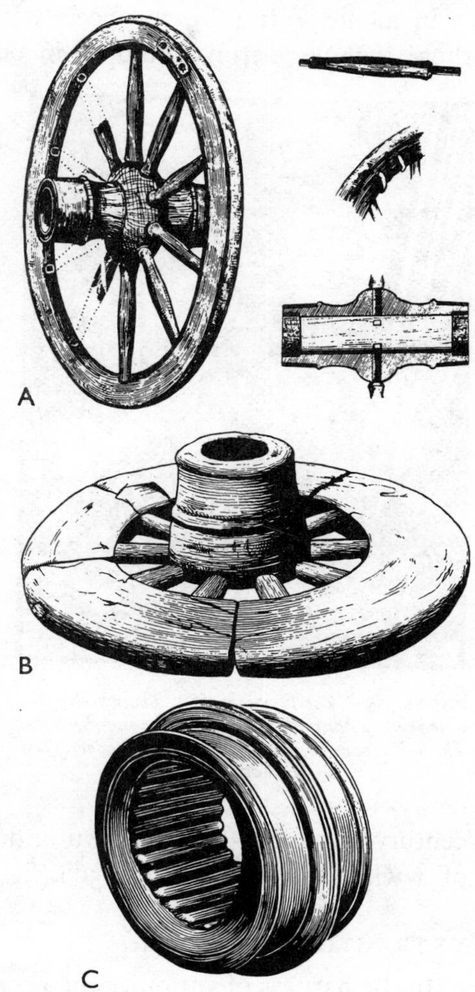

FIGURE 504—(A) *Wheel with a felloe in one piece;* (right) *spoke fitted into the felloe with a round tenon, into the hub with a square one; joint of the felloe; section of the hub. From the Roman fort at Newstead, Melrose. Diameter 3 ft. Second century.* (B) *Wheel found in a bog at Sodermanland. Probably medieval.* (C) *Hub of a wheel of the Dejbjerg wagon, showing grooves for bearing-rollers. First century* B.C.

The felloe of a Celtic wheel, as of those of the Thebes chariot (vol I, figure 525), was a single piece of ash, bent circular. The Celtic wheels were bound with iron tires skilfully welded into a full circle out of several pieces of metal. Those at Llyn Cerrig in Anglesey are of a high quality carbon steel (figure 482). The absence of nail-holes in Celtic tires shows that they were placed red-hot over the wooden wheel and then shrunk on, the quenching of the red-hot metal with water hardening the steel.

A few early medieval wheels have been found, notably in Scandinavian bogs. They are heavy, and have six or seven broad felloes with one spoke to each (figure 504B). Medieval illustrations also often reveal such a construction. Wheel-rims were being built up of a number of separate felloes in Roman times, as sometimes in Greece; the later (second century A.D.) heavier wheels from the fort of Newstead in southern Scotland had twelve spokes, and the felloes were in six sections, each attached to the next by dowels.

In medieval times cart-wheels were used both with and without iron tires, those that were iron-bound often being forbidden in towns because of their effect on roads (p 533). Even more destructive of surfaces must have been the nails frequently seen as a series of projecting studs round the wheels (figures 499, 500). In the larger wheel from Newstead the dowels of the spokes passed right through the felloes, and took the wear in the same way as the cart-nails. Ready-made pairs of wheels could be bought in English market-towns in the Middle Ages; they were manufactured on a considerable scale in the woodland areas, such as the Chilterns.

FIGURE 505—*Wheelwright's shop. The wheelwright is making a 'dished' wheel, drilling out the hubs. The other man is splitting timber for spokes. Jost Amman, 1568.*

The final development of the wheelwright's art is the construction of a wheel not in one plane but as a flat cone. This 'dished' wheel is designed for strength against the sideways thrust inevitable with the swaying of heavy loads drawn over erratic surfaces. Such wheels may be seen in illustrations from the fifteenth century onwards. One is shown under construction in Jost Amman's woodcut of a wheelwright's shop in 1568 (figure 505).

V. TRACTION-HARNESS

In the harness of antiquity the breast- and girth-bands met on the yoke at the withers. On a horse, if a breast-band dragged back at the nape, it inevitably slid up the throat, and exerted a choking pressure on the windpipe[1] when the animal pulled hard. The ancient peoples were aware of this trouble, and it is remarkable that improvements were so long in coming. The modern type of horse traction-

[1] It must be remembered that the blood-pressure of the veins in the neck is very low, and the vein-walls have little ability to resist compression, so that venous congestion in the head is easily developed. Indirect pressure on the vagus nerves would also cause some discomfort and vascular disturbances, such as slowing of the heart-beat. A strap from the breast-band passing between the legs to the girth-band kept the former down (figure 486), but chafing must have been considerable.

harness has been usual in Europe since the twelfth century and was thus a very great advance in utilization of power-resources. As well as economizing draught-horses and enhancing bulk-transport capacity, it facilitated rapid long-distance travel in more comfortable and heavier vehicles, which developed continuously until the railway age.

Improvements in traction-harness were attempted in several ways. For instance, the breast-bands, through which all the horizontal pull of the horse was transmitted, were set horizontally instead of diagonally (figure 506). They were joined to the middle of the girth-band, from which the pull was still taken at the withers. Such straps must have caused chafing even when well placed. This system appeared in the third century A.D. in Han China and in Persia, and in the west (Byzantium) by the ninth century (figure 507); it was, however, not the forerunner of modern harness.

FIGURE 506—*Roman relief showing a four-wheeled travelling-wagon or funeral-car. Note the horizontal strap-harness leading up to what is virtually a collar at the horses' necks. The girth-band passes over a saddle cloth. From Vaison, southern France.*

Shafts. The substitution of a pair of shafts for the single draught-pole was inevitable if animals were to be used singly or in tandem. The shafts could be attached well down the breast-band so that the pull did not make it ride up the throat. They became common in China from about the time of Christ, though they were sometimes combined with a yoke and used with oxen, and are to be seen occasionally in Imperial Roman sources (figure 489). Shafts are not known in any intermediate sources at this period and possibly were used independently

FIGURE 507—*War-chariots on a bone carving from a Byzantine casket. The traction-harness is composed of horizontal breast-band and collar. Ninth century A.D.*

in the Far East and the west. The Roman world, however, does not seem to have appreciated the advantages of shafts, and they were not generally employed in the west until the Middle Ages.

Traces from the vehicle could be attached well down the breast-band in the same way as shafts. The Greeks and Celts used traces (figure 483), but those on Greek vehicles were attached to the withers of the outer horses of a four-horse team (figure 486). Though Celtic chariots were usually drawn by two horses, there is little evidence as to how the traces were attached to the animals. Traces to the middle of the breast-band do not appear until about the tenth century (figure 508), and in the Middle Ages are commonly shown—usually as ropes—on both two- and four-wheeled vehicles (figure 499). Their use enabled a team of draught animals to pull in file more effectively than in parallel. This draught-train is seen in Roman times, but does not become usual until the Middle Ages, when it was also used in the east.

FIGURE 508—*Tenth-century harnessing by shafts and collar. Illustration in a Frankish manuscript.*

Horse-collar. The girth-band became superfluous when shafts or traces were attached to the middle of the breast-band; the latter thus became a collar (figure 506) and the modern type of harness was born. In western sources this is seen clearly in the tenth century A.D. (figure 508), though there is more than a suspicion of it in Roman times (see figure 489, where the breast-band, though badly placed, is really a collar).

The final step in the evolution of modern traction-harness was the padding of the collar. Small, apparently padded, shoulder-collars are sometimes to be seen hanging from the yokes of Roman vehicles (figure 493), but it is not until the twelfth or thirteenth centuries that the good padded shoulder-and-breast horse-collar of modern type appears in European illustrations (figures 497, 499, 500); from then on it is common. The tables are now turned, for in medieval illustrations oxen are sometimes seen in padded collars, that is with the harness suited to horses.

Cart-saddle and postillion. In Roman and medieval representations it is evident that the pressure of the strap holding the weight of the shafts was often distributed over the horse's back by a cart-saddle, to reduce chafing. Another saddle might also be added to one horse, usually that nearest the vehicle, upon which the postillion rode. The postillion riding a draught-horse does not emerge definitely until the twelfth century (Bernard of Clairvaux), but from the thirteenth century onwards he is a frequent figure in illustrations, brandishing a whip or a multiple-strand flagellator (figure 497).

Conclusion. The evolution of the modern type of horse traction-harness, usual in Europe from about the twelfth century onwards and one of the more significant developments of this period, presents a complex picture of interaction between east and west. A definitive account of its progress can scarcely yet be given. The less significant horizontal breast-strap seems of oriental origin, as do shafts which, although they may have been developed independently in the east and in the west, were not fully appreciated in the Roman west. The horse-collar and the development of padding may well have been Roman and Byzantine contributions, as was probably also the use of side-traces. In Persia the Sarmatians, the Parthians, and their successors from A.D. 226, the Sassanids, must have played their part in transmitting these ideas between east and west. But in the west it was the transport requirements of the Roman Empire that called forth such improvements in traction-harness.

FIGURE 509—*Diagram of a Scythian saddle and stirrup, interpreting a scene on the Chertomlyk vase (cf figure 431). The hanging loops apparently of leather may serve only for mounting. On the vase the loop at the back is hidden by the horse.* c *380 B.C.*

VI. HORSE-RIDING

The effective use of the horse as a riding-animal was largely developed by the nomadic peoples of the steppe-lands of Asia during the last millennium B.C. The practice was diffusing into eastern Europe from about 800 B.C., adding a considerable element to the offensive arms of warlike peoples. Horse-riding seems, however, to have been a rapid development in Greece, being recorded on vases from the late eighth century. By classical times cavalry had superseded chariotry as an instrument of war.

The earlier steppe nomads must, like the earlier Greeks, have ridden bareback. Horse-riding was not fully effective until some form of firm saddle and stirrup gave the horseman a secure seat, especially important to the warrior. Both these improvements arose on the steppes, whence they were transmitted eastward and westward, probably by the semi-nomadic horse-riding peoples. Other important items of riding-equipment, the shoeing of horses, the curb-bit, and the spur, seem to have developed first in the west, the last two not reaching the Far

East until modern times. Thus the full development of horse-riding equipment resulted from an interaction of west and east, just as did that of horse-traction.

Horse-cloth and saddle. The horse-cloth, perhaps originally an animal-skin, was used by Scythians on the steppe-lands (sixth to third centuries B.C.). That seen on the Chertomlyk vase (figure 509) of the fourth century B.C. has developed more nearly to a true saddle, for it is strapped on to the horse by breast- and girth-straps and seems to have padded bows. Bareback riding was common in Greece, though the horse-cloth was used in Assyria from the ninth century B.C. and by the Greeks from the sixth century B.C. It had become a usual part of a Greek soldier's equipment by the fourth century B.C., and was in general use in the Roman world by the first century B.C. The Germanic tribes considered a horse-cloth a mark of laziness.

FIGURE 510—*Roman saddle with padded bows. From a relief on Theodosius's column at Constantinople (379–95). Similar saddles are seen on monuments in the west. After a seventeenth-century engraving.*

There is no evidence that the Romans used padded saddles until the fourth century A.D. (figure 510). The true riding-saddle was evidently introduced to the later Roman world by oriental elements, probably in the Imperial cavalry. Such a padded saddle was placed over the horse-cloth, and was known as a *sella* or chair. The padded saddle had been used in China in Han times (A.D. 25–220).

Stirrups. Straps hanging from horses' girths and apparently ending in loops may be seen on the fourth century B.C. Chertomlyk vase (figure 509) and on carvings of the second century B.C. at Sanchi and elsewhere in India. Such a loop may, however, have been merely an aid to mounting, for while actually riding a man might be in danger if he could not withdraw his foot quickly from a pliant loop. It was the rigid metal loop that made the use of the stirrup an effective advance in horse-riding. In western Asiatic and European sources no horseman is depicted engaged in vigorous riding with his feet in stirrups until well on in post-Roman times.[1] The use of stirrups (figure 511) is mentioned in China about A.D. 477; stirrup-rings may be seen on Chinese monuments from the sixth century, and the earliest metal stirrup-rings in the Far East are of this period.[2] The idea of

[1] The ledge seen under the feet of Shalmeneser III (859–825 B.C.) on the bronze reliefs is only a foot-rest for a royal personage riding sedately, but perhaps the germ of the stirrup idea is there. It is an isolated example.

[2] NEEDHAM, J. ('Science and Civilisation in China', I (Cambridge, 1954), p. 167) reproduces a silhouette from an

these stirrup-rings, like that of the leather loop, was probably developed among the steppe horse-riding peoples, but the earliest stirrup-rings are of cast iron or bronze, and were probably made under the influence of the technical skill of Chinese craftsmen. In the west, stirrups—the cast iron variety at that—are first found in Hungary, in burials apparently of Avars, horse-riders who came from the central Asian steppe about A.D. 560. Their use is recommended for Byzantine cavalry soon after this. In north-west Europe stirrups are first found in eighth-century graves of Vikings (grave III at Vendel in Sweden, for instance), and were evidently adopted by them as a result of their eastward expeditions towards the Black Sea at that time. It is quite likely that it was largely through the Viking horsemen that the stirrups of medieval Europe were introduced, though some development in the south from Byzantine usage cannot be discounted. The earliest stirrups recognizable as such were iron 'rings', varying in shape from round to triangular, and hanging from leather straps. From the earliest stirrups onwards there has naturally been a tendency to give the foot a flat rest, and the shape of the foot has dictated that of both medieval and modern stirrups (figure 512). From Viking times onwards, stirrups were often decorated with patterns of inlaid silver or bronze.

FIGURE 511—*Saddled Chinese horse. Note the stirrup. T'ang period, seventh century A.D.*

Spurs. Spurs were not used by the earlier Greeks. They imply boots for riding, which Xenophon (fl *c* 400 B.C.) recommends [5]. Though he does not mention spurs, one appears on the foot of an Amazon on a vase of this date, and bronze spurs have been found at Dodona. Spurs have been found on Celtic sites such as La Tène from about the fourth or third centuries B.C. (figure 513 A), and the Romans used them from that time onwards, if not earlier. They do not appear to have reached the Far East until recent times.

These early prick-spurs are all stumpy, with a short prick, and with studs or buttons at the open ends for attachment of the strap (figure 513 A). The prick-spur was universal from then on over the whole western world, the prick and the arms being increased in length. In the eleventh and twelfth centuries there

early (1821) rubbing of a tomb-shrine relief of A.D. 147 depicting the 'flying gallop', showing the rider with his foot in a stirrup, but interpretation of such details from these old rubbings is uncertain.

was a tendency to check the effectiveness of the long prick by some sort of stop-mould, or by reducing it to a cube set diagonally (figure 513 B, C). The straight arms gave way in the thirteenth century to those curved to the shape of the foot. From the thirteenth century, prick-spurs were superseded in the west by rowel-spurs (figure 513 D, E), which first appear in a Spanish manuscript of about 840. In the later Middle Ages and afterwards, the spiked wheels tend to be set out on extravagantly long shafts from the back of the boot.

These post-Roman spurs are of iron, and in Britain at any rate they had a coating of tin flashed over. This tinning—which is not found on stirrups—can be traced as a continuous tradition of the spur-maker in Britain from Saxon times to the seventeenth century, and illustrates the tenacious continuity of workshop traditions.

Bridle-bits. The principle of controlling a horse by reins attached to a bit through its mouth was well developed in the ancient world. Variation on the basic design of this simple *snaffle-bit* (figure 514 A) has continued unceasingly into modern times. Most earlier bits were of two links (vol I, figure 522), but the three-link bit, with a small centre link in the mouth (sometimes made knobby) (figure 514 B) rendering it more flexible so that it was less easy for the horse to gain control by biting, was used in Greece and became common among the Celts. Two-link bits nevertheless continued in use through the Middle Ages and later.

Xenophon recommends the use of flexible mouthpieces, and says that a man must have two bits, a rough one for training horses, and a smooth one for use thereafter [6]. Nevertheless, rough and spiky bits were probably used in antiquity for more than mere training.

Cheek-pieces, nose-bands, and cheek-straps were introduced in Assyria (vol I, figure 523), whence the Greeks took them. Later cheek-pieces are only stylistic variants; there is no change in principle.

As well as the snaffle-bit with simple mouthpiece used with a single rein there appeared in early Roman times the *curb-bit* (figure 514 C). This is the sort of device one would expect to have been brought to the west by some oriental element in the Roman armies. However, though curb-bits are to be seen in the third century A.D. in the Sassanid kingdom, they appeared in the west as much as five or six hundred years earlier, having been found in a Celtic grave of the third century B.C. at Canosa, near Bari, Italy, as well as on Roman sites. It must have been the development of closely organized cavalry that made some device for more severe control of the horse imperative.

The function of the curb-bit is to halt the horse by forcing its head upwards; horses run with their heads forward. The fully developed curb-bit

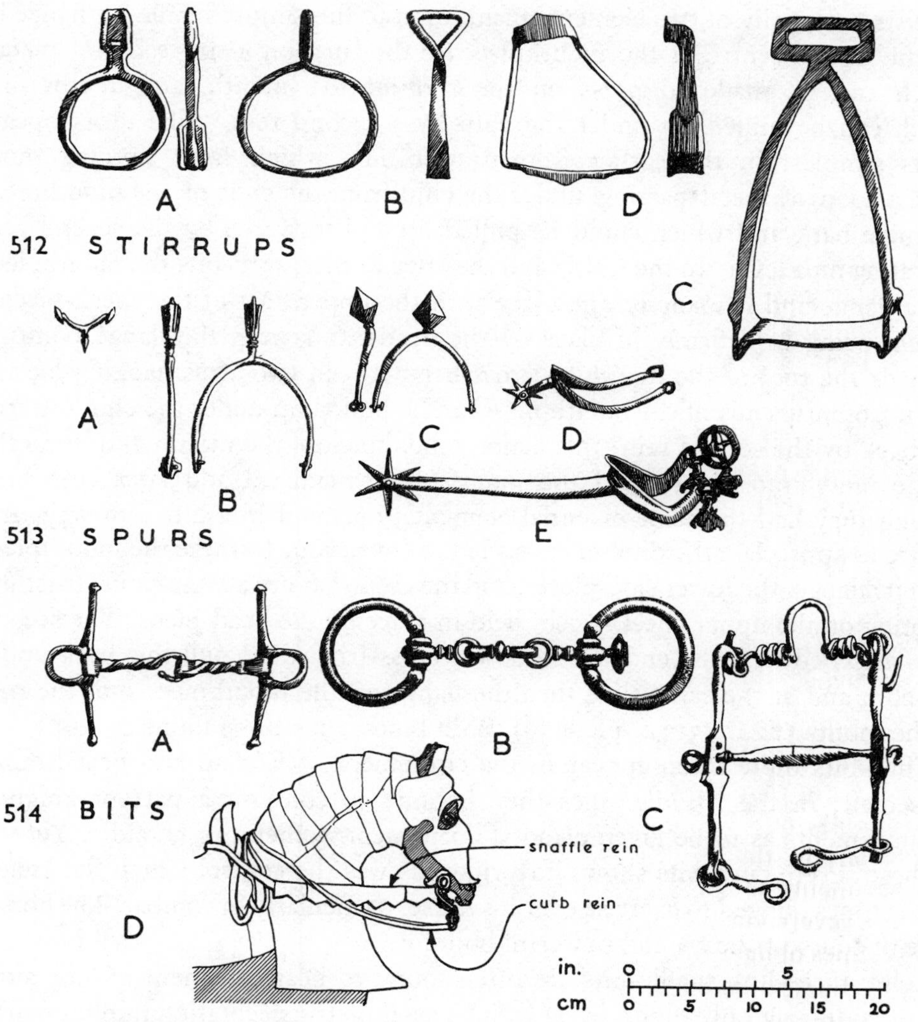

FIGURE 512—*Stirrups:* (A) *The earliest type in Europe, probably brought in by the Avars from the east, cast iron. Hungary, sixth to seventh century* A.D. (B) *Simple bent metal stirrup, one of the earliest in the north. From Vendel, Sweden. Eighth century* A.D. (C) *Eleventh to twelfth century stirrup from Dolkheim, East Prussia.* (D) *Asymmetrical stirrup of early fourteenth-century type. From London.*

FIGURE 513—*Spurs:* (A) *Early Iron Age prick-spur from Stradonitz, Bohemia.* c *second century* B.C. (B) *Eleventh-century prick-spur from Oxfordshire.* (C) *Twelfth-century prick-spur from London.* (D) *and* (E) *Rowel-spurs from London;* (D) *early fourteenth century,* (E) *fifteenth century.*

FIGURE 514—*Bits:* (A) *Simple medieval iron two-link bit. From London Wall.* (B) *Three-link bit with decorative boss on one side only, implying use in paired draught. Anglesey. First century* B.C. (C) *Iron curb-bit from Canosa, Italy. Third century* B.C. (D) *The front part of a horse-panoply showing a curb-bit with a curb-rein in addition to the ordinary rein which is decorated with trappings. From the Imperial Armoury, Vienna. 1502.*

consists essentially of two elements in addition to the simple snaffle—a flange or tongue in the centre of the mouthpiece (in the curb-bit a single bar of metal) which can be made to press on the roof of the mouth, and a cross-bar which can be pulled up under the chin by a second rein.[1] The idea appears at its simplest in the earlier Roman curb-bits, which have nothing more than a loop of metal passing under the chin from the ends of the mouthpiece (a single bar), and which could be pulled up and forwards by the second rein operating on a lever. In the Celtic and the later Roman curb-bits the mouthpiece has a flange and is made in one piece with the upper part of the cheek-pieces, which, being held firmly in place by the head-band, keep the flange pointing towards the roof of the mouth. A cross-bar between two arms hanging loosely down from the ends of the mouthpiece can be pulled up under the chin towards the back by the second rein; this action tends to close the mouth and press the flange more into the roof of the mouth. The medieval and later curb-bits, though they had the same essential elements, operated in a different way, and failure to appreciate the difference has led to confusion. In these the mouthpiece with its flange, the lower side-pieces, and the cross-bar are all one rigid structure, pivoting on the upper cheek-pieces held in place by the head-piece. The second rein, attached near the ends of the lower cross-bar, could pull this back under the chin and at the same time turn the flange on the mouthpiece into the roof of the mouth (figure 514D, plate 36). Both types can still be found in use.

Curb-bits quite often appear in the equipment of Roman and post-Roman horsemen; in the Middle Ages they became so constant a part of knightly accoutrements as to be given elaborate ornament, sometimes heraldic. Yet use of these severe curb-bits shows how rigorous was the control which the knight was at times obliged to exercise over his horse, particularly in combat. The horses were of course of heavy and powerful build.

Celtic three-link snaffle-bits are often found to bear ornament on one side-link or side-ring only (figure 514B). Such asymmetric decoration implies paired draught, and in some regions, such as Ireland, may constitute almost our only surviving archaeological evidence for the use of wheeled vehicles.

These bits were of cast bronze, or iron coated with thin sheet bronze, or with bronze or tin by flashing on. All the rest of the bridle was normally composed of leather straps, with occasional metal rings, buckles, or strap-ends, frequently ornamented with enamelled metal or openwork phalerae (metal disks or bosses,

[1] Curb-bits often enough appear in sculptures and illustrations without a double rein. They are used nowadays with a single rein when a very sharp stopping control is needed, but some sensitivity of control is thereby sacrificed.

vol I, figure 93). Complete iron nose-bands are known from Roman contexts, presumably for use with curb-bits.

Hanging down from the yoke of the Egyptian Thebes chariot (vol I, figure 525) are two wooden Y-shaped pieces. These perhaps hung from the bridle-bit, for leading the draught-horses in procession. Pieces with this function, but of iron and more rectilinear design, are known from Celtic contexts, and are found in the early first century A.D. in Roman forts on the Germanic frontier. Cast bronze pieces shaped like a U with a handle hanging down from the centre are peculiar to Ireland, and, as with horse-bits, decoration down one side only shows that they were for leading a draught-pair of horses.

Horse-shoes. Shoeing (p 515) considerably increases the efficiency of the horse. Not only are the hooves protected, but they have a better grip on the ground, an advantage both in riding and in driving. Horseshoes (see p 515) are common enough on Roman sites from the first century B.C., and a mule's loss of its shoe is mentioned by Catullus (*c* 84–54 B.C.) [7]. The usual type was comparatively light, of iron, and had nail-holes stamped through from one side, resulting often in a wavy outline to the outer edge. The free ends were turned over to form a calkin, which, with the nail-heads when they projected as on medieval horse-shoes, gave a good grip on the ground. This type persisted well into the Middle Ages, but plainer, heavier shoes were also used from Roman times onwards, and these became the usual fashion in the later Middle Ages. In the Roman world the curious 'hipposandal' is also commonly found (figure 469). It is a smooth iron plate bent over to form loops at each end, and was evidently tied on to the horse's hoof, to which it undoubtedly gave protection: a few are known with a spike to give extra grip. Probably it was for use on cobbled or other rough road surfaces, as it persisted rather too long to be merely a transition from the esparto shoes (*soleae sparteae*) used on oxen to the shoe with a true grip. Shoes were by no means universally used, however, and numerous beasts must have gone unshod.

REFERENCES

[1] CAESAR *De bello gallico*, V, xv. (Loeb ed., p. 252, 1917.)
[2] *Idem Ibid.*, V, xix. (Loeb ed., p. 258, 1917.)
[3] HOMER Odyssey, X, lines 103–4. (Loeb ed., Vol. 1, p. 352, 1919.)
[4] HESIOD *Opera et dies*, line 425. (Loeb ed., p. 34, 1914.)
[5] XENOPHON On the Art of Horsemanship, XII, 10. (Loeb ed., *Scripta minora*, p. 360, 1925.)
[6] *Idem Ibid.*, X, 8–11. (Loeb ed., *Scripta minora*, pp. 348 ff., 1925.)
[7] CATULLUS XVII, line 26. (Loeb ed., p. 24, 1912.)

BIBLIOGRAPHY

ARENDT, W. W. "Sur l'apparition de l'étrier chez les Scythes." *Eurasia Septentrionalis Antiqua*, **9**, 206–8, 1934.

BERG, G. 'Sledges and Wheeled Vehicles.' Nordiska Museets Handlingar, no. 4. Stockholm, Copenhagen. 1935.

BIVAR, A. D. H. "The Stirrup and its Origins." *Oriental Art*, new series **1**, ii, 1955.

CHILDE, V. G. 'Prehistoric Migrations in Europe.' Kegan Paul, Trench, Trubner, London. 1950.

FORBES, R. J. 'Bibliographia Antiqua. Philosophia Naturalis', chapter 10. Nederlandsch Instituut voor het nabije Oosten, Leiden. 1950.

FOX, SIR CYRIL F. 'A Find of the Early Iron Age from Llyn Cerrig Bach, Anglesey.' Nat. Mus. Wales, Cardiff. 1946.

Idem. "Sleds, Carts and Waggons." *Antiquity*, **5**, 185–99, 1931.

GHIRSHMAN, R. "S. V. KISSELEV, Histoire de la Sibérie du Sud. Matériaux et recherches archéologiques en U.S.S.R." *Artibus Asiae*, **14**, 168–89, 1951.

JACOBSTHAL, P. 'Early Celtic Art' (2 vols). Clarendon Press, Oxford. 1944.

KLINDT-JENSEN, O. "Foreign Influences in Denmark's Early Iron Age." *Acta Archaeologica*, **20**, esp. 87–108, 1950.

LANE, R. H. "Waggons and their Ancestors." *Antiquity*, **9**, 140–50, 1935.

LEFEBVRE DES NOËTTES, R. J. E. C. 'L'attelage et le cheval de selle à travers les âges.' Picard, Paris. 1931.

LONDON MUSEUM. 'Medieval Catalogue', by J. B. WARD PERKINS. London Museum Catalogues, no. 7. London. 1940.

LORIMER, HILDA L. "The Country Cart of Ancient Greece." *J. Hell. Stud.*, **23**, 132–51, 1903.

Idem. 'Homer and the Monuments', esp. pp. 307–28, on chariots, and pp. 154, 490, 504, on horse-riding in Greece. Macmillan, London. 1950.

MINNS, SIR ELLIS H. 'The Scythians and Greeks.' University Press, Cambridge. 1913.

RIDGEWAY, SIR WILLIAM. 'The Origin and Influence of the Thoroughbred Horse.' University Press, Cambridge. 1893.

STURT, G. 'The Wheelwright's Shop.' University Press, Cambridge. 1923.

Mining-truck running on wooden rails (pp 548, 655).
After Sebastian Münster's Cosmographia universalis. *1550.*

16

SHIPBUILDING

T. C. LETHBRIDGE

THE sources of our knowledge of ancient shipbuilding are of three kinds; first, a very few examples of the remains of ancient vessels; secondly, written accounts; thirdly, contemporary illustrations.

The first class is naturally much the most important, but it is so small that there is a danger of building too large a structure on the evidence obtained from a single specimen. When a complete vessel is discovered and its qualities are obviously good, there is a tendency to regard it as the height of perfection everywhere in its day. The ships of the Vikings buried with the Norse chieftains are in this class. Much larger vessels had been built in the Mediterranean a thousand years earlier, and already embodied principles of construction unknown in the north but fundamental to all big wooden ships down to the present century.

Literary evidence depends largely upon the powers of observation of the writer. If he was not used to nautical matters it may be very misleading. In any case it must almost always be picked out sentence by sentence from works dealing with other subjects.

Pictorial evidence is often unreliable. An artist, unless he has been a seaman himself, seldom knows what he is trying to draw when depicting a ship. He does not understand the purpose of the gear he sees and so leaves out many things of importance and exaggerates others to suit his composition. He tends to shorten all drawings of the hulls of vessels and to accentuate all curves. We therefore see many pictures of ships looking like slices of melon, surmounted with a skewer adorned with a piece of notepaper and a few bits of string.

All these things have to be considered from every angle. The resulting picture will of necessity vary with the outlook of the historian and can seldom be accepted as more than a personal opinion. With these warnings, we turn to Homer.

The days have passed in which critics wasted their lives trying to reduce the 'Iliad' and 'Odyssey' into the work of fifteen or twenty different poets. It is now assumed that these great poems give a reasonable picture of life in one cultural period which, archaeologically, was the Late Bronze Age. It is clear that in the society described by Homer there was already a sharp distinction between the ship-of-war and the merchantman (figure 515). The difference is so sharp that it was evidently the product of many centuries of evolution.

The method of building the Homeric ships was almost the same as that employed for most small wooden ships today. The vessels had keels, stem- and stern-posts, and ribs (frames or timbers) covered with outside planking. The construction was fastened together with wooden pegs (tree-nails), a method which has only recently been superseded by metal fastenings. These ships were carvel-built, that is, their planking met edge to edge and did not overlap as in the northern clinker-built ships of later times. It is interesting to note that the keel, the most important element of longitudinal strength, was already fully developed. It was many centuries before the shipwrights of the north employed

FIGURE 515—*Greek sailing merchant-ship* (left) *and a war-galley. From a black-figured kylix, sixth century B.C. The galley is apparently a bireme, the lower oars working through round ports.*

the keel. In ancient Egypt it was apparently unknown. Ancient Egyptian wooden vessels were simply composite copies of early reed boats, with no elaborate constructional plan (vol I, pp 734–6). Modern examples may still be seen on the Nile.

The Homeric warship was a long narrow vessel, light enough to be hauled up on the beach by her crew. In this respect she resembled a Norse longship, whose crews could normally carry their vessels up from the water. Homer gives these vessels a crew of fifty men, which implies that their length must have been in the neighbourhood of 125 ft. They had a small forecastle raised above their ram-shaped bow, a small poop with a high curving stern-post, and a raised gangway running the whole length of the ship. There are representations of vessels of this kind on vases painted in the so-called Greek geometric style, which dates from the beginning of the Iron Age (figure 516). Their construction was undoubtedly light. They were used only in summer, and in later warfare we often hear of such a vessel opening out when it rammed another.

These Homeric ships were fitted with a mast, which apparently was lowered by a forestay and had its foot in some kind of tabernacle. The square-sail set on this mast was apparently used only for running before the wind, and the high

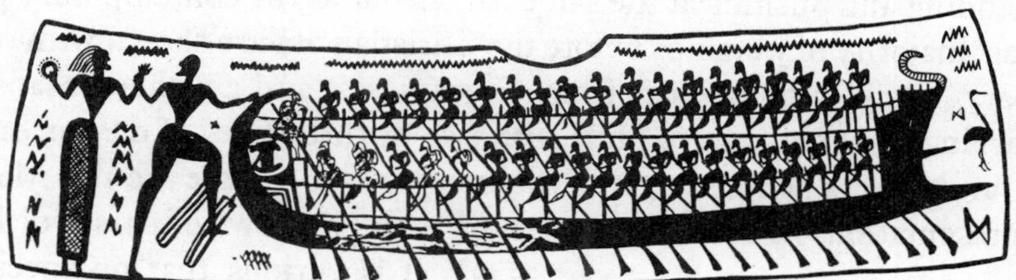

FIGURE 516—*A 'Homeric' war-galley; Paris is leading Helen aboard ship. From a geometric vase, eighth century* B.C. *Note the two steering-oars. The right-hand portion of the ship clearly shows that the painter intended to show one row of oarsmen on each side.*

poop was designed as a protection against following seas. The vessels were steered by a rudder hung on each quarter. They were very low in the waist, and the oarsmen rowed with the oars resting on a raised rail. This may have been soon replaced by a built-up side, for the fine paintings on many vases of about 500 B.C. show vessels with completely built-up sides and the oars operating through rowing-ports (figure 517).

There has been much controversy on the methods of rowing these early warships. Such terms as bireme, trireme, and quinquereme are taken by some authorities to indicate that the vessels in question were rowed by oars operated at two, three, and five different heights from the water. Another school of thought believes that these terms refer to the number of men employed on each of a single row of oars, in the same way that the modern term double-banked oars implies two men to each oar. The second school can point to the latest war-galleys of western Europe, with five men to each oar, and to the still existing Portuguese *saveiro*, where the number is nine or even ten. The ratio of one man to one oar becomes almost incredible for the quinquereme, in which the length and weight of oars would seem beyond the power of one man.

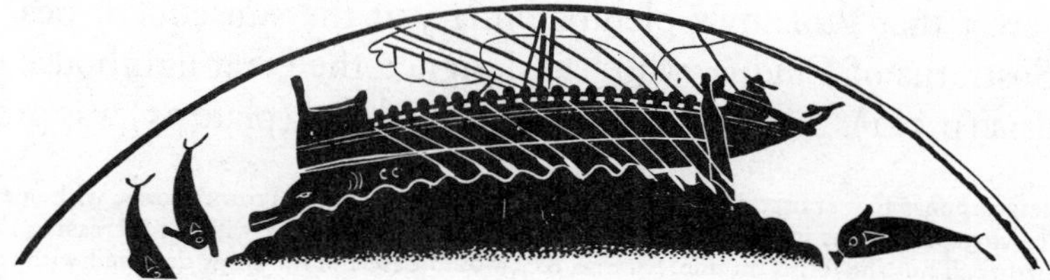

FIGURE 517—*Greek galley of later type, with raised sides. From a black-figured kylix, sixth century* B.C.

In the Homeric vessels there seems to have been only one man to each oar. In later times the literature certainly implies men at different levels, but this can be taken in two ways. For instance one man may be sitting and pulling while another can be standing and pushing at the same oar. As far as the contemporary pictures go, the vast majority of warships before the Christian era are shown with only one line of oars. Yet it is quite clear from Greek vases, and a rather unsatisfactory Greek carving, that ships were built with two or three lines of oars on each side, the ports and rowers being staggered (figures 515, 518); there are also Mesopotamian representations of sea-going warships with two rows of oars (vol I, figure 543). It is curious that no special term seems to have been used to describe the far more commonly represented ships with a single tier of oars.

FIGURE 518—*Greek trireme, from a carving at the acropolis, Athens. The three rows of oars appear to work upon thole-pins or in ports in the three thick horizontal bands. The uppermost of these bands probably represents an outrigger projecting from the side of the vessel, and the second band (carrying the middle row of oars) the edge of the hull. The third and lowest row of oars is set low down in the hull.*

By about 250 B.C. large fleets of war-galleys were possessed by several eastern Mediterranean powers. Most of them were quinqueremes. However, the war between Antigonus II of Macedon and Ptolemy II of Egypt, which culminated in the battle of Cos (*c* 258 B.C.), introduces us to a new nautical term and perhaps to an entirely different type of vessel. This was the cataphract, which seems to have originated at Corinth. It is said that the great cataphract of Antigonus had in her construction enough timber to build fifteen quadriremes, and that her rowing-power was as nine to four. From this we may conjecture that, whereas a quadrireme had four men to an oar, this cataphract had nine.[1] The term cataphract was used not only for vessels but for heavy mailed cavalry; it refers to the armour. It is thought that cataphracts were protected from ramming by the thickness of their side-planking, and that there were decks above the rowers for their protection, and for the use of the boarding-parties on whom the chief reliance was placed.

The victory of Antigonus at Cos brought the cataphract into great prominence. It is of interest that Ptolemy's plenipotentiary at the subsequent peace conference was Sostratus of Cnidus, who had designed the great lighthouse or pharos at Alexandria (p 521). The famous 'Winged Victory' (plate 35) was probably set

[1] Some writers suppose that at first the Greek ship developed three (or five) rows of oars, with one man to each, but that later rowing-power was increased by applying more men to each oar without increasing the number of rows of oars further. Thus the terms bireme, trireme, &c., would describe the ships depicted with more than one row of oars and one man to each; the larger terms (15-reme, 40-reme, &c.) similar ships having five men to each oar in three banks, or eight men to each oar in five banks. But the evidence for the existence of such vessels is doubtful.

up by Antigonus as a memorial of the battle of Cos. She is standing on a model of the stem of a cataphract, or *corinthia* as the type was also called. Similar victories appear on coins of an earlier date (figure 519).

By the middle of the third century B.C. there had developed a sharp division into two types of war-vessels, which remained with little change for centuries: the light galley and the heavy galleon.

The Homeric description of the preparations of Odysseus for his escape from Calypso's isle, gives what must be an eye-witness's account of the method of construction of a broad-bottomed trading-vessel. We learn that the whole construction was pegged together with tree-nails, which passed through the planking into the numerous internal ribs. The vessel was partly decked, and so low in the waist that it needed hurdle-work backed with brushwood to serve instead of the waist-cloths still rigged in small Levantine traders. It was fitted with a single quarter-rudder and set a sail, presumably square, with a yard, halyards, braces, and sheets. There is no mention of standing rigging. However, the vessel was a sailing-ship and not a pulling-boat. From this time onward the merchantman relied on sail and the warship on oars.

FIGURE 519—*'Winged Victory' mounted on the prow of a galley. From a coin probably issued in celebration of the naval victory of Demetrius I of Macedonia over Ptolemy I at (Cyprian) Salamis, 306 B.C. The projection of the outrigger below the plinth carrying the figure is clearly shown, as it is on the (incomplete) statue in the Louvre (plate 35).*

This was an economic necessity, for the merchant could not pay the wages of large crews. The great developments of sail were due to the ingenuity of merchant skippers. Warships, with relatively huge crews, could neglect labour-saving devices. This can be seen in the history of our own naval development, where huge topsails and 'man-killing' jibs remained in use in the Royal Navy long after they had been split up into smaller units by the traders.

Few reliable descriptions or pictures remain of the early merchantmen, but there is one fine drawing of a Greek trader on a vase of about 500 B.C. (figure 515, left). It shows a well designed hull, with a bow which in later years was to be termed a fiddle or clipper bow. The ship, with a single mast and square-sail, its only source of power, appears to be of no great size. The mast is supported by a single shroud on either side. The hurdle-work in the waist is shown.

From very early times merchant vessels sailed the whole length of the Mediterranean. Biblical evidence, dating from about 400 B.C., describes voyages from Palestine to Tarshish (*Tartessus*)[1] in south-western Spain in the time of Solomon (*c* 950 B.C.). Although the evidence is not contemporary, such biblical traditions have often proved correct. Early Phoenician ivories, carved in Palestine about the tenth century B.C. (vol I, p 672), have been found in Spain, which they presumably reached by sea through the 'ships of Tarshish', to use the biblical term. Irish gold-work has also been found in Spain. It is clear that the trade of Tartessus was most important. It is probable that its people were not only in constant touch with the eastern Mediterranean and the British Isles, but that to them was due the discovery of Madeira and possibly the Azores. Much would be learnt were the site of Tartessus located and excavated.

Tartessus was in touch with the Greek world, and Herodotus states that Colaeus of Samos brought a rich cargo thence about 650 B.C. Carthage, a colonial foundation of Phoenician Tyre (814 B.C.) furnished the greatest merchant traders of the old world. They not only destroyed Tartessus (about 500 B.C.), but closed the Straits of Gibraltar to Greek and Etruscan shipping. Their shipbuilding must have been of a relatively high order, for their coins have been found at Corvo in the Azores in mid-Atlantic. They traded far up the western European seaboard. Little is left, however, to give an idea of their vessels. Some of their coins seem to show war-vessels of the Greek type, but of their merchantmen we have hardly a trace—as serious a gap in knowledge as that of the 'ships of Tarshish'. Yet before 500 B.C. Hanno the Carthaginian sailed so far down the west coast of Africa that he was able to return with heads of gorillas set on his stems. At about the same time, Himilco voyaged to the tin lands in the north, presumably Britain and Ireland.

It must be remembered that, despite our great mass of information about the ancient world, this is nevertheless only a minute fragment of what was once known. Where the account of one voyage survives today, hundreds of such voyages had been made; where one foreign object is found in an excavation, very many similar things once existed in that same area. Unless a people as a whole is given to representing its ships in pictures, it is most unlikely that representations of them will survive. Many Irishmen made prodigious voyages in the early centuries of the Christian era, yet there is no single illustration of what their ships were like. Nothing survives but some brief descriptions of what they took with them and how many men some vessels contained. The ruins of Tartessus must conceal many secrets, but there are very few Carthaginian representations

[1] *Tarshish* is a Phoenician word meaning smelting-works. Vessels engaged in transporting metals or ores were known as *tarshish*-ships. Tartessus was presumably the site of a particularly important smeltery.

and none have come to light in the Greek trading-colonies in southern France and Spain.

It is thus a matter of great interest that the wreck of a Greek merchantman was located in 1954 off the island of Grand Congloue near Marseilles. First reports, furnished by divers, stated that the vessel was 100 ft long and sheathed with lead. A cargo of wine-jars and platters was raised (figure 100). Trade marks on the wine-jar stoppers are thought to indicate that the ship was of about 250 B.C. and carried a cargo of wine and pottery from Rhodes. The lead sheathing is of great interest. It shows that the Greek traders had already mastered the scourge of the shipworm (*Teredo navalis*, L), a boring bivalve mollusc which often made English and Dutch eighteenth-century ships voyaging to the Mediterranean unseaworthy after a very few years.

It has been long known that lead sheathing was in use in early Roman times, for this method of protecting the hull was observed on Caligula's vessels in Lake Nemi (p 572). However, its use as early as 250 B.C. shows the high standard that wooden ship-construction had already reached (figure 522).

The skill involved was no new thing. At least as early as 2000 B.C. men were passing by sea from the east end of the Mediterranean to Spain, and from Spain to the north of Scotland. About 300 B.C., owing to some relaxation of the Carthaginian control of the Straits of Gibraltar, a Greek from Marseilles, Pytheas, could voyage thence to Iceland and back. He returned with the first reliable geographical reports of Britain that have survived. A vessel such as the merchantman wrecked off the Grand Congloue was perfectly fit for a passage from Britain to Iceland, which Irish monks often undertook in later centuries in skin boats. With such vessels the Carthaginians or Greeks could reach not only the Azores but the Americas. It was recognized that the world was spherical, and they may well have tried to sail round it. The idea that ancient seamen never ventured out of sight of land is false. In 1500 B.C. men were sailing direct between Crete and Egypt.

The war that broke out between Rome and Carthage was to have a remarkable effect on maritime affairs, for it led the Romans, an essentially agricultural people, to become masters of the Mediterranean and all the western seas. The Byzantine historian Procopius (d *c* 565) relates that it was a Carthaginian warship, blown ashore on the coast of Italy, which served as a model for the future Roman fleet. Judging by the Roman tactics in subsequent naval actions, this vessel must have been a cataphract.

The Romans did not waste time on evolving elaborate manœuvres with light ships but relied entirely on boarding. For this purpose they evolved a hinged gangway, with a spike at its outer end. This gangway was kept raised till the enemy ship was alongside. It was then let go with a run. The spike pierced the enemy's deck and not only grappled the two vessels together but provided a boarding-bridge for the Roman marines. Although the Romans were very successful in their naval battles, it was a long time before their sea-captains mastered the mariner's art. Frequent disasters on a large scale occurred when their fleets were caught on lee shores and blown aground.

A very fine carving at Rome (figure 520), shows the bow of a Roman warship at a slightly later period. The oars are in two tiers—these are best not called banks, a term liable to lead to confusion. The rowing-ports are made watertight by leather collars on the looms of the oars, secured to the ship's side. The vessel is fitted with a strengthened ram and has a high, curving stem. A heavy, carved baulk of timber is used as a cathead for the anchor. A raised turret on the forecastle head serves as a bridge for conning the ship and for the look-out man. A small mast is stepped like a bowsprit right forward. On this a sail, like the spritsail of Elizabethan times, was set to pull the ship out of action if too many of the oars were disabled. The entry-port is beside the cathead, and aft of the entry-port a raised and projecting gangway runs the whole length of the ship's side, above the oars, to accommodate the marines in action. There seems little doubt that this vessel, usually spoken of as a bireme, is a cataphract. It is possible that both terms could be applied to it.

FIGURE 520—*Roman warship. From a carving in the Temple of Fortune at Praeneste, c 30 B.C. The lower row of oars emerges from the side of the ship below the outrigger; the upper row from the lower part of the outrigger. It is doubtful whether the carving on the upper part of the outrigger, on which the marines stand, represents ports for a third row of oars or not.*

Among the wall-paintings at Pompeii (destroyed A.D. 79) are several good pictures of war-galleys (figure 521). These are all shown with a single tier of oars. The two quarter-rudders project through ports in the ends of the fighting-gangways mentioned above. The sterns, if these gangways were absent, would be somewhat like those of Norse longships at a later date. The vessels are, however, fitted with ram bows, which were rare in the north. Only one mast is shown. This is stepped rather forward of the centre, with its head raked sharply forward. It has a fore-stay and a square-sail controlled by braces. The officers and perhaps the helmsmen were protected from sun and weather by an arched canopy over the poop, from which no doubt the canvas or other covering was stripped before action.

This general form of war-galley had not changed very greatly fifteen hundred years later. When galleys finally went out of use at the close of the eighteenth century, the following general changes might have been noted by a Georgian seaman who had read his classics. The two quarter-rudders had given place to a single rudder hung from pintles on the stern-post; the fulcrum of the oars had become fixed on the rail of what had once been the marines' gangway; the ram had been raised clear of the water, forming a pronounced beak; and lastly the single square-sail mast had given place to two or more masts setting triangular

FIGURE 521—*Pictures of war-galleys from Pompeii.* (Above) *with oars only;* (below) *with single mast and square-sail.*

lateen sails. The rudder-development had long been common to all vessels. The boarding-force of marines had been replaced by two 24-pounder guns in the ship's forecastle, and the gangway could thus be used as an outrigger for oars of greater length. The general idea, however, had survived. These last galleys were not the children of the heavier, protected cataphract, but of the lighter, swifter bireme or trireme of the Greeks. From the known fact that some of them had five men pulling each oar, we may perhaps be justified in speaking of them as quinqueremes.

The heavy vessel of the cataphract type was not always the victor in naval actions. The battle of Actium in 31 B.C. is said to have been won by the light galleys of Octavian (who later became Emperor as Augustus), which over-

FIGURE 522—*Roman pleasure-galley from Lake Nemi, showing the lead sheathing and supports for the outriggers. c A.D. 30.*

whelmed the heavy ships of Mark Antony. We have little idea of the size to which the largest ships of the cataphract type grew, but that they were not much, if any, smaller than ships of the line of the Napoleonic era seems probable from the pleasure-ships of Caligula (A.D. 37–41) found in Lake Nemi. That emperor spent vast sums on building such craft. He used to be rowed along the Italian coast in a vessel with ten banks of oars. It is impossible to say how so many banks (if that is the sense of the phrase) could be arranged in tiers without compromising stability, but the ships certainly must have been of great size.

That remains of large wooden ships exist in Lake Nemi has been known since the Middle Ages. In 1905 divers prepared plans of two of the vessels. The larger was supposed to be about 450 ft long, 192 ft in beam, and with an overall depth of 51 ft. When Mussolini had Lake Nemi drained so that the ships could be examined, little remained in position above the floor-timbers of the hulls. The main features of the construction were, however, clear (figure 522).

These ships of Nemi were modelled on sea-going types, though built for use on a small lake. They were strongly constructed of carvel-build, with larch-planking pegged with tree-nails. Outside the planking was a thick padding of cloth smeared with pitch, and outside the cloth was a sheathing of lead fastened to the planking with bronze studs. We are not concerned here with the ostentation displayed in their internal fittings, their heavy mosaic pavements laid on decks, their statues set up in gardens with ponds fed through lead piping, and so on. The important fact for our purpose is that shipbuilding had reached so high a level by A.D. 30–40 that vessels could be built larger than any wooden ship of the line at the time of the Crimean War.[1] H.M.S.

FIGURE 523—*Roman merchant ship, from a carving on a tomb at Pompeii. Sailors are handling the square mainsail. The fore part of the ship is imperfect but the raked mast for the small foresail is visible.*

[1] Ships of the line, however, were intended to bear the stress of storm at sea, and were not built for the smooth waters of a lake.

Duke of Wellington (131 guns), launched in 1852, was about 240 ft on the gun-deck and 202 ft on the keel, with a beam of 60 ft and a depth of 24 ft. Her tonnage was 3771. Making due allowance for Caligula's lunatic extravagance, it is fair to say that the cataphracts of Mark Antony at the battle of Actium must have been comparable in size, and to some extent in under-water body, with Nelson's ships at Trafalgar. But they did not need the immensely strong

FIGURE 524—*Roman merchant ship in port. From a carving found at Ostia, showing the quarter-rudder, raked foremast, rigging, and triangular topsail above the yard. Note lighthouse in background.* c A.D. 200.

construction of eighteenth-century warships. The catapults on the decks of Mark Antony's ships, on which he is said to have relied for victory, could never have driven missiles with anything approaching the force of gunpowder.

It is probable that the building of great vessels reached its peak in the days of Caligula. No enemy remained at sea to challenge Rome on equal terms. When Saxon pirates began to infest the northern seas, no cataphract was required to deal with them. The need was then for light, fast cruisers and patrol-craft. So the first age of the great ship came to an end; not through any failure in the shipwright's art, but simply because there was no necessity for going on with it.

The merchant vessel has an entirely different story. Rome required a constant stream of sea-borne imports, especially corn. Piracy had been suppressed in the Mediterranean, where the only danger to traders was the stress of weather. From all corners of the known world—Spain and Tunisia, Britain and northern Gaul,

Egypt and Asia Minor, India and Ceylon, the Baltic and Black Seas—metals, textiles, pelts, precious stones, ivory, wine, oil, and above all grain flowed into Italy. Most of this vast trade was carried on by sea. It would be absurd to think that the shipping of the time was not fit to carry it. Where the shipping was handicapped was not in the construction of the vessels themselves, but in the lack of a suitable rig to enable them to use for their purpose winds of any direction.

It is almost certainly incorrect to think that ships of this Roman age could do nothing more than run before a following wind, but it is clear that they could never beat to windward in anything but the most limited sense. Most vessels therefore waited for favourable winds before starting on their voyage, in the same way that traders in the Indian Ocean used to wait for the monsoon.

There are several surviving carvings on stone that give a good idea of the typical Roman merchant vessels of the first and second centuries of our era (figures 523, 524). Remembering that almost every artist makes his ships appear too short and the line of the sheer (the curve from bow to stern) too great, we yet cannot fail to get a very favourable impression from these ancient carvings. The hulls are round to the extent that most modern sailing fishing-vessels were round. The beam probably went about three-and-a-half or four times into the length. The sterns are beautifully designed for lifting to a following sea. The two quarter-rudders are still there, and the single stern-post rudder, difficult to control, has not yet interfered with the old Greek ideas of construction. The hull is very strongly built, with three or more great wales to protect the planking from chafing against the quayside, and to add rigidity to the ribs when pitching in a seaway. The bow has a beak above the waterline for the convenience of the men handling the foresail. There are catheads for the anchors, deck-houses, and a stern gallery with a shrine at the outer end for the vessel's tutelary deity. There are a great mainmast, stepped amidships but raked forward by a stout forestay, and a foremast, even more sharply raked, in the bows. A square-sail is set on both these masts and is controlled by braces in the modern manner. A triangular top-sail, or perhaps a pair of top-sails, one on each yard-arm, is set above the square-sail on the mainmast. The mainmast, which in at least one carving appears to have been so large that it was built of several spars scarfed together, is supported by several shrouds. These shrouds are set up with lanyards and dead-eyes, but instead of being made fast to chain-plates bolted to the ribs, as in recent vessels, the dead-eyes are secured to ring-bolts bolted through one of the great lateral wales. The method is, however, essentially modern. A very slight alteration in the rake of the masts, coupled with the universal change from square to fore-and-aft rig, gives one the modern Italian *navicello*. From this it is but a step to the favourite yacht-rig of America, the two-masted stay-sail schooner. The Roman vessels look low in the bow and high in the stern compared with these modern types, but this is because they were designed for running before the wind rather than for punching into a head sea when beating against it. More than anything else this shows the limitations of the Roman traders. The fault persisted far into the modern age and may be observed in most pictures of Elizabethan vessels.

This was not the only type of merchant ship. Many mosaics from the Mediterranean region illustrate sailing-vessels and even pulling-boats built with a ram bow, in some cases a projection of the keel beyond the foot of the stem-post. This would obviously protect that post if the ship were involved in a collision with another vessel. It is reasonable to suppose that this form of bow is of very great antiquity, evolved when early boats were being constructed with a keel—which was in effect the original dug-out hull from which all carvel-built vessels are descended. The stem- and stern-posts (to which the first strakes—side-planks—were secured at their ends) could be placed either inside the dug-out body, or at its extremities. From the type of boat with an internal stem-post the vessel with a ram was derived; from the other type all recent carvel-built vessels are derived.

FIGURE 525—*Bronze model of the hull of a merchant sailing-ship, from ancient Beth-Marē, Litani valley, Lebanon.* A.D. *121–2. Scale* c *1/7.*

It is of the greatest interest, therefore, that a very fine model of a merchant vessel of the ram type, apparently to scale, has recently been discovered in Syria (figure 525). It is of bronze, 15 in long, and had been adapted for use as a lamp. The type of ship represented by the model is believed to date from the time of Hadrian (A.D. 117–38). Its shape is most remarkable, amazingly round by modern standards, so that its length is but two-and-a-half times its beam. Many of the constructional features of this type of vessel have already been noted: the two great strengthening wales on the outside of the hull; the pair of quarter-rudders; the two masts; and so on. The original of the model, however, seems to have been pierced for rowing-ports and also to have had a much higher bulwark forward than did ships without the ram. The weight of the ram may have caused the vessel to pitch more heavily than the other type. The survival of this feature, long after the seas had been swept clear of pirates by the Roman navy, can be attributed only to the conservatism of shipbuilders, who were unwilling to fit the stem-post in any other way than that to which they were accustomed. Some vessels in the north, beyond the reach of Rome, retained a vestigial ram even longer.

In later Roman mosaics ships can be seen without beaks and without galleries in the stern. Both stem- and stern-posts are carried up high. Except for the more pronounced forward rake of the stem-post one might think that the artist had been attempting to show a vessel of the later Viking type.

The only discovery recorded of a Roman merchant ship was in the mud beside Westminster Bridge, London, in 1910. Little remained but the floor-timbers and part of the planking (figure 526). From coins wedged between members of the hull, it was thought that the ship was of *c* 280–300. There was too little remaining to yield a clear picture, but the ship was more than 60 ft long and was carvel-built, being pegged together with tree-nails. It was not sheathed with lead. The wales along the sides were still to be seen. These, unlike the rubbing-strakes of modern fishermen, were pegged directly on to the ribs between the strakes of planking.

FIGURE 526—*Remains of a Roman ship found in the Thames near Westminster Bridge. The ribs are 5 in wide and placed 10 to 20 in apart. The fragment is 38 ft long and 18 ft wide; it probably represents a sailing sea-going vessel.* c *A.D. 300.*

Other Roman vessels must have been found from time to time, but not recognized owing to the close resemblance between them and more recent types.

The river-systems of the ancient world were of great importance in providing main routes of communication, for water-transport (even with Roman roads) was always cheaper than transport by cart or pack-animal. The oldest river-navigations are those of the Nile and the Euphrates, but as early as *c* 500 B.C. the French rivers were extensively used for the carrying of heavy goods across country from Marseilles. Strabo describes the wine-barges on the Douro, and the *barco rabelo* engaged in this trade today preserves some very ancient features. Not only does it set a square-sail which might have been patterned on an ancient Greek picture, but it is steered by a sweep over the stern in the manner of some early Egyptian vessels.

Inland barge-traffic was of considerable economic significance under the Roman Empire. Pictures of the boats are found on monuments erected for members of the powerful bargemen's guild in Gaul. Some depict boats with several oars a side and with the ram bow and curving stern of sea-going ships. Others were apparently double-ended, with high curved stem- and stern-posts (figures 102, 103).

The barge type seems to have been more or less standardized all over the ancient world. Even today there is little distinction between the old barge types of England and Italy. It is not to be supposed that the apparent clinker-build of our English barges indicates a Viking origin for this kind of vessel. Although they appear to be clinker-built, they are really pegged together with tree-nails. Their general form, with curved stems and sterns, and their paintwork indicate their southern origin.

The characteristic river-barge is built without a keel. In place of it there is a wide and heavy plank, which with five or more similar planks forms the flat floor. The floor-timbers, or ribs, are very heavy and pegged to large knees, which in turn are fastened to a heavy stringer (that is, an inside strake of planking) above water-level. It is this stringer that gives the barge its longitudinal strength. A vessel presumed to be an Anglo-Saxon barge was found in 1900 at Walthamstow, Essex. It had the flat floor of a barge, with a length of about 45 ft and a beam of 7 ft. The planking, clinch[1]-fastened, must be regarded as of a hybrid type. The boat had been pulled out of the water and placed bottom-up over the body of a Viking of about the ninth century; the boat itself was English.

Later, some barges tended to develop into sea-going vessels, though long retaining their original single mast and square-sail. In the end, however, most of them, except the Humber keel, adopted the sprit-sail rig. Their highest development was undoubtedly the Thames barge, but this was not unique. The Humber keel was as well known, though more primitive in appearance. The Severn trows, like the Thames barge, adopted a form of transom in the stern. All, however, were derived from the same general type of flat-bottomed river-barge (as were the granite-barges of Falmouth and the stone-barges of Caen), which was in all probability very ancient even when the Romans employed it on the rivers of Gaul and Britain.

With the shifting of the centre of power from Rome to Byzantium, a most serious gap appears in our knowledge of Mediterranean shipping, which has given an erroneous impression of the importance of northern vessels in the general trend of the shipwright's art. With the exception of the building of great warships, it is improbable that any of the constructional skill of the Mediterranean shipbuilders was ever lost. Unfortunately the Byzantine Greeks were not

[1] To clinch a nail is to fasten it by hammering the point down over a rove or washer.

marine artists and the information locked up in their contemporary documents has not been completely examined.

It is clear that for many centuries the Byzantine navy was kept at a high pitch of efficiency, and that free, paid oarsmen took the place of the slaves of the Roman period. It is not so clear what the vessels were, although many were evidently of the lighter galley type. They are completely overshadowed today by the actual survivals of northern war-vessels.

Northern shipbuilding diverged at an early period from that of the Mediterranean. It is probable that this divergence took place many centuries before the time of Homer, and that early types which passed up the great eastern European waterways became established in the Baltic area for hundreds of years. From these primitive northern vessels the clinker-build was evolved. Not until those northern boat-builders had been in touch with the work of southern shipwrights in Gaul, in Britain, or even in the Baltic itself, did they evolve a seaworthy type of war-vessel of their own. All their early designs lacked that essential of sound construction, the keel.

Two pagan customs, surviving in the north long after they had been abandoned in southern lands, have provided us with actual examples of the northern ships ranging over a period of perhaps a thousand years. These customs were, first, the practice of devoting the spoils of war to some water-god, and secondly, the burial of a man or woman in his or her vessel.

Of the first type only a few examples have been found. The second rite was common to many of the northern lands. Examples of it have been found, for instance, in Sweden, Norway, Scotland, England, and northern France. The votive ships discovered are relatively earlier than those of ship-burials. The Hjortspring boat found near Als, Jutland, is little more than a composite dug-out canoe; it is composed of five adzed planks (strakes), which were fastened to the ribs (timbers) by lashings passing through lugs or cleats left by the adze-work on the inner surface of each plank. This extremely primitive round-bottomed boat shows no trace of contact with southern models (figure 527 A).

By the fourth century A.D. a great change had come over northern shipbuilding. In the preceding five or six hundred years, contact had evidently been made with southern vessels. Two vessels were found in a bog at Nydam in Schleswig, one nearly complete. Here the primitive methods employed in the Hjortspring boat are still used, but the ends of the vessels are swept up to resemble southern craft (figure 527 B). There is as yet no true keel, but the bottom plank has been reduced in width and thickened, and ten planks are now employed instead of five. There is no mast. This vessel is of true clinker-build, for the edges of the overlapping planks are fastened together with clinched iron rivets and not sewn as in earlier craft. The ship was rowed by fifteen pairs of oars, fastened by means of raw-hide grommets to horned wooden tholes on the gunwale, and steered by a single quarter-rudder. Thwarts for the oarsmen were fixed just below the gunwale.

Although this vessel is a great advance on that from Hjortspring, it is still extremely primitive, and the Anglo-Saxon invaders of Britain at this time could not have hoped to compete with the well constructed galleys of Rome. The location of the Roman defence forts strongly suggests that these pirates never crossed

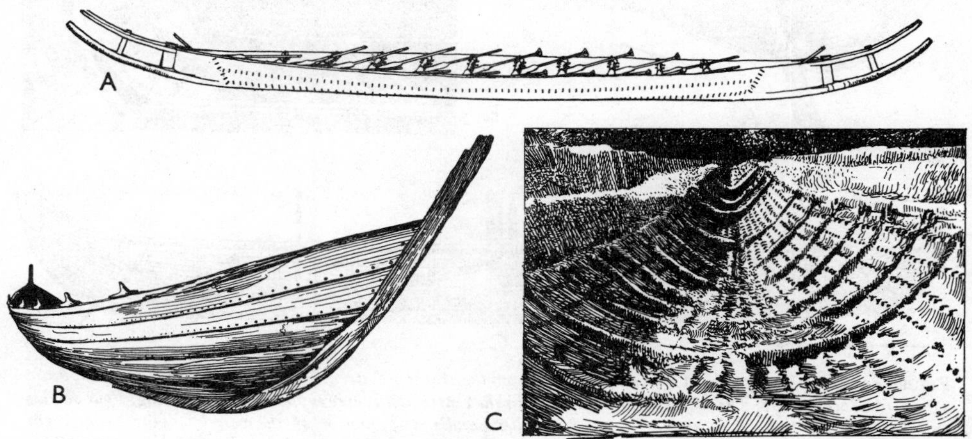

FIGURE 527—(A) *Model of the early Viking boat found at Hjortspring, south Jutland. It accommodated twenty oarsmen seated in a double row. Early Iron Age.* (B) *Viking vessel found at Nydam: late fourth century A.D. Length 76 ft over all, beam 11 ft.* (C) *Interior view of the stern of the Sutton Hoo ship, excavated in the ground. The form of ribs and planking is shown by 'replacement soil' in which the iron nails remain while the timber has perished. The vessel is an open rowing-boat, about 80 ft long, 14 ft wide, and 4·5 ft deep. The strakes are joined together, and are not single timbers as in the Nydam ship.* c *A.D. 660.*

the North Sea, but hugged the coast from Germany almost to the Straits of Dover before attempting the crossing. Except for its larger size, the galley found in the Anglo-Saxon burial mound at Sutton Hoo is much the same as the Nydam vessel (figure 527 C).

The pirate galleys were so long, narrow, and longitudinally weak that much had to be improved before the seaworthy longships of the Norsemen could pass at will from Norway to Ireland, and from Ireland to Spain. More strength was needed in beam and longitudinally. When this had been obtained by changing the rake of the stem- and stern-posts, it was further necessary to build up the sides. A true keel was also fitted. The Gokstad (Norway) ship (figure 528), which may date from about the beginning of the ninth century, was preserved almost complete in a burial mound believed to be that of the local king Olaf Geirstad-Alv, who died soon after 800. It set a square-sail on a single mast amidships, had rowing-ports cut in the heightened sides, and was pulled by sixteen oars a side.

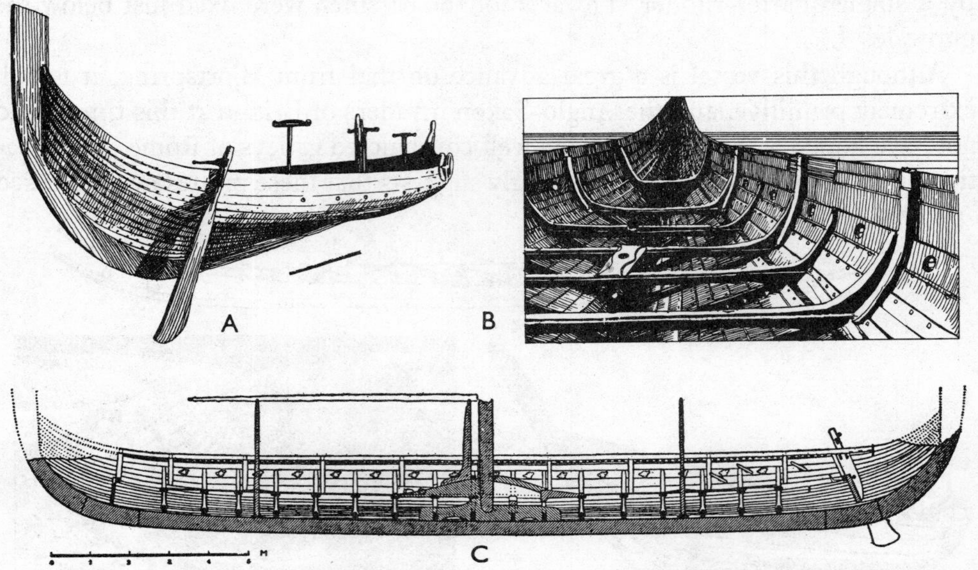

FIGURE 528—(A) *The restored Gokstad ship, seen from the starboard stern.* c *800. Length 78 ft over all, beam 16·75 ft, depth 6·75 ft. The stump of the mast rises in the centre, with three supports for an awning. Four shields still hang on the bows. The steering-oar has a tiller handle.* (B) *Interior of the vessel looking towards the stern, showing the step for an awning-support. A boat similar to this was sailed across the Atlantic in 1892.* (C) *Longitudinal section of the Gokstad vessel, showing the method of stepping the mast.*

It was nearly 80 ft long over all and more than 16 ft in beam. Its displacement was of the order of 20 tons.

Some archaeologists, however, have suggested that the Gokstad ship may be nearly two hundred years younger than the date suggested above. Should this suggestion be generally accepted, it would be necessary to conclude that the early Norse pirate fleets were made up of vessels whose development was far less advanced than has hitherto been believed. There are records of improvements in warship-design in the time of King Alfred (871–99). Thus the Anglo-Saxon Chronicle relates (A.D. 897):

> Then King Alfred commanded long-ships to be built to oppose the esks [pirate galleys]; they were full-nigh twice as long as the others; some had sixty oars, and some had more; they were both swifter and steadier, and also higher than the others. They were shaped neither like the Frisian nor the Danish, but so as it seemed to him they would be most efficient.

This one quotation is enough to show how great are the gaps that exist in our knowledge of shipbuilding. No one has any idea what kind of warships were built by the Frisians or the Danes. It is known only that the chieftains of Norway were buried in vessels of a certain type, some of which are still preserved, and it is only assumed that they were the finest examples of the craft of their day. Alfred, with his extensive study of the Roman plans of defence against the Saxons, was perhaps responsible for the reintroduction of some of the Roman knowledge of shipbuilding. His galleys approached the limit of size for an open, clinker-built wooden ship. It may be, however, that he was not bound by any northern design. So much writing was deliberately destroyed between Alfred's time and ours that detailed accounts of Roman shipbuilding available to him may have been lost.

This was only the first stage in the great age of Norse shipbuilding. The ships were to grow in size until they had at least thirty and perhaps sixty oars a side. They were light, buoyant vessels, but it is a mistake to think that they compared in seaworthiness with the traders of the time. In the Sagas (twelfth and thirteenth centuries), for example, we are told that it was not thought safe to go to the Faeroe Islands in a longship, a voyage to be undertaken only in a merchant vessel.

It was probably due to the Carolingian Franks that many new ideas came to the north; either from Gaul or from Britain the Vikings learnt new ideas of ship-construction and of navigation. It is, however, possible that this knowledge was obtained in Spain, either from the Moorish invaders or from the Spaniards themselves, who kept an ambassador at the Norse court in Dublin. Wherever it came from, it gave the northern peoples a freedom of movement over wide areas of open ocean, which had apparently been enjoyed by Irish navigators shortly before the Viking age. Parts of France, England, Ireland, and Scotland were colonized, together with Iceland and Greenland. America was accidentally discovered and partially explored. Were it not for the Norse custom of handing down stories of the lives of some of their prominent men, and their habit of burying their chieftains in ships, we should probably know no more about their widespread voyages than we do of those of their Irish predecessors. It seems probable that the Irish reached at least as far as Iceland, Greenland, Newfoundland, and the Azores, whither they voyaged in large, skin-covered boats holding twenty to thirty men apiece. St Brendan (484–577) was the most famous of these Irish explorers.

Little is known about the boat-construction of the ancient Irish. While it is clear that most of their boats were capable of holding only three men, others were of a much larger size. A gold model of a boat, probably of the larger kind,

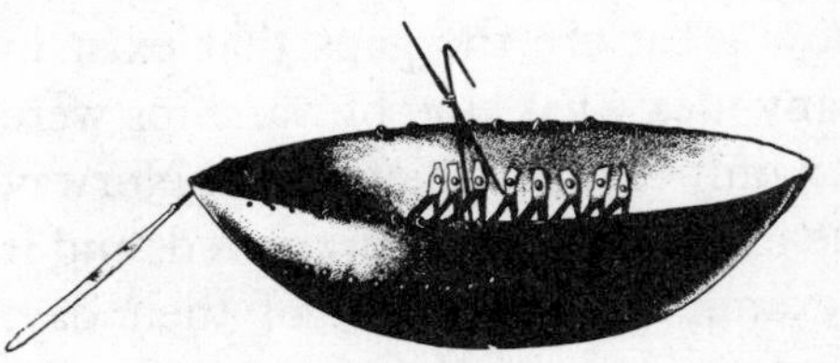

FIGURE 529—*Gold model of a boat, from Lough Foyle, Ireland. Date* c *A.D. 1.*

dating from about A.D. 1, was found beside Lough Foyle (figure 529). The vessel on which the model was based had nine oars a side, passed through grommets in the manner of the Anglo-Saxon boats. As there is no evidence of tholes the grommet may have passed directly round the wooden rail which served as a gunwale, as in the Eskimo *umyak*. The boat is of the same construction at bow and stern. It has a steering-oar in a grommet on the port side (whereas the Teutonic peoples steered on the starboard side) and a small mast amidships for a square-sail. Although it is not certain that the model represents a skin-boat, it is most probable that it does. In the Pepysian library at Magdalene College, Cambridge, there is preserved a seventeenth-century drawing of a large skin-boat which is by no means unlike the model, apart from its transom stern. There is no reason to doubt that the Irish could build such vessels carrying twenty men, or more. Since the Eskimo *umyak* of a comparable build can convey up to forty persons, St Brendan's twenty or thirty is by no means an implausible number.

FIGURE 530—(Above) *Two-masted ship and a light-house, from a relief on the campanile at Pisa; twelfth century. Apparently both masts carry square sails; the foremast is strongly raked. The vessel has fore- and after-castles and a starboard steering-oar. The main strakes are strongly marked.* (Below) *Single-masted ship, with fore- and after-castles, on the late thirteenth-century seal of Winchelsea. The starboard rudder is clearly shown. Two men (aft) raise the anchor with a windlass, assisted by two others (forward) hauling on the cable.*

The ships of the Middle Ages are far harder to understand than those of classical times. A whole new nomenclature appears, difficult to correlate reliably with particular types of vessel. Illustrations are mostly poor—so bad in most cases as to suggest that the artists had little first-hand knowledge of the ships they depicted.

In the north a large, round, clinker-built merchant ship setting a single square-sail on a mast stepped amidships was known as a cog. The cog was probably derived not from the northern warship but from an enlargement of the round ship used for trading by all peoples bordering on the North Sea. Since piracy was commonplace everywhere, all ships had to be capable of defence; thus the distinction between warship and merchantman became obscured. Cog fought cog, as galley fought galley.

Raised fighting-platforms (castles) were erected at either end of such ships (hence the term forecastle). The notion of the castle may have been introduced to northern waters from the Mediterranean by crusaders, but the structures were much cruder than those on Roman ships (figure 530).

FIGURE 531—*Single-masted ship with stern-post rudder on the seal of Elbing, Poland.* c *1242.*

The northerners learnt, in the end, to sail to windward, a purpose for which lateen sails were widely used by Roman boats. It may be that Raud the Strong, who was tortured to death by the Norwegian king Olaf Trygvasson in the tenth century because he had 'wind at will', had learnt this art. It was certainly well known in the fourteenth century, for Froissart states that after the battle of Aljubarrota in 1385 the Portuguese ambassadors chartered 'a vessel called a "lin" which keeps nearer the wind than any other'. On their passage from Portugal to England they were three days at sea and made a Cornish landfall on the fourth, showing that the sailors of this period had no hesitation in sailing across the open ocean.

FIGURE 532—*Pictures of boats with lateen sails, from a Byzantine manuscript of* c *886. The halyards and pulleys controlling the sail are clearly shown.*

The whole of maritime history is unfortunately riddled by extensive gaps. It is not known for certain when the invention of the stern-post rudder came into general use, or where it was evolved. There is evidence that this type of rudder was known in China some centuries earlier than in Europe (p 771), and it is at least probable that it was used in Byzantine vessels before it was fitted in those of the north, where it is found in the thirteenth century (figure 531). To the Graeco-Romans also we must probably give the credit for the invention of the lateen sail.

This sail, which is no more than the old Mediterranean square-sail with the after end of the yard tilted up into the air and the leading edge of the sail cut short, made beating into the wind easy. How early this first fore-and-aft sail, typical of the eastern Mediterranean and the Red Sea, was invented nobody knows (figure 532). 'Lateen' is a phonetic rendering of the French (*voile*) *latine*, Latin (sail), and we now know that this type of sail came from the Roman Empire of the east, together with the sprit-sail.

FIGURE 533—*Castilian merchantman of the thirteenth century, with two lateen sails. From the* Lapidario *of Alfonso the Wise. Thirteenth century.*

The stern-post rudder could not be handled in the same way as a rudder hung on the quarter of the vessel. For untold ages ships had been built with high stern-posts designed for the lift of the vessel to a following sea. When the rudder was hung right under this type of stern, some means had to be devised for controlling it. One method, adopted in the north and still to be seen fitted to small Norwegian and Faeroese boats to this day, was to fit a lateral bar at right-angles to the head of the rudder and to extend this again at right-angles with a tiller, which thus came to the hand of the helmsman sitting by the ship's quarter-rail. This had obvious weaknesses, and before long most shipbuilders decided that it

FIGURE 534—*Sailing-galley with two lateen sails. From a woodcut, 1486.*

was not good enough. The high stern-post had to go, and in its place a port was cut right in the stern, through which the tiller passed direct from the head of the rudder to the helmsman's hand. This was almost certainly a Mediterranean device, for illustrations of northern vessels still show the high stern-post and outboard steering-arm as late as the sixteenth century. A fourteenth-century ship of the Byzantine emperor John Palaeologus V (1354–91) is shown with the fully developed rudder. It is interesting to note that this vessel sets lateen sails on two masts and has two tiers of oars; one tier is rowed on the rail of the old *katastroma*, or fighting-deck, and one beneath it.

FIGURE 535—*An early 'carrack', with mainmast and a small mizzen-mast, from King Henry VI's psalter, c 1430. The stern-post rudder and its tiller are clearly shown.*

By the thirteenth century the volume of trade in the western Mediterranean was very considerable, much of it passing in the ships of Venice, Genoa, and Pisa. These, whether fitted with oars or not, carried two masts (the foremast being the taller) with lateen sails (figure 533). Such vessels were hired by St Louis for his crusade (1268) and were capable of sailing to the English Channel. About the beginning of the fourteenth century, however, the northern type of ship (cog) with a single mast, square sail, and stern-post rudder was adopted by the Italian merchants, although the galley continued to be rigged with lateen sails for at least two centuries longer (figure 534). Somewhat later, when the round sailing-ship had developed three masts, the mizzen was often rigged with a lateen sail (figure 536).[1]

The round merchant sailing-ship, clinker-built in the north but carvel-built in the south like Roman vessels, was rapidly improved during the first half of the fifteenth century, particularly in the Mediterranean. The first stage was the stepping of a second mast, the small foremast characteristic of early 'carracks', whose build was soon imitated in the north (figure 535). Then, owing to the pressure on the rudder exerted by the fore-sail when beating, it was found necessary to provide some balance farther aft. This led to the stepping of a third

[1] A late fifteenth-century vessel of Portuguese origin, the 'caravel', had three masts fitted with lateen sails.

mast (mizzen) on the poop. This rig became the standard for the later carracks, which were to open the seas of the world to navigation.

The first decisively dated picture of a three-masted carrack is on the seal

FIGURE 536—*A three-masted merchantman* (kraeck = *carrack*) *with a lateen sail on the mizzen-mast. From a late fifteenth-ceetury Flemish engraving on copper* (see also *plate* 9).

of Louis de Bourbon, 1466; soon afterwards illustrations of this type of ship become common. They show a very strongly built round ship, with numerous wales along the sides (figure 536). The bow is swept up high with a forecastle-head or deck. The waist of the ship is relatively low and the aftercastle, or poop, is built up with two or more decks. The tiller passes into the hull of the ship through a port, above what must be the level of the main deck. There is a curved projection forming a short transom above the head of the rudder, by no means unlike that of the Hastings fishing-luggers only recently extinct. In the following years the transom was to be carried down below water level; but at this stage the stern is still rounded as in the Roman vessels. At least two galleries were built above the transom and projected aft well out beyond the rudder. Everything gives the impression of strength rather than speed.

There are three masts. The mainmast, stepped more or less amidships, is a large spar and appears in some pictures to be composite. Its great square-sail yard is formed of two spars, scarfed together in the middle like the lateen-yard of an Arab dhow today. Above the halyards and lifts for the yard is a round top surmounted by a light topmast, which sometimes shows a light topsail-yard crossed. The mast is stayed with a large number of shrouds with dead-eyes and ratlines. Ratlines for ease in mounting the shrouds were apparently not known in Roman times. The foremast is short compared with the mainmast. It is the old *artemon* or foresail-mast, but is no longer raked forward. There is a bowsprit, more properly perhaps a bumkin or short beam, projecting from the forecastle, which may have been used to extend the foot of the foresail on a wind. The third mast, which we should now call a mizzen, sets a lateen sail. The ship has therefore that combination of fore-and-aft and square sails which was to become indispensable for all larger sailing-vessels down to the end of the age of sail.

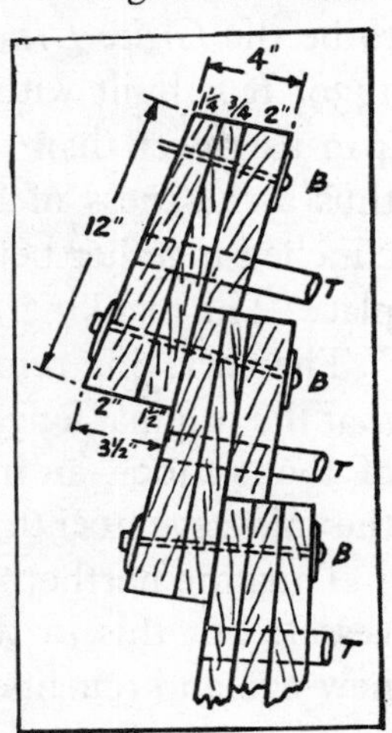

FIGURE 537—*Section of planking of the* Grâce Dieu, *1418, showing the method of clinker-build and the use of tree-nails* (T) *passing through three thicknesses of plank to the ribs, and iron bolts* (B) *holding together the five thicknesses at the overlap.*

When we look at this type of ship, it is clear that we are at the end of one age and the beginning of another. This is essentially a modern ship—and yet one of the greatest antiquity. It owes much to Greece and Rome, yet combines certain entirely new features which have greatly increased its efficiency. Where the ships of Rome had to wait for a favourable wind, this ship could make use of almost any wind. Where the Roman ship was at the mercy of a lee shore, as is shown in the account of St Paul's shipwreck, this vessel could at least make an attempt to claw off the land. It would be a slow and painful business, but it could be tried.

The decline of the Moorish kingdoms in Spain opened the doors to the Mediterranean once more. It is no accident that the fall of Granada (1492) coincided with Columbus's first voyage to the west, for, with no immediate menace close at hand, Spain and Portugal were able to make use of the shipbuilding and navigational knowledge of the other European peoples.

The largest example of a clinker-built ship to be identified with any certainty is the wreck in the Hamble river above Bursledon bridge. This vessel is believed to be the *Grâce Dieu* built at Southampton for Henry V in 1418 and burnt in 1439. It is built with three layers of oak planking, of which the inner plank is 4 in narrower than the outer two, which are 12 in wide (figure 537). There is thus a thickness of five planks at each seam. The whole skin was apparently clinched together before the ribs were inserted within the hull and wedged into place. The planking was secured to these ribs by tree-nails.

The surviving remains are 135 ft in length and $37\frac{1}{2}$ ft in beam. It is believed that the original length was 200 ft, and the beam 50 ft, with a tonnage, therefore, of about 1400—an incredible figure for the time and not attained again before the mid-seventeenth century.

Thus the northern shipbuilders pushed to its maximum size the clinker-built vessel; but this in turn was to give way to carvel-built ships, which began a new age and remained familiar for centuries in the south.

BIBLIOGRAPHY

ABELL, SIR WESTCOTT (STILE). 'The Shipwright's Trade.' University Press, Cambridge. 1948.

ANDERSON, ROMOLA and ANDERSON, R. C. 'The Sailing-Ship. Six Thousand Years of History.' Harrap, London. 1947.

BOWEN, F. C. 'The Sea. Its History and Romance', Vol. 1. Halton and Truscott Smith, London. 1924.

BRØGGER, A. W. and SHETELIG, H. 'The Viking Ships, their Ancestry and Evolution' (trans., rev., and abridged by KATHERINE JOHN). Dreyers Forlag, Oslo. 1953.

CASSON, L. "Fore-and-aft Sails in the Ancient World." *The Mariner's Mirror*, **42**, 3–5, 1956.

CHATTERTON, E. K. 'Sailing Ships and their Story' (new ed.). Sidgwick and Jackson, London. 1923.

HOLMES, SIR GEORGE (CHARLES VINCENT). 'Ancient and Modern Ships.' Part I: 'Wooden Sailing-Ships.' Chapman and Hall, London. 1900.

HORNELL, J. 'Water Transport.' University Press, Cambridge. 1946.

LA ROERIE, G. and VIVIELLE, L. 'Navires et marins de la rame à l'hélice', Vol. 1. Duchartre and van Buggenhoudt, Paris. 1930.

MOLL, F. 'Das Schiff in der bildenden Kunst' (2 pts). Schroeder, Bonn. 1929.

SMYTH, H. W. 'Mast and Sail in Europe and Asia' (new ed.). Blackwood, Edinburgh. 1929.

The Mariner's Mirror (Quart. J. Soc. Naut. Res., London), **4**, 1914. (For further references).

17

POWER

R. J. FORBES

I. FIVE KINDS OF PRIME-MOVERS

IN each epoch of history many factors, such as the materials available, the accumulated skill and experience of craftsmen, economic and social conditions, religious or ethical tenets, and even philosophical doctrines, interact to determine the nature of technology. At each stage craftsmen command certain tools and machines for the various operations needed to transform raw materials into finished products. Such things as the hammer and chisel or the rotary quern are sometimes called 'direct actors' to distinguish them from prime-movers, which are machines that provide motive power for other tools or machinery. These last, which convert the power of animal muscles, running water, wind, or heat into a convenient form of mechanical energy, serve as measures of man's ability to harness the forces of nature.

Thus, after a certain cultural stage has been reached, the availability of prime-movers is the conditioning factor in technology. It is they that determine the size of the units of metal, of wood, or of other material with which the craftsman works, as well as the nature and dimensions of the machines, tools, and other products that can finally be made from the units thus fashioned [1]. The introduction of a new prime-mover generally renders energy available in a more concentrated form, and permits a new level of production. From this point of view there are five stages in the history of technology. In the first stage men had muscle-power only. During the Neolithic revolution the domestication of animals increased the total amount of tractive power available, but without raising the level of energy-production (vol I, ch 13). Even the use of the horse and the dromedary during the second millennium B.C. did not change this situation. The second stage is, therefore, important mainly for this increase in total man- and animal-power rather than for a higher or more concentrated level of energy-production. Nevertheless, its results are manifest in the greater variety and larger number of tools produced from the time of the ancient empires onwards.

The third stage opens with the introduction of the water-mill during the late Roman Empire. Its earliest form, the Norse mill (figure 538), again implies a mere shift of the motive power, from animal muscles to a machine moved by running water, rather than a new level of energy-production. The querns formerly rotated by two slaves, and donkey-mills (figures 76–78) of 0·4–0·5 hp, were now sometimes moved by a primitive water-mill the energy-output of which was of about the same magnitude. This machine added to the total energy available to mankind but did not supply larger quantities of power per unit.

When, however, Roman engineers converted this primitive mill into the Vitruvian mill (p 595 and figure 540) they created a prime-mover which, even in its most primitive form, yielded some 3 hp. For various reasons (pp 601 ff) this water-mill did not come into general use in the Mediterranean world. Western Europe realized its great importance during the early Middle Ages, and its rapid technical development and dissemination (pp 608 ff) then placed in the hands of mankind a prime-mover capable of yielding some 40–60 hp. The windmill, introduced during the same period, was another source of an equivalent power (pp 617–20, 623 ff). Water-mill and windmill dominated technology until the end of the eighteenth century, and their capacity determined the range of the machinery, processes, and products used during that period.

FIGURE 538—*A water-mill of the Norse type used in Shetland until recently. The blades were lifted clear of the stream by the rope.*

The fourth stage was initiated by the steam-engine, which by 1850 had become a prime-mover capable of yielding more energy than either the water-mill or the windmill. At the present time we are on the verge of an unpredictable fifth stage, that of atomic power. Here we shall discuss the third stage, the introduction of the water-mill into the classical world and the changes that it wrought in classical technology.

II. SLAVES AND HARNESSED ANIMALS

During almost the whole of antiquity the only prime-movers were men and animals. Even when the more economical form of water-mill was devised about the beginning of the Christian era its penetration was so slow that the situation remained practically unchanged for another 400 years.

An acute shortage of labour, which incidentally contributed to the spread of the water-mill and other machinery, was first encountered during the later

Roman Empire. Generally in antiquity the state had at its disposal concentrations of man-power, in the form of statute-labour or slaves, for the construction of great public works or the erection of monuments. Concentration of energy was obtained by the use of large gangs. Massive machines, such as the cranes used in architecture (figure 603), and water-wheels for draining mines, were worked by them or by animals. The tread-mill was often employed in machines for converting muscle-power into rotary motion. Gangs of soldiers or galley-slaves provided motive power for ships, and for constructing roads and aqueducts.

Slavery was an accepted social institution throughout antiquity, but there is no basis for the sweeping assertion that it impeded the use and evolution of machinery and engineering. During 3500 years before our era the customs and consequences of slavery were many, and there were great local and regional variations. In spite of the common view, slavery played only a minor part in the economy of Egypt before the Hellenistic period, and certainly did not hinder the evolution of crafts [2]. In the Near East slavery was part of the economic pattern from the beginning, but apart from temple-slaves and state-slaves the proportion of the bound population in every country and at almost every time was insignificant. There were no private estates or industries employing masses of slaves, large landowners preferring free tenants. The basis of Near Eastern economy was formed by the free tenant-farmer and cropper in agriculture, and the free artisan and day-labourer in industry. The slave had his place in the household, but always remained half free and half chattel. The Near East also knew the bondman, so frequently mentioned in the Bible. He worked as a kind of contract-coolie to pay off a debt or to earn a lump sum of money. He was a paid servant and not truly a slave [3].

In the free Greek cities the craftsman was in competition with household handicrafts. In the craft-shops free men and slaves worked side by side, and the rate of pay for workmen of each status was equal, so far as our sources indicate. This Greek system of employing slave-labour alongside free craftsmen was uncommon in Italy before 350 B.C., and then it spread only slowly as the Romans conquered the world.

There was an increase in the employment of slaves in workshops during the second half of the third century B.C. [4]. From then onwards slaves were employed on the latifundia or great estates as a dependable and permanent group to carry on the routine work of an elaborate mixed cultivation. The growing of wheat and barley, however, remained largely in the hands of free farmers, while handicrafts, business, and domestic service absorbed most of the slaves, apart from public works and state-owned mines. This central period of the extension of the use in Italy of skilled slave-labour together with that of free artisans (*c* 200 B.C.–*c* A.D. 100) witnessed the growth of manumission of skilled slaves and of their influence in business and politics. The humanitarian doctrines of the Stoic philosophers were generally absorbed about the turn of our era, as they became more fundamental and outspoken.

But why did not harnessed animals replace human labour more extensively in antiquity? The answer seems to be, because of the disastrous results of ignorance of animal anatomy. Thus ox-harness was used on asses, mules, and horses (pp 538, 544). The ox had been used for many centuries for ploughing, and to a certain extent to pull heavy loads and carts in the country-side. Beyond the farm the ox was too slow for land-transport, and it could not be employed in urban workshops and mills. In applying the ox-harness to the donkey and horse the ancients robbed these animals of most of their natural tractive power (p 522) [5]. Though the horse when used as a pack-animal could carry four times as much as a man, the harness prevented it, when used for draught, from exerting more than part of its available strength. The harness was so unsuitable that it choked the animal when pulling hard. Thus instead of pulling fifteen times as much as a man, the horse pulled barely four times this amount. If one horse was equivalent to four slaves, the shrewd Roman agronomists, comparing this ratio with that of the amount of food consumed by the horse and slave, which also happens to be 4: 1, realized that the question whether it would be economic on the farm to use slaves or the less adaptable horse depended on whether man-power was plentiful or not.

Indiscriminate use of the throat-and-girth harness hampered the more frequent use of draught-horses. The flour-mills of ancient Rome were driven either by two men or by a donkey, and there was no economic incentive to prefer the draught-animal. In the early Middle Ages the breast-strap or postilion harness was acquired from the steppe tribes who invaded Europe. They had also known the horse-collar from the seventh century, and by the twelfth were using it almost exclusively for draught (figure 508) [6]. Between the tenth and the twelfth centuries collar-harness was generally adopted in western Europe, enabling the horse to displace the ox in drawing ploughs and farming-equipment, even though it consumed more expensive food (figure 57).

We know from various sources that a horse attached to a vehicle by the ancient harness could not pull more than 62 kg. This made land-transport very expensive, and prohibitively so for mass-products. Nor did the ancients know a more efficient way of yoking many animals to a wagon than by an increase in the number of yokes. A much better energy-output was obtained when the arrangement in tandem was adopted in the early Middle Ages (figures 486, 499).

Finally, the ancients had no proper horse-shoes. Metal, leather, or straw *solae* were temporarily attached to the feet of horses, mules, or camels when the ground was hard or slippery, but the true iron horse-shoe, permanently attached to the hoofs with nails, seems to have been adopted from the nomads of the steppes by the Germanic tribes about the second century B.C. (figure 469). In the northern provinces of the Empire, however, the horse-shoe penetrated very slowly until its general adoption in the eighth century A.D. By then there had been evolved a new type of horse-shoe with a better grip on the hoof, the ancestor of the horse-shoe as we know it [7].

III. GREEK OR NORSE MILL AND WATER-TURBINE

Lack of suitable harness and of proper methods of protecting the hoofs of draught-animals left the ancient world with man as the main prime-mover until the advent of the water-mill. Mills of this type are first mentioned in a poem by one Antipater of Thessalonica (*c* first century B.C.), which has been rendered as follows:

> Cease from grinding, ye women who toil at the mill; sleep late, even if the crowing cocks announce the dawn. For Demeter has ordered the nymphs to perform the work of your hands, and they, leaping down on the top of the wheel, turn its axle which, with its revolving spokes, turns the heavy concave Nisyrian mill-stones [8].

Clearly these lines do not refer specifically to an overshot water-mill, as some claim, but to a water-mill in general. Strabo (d *c* A.D. 21) tells of the water-mill at Cabeira in Pontus that Mithridates erected near his new palace in 65 B.C. [9]. These, the oldest water-mills of which we know, were for the grinding of corn—a constantly recurring burden in every ancient household, and one that provided a particularly strong incentive to mechanize the corn-mill.

The most primitive water-mill was the Greek form, often called the Norse mill, in which a vertical shaft or axle bore at its lower end a small horizontal 'wheel' composed of a number of scoops (figures 538, 540 A). The shaft passed upwards through the lower mill-stone and was fixed to the upper stone by a cross-bar spanning the aperture or 'eye' of the stone. Such a mill is also called the horizontal water-mill [10]. It requires a running stream to work satisfactorily, but does not necessarily need a mill-pond and other means of regulating the water-supply, as later types do.

We do not know the origin of this simple water-mill, which was essentially a feature of peasant culture. It was always limited to mountainous regions, and was often made more effective by a mill-race and chute. Its low efficiency rendered it unsuitable for the commercial production of flour, each mill serving the limited needs of a single farmer. The stones were small and revolved slowly, once with each revolution of the water-wheel itself, and were usually incapable of grinding more than a sack of corn at a time. The Norse mill probably originated in the hilly region of the Near East [11], whence it spread westward and eastward. It was never found in Egypt or Mesopotamia, probably because of the great rise and fall in the level of the big rivers. It could be made to work only with the small volumes of water moving at high velocity common in hilly districts, whereas the river-valleys offered only large volumes of low-velocity water.

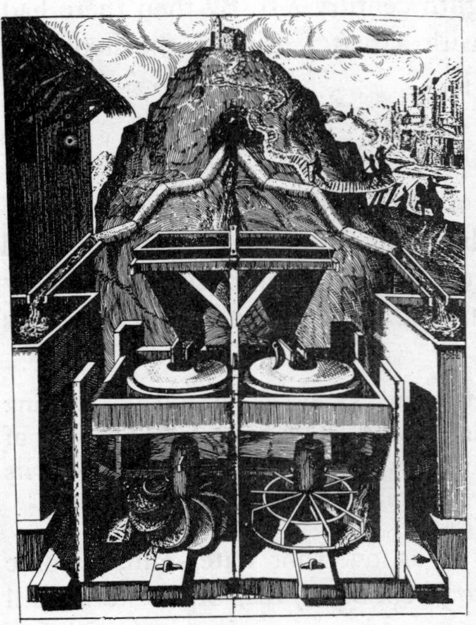

FIGURE 539—*A late illustration of a horizontal water-mill, also known as the Greek or Norse mill, showing two forms of wheel. This type of mill needs a steady stream of water and was often served by an aqueduct. 1617.*

The Norse mill may well have been known to Pliny. He pointed out that 'in the greater part of Italy a bare pestle is used [to grind corn], and wheels also that water turns round as it flows over them and the mill-stone' [12]. He probably refers to north Italy.

Norse mills gradually spread, reaching Ireland and China by the third and fourth centuries A.D. Recent excavations seem to prove that they were known and built in Denmark by the birth of Christ and rapidly increased in number in the following centuries [13]. They were very popular until the late Middle Ages, and were in use on the Garonne (France) as late as 1588 (figure 539). The Norse mill has survived into recent times in the Orkney, Shetland (figures 538, 540 A), and Faeroe Islands, Norway, Rumania, the Lebanon, and many parts of central Asia [14].

Though the Norse mill provides little power and only a slow rotation of the mill-stones, sufficing to grind a small amount of flour for domestic consumption, it is the precursor of the water-turbine. Earlier forms of these water-turbines seem to go back to the Middle Ages, and a kind of Pelton wheel sketched about 1430 was said to have been invented by a pope and to have been built near the

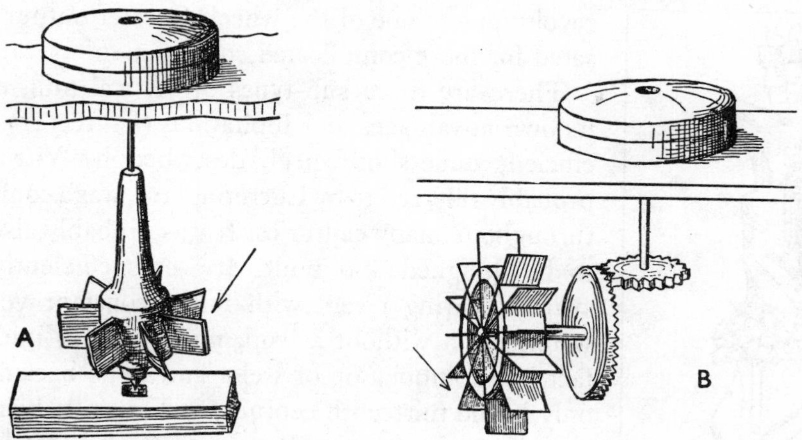

FIGURE 540—(A) *A 'horizontal' water-mill, the wheel being horizontal. The upper end of the shaft is fixed to the upper mill-stone. A small but rapid stream was directed against the wheel by a chute.* (B) *A 'vertical' mill, the wheel being vertical. It could be undershot, overshot, or a breast-wheel (cf figure 541), and the drive was communicated to the mill-stone by means of gears.*

monastery of St Georgenberg (Inn valley) [15]. Such mills continued in use till the eighteenth century in Provence and the Dauphiné, and in Umbria and other regions of Italy.

IV. VITRUVIAN WATER-MILLS

The Greek mill, though mechanizing domestic corn-milling, would have had little effect had it not inspired a Roman engineer of the first century B.C. to construct the more efficient vertical or Vitruvian mill. Using his knowledge of rudimentary gear-wheels and other products of Hellenistic mechanical skill, the inventor transformed the old Greek mill into a much more useful machine (figure 540 B). The water-wheel was now placed in a vertical position. In order to move the mill-stones its horizontal shaft was geared to the vertical shaft turning the upper of these stones. In Roman water-mills the mill-stones usually made five

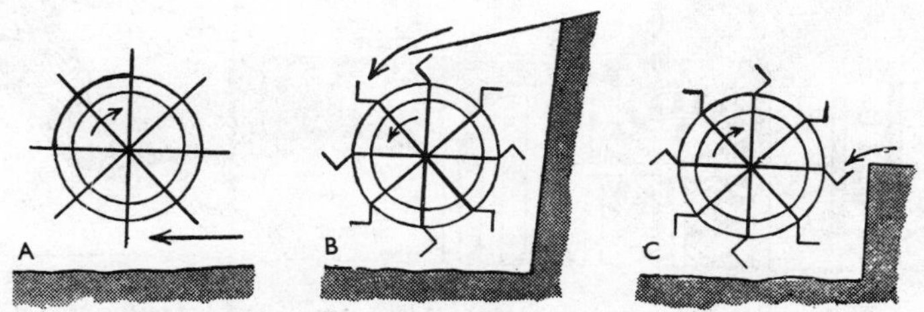

FIGURE 541—*Diagram of:* (A) *undershot wheel;* (B) *overshot wheel;* (C) *breast-wheel.*

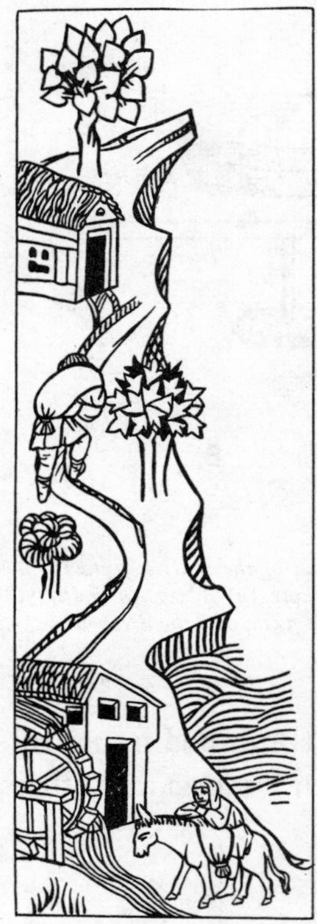

FIGURE 542—*A mill with overshot wheel supplied by a chute. From a woodcut, 1423.*

revolutions to one of the wheel. Greater output compensated for more complicated construction.

There are three sub-types of vertical mill, each with its own advantages and limitations (figure 541). The less efficient undershot wheel, described by Vitruvius and probably referred to by Lucretius [16], was a common type throughout many centuries. It was probably also the first to be designed and built. It works efficiently only in swiftly flowing rivers with fairly constant volumes of water, often without a dam, mill-pond, or mill-race. In fact, the combination of weirs and mills became general only in the thirteenth century, earlier mills being fed by aqueducts if necessary. It was later found that in practice a greater efficiency could be obtained by causing the water to impinge on the top of the wheel and to turn it by its descending weight (figures 541 B, 542). This was the principle of the overshot wheel, which needs a well directed and regulated water-supply, commonly collected in a mill-pond from rivers and springs, then delivered through a sluice to a mill-race and chute properly set for the correct impact of the water on the wheel. Sluggish rivers, or rivers of variable volume, can thus be put to advantage, the river-level being raised by a weir or dam to force its overflow or part of its waters into the mill-pond (figure 543). Aqueducts are equally suitable for the supply of water (figure 544).

This most efficient type of water-mill was not very common in classical antiquity. The earliest known example, in the *agora* at Athens, was built about

FIGURE 543—*The stream, with fish-traps, is dammed to make a mill-pond and raise a head of water to supply the overshot wheel by a chute. Luttrell Psalter,* c·*1338.*

A.D. 470 [17]. Even in the Arab world it did not predominate, for Al-Qazwīnī considers worthy of mention the fact that Al-Adri (the Abderan) reports an overshot wheel near Waluta (Majorca) fed by a chute 80 cm in diameter and

FIGURE 544—*An ore-crushing machine worked by an overshot wheel, which is itself driven by water from an aqueduct. Agricola, 1556.*

350 m in length. It was built because 'if the water falls the common mills cease to turn' [18].

Venantius Fortunatus, as late as the sixth century, calls the water-mill simply *mola*, the ancient term used for any grinding-device. Ausonius (fourth century) describes it poetically as *cerealia saxa* [19]. Vitruvius [16] and Strabo [9] call it *hydraleta hydraletēs*. The latter term was generally attached specifically to the later vertical water-mill. Remains of three of these early mills, all built during the Roman Empire, are known.

At Venafro on the Tuliverno, near Naples, the mill-race and remains of an aqueduct were discovered, together with the imprints of a water-mill in tufa-sediments (figure 545) [20]. The heavy nave (diameter 74 cm = 10 Roman digits) carried 18 paddles each formed like a spade and joined by two parallel rims, thus forming a wheel of diameter 1·85 m. Fed by an aqueduct this undershot wheel would have an estimated output of about three hp, and the mill-stones, at 46 revolutions per minute, could grind 150 kg of corn an hour. If we compare this with the usual seven kg an hour ground by two slaves in hand-querns, or with the output of a donkey-mill, we see that this Vitruvian mill meant a revolution in the grinding of corn. Its output of energy was much higher than that of any machine driven by man or beast, and indeed of any other power-resource of antiquity. Once it was realized that the water-mill could not only grind corn but supply power to other machines, possibilities of technological operations at a new level were disclosed.

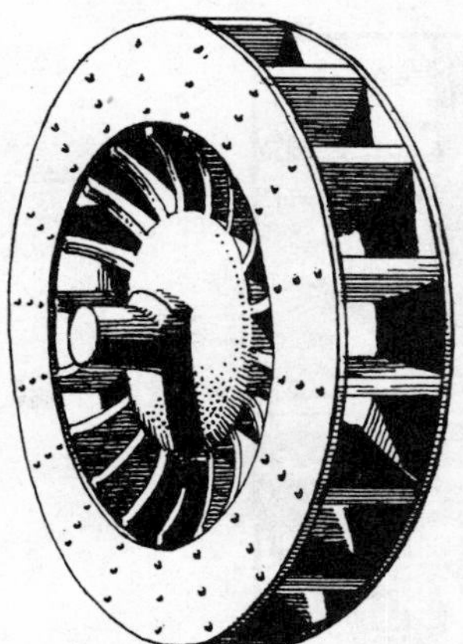

FIGURE 545—*Model of an undershot Roman water-wheel from Venafro, near Cassino, Italy. It is reconstructed from impressions left in lava that had flowed over the mill. Scale 1/30.*

Industrial production was already dependent on water-power in the Roman flour-factory at Barbegal, six miles from Arles (figures 546, 547). Alongside the aqueduct of Saint Rémy, built in the time of Agrippa (63–12 B.C.), which conducts water to Arles, there is another collecting the waters of the Arcoule, the valley of Les Baux, and the well of Manville. This aqueduct of Les Baux, with a channel 2 m wide and 5·6 m deep, is built of bricks and concrete, and dates from after the reign of Hadrian. At Barbegal the channel is doubled and made to slope at an angle of 30°, with a fall in level of 18·60 m. Two sets of eight overshot wheels were built into this sloping conduit,

FIGURE 546—*Conjectural reconstruction of the mills at Barbegal, near Arles.* A.D. *308–16.*

each having a diameter of 220 cm and a width of 70 cm. Wooden gear-wheels were fixed with lead to the iron axles and the gearing was housed in chambers under the mill-stones, of which there was one pair to each wheel and not two as at Athens. Thus the stones could be made heavier and more efficient. The lower stone was 45 cm thick and its diameter was 90 cm. The upper stone had a funnel for feeding the corn to the milling surface [21].

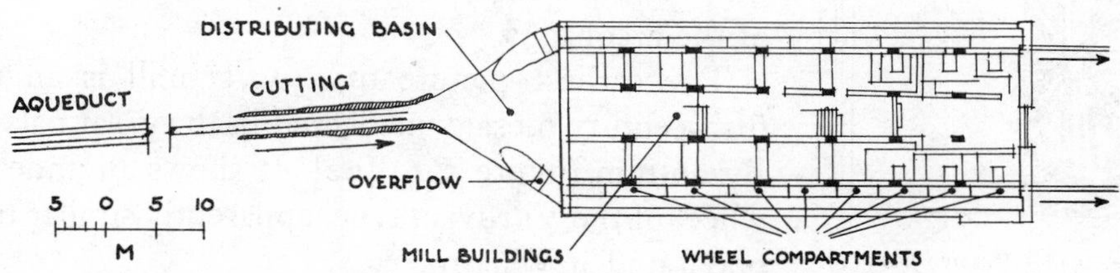

FIGURE 547—*Ground-plan of the mills at Barbegal.*

Comparison with older types of corn-mills has revealed that the hourly capacity was 15–20 kg of corn per set of stones, that is, a total production of 240–320 kg of flour an hour, or 2·8 tons in a ten-hour day. This would be sufficient to feed a population of 80 000, but Arles numbered hardly more than 10 000 inhabitants in the third century A.D.

Arles was an important harbour supplying corn to Rome, and the origins of this factory may go back to the second century when the food-controller for the province of Narbonne settled there. It is supposed that the Barbegal flour-mill was built by Q. Candidius Benignus, the famous local engineer who, according to an inscription, 'was skilled as none other; none surpassed him in the construction of machines and the building of water-conduits'.

In its present state the mill seems to have been constructed in the time of Constantine I, who resided at Arles from A.D. 308 to 316. At that period Arles had important warehouses for the storage of corn, and the flour-mill may have supplied local demand and the army of the province of Narbonne. We have archaeological evidence for a similar flour-mill at Préty (*Pistriacum*), near Tournus, Burgundy, which ground the corn of the Saône valley and supplied the army of northern Gaul [21].

A third example of an early vertical water-mill, in the Athenian *agora*, had an overshot wheel with a diameter of 324 cm, and two sets of mill-stones [17]. The wooden axle was made of a single beam, some 350 cm long, and turned in wooden bearings. The mill was built, according to the archaeological evidence, in the long reign of Leo I (457–74), and was probably destroyed by fire at the

FIGURE 548—*Mosaic from the great palace of Byzantium, representing a water-mill. Early fifth century.*

time of the great Slav invasion (582). This mill and its race with a single trap imply a copious and steady stream. The water-supply from Hymettos and Pentelikon brought in by the Peisistratids was relatively modest. However, excavations of the *agora* have shown a series of springs used from the earliest times and probably originating from the north slope of the areopagus.

The earliest picture of a water-mill is an early fifth-century mosaic uncovered in the great palace in Byzantium (figure 548) [22]. It shows an undershot wheel of the Vitruvian type, apparently similar to the one found at Venafro.

V. INTRODUCTION OF THE CORN-MILL

The conjunction of the paucity of archaeological material with the absence of records of disputes over mill-rights and yields of flour (both so common in medieval texts) is impressive. When, therefore, in the fourth century A.D. there is evidence of a rapid increase in the number of water-mills; when they come into use on the confines of the Roman Empire as testified by Ausonius [19], who refers to water-mills for cutting and polishing marble on the Ruwer (a tributary of the Moselle) in 379; when we find an epitaph at Sardes [23] mentioning a *manganareios hydraleta* (engineer of the water-mill); and when tradition relates that the Hellenized Persian Metrodorus introduced the water-mill into India during the reign of Constantine the Great (311–37); then the reasons for the sudden exploitation of this new source of power, after so many centuries during which the water-mill remained comparatively unimportant, are worthy of examination.

They can be best understood by considering the production of flour in the imperial city of Rome itself. Under the Empire corn-grinding had become a specialized trade, as had the baking of bread at an earlier period, centralizing an operation formerly part of the everyday routine of the Roman household and relieving the baker who had formerly produced both the flour and the bread. The centre of these bakeries and corn-mills, which were turned by horses and donkeys, was the Janiculum, a hilly quarter on the far side of the Tiber. When Caligula (A.D. 37–41) confiscated the horses in the bakeries of Rome a bread famine followed.

Some of these mills were even man-driven, slaves and paupers supplying the necessary labour. During the early Empire unemployment was still a serious problem. Vespasian (A.D. 69–79) is said to have refused the building of a water-driven hoist 'lest the poor have no work'. Constantine the Great, who had to cope with a shortage of labour, employed criminals in the flour-mills; most of the mills, however, were driven by horses or donkeys by this time (figure 78).

The acute shortage of labour in the fourth century considerably promoted the use of the water-mill for grinding corn. It suggested to Palladius (? fourth century A.D.) that water-mills would ease the burden of men and animals [24]. He mentions corn-mills driven by the water of public baths and aqueducts. We know, indeed, that the overflow of the aqueduct of Trajan was used for manufactures sited in the Janiculum, and that water-mills were constructed in the baths of Caracalla dedicated in A.D. 216 [25]. When water-mills invaded the Janiculum, despite the opposition of vested interests and the owners of the old type of horse-mills, water was often diverted from the old aqueduct of Trajan. In an edict of 395 Honorius and Arcadius forbade this misuse of water [26]. According to Procopius (sixth century) all the water-mills were found from of old on the slope of Janiculum, because much water was conveyed by an aqueduct to the crest of the hill, whence it rushed down the declivity with great force [27].

Numerous and repeated edicts forbidding millers to divert water from the aqueducts to their mills show the increase of their number up to and including the sixth century [28]. Later Visigoth laws dealing with the destruction of mills and sluices declared that 'anyone violently injuring a mill shall repair the injury within thirty days; and the same with regard to pools and sluices attached to mills' [29].

The evolution of the water-driven corn-mill gave rise to a new class of craftsmen, the *molitores* or *molendarii*, millers generally owning the mill, grinding corn, and sometimes even preparing malt or baking bread. The latter craft, however, largely remained in the hands of the bakers, whose guild had been founded by the Emperor Trajan about A.D. 110.

VI. OBSTACLES TO INTRODUCTION OF THE WATER-MILL

It is obvious from the above facts that the Roman engineers of the first century B.C. had transformed the primitive Norse mill into a real prime-mover. Why was its introduction delayed until well into the third century A.D. and even then restricted to a few centres only?

Several factors combined in enforcing delay, one springing directly from the physical geography of the Mediterranean basin. Most rivers in this centre of the Roman Empire carry widely varying quantities of water in different seasons;

this greatly hampered the introduction of undershot water-mills constructed on a masonry foundation, since during a large part of the year the water-supply would be insufficient to turn the wheel. Hence the building of water-mills was restricted to those few sites where a constant and liberal water-supply was available from river or aqueduct, and where there was also a heavy demand for flour and a lack of cheap labour. It is significant that all the water-mills excavated (at Venafro, Barbegal, Tournus, Rome, and Athens) were fed artificially by aqueducts in order to procure a constant flow. A system of aqueducts to which mills might be connected represented a heavy capital investment, possible only for the great centres of the Roman world such as Rome, Athens, or Byzantium. In a few cases such military centres as Barbegal or Tournus would qualify by their size for the erection of flour-factories. Labour, too, was plentiful and cheap until the end of the third century, precisely the period when the donkey-mills or slave-driven mills of Rome were gradually ousted by the water-mill. In western Europe the geographical conditions were different. There were many mountain streams and brooks whose constant water-supply favoured the construction of 'a mill on every manor that had a stream to turn it' during the Middle Ages.[1]

Nevertheless, this does not fully explain why the Roman engineers applied the water-mill to the production of flour only, and why they did not employ it as a prime-mover except in a few instances at a very late date. It may be true that the Roman engineers were not particularly ingenious, and that most of their skill lay in applying usefully the inventions of Greek mechanicians, but they did possess a body of practical experience sufficient to enable them to build machines moved by water-wheels. They were familiar with all the elementary principles of mechanical engineering, and in fact their practical knowledge of machinery was not surpassed until the eighteenth century [30]. Many examples prove that this skill could be translated into practical results. When capital was available and the state exerted its influence, mechanization in antiquity could go very far indeed, as in the arts of war. From Hellenistic times onwards mechanical missile-weapons based on the bow or sling were developed and built in large quantities. Tests on such machines have proved their efficiency, and the number of such ballistas, catapults, and other siege-engines in the Roman army indicates a degree of mechanization that was not exceeded before modern times (ch 20).

For building-purposes large cranes, driven by tread-mills, were constructed to raise heavy blocks (figures 578, 603). Hoisting-machines driven by water-power

[1] Especially as the lord of the manor derived a toll from the obligatory use of the manorial mill by the villeinage.

were not introduced either for building or for raising water out of mines (in which human tread-mills were employed also), apparently because the state was for long concerned to provide useful labour for the indigent, who were largely dependent on the corn-doles. Only in rare instances did public works or services encourage mechanization. Otherwise the arts and crafts were left to themselves to work out their own destinies. Mainly household handicrafts, like spinning and weaving, show little progress. The classical potter produced excellent thin-walled ware, but progress was artistic rather than technological in character and there was no mechanized mass-production (ch 8, pt I). Lack of stimulus to industrialize left technology practically stagnant during the Roman Empire. A tendency towards specialization and mass-production [31] in a few instances (e.g. in pottery, shoe-making, and jewelry-manufacture) was satisfied by concentrations of man-power using unchanged handicraft methods; thus the social structure and the cheapness of labour frustrated any inventiveness on the part of engineers. Their efforts were only too often wasted on show-pieces meant to impress the masses at public festivals.

Standards of economic efficiency in the ancient world were very different from those of modern industrial society and, above all, the fertile interplay between science and technology was almost totally absent. The ancient engineer was sometimes aware of its necessity, for as Pappus (third to fourth century) remarked:

> The mechanicians of Hero's school say that mechanics can be divided into a theoretical and a manual part; the theoretical part is composed of geometry, arithmetic, astronomy, and physics; the manual of work in metals, architecture, carpentry, and the art of painting, and the manual execution of these things. The man who has been trained from his youth in the aforesaid arts and in addition has a versatile mind, will be, they say, the best architect and inventor of mechanical devices. As it is not possible for the same man to excel in so many academic studies and at the same time to learn the aforesaid crafts, they advise one who wishes to undertake 'mechanical work' to use such crafts as he already possesses in the tasks to be performed in each particular case [32].

Only in the late Middle Ages did such superior craftsmen succeed in making their views effective; in antiquity science ignored them.

The ancients had no word to describe the scientist: they called him merely a philosopher. Natural philosophers or *physiologoi*, students of 'the natural growth of things', observed nature and took many of their images from the crafts of the Greek towns. Mechanistic thought-patterns were used almost exclusively to explain certain natural phenomena, but for the most part the observations served merely to illustrate philosophical theories [33]. The philosopher attached no importance to manual labour or the works of technology and engineering: even physicians were considered *technitai* or technicians. Through study of nature they sought the wisdom and intellectual satisfaction which each free citizen should endeavour to acquire in his leisure (*otium*). Application to the mechanical arts (*neg-otium*) was definitely inferior; by exercising their trade the *banausoi* (craftsmen) killed the spirit. Plutarch calls the machines Archimedes (p 632) made when the Romans besieged Syracuse the by-products of geometry, and says that 'the construction of engines and in general every trade that is exercised for its practical value is lowly and base' [34].

This typically Greek contrast between the liberal arts and the *artes mechanicae* precluded all efficient co-operation between science and technology. The classical scientist at his best had a clear apprehension of scientific method; he was aware of the need for deduction and verification, and he conducted experiments. But he was too naïvely empirical and scarcely appreciated the difficulties involved in adequate observation. The effect of craft-knowledge on his science seems extremely small and doubtful [35]: certainly classical science yielded no conclusions helpful to the engineer and the craftsman, who were left to travel the hard road of trial and error alone. The Baconian notions of utility and progress had no place in this world; as Bacon himself phrased it: '[Great technical discoveries] were more ancient than philosophy and the intellectual arts; so that, to speak the truth, when contemplation and doctrinal science began, the discovery of useful works ceased' [36].

Nowhere in antiquity was there an impulse towards industrialization demanding a more concentrated output of energy from prime-movers and leading to specialization, mechanization, and standardization. Food-products were manufactured on a large scale, partly with the use of improved crushing- and grinding-machines, but only manual labour was used. The larger grouping of craftsmen in an *ergastērion* or *officina* hardly constituted a factory in our sense. The Athenian *ergastēria* with 10–30 craftsmen were not the result of heavy capital investment for the purchase of machinery. They housed a number of craftsmen of the same trade and corresponded rather to the *ateliers* or *manufactures* of the eighteenth century. The net profits of these workshops were not reinvested to increase their output, but were consumed by the craftsmen themselves [37]. No large sums were available for the development and construction of machinery to replace human or animal power, cheap enough in a world which produced the bare necessities of life for the masses and a limited range of luxuries for the few. Mass-production of goods would in fact have meant over-production in the ancient world, which invested its capital in slaves and land. Applied science had already completed its task, so the ancients thought, and this may have been one of the causes of the failure of the Roman Empire, as it was of the Hellenistic world of the east [38].

Only in the larger centres did the mechanical production of flour, one of the necessities of the masses, become urgent by the fourth century, when the expansion of the Roman Empire had long since stopped and labour grew more and more scarce, even in the form of slaves. By then Roman authors began to complain of the neglect of applied science and, not without reason, to praise the inventiveness of the barbarians. 'This is a quality which we see granted without respect of persons; for although the barbarian peoples derive no power from eloquence and no illustrious rank from office, yet they are by no means considered strangers to mechanical inventiveness, where nature comes to their assistance', says the unknown author of the *De rebus bellicis*, about 370, whose inventions were specially designed to save man-power in the army [39]. Once this widespread social demand for cheaper bread existed, the vested interests in slave- or donkey-driven mills could easily be overcome by introducing the mechanical devices already known for centuries.

By the time the water-wheel was introduced as a prime-mover a spiritual revolution had swept away the classical views on the degrading character of manual labour. Slavery had always depressed the social and economic conditions of the free craftsmen, kept their wages low, and subjected them to the contempt of the intellectual classes. Combined with restricted demand it had frustrated mechanical inventiveness and the organization of efficient methods of production of cheap goods for all. Change came in the later Roman Empire, when slavery was already on its ebb, through a new attitude towards the poor. The ruthless suppression of the weak in classical antiquity (notably between 200 B.C. and A.D. 100) showed its worst face in the treatment of slaves, which in its turn led to frightful revolutions, primarily based not on social programmes but on the elemental will to survive. The poor were regarded as victims of the gods in the ancient Near East, and were recommended by religion to the sympathy of their wealthier brethren. Only the Old Testament rose to the moral concept of the inherent brotherhood of man (Job xxxi. 15) and to the denial of the right to own a man in perpetuity. In the classical world several schools of philosophers, notably the Stoics, preached this human brotherhood. They tried to instil philanthropy and humanity into the hearts of their fellow-countrymen, who should no longer treat their poor and weak brethren as chattels but as human beings. The advent of Christianity introduced a more basic concept, for which the new term *caritas* (charity) was coined; this was more than mere friendliness and hospitality.

Christianity not only changed the attitude of the citizen towards the poor and the slave: it radically attacked the classical views and extolled the value of manual labour. In this way slavery was doomed to disappear in the long run, and the craftsmen were to gain an honourable place in society. The door was opened to the view that nature should be used to serve mankind, that its forces should be captured and trained to ease his task and his life [40].

The ancients had always believed in an animated nature. To most of them it was a work of art of beautiful suitability in which Form had not always completely subjected Matter. Others, such as the atomists, considered that nature had achieved its structure by boundless waste. None of them would have accepted the Newtonian world-machine created by God and left to run its course, the concept so dear to the eighteenth-century scientist. Even Democritus, Epicurus, or Lucretius would have shuddered before such blasphemies. In reality the ancients always based science on animism. Thales says that 'all things are full of gods' and that 'the magnet is proved to have a soul since it can move iron'. This animism is still an essential element in the final philosophy of antiquity, Neoplatonism. The poet's 'water-nymphs moving the axle of the water-wheel' were real and not just an image. The ancient world did not dream of man's harnessing these supernatural powers until Christianity, by its opposition to animism, opened the door to a rational use of the forces of nature. The last obstacles to the introduction of prime-movers had fallen, when by the fourth century A.D. the Roman Empire became officially Christian.

FIGURE 549—*Paddle-driven warship; the capstans are rotated by oxen. From a manuscript of* c *1436, probably based on a late Roman picture.*

VII. PADDLE-WHEEL AND FLOATING MILL

The invention of the ship propelled by paddle-wheels (*c* A.D. 370) was inspired by the water-mill, for in principle the paddle-wheel is simply a water-mill inverted by applying power to its axle. Its probable inventor was the unknown Latin author of the *De rebus bellicis* (p 605) [41]. In his warship three sets of oxen turn capstans driving three paddle-wheels on the outside of the ship (figure 549; cf plate 37 B). This invention precedes by several centuries the man-moved paddle-wheels claimed by the Chinese to have been used in a naval battle of the twelfth century.

FIGURE 550—*A floating mill. Two boats tied in the fastest part of the river were each fitted with mill-stones and between them was fastened a water-wheel to turn the mills. 1595.*

The displacement of older types of mill by the water-mill within 150 years is evident from the attempt of the Goths, when besieging Rome in 537, to starve the city by cutting the aqueducts. This prompted the invention of the floating mill by the military commander Belisarius and his engineers. Procopius relates:

FIGURE 551—*A corner of the harbour of Cologne with floating mills and what is possibly a small dredger for removing mud and silt. The incomplete cathedral in the background has a crane on one of its towers. Woodcut of 1499.*

Belisarius hit upon the following device. Just below the bridge connected with the circuit-wall, he fastened ropes from the two banks of the river and stretched them as tight as he could, and then attached to them two boats side by side and two feet (*sic*) apart, where the flow of the water comes down from the arch of the bridge with the greatest force, and placing two mills on either boat, he hung between them the mechanism by which mills are customarily turned. And below these he fastened other boats, each attached to the one next behind in order, and he set the water-wheels between them in the same manner for a great distance. So by the force of the flowing water all the wheels, one after the other, were made to revolve independently, and thus they worked the mills with which they were connected and ground sufficient flour for the city [42].

This floating mill was to spread east and west in succeeding centuries (figures 550, 551). In the tenth century floating paper-mills were found on the Tigris near Baghdad, and we hear of 'boat-mills' near Venice. In the Seine, floating mills were built under the Grand Pont in Paris during the reign of Louis VII (1137–80) and destroyed with this bridge in 1296 (figure 552). They are noted on the Garonne (1290) and the Loire near Orleans (1306), but attempts to build them on the Thames failed twice (in 1525 and in the eighteenth century). In many parts of Europe, however, including the Po valley, they have survived until recent times.

FIGURE 552—*Floating mills under the arches of a bridge over the Seine in Paris. From a French manuscript of 1317.*

VIII. SPECIALIZATION OF THE WATER-MILL

By the fourth century men had discovered the real importance of the water-mill. Just as, many centuries later, the steam-engine which began as a steam-pump in the mines developed into a prime-mover, so, much more gradually, the water-mill was recognized as a prime-mover or power-resource rather than as a mechanized corn-mill. This is evident from the 'marble mills' recorded near the Moselle in A.D. 379 [19] and the very ancient French saw-mills mentioned by Bélidor (1693–1761) [43]. Authors began to suggest timidly all kinds of applications for water-power. This trend was stimulated by the growing conviction of the Christian community that the use of human or animal power should be avoided if nature could be made to do the work.

Though the troubled era in Europe following the disintegration of the Roman Empire was hardly favourable to technological progress, there is evidence that the number of water-mills steadily increased between the tenth and twelfth centuries [44]. In western Europe Gregory of Tours (*c* 540–94) reports water-mills near Dijon and on the Indre [45]. In Merovingian times they were sufficiently important to be protected in the Salic laws [46]. Water-mills appeared as a source of fiscal revenue in Charlemagne's *Capitulare de villis* (800) [47]. The eleventh and twelfth centuries show the great progress of the medieval industrial revolution, a strong advance in mechanization in different industries followed by a period of stability until the fifteenth century. The rapid spread of water-mills may be illustrated from some local figures. On the banks of a minor tributary of the Seine at Rouen there were 2 mills in the tenth century, 5 in the twelfth, 10 in the thirteenth, and 12 about 1300. In the district of Forez there was only one early in the twelfth century, but there were nearly 80 in the thirteenth. In Aube 14 mills in the eleventh century became 60 in the twelfth, and nearly 200 in the early thirteenth. The townsmen of Troyes built 11 water-mills on the Seine and the Meldanson in the period 1157–91. By 1493 the district had 20 corn-mills, 14 paper-mills, 2 tanning-mills, 4 fulling-mills (figures 187, 553), and 1 cloth-mill: a total of 41.

FIGURE 553—*An undershot water-wheel working a fulling-mill. Strada, 1617.*

During the same period there was an extension of the use of water-power, as exemplified in its application to water-lifting and irrigation-wheels (abbey of St Bertin, 1095–1123), oil-mills (eleventh century in Graisivaudan), mills grinding pigments (Péronne, 1376), malt-mills (Béthune, 1138), wood-turning-mills (Vizille, 1347), and grinding-mills for cutlery (Forez district, 1257). Other mills had to use a variety of cams and gear-wheels to move their machinery; such were fulling-mills (Dauphiné, 1050; Forez, 1066; Champagne, 1101), tanning-mills (Graisivaudan, eleventh century; Issoudun, 1116), hemp-mills (Graisivaudan, 1040), iron mills (1116), saw-mills (illustrated in Villard de Honnecourt's sketchbook of about 1250 (figure 584)), and paper-mills (thirteenth century). In the south of France there was an intimate connexion between irrigation-canals and the establishment of water-mills. After the building of the Vaucluse canal (1101) licences were given for the building of water-mills and wind-mills, and mills for grinding corn, pressing olives, and fulling cloth. The large monasteries were particularly interested in the building of water-mills. In 1020 the abbey of St Victor, near Marseilles, possessed a water-mill on the Jarret near Huveaune, and a branch of the Charente turned the mills belonging to the abbey of Saintes. The Cistercian monks, indefatigable reclaimers of barren and waste lands, were equally prominent in the diffusion of the water-mill. Their abbey in southern Champagne had fulling-, paper-, and hammer-mills together with water-wheels in the thirteenth century. They were particularly involved in the building of water-driven hammer-forges (figure 554) and they pioneered the use of water-mills in iron metallurgy not only in France but in Germany, Denmark, and Britain. Out of thirty French documents of the twelfth century concerned with hammer-forges and iron metallurgy twenty-five were drawn up by Cistercian monks as producers. Some of their abbeys even had water-driven workshops combining very different crafts under one roof (p 650).

Tide-mills, though rare, appeared as early as 1125–33 near the mouth of the Adour. They are known to have been used on the Italian coast near Venice a century later, and a drawing of tide-mills made in 1438 by the engineer Jacopo Mariano still exists. Dover harbour had an early one dating from before the Conquest. Most tidal mills, however, date from the eighteenth century; they never played any important part in industry.

In Britain there was a similar spectacular rise in the number of water-mills, accompanying an increase of population, the development of towns, and the growing mechanization of several crafts. The earliest reliable allusion to a corn-mill in England occurs in a charter granted by Ethelbert II in 762 to the owners of a monastic mill situated east of Dover. Gradually, corn-grinding was concentrated in horse- and water-mills, without, however, completely displacing the household quern. By the tenth century the water-mill had invaded Ireland. A century later Domesday Book (1086) reveals the existence of no fewer than 5624 water-mills in 3000 communities south of the Trent and the Severn [48]. Some of these mills may have been of the horizontal or Norse type, though this was more typical of the Celtic and Norse settlements in the hills. The vertical or Vitruvian wheel prevailed in the long run.

Water-mills eased the housewife's task, and introduced the miller. They were sources of profit to those who erected them upon their land, and a mill was attached to every manor that had a stream to turn it—that is, to about one-third of the Domesday manors. Mills are recorded in clusters, with the greatest density around the central highland and in the east facing the European continent. On the average there was one mill for every fifty households. This distribution supports the notion that the use of water-power was introduced from the continent, first into Kent then into Lincolnshire and Norfolk.

The corn-mill was soon followed by industrial mills. Domesday Book itself mentions a few stamping-mills for crushing ore, and hammer-mills. Later we hear of fulling-mills (1168), tanning-mills (1217), paint-mills (1361), and saw-mills (1376). By the fourteenth century every village of moderate size had a mill of the new type, rented by its miller from the manorial lord; conflicts between lord and tenant over mills figure largely in legal documents. Mills in the river valleys played a large part in technological developments, such as the thirteenth-century substitution of fulling with water-driven hammers for the old hand- or foot-fulling. This caused a migration of the fulling-trade to the country, freeing the craftsmen from restrictive local regulations [49].

We find a similar picture in central Europe. By the eighth century water-mills were well established in Thuringia, while the Bohemian chronicle of Václav Hajek (Wenceslaus Hagecius, d 1552) claims that the first water-mill in Bohemia was built in A.D. 718—though this is now generally discredited [50]. The Emperor Henry I is said to have founded Goslar in 922 on a spot where once a water-mill stood. The town of Augsburg built its *Lechwehr*, a weir to provide water to its mills, about 1000. A water-mill with a mill-pond, a water-race, and an artificial conduit cut into the rock was built near the abbey of Viecht in the Inn valley in 1097 [15]. By the twelfth century water-mills had penetrated into Scandinavia; they reached Iceland and Poland about 1200. In 1242 mills were built on the Oder near Breslau and in 1272 the citizens obtained a privilege concerning the conveyance and supply of water. The first German paper-mill was built near Nuremberg by Italians in 1389.

The water-mill was early applied to metallurgy in central Europe, powerful water-driven hammer-forges and bellows (figures 554, 555) contributing materially to the increase in size and quality of wrought iron objects, and to the new

FIGURE 554—*Water-driven hammer-forge. Strada, 1617.*

process that produced cast iron in large quantities for the first time. Thus water-power became the basis of mining and metallurgy. In this role it penetrated to regions still unconquered. By the twelfth century it was used in the copper-mines of the Harz and the silver-mines of Trient. Water-power followed silver- and copper-mining into the Alps and Scandinavia (sixteenth century). Water-driven forge-hammers were common in the thirteenth century; early in the fourteenth water-mills were used in the production of drawn wires (figure 39); water-wheels operated mining-hoists and grinding-stones from the thirteenth century; and in the fifteenth there were water-driven machines for boring gun-barrels.

FIGURE 555—*Water-driven bellows. Ramelli, 1588.*

This widespread use of the water-wheel in industry encouraged the improvement of gearing, and of practical mechanics generally, as is shown by the engineers' handbooks of the early sixteenth century, but unfortunately little is known of the construction of medieval water-driven machines. Transport also was made easier, as cities and industries tended to grow where there was running water and therefore a supply of energy.

We have only disconnected information on the spread of the water-mill in the east, whither it certainly travelled during the later Roman and early Byzantine periods. It was well known in China in the seventh century and was introduced into Japan for irrigation about 800. The Talmud refers to water-mills, which seem to have been brought into Palestine during the Roman domination. In Islam there was a very strong tradition of Hellenistic engineering skill, and the works of Philo, Vitruvius, Hero, and others were early available in Arabic. A Greek emissary is said to have built a large water-mill at Baghdad in 751, though another, more probable, account places the arrival there at about 790. In the main the water-mill was used for irrigation, but the Muslims improved the older methods of construction and encouraged the use of water-power. Many cities on the Euphrates and the Tigris had floating mills, and the coastal town of Basra on the Persian Gulf even had tide-mills, mostly for grinding corn. Syria was very famous for its water-wheels, notably the big wheels on the Orontes at Antioch and at Hama, where there were no fewer than thirty-two about the year 1300. Qaisar ibn Abu'l-Qāsim (d 1251) was an especially celebrated builder of water-wheels. It was formerly supposed that the crusaders, impressed by these machines, had introduced them into western Europe. But though this eastern revival of Hellenistic mechanics coincided with the beginning of the fuller utilization of water-power in the west, the trend of development was very different. When the water-mill had become a major source of power in the west, Al-Qazwīnī (1203–83) could describe only various water-wheels in his 'Geography', and Al-Jazarī in 1205 mentions a water-wheel as among the many amusing automata one could make. Apart from its use in irrigation, the water-wheel never became a principal prime-mover in the Muslim world, as economic conditions did not stimulate further developments.

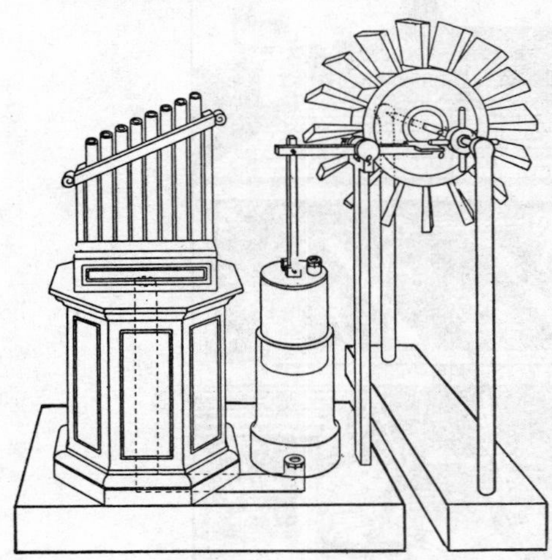

FIGURE 556—*The wind-organ of Hero, as reconstructed in the sixteenth century on the basis of his texts.*

IX. WINDMILLS OF THE EAST

It is generally agreed that the windmill was unknown to the Greeks and Romans [51]. Hero of Alexandria describes a wind-organ, the piston providing the air for the organ pipes being moved by a wheel 'which has oar-like scoops as in the so-called wind-motors' (figure 556). The Greek word is *anemourion* (wind-vane), a term which, except in this text, occurs only in the writings of a twelfth-century bishop. Some of the drawings in the Hero manuscripts show a kind of windmill moving the piston, but they are later additions, to be dismissed as interpretations by Christian or Muslim scribes who knew the windmill proper. As no other classical engineer mentions anything like a windmill we must conclude that it was unknown, despite Hero's *anemourion* [52].

The ancestor of the eastern windmill may be the prayer-wheel, in which scoops catching the impact of the wind rotate a vertical axle. Chinese travellers found this in use in central Asia about A.D. 400. The first real windmill is reported from Islam. During the reign of the Caliph Omar I (634–44) a certain Persian claimed to be able 'to build a mill that is rotated by the wind', and when he affirmed this claim on interrogation the Caliph made him build one [53].

FIGURE 557—*The horizontal windmill described by Al-Dimashqī, c 1300. The sails below turn a vertical shaft; this rotates the upper stone, above which is a hopper for supplying grain to the mill-stones.*

Our next information comes from two Persian geographers, the earliest writing about 950, who speak of windmills in the Persian province of Seistan:

> The region consists of evaporated salt-lakes and sand, and is hot. In it we find clusters of palm-trees and no snow falls there. The land is flat and one sees no mountain. There strong winds prevail, so that, because of them, mills were built rotated by the wind.
>
> The masses of sand in this region drift hither and thither, and if no measures had been taken they would have overwhelmed communities and cities. I have heard that, when they want to remove sand from some place without spreading it over the adjoining fields, they enclose the sand with a structure like a fence, of timber, thorns, and the like, higher than the sand-dune. In the lower part they make a door. Through this the wind enters and blows away the upper levels of the sand like a whirlwind, so that it falls harmlessly far away [54].

The second paragraph, on the control of drifting sand, has been cited to show that the inhabitants of Seistan seem to have been quite capable of directing air-currents at this early date. They applied their skill chiefly to the building of windmills. These are first mentioned by the Arab geographer Al-Masʿūdī (A.D. 947), who says: 'Seistan is a land of wind and sand; the region is characterized by the fact that there the wind turns mills which pump water from wells to irrigate the gardens. There is no place on earth where people make more use of the wind' [55]. About three centuries later the Persian scholar Al-Qazwīnī (d 1283) relates of Seistan: 'There the wind never rests, and they build their mills to make use of it. They grind corn only with these mills. It is a hot country and it has mills using the wind' [56]. He goes on to describe the inhabitants' control of drifting sand in terms that are repeated by other Arabic geographers [57]. Thus apart from this local use of the wind to shift sand we find windmills in Seistan grinding corn and pumping water.

It would seem that Seistan was the home of the eastern windmill and that its invention should be placed in early Muslim or even pre-Muslim times. The first description of its construction is given (together with a drawing) by the Syrian cosmographer Al-Dimashqī (1256/7–1326/7) (figure 557):

> When building mills that rotate by the wind [the Seistanis] proceed as follows. They erect a high building like a minaret, or they take the top of a high mountain or hill, or a tower of a castle. At the top they build a two-storey structure. The upper part contains the mill that turns and grinds; the lower one contains a wheel rotated by the enclosed wind. When the lower wheel turns the upper mill-stone turns too. Whatever wind may blow, the mills rotate, though only one stone is moved.
>
> After they have completed the double structure, as shown in the drawing, they make four slits or embrasures [in the walls] like loopholes, but reversed, for the wider part opens outward and the narrow slit is inside, forming ducts for the air, in such a way that the wind penetrates into the interior with force, as from the goldsmith's bellows, . . . from whatever direction the wind may blow. When the wind has entered this structure through the entrance prepared for it, it finds in its way a 'reel' like that on which the weavers wind one thread over another. This device has twelve arms or one could use six. Upon these fabric has been nailed, like the covering of a lantern, only in this case the fabric is divided over the different arms so that each is covered individually. The fabric has a bulge which the air fills and by which the arms are pushed forward. Then the air fills the next one and pushes it on, then it fills the third. The reel then turns; and its rotation moves the mill-stone and grinds the corn. Such mills are required on high castles and in regions which have no water but a lively movement of the air [58].

A later historian tells us: 'In Afghanistan all windmills and water-raising wheels are driven by the north wind, to which they are orientated. Attached to the windmills there are series of shutters which are closed to shut out, and opened to admit, the wind. For if the wind is too strong the flour burns and becomes black; even the millstone itself may grow hot and disintegrate' [59].

This type of windmill appears to be directly related to the Greek or Norse mill, which originated in the mountain range of north-west Persia. The windmill seems to be a local Persian adaptation of the Greek mill to a region where there is no water but where steady winds prevail. After having been confined to Persia and Afghanistan the invention subsequently spread throughout Islam, and beyond it to the Far East in the twelfth century. In Islam, India, and China the primitive windmill became an important prime-mover, used to grind corn, to pump water, to crush sugar-cane, and so on. From the Egyptian sugar-cane industry they were, centuries later, introduced into the West Indies when experts from Egypt went there to help in establishing the first sugar-plantations. Owing to the devastation of irrigation-works by Mongols and Turks the windmill has now practically disappeared from Egypt and the Near East, except in Persia.

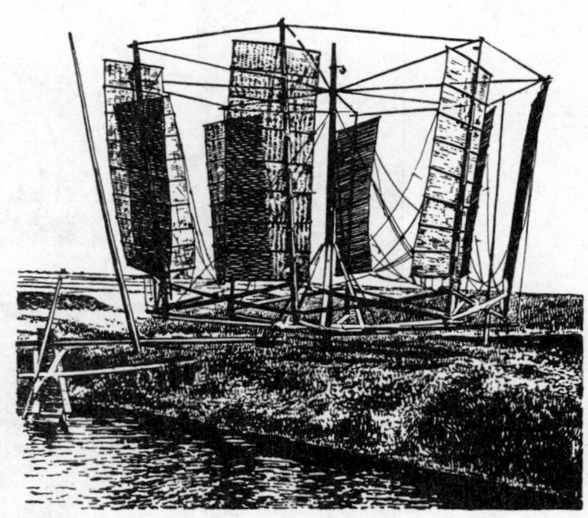

FIGURE 558—*Modern Chinese windmill used with a chain-pump for irrigation.*

For irrigation and for the pumping of brine from deep wells, the Chinese still use a windmill derived from the Persian type, in conjunction with a chain-pump (figure 558). It is not built inside a mill-house but stands unprotected in the open, carrying on its frame sails adapted from those of Chinese fishing-boats [60]. This form was first described in 1655, but apparently is much older.

X. THE WINDMILL AS PRIME-MOVER

There has been much speculation on the origin of the windmill in the west. Some have contended that it passed from Persia to north-western Europe by way of the trade-routes through Russia and Scandinavia. The order of the occurrence of the first windmills in the different European countries suggests rather that it might have come from Islam by way of Morocco and Spain. On the other hand, the construction of the western mill (which has, from the earliest post-mill onwards, four vertical sails attached to a horizontal shaft) is entirely different from that of its eastern counterpart.[1] Thus it does not seem to derive from the Greek

[1] Aerodynamically, the western mill—a kind of air-screw reversed—is far more advanced and ingenious than the eastern. In the former the wind impinges on the whole of the sail-area all the time; in the latter on only a

water-mill and may have some completely different origin. Some writers have accordingly derived it from the Vitruvian water-mill. As a wind-vane of a sort was known to Hero in antiquity (p 615), it is just possible that this idea finally emerged in the western windmill. The mechanical arrangements of the post-mill

FIGURE 559—*Modern Aegean windmills showing the triangular sails, spread out like the sails on a boat.*

resemble those of a Vitruvian mill with the drive inverted; that is, down from the sails rather than up from the water-wheel. A mill-wright familiar with the Vitruvian water-mill could have constructed the mechanism of the post-mill, if he had known how to shape and trim the sails to obtain continuous rotation. This new aerodynamic device was the essential and novel invention in the western windmill (pp 623–8).

We should like to know more of the history of the quaint windmills of Crete and the Aegean islands (figure 559), a few examples of which can be found as far west as Portugal. They have sails set like those of a boat and not spread on a lattice framework. The shaft has eight or twelve arms, slender round poles supported by wire stays. The sails are triangular, with their base towards the circumference and their apex pointing to the centre. One edge is attached to the end of the pole, the other to a rope. While the mill is idle, the sails are twisted along the poles; when in use they are unfurled to a greater or less extent. The

part of the sail-area. Thus the problem of obtaining continuous rotary motion from the constant pressure of wind on a plane surface is solved far more efficiently by the western type of sail, the differentiation of which in principle from the eastern is so great that it cannot be called a mere adaptation.

top of the mill can be turned towards the wind by using a long pole as a lever to move the conical cap, which is supported on a ring of hard wood resting on the tower itself. It has been suggested that the root idea of the western mill is to be found in both the wind-motor and the sails of a ship. But these Cretan windmills

FIGURE 560—*Post-mills and tower-mills in the Low Countries. Stradanus, c 1600.*

are tower-mills, and we know of no tower-mill antedating the post-mill. Thirteenth- and fourteenth-century manuscripts do not show jib sails, which are found only in the Mediterranean area and the Iberian peninsula. These mills are therefore probably later adaptations of the western tower-mill to Mediterranean conditions [61].

By the early thirteenth century the windmill became the typical prime-mover of the plains stretching from eastern England across the Low Countries (figure 560) and north Germany into Latvia and Russia. In its centre, in the Low Countries, the windmill is now primarily used as a prime-mover for water-lifting or pumping-machinery [62]. It has been claimed that the Cistercian monks introduced it during the large-scale drainage of the lakes and fens in these regions. The production of peat for the larger settlements here formed shallow lakes, which grew larger in the period of 1000–1400. The systematic drainage of these lakes and fens started about 1300, after the encroachment of lakes and sea had been temporarily stopped by dikes and dams. Contrary to expectation, however, the first windmills of the Low Countries were corn-mills. The oldest Dutch document is that of Count Floris V granting the burghers of Haarlem the privilege of paying six shillings tax for a windmill and three for a horse-mill, whereas non-citizens had to pay twenty shillings (1274). In the year 1299 the windmill of the former monastery Koningsveld at Delft was built, and accounts of the Count of Gelre for the year 1294 show the building of a corn-mill at Logchem. Only in 1430 do we hear of the first marsh-mill driven by the wind (called *wipmolen*, p 625) and not until about 1600 did it become a common feature of the Dutch drainage-system (p 689).

In 1222 the town of Cologne built a mill on its town wall, a practice common in late-medieval towns. The Dutch must have quickly become notable mill-wrights, for a mill at Cologne of 1392 was built by Dutchmen, as was one at Speyer a year later. In 1237 we hear for the first time of an Italian windmill, at Siena; in 1337 another is mentioned near Venice. Through central Europe and southern Scandinavia the windmill reached Finland about 1400.

Some little-known suggestions for the use of the windmill indicate appreciation of its unusual adaptability. Thus a Munich manuscript of 1430 proposes to use a wind-shaft with sails to act as a brake when lowering heavy loads, such as blocks of natural stone. In the book dedicated by Walter de Millinate to Edward II of England in 1326 the author illustrates the use of a post-mill for flinging beehives into a besieged town. Lastly, the Italian physician Guido da Vigevano sketched in 1335 a design for a fighting-car or 'tank' to be driven by a wind-shaft with sails (figure 659). A similar scheme was put forward in the fifteenth century by the military engineer Roberto Valturio (d 1484).

REFERENCES

[1] Cranstone, B. A. L. *Man*, **51**, 48, 1951.
[2] Forbes, R. J. *Arch. int. Hist. Sci.*, no. 12, 599, 1950.
[3] Mendelsohn, I. 'Slavery in the Ancient Near East.' University Press, Oxford. 1949.
[4] Westermann, W. L. *J. econ. Hist.*, **2**, 149, 1942.
[5] Lefebvre des Noëttes, R. J. E. C. 'L'attelage et le cheval de selle à travers les âges' (2 vols). Picard, Paris. 1931.
[6] Needham, J. *J. R. cent. Asian Soc.*, **36**, 139–41, 1949.
[7] Carnat, G. 'Le fer à cheval à travers l'histoire.' Vigot, Paris. 1952.
[8] *Anthologia Graeca*, IX, no. 418. (Loeb ed. Vol. 3, p. 232, 1918.)
[9] Strabo XII, C 556. (Loeb ed. Vol. 5, p. 428, 1928.)
[10] O'Reilly, J. P. *Proc. R. Irish Acad.*, **29**, sect. C, 55, 1902 (1904).
Curwen, E. C. *Antiquity*, **18**, 130, 1944.
[11] Mayence, F. *Bull. Mus. R. Art Hist.*, **1**, 5 and fig. 5, 1933.
[12] Pliny *Nat. hist.*, XVIII, xxiii, 97. (Loeb ed. Vol. 5, p. 250, 1950.)
[13] Steensberg, A. 'Farms and Water-mills in Denmark through 2000 Years', p. 294. Nationalmuseet, 3. Afd. Arkaeol. Landsbyundersøgelser, no. 1. Copenhagen. 1952.
[14] Bennett, R. and Elton, J. 'History of Corn Milling', Vol. 2, chap. 3. Simpkin, Marshall, London; Howell, Liverpool. 1899.

[15] REINDL, C. *Wasserkraftjb.*, **1**, 1, 1924 (1925).
[16] VITRUVIUS *De architectura*, X, v. (Loeb ed. Vol. 2, p. 304, 1934.)
LUCRETIUS *De rerum natura*, V, line 515 f. (Loeb ed, p. 378, 1924.)
[17] PARSONS, A. W. *Hesperia*, **5**, 70, 1936.
[18] WIEDEMANN, E. *Mitt. Gesch. Med. Naturw.*, **15**, 368, 1916.
[19] AUSONIUS *Mosella*, lines 362–4. (Loeb ed. Vol. 1, p. 252, 1919.)
FORTUNATUS *Carmina*, III, no. 12, lines 37–38. Ed. by I. LEO. Mon. Germ. hist., *Auctores antiquissimi*, Vol. 4, Pt I, p. 65. Weidmann, Berlin. 1881.
[20] JACONO, L. *Ann. Lavori pubbl.*, **77**, 217, 1939.
REINDL, C. *Wasserkraft, Münch.*, **34**, 142, 1939.
[21] BENOIT, F. *Rev. archéol.*, sixième série, **15**, 19, 1940.
SAGUI, C. L. *Isis*, **38**, iii and iv, 225, 1948.
[22] BRETT, G. *Antiquity*, **13**, 354, 1939.
[23] BUCKLER, W. H. and ROBINSON, D. M. 'Sardis', Vol. 7, Pt I, no. 169. Publications of the American Society for the Excavation of Sardis. Brill, Leyden. 1932.
[24] PALLADIUS *De re rustica*, I, xli (xlii).
[25] ASHBY, T. 'The Aqueducts of Ancient Rome' (ed. by I. A. RICHMOND), p. 46. Clarendon Press, Oxford. 1935.
[26] CODEX THEODOSIANUS, XV, 2, 6. (*Theodosiani Libri XVI*, ed. by T. MOMMSEN and P. M. MEYER, Vol. I, Pt II, p. 816. Weidmann, Berlin. 1905.)
[27] PROCOPIUS History of the Wars, Bk. V (The Gothic War), xix, 8. (Loeb ed. Vol. 3, p. 186, 1919.)
[28] CORPUS JURIS CIVILIS. *Codex Justiniani*, XI, 43, 10.
CASSIODORUS *Variae*, III, xxxi, in MIGNE *Pat. lat.*, Vol. 69, col. 593. For an Eng. trans. see HODGKIN, T. 'The Letters of Cassiodorus', p. 213. Frowde, London. 1886.
[29] *Lex Visigothorum*, VIII, 4, 30. *Mon. Germ. hist.*, *Leges*, sect. I, Vol. 1. Hahn, Hanover, Leipzig. 1902.
[30] DRACHMANN, A. G. 'Ktesibios, Philon and Heron.' *Acta historica scientiarum naturalium et medicinalium*, Vol. 4. Bibliotheca Univ. Havniensis, Copenhagen. 1948.
REHM, A. *Arch. Kulturgesch.*, **38**, 135, 1938.
[31] PLINY *Nat. hist.*, XXXIV, vi, 11. (Loeb ed. Vol. 9, p. 134, 1952.)
SAINT AUGUSTINE *De civitate Dei*, VII, 4. (Trans. by M. DODDS, Vol. 1, pp. 264 f. Hafner, New York. 1948.)
[32] PAPPUS of Alexandria *Collectiones mathematicae*, VIII, 1. (French trans. by P. VER EECKE. Vol. 2, p. 810. Desclée de Brouwer, Paris, Bruges. 1933.)
[33] TORREY, H. B. *Amer. Nat.*, **72**, 293, 1938.
GOMPERTZ, H. *J. hist. Ideas*, **4**, 161, 1943.
[34] PLUTARCH *Vitae parallelae: Marcellus*, xiv, 4. (Loeb ed. Vol. 5, pp. 470 ff., 1917.)
[35] DIJKSTERHUIS, E. J. *Hermeneus*, **18**, 23, 1946.
[36] BACON, FRANCIS. *Novum organum scientiarum*, I, v, Aphorism 85.
[37] OERTEL, F. *Rheinisches Museum*, new series, **79**, 230, 1930.
FORBES, R. J. *Arch. int. Hist. Sci.*, no. 8, 919, 1949.
[38] ROSTOVTZEFF, M. 'Social and Economic History of the Hellenistic World' (3 vols). Clarendon Press, Oxford. 1941.
[39] THOMPSON, E. A. (Ed. and Trans.). 'A Roman Reformer and Inventor. Being a new text of the Treatise *De rebus bellicis*', p. 107. Clarendon Press, Oxford. 1952.
[40] GEOGHEGAN, A. T. 'The Attitude towards Labor in Early Christianity and Ancient Culture.' Studies in Christian Antiquity No. 6. Catholic University of America, Washington. 1945.

[41] THOMPSON, E. A. See ref. [39], pp. 50 ff.
[42] PROCOPIUS. See ref. [27], xix, 19–20. (Loeb ed. Vol. 3, p. 190, 1919.)
[43] BÉLIDOR, B., FOREST DE. 'Architecture hydraulique, ou l'art de conduire, d'élever et de ménager les eaux pour les différens besoins de la vie', Vol. 1, chap. 2: "Moulins à scier le bois, le marbre, et à percer des tuyaux", p. 321. Jombert, Paris. 1737.
[44] GILLE, B. 'Esprit et civilisation technique au Moyen Âge.' *Conf. du Palais de la Découverte, Janvier 1952*. Paris.
[45] GREGORY of Tours. *Historia Francorum*, III, xix; *Vitae patrum*, XVIII, 2 (ed. by W. ARNDT and B. KRUSCH), *Mon. Germ. hist.*, *Scriptores rerum Merovingicarum*, Vol. 1, pp. 129 and 734 f. Hahn, Hanover. 1885.
[46] *Lex Salica* ed. by J. F. BEHREND (2nd ed.), XXII, 1–3, pp. 39 f. Böhlau, Weimar. 1897; ed. with German trans. by K. A. ECKHARDT, XXIX–XXX. Germanenrechte, Neue Folge, Abt. Westgermanisches Recht, Böhlau, Weimar. 1953.
[47] *Capitularia regum Francorum* ed. by A. BORETIUS, no. 32 (*Capitulare de villis*), p. 89. *Mon. Germ. hist.*, *Leges*, sect. II, Vol. 1, Hahn, Hanover. 1883.
[48] HODGEN, M. T. *Antiquity*, **13**, 261, 1939.
[49] CARUS-WILSON, ELEANORA M. *Econ. Hist. Rev.*, **11**, 39, 1941.
[50] HAJEK, VÁCLAV. 'Kronika Czeska', fol. VIII[v]. Jan Severýn the younger and Ondře Kubes, Prague. 1541; German trans. by J. SANDEL. 'Böhmische Chronica Wenceslai Hagecii', fol. 10[v]. Weidlich, Prague. 1596.
[51] TITLEY, A. *Trans. Newcomen Soc.*, **3**, 41, 1924.
VOWLES, H. P. *Ibid.*, **11**, 1, 1930–1.
[52] HERO of Alexandria *Pneumatica*, I, 34 (ed. by W. SCHMIDT, Vol. 1, pp. 204–7. Teubner, Leipzig. 1899. Eng. trans. by B. WOODCROFT, p. 108. Whittingham, London. 1851.)
[53] AL-MAS'ŪDĪ, 'ALĪ IBN ḤUSAIN. 'Les prairies d'or' (ed. and trans. by BARBIER DE MEYNARD and PAVET DE COURTEILLE), Vol. 4, p. 227. Société Asiatique, Paris. 1865.
MUIR, SIR WILLIAM. 'The Caliphate' (new rev. ed. by T. H. WEIR), p. 187. Grant, Edinburgh. 1915 (Reprinted 1924).
[54] AL-IṢṬAKHRĪ, IBRĀHĪM IBN MUḤAMMAD AL-FĀRISĪ. 'Das Buch der Länder' (German trans. by A. D. MORDTMANN), p. 110. Akademie von Hamburg, Schriften. Vol. 1, Pt II, p. 134. Dieterich, Göttingen. 1845.
IBN HAUQAL, MUḤAMMAD ABU'L-QĀSIM. *Viae et regna* (ed. by M. J. DE GOEJE), p. 299. Brill, Leyden. 1873.
[55] AL-MAS'ŪDĪ, 'ALĪ IBN ḤUSAIN. See ref. [53], Vol. 2, p. 80. Paris. 1863.
[56] AL-QAZWĪNĪ, ZAKARĪYA IBN MUḤAMMAD. 'Kosmographie' (ed. by F. WÜSTENFELD), Pt II, p. 134. Dieterich, Göttingen. 1848.
[57] ABU'L-FIDĀ', ISMĀ'ĪL IBN 'ALĪ 'IMĀD AL-DĪN. 'Géographie d'Aboulféda' (French trans. by M. REINAUD and S. GUYARD), Vol. 2, Pt II, p. 105. Imprimerie Nationale, Paris. 1883.
[58] AL-DIMASHQĪ, MUḤAMMAD IBN ABĪ TĀLIB. 'Manuel de la Cosmographie du Moyen Âge' (French trans. by A. F. MEHREN), p. 247. Leroux, Copenhagen, Paris. 1874.
[59] MEZ, A. 'Die Renaissance des Islams' (ed. by H. RECHENDORF), p. 439. Winter, Heidelberg. 1922.
[60] VOWLES H. P. and VOWLES, M. W. 'The Quest for Power', p. 124. Chapman, London. 1931.
[61] COBBETT, L. *Antiquity*, **13**, 458, 1939.
BATHE, G. 'Horizontal Windmills.' Published by the author, Philadelphia. 1948.
[62] BOONENBURG, K. 'Onze Windmolens.' Heemschutserie no. 69. De Lange, Amsterdam. 1949.
Idem. 'Windmills in Holland.' Netherlands Government Information Office, The Hague. 1951.

A NOTE ON WINDMILLS

REX WAILES

THE earliest authentic mention of a windmill in western Europe yet discovered is in a deed, reliably dated as *c* 1180, recording the gift of land near a windmill to the abbey of St Sauvère de Vicomte in Normandy [1]. Research has multiplied the number of later references, increasing in frequency with the passage of the centuries, without disclosing any of earlier date.

The following century provides the first known illustration of a windmill, in an initial letter of the 'Windmill Psalter' written at Canterbury *c* 1270, and now in New York (figure 561) [2]. This shows quite clearly a post-mill of orthodox construction which was, it may safely be assumed, a feature of the landscape familiar to the artist who painted it. The fact that the windmill was becoming fairly commonplace in the late thirteenth century is consistent with a first reference to it about a century earlier. Early references describe a windmill specifically as *molendinum ad ventum*, other mills being known simply as *molendina*.

The post-mill consists of a box-like wooden body, carrying the sails and containing the machinery, mounted on a suitably braced upright post on which it can be turned so that the sails may face the wind (figures 561–3). The timber sub-structure supporting the post rests on brick or stone piers, but is not fixed to them in any way, the mill being so constructed as to make it inherently stable on the piers (figure 592).

One might well suppose that, in the transition from the Vitruvian water-mill to the

FIGURE 561—*Post-mill. From the illuminated initial on the first page of an English psalter. This is said to be the earliest representation of a windmill in a book. 1270.*

FIGURE 562—*Post-mill showing the wooden body of the mill and the long pole with which the whole mill was turned to face the wind. From an Italian manuscript. Early fourteenth century.*

post-windmill as we know it, there would have been an intermediate stage with the stones placed above the main driving-shaft; in effect a water-mill on stilts with windmill-sails substituted for the water-wheel. No evidence for such an intermediate stage has, however, been discovered, which suggests that the twelfth-century mill-wrights possessed sufficient originality and initiative to make the advance at a single jump.

In the U.S.S.R., however, a post-mill of this intermediate form was photographed in

FIGURE 563—*Post-mill with beam illustrating the 'Physics' of Aristotle. From an initial in a French manuscript. Fourteenth century.*

FIGURE 564—*Tower-mill from a fifteenth-century stained glass window at Stoke-by-Clare, Suffolk.*

the 1920's by some English travellers. Although, judging by its appearance, this mill was of no very great age, it is of interest as showing that the construction suggested above is possible, though clumsy. It is also inefficient, since (*a*) the sails are too low to catch the wind satisfactorily; (*b*) there is little or no storage space at the top to feed grain into the hoppers of the stones by gravity; (*c*) the stones are too high up in the mill for convenience in working; and (*d*) the main drive has to be passed to obtain access to the stones.

Another type is the sunk post-mill. The maritime 'Judgements of Oléron', said to have been established by Eleanor of Aquitaine about the middle of the twelfth century and adopted in England *c* 1314, contain the earliest reference to this type of mill:

> And likewise with regard to windmills, some of which are altogether held above the ground, and have a high ladder, and some have their foot fixed in the ground, being as men say, well affixed; and accordingly they are not movable, for they cannot be detached from the ground nor removed without damage to their original structure [3].

Remains of this type of mill have been excavated in England at various places, the most recent being at Sandon Mount, Hertfordshire [4]; they show that the whole substructure of the mill, including the post and its supporting timbers, was buried in the ground. The idea seems to have been to avoid any possibility of the mill being blown over forwards if taken aback in a high wind; the disadvantages were that its effective height was lowered and that it therefore caught the wind less readily, while in addition the timber-work rotted easily.

Turning the body of the mill, complete with sails and machinery, to face the wind demanded considerable effort, and it is not surprising that the tower-mill is illustrated as early as 1390. Its invention has been erroneously ascribed to the mid-sixteenth century, but it seems logical, in view of this illustration, to put it at the beginning of the fourteenth. In this type of mill a tower of brick or stone contains the machinery, and only the top or cap carrying the sails is turned to face the wind (figures 564, 565).

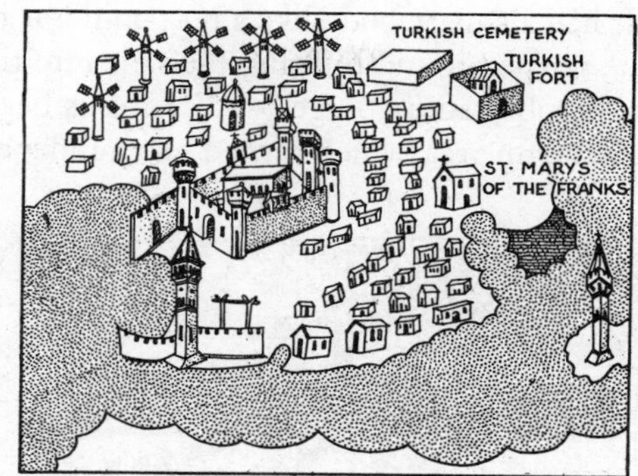

FIGURE 565—*Byzantine tower-mills at Gallipoli. Pen-drawing from a traveller's account. 1420.*

Further incentives to the use of tower-mills may well have been the scarcity of sufficiently massive timber for the construction of post-mills, and the desirability of erecting windmills on the walls of towers and castles. In such positions a wooden post-mill would have been very vulnerable to enemy attacks which a stone tower-mill could withstand (figure 566).

Up to this time windmills were used solely for grinding corn, but in 1430 the Dutch invented the *wipmolen* or hollow post-mill and applied it first to drainage (p 689) and then to other purposes. It had a small post-mill body with a shaft passing down through the centre of the post to operate machinery below.

In 1592 the first wind saw-mill was built at Uitgeest, Holland, by Cornelis Cornelisz.

FIGURE 566—*Tower-mills on the Island of Rhodes. Many more are standing on the outer harbour-wall. Part of a woodcut, 1486.*

It had a square body like a post-mill but higher, and was mounted on a raft which could be warped round to bring the sails into the wind. Logs were delivered and sawn timber was collected by barge [5]. From this beginning sprang the fine *paltrok* saw-mills of the Zaandam area, which turned on an independent ring of rollers running on a track laid on

FIGURE 567—*A tower-mill showing the internal construction of the mill and the method by which the top only of the mill was turned. The piece of pipe shows the very simple but very efficient ball-and-chain method of raising water. Ramelli, 1588.*

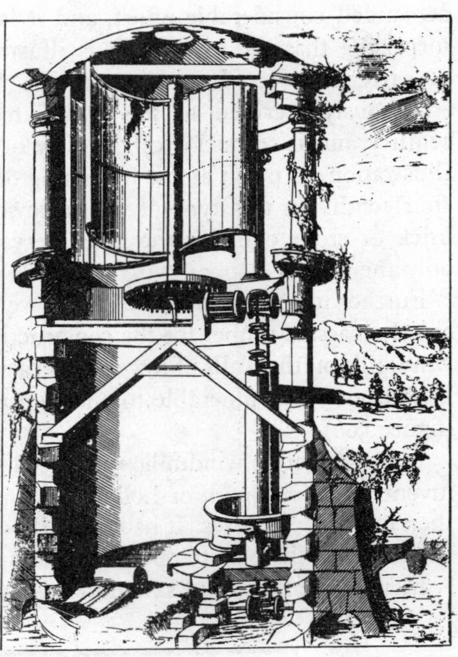

FIGURE 568—*A tower-mill with canvas sails under a dome. This is the earliest known illustration of a horizontal windmill. Besson, 1578.*

a low brick base. This would seem to be a very early example of an independent roller-bearing.

The mechanician Ramelli in 1588 [6] is the first to depict the internal working-parts of windmills; he shows both post- and tower-mills for grinding corn. While the body of a tower-mill could be, and later was, enlarged to contain more stones and machinery than that of a post-mill, this development does not appear to have been made at that period. The limiting factor was probably the efficiency of the sails, which, judging by their design, would not have generated sufficient power in an ordinary good wind to drive more than one pair of stones and the sack-hoist.

Ramelli also shows a tower-mill operating a chain-pump (figure 567). A chain-pump is again shown by Jacques Besson in the earliest known illustration of a horizontal windmill. His book *Théâtre des Instrumens Mathématiques* [7] was written before 1569 and published posthumously in 1578. The horizontal mill has curved sails made of canvas mounted under a dome at the top of a tower (figure 568).

Veranzio [8] in 1595 illustrates five different types of horizontal mill. One resembles Besson's, but with straight wooden sails and fixed curved guide-vanes surrounding them; another has fixed, straight windshields with gaps between them, and provision for detachable doors to close the gaps according to the direction of the wind. A third is like a huge chimney-cowl with a conical roof and is made of timber (figure 570); a fourth has four timber sails like long horizontal V's; while a fifth has canvas-spread hinged and feathering shutters in the sails (figure 569).

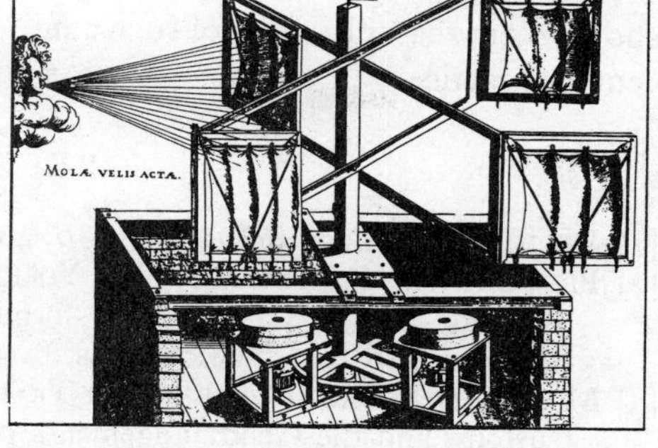

FIGURE 569—*Hinged sails of canvas are shown in this section of a tower-mill, as well as the clasp-arm construction of the gear-wheels. Veranzio, 1595.*

Three out of the five illustrations show gear-wheels of clasp-arm construction, with four arms crossing each other to form at the centre a square through which the main shaft passes (figures 569, 570). These illustrations of Veranzio are the earliest examples of the improved clasp-arm wheels in windmills. The earlier compass-arm wheel construction entailed the mortising of the arms right through the shaft, thus weakening it where strength is most needed.

How many of these early designs were actually put into practice is not known, but it must not be assumed that all those depicted in the many books describing improved machines had been tested or were indeed even practicable. Many were no more than ingenious suggestions or pious hopes.

The dates inscribed on continental windmills are often equally misleading. In the eighteenth century the figure 7, as then carved, much resembled the modern continental figure 1 with a serif, and the modern continental 7 with the crossed down stroke was unknown. As a result honest and enthusiastic, but uncritical, antiquarians have accepted without question the assertions of proud owners that their windmills date from

FIGURE 570—*A tower-mill with sails of wood under a conical roof. Two sets of mill-stones are shown being worked simultaneously; with a less powerful wind only one set would be turned. Veranzio, 1595.*

the twelfth century (for example, 1769 misread as 1169) when a little reflection would have shown them that the design of such a mill was an example of eighteenth- and not twelfth-century practice.

REFERENCES

[1] DELISLE, L. *J. Brit. archaeol. Ass.*, **6**, 403, 1850.

[2] PIERPONT MORGAN LIBRARY, NEW YORK. 'Catalogue of Manuscripts from the Libraries of William Morris, [*et al.*] now forming a portion of the Library of J. Pierpont Morgan', MS 19. Chiswick Press, London. 1907.

[3] MONUMENTA JURIDICA (ed. by SIR TRAVERS TWISS). "The Good Usages and the Good Customs and the Good Judgements of the Commune of Oleron", chap. xci in Vol. 2, p. 386. Rolls Series, London. 1873.

[4] WESTELL, W. P. *Trans. St. Albans and Herts. archit. and archaeol. Soc.*, new series, **4**, 173–83, 1934.

[5] VAN NATRUS, L., POLLEY, J., and VAN VUUREN, C. 'Groot Volkomen Moolenboek' (2 vols). Johannes Covens and Cornelius Mortier, Amsterdam. 1734; 1736.

[6] RAMELLI, A. 'Le diverse et artificiose machine.' Published by the author, Paris. 1588.

[7] BESSON, J. 'Théâtre des instrumens mathématiques.' B. Vincent, Lyons. 1579.

[8] VERANTIUS, F. *Machinae novae*, Pls XIII, XII, XI, IX, VIII. Venice. 1620 (?).

'The Mills of Babylon.' A black and white timber building was built on piles over the river, and under the arches of this bridge-like structure were fixed the three water-wheels. From a fifteenth-century French manuscript.

18

MACHINES

BERTRAND GILLE

I. ORIGIN OF THE MACHINE

THERE was much vagueness in the use of the word *mēchanē* by the Greeks and of the cognate word *machina* among the Romans. These words were generally used to denote any ingenious invention. There was, however, another and more precise use. Thus the Greeks differentiated between two types of machines: *simple machines*, of which there were five, namely lever, wedge, screw, pulley, and winch, to which a sixth, the inclined plane, was sometimes added; and *complex machines*, which were combinations of simple machines.

In practice the word machine seems to have been applied to any instrument interposing a certain number of moving parts between driving-force and acting-part. The winch intervenes between the man turning it and the tense rope. An inclined plane intervenes between the elevation to be attained and the cart to be raised. Nowadays we should say that some such simple elements are merely parts of machines; but the conscious distinction of these basic parts was a necessary step toward the improvement and application of more complicated devices.

It is obvious that nature has included the elements of machines in the structure of animals. There are no tribes, however primitive, that do not systematically use some forms of lever. Combined levers are also very frequent, either in the form of pincers, finally developed only when metal makes its appearance; or in the more primitive form of tongs, of which the earliest, the bent stick, is in effect two levers joined at one end and forced open by a spring. Eye-brow tweezers were widely used in the ancient civilizations and are known from the Bronze Age in Sumeria, the Indus valley, Egypt, western Europe, Africa, and eastern Asia; their use is even more widespread than that of the razor. Most agricultural implements are special adaptations of levers.

The lever is the basis of all machines, though for extensive use it must be combined with other mechanical elements. For long ages men could use it only in its simplest applications. Elaboration began with one other device, the spring, which strictly speaking is not a mechanical element but rather a means of storing energy. Springs also exist everywhere in nature—all that was needed was to imitate them. Bivalve shells may well have been employed as tongs, the two parts acting as levers and the joint as a spring. Most traps made by primitive peoples are mere combinations of levers and springs.

A great step forward in machinery was made with the introduction of the wheel. Its origin and early history have already been discussed (vol I, ch 9). There it was seen that from the wheel two essential machines were very early evolved, namely vehicles and the potter's wheel.

The pulley is a special form of wheel, the early history of which is very obscure. It brings in another element, namely the rope used merely as a flexible link; the driving-belt came much later. Assyrian representations of the pulley are doubtful, and it seems to have been unknown in ancient Egypt. It is mentioned in the work *Mechanica*, falsely ascribed to Aristotle (d 322 B.C.) but of his school and not much later than he. The pulley was known to Vitruvius (first century B.C.) and was in wide use in the early centuries of the Roman Empire (pp 658 ff). Of the simple machines the pulley and wedge, but hardly the winch or the screw, have perhaps analogues in anatomy.

II. GREEK MACHINERY

Among the Greeks before the sixth century B.C. mechanical devices were still much as in the ancient empires, where the introduction of metal-working brought about the improvement of only a few simple machines such as tongs and pincers. From the sixth to the fourth century the Greeks made remarkable progress in mechanization. This was connected with the evolution of two techniques characteristic of their culture, namely navigation and the theatre. The stage and the ship used certain very similar machines. The pulley and its indispensable complement, the winch, characterize both. The pulley is a grooved wheel around which a rope passes, while the winch is a rotating drum to which lever-arms are fitted. The winch may moreover be used in various forms with either a horizontal or a vertical axle. Certain plays, notably those of Euripides (480?–406? B.C.) require extensive machinery, for in them a god is often let down on to the stage, as though from heaven, to clear up a human entanglement. This is the *theos ek mēchanēs*, *deus ex machina*, 'the god from the machine', in the sarcastic phrase of Lucian (A.D. *c* 120–*c* 180). It seems that machinery was first used in drama in 427 B.C., but it is impossible to decide whether the winch and pulley were employed first in the theatre or in ships and ship-building. There are Greek surgical works of the fifth century B.C. that describe ways of straightening broken limbs by extension. Since these involve winches but not pulleys it seems likely that the former device preceded the latter.

The evolution of ship-building at this period proves that machinery for working timber had well advanced (pp 563–7). The wood-turning lathe appears from certain texts to have been known at the end of the sixth century B.C. The auger, a boring-tool turned by a lever-handle, of which medieval examples are shown in figure 359, has been attributed to the fifth century B.C. The bow-drill (figures 571, 204, 206), in which the drill-bit is driven by friction of the bow-string on the axle holding it, appeared in Greece at this time. There can be little doubt that this implement derives from Egypt, for it is represented there both in mural decorations and by actual examples (vol I, figures 112, 487–9). The motion transmitted is an alternating one, but this does not affect its purpose.

FIGURE 571—*Greek cabinet-maker using a bow-drill to pierce a hole in the lid of a chest. From an Attic vase, fifth century B.C.*

The auger, translating a circular motion to a linear motion along its axis of rotation, is related to the screw, which was certainly known before Archimedes (*c* 287–212 B.C.), to whom it has been falsely ascribed. It may have been invented by Archytas of Tarentum, a Pythagorean philosopher and mathematician (*c* 400 B.C.), though the evidence is unreliable (p 677).

Apart from the origin and history of screws is the problem of how the early examples were made. The spiral itself is a very ancient motif and is found in the flat on Neolithic monuments, as in Malta (*c* 2000 B.C.). Such spirals were doubtless traced by a piece of carbon attached to a thong, the other end of which was fastened to a stick projecting from the surface to be inscribed. Figures of this kind may then have suggested the spiral pointing of augers, for which the shells of many molluscs furnish a model. Such a spiral is different, however, from the screws of ancient machines, which were cut on cylinders, not on cones.

The earliest screws of cylinder-type were doubtless of wood, as were those of the presses found and figured at Pompeii (figure 193), and most of those discussed by Hero (figures 87 and 572). Screws of metal were, however, known in classical antiquity. For instruments of precision, such as Hero's *dioptra*, the cutting had to be fine and exact. The method of attaining this precision is known. A sheet of soft metal in the form of a right-angled triangle was wound round the cylinder so that one arm of the right-angle was parallel to the axis. The hypotenuse would then trace a spiral on the surface of the cylinder. This metal spiral was used as a guide to the file, chisel, or gouge cutting the spiral groove. One section of the spiral having been cut, the metal triangle could be moved along the cylinder and the spiral further extended. Spirals could also be multiplied, so that a screw might contain two or more worms. By the same means it was possible to cut the cogs of wheels obliquely. Such cog-wheels were in effect short screws with many worms.

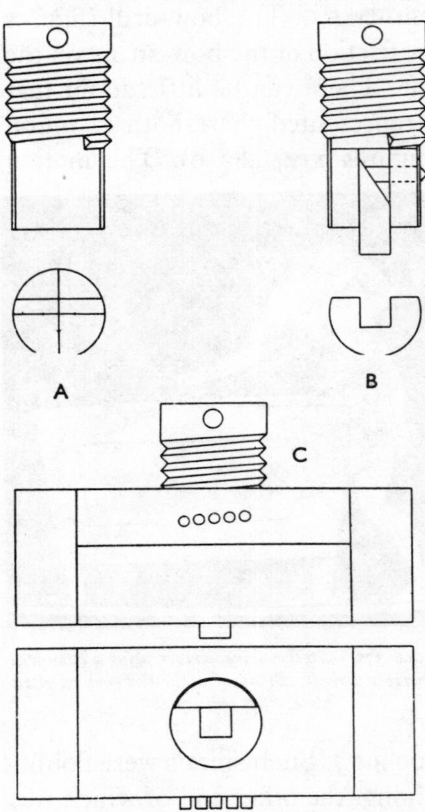

FIGURE 572—*Reconstruction of Hero's screw-cutter.* (A) *Side and bottom views of a male screw. It has a hole for a handle at its upper end, and a smooth peg at its lower end.* (B, above) *A slot has been cut into the peg, and an iron cutter—note the oblique cutting-edge—which is to cut the screw-thread of the female screw in a continuous line has been pegged into it.* (B, below) *The end of the peg seen from below.* (C, above) *The male screw with its peg thus prepared is here inserted into a hole in a plank. Five small pegs in the side of the plank form a provisional screw-thread to guide the cutter. The female screw is cut in the second plank, which also has had a hole bored through it.* (C, below) *Bottom view of the screw-cutter in action. Note that the third of the wedge sawn off in making the slot has been replaced.*

Among the Greeks of the fourth century B.C. machines had reached the point at which mathematical analysis could begin to take shape. According to Diogenes Laertius (third century A.D.) Archytas, the alleged inventor of the screw, was the first to apply geometry to mechanics. Plato indeed reproaches this early Pythagorean (fl *c* 400 B.C.) for seeking to solve certain problems of geometry with mechanical instruments. The Pythagoreans in fact were among the Greek thinkers closest to reality and experience; by losing touch with experience ancient learning was misled into formalism. The mathematical theory of the lever originated in the pseudo-Aristotelian *Mechanica* of the fourth (? third) century B.C. (p 630). This work consists of thirty-five discussions on problems related to the lever, many of which are not theoretical but eminently practical matters.

Archimedes of Syracuse (*c* 287–212 B.C.), greatest of all the ancient exponents of mechanics and one of the greatest mathematicians of all time, was the first of the long line of those who have promoted science by propounding practical and definite problems. Like Pythagoras and

Archytas, like the Egyptian and Babylonian surveyors before him, and like Leonardo and Galileo after him, Archimedes was a geometer because geometry is a technician's science. His researches on statics revealed the fundamental principles relating to the lever and the centre of gravity. His studies were of great assistance to those who sought to construct purposeful machines. These geometrical investigations of Archimedes were important, since, if we may believe Plutarch (*c* A.D. 46–*c* 125), they enabled him to calculate, for example, the number of pulleys needed for lifting a given weight with a given force. He knew how to calculate mechanical advantage, than which there is nothing more important for all lifting-devices.

According to legend, Archimedes invented and constructed many machines, especially for military purposes (pp 604, 699, 714), but on the nature of these devices the evidence is vague. The only machine which is both associated with his name and precisely known is the 'Archimedean screw' for irrigation (p 677). Here it seems that legend has incorrectly attributed to him a contrivance that had long been in use in Egypt.

The lead given by Archytas and Archimedes towards the more accurate study of machines was followed by others, particularly at Alexandria. The school there was active in the field of mechanics between the third and the first centuries B.C. Its members were, however, theoreticians rather than practical men; they certainly applied themselves to problems put forward by technicians but seem to have taken little part in actual machine-construction. Rarely do they present their devices as their own inventions: some of them had therefore been invented earlier, though it is impossible to say when. These mechanisms were mostly applied only to useless scientific toys, such as the aeolipile, sometimes said to be the remote ancestor of the steam turbine. Here steam was led into a pivoted spherical container through its axle, and the sphere pierced by two tubes twisted so that the jets of steam from them caused it to revolve. Nevertheless, apart from an occasional exception, such as Philo of Byzantium (see p 708), the Alexandrian mathematicians and technicians are less interesting for their theoretical ideas than for the mechanisms they brought into use.

An early Alexandrian mechanician was Ctesibius (see p 708), to whom the discovery of the elasticity of air is attributed. He appears to have invented compressed-air weapons and suction- and force-pumps, which were in fact usefully applied in the next century. It is also claimed that he devised two special instruments, the clepsydra (vol I, figure 48) and the hydraulic organ (figure 573, p 634).

The use of compressed air for blowing a fire was known to the Egyptians (vol I, figure 383), but the passage from such rudimentary instruments as theirs to metallic cylinder- and piston-pumps would mean a tremendous advance. And in fact such pumps, though well attested for antiquity (figure 574), are not often mentioned or frequently found, and they do not appear to have been common even during the Middle Ages. The application of the piston and cylinder to pumps could not come into effective general use until the problem of transforming certain movements had been solved.

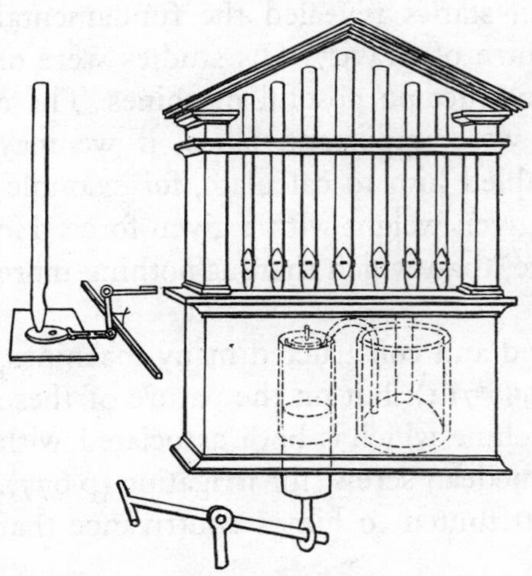

FIGURE 573—*Hero's hydraulic organ, based on that of Ctesibius. Water is forced into the right-hand compartment by the manually operated piston-pump, displacing air which is admitted to the pipes by the key-valve shown on the left.*

Both force- and suction-pumps must have been fitted with inlet and delivery valves. Ancient technical skill was not equal to fashioning the improved types of valve now employed, so craftsmen were limited to more or less ingenious expedients. They used a leather elbow-pipe, such as was later adopted by Papin (1647–1712), but the most common practice was to fit a small metal plate sliding on four pins at the base of the cylinder of the pump. The pins had flattened heads to prevent the plate from sliding completely off. Pressure inside the cylinder forced the plate against the inlet-hole, so sealing it, while pressure from without forced the plate back along the pins to allow free entry. This valve could, of course, easily be reversed to work in the opposite sense.

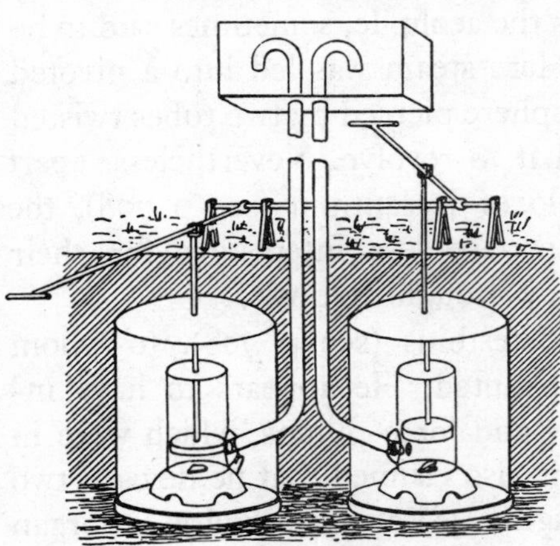

FIGURE 574—*Philo's combined suction- and force-pump, with inlet and outlet valves.*

A mechanician comparable with Ctesibius was Hero of Alexandria (see p 708), many of whose works survive. Like his predecessors, Hero applied himself to theoretical problems that arose in connexion with machines. Carrying on the work of Archimedes he studied the theory of the centre of gravity, and the general theory and conditions of equilibrium and motion of the five simple machines (pp 629, 660). Moreover he reduced these to the principle of the lever or of the wheel, or to a combination of the two. He noted the mechanical advantage to be derived from combinations of wheels, particularly of cog-wheels and gears. He attacked many problems that Archimedes had put forward, such as those involving cog-wheels and other devices for moving a given weight by a given force, and produced the simplest solutions.

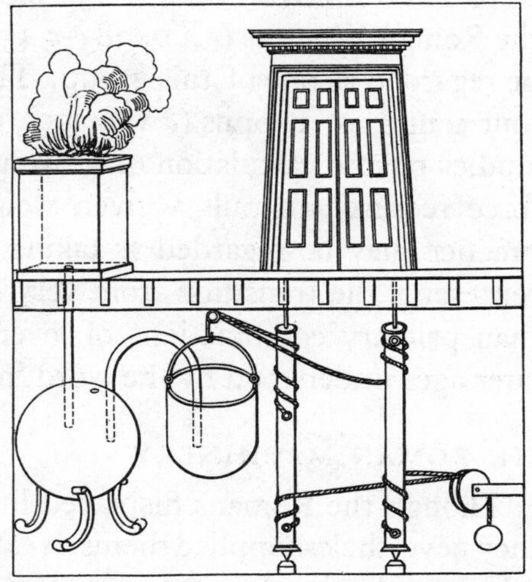

FIGURE 575—*Hero's device for opening the doors of a temple when a fire is lighted upon the altar. Expanding hot air in the altar drives water from the container into a bucket, which in descending turns the door-spindles by means of a rope, raising the counter-weight.*

The treatise of Hero on these topics was translated into Arabic by order of the ʿAbbāsid Caliph Ahmad ibn Muʿtasim (833–42). Other works by Hero concern engines of war, clocks, compressed-air engines, and screw-cutting (figure 572). One is devoted to automata. Such toys either worked separately or were combined for theatrical purposes (figures 573, 556). The accounts of these automata are interesting as introducing for the first time certain mechanisms that formed the basis of later machines. Among them are the crank, the cam-shaft, and systems of rotation with counter-weights (figure 575). Some of the devices of Hero of Alexandria involved the principles of the vacuum and of the incompressibility of water (figure 576).

It seems reasonable to suppose that Hero had a determining influence on later practitioners. Theatrical automata, for instance, were widespread in the Hellenistic and Roman world. The popularity of Hero's work is shown by the large number of extant medieval copies of them. It is symptomatic that these survivors are the practical texts, while the manuscripts containing the theoretical treatises have perished. Many of Hero's ideas were used by the Arabs, among whom a flourishing guild of automaton-makers existed until the sixteenth century.

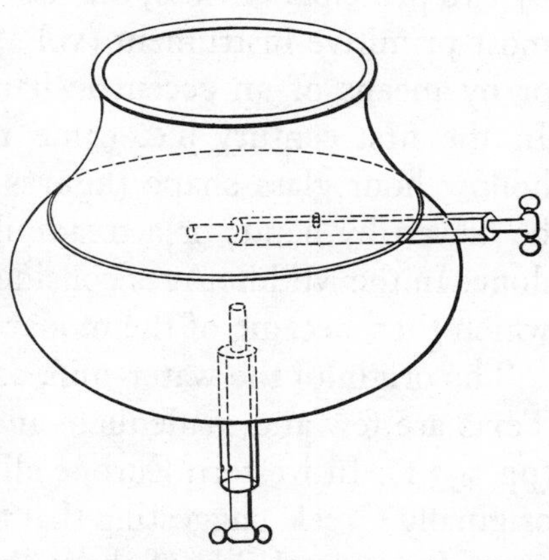

FIGURE 576—*Hero's cupping-glass, for bleeding, applying the principle of the syringe.*

Hero of Alexandria was but one of many to use various mechanical ideas (figures 639–45). Constructors of war-engines were the most numerous. Such were Philo of Byzantium (*c* 150 B.C.), Apollodorus (second century A.D.), and the Roman Vegetius (*c* A.D. 383–*c* 450). Vitruvius (fl *c* 27 B.C.–A.D. 14) also may be regarded as one of this group. The school of Alexandria was to continue for four centuries. Pappus (*c* A.D. 300), one of its later members, was the author of studies on the calculation of gearing, and he also furnished information on the force required to pull a given weight on inclined planes. Modern machine-practice may be regarded as taking its rise at Alexandria, for the school there represents the transition from very simple mechanisms, involving hardly more than primary combinations of levers, to the far more complex devices which later ages understood by the word 'machines'.

III. ROMAN MACHINERY

Though the Romans displayed little inventive genius in mechanical matters, they nevertheless applied themselves to perfecting some of the techniques they had inherited. Under their domination, during the few centuries around the beginning of our era, machinery underwent a considerable development that cumulatively amounted to a mechanical revolution. The chief element in this progress was the wide adoption of rotary motion, as in mill-wheels and treadmills (figures 540, 541, 578, 603). To make wider use of rotary motion it was necessary to adapt the instruments hitherto employed, and this could be accomplished only slowly.

The most spectacular example of early application of rotary motion, apart from transport, is the milling of corn. Till the first century B.C. corn was ground by two principal devices, the roller and the disk. The ordinary stone roller is the most primitive instrument (vol I, figure 176); the stone disk revolved by hand or by means of an eccentric handle was a more complex device (figure 75). In the first century B.C. came the introduction of the Roman mill-stone of hollow hour-glass shape (figures 76–78). A small example of this type could be driven by hand; or a treadmill could be fixed to it, and this was very soon done. In the Middle Ages considerable improvements were made to mill-stones, which then became of the modern form.

The origin of the water-mill, as of most early machines, is still far from clear. Texts are few and inadequate and early archaeological evidence is non-existent (pp 593 f). In western Europe all the terms for describing the water-mill were originally Greek, suggesting that the invention came from the eastern Mediterranean. It was probably of about the same date as the treadmill.

Another instrument widely used in the Mediterranean districts, the olive-press, was similarly transformed into a rotary machine. At first it was indistinguishable from the grape-press (figure 81) and exploited the principle of the lever. Later, for pressing olives a large round stone rolling edgewise in a circular basin was devised (figure 80). It was fitted with an axle pivoting round a centre-post which could be coupled to an external treadmill. In Europe it was commonly adapted to a water-mill. This type of press is still in use over all north Africa; in Normandy, where it is used for crushing cider-apples; and elsewhere, for grinding corn.

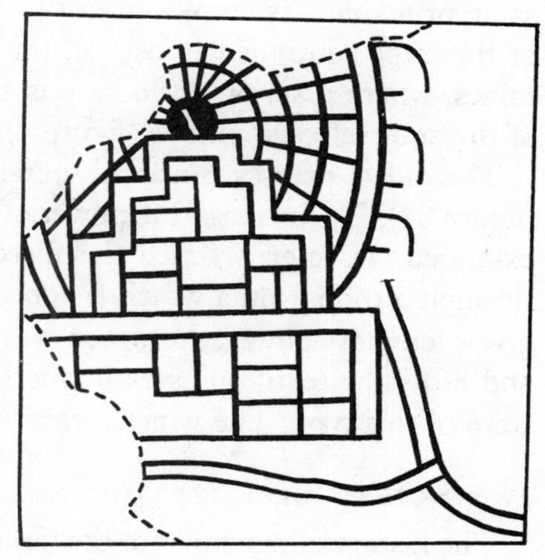

FIGURE 577—*Roman water-raising scoop-wheel, from a mosaic found at Apamea. Second century.*

A third application of the treadmill was for the raising of water (p 603). In Egypt the shaduf (vol I, figures 344–7), and later the Archimedean screw (figures 617–19), have done duty for ages. Where water had to be raised from great depths, as from mines, the Greeks and Romans used ropes, pulleys, and buckets in the new machines employing rotary motion (figure 616, 2): these were worked by drums on the surface, driven through gearing by treadmills. Thus it became possible to construct powerful machines supplying a network of irrigation-canals. It is in these great schemes for public services that the genius of the Romans was at its best. They undertook, particularly in north Africa and Spain, systems of irrigation-canals on a magnificent scale (pp 677 f), restored and used by the Arabs centuries later. The Arabs also rebuilt the ruined water-raising apparatus, to which they gave the name *nāʿūrah*—westernized to noria (p 671).

Water-raising wheels are often mentioned in classical Latin literature. One of the earliest representations is on a mosaic from Apamea, Turkey (figure 577). Some were worked by the squirrel-cage method. These continuously rotating machines became widespread in the Empire, but they were by no means the sole manifestation of the Roman genius

FIGURE 578—*Roman hoist for raising blocks of stone, operated by a man in a squirrel-cage. From a relief at Syracuse. First century.*

for mechanical engineering. The cages, revolved by the weight of men or beasts inside them, again involved the principle of the lever. The device was also combined into a crane for hoisting building-materials (figures 578, 603), and the same principle was employed in the many medieval cranes (plate 38 B). Cranes of this type continued in use till the nineteenth century; they were adapted to mines, where a series of floors was built, on each of which there worked one of the squirrel-cage pumps (figure 5).

The oldest presses consisted merely of a simple lever depressed by a weight (figure 81). To increase the power of the machine the arm of the lever had to be extended. In later practice the force upon the arm of the lever was exerted through a rope from a winch fastened to the arm (figures 82, 83). Later still, to give even more power, the winch was sometimes replaced by a screw (figures 85 and 86). The textile-presses of the Byzantines and the oil-presses in Pompeii were of this type. The winch-press was, however, the more frequently used.

IV. ANCIENT CIVILIZATION AND MECHANIZATION

The history of machines in the Graeco-Roman world raises problems that cannot yet be resolved. The disdain of technical problems evinced in ancient writings was peculiar to the literary class and was not shared by the population as a whole. This is proved by the importance attached to these devices in legislative texts, particularly those from Athens. They show that manual dexterity and skilled labour were in actual fact held in great regard.

The question of the attitude of the ancients in general towards that more advanced technological level represented by machines is partly also a matter of period. In the days of Plato and Aristotle machinery had not developed enough to arouse serious social problems. Winches, pulleys, and bow-drills are not impressive machines. The theatrical automata of which Hero (see note, p 708) gives detailed descriptions (p 635) may have existed in a less perfected form a few centuries before him. It may be that the mysterious and unexplained motions of their figures caused an unsophisticated public to think that one day the work of machines might replace that of men. In one of the scenes described by Hero there is a naval dockyard where puppets cut and saw timber. If an automaton could do work on the stage, why not machinery in real life?

Many modern economic writers suggest that ample and cheap slave-labour may have obviated the incentive to the improvement of machines. Aristotle (d 322 B.C.) at the beginning of his 'Politics' says that slaves would become redundant if shuttles and plectra could move automatically, but he did not suggest that slavery prevented plectra and shuttles from being mechanically improved.

And with slavery should we not include manual labour in general? The argument attributing the disappearance of slavery to the change from the yoke to the horse-collar, which rendered animal-power more economic, is also unconvincing; in classical antiquity human traction was used only for enormous loads and not for ordinary transport. In the Middle Ages, and for long afterwards, transport was mainly by pack-animals. For long-distance transport, even with the horse-collar, the tracks remained unsuited until the road-system itself was reformed (ch 14).

The economic conditions of slavery have been too little studied, but partial mechanization, by reducing the number of workers, certainly did produce a nucleus of industrial concentration. We can perhaps here disregard the large workshops for weapons and other military supplies created between the first and third centuries A.D., for these were the response to the needs of the army and were perhaps unrelated to technical or economic changes. There were, however, other and more truly industrial factories which adopted the technical improvements discussed above. Thus at Barbegal, near Arles, excavations have revealed sixteen mills on eight symmetrical floors (figure 546). This important milling-site has been related to the economic revolution brought about by the reforms of Diocletian (284–305) and Constantine (311–37). State-control of both distribution and industrial production was now taking the place of the domestic economy and free-labour system which had served the earlier Empire.

Other large factories besides that at Barbegal are known from the later Empire. There were in north Africa oil-manufacturing plants on a comparable scale. At Djemila, in Tunisia, there are the remains of a factory equipped with so many presses that it was clearly working on an industrial scale. At Kherbet-Agoub in Algeria the oil-works included twenty-one platforms for presses. Not only was the whole organization of these works designed for large-scale production but a town had sprung up around the factory. The same was the case with the Bir Sgaoum oil-works in Algeria.

Moreover a profound economic change as a result of mechanization is evidenced by the numerous enterprises of less importance that appeared in all Roman towns. The mills still to be seen in the ruins of Pompeii (figure 77) and Ostia, as well as many urban oil-works in north Africa, give evidence of this semi-mechanized stage.

V. THE EARLY MIDDLE AGES

What became of this mechanization between the Roman period and the later Middle Ages, that is, between the fifth and tenth centuries? Times of trouble and invasion are not propitious for inventions or for the development of the techniques of heavier industry. Insecurity and barbarian invasion ruined industrial urban centres. Large factories disappeared in the turmoil. In north Africa and Spain invasion and war caused the neglect and ultimate disintegration of Roman irrigation-works, which no longer received the necessary minimum of management and control for effective functioning. The decline of urban manufacture encouraged industrial production in rural districts. The water-mill gave way to the hand-worked mill adapted to the family unit. Large oil-works also disappeared and were replaced by small private presses. Domestic manufacture required neither advanced technique nor much labour. It is the division of labour, particularly in urban centres, that necessitates advanced mechanization and its consequent mass-production.

Documents from the ninth and tenth centuries yield information on these small domestic industries. Dues in kind indicate that most industrial products were manufactured by peasants during the slack seasons. The women spun and wove both flax and wool; the men were employed on such tasks as extracting iron from its ores and making farm implements, either by using this new metal or by re-forging old iron. Grain, malt, and oil were treated in hand-worked machines. Only very rudimentary installations, with few and cheap implements, were required, and little transport was involved in carrying raw materials and finished products within the frontiers of the manor.

Between the tenth and the twelfth centuries there were great economic, demographic, and technical changes. Population was increasing rapidly all over western Europe, and was becoming redistributed. Towns were rising again. Monetary circulation increased, the minting of gold was resumed, and the trade balance with the east became favourable. Renewed accumulation of capital led to investment in the more technically advanced industries; important investors were the large monasteries and the Italian merchants and bankers. In rural districts the same evolution was manifested by the extension of cultivation. By the gradual disappearance of villeinage, the system that tied a man to the place of his birth, a greater labour force became available for the towns. This movement was complicated, but not halted, by political and intellectual changes.

VI. LATER MIDDLE AGES. CONVERSION OF MOTION

The Middle Ages added few basic improvements to the machines inherited from antiquity, though they extended this primitive mechanization to its limits. If we are surprised that obvious improvements were neglected, it should be remembered that the same main difficulty in machinery-construction was encountered during the Middle Ages as in antiquity, namely the want of adequate materials. Most machines remained of wood, which lacks strength and resistance to frictional wear.

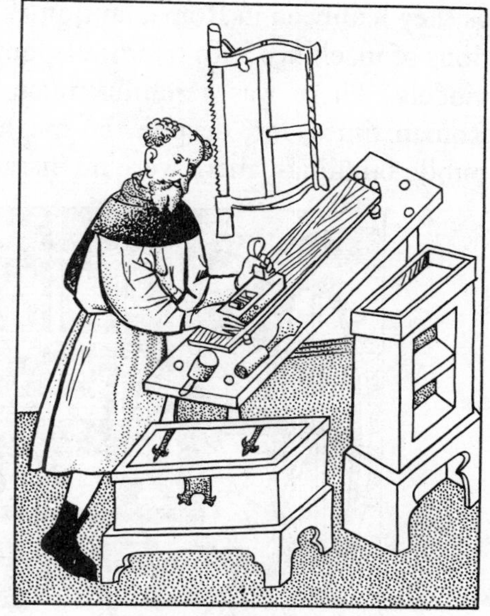

FIGURE 579—*Carpenter using a plane. His chisels, frame-saw, and completed work are also shown. Note the stops holding the plank, and the handle at the fore end of the plane which is used in the modern manner. From the* Mendel Brüderbuch, *Nuremberg. c 1404.*

The tools of medieval craftsmen for cutting and working wood were also very inadequate, but there are a few valuable ones that were in a sense reintroduced. Thus the plane was known in antiquity (figure 206 P), but apparently became rare or forgotten, though indispensable for working-parts that have to be fitted together. It seems to have reappeared in the twelfth century. John of Garland mentions it about 1220, and it is apparently represented in two Catalonian sculptures of the end of the twelfth century at Gerona and San Cugat, near Barcelona. It appears unmistakably in the thirteenth century and from that time is well represented (figures 358, 579). Advances in the manufacture of steel in the Middle Ages certainly also contributed to the improvement of tools.

Nevertheless in major mechanical apparatus little progress was shown until towards the end of the period. The hoisting-devices used in building remained

FIGURE 580—*Building operations in progress.* (Left to right) *The master-builder with his royal patron; carrying stone up a ramp in a hod; wheel-barrow; plummet-level on wall; simple hoist; stone-masons at work; plumb-line used on face of wall; boring hole with auger in scaffold-pole. From a mid-thirteenth-century manuscript.*

as they had been in Roman antiquity. There are numerous medieval representations of machines with treadmills, but they reveal no difference from the Roman models. There was certainly much less building on the grand scale than in Roman times, for, despite the magnificence of the cathedrals and a few large public buildings, there were no massive aqueducts, no theatres, no considerable

FIGURE 581—*Water-driven stamp-mill: the heavy stamps are raised by cams on the shaft.* c *1480.*

number of important bridges. Though a few houses in the later Middle Ages were of stone (p 446), town houses were usually of timber and rammed earth and their construction needed little labour or machinery. Even in building cathedrals little scaffolding was used, the fabric itself supplying or supporting working-platforms as it went up (figure 347). Simple winches and treadmill-winches are almost the only machines to be found in the many miniatures representing the construction of buildings (figure 580), though the slewing-crane appears early (figure 600 B).

The wheelbarrow was the one distinctively medieval device that facilitated the movement of earth, as in digging foundations. It is more easily handled than a basket, though its capacity may not be any greater; its chief advantage is that a man pushing a wheelbarrow can walk much more quickly than one bearing a basket with the same load. The first figures of the wheelbarrow are of the thirteenth century (figure 580), but it had long been known in China (table, p 770), and news of its simple construction may have come through Islam or Byzantium.

Mechanization is directly conditioned by the use of certain forces. Apart from man-power, antiquity had known and employed only treadmills for yielding continuous rotary motion. Hydraulic power, widely used in the Middle Ages for milling corn, was of the same kind. Treadmills and water-mills could be coupled directly to all machines requiring continuous rotary motion.

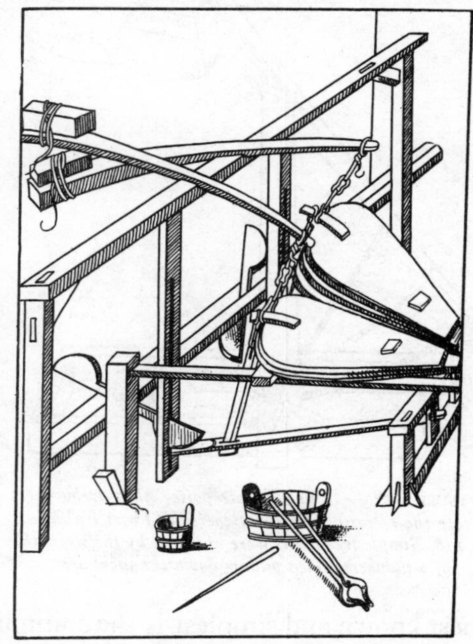

FIGURE 582—*Bellows operated by cams. Counterbalances are used to open the bellows after they have been compressed by the cams.* c *1480.*

For hammers or crushers, it was possible to use the cam-shaft, known to Hero of Alexandria (p 635). The cam acted upon the shaft of the hammer or crusher, which fell back again under its own weight (figure 581). The small iron or copper hammers used in tanneries, paper-mills, and fulling-mills were thus operated (figures 187, 553). Later the principle of the cam was applied also to bellows and heavy hammers in the metallurgical industries (figures 554, 582). In this device the weight of the implement sufficed to effect the restoring-movement which did the work. A heavy counterpoise was also used to operate the trebuchet (figure 583, p 724).

Another problem arose when the weight of the tool could not be used to complete the cycle. This was the case with saws, where a cam could pull the saw but another agent was needed for the reverse motion. In the Middle Ages the power of springs was used for such purposes in many machines. Thus in the machine-saw of Villard de Honnecourt (fl *c* 1250), the return movement is effected by the relaxation of a pole previously bent by the forward stroke of the saw (figure 584). Spring-mechanisms were often used to derive either a reciprocating motion from a continuous rotation, or a continuous rotary motion from a reciprocating motion.

The same idea was applied to lathes for wood-turning, which were based on a very simple principle. A rope wound around the shaft bearing the lathe-chuck and attached to a treadle caused the chuck to revolve when the treadle was depressed. For proper working it was necessary to cause the chuck to revolve continuously a certain number of times, and then to rotate the shaft in the

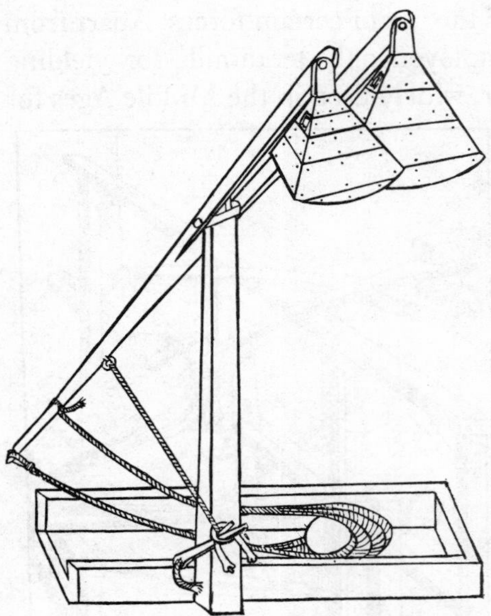

FIGURE 583—*Double-counterpoise sling-trebuchet. Note the releasing-device at the side. From Valturio, 1478. Simple trebuchets were operated by the strength of a number of men pulling down the short arm.*

opposite sense, to wind the cord again ready for a second impulse on the treadle. To effect this reversal the simplest means was to join the ends of the rope to two treadles operated alternately, but the amplitude of the two treadles was limited. A spring to make the shaft of the lathe revolve in the restoring direction was more popular. The spring could be an ordinary pole, as with Villard's saw, or the relaxation of a bow-like device arranged in various ways could be used. We have representations of all these instruments from the thirteenth century (figures 585, 586).

Improvements to small machinery and small elements in machines were also but slight. Yet new devices and designs give proof of some medieval inventive spirit. Among these new inventions the best known and simplest is the spinning-wheel (figures 168, 587). In early times spinning was done with spindle and distaff. The transition to the spinning-wheel is again obscure, and even the date of the first examples is in debate. It is possible that their design followed from that of the spool, which had long been in use. Many spinning-wheels on stained glass or in miniatures, such as in those of the thirteenth century at Chartres, are in reality spools. In the west the spinning-wheel is definitely mentioned in the first half of the thirteenth century, and very clearly in the statutes of the Drapers' Guild at Spires in 1280, and of that of Abbeville in 1288. The spinning-wheel remained a rare implement almost to the close of the Middle Ages and was still of primitive construction—scarcely more than a spindle mounted on a revolving axle.

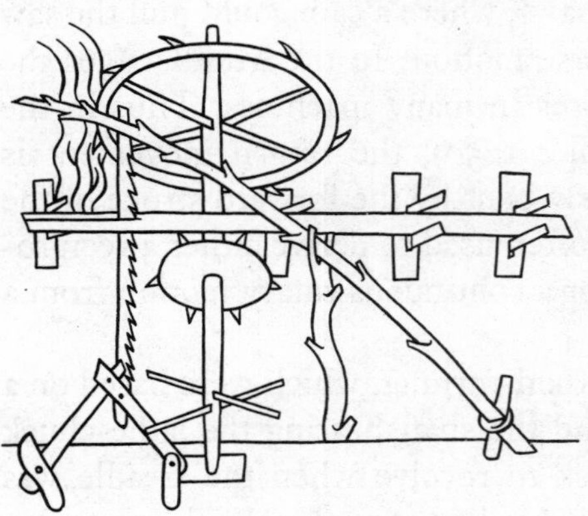

FIGURE 584—*Villard de Honnecourt's sketch of a water-powered saw. c 1250.*

The final operation of twisting silk thread is very similar to spinning with a spinning-wheel. It was owing to this fact that mechanization first appeared in the industry (ch 6). Important technical improvements were introduced in thirteenth-century Italy into the twisting and winding of silk. These operations must have been carried out more or less mechanically since they were done in water-mills, the so-called 'Bolognese' mills. It is difficult to say precisely what the improvements were, since we learn about the twisting of silk fibres only from much later writers, such as Vittorio Zonca (1607). Weaving-operations themselves were hardly affected by further mechanization in the Middle Ages.

FIGURE 585—*Turning as represented in a thirteenth-century painted glass window in the cathedral at Chartres. The work revolves between two pivots: it is turned by the rope fastened to the treadle, which is probably pulled up by a springy pole at the upper end.*

Of the equipment used there is little to be said, for it hardly changed between antiquity and the eighteenth century. There was, however, a certain amount of specialization of tools (ch 11).

VII. LATER MIDDLE AGES. CONVERSION OF POWER

Besides the conversion of motion there is the troublesome problem of the conversion of power. How can a force or a velocity be increased or reduced to the required degree? Antiquity had bequeathed to the Middle Ages a wonderfully fruitful instrument in the gear. From the time of Pappus (*c* A.D. 300) the method of calculating gear-ratios was known, but gearing was mainly intended for increasing speed of rotation, and this was accompanied by a corresponding loss of power.

With certain devices, a moderate control of speed and power was attained during the Middle Ages. A good example is the hydraulic hammer used for forging (figure 554). Here, regulation was first effected by graduating the width of opening of the sluice-gate supplying the water to the paddle-wheel. The operation was limited

FIGURE 586—*Turning a wooden bowl on a pole-lathe. From a fourteenth-century French manuscript.*

FIGURE 587—*Crank-rotated machine for spinning cord for snares. From a fifteenth-century French manuscript.*

by the maximum flow of the sluice and was useless if the operator had no control over the head of water. An additional system came into use when the cams that acted on the shafts of the hammers were fixed to a ring which in turn was wedged on to the driving-shaft. For each hammer a variety of rings fitted with a variable number of cams could be provided, and to vary the strength of the blow it was necessary merely to change the ring striking upon the shaft of the hammer. This was fairly easy, since the rings were not rigidly fixed but wedged to their shafts.

It was more difficult to invent devices to increase power and reduce speed. One of the first of these applied the principle of the lever directly to winches by increasing the relative length of the lever-arm of the drum. Machines of this type were used for lifting large blocks of stone (figure 588). Recesses were cut round the circumference of the horizontal winch, and ratchets were fitted inside them. The winch was revolved by a pivoted lever engaging with these ratchets and pressing upon them. During the reverse movement of the lever it passed over the teeth of the ratchets as these slipped back into the recesses made for them. The operator could put his full weight on the lever and was, moreover, assisted by the steps upon which he could pull with his arms. A second ratchet prevented the winch from unwinding. Sometimes there would be two lever-arms joined at the end by a plank which could accommodate several men, thus increasing the effective power considerably.

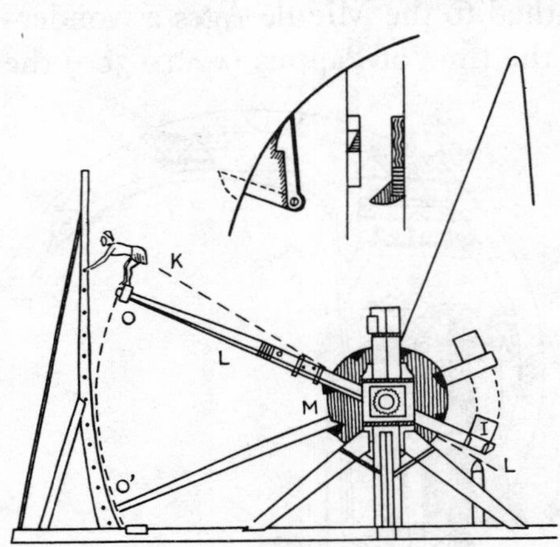

FIGURE 588—*Reconstruction of a medieval winch for raising heavy burdens by means of an oscillating lever and ratchets.*

Multiplication of forces by pulleys had been known since the invention of pulley-blocks in antiquity (p 658), and pulleys remained in common use in the Middle Ages. Reduction of velocity by the screw was, however, probably a medieval invention. The earliest known record is in the chronicle of Gervais the monk (*c* 1200), who mentions the use of screws for lifting loads. Villard de Honnecourt in the thirteenth century portrays the screw-jack for lifting very heavy burdens (figure 589). His astonishment at the power of this machine suggests that it was not yet common and that he had only recently learnt of it.

With the screw-jack and the ratchet it was possible to construct many types of hand-winch, using levers, racks-and-pinions, and so forth. Such machines could lift very heavy loads by repetition of a series of small movements. We have no pictures of the small hand-winch earlier than the fifteenth century (figure 349), but by then it was commonly used in bending the cross-bow (figure 655 C).

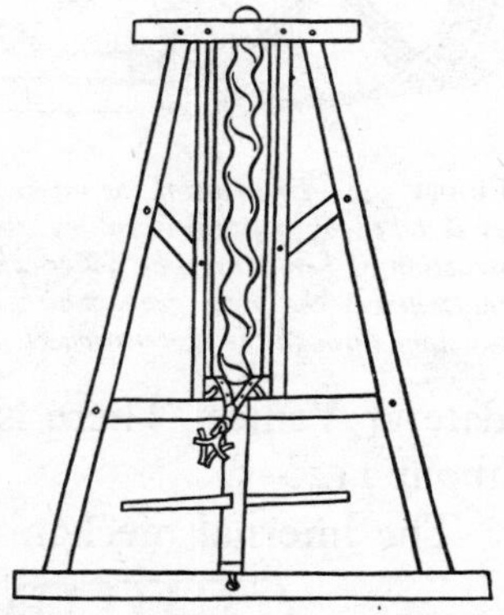

FIGURE 589—*Villard de Honnecourt's sketch of a screw-jack. c 1250.*

Reduction-gearing was used also in presses. Screw-presses almost completely replaced winch-presses during the Middle Ages. The screw was placed at the end of a huge lever, often more than 10 m long (figure 84). Notable extant examples are the presses of the Dukes of Burgundy at Clos de Vougeot and Chenôve, of the thirteenth century. Though tracing screw-threads requires much skill, the screws of oil-presses in the south of France were early standardized; Villard de Honnecourt, however, gives no indication of how the tracing was carried out.

The examples described show us that during the Middle Ages the problems of converting motions and power found solutions adequate for the practical needs of the time. Building with enormous blocks of stone in the Roman fashion was seldom attempted during this period. Notable exceptions are the monolithic columns of the chancel at Vézelay, Yonne, and in the cathedral at Langres.

Good use was made of hydraulic power during the Middle Ages, but were attempts made to improve the efficiency of machines? It is to be remembered that the very concept of efficiency in the mechanical sense is essentially modern. Nevertheless in the Middle Ages a man would doubtless consider whether his mill-wheel revolved better or worse than that of his neighbour. By the fourteenth century there were in use three positions of paddle-wheel in relation to the stream of water: the undershot wheel, the overshot wheel, and the horizontal wheel. Some were mounted on boats, to catch the rapid central current; these, of course, had undershot wheels (ch 17). It is difficult to say whether slanting of the paddles to obtain better efficiency was ever tried, for drawings of the period are not good enough to warrant deductions from them.

Experimental efforts at the improvement of machines were certainly made;

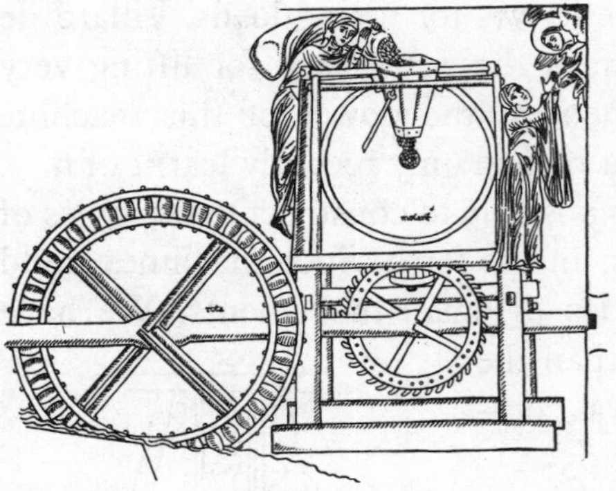

FIGURE 590—*Drawing of the interior of a corn-mill, as sketched by a twelfth-century abbess for the instruction of her nuns. The water-wheel on its thick shaft drives the main gear, which turns the upper stone through the lantern-wheel and its shaft.*

there are, for instance, many traces of them in the notebooks of Leonardo da Vinci. Thus he investigated the problems of mill-wheels—the position of the wheel, the form of its paddles, the angle of the water's impact. These studies led him to a conception of the principle of the hydraulic turbine, which, even if it originated in antiquity with Hero, had later been forgotten. A special type of mill was the tide-mill, of which the oldest examples, those at Dover, were built in the time of William the Conqueror. There were others of about the same date at Venice. There is documentary evidence of yet others at Bayonne at about 1120–5.

The internal mechanism of mills—toothed wheels, transmission-shafts, and gears—does not seem to have changed in the course of centuries (figures 590, 591). Eleventh-century pictures show this characteristic and long-lived mechanism. The main shaft, resting on a lead-covered wooden bearing, was held in place by an iron strap. In the windmill it was a timber of huge dimensions. The large gears—in which each tooth was composed of a separate block of wood—and the pinions or rundles were all of wood. Little metal was used, so that the various parts, especially the gear-teeth, wore quickly. Leonardo da Vinci studied the effect of varying the angle at which the gear-teeth meshed.

FIGURE 591—*Water-mill depicted on a Spanish reliquary Second half of the thirteenth century.*

The Middle Ages are characterized less by a general progress towards mechanization than by the development of a small group of machines, of which the mill was the most important (figure 592). There were too many social obstacles and too many technical difficulties for any general mechanization. Nevertheless the progress achieved was far from negligible and marks a considerable advance on the machinery of the ancient world.

The development of mechanization was retarded by many factors that are still acting. First, there was traditionalism, strengthened both by the simple resistance to a new device and by the isolation of some regions with a corresponding perpetuation of old tools and methods. Subject peoples, depressed classes, and ethnic minorities were and are notoriously unwilling to accept technical improvements. Another manifestation of traditionalism is a contempt for the products of new machines and methods; examples are the refusal of weavers to accept thread from spinning-wheels at Abbeville in 1288, and many drapers' regulations in the following centuries. When the guilds had at last to yield they allowed the thread spun on a wheel to be used only for the warp. There was a similar objection to the fulling of cloth in a mill. In the cloth-making towns of the north this was forbidden from the beginning for cloth of the better qualities. Elsewhere the prohibition of fulling-mills for such cloth came in later: at Blois in 1292, at Nogent-le-Roi in 1403, and at Chartres in 1444.

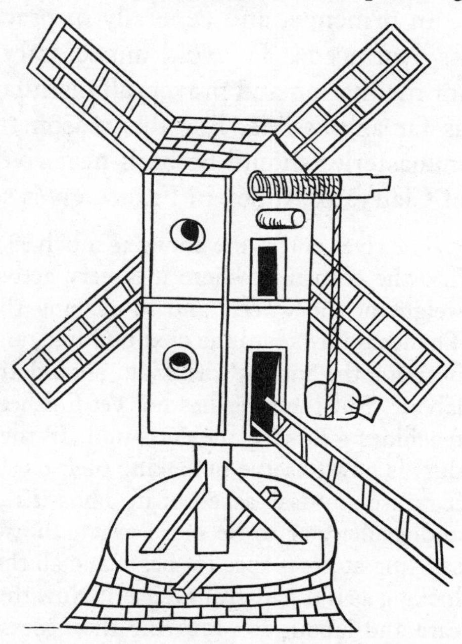

FIGURE 592—*Post-mill with an auxiliary drive to a drum, by which sacks of grain are lifted from the ground. The sails are four wooden frames, over which cloth was stretched. From a fifteenth-century German manuscript.* (*See also p 623 and figures 561–3.*)

Secondly, the new machines often aroused opposition because they were subject to monopolies or the imposition of extra taxes. The feudal rights of the lord of the manor on which the new mill was built were generally carefully preserved, as with many flour-mills, iron-mills, barley-mills, tan-mills, and fulling-mills, thus increasing the burdens on those who might use them. To avoid such dues the people retained their old tools and equipment, perhaps with a certain inner satisfaction at continuing the methods of their fathers.

VIII. MONASTERIES AND MECHANIZATION

The manorial system never aimed at complete self-sufficiency save perhaps for a short period after the barbarian invasions. There was, however, one important type of concentration of manufacture that favoured mechanization. This is exemplified by the monasteries, particularly those of the Cistercian order, which formed the sole economic autarkies of the Middle Ages. They played a major part in the development of machinery and in particular of the various applications of the water-mill.

In principle, and generally in practice, each monastery had itself to meet all its own needs. To avoid unnecessary labour, which reduced the time available for meditation and prayer, all manufactures in monastic hands were mechanized as far as possible. For this reason the Cistercian regulations recommend that monasteries should be built near rivers that could supply power. A description of Clairvaux Abbey in France gives some idea of this mechanization:

> The river enters the abbey as much as the wall acting as a check allows. It gushes first into the corn-mill where it is very actively employed in grinding the grain under the weight of the wheels and in shaking the fine sieve which separates flour from bran. Thence it flows into the next building, and fills the boiler in which it is heated to prepare beer for the monks' drinking, should the vine's fruitfulness not reward the vintner's labour. But the river has not yet finished its work, for it is now drawn into the fulling-machines following the corn-mill. In the mill it has prepared the brothers' food and its duty is now to serve in making their clothing. This the river does not withhold, nor does it refuse any task asked of it. Thus it raises and lowers alternately the heavy hammers and mallets, or to be more exact, the wooden feet of the fulling-machines. When by swirling at great speed it has made all these wheels revolve swiftly it issues foaming and looking as if it had ground itself. Now the river enters the tannery where it devotes much care and labour to preparing the necessary materials for the monks' footwear; then it divides into many small branches and, in its busy course, passes through the various departments, seeking everywhere for those who require its services for any purpose whatsoever, whether for cooking, rotating, crushing, watering, washing, or grinding, always offering its help and never refusing. At last, to earn full thanks and to leave nothing undone, it carries away the refuse and leaves all clean [1].

Most early abbeys had an extensive water-system of this type. In some of them the various hydraulic workshops were concentrated in a single factory, as, for example, at the French abbey of Royaumont, near Paris. There the 'works' was built over the river itself, which passed along the axis of the building through a high and narrow tunnel 32 m long and 2·35 m wide. The workshops supplying the material needs of the abbey, such as the grain-, tanning-, fulling-, and iron-mills, were situated on each side of this tunnel. At the abbey of Vaux de Cernay, also near Paris, there was a similar works. The abbey of Fontenay in Burgundy still has its factory, a huge structure with four rooms built at the end of the twelfth century and measuring 53 m long by 13·50 m wide. The forge was in the second room. The river passed alongside the building and the grain-mill was at the end and built over the river.

Many other Cistercian abbeys had similar buildings where all the equipment used for their work was concentrated. In the twelfth-century Fountains Abbey, Yorkshire, an underground river passes through a series of tunnels feeding a brewery, a corn-mill, and various workshops. There were also private factories in the Limousin district in France; the Materre mills included those for corn, hemp, and tanning. In the same district there was a single mill which dealt with flour-grinding, cloth-fulling, and tanning.

FIGURE 593—*Sharpening a sword on a grindstone revolved by a crank handle. From the Utrecht psalter. Ninth century.*

IX. FIRST APPEARANCE OF MODERN MACHINERY

The first medieval technical revolution occurred in the twelfth and thirteenth centuries. The fourteenth century was a period of adaptation, when the disastrous Hundred Years War (1338–1453) between England and France, and the Black Death (1348–50), contributed not a little to a slackening of progress. This was to reappear with added vigour in the second half of the fifteenth century.

The new phase of technical advancement occurred in a region distant from the protracted war. Northern Italy, southern Germany, and the Rhine valley enjoyed a very rapid cultural evolution—the beginnings of a renaissance in art, literature, science, and technology. Now appeared the first of the long line of engineers who were later to form or to train the various groups of technicians—men who were at the same time architects, mechanical engineers, and military engineers.

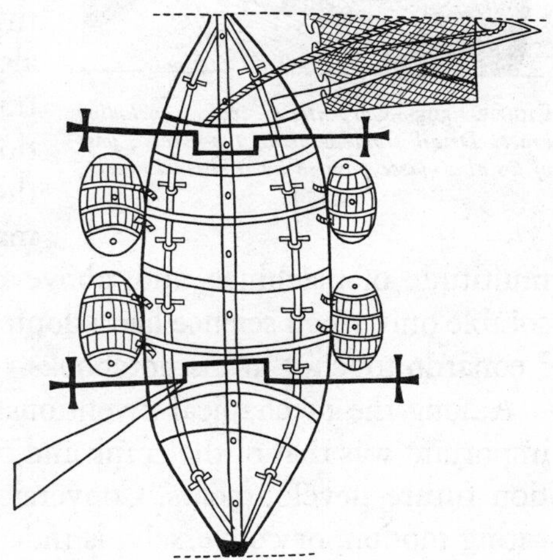

FIGURE 594—*Crank-propelled paddle-boat (with buoyancy tanks!). From a manuscript of 1335.*

Leonardo da Vinci's description of himself when writing to the Duke of Milan corresponds perfectly to this new definition of the 'compleat technician'. He says that he has a process for the construction of very light bridges, capable of easy transport. He knows how to drain moats and make scaling-ladders. He knows how to make light cannon easy of transport, capable of ejecting inflammable matter. By narrow and tortuous subterranean tunnels, he can create a passage to inaccessible places, even under rivers. He can make cannon, mortars, and engines of fire different from those now in use, and can replace them by catapults and other projectile weapons at present unknown. He can compete with anyone in architecture, and in the building of canals.

FIGURE 595—*Carpenter's tools, including brace. Detail from 'Bearing the Cross', part of an altar-piece painted by Meister Francke, 1424.*

This superbly gifted technician had of course progressed far beyond his predecessors. He had a better knowledge of science than they, and was more given to conscious thought on the development of the means at his disposal. The development of machinery now required comprehensive solutions of problems. Interposed between prime-mover and operative tools there must be mechanisms that do not belong specifically to any one of the five simple machines (p 629). There are also such general problems as friction, the transformation of motion, and the reduction and augmentation of power. Moreover, there are problems of stresses and strains in materials, problems of mechanization in a multitude of machines, and, above all, strictly scientific problems that became soluble only when science had adopted the experimental method. On all of these Leonardo touches in his notebooks.

Among the mechanical inventions at the close of the Middle Ages, the most important was that of the crank and connecting-rod, which was largely to condition future developments. Conversion of continuous rotary motion to reciprocating motion, or conversely, is indeed found in many machines, but the only way of effecting these conversions known to the Middle Ages involved springs (p 644), a method with many drawbacks.

The invention of the combined crank and connecting-rod was stimulated by two other inventions. First was the simple crank (figures 593, 594) which had been known since late antiquity. It was a natural development from the attachment of a handle perpendicular to a wheel. Following this was the carpenter's brace, the history of which is obscure (p 230). In John of Garland's dictionary (thirteenth century) there is possibly a reference to it, but the first illustration is in a painting of 1424 (figure 595). Another early example of it is in a fifteenth-century picture by Robert Campin (plate 12).

The difficulty of assembling the moving parts no doubt long delayed the appearance of the crank and connecting-rod system. Friction had a very detrimental effect when the parts were of wood. The earliest depicted example of this system is in a German manuscript, at Munich, of 1421–34 (figures 596, 597). It is a mill probably operated by hand or foot, and the main parts seem to be of metal. This type of mill was still in use in southern Germany in the nineteenth century. Another drawing from the same manuscript shows a machine of better

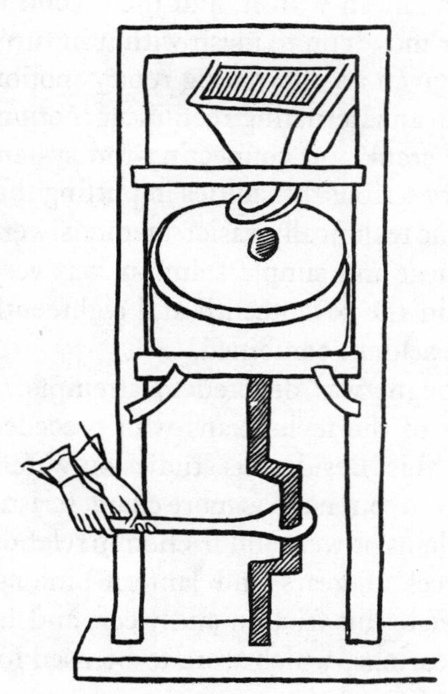

FIGURE 596—*Hand-mill worked by crank and connecting-rod, allowing a simple to-and-fro motion of the hands. From a manuscript of* c *1430.*

FIGURE 597—*Mill operated by double-throw crank through gear-drive. The two connecting-rods are apparently supposed to be linked to a pair of treadles, but this part of the machine is not clearly represented. From a manuscript of* c *1430.*

design. There are very few other examples of the crank and connecting-rod system before the end of the fifteenth century.

The problem of the conversion of a continuous rotary to a reciprocating motion greatly interested Leonardo. It arose naturally in considering suction- and force-pumps. Leonardo produced several solutions. The first used a cylinder round which a double helical groove was cut. A peg fixed to a piston-rod fits into this groove. Rotation of the shaft carrying the cylinder imparts an alternating motion to the piston. A second scheme evades the difficulty without actually solving it. A fly-wheel operated another wheel to which two piston rods were attached, with an alternating rotation. Here there was no transformation of motion but merely a transmission of an initial alternating motion. His third device was that adopted by all the constructors of machines in the sixteenth century. The continuous rotary motion is transmitted to a disk with teeth on only half of its periphery. These teeth engage with two lantern-pinions[1] mounted on a single shaft, which is at right-angles to that carrying the toothed disk. The two pinions are so placed at each end of the diameter of the disk that the first is driven in one direction as the teeth of the disk mesh with it, and the second is driven in the reverse direction as the half-gear moves on to mesh with it in turn. The shaft carrying the two pinions is thus given an alternating rotary motion (figure 169). It is not difficult to convert this to an alternating rectilinear motion. Leonardo's drawings also include a genuine crank and connecting-rod system applied to a saw. But there must have been very serious difficulties in putting this system into practice, and therefore inferior but technically easier methods were preferred. That employing the half-gear, a neat and simple solution, was very widely used in the sixteenth century. Even in the seventeenth and eighteenth centuries the crank and connecting-rod were seldom combined.

Some of Leonardo's most penetrating thought was devoted to attempts to rationalize basic mechanical problems. None of the technicians who preceded him is known to have attempted anything like this. Besides his studies on water-mills (p 648) many other examples could be given, but none is more characteristic than his study of gears. He examined the problems of wear and friction in relation to the angles formed by the engaging surfaces of gears and lantern-pinions. He sought a method of gearing that would obviate this friction and wear, and he discovered the basic principles of irregular gearings, which were to be used for the first time in the following century.

In mining, mechanization appears to have made some headway (ch 1). Pumps

[1] A form of pinion consisting of two circular disks connected by round pins, which act as the teeth in a modern gear-wheel.

operated by crank and connecting-rod must considerably antedate Agricola's description of 1556 (pp 16 f). Small trucks on rails appeared for the first time at the end of the fifteenth century (p 548) [2].

Machinery for working metals was greatly improved during the fifteenth century. First there was a tendency to increase the capacity of furnaces for the reduction of ores and thus to increase the amount of metal available. A time arrived when natural draught and fans no longer sufficed to supply enough air. Huge bellows were then used, which very soon came to be operated by hydraulic wheels and counterpoises (figure 582). The first blast-furnaces with water-driven bellows were probably those in the Liége district or on the banks of the Rhine early in the fifteenth century (figure 36). No further great advances in blast furnaces were made until the sixteenth century.

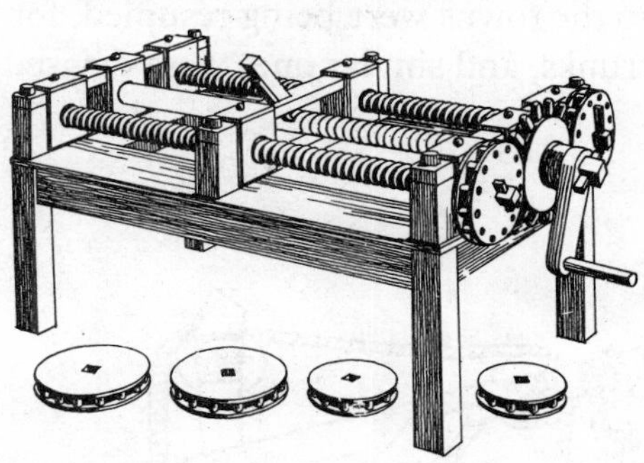

FIGURE 598—*Leonardo's screw-cutting machine. The rod to be cut is mounted in the centre and turned by the crank. A carriage bearing the tool* (centre) *is advanced by two lead-screws, rotated through the gears by the same crank. Varying the gear-ratios varies the pitch of the thread cut on the rod. From a model in the Science Museum, South Kensington.*

Many machines for working metals are illustrated by Leonardo, though it is uncertain whether they represent his own ideas or apparatus already known. Thus there are machines for making iron bars of a given sectional form. One method that was hardly practicable was to pass the iron bar, presumably hot, through a matrix to give it the required shape. The rod was generally depicted as drawn by a screw operated by water-power. In the following century foundries equipped with rolling-mills produced the same products much more easily. On the other hand this method is of proven value for making wire, and hydraulic wire-drawing works were known in southern Germany by the end of the fifteenth century. At the same period the manufacture of tin-plate began in Nuremberg. This industry implies suitable machines for making sheet-metal. One of the most original of Leonardo's machines is the screw-cutter (figure 598).

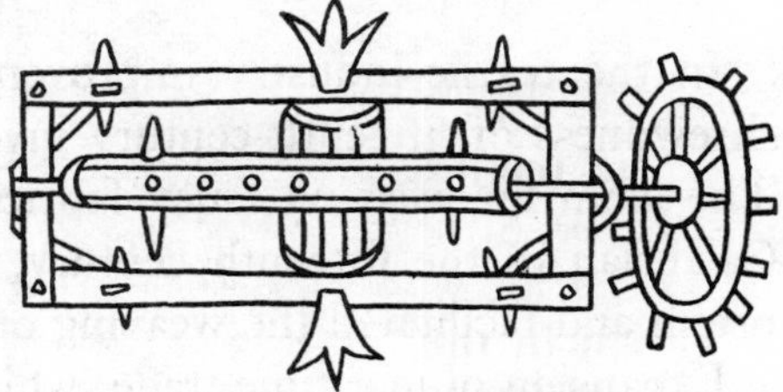

FIGURE 599—*Machine for boring out tree-trunks for use as water-pipes; possibly the wheel shown on the shaft bearing the cutting-tool is intended to be a water-wheel. The barrel-like piece is presumably intended to thrust the trunk forward as the work proceeds. From a manuscript of* c *1430.*

Many examples of boring-apparatus appear in fifteenth-century manuscripts, such as a machine for boring timbers seemingly driven by a water-wheel (figure 599). Leonardo depicts a similar but vertical machine. Such machines were important to an age when water-supplies to the towns were being resumed, for the conduits were mostly of hollowed tree-trunks, and similar ones were devised to ream out the bores of cannon.

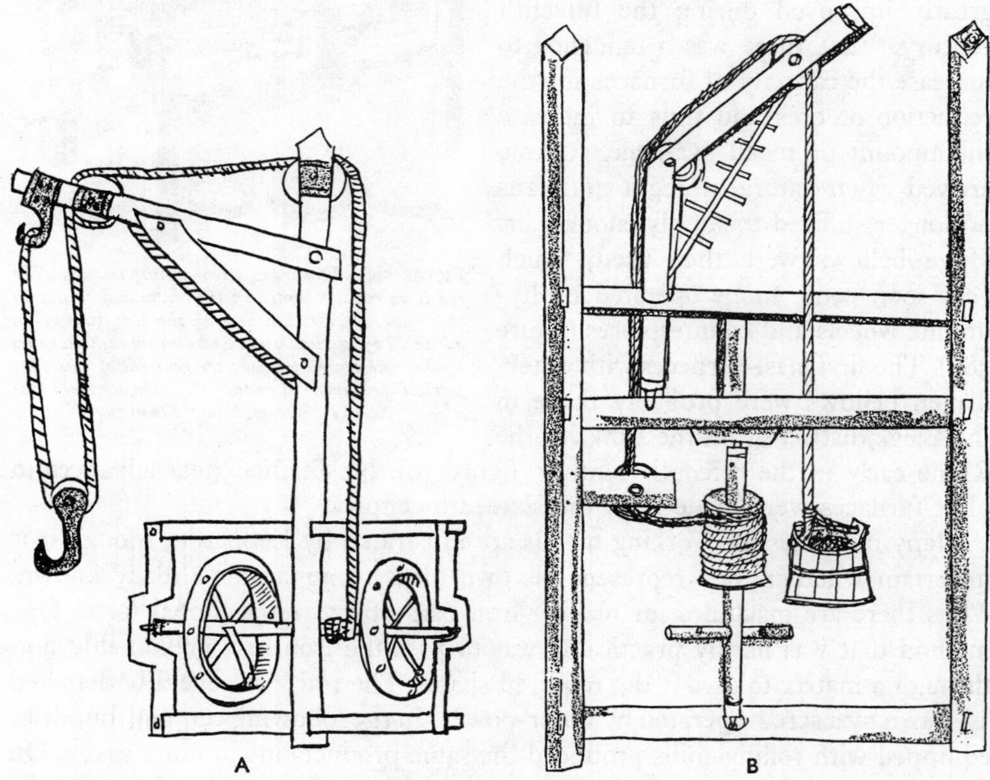

FIGURE 600—(A) *Simple form of crane.* (B) *Early example of slewing-crane. Both from a manuscript of c 1430.*

In the textile-industry improvements of the spinning-wheel manifest the shrewdness of fifteenth-century invention. Weaving also made some progress. The loom for weaving fancy fabrics, said to have been invented by John the Calabrian in the fifteenth century, was an advance on the looms of Chinese origin and facilitated the weaving of silk fabrics of great beauty.

Expansion of maritime trade, which became very considerable with the great geographical discoveries, brought with it the problem of ports and in particular of dredging. As early as 1435, the Dutch had built at Middelburg a ship which they called the 'scraper'. It was fitted with a kind of huge rake which loosened the bottom of the harbour channel. The current then carried away the silt and sand. In many places more primitive methods continued, as at Honfleur, where the harbour channels were cleared at the spring tides by ploughs and wheelbarrows. At other times the water was penned up at high tide in a basin whence it was subsequently released at low water to scour the harbour.

The invention of movable lock-gates (figure 625) made possible the active extension of canal-building for inland navigation. Lock-gates of some kind had

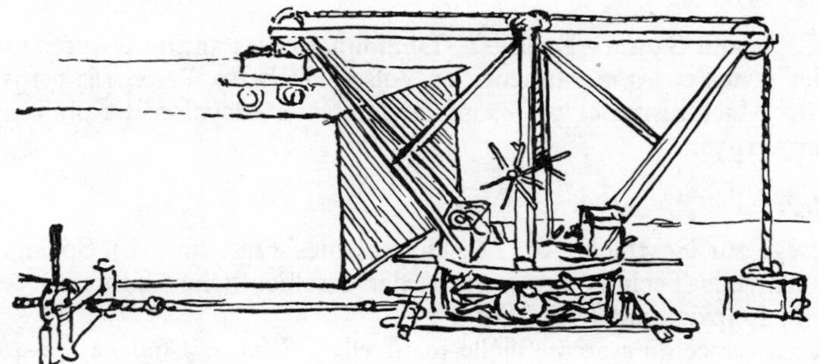

FIGURE 601—*Double swivelling-crane sketched by Leonardo, designed to raise and transport blocks of stone. The winch on the left is apparently intended for dragging the whole machine forward on its foundation.*

long been known, and had been used since 1180 at Damme on the canal from Bruges to the sea. The Duc de Berry also had a lock constructed in 1394 on the canal going from Njort to the ocean. In the fifteenth century the lock-gate system was much improved. From then on, paddles were used, as today, to control the flow of water in and out of the lock-chamber through sluices cut in the gates or sides of the lock. Filippo Visconti of Milan used this system from 1440 onwards, the Venetians used it on the Piovego in 1481, and it was described by Alberti in his treatise on architecture in 1450.

Thus the fifteenth century appears as particularly fruitful for the development of mechanization in many different fields. The German and Italian manuscripts that have been preserved prove that their authors made serious efforts to find solutions to the various problems encountered in the evolution of technology. They often combined devices that had long been known and made them into machines adapted to the needs of their times. By combining the Archimedean screw with the windmill, a combination first mentioned in 1404, an engine was designed which made it possible to drain land in Holland (p 689). Cranes were made to swivel (figure 600), and implements illustrated in fifteenth-century miniatures and sixteenth-century paintings formed the principal hoisting-apparatus in ports until the middle of the nineteenth century. Leonardo da Vinci even invented cranes that moved on a sort of railway, probably wooden (figure 601).

REFERENCES

[1] *Bernardi Vita*, II, ch. 5, no. 31, in MIGNE, *Pat. lat.*, Vol. 185, cols. 570–2.
[2] WOLF, F. 'History of Science, Technology, and Philosophy in the Sixteenth and Seventeenth Centuries' (2nd ed.), p. 511, fig. 265. Allen and Unwin, London. 1950.

BIBLIOGRAPHY

Antiquity

DAREMBERG, C. V. and SAGLIO, E. (Eds). 'Dictionnaire des antiquités grecques et romaines. D'après les textes et les monuments' (5 Vols.). Hachette, Paris. 1873–1919.
SCHUHL, P. M. 'Machinisme et philosophie.' Nouvelle encyclopédie philosophique, no. 16. Alcan, Paris. 1938.

The Middle Ages

BECK, T. 'Beiträge zur Geschichte des Maschinenbaues' (2nd enl. ed.). Springer, Berlin. 1900.
FELDHAUS, F. M. 'Die Technik der Vorzeit, der geschichtlichen Zeit und der Naturvölker.' Engelmann, Leipzig and Berlin. 1914.
GILLE, B. "La naissance du système bielle-manivelle." *Tech. et Civil.*, **2**, 42–46, 1952.
Idem. "Léonard de Vinci et son temps." *Tech. et Civil.*, **2**, 69–84, 1952.
Idem. "Le machinisme au moyen âge." *Arch. int. Hist. Sci.*, no. 23/4, 281–86, 1953.
Idem. "Le moulin à eau." *Tech. et Civil.*, **3**, 1–15, 1954.

A NOTE ON ANCIENT CRANES

A. G. DRACHMANN

THE pulley and the windlass were known to Aristotle, who notes in his 'Mechanics' that 'in building construction they can easily raise great weights; for they shift from one pulley to the other, and again from that to capstans and levers; and this is equivalent to making many pulleys' [1].

We might almost suspect that he knew the compound pulley, if it were not directly stated by several authors that Archimedes was the inventor of both the triple and the compound pulley, which he used in his famous demonstration of dragging a fully loaded three-masted ship on dry land. He did this feat single-handed, using a windlass with an endless screw [2]. It was the prospect of the almost unlimited force available in this combination that impelled him to offer to move the Earth itself if he were only given a fixed place to stand.[1] Among Archimedes' many inventions for the defence of Syracuse against the Romans (p 714) this powerful crane was very active and very much respected by the enemies. After his death it fell into disuse, and while the triple pulley was adopted everywhere, the endless screw disappeared.

Vitruvius has much to say about the raising of heavy burdens [3]. The most common

[1] Some, however, suppose that Archimedes was referring to the lever. See p 633.

form of crane consisted of two strong beams, joined at the top by a clamp, but straddled apart on the ground, like the modern sheerlegs (figure 602). The burden was lifted by means of a triple pulley; a windlass was mounted on bearings fixed to the legs. The crane was held in position by means of three or four stays fastened to stakes. We are not told how the burden was moved in a horizontal direction; a two-legged crane could be moved in only one plane by means of its stays. Instead of the triple pulley, a compound pulley with five sheaves could be used.

If a greater burden had to be lifted, so that a heavier crane had to be used, it was set up by means of its own windlass, by hauling on a back-stay fastened to a stake far back and running over blocks at the stake and at the top of the crane. When the windlass is turned, says Vitruvius, 'it will of itself raise the engine without danger', but he does not observe that this cannot be done unless the crane is first raised a little way from the ground. The danger would begin when the top of the crane was above the heads of the men trying to lift it. Once set up, it was made secure by ropes and stays tied to stakes.

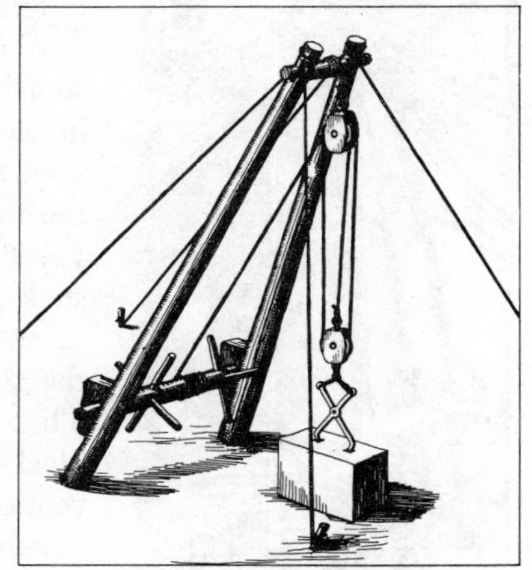

FIGURE 602—*Reconstruction of Vitruvius's crane.*

If the burden was too heavy for this arrangement, a sort of geared windlass was used. Instead of the windlass on the foot of the crane a drum was mounted on an axle passing through the bearings. The middle of the rope was made fast to the ring in the lower block, and each of its ends was passed through the upper and lower blocks the requisite number of times, and then fastened to the axle, one at each side of the drum. A rope round the drum went to a windlass and when that was turned, the burden was lifted 'softly and without danger'. Another way was to exchange the windlass for a 'larger wheel' placed either in the middle or at one end, and to lift the burden more quickly by means of 'tramping men'.

Two reliefs, one in the Lateran Museum [4], the other in the museum at Capua [5], show us what is meant by the 'larger wheel' and the 'tramping men'. In the Lateran relief we see a single mast, held by stays with triple or compound pulleys; on the side of the mast is an enormous treadmill, which five men are working (figure 603). In the Capua relief the wheel seems to be placed apart from the mast; only two men are treading it (figure 578).

There is still another way, says Vitruvius, for lifting heavy burdens quickly, but only an expert should tackle it. Three blocks are used on the burden and on the top of the mast; the three ropes are pulled by three gangs of men, and it is clear that discipline and command are necessary to avoid mishaps. Only one mast is used, and this, says Vitruvius, is useful because it is possible to put down the burden either to the right or the left. He adds that all these ways are used not only for building, but for loading and unloading ships; here are also used horizontal beams swinging in bolts. Pulleys are further used for hauling ships over dry ground.

After the cranes Vitruvius relates some ingenious methods of overcoming special difficulties. The shafts of the great columns for the temple of Diana at Ephesus had to be transported along a flat but soft stretch of ground where wagons could not go. So the contractor, Chersiphron, put iron staples into the ends of the columns, made a frame long enough to reach from end to end, and allowed the columns to be dragged by oxen, like enormous garden rollers.

FIGURE 603—*Part of a relief from a Roman sepulchral monument of c A.D. 100. Five 'tramping men' are seen inside an enormous treadmill, which is being used in building the monument. The compound pulley and the stays for moving the mast are clearly shown.*

When it came to the capitals, his son Metagenes went a step further. He embedded the square ends of the capitals in wheels 12 ft high and dragged them along in the same way. Later on, one Paconius lost his money trying to do the same thing in another way. He had to move a block, 12×8×6 ft. He included it in wheels of 15 ft, put round sticks, 2 digits thick, less than a foot apart all round it, wound a rope round the sticks, and let oxen draw the rope; but the contrivance would not run straight. These were not the normal ways of moving heavy blocks; Vitruvius merely describes them as clever inventions.

After Vitruvius, who lived at the time of Augustus, we come to Hero, who lived about A.D. 70. In the second book of his 'Mechanics' [6] he describes the five 'simple powers', that is the windlass, the lever, the pulley, the wedge, and the screw. These are the five mechanical elements that allow us to transform a small force acting over a long distance into a large force acting over a short distance. Hero shows how they are used for lifting heavy burdens. In the third book [7] he gives some examples of engines in actual use for this purpose. A crane, we learn, may have one mast, or two, or three, or four. A crane of one mast consists of a strong beam, somewhat taller than the height through which the burden is to be lifted; it is placed upright on a platform of wood and held in place by three or four stays. To strengthen it a rope is wound round it with many turns placed some 12 in apart. These also act as footholds for those who have to climb the pole. A drawing shows these windings forming a single spiral round the mast; the text appears to indicate separate lashings, like those on a fishing-rod. A pulley is fastened to the top of the mast, another to the burden; the rope is pulled by hand or by a capstan, which is placed apart from the mast. When the stone has been lifted high enough, one of the stays is loosened and the mast is allowed to lean over till the stone can be lowered to its place on the wall. If the mast cannot slope far enough, the stone is put on rollers and levered to its place. A rope is then attached to the platform and the mast dragged along to another place, to lift another stone, the stays being slackened or tightened as required.

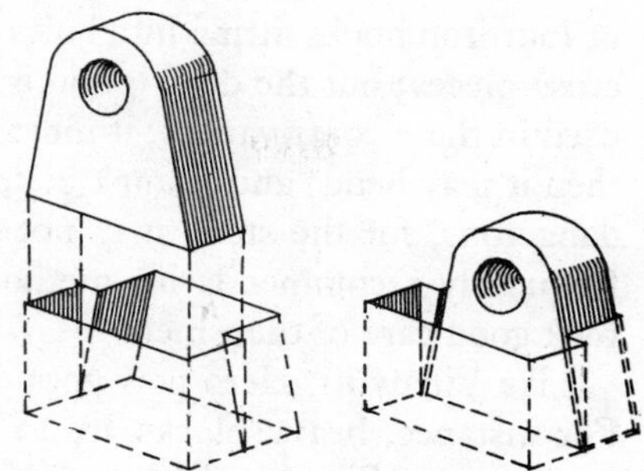

FIGURE 604—*Hero's 'hanger' for lifting a stone. When the hanger had been placed in the undercut hole, a block of wood was inserted beside it to keep it in place.*

If the crane has to have two masts, they are placed on 'the instrument called threshold', as the Arabic text has it, using the Greek word for threshold. The two masts slope inwards to the extent of one-fifth of their height, and are joined at the top by a piece of timber, to which the pulley is fastened. Their bases are made fast to the threshold, and the rope is drawn either by hand or by animals. The masts are held up by stays.

Cranes with three masts consist of three posts, joined at the top and standing on the ground. The pulley is fixed at the top. Such a crane is solid and stable, but it can be used only where it can stand across the place where the burden is to be placed.

For very great burdens four masts are used, placed vertically and forming a square Their tops are joined by shorter pieces of timber, and by two diagonal beams that carry the pulley at their intersection.

Then Hero makes a very interesting remark: in setting up cranes the use of nails and plugs should be avoided; the greater the burden, the more we should content ourselves with ropes and strings, and not try to nail the timbers together.

The burdens to be lifted are massive stone blocks for buildings. Burdens could be slung from ropes, and Hero describes several ways of lifting one without placing the ropes beneath it: it was then possible to place it on the right spot at once. One of these 'hangers', as he calls them, was a solid ring of iron on a foot fitting into an undercut hole in the upper surface of the block. The foot entered the hole from the side, and a piece of wood was driven in beside it to keep it in place (figure 604). Another construction did away with the hole beside the undercut hole. Here there are two eye-bolts fitting each into one of the undercuts; a third bolt with straight sides is pushed down between them to hold them in place, and an iron rod passes through the eyes of all the bolts and serves as a handle for the lifting (figure 605). There is also a description of a 'crayfish' with three

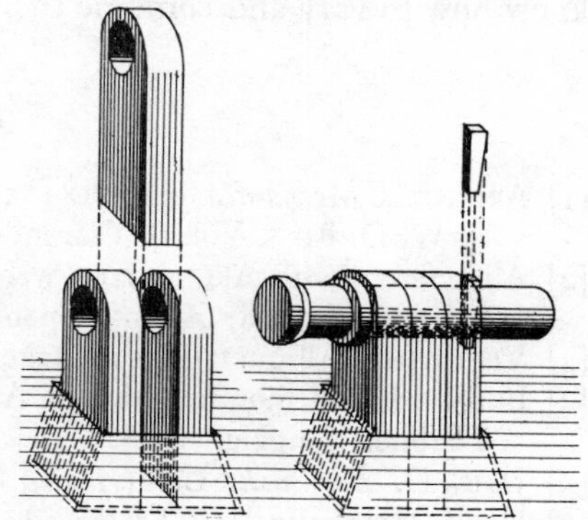

FIGURE 605—*Hero's triple 'hanger'. When the flat, middle eye-bolt is in position, the two outer bolts are safe in their undercut hole. The cross-bolt is used for the lifting. The head and the wedge are not mentioned in the text, but they seem to be a necessary precaution.*

or four iron hooks fitting into holes in the sides of the stone, and kept in place by wooden cross-pieces; but the description is not clear. Hero advises care in the choice of the iron used in these contrivances; it must not be too hard, for then it is brittle, nor too soft, for then it may bend; and it must be free from folds, warpings, and fissures. These are very dangerous, for the stone may not only fall, but may hit the workmen. Vitruvius, too, frequently recommends his methods as 'safe': evidently the contractors of those days took good care of their men.

Like Vitruvius, Hero now goes on to describe the solution of some special problems. For instance, heavy blocks are to be taken down a steep way from the quarry to the stone-cutter. Two roads are made, and two carts, each of four wheels, are used. One below is loaded with waste material, the other, above, with the block. A long rope, running over pulleys, connects the two carts, and the counterpoise is dragged upwards by oxen. To lift a heavy column from its base in the upright position, a rope is passed over two pulleys, with a container as a counterpoise. The container is filled with stones till it lifts the column. How the whole thing is moved to another base is not explained. One may suspect that the text is corrupt, and that the problem was to repair the base without having to take down the column. To get heavy weights aboard a float for water-transport the float might be placed on sacks full of sand. When the burden was safe on the float, the sand was let out of the bags, and the float with its burden became waterborne. There is also a contrivance for restoring a wall to the vertical, if an earthquake has caused it to lean. A strong piece of timber is placed in a ditch dug parallel to the foot of the wall, on the side towards which the wall is leaning; strong beams connected by cross-pieces are planted against this timber, and a windlass with a pulley, from the other side of the wall, will drag the wall back to its proper position.

As has been remarked elsewhere, no toothed wheels were used; only the three simple powers: the windlass, the lever, and the pulley. But the ancient technicians certainly knew how to vary and combine them for many different purposes.

REFERENCES

[1] ARISTOTLE *Mechanica*, chap. 18 (853^{a-b}), trans. by E. S. FORSTER in 'Works of Aristotle', ed. by W. D. ROSS, Vol. 6. Clarendon Press, Oxford. 1913.

[2] All references to ARCHIMEDES were collected by J. L. HEIBERG *Quaestiones Archimedeae*, chap. 3, pp. 35 ff. Klein, Copenhagen. 1879.

[3] VITRUVIUS. All quotations are taken from X, ii, 1–14. (Loeb ed. Vol. 2, pp. 278 ff.)

[4] BRUNN, H. "I monumenti degli Aterii" in 'Kleine Schriften', Vol. 1, pp. 85 ff., and fig. 27. Teubner, Leipzig. 1898.

[5] JAHN, O. *Ber. sächs. Ges.* (*Akad.*) *Wiss.*, phil.-hist. Kl., **13**, 302, and Pl. IX, 2, 1861.

[6] HERO *Mechanica*, II. (Arabic ed. and French trans. by CARRA DE VAUX. *J. asiatique*, neuvième série, **2**, pp. 227 ff., 1893.)

[7] *Idem Ibid.*, III, 2–12. (*Ibid.*, pp. 484 ff.)

19

HYDRAULIC ENGINEERING AND SANITATION

R. J. FORBES

I. EARLY WELLS AND CISTERNS

THE earlier stages of organized water-supply reach back very far. In the Early Bronze Age the springs that served the communities of Europe were often encased in wood. Rain-water was occasionally collected, and timber-framed shafts were cut into strata of impermeable clay to provide for storage. In the ancient Near East almost every hill-city depended on springs at the foot of the town-mound (vol I, p 525). There is frequent confusion between the natural spring and the man-made well. Here we shall confine the word 'well' to describe shafts for obtaining water vertically below the spot where it is required, when it is not obvious at the surface. Many so-called holy wells were but enclosed springs, often deepened later. Wells are characteristic of life in cities and on irrigated farms, where larger water-supplies are needed.

Observation of nature and experience in mining and tunnelling gradually gave rise to practical methods of water-finding. The Roman author Vitruvius gives a few scientific indications for locating a good supply [1], such as noting mist close to the ground, examining the type of soil, and observing the vegetation. These should be followed by trials, in which metal vessels or unbaked clay pots are buried for a short time and inspected for moisture.

Ancient wells were never drilled but were dug by hand. Most were circular and all were lined, usually with wood, stone, or brick. Sometimes pottery rings took the place of brick linings, as in Crete and in Mycenaean sites on the Greek mainland. The acropolis of Athens had two stone-lined wells 18 m deep. The Romans preferred brick or concrete linings, but in military camps and other temporary settlements they used wooden barrels, or plankings between four solid corner-posts, or mortised plank constructions [2]. Such wooden casings were later adopted in other parts of Europe (figure 606).

For drawing water, gourds, shell dippers, pottery-jars, and buckets have been in use for millennia. For deep wells relays of men and women passed the water-jars hand to hand from the water's surface (figure 607). With the evolution of water-lifting machinery (pp 675 ff) new methods were introduced. Both Greeks and Romans used machinery for water-supply in towns. Thus the water for the warm baths near Pompeii came from a well (section 2 × 3 m, 25 m deep) adjoining a basin (1·5 × 2·0 m) in a little building that also contained a long narrow room with a treadmill. Worked by two slaves, this circulated two sets of chains of bailers hanging in the well. The output would be about 440–800 gallons an hour [3].

Apart from hewing cisterns in rock the Greeks and Romans built them of masonry or concrete with barrel-vaults and pillars, and often added smaller settling-basins to them. In Hellenistic and Imperial Roman times the large cities, such as Alexandria and Byzantium, built huge cisterns (figure 608). In the latter city the largest measures 141 by 73 m and has 420 columns. The immense cisterns of Valens (364–78) and Justinian (527–65) still serve Istanbul [4].

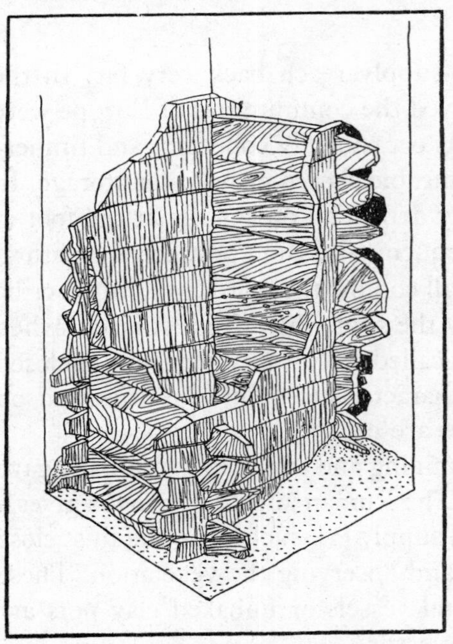

FIGURE 606—*Medieval wooden-lined well found near Happisburgh in Norfolk. The shaft is 5 ft 9 inches square, and was originally 22 ft deep. Twelfth century.*

II. CONDUITS, PIPES, AND SEWERS

In prehistoric Europe timber was easily available, and plank channels or hollow trunks formed the basic unit for conduits. An enclosed spring of mineral water at St Moritz is one of the oldest in Europe (figure 609). Water from this spring was conveyed in two lines of large hollow tree-trunks [5]. Wooden pipes were also used by the Romans (who sometimes applied iron collars to strengthen the joints) and were very common in medieval Europe when they were usually about 6 m long and 8–12 cm in diameter. Such medieval water-mains could withstand pressures up to $3\frac{1}{2}$ atmospheres. Some towns, such as Nuremberg, with an elaborate system of wooden pipes, devised special machines for boring them (figure 599).

Properly sealed earthenware pipes can work at pressures of up to 50 atmospheres, and may be strengthened by embedding in concrete, as was the Roman practice. Open terracotta conduits were first used on the mainland of Greece by the Mycenaeans, though the Cretans had covered their ducts to avoid the accumulation of silt (vol I, figure 349) [6]. Earthenware or stone pipes were very early preferred for small quantities of water and for drains. Such pipes were used in the Indus cities before 1500 B.C., and the earthenware pipes, masonry sewers, water-closets, and drains of some Mesopotamian cities of comparable date are still in working order. The Greeks made their earthenware pipes in curved sections as well as straight, and of a tapering shape so that each fitted into the next, a method adopted by the Romans.

FIGURE 607—*Conjectural reconstruction of a Mycenaean spring-approach at Athens.*

Very ancient metal pipes are also known, as at the temple of Sahure at Abusir in Egypt, where they are of hammered copper 1·4 mm thick. These pipes were cemented into grooves cut in solid flagstones. Bronze pipes conveyed water from the mainland to the Phoenician island-city of Tyre; pipes of lead and of bronze were employed by the Greeks, but the most extensive use of lead pipes for the local distribution of water came with the Romans, though they were well aware of the danger of lead poisoning [7]. Strips were cut from cast sheets about ten feet wide, bent round a wooden former and joined by solder. The ratio of tin to lead in the solder was carefully defined (p 47) but leaks often occurred. The lead pipes of the Roman water-board are notable as the first known series of industrial products to be standardized, on the basis of the *quinaria*—a pipe formed from a strip five digits (9·2 cm) wide.

Bronze pipes were too costly for common use, but they are occasionally found in the villas of the wealthy, as at Rome, Pompeii, and Baiae. Fittings, stops, and taps, of which the Roman world knew many varieties, were, however, always of bronze. Some private houses at Pompeii had as many as thirty taps.

Besides wood, pottery and lead pipes were widely used in the Middle Ages. Lead pipes were soldered with tin solder and held well, though Vincent of

FIGURE 608—*Cisterns at Carthage hewn out of the rock.*

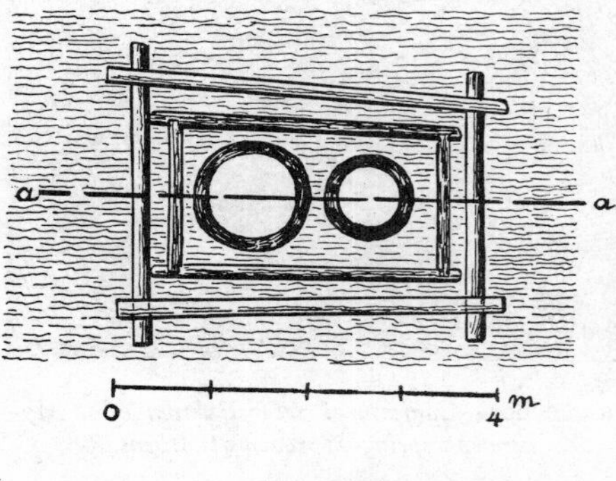

FIGURE 609—*Plan and section of an enclosed spring found at St Moritz, Engadine, probably of the Late Bronze Age.*

Beauvais in his *Speculum* (1254) thought that tin solder facilitated corrosion. By the fifteenth century pipes of cast iron were very occasionally used, as for the water-supply to a public fountain in Augsburg in 1412.

III. THE ORIGIN OF AQUEDUCTS

The *qanaat*, already widely spread in antiquity (vol I, pp 532–4), is essentially the same as the adit driven horizontally into the mountain-side by the ancient miners. Armenia is one of the oldest mining and metallurgical centres in the Near East and, for this and other reasons, is probably the original source of the *qanaat*. They were already widely used in Persia in the period of the Achaemenian kings (sixth century B.C.–331 B.C.). Megasthenes found that they were so frequent in northern India about 300 B.C. as to need a staff of 'officials who inspect the closed canals from which the water is distributed into conduits, that all may have equal use of it' [8]. The *qanaat* had now spread to Trans-Caucasia, Afghanistan, Kashgar, and Chinese Turkestan; its construction was the privilege of century-old guilds of specialists [9]. The Persians introduced this technique to Egypt, and more particularly to Kharga oasis [10], probably during the reign of Darius I (521–485 B.C.). The technique spread across the Sahara as far as southern Morocco, and was used at Roman Garama, Fezzan. The water-tunnels of the Graeco-Roman world are often very similar.

The earliest Greek medical works claim that good water is essential for health (figure 610) [11]. Thus the 'Hippocratic Collection', a work on 'Airs, Waters, and Places', of *c* 400 B.C., has many shrewd observations on the quality of water

FIGURE 610—*Greek women taking a shower-bath. From a black-figured vase. Sixth century B.C.*

and general health. To obtain good water the Greeks freely applied the methods of the Near East. The aqueduct at Athens which brought water from Mount Pentelicus was of the *qanaat* type, its underground channel having a vertical air-shaft about every 15 m.

A second type very common in ancient Greece was the pipe-line supported by stones, which boldly followed the shortest way from its source through valleys and tunnels. Its oldest example may be the aqueduct of Samos, probably of the

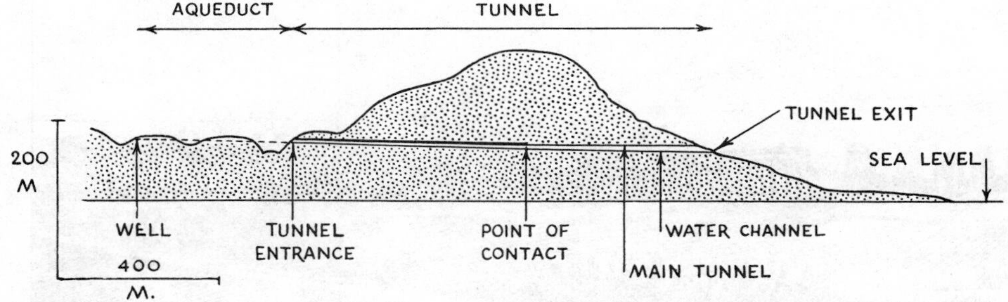

FIGURE 611—*Diagrammatic section of the aqueduct of Samos, probably sixth century B.C. The water was taken by the shortest route, through a pipe-line supported on stone pillars, and a tunnel. The tunnel was cut from both ends, but by a miscalculation a large bend was required to enable the sections to meet.*

sixth century B.C., which Herodotus regarded as one of the three greatest works in any Greek land (figure 611) [12]. He relates that the designer was Eupalinus of Megara, who is thus the first hydraulic engineer whose name has reached us.

The Samos tunnel is just over 1000 m long, and was bored from both ends.

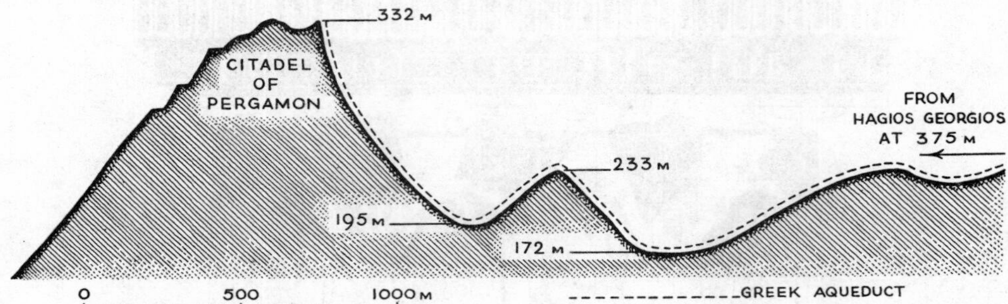

FIGURE 612—*Diagrammatic section of the aqueduct of Pergamon, showing the pipe-line siphon crossing two valleys and the intervening ridge to rise to the level of the citadel, almost in a straight line.* c *180* B.C.

The two passages failed to meet by about 5 m, so that the one from the source had to be curved sharply to make contact. The tunnel is of about 1·75 m square in section, but the water ran in a trench cut in its floor. This trench has a greater slope than the tunnel from which it was built; at the entrance it is about 2·5 m below the floor of the tunnel, and at the exit about 8·3 m. Clay pipes were laid in the trench, which was then filled in, leaving communicating shafts at intervals. Some of the surveying-devices that the Samians may have used for the construction of this aqueduct are mentioned on p 672.

In Asia Minor and elsewhere the Greeks introduced a daring new hydraulic

FIGURE 613—*Ruins of the* Claudia *and* Anio Novus *aqueducts; see figure 614.*

device, the siphon. It avoided the expensive tunnelling or the long tortuous windings of a gravity-flow duct. Though small siphons had been used for centuries, the first attempt to explain their action was that of Hero (fl A.D. 65). Using the siphon as a short cut across a deep valley entailed high pressures, and corresponding strength was needed in the lower part of the siphon. The aqueduct

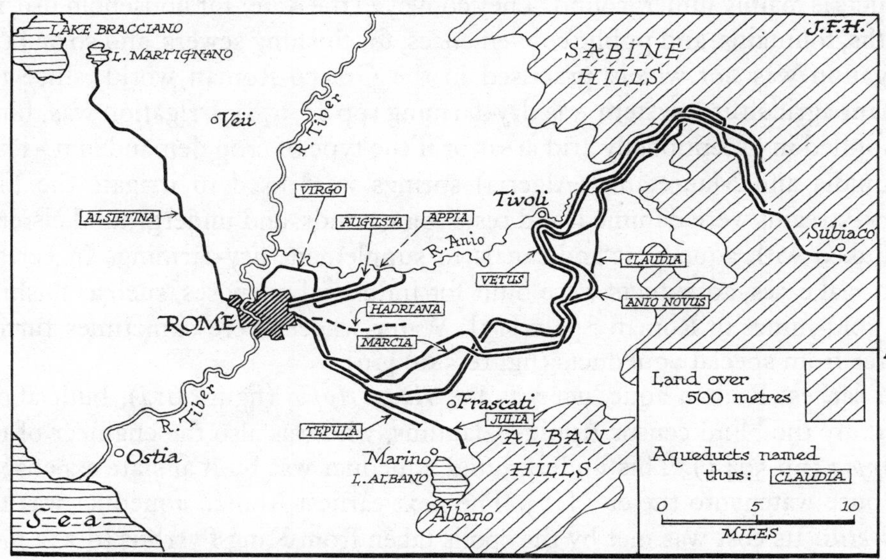

FIGURE 614—*Map showing the routes of the Roman aqueducts.*

of Pergamon, built *c* 180 B.C., is a typical example of the use of such a siphon (figure 612) [13]. The water-supply was conducted by gravity-flow from a spring in the mountains (ht 1174 m) to two settling tanks at Hagios Georgios (ht 375 m) east of the city. From these tanks a pressure-line brought the water across two valleys (hts 172 m and 195 m) and over the intervening ridge (ht 233 m) to the citadel (ht 332 m). Pressures up to 20 atmospheres had to be overcome in the lower valley. The course of the pipes is traceable by the stone slabs, 1·2 m apart, which supported them. They were probably of metal, and were removed when the siphon was replaced in Roman times by a system of aqueducts, which was more reliable. These bridged the valleys and supplied only the newer parts of the city at the base of the hill.

Siphons and bold tunnelling are typical of the water-supply systems built in the many Greek and Hellenistic city-states from Sicily and southern Italy to Asia Minor. The Romans seldom used the siphon, probably because of leakages and the poor materials available for high-pressure piping.

IV. ROMAN AQUEDUCTS

The Roman aqueducts are widely known for their superb arches, which span many valleys of Spain, France, and Italy, and notably the plains of the Campagna around Rome (figure 613). It is sometimes not realized that the course of these aqueducts is mainly underground. They conveyed the water for household use, for the baths, fountains, and public conveniences, for flushing sewers, and so on [14].

Irrigation was not widely practised in the Graeco-Roman world, since the dominant agricultural system was dry-farming (pp 81, 678). Irrigation was, however, applied in exceptionally arid areas or if the type of crop demanded it. Thus in Numidia and Mauretania (Algeria) springs were used to irrigate the hill-country, streams were dammed, and reservoirs, ponds, and underground cisterns were linked with aqueducts and canals to supplement dry-farming. In certain exceptional cases, aqueducts were built for industrial purposes, such as flushing at the gold-mines in Roman Spain [15]. Water-wheels were sometimes turned by water from special aqueducts (figure 546) [16].

The earliest Roman aqueduct was the *Aqua Appia* (figure 614), built about 300 B.C. by the blind censor Appius Claudius, who was also the engineer of the *Via Appia* (pp 504 f). This underground aqueduct was built at state expense to bring pure water into the city [17]. The next earliest Roman aqueduct was the *Anio Vetus*. Its cost was met by the spoils taken from King Pyrrhus (d 272 B.C.) just as the *Aqua Marcia* (figure 614, plate 39) is said to have been financed about 146 B.C. from the booty taken at Corinth and Carthage [18]. Early Republican times saw the extension of Rome's water-supply as well as the construction of some local aqueducts [19]. By 125 B.C. the water-supply of Rome had been doubled, to meet the city's rapid expansion.

In the troubled days of the later Republic (first century B.C.) the aqueducts were neglected; the Emperor Augustus (27 B.C.–A.D. 14) had to repair many channels and ducts that had fallen into decay, and to build others [20]. The Imperial period was a great era for building, and several emperors increased the water-supply of their main cities. The huge aqueduct of Claudius took fourteen years to build. His engineers had to grapple with the problem of quarrying blocks of tufa from the banks of the Anio river and hauling them several miles to the building-site. No fewer than 40 000 wagon-loads were transported every year [21]. Hadrian (117–38) built many aqueducts in various parts of the Empire [22] and later emperors, up to Justinian, continued this policy [23].

In Egypt Augustus built an aqueduct for Alexandria. In the rest of the country the older forms of irrigation-machine were also used for supplying the towns.

Thus at a military camp near Memphis, wrote Strabo, 'the water is conducted up from the river by wheels and screws. 150 prisoners are employed on the work' [24]. In A.D. 113 a town in the Fayum built a water-wheel that lifted water into reservoirs to feed the town fountains, a bath, a brewery, and two synagogues [25].

In Asia Minor, the old Greek initiative was reawakened by Roman rule [26]. Sardis, Ephesus, Pergamon, Smyrna, Miletus, and Nyssa received new aqueducts. In Spain the large aqueducts of Merida and Segovia still stand and are well known. The bridge of the latter is over 1000 m long and 50 m high. In France the aqueducts of Nîmes (19 B.C.) (figure 371) and Arles are famous. Lyons had several great water-channels, including a siphon with seven lead pipelines of 27-cm diameter. For this siphon some 10 000 tons of lead were used [27]. In Roman Germany there were great aqueducts at Cologne, Bonn, Mainz, and Treves [28].

In Britain there were Roman aqueducts at Lincoln, Dorchester, and elsewhere. That of Wroxeter supplied private houses, through separate sluices for each house; in Silchester, the water arrived below the town level and was raised by force-pumps. An 8-mile aqueduct supplied gold-workings in Carmarthenshire with water to wash the ore.

V. ORGANIZATION OF ROMAN WATER-SUPPLY

Vitruvius devotes one of his ten books to water [29], in which he discusses the construction of aqueducts, wells, and cisterns; but a better source of technical information on Roman water-supply is a work of Sextus Julius Frontinus [30], water-commissioner of Rome from A.D. 97 to 103 or 104. 'With such an array of indispensable structures carrying so many waters,' he enthusiastically explains, 'compare the idle Pyramids or the useless, though famous, works of the Greeks' (figure 614) [31].

Originally the water-supply of Rome was controlled by municipal magistrates. Augustus centralized their responsibilities in his friend Marcus Vipsanius Agrippa. A law of 9 B.C. entrusted these duties to a board consisting of a water-commissioner of consular rank, as chairman, and two technical advisers. Frontinus was the seventeenth of such commissioners. During the reign of Claudius a reorganization (A.D. 52) transferred to the commissioner the entire responsibility for water-supply and the necessary funds for upkeep and construction.

The water-commissioner had a staff that included an architect, administrators, and a whole series of graded inferior officers. In addition to these, Agrippa, the first water-commissioner, had at his disposal a band of 240 slaves, whom he trained and left to the state at his death. There was a second gang belonging to Caesar. These slaves, with further specialists and hired free labour, formed the regular technical staff. A large body of men was certainly needed to maintain the eight main aqueducts that served the capital [32].

Of these eight aqueducts (figure 614) five drew their waters from springs and artesian wells, two from rivers, and one from the sea. The older aqueducts were mainly subterranean, partly as a precaution against attack. The five built later reached the city at the highest possible level (plate 39) and thus risked being cut, as in fact happened during the siege of Rome by the Goths. The sum of the cross-sections of the eight ducts was about 7·5 sq m, and their total length was 350 km, of which only 47 km lay above ground. Delivery has been calculated as about 1 010 600 cu m per day. In the fourth century A.D. Rome had 11 public baths, 856 private smaller baths, and 1352 fountains and cisterns.

The source of the water being chosen, plans for its course were carefully made. Underground ducts were preferred, and the beautiful ducts on tiers of arches were avoided, if possible, because of their high cost. Frontinus mentions that he had drawn plans of all existing aqueducts. The course chosen usually had a gradient of 1:200–1:1000. Every effort was made to avoid a pressure-supply, for siphons were expensive to build and repair. The siphon was, however, occasionally used to cross steep valleys (figure 612).

The gradient of the aqueducts was not always evenly maintained, for the only levelling-instruments available were very simple. The ancient Near East used a plumb-bob level in the form of a capital A with a plumb-line hanging from the apex. The Greeks and Romans had a simple water-level or *chorobates*, the *dioptra*, and the *groma* to trace right angles. When these instruments were used skilfully, it was possible to maintain a gradient of 1:2000 very closely. Even reasonably accurate tunnelling was possible with them, and we seldom hear complaints of careless and negligent work.

From spring or river the water was led to settling-tanks, which usually had two compartments with sloping floors to facilitate cleaning. The water was then conducted to the duct proper, often 15 m underground, which had air-shafts every 40 to 50 m to prevent air-locks and to allow inspection and cleaning. The duct varied from 0·5 to 3·0 sq m in cross-section. It consisted of a concrete lining 50–60 cm thick, enclosed and carried by a mass of masonry. As many of the springs tapped were fairly hot, the Roman aqueducts suffered from serious incrustations of calcium carbonate and had to be cleaned frequently. The waters of only the *Aqua Virgo* and *Aqua Marcia* were so pure that they did not need settlement [33]; that of the *Aqua Alsietina* was undrinkable, and with part of the supply of the *Aqua Traiana* served for water-mills, fountains, and sewer-flushing. The water of the aqueducts arrived at distribution-tanks; these were not reservoirs for they had only little storage capacity. The supply worked on the principle of constant off-take. Each aqueduct branched to a number of tanks; the *Aqua Appia* served 20 and the *Anio Vetus* 92, the total for Rome being 247.

VI. WATER INSTALLATION AND DISTRIBUTION IN ROME

Frontinus relates [34] that there were three main groups of water-supply:

		Per cent
(*a*) At the Emperor's disposal		17·1
(*b*) To private persons (houses and industries)		38·6
(*c*) Public supplies:		
19 military barracks	2·9	
95 official buildings	24·1	
39 public buildings and theatres	3·9	
591 cisterns and fountains	13·4	
		44·3
		100·0

From the bottom of each tank three mains tapped water for the fountains, baths, and official buildings. Ten higher mains delivered water to private consumers, blocks of houses, industries, and so forth. Tapping from the main was illegal, but the overflow from reservoirs, fountains, and public baths was often free, though the bulk of it was used for flushing drains or for industrial purposes. The basement of the baths of Caracalla contained mills driven by this surplus. Some buildings had secondary reservoirs in which the water supplied was pumped by water-wheels or force-pumps (figure 342). In the provincial towns supplies were not always ample; in Pompeii, for example, the authorities were often forced to ration the water to private persons during certain periods of the day to ensure supplies for the baths and public buildings.

The Emperor could grant any syndicate or person the right, even for life, to tap the mains for his own use, but generally the officers in charge of reception tanks charged consumers according to the cross-section of their delivery-pipes. These private delivery-pipes were calibrated on the *quinaria* as unit (p 665). The head of water was unspecified but tacitly assumed to be constant. Frontinus discusses the set of standard pipes based on the *quinaria*, and details a series of 25 gauges ranging from one *quinaria* to a pipe of 120-*quinariae* capacity [35]. There were officers whose duty it was systematically to inspect and stamp them. Regulations for this process survive.

Frontinus not only surveyed and restored the whole water-supply of Rome, but tried to establish correct figures on which his bureau could work. He was not satisfied with the total intake of the aqueducts, and by measuring cross-sections of ducts, but ignoring velocities, he attempted to calculate the quantities of water delivered from the sources, at the reservoirs, and finally to the consumers. His ignorance of hydraulic principles led him to blame the loss entirely on 'the dishonesty of the water-men, whom we have detected diverting water from the public conduits for private use. But a large number of landed proprietors also, past whose fields the aqueducts run, tap the conduits; whence it comes that the public water-courses are actually brought to a standstill by private citizens, just to water their gardens' [36]. He sought to correct these faults and awards himself the epitaph: 'The expense of a monument is superfluous; my memory will endure if my actions deserve it.'

VII. TESTING AND PURIFYING WATER IN ROME

Ancient physicians and engineers alike stress the need for purity in water. Galen, Vitruvius, and many more denounced the use of lead for cistern-linings or pipes, but lead-poisoning continued to occur (p 419) [37], though the extensive incrustation of calcium and lead carbonates on pipes may have effectively reduced it. Vitruvius recommends the following methods for testing the purity of water: 'The water, being sprinkled over a vessel of Corinthian bronze [a gold–silver–copper alloy] or any other good bronze, should leave no trace. Or if water be boiled in a copper vessel and allowed to stand and then poured off, it will also pass the test if no sand or mud be found in the bottom of the copper vessel. Again, if vegetables boiled in the vessel are soon cooked they will indicate that the water is good and wholesome' [38]. Several ancient texts mention that certain waters 'can bear a little wine', and it is possible that by dosing water drop by drop with a strongly coloured wine the Romans were estimating roughly the lime-content of their water [39].

To purify water, ancient recommendations range from mere exposure in the sun and air to filtration. Herodotus notes with approval that the water of the Karscheh (*Choaspes*), which flows by Susa, was boiled and stored in silver flagons for the Persian kings [40]. Porous filters of tufa have been found, and filtration through wool or lengths of wick was well known. Athenaeus of Attalia wrote a work 'On the Purification of Water' (*c* A.D. 50) in which filtration and percolation are discussed. Percolation through layers of sand is also recommended by Vitruvius [41]. Ancient authors suggest that water may be purified or rendered wholesome by the addition of various substances, the most common and effective of which was wine.

VIII. METHODS OF WATER-LIFTING

Most of the simple discontinuous methods of lifting water still to be seen in many parts of the east, such as the shaduf, made their first appearance before classical times (vol I, ch 19). The next advance in mechanical water-lifting was

FIGURE 615—*Modern* sāqiya *(wheel of pots) turned by a camel, drawing water from the Nile. Note the very crude construction.*

the provision of a continuous supply of water. This was achieved by the application of two inventions of the Greeks—the toothed wheel and the screw (pp 116 f, 631 f). The Greek achievements in water-lifting are perhaps related to the new ability of their philosophers to apply the results of their study of natural phenomena to daily life. Babylonian astronomers were adroit in their mathematical tabulation of celestial events. The Greeks, however, possibly for the first time, cast their observations into geometrical schemes that were intended to explain the phenomena. Thus they conceived systems of the mechanism of the heavens in complex rotary movements, which suggested practical applications of these movements.

'All machinery', says Vitruvius, 'is generated by Nature and the revolution of the universe guides and controls.... Since, then, our fathers had observed this to be so, they took precedents from Nature; imitating them and led on by what is divine, they developed the comforts of life by their inventions. And so, they rendered some things more convenient by machines and their revolutions and other things by handy implements' [42].

The invention of the toothed wheel is sometimes ascribed to Archimedes (d 212 B.C.) (p 632); it certainly cannot have appeared much earlier. It enables a vertical wheel carrying pots to be rotated by gearing it to a horizontal wheel turned by man or beast. This is the 'Persian wheel' or *sāqiya*. In practice, to

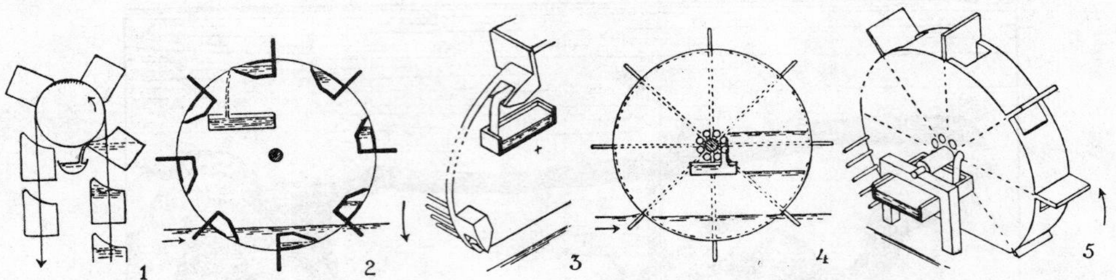

FIGURE 616—*Three types of water-lifting machine described by Vitruvius:* (1) *chain of buckets passing round two pulleys;* (2, 3) *scoop-wheeling;* (4, 5) tympanum *of eight sections. The paddles enable it to be rotated by the flow of the stream.*

prevent too serious a loss of water from the pots, the collecting sluice must be placed well below the top of the wheel, so that the lift cannot be much greater than the radius of the wheel (figure 615). A variant, called the *tympanum* by Vitruvius, allowed a larger quantity of water to be raised through a height equal to the radius, with less wastage (figure 616). It consisted of a drum with a horizontal axle, divided into eight compartments by planks radiating from the axle. An opening in each compartment, six inches wide, admitted water from the lower level at the circumference of the drum. As the compartments were raised by the revolution of the drum the water ran out of holes cut into each of them near the axle and was collected in a trough [43].

In Egypt rotary devices were slowly introduced in Hellenistic times. The *sāqiya* is first mentioned in the second century B.C. By Roman times it had begun to be replaced there by the *tympanum*, which, turned by a treadmill, provided an abundant supply of water for irrigating gardens or dissolving salt in salt pits. One of the mechanisms was in use in the Fayum about 20 B.C., and Strabo (? 63 B.C.–A.D. 21) writes of a ridge of fertile land in Egypt to which water was raised from the river by wheels and

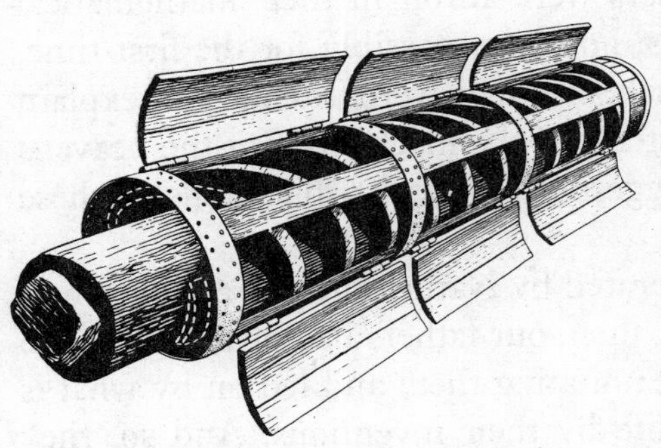

FIGURE 617—*Archimedean screw of oak from a mine at Sotiel, Spain. The axle is 24 inches in diameter, and the screw projects 2·4 in from it.*

screws, and where 150 prisoners were employed in the work [24]. Hellenistic times also saw the introduction of water-lifting machinery into Palestine and Mesopotamia.

Though the invention of the screw is ascribed by tradition to Archytas of Tarentum (fl 400 B.C.), our earliest definite information on it concerns Archimedes and his famous water-screw. Diodorus (first century B.C.) claims that the Nile delta was irrigated by machines invented and introduced by Archimedes, which, from their form, were called *cochleae* (snails) [44], though elsewhere he says that there were devices in mines and wells before Archimedes introduced them on his visit to the court of the Ptolemies (p 633) [45].

FIGURE 618—*Archimedean screw for water-raising, from a mural at Pompeii, showing the method of turning the screw by treading with the feet. It should of course be shown sloping, not horizontal as it appears in the picture.*

From the remains of an ancient water-screw found in Egypt [46], from pottery models, and from descriptions, it is clear that water-screws were worked by some form of treadmill and not by the crank, which was unknown to ancient engineers (figures 617–19). The water-screw is still common in certain parts of Upper Egypt and of the Arab world for small lifts. It was introduced into Morocco in Byzantine times, and from there spread to western Europe.

The energy of running water was harnessed during the first century B.C. or a little earlier, and led to the last important stage in the evolution of water-raising machinery—the employment of the water-wheel as prime mover (ch 17). In the water systems of Imperial Rome the water-wheel was important for obtaining a good head of water when the supply was delivered at a low level [47]. This was the case at Ostia, the port of Rome. At the main baths were storage-tanks from which water was raised by wheel to a tank on an upper floor, whence

FIGURE 619—*Negro slave working an Archimedean screw with his feet. From a late Roman–Egyptian terracotta relief.*

an adequate supply at fair pressure could be obtained. A similar method was adopted in Roman Silchester and other sites. It remained in vogue in Europe until the last century.

IX. GREEK AND ROMAN IRRIGATION AND DRAINAGE

Water-conservation was the principle at work in the classical technique of cultivation. The dry-farming technique (pp 81, 670), propagated by the great weight of classical agricultural doctrine, held for Greece and Italy, but it failed in the wetter climate of north and west Europe. Irrigation was applied only if the climate was arid enough to demand it, though supplies of water were provided for types of crop that needed large quantities, such as onions, cabbages, lettuces, and pears. In certain cases vineyards also were irrigated. Irrigation-water was generally supplied from aqueducts. There is an inscription with a plan of an aqueduct, the *Aqua Crabra* near Tusculum, which gives the names of the properties irrigated, the number of pipes supplied, and the hours when they could be opened.

In north Africa, where climate and conditions were more like those of the Near East, irrigation was much more common [48]. Springs were used to irrigate small hill-plots in Numidia. Everywhere in Algeria are vestiges of Roman hydraulic works. Streams were dammed, and water was stored in large cisterns, ponds, and underground tanks, and distributed by canals and aqueducts. In Italy itself this system of irrigation was usually unsuccessful, as the technique was not correct for that terrain. Only after the eighth century A.D., when the Arabs introduced such crops as rice into the western Mediterranean, did irrigation-techniques work successfully in Italy.

In the ancient Greek world the cultivated area was restricted, for the area of potentially fertile land was largely broken by shallow lakes, swamps, and marshes. Attempts at drainage were made from the earliest times, particularly in Boeotia [49]. Lake Copais, now drained, was a vast reed-swamp. The outlets of this depression were a number of natural subterranean crevices which were sometimes blocked by earthquakes. The ancients often attempted to make new outlets, which constantly fell into disrepair. Efforts to drain the lake continued to be made until well into Christian times. Empedocles (fifth century B.C.) is said to have drained the swamps near Selinus and to have improved the climate and harvests of Agrigentum, Sicily, by blocking a windy rift in the hills [50]. An inscription of about 320 B.C. records a contract by a city in Euboea to reclaim fertile lands by hydraulic works.

The Romans were far more active in drainage than the Greeks [51]. Many

Italian rivers like the Arno and the Tiber cannot always contain the torrential spring floods, and form swamps near their mouths or flood the fields along their banks. In Rome itself the Tiber was regulated towards the end of the third century B.C. by building a stone embankment in three stepped stages; this in fact proved better than the modern straight quay.

The drainage of several regions seems to date back to the early Republic. It was designed to improve farm-lands over large areas, and indicates the presence of great landlords controlling large resources of labour. This drainage was achieved by water-tunnels, often subterranean, with air- and construction-shafts at regular intervals. These occur in the Pontine marshes, south of Rome, and at several places in Etruria. Lake Velinus and the plain of Raete were drained by an 800-m canal partly cut in the rocks. The drainage of the large Lake Fucinus was undertaken by the Emperor Claudius I in A.D. 41 and completed by 30000 men in eleven years [52]. A 3½-mile tunnel, with air-shafts at regular intervals, had a cross-section of 16 sq yds. It added 50000 acres to the Imperial estates.

There were several attempts at draining the Pontine marshes, but without definite results. The Roman drainage of the Po valley was much more successful. The project formed part of the policy of settling discharged soldiers in this newly conquered area. Thus the consul Marcus Aemilius Scaurus in 109 B.C. drained the plains by running navigable canals from the Po as far as Parma [53], before constructing the *Via Aemilia* connecting this district with Rome (p 505), and provided good farm land in Cisalpine Gaul. Other engineers drained the district round Bologna, Piacenza, and Cremona, and canalized the Adige between Ferrara and Padua. Claudius I (41–54) finally crowned the work by drainage north of the Po, and by building the *Via Altinata*. These projects also included the drainage of the marshes of Ravenna, which town became increasingly important when the Adriatic fleet was stationed there [54].

Roman drainage and river-regulation projects were not confined to Italy. The army canalized one of the mouths of the Rhône to safeguard the regular oversea supplies of the army [55]. This canal (*fossa marina*) silted up in the first century A.D., but Arles remained an important garrison town with good communications for inland shipping. In the fourth century A.D. there was general complaint of the neglect of drainage in the lower parts of Gaul [56].

In Britain the Romans tried to reclaim the Fens. They built an embankment to protect the flat lands of the Wash from the sea, and other dikes enclosing marshy land at the mouth of the Usk, Monmouthshire. They drove an 8-mile canal connecting the Cam below Cambridge with the Ouse half-way between St Ives and Ely. Another canal 25 miles long and 60 ft wide was made from Peterborough on the Nene to the Witham, 3 miles below Lincoln. The Fosse Dike connected the Witham and the Trent. In connexion with these drainage-canals a 60-ft causeway was built along some 30 miles of their banks [57].

In Holland, Drusus in A.D. 12 laid a mole in the Rhine west of Cleves, to keep excess water from flowing through the Waal and flooding the southern part of the district between the Rhine and Waal. He dug a canal between the rivers Rhine and Yssel to relieve the Rhine of surplus water [58]. About A.D. 45 Corbulo

FIGURE 620—*Shaduf on a Flemish farm, from an engraving by Pieter Brueghel, c 1550.*

connected the Rhine and Meuse with a 23-mile canal running along the site later occupied by The Hague [59], providing his ships with a safe passage that avoided the sea.

X. MEDIEVAL IRRIGATION AND DRAINAGE

The Arabs greatly extended the irrigated area in Spain and Sicily. The Moors of Spain knew how to drain rivers and irrigate their fields by branched channels efficiently distributing the available water. The big lifting-wheels at Toledo may date back to their time. This technique was taken over by the Christians at their reconquest of the city in 1085, and the irrigated area in the Ebro valley was materially increased in the thirteenth century. In the Roussillon[1] (south-east France), where Arabic influence was strong, irrigation methods and crops were

[1] Corresponding largely to the modern department of the Pyrénées Orientales.

introduced in the eleventh and twelfth centuries, and by the thirteenth a system of canals, water-wheels, and water-mills for irrigating crops had developed there. In Italy irrigation-works appear in Lombardy in the twelfth century and some hundred years later in Emilia. Near Milan, the Cistercians introduced the use of city refuse and sewer-water as fertilizers on their land about 1150, and the irrigated meadows there are of the fourteenth century.

The Arabs were also responsible for the spread of various forms of the water-wheel, sometimes known to the west by their Arabic or Persian names. The early western terms for this machinery are a particularly apt illustration of classical tradition being overrun by Egyptian, Arab, Berber, and Persian elements, which spread west and north by way of Sicily and Spain. The shaduf was used by Spanish gardeners of the sixth century A.D. It spread slowly north and is found in Germany and Flanders in the fourteenth century (figure 620).

The farm-lands of France were greatly extended by the deforestation in Normandy and Maine that began in the tenth century, but the next two centuries, culminating in the reign of St Louis (1226–70), form the real age of reclamation. Throughout Europe, however, the most important gains of arable land were not from forests but from marshes and rivers, where drainage was the dominant factor.

The main development of marshland drainage was in the Low Countries, at the mouths of the Scheldt, Meuse, and Rhine. The techniques spread eastwards in the twelfth century, first to the lowlands west of the Elbe and then east of this river. The evolution of drainage-techniques can best be studied in Flanders and the four sea provinces of the Netherlands: Zeeland, Holland (originally the name of a district near Dordrecht), Friesland, and Groningen [60]. The situation in Roman times is well described by Pliny, who visited the country:

> In this land twice each day and night the ocean with its vast tide sweeps in a flood over a measureless expanse, covering up Nature's age-long controversy and the region in dispute between land and sea. There this miserable race occupy elevated patches of ground or hand-built platforms above the level of the highest tide, living in huts on the chosen sites like sailors in ships when the water covers the surrounding land [61].

This is only part of the picture, for in several regions of the Low Countries the sea encroached less on the land during the fairly dry period from about 1500 B.C. to A.D. 500. Many of the mounds mentioned by Pliny grew up from the refuse collected in inhabited sites. When from A.D. 500 the climate grew moister, and the land sank some 10 cm per century, the situation worsened. The mounds or *terpen* characteristic of the region were artificially raised against floods (vol I, p 322). There are some 1260 of them, varying in area from 5 to 40 acres, in a stretch of 60 by 12 miles. As the sea encroached farther, floods became more frequent. Dikes on the danger-spots at first sufficed. Some were built as early as the seventh century, and by A.D. 1000 Friesland was relatively safeguarded by them. In Flanders and Zeeland land was gradually wrested from the sea. The saltings and mud-flats were first naturally consolidated by seaweed, then marsh samphire (*Salicornia* spp) and marram grass (*Ammophila arenaria* L) gradually raised the beaches. The grass beaches suffered less and less from periodical tidal

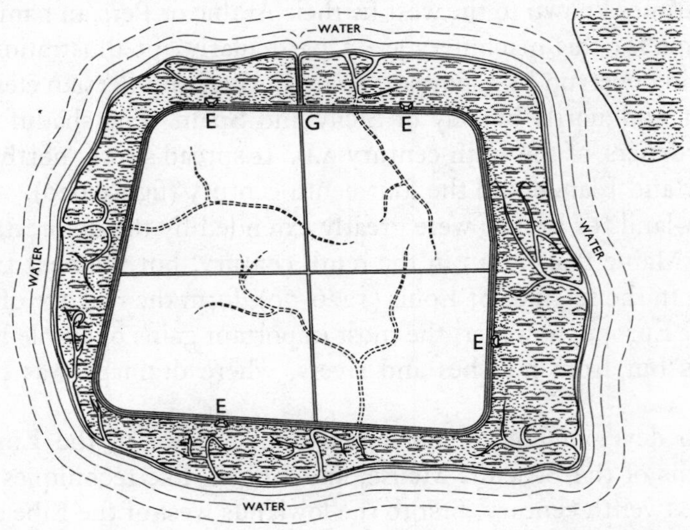

FIGURE 621—*Sketch of a simple polder. Two straight ditches are cut at right angles across the reclaimed area surrounded by an embankment.* G *is a lift-gate through which water flows from the polder at low tide. At* E *are heaps of earth and clay for repairing the dikes. Based on Andries Vierlingh's manuscript* Tractaet van Dyckagie, *1579. Scale* c *1/15,000.*

floodings and could be used as grazing for sheep. This land was gradually drained properly, and either ploughed or used for horses and cattle. The lush grass growing there was ideal for strong heavy breeds. This reclamation became a basic policy of the Counts of Flanders who, from the twelfth century, granted to abbeys and chapters in the maritime region new lands still exposed to the sea.

As here, so in the rest of Europe, land-reclamation was started and led by religious houses. Thus the Benedictines worked on the irrigation of Roussillon, Saintonge, Maine, Île de France, and Bavaria. More to the north the Premonstratensians (White Canons) and Cistercians cultivated the waste-land. Their main works were in Germany between the Rhine and Elbe, in Saxony and Thuringia, and finally in Lusatia and Bavaria.

By the tenth century the Netherlanders had gained considerable experience in land-reclamation. Their skill was recognized beyond their homeland, and as early as 1106 the Archbishop Frederick of Bremen called in colonists from Holland to reclaim bogs north-east of his see. It should, however, be realized that until 1500, despite such efforts, more land was lost than gained, and this is particularly true for the earlier period. The sea, owing to the subsidence of the land, encroached more and more so that the Zuider Zee, for example, reached its largest extent about 1300. During a storm on 14 December 1287 over 50 000 people were drowned in the area, and it was several years before people ventured there again.

By gradually encircling threatened land by dikes, polders were made (figure 621). These areas of reclaimed land discharged their water into the sea at ebb-tide through gates or sluices which could be closed when the tide began to flow. The main principle of this drainage was to shorten the coast and close the bays, creeks, and gaps. Early in the thirteenth century certain bays in the north of Holland had already been dammed. The old sea-dikes now became inner dikes. Soon after 1300 the great dike encircling West Friesland was ready. By the end of that century most of southern Holland was protected by dikes, and land was gained in the same way in Zeeland. Between 1200 and 1500 over 285 000 acres were gained and held.

The sea was not the only enemy of the drainage engineer. With the rise of the cities peat was dug in the low country behind the coastal dunes. The shallow lakes thus formed encroached on the land. They too had to be diked, drained, and recovered as farmland. The retaining dams became dikes of a new type of polder. In this region the villages are built on the dikes or beyond the ancient lakes. In the polders reclaimed after 1500, which were lake-bottoms originally, the villages lie in the polder itself.

Until the fifteenth century salt was made by incinerating the halophytic peat and leaching the ashes. The practice dated from long before the Carolingian period, and some of the towns in Zeeland and elsewhere on the North Sea coast as far east as Schleswig grew rich on the salt trade. Peat-burning was restricted during the fifteenth century, and forbidden by decree in 1515. Peat-cutting destroyed good land and often endangered the polders by inducing a higher water-table. The cutting and destruction of former sea-dikes, now inland, also endangered the land. This was prohibited in 1595. Finally, shifting currents and especially very low ebb-tides often endangered dikes that were not solid enough. The dikes, originally private ventures, thus became planned and organized attempts to keep the sea at bay.

The construction and maintenance of the dikes and polders had now become a task of the community, for an apportioned responsibility on the dike went with each share of land. Gradually more and more polders sprang up, as men co-operated in draining and discharging the polder-water, either through a com-

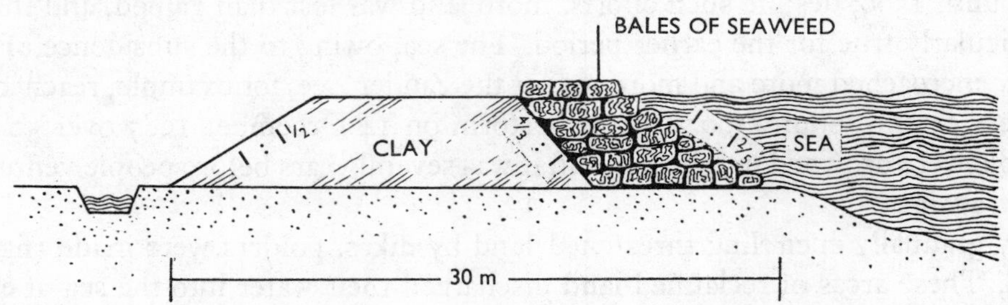

FIGURE 622—*Typical dike-construction of the Middle Ages in Zeeland. The core was earth and clay with bundles of seaweed piled along the sea-edge. This gradually formed into a solid mass protecting the core. Based on Andries Vierlingh, 1579.*

mon gate directly into the sea, or first into a dammed bay or stream and thence to the sea. Brotherhoods or companies were formed to expedite the work, which gradually won the support of the central authorities. By 1500 the attack on the sea had gained sufficient impetus for town councils to clamour for charters to build dikes and reclaim land. When the land had practically been safeguarded, the inland waters formed by peat-delving activities became more immediate dangers, and merchants formed companies to reclaim them and exploit the farm-lands thus won. A more systematic planning of the reclamation of the lakes of northern Holland was started by Lamoral, Count of Egmont (1522–68). His success was possible because of the accumulated knowledge of dike-construction and the

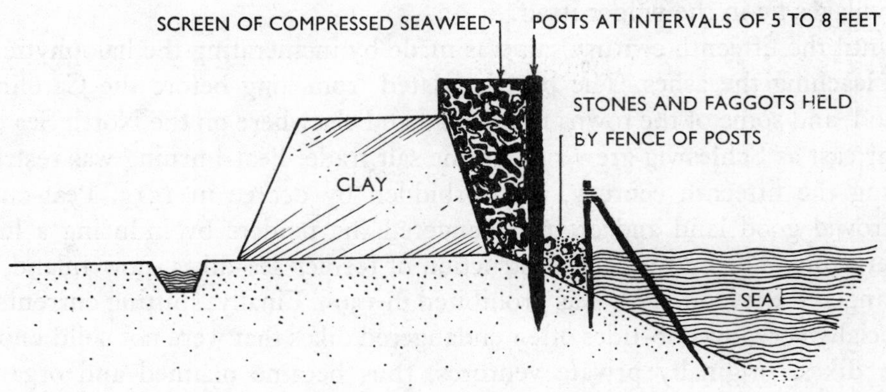

FIGURE 623—*Another type of medieval dike, in which the core is protected first by seaweed compressed behind piles and secondly by a wall of stones and faggots also retained by piles. Based on Vierlingh, 1579.*

introduction of the windmill-driven scoop-wheels to pump the rain- and seepage-water of these deep polders.

XI. DIKES, WINDMILLS, AND SLUICES

The oldest medieval dikes were simply formed from lumps of boulder-clay consolidated by treading with oxen or horses. The foundation was usually carefully denuded of vegetation, and in front of the dike there was generally a stretch

FIGURE 624—*A lift for transferring boats from one level to another in a canal. They were hauled on to a platform* A, *which was then moved up or down the inclined plane* B *to the new level with the aid of a horse-winch. Zonca, 1607.*

of foreshore. Some dikes, mostly in Zeeland, had an earth core plastered with clay on the slopes; some had stepped slopes, and the top slope of 1·5 in 1 was plastered with clay and planted with grass. The slope facing the sea needed special protection. This was provided in the fourteenth century by bundles of seaweed piled step-wise against the vertical front of the dike as far as the top and over it (figure 622). The seaweed was compacted by its own weight, and the heat evolved as it putrefied gradually formed it into a solid mass. In Zeeland and elsewhere where seaweed was not available reed was used. The reed-dike had a frontage of reeds piled up with the roots seaward. The strength of this protective layer was determined by the length of the reeds. There was no compaction and the reed layer was liable to rot; it had to be renewed every five years or so. During the fifteenth century more lasting forms of protecting the dike were introduced, such as piling. In 1440 the dike from Amsterdam to Muiden was thus protected; others soon followed. Piles were also used as breakwaters. The piles were generally 7–10 m long and 30 cm in diameter. They were held together by double girders and iron bolts. At the end of the fifteenth century a new form of dike-protection came into use. Two rows of short piles were driven,

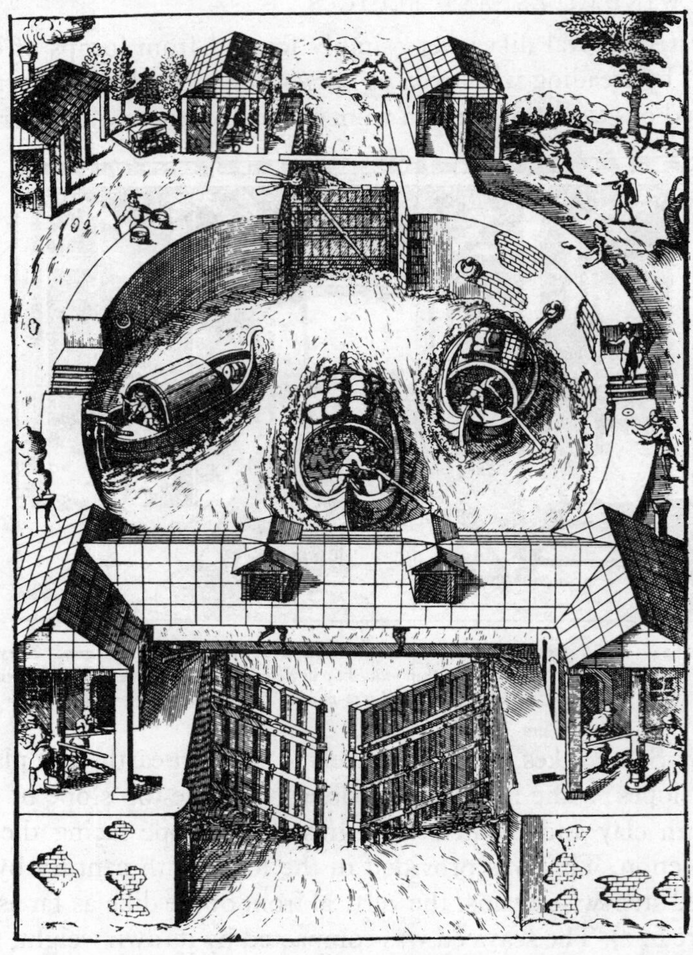

FIGURE 625—*A lock-chamber for transferring barges from one level to another in a river. The down-stream gate (foreground) is shown as opening in the wrong direction. Note the windlasses for moving the gates. There are no sluices for filling and emptying the chamber. Zonca, 1607.*

a few feet apart, the space between the rows being filled with bundles of faggots held down by stones (figure 623). This was the ancestor of the later fascine work.

Once the attempts to keep the sea out had succeeded, a new task began. Rain and seepage-water had to be drained at regular intervals. In early times the canals collecting the surplus water from the ditches discharged it to the sea at ebb-tide or in stages, first into the diked bay or canal-system and thence into the sea. This was effected by sluice-gates or by weirs built into canals or dikes. Such weirs in canals were of course a hindrance to shipping, which had to be towed

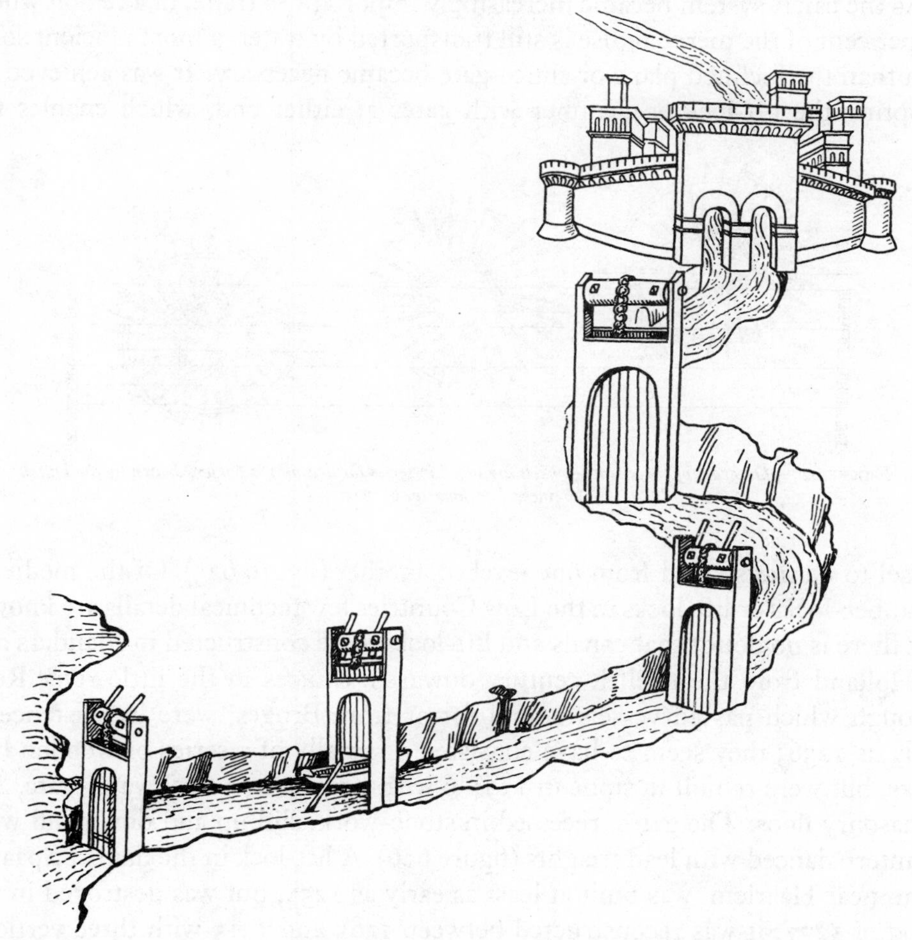

FIGURE 626—*A river divided into level 'pounds' by a series of vertical sluice-gates. When one gate was raised, the boat had to be forced through against the rush of water. From a fifteenth-century Italian manuscript on water-engineering.*

over them with windlasses or by man-power. Often the goods had to be transshipped from one barge to another. Lifts for ships sometimes took the form of inclined planes over which they were hauled (figure 624); examples still exist in some parts of the Netherlands. In the case of polders the excess water was often discharged through the dike by a conduit which could be closed by a movable water-tight gate, moved up and down with chains. Such gates, which first moved between wooden posts encased in the dike, were mounted in masonry in the early Middle Ages.

As the canal-system became increasingly important to trade, in a region where 85 per cent of the merchandise is still transported by water, a more efficient solution than the inclined plane or sluice-gate became necessary. It was achieved by adopting the lift-lock, a chamber with gates at either end, which enables the

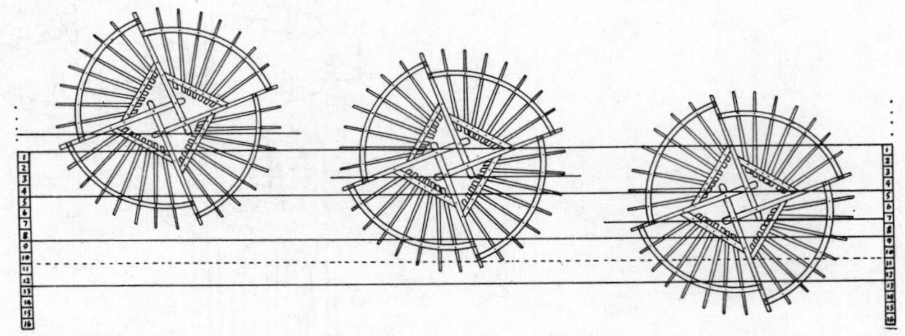

FIGURE 627—*Diagram of three scoop-wheels arranged in series for draining a polder, drawn by the Dutch engineer Leeghwater.* c *1610.*

vessel to be transferred from one level to another (figure 625). Of the medieval chamber-locks or lift-locks in the Low Countries few technical details are known, but there is no doubt that canals and lift-locks were constructed in Flanders and in Holland from the twelfth century onwards. Sluices in the little river Reie, through which passed vessels bound for or from Bruges, were in existence as early as 1236; they seem to have consisted originally of a series of wooden lift-locks, but were rebuilt in stone in 1394–6 with a sea-gate, a freshwater-gate, and a masonry floor. The gates, recessed in stone-work, slid up and down and were counterbalanced with lead weights (figure 626). A lift-lock in the dam at Spaarndam, near Haarlem, was built at least as early as 1253, but was destroyed in the flood of 1277; it was reconstructed between 1285 and 1315 with three vertical-lift gates, which were replaced in 1567 by three sets of mitred gates. This lock was about 24 ft wide, and provided for 'the easy passage of sea-going ships'. The use of sluices and locks was much extended from the fourteenth century, when canals came to be used more frequently for shipping.

Sluices and locks were efficient as long as the polders represented land won from the sea. From lower levels, such as the lakes left in the ancient polders, discharge by natural flow at ebb-tide was not possible and pumps were needed. Sometimes the new polders were so deep that the water was pumped in stages by a series of windmills (figure 627). Until steam-engines were introduced (1787), windmills were always used to drive the pumping-machinery, to which they had been applied in the fourteenth century. By 1400 Holland had many marsh-mills, but even in 1514 there were but eight north of Amsterdam, where the drainage of the lakes had not yet begun. The new marsh-mill was a recasting of the old post-mill built to grind corn (figure 562), and one of the earliest of them was constructed in the polder Bonrepas near Schoonhoven. This mill proved so efficient that by the end of that century over 17 marsh-mills pumped their water into the Vlist.

The marsh-mill turned on a post, the wind-shaft and sails on one side and the tail-pole and stairs on the other balancing each other, to facilitate setting to face the wind. It is not known whether horse-driven scoop-wheels preceded the marsh-mill. Before the use of wind-power there were, however, a few horse-driven Archimedean screws in the Low Countries. In the fifteenth century these were geared to marsh-mills. In the days of Stevinus, at the end of the sixteenth century, the screw was usually at an angle of about 30°, its diameter being 1·50–1·80 m. With the usual 40–50 revolutions a minute it raised water 4–5 m and would pump about 40 cu m a minute. The sixteenth century was to see the start of a great reconquest of drowned and swamped land.

The relation between the evolution of locks in the Low Countries and that in Italy (p 657) is still a problem, but certain differences in their construction correspond to differences in the parts played by the canals of the two countries.

XII. MEDIEVAL WATER-SUPPLY

With the fall of the Empire the lack of central authority and consequently of adequate public funds led to the decline of all public services. Their organization was left to private citizens or to municipal authorities. Only in certain large urban centres did even remnants of the Roman system survive. In the early seventh century, charge of aqueducts was commonly undertaken by the bishops. In most cases, however, towns were forced to rely on local wells, springs, and rivers. The archbishops of Salzburg, for instance, obtained their daily supply of fresh spring-water by messenger. Ordinary citizens had private wells or cisterns, or drew on the town wells and fountains. The water from public fountains and wells was often distributed by professional water-carriers (tailpiece, p 187). Town wells worked either with a bucket and windlass or, if the water was near the surface, by a bucket with a counterweight. Many towns changed from wooden to copper buckets, which could be more easily purified.

As the efficient superintendence of sewers and street-cleaning declined [62], hygienic conditions fell far below the standard of Imperial Rome, where pure water had been available even far beyond the confines of the city. Wells, being also used for the storage of rain-water, were often in low-lying ground and close

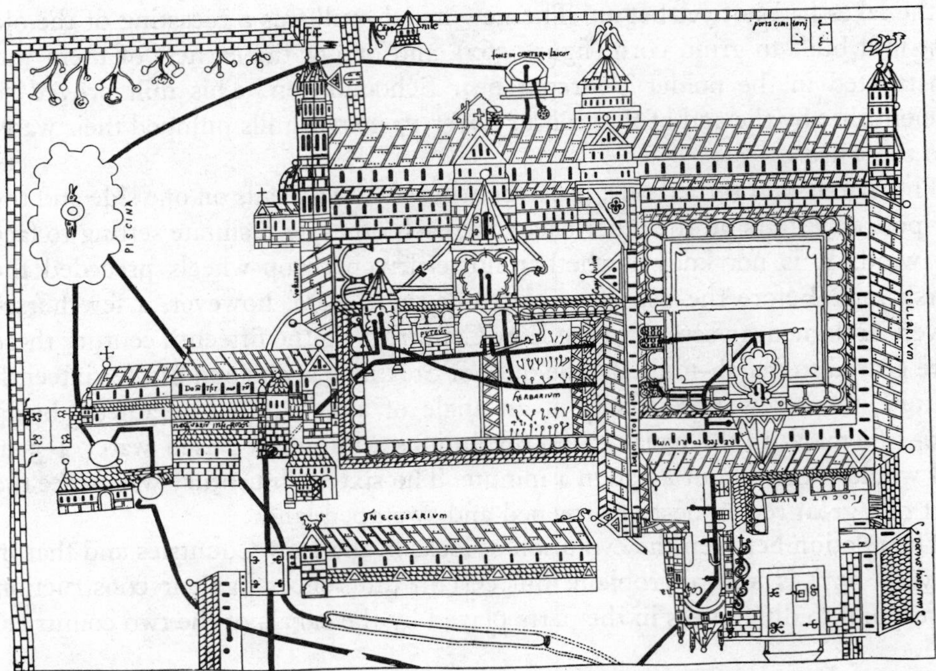

FIGURE 628—*Part of a twelfth-century plan of the cathedral and priory at Canterbury with the systems of water-supply and drainage. The plan is orientated with south at the top and north at the bottom. The part not represented extends to the north town-wall, and beyond it to the source three-quarters of a mile distant from the central tower of the church (seen in the plan). Thence the water was conveyed through a conduit-house, and by leaden pipes, passing five settling-tanks, through the city-wall into the precincts of the cathedral. The black lines in the illustration indicate the course of the feeding-pipes inside the greater part of the precincts. Note the reservoirs (e.g. the octafoil basin in the centre with its pair of stand-pipes, the similar smaller one in the cloisters to the right, and the* fons *near the central tower) where part of the water coming from the pipes was stored. Part of the water that fed the fish-pond* (piscina) *was used to flush the latrines* (necessaria) *and was eventually carried into the town ditch. Note the natural wells in the centre cloister* (puteus) *and near the* fons *or source at the top. They were used when the main system failed. The drainage-system is represented by simple double lines. The rain-water was conducted to this system; gutters can be seen around the cloisters on the right and protruding from the main church-building. It is believed that the plan was drawn by the engineer of the system shortly after its completion in 1153.*

to cess-pools and latrines. Thus epidemics quickly spread. Open gutters in the middle of the street carried refuse, while rain-water pouring from the roofs was not properly drained. The streets, seldom paved, were often mud-pools from which the excreta of pigs and other animals leaked into wells and private plots [63].

In Paris the situation started to improve when Philip Augustus (1180–1223) enclosed the city within walls about 1190. At that time the abbey of Saint-Laurent had a reservoir at Pré-Saint-Gervais which held water from springs at Romainville. This was now piped to Paris in lead pipes [64]. The abbey of Saint-Martin-des-Champs had further repaired 1100 m of the masonry of the old aqueduct of Belleville, and mains were being laid to the public fountains. The convent of the Filles-Dieu was the first to get a concession to pipe water from the public main to its establishment (1265). Many nobles and merchants followed suit, but illicit tapping became so frequent that a decree (1392) had to be enacted against it. In 1404 Charles VI took steps to prevent pollution of the Seine. Citizens now began to take an interest in an ample and wholesome water-supply, and in 1457 the Provost of the Merchants paid for the repairs of about 200 m of the Belleville aqueduct.

These developments were typical of those in many medieval towns. At Canterbury the priory of the cathedral had an elaborate water system (figure 628). Certain towns were able to keep the old Roman aqueducts in some sort of repair. Others, like Plymouth, built simple aqueducts carrying water from spring or river into the town along open ducts or, like Constance, piped their water from springs. The advent of pumps in the fifteenth and sixteenth centuries changed the picture completely (p 654). Attempts at pumping had started earlier; they failed because of their undeveloped machinery, but by the end of the fifteenth century piston-pumps driven by undershot water-wheels, some 5–6·5 m in diameter, were being introduced in south German towns.

In later medieval times municipal authorities or groups of citizens replaced ecclesiastics and monasteries in undertaking water-supply. The cost of driving wells was borne either by the whole town or by groups of 'well-brethren'. The latter usually catered for their members only, and left to the town the problem of supplying water for the poor, complaining when money was extracted from them for some public service. The actual construction and repair of the mains was often in the hands of professional well-diggers quite unskilled in the planning of water-supply.

During the fourteenth century the towns in Flanders, the wealthiest part of the western world, set a good example by enforcing laws on the cleanliness of water and food (tailpiece). The Black Death of 1348 and the following years stimulated such measures. Its memory resulted in the first British urban sanitary act (1388), which forbade the throwing of filth and garbage into ditches, rivers, and waters.

Towards the end of the Middle Ages scientific works on water-supply begin to appear. Konrad Kyeser devoted the whole third book of his *Bellifortis* (1405) to hydraulics and water-supply. His contemporary Giovanni da Fontana not only wrote on aqueducts (1420) but actually proposed a hot-air engine to raise water from a well. They were heralds of the change from a gravity-flow system to a pressure-system.

REFERENCES

[1] VITRUVIUS *De architectura*, VIII, i, ii. (Loeb ed. Vol. 2, pp. 136 ff., 1934.)

[2] JACOBI, H. *Saalburgjb.*, **8,** 32–60 (esp. 36), 1934.

[3] POMP, R. *Technikgeschichte*, **28,** 159, 1939.

[4] FORCHHEIMER, P. and STRZYGOWSKI, J. 'Die byzantinischen Wasserbehälter von Konstantinopel.' Byzantinische Denkmäler (ed. by STRZYGOWSKI, J.), Vol. 2. Mechitharisten-Congregation, Vienna. 1893.

[5] HEIERLI, J. *Anz. Schweiz. Altertumskunde*, new series, **9,** 265, 1907.

[6] EVANS, SIR ARTHUR. 'The Palace of Minos', Vol. 2, Pt. II, pp. 462 f.; Vol. 3, p. 253, fig. 173. Macmillan, London. 1928; 1930.

[7] VITRUVIUS *De architectura*, VIII, vi, 10–11. (Loeb ed. Vol. 2, p. 188, 1934.)

[8] STRABO XV, C 707. (Loeb ed. Vol. 7, p. 82, 1930.)

[9] FEILBERG, C. G. in *Øst og Vest* (Festschrift for A. Christensen), 105, Munksgaard, Copenhagen, 1945.

[10] CATON-THOMPSON, GERTRUDE and GARDNER, ELINOR, W. *Geogr. J.*, **80,** 382, 1932.

[11] SCHMASSMANN, H. W. *Ciba Rdsch.*, **91,** 3357, 3375, 1950.

[12] HERODOTUS III, lx. (Loeb ed. Vol. 2, p. 76, 1921.)

[13] WIEGAND, E. *Gas- u. Wasserfach*, **76,** 513, 1933.

[14] BERNARDI, M. 'L'igiene nella vita pubblica e nella vita privata di Roma.' Del Bianco, Udine. 1941.

[15] STRABO III, C 146. (Loeb ed. Vol. 2, p. 40, 1923.)
PLINY *Nat. hist.*, XXXIII, xxi, 74–76. (Loeb ed. Vol. 9, pp. 56 ff., 1952.)

[16] BENOÎT, F. *Rev. archéol.*, sixième série, **15,** 19, 1940.

[17] DIODORUS XX, xxxvi, 1–4. (Loeb ed. Vol. 10, p. 236, 1954.)

[18] FRONTINUS *De aquae ductu*, I, 6, 7. (Loeb ed., pp. 342 ff., 1925.)

[19] LIVY XL, li, 7; XLI, xxvii, 11–12. (Loeb ed. Vol. 12, pp. 28, 158, 1938.)

[20] *Res gestae divi Augusti*, iv, 20. (Loeb ed., p. 376, 1924.)

[21] PLINY XXXVI, xxiv, 122.
SUETONIUS *Vitae XII Caesarum*, V: *Claudius*, xx, 1. (Loeb ed. Vol. 2, pp. 36 ff., 1914.)
FRONTINUS *De aquae ductu*, I, 14. (Loeb ed., pp. 354 ff., 1925.)

[22] *Corpus inscriptionum latinarum*, III, 549, 709, 1446; IX, 5681; XIV, 2797.
PAUSANIAS VIII, xxii, 3. (Loeb ed. Vol. 4, p. 4, 1935.)

[23] DALMAN, K. O. 'Der Valens-Aquädukt in Konstantinopel.' Istanbuler Forschungen, 3. Archäol. Inst. dtsch Reich, Abt. Istanbul, Berlin. 1933.

[24] STRABO XVII, C 807. (Loeb ed. Vol. 8, p. 86, 1932.)

[25] *Papyrus Londinensis* 1177 in 'Greek Papyri in the British Museum' (ed. by SIR FREDERICK G. KENYON and SIR HAROLD I. BELL), Vol. 3, pp. 180 ff. Trustees of the British Museum, London. 1907.

[26] WEBER, G. *Jb. dtsch. Archäol. Inst.*, **19,** 86, 1904; **20,** 202, 1905.

[27] STÜBINGER, O. "Die römischen Wasserleitungen von Nîmes und Arles." *Z. Gesch. Archit.*, 3. Beiheft. Heidelberg. 1910.

[28] SAMENREUTHER, E. 'Römische Wasserleitungen in den Rheinlanden.' 26. Bericht der Römisch-Germanischen Kommission. 1938.

[29] VITRUVIUS *De architectura*, VIII. (Loeb ed. Vol. 2, pp. 132 ff., 1934.)
[30] HERSCHEL, C. 'The Two Books on the Water-supply of the City of Rome by Sextus Julius Frontinus' (2nd ed.). Longmans, Green and Co., London. 1913.
[31] FRONTINUS *De aquae ductu*, I, 16. (Loeb ed. Vol. 2, pp. 356 ff., 1934.)
[32] LÉGER, A. 'Les travaux publics, les mines et la métallurgie aux temps des Romains', pp. 551–676: "Aqueducs." Librairie Polytechnique, Paris. 1875.
[33] PLINY *Nat. hist.*, XXXI, xxiv–xxv, 41–42.
STATIUS *Silvae*, I, v, lines 23 ff. (Loeb ed. Vol. 1, p. 60, 1928.)
[34] FRONTINUS *De aquae ductu*, II, 78–87. (Loeb ed., pp. 406 ff., 1925.)
[35] *Idem Ibid.*, I, 24–63. (Loeb ed., pp. 364 ff., 1925.)
[36] *Idem Ibid.*, II, 75, 112–16. (Loeb ed., pp. 404, 442 ff., 1925.)
[37] KOBERT, R. "Chronische Bleivergiftung im klassischen Altertum" in DIERGART, P. (Ed.). 'Beiträge aus der Geschichte der Chemie' (Memorial Volume for GEORG W. A. KAHLBAUM), pp. 103–19. Deuticke, Leipzig, Vienna. 1909.
[38] VITRUVIUS *De architectura*, VIII, iv, 1–2. (Loeb ed. Vol. 2, pp. 176 ff., 1934.)
[39] TRILLAT, A. *C.R. Acad. Sci.*, **162,** 486, 1916.
[40] HERODOTUS I, clxxxviii. (Loeb ed. Vol. 1, p. 352, 1920.)
[41] VITRUVIUS *De architectura*, VIII, vi, 15. (Loeb ed. Vol. 2, p. 192, 1934.)
[42] *Idem Ibid.*, X, i, 4. (Loeb ed. Vol. 2, p. 276, 1934.)
[43] *Idem Ibid.*, X, iv, 1–2. (Loeb ed. Vol. 2, p. 302, 1934.)
[44] DIODORUS I, xxxiv, 2. (Loeb ed. Vol. 1, pp. 112 ff., 1933.)
VITRUVIUS *De architectura*, X, vi, 1–4. (Loeb ed. Vol. 2, pp. 306 ff., 1934.)
[45] DIODORUS V, xxxvii, 3–4. (Loeb ed. Vol. 3, pp. 196 ff., 1939.)
[46] PRICE, F. G. H. *Proc. Soc. Antiq. London*, second series, **16,** 277, 1897.
[47] ASHBY, T. 'The Aqueducts of Rome', pp. 46–47. Clarendon Press, Oxford. 1935.
[48] BRUNHES, J. B. 'L'irrigation, ses conditions géographiques, ses modes et son organisation dans la Péninsule Ibérique et dans l'Afrique du nord.' Naud, Paris. 1902.
[49] KENNY, E. J. A. *Ann. Archaeol. Anthrop.*, **22,** 189, 1935.
[50] DIOGENES LAERTIUS VIII, lix–lx, lxx. (Loeb ed. Vol. 2, pp. 374, 384, 1925.)
[51] LÉGER, A. See ref. [32], pp. 415–25.
[52] TACITUS *Annales*, XII, lvi. (Loeb ed. Vol. 3, pp. 396 ff., 1937.)
SUETONIUS *Vitae XII Caesarum*, V: *Claudius*, xx, 1–2; xxi, 6. (Loeb ed. Vol. 2, pp. 36 ff., 43 ff., 1914.)
[53] STRABO V, C 217. (Loeb ed. Vol. 2, pp. 328 ff., 1923.)
[54] *Idem* V, C 213. (Loeb ed. Vol. 2, p. 312, 1923.)
[55] PLUTARCH *Vitae parallelae: Caius Marius*, xv, 1–4. (Loeb ed. Vol. 9, pp. 500 ff., 1920.)
[56] *Panegyrici* V (VIII), vi. ('Panégyriques Latins' ed. with French trans. by E. GALLETIER, Vol. 2, p. 94. Les Belles Lettres, Paris. 1952.)
[57] TACITUS *Annales*, II, viii. (Loeb ed. Vol. 2, p. 394, 1931.)
Idem Historiae, V, xix. (Loeb ed. Vol. 2, pp. 206 ff., 1931.)
SUETONIUS *Vitae XII Caesarum*, V: *Claudius*, i, 1. (Loeb ed. Vol. 2, p. 2, 1914.)
[58] HETTEMA, H. 'De nederlandsche Wateren en Plaatsen in der romeinschen Tijd', p. 105. Nijhoff, The Hague. 1938.
VOLLGRAFF, C. W. *Meded. vlaamsche Acad. Kl. Wet.*, **1,** 12, 555, 1938; **2,** 6, 141, 1939.
[59] TACITUS *Annales*, XI, xx. (Loeb ed. Vol. 3, p. 280, 1937.)
[60] BIJL, J. G. (Ed.). *Congr. int. Geogr., Amst.*, 1938.
COOLS, R. H. A. 'Strijd om den grond in het lage Nederland.' Nijgh, Van Ditmar, The Hague. 1948.

DIBBITS, H. A. M. C. 'Nederland-Waterland.' Oosthoek, Utrecht. 1950.

HÉRUBAL, M. 'L'homme et la côte.' Coll. Géographie humaine, no. 10. *Nouv. Rev. franç.*, Paris. 1937.

VAN VEEN, J. 'Dredge, Drain, Reclaim. The Art of a Nation' (4th ed.). Nijhoff, The Hague. 1955.

[61] PLINY *Nat. hist.*, XVI, i, 2–3. (Loeb ed. Vol. 4, pp. 386 ff., 1945.)

[62] SABINE, E. L. *Speculum*, **9**, 303, 1934; **12**, 19, 1937.

[63] THORNDIKE, LYNN. *Ibid.*, **3**, 192, 1928.

[64] PARSONS, W. B. 'Engineers and Engineering in the Renaissance', pp. 240 ff. Williams and Wilkins, Baltimore. 1939.

BIBLIOGRAPHY

ASHBY, T. 'The Aqueducts of Ancient Rome' (ed. by I. A. RICHMOND). Clarendon Press, Oxford. 1935.

BROMEHEAD, C. E. N. "The Early History of Water Supply." *Geogr. J.*, **99**, 142–51, 183–96, 1924.

BUFFET, B. and EVRARD, R. 'L'eau potable à travers les âges.' Solédi, Liége. 1950.

CALDERINI, A. "Macchine idrofore secondo i papiri greci." *R.C. Ist. Lombardo Sci. Lett.*, seconda serie, **53**, 620–31, 1920.

Ciba Z., **9**, no. 107, 'Das Wasser', 1947.

CLARK, J. G. D. "Water in Antiquity." *Antiquity*, **18**, 1–15, 1944.

DUPONT, G. 'L'eau dans l'antiquité.' Geuthner, Paris. 1938.

GERMAIN DE MONTAUZON, CAMILLE. 'Les aqueducs de Lyon.' Leroux, Paris. 1909.

GRIMAL, P. "Vitruve et la technique des aqueducs." *Rev. Philol.*, troisième série, **19**, 162–74, 1945.

JACOBSEN, T. and LLOYD, H. F. SETON. 'Sennacherib's Aqueduct at Jerwan.' Univ. of Chicago Orient. Inst., Publ. 24. Chicago. 1935.

LAESSØE, J. "The Irrigation System at Ulḫu, Eighth Century B.C." *J. Cuneiform Stud.*, **5**, no. 1, 21–32, 1951.

LUCKENBILL, D. D. (Trans.). 'The Annals of Sennacherib', col. VIII, lines 22–48. Univ. of Chicago Orient. Inst., Publ. 2. Chicago. 1924.

MAHUL, J. "Les tuyaux de plomb, histoire et progrès de leur fabrication." *Nature, Paris*, **65**, no. 3014, 503–10, 1937.

ROBINS, F. W. 'The Story of Water Supply.' Oxford University Press, London. 1946.

VAN DEMAN, ESTHER B. 'The Building of Roman Aqueducts.' Carnegie Institution, Publ. 423. Washington. 1934.

Clearing out a water-course. From a fifteenth-century manuscript.

20

MILITARY TECHNOLOGY

A. R. HALL

THE object of this chapter is to describe briefly the application of a variety of techniques of metal-working, of machine-making, and of building construction for purposes of war. The states that rose and fell in the Mediterranean and west European areas during the millennium covered by this volume were essentially fighting states; their prosperity depended upon their ability to subdue their neighbours, to defend their civilization against barbarian invaders, and to protect their trade. When they ceased to be able to do these things they collapsed. So the power of Persia passed to Macedonia, that of Macedonia to Rome, and that of Rome (in the west) to the Germanic tribes; in each case the transfer of power followed the achievement of a military ascendancy on the one hand and a military decline on the other.

In determining the fortunes of war, the field-equipment and the technical skill of the different armies which at various times attained a tactical supremacy were naturally of high importance; but we must not forget that generalship, the skill with which equipment and techniques were employed, the morale of the fighting men, and the resourcefulness with which they were strategically and tactically directed were together more important still. Hence we cannot understand the changes in weapons, armour, and fortification without knowing something of the development of methods of attack and of defence, of the art of war in the larger sense. On the field of battle, the medieval knight was a very different figure from the Roman legionary, not so much because of changes in metallurgical technology as because of changes in methods of fighting.

It will be convenient to divide the period into three sections, each having its own characteristics: (I) The Greek and Macedonian period; (II) the Roman period; (III) the Middle Ages. Within each of these sections we shall discuss firstly hand-weapons and armour, then projectile weapons, and finally fortifications and siege-warfare.

I. THE GREEK AND MACEDONIAN PERIOD (*c* 500–200 B.C.)

The hand-weapons of this period—the spear, the sword, and the dagger—had long been familiar (vol I, ch 22). Bronze weapons, such as spear-heads, had been cast in piece-moulds, but iron demanded the blacksmith's art of forging. The crude bloom obtained from the furnace, containing much cinder and slag, was reheated and hammered until the metal was purified and consolidated. The weapon was then brought to the required shape by cutting up the bloom with chisels and further hammering upon the anvil. It could then be polished with abrasive stone, and the cutting-edge ground sharp. Finally, the iron was hardened by carbonizing (vol I, p 596), quenched, and tempered to give a suitable combination of hardness and toughness.

FIGURE 629—*Early Greek armour: the death of Achilles. From a black-figured amphora, sixth century* B.C.

A revolution in the art of war had shaken Asia about the middle of the second millennium B.C., when horses were applied to light chariots; armies thus gained a new mobility and impetus. Chariotry, and rather later cavalry, formed the *élite* of Asiatic armies for well over a thousand years. But in Greece, where there were small city-states instead of vast empires, where the terrain was almost everywhere difficult, and where (at first) the professional soldier did not exist, conditions were different. The core of the Hellenic army was the line of heavily armed infantry or hoplites, drawn up eight deep, which advanced steadily to its front with the object of pushing its way through the enemy. The hoplite wore a helmet with nose and cheek-pieces, a breast-plate, and greaves to protect the legs, all of bronze. He carried a heavy bronze shield of elliptical form, secured to his left arm and hand by leather bands. His weapons were a short, straight iron sword, and, principally, a nine-foot iron-tipped spear (figure 629).

Flanking the line of hoplites, and useful in reconnaissance or pursuit, were the lightly armed peltasts (originally the poorer males of the Greek city-states, who could not afford body-armour); these carried javelins as missile weapons, and a light shield for defence. The peltast did not receive as much pay as the hoplite; he did not enjoy the same standing in the army, and was more often a mercenary. Besides these two classes of infantryman, the Greek cities produced a few nobles on horseback—who never constituted an organized cavalry force—and a few archers and slingers. The bow—especially the short composite bow—was an Asiatic weapon, never popular among the Greeks or Romans. Ulysses, when he went to the Trojan war, left his great bow, which no man could bend, in the care of his wife Penelope. The Cretans and Rhodians, like the Balearic islanders later, were skilled in the use of bow and sling, but there were not many such in a typical Greek army, which was therefore always liable to annoyance from an enemy who refused to stand and await the hoplites' steady march. The success of the line of hoplites depended on its ability to resist attrition from cavalry and missile weapons until it was able to drive the enemy from his chosen position.

These characteristics of Greek military technique are all demonstrated in Xenophon's story of the attempt of Cyrus the Younger to win the Persian Empire from his brother Artaxerxes (401 B.C.). His army included, besides Asiatics, some 9600 Greek hoplites and 2300 Greek and Thracian peltasts; he had only 2600 cavalry in all. In the decisive battle at Cunaxa the hoplites easily routed the more numerous Persian infantry, while the 6000 Persian cavalry (some of whom were horse-archers) drove the remainder of Cyrus' forces from the field, killed him, and thus left the Greek infantry utterly isolated. In their famous retreat the Ten Thousand were much harassed by the Persian cavalry and bowmen, though no attempt was made to break their ranks in pitched battle. The best the Greeks could do was to organize a small troop of about fifty horse and equip some Rhodian slingers, who used lead bullets; 'the one lesson taught by Cyrus' expedition was that no one need hope to conquer Persia without a cavalry force very different from any which Greece had yet envisaged' [1].

The lesson was applied, not by the Greeks themselves, but by their Macedonian neighbours to the north, who, with the Thessalians, were the only civilized peoples in Europe to make much use of the horse in war. The Macedonian army was organized by Philip II, probably at the beginning of his reign (359–336 B.C.). The core of its strength was the 'Companion' cavalry, originally recruited from the land-owning class. This body, up to 2000 strong, was usually led by the great conqueror Alexander, Philip's son, in person. Besides the Companion cavalry, Alexander's army of conquest numbered as many as 6000 other horse drawn from other peoples, including the Asiatics whom he had earlier defeated. The infantry—up to 25 000 men—were divided into three main groups. The premier corps was the phalanx, highly trained, fast marching, and capable of orderly manœuvre on the battlefield. The phalanx was twice as deep as the older hoplite formation, and less densely packed, so as to give greater freedom in manœuvre and weapon-play. The second corps, the hypaspists, do not seem to have differed in any marked way in equipment or military function from the phalangists. Thirdly, Alexander employed smaller numbers of lightly armed men equipped with javelins, bows, and slings. These were found particularly effective against chariotry and elephants.

The Companion cavalry wore heavier body-armour than the infantry, and a helmet; their weapons were a sword and a thrusting-spear. They had no shield. Light cavalry wore leather armour instead of bronze and carried a longer spear. The heavy infantry wore helmets and greaves and a leather jerkin protected by metal plates. They carried a small shield and a massive thrusting-spear (*sarissa*), 13 or more feet long.

With these troops Alexander conquered the known world east and south of Greece. He could march 400 miles in two weeks, or when pressed 135 miles in three consecutive days. He defeated armies superior in number, chariotry, and elephants; he stormed fortified places both by sea and by land. The cavalry was his major striking-force in battle, the main function of the infantry being to hold their ground and withstand the charges of the enemy horse. But Alexander's victories were won by thought rather than by brute force, and as his army combined all arms he was able to use each in any situation that seemed to demand it.

Missile weapons depending on the unaided strength of a man's arm—the javelin, bow, and sling—never had a prominent place in Greek military technique. They rarely had much effect on the course of battle, for the peltasts armed with javelins were skirmishers rather than infantry of the line. Yet the Persians had made themselves masters of western Asia by using their archers to throw the enemy into disorder, and then charging with cavalry. Only when they entered Greece were they decisively checked by the armoured hoplite, who could stand up to a flight of arrows, and upon whose spears it was useless for cavalry to charge. With its flanks protected, as at the battle of Plataea (479 B.C.), the hoplite line was invincible. Thereafter, for a time, the bow and other missile weapons were incidental to the spear.

At the same time, however, great development was taking place in the construction of more powerful missile engines. This was largely due to the ingenuity of Greeks in the service of Macedonia and the Successor states which divided Alexander's vast empire. Diades, Alexander's own engineer, was a Thessalian. The greatest military engineer of all antiquity was the scientist and mathematician Archimedes (*c* 287–212 B.C.), whose devices protracted the defence of Syracuse against the Romans; he was slain by a Roman storming-party. Except in relation to war, the earlier Greeks turned little of their intellectual energy to the improvement of engineering and technology, but their brilliance in this one direction shows that, considering the means and resources open to them, they were far from lacking in mechanical dexterity. Greek war-machines were adopted, without important modification, by the Roman military command, and indeed provided the basic pattern for all war-machines before the invention of gunpowder—and even later, as the drawings of Leonardo da Vinci show. We shall now trace the history of these war-like engines, postponing a fuller description of their construction to the Roman period, for which the sources are more informative.

The first missile engines resulted from the endeavour to increase the penetrative force generated by the bow. There were two obvious defences against the ordinary archer, a heavier and larger shield, and more massive body-armour. The archer's reply to these defences was limited by the strength of his arms, unless he was able to bring into action the more powerful muscles of his back and legs. This involved the transformation of the bow into a more complex machine—the addition to it of a stock, and in the stock some contrivance or lock, so that the bowstring could be held under tension while the arrow was fitted and aim taken. The Greeks knew of such a cross-bow, called *gastraphetes* (that is, 'stomach bow'), said to have been invented by Zopyrus of Tarentum, but apparently it was not a common weapon. However, Archimedes at the siege of Syracuse made loopholes in the ramparts about 3 in wide at the height of a man, through which *skorpidia* ('scorpions') were discharged at the enemy. Roman writers also used the term 'scorpion' as though it applied to a small hand-weapon, and it would seem reasonable to suppose that the 2000 scorpions handed over by Carthage in 149 B.C. were actually cross-bows of some kind.

The principle could be applied even more effectively on a larger scale. The Greeks, unlike medieval soldiers, never used a hand cross-bow in which mechanical advantage was obtained by a winch or lever, but they did apply the same idea to larger engines that stood independently on the ground. The first of these catapults were probably giant cross-bows having a composite elastic element made of wood, horn, and sinew. These, rather than the torsion catapults to be mentioned later, were perhaps the engines said to have been invented at Syracuse for the tyrant Dionysius about 400 B.C. In his wars with Carthage, then the great mistress of siege-craft, Dionysius proved himself an original commander. His catapults were said (doubtless with much exaggeration) to be capable of battering walls from a range of three or four hundred yards. A catapult arrow is heard of at Sparta about 370 B.C., when it was regarded as a great curiosity, and there were two catapults at Athens about 358 B.C. [2]. There seem to be no allusions to the catapult in oriental literature before that in 2 Chron. xxvi. 14, 15:

> Uzziah prepared for all the host shields and spears, and helmets, and coats of mail and bows and stones for slinging. And he made in Jerusalem engines, invented by cunning men, to be on the towers and battlements, to shoot arrows and great stones withal.

Uzziah lived in the eighth century B.C., but the 'Book of Chronicles' was probably not put together before 300 B.C., so the passage is an instructive anachronism. In Europe, Philip of Macedon had plenty of tension-catapults, and the first torsion-engine is heard of in Alexander's siege of Tyre (332 B.C.), where he had stone-throwers which must have been of this type. According to Pliny, 'among missile engines the scorpion [was invented] by Pisaeus, the catapult by the Cretans, the ballista and sling by the Phoenicians of Syria' [3]. While Pliny's ascriptions of inventions are usually worthless, the association of the torsion-catapult (ballista) with the Phoenicians is not implausible, for Alexander employed Phoenician engineers. However this may be, it was the Greeks who brought such missile engines into warfare as standard equipment.

FIGURE 630—*A stick-slinger.*

The power of the tension-catapult or giant cross-bow was limited by the elastic strength of the bow itself. The ancients had no steel springs sufficiently massive to be used for this purpose, and they found that better results could be obtained by employing the force exerted by thick twisted cords. There were three principal types of these machines—(*a*) the two-armed, arrow-shooting ballista, (*b*) the two-armed, stone-shooting *pẹtrobolos*, and (*c*) the single-armed, stone-throwing *monagkon* or onager, which was often fitted with a sling to increase the velocity of the projectile. This last form was a mechanized version of the stick-sling (*sphendone*) swung with two hands (figure 630) [4]. Although the technical descriptions of these machines (to which we turn later) come from Roman times, they were certainly all in use before 200 B.C. They were early associated not only with arrows and stones but with pyrotechnical preparations. On the last night of the siege of Rhodes by Demetrius in 305 B.C., the Rhodians are said to have directed 800 flame-carriers at his wooden siege-engines. These were probably earthenware pots containing naphtha, pitch, and similar inflammable materials. Such compositions were the precursors of the notorious 'Greek fire' of later times, which also was projected from catapults (p 375).

Despite their cumbersome construction and slow rate of fire, ancient military experts clearly placed considerable faith in their machines. Alexander carried arrow-shooting catapults (or rather, the vital parts for constructing them) on his Asian campaigns; at Mount Aornos (327 B.C.) he went to considerable pains to bring them into action [5]. The vital materials for the twisted cords—women's hair and animal sinews—were prized commodities; Rhodes sent 300 talents of hair and 100 of sinews to Sinope in 250 B.C., and in 225 Seleucus (another of the Successors) sent Rhodes a present of several tons [6]. These strange weapons were taken very seriously, and indeed their performance was not contemptible.

FIGURE 631—*East wall of Troy VI (to right) and north side of a tower. The tower projects 7 metres from the line of the wall. 1500–1300 B.C.*

The missile engine, in its original form of the hand cross-bow, was directed against armoured men, but it soon found its major role in siege-warfare. It was useful for breaking down gates, palisades, portable towers, and other wooden works of attack or defence. It could clear the defenders from a parapet or tower, or alternatively hamper the attackers in advancing their assaulting-gear. It could carry fire to houses, tents, or wood-work. It was even capable of effecting a breach in walls. Like the modern commander with his guns and mortars, the ancient general could employ either the direct (ballista) or the indirect (*monagkon*) trajectory. Too slow and immobile for work in the field, in the more leisurely conduct of a siege the catapult provided an efficient artillery. This had its inevitable effect on the design of fortifications.

A great deal had already been learnt, in the course of human history, about the design of strong places, whether the Iron Age camp like Maiden Castle with its circles of ramparts and ditches, or the cities of Mesopotamia with high walls of mud-brick, broken by regular series of massive towers. The Greeks also were accustomed to elaborate fortifications from an early period in their history. The various cities of Troy (Hissarlik) were protected by walls of dry masonry (figure 631) or mud-brick, and the Homeric besiegers built for themselves a fortified camp surrounded by a ditch and an earthen wall reinforced with towers. The construction of a glacis in front of a way running before the walls, from which the defenders could meet the attackers hand-to-hand while under the cover of arrows or other missiles shot from the walls, was another ancient device (figure 632).

We need not here enter into architectural details; the one significant change, enforced by the use of more efficient siege equipment, was the replacement of wood and brick by stone, with which all sizeable cities were fortified from the fourth century B.C. onwards. Otherwise the defence relied upon the strengthening of well tried methods, such as the reduplication of ditches within bow-shot (as at Syracuse) and the multiplication of missile weapons. Some places were almost impregnable; the legendary ten years' siege of Troy (which ultimately fell to Ulysses' stratagem) is matched by Nebuchadrezzar's thirteen years' siege of Tyre, ending in failure—though the city fell to Alexander, who succeeded in building a mole from the mainland to its island walls. Rhodes resisted Demetrius, and Syracuse might have resisted the Romans had its defenders been as resolute as Archimedes was ingenious.

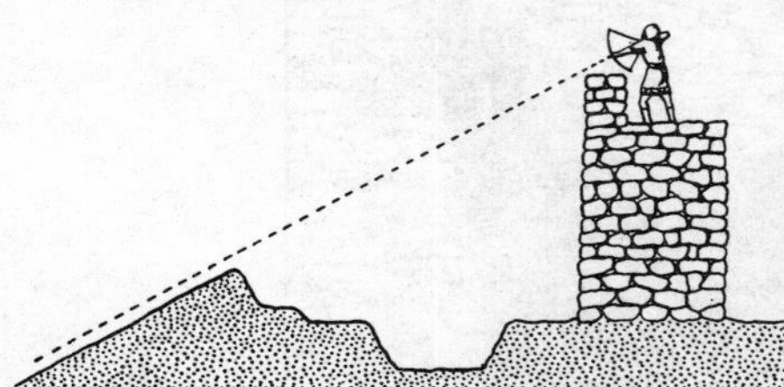

FIGURE 632—*Glacis and covered way.*

Those who sought to possess themselves of a fortified place had many resources at their disposal. Treachery often delivered it into their hands. An unexpected or night attack might carry the walls. Investment could cut a city off from the surrounding country-side and force its surrender from starvation, to which plague was often joined. If other methods failed, time pressed, and it was determined to force a breach, two methods offered. The first was mining: a tunnel was driven beneath a portion of the curtain or a tower, while the masonry was temporarily supported on wooden props. The miners then fired the props and retired to await the collapse of the masonry, whereupon the confused defenders were charged. The second was the use of the battering-ram, by which a breach was gradually hammered out of the wall. Direct assault, and either of the methods of forcing a breach, assumed that the assailants were capable of holding the ground immediately below the wall. For this it was necessary to fill the ditch or ditches, and climb the outer ramparts or glacis; at this stage the attackers were most vulnerable, and the covering-fire provided by missile engines was most valuable. In approaching the wall, and in their work upon or beneath it, the attacking forces made use of a variety of protective devices. These, like the missile engines, we shall describe in the following section.

II. THE ROMAN PERIOD (*c* 200 B.C.–A.D. 400)

The Roman soldier was originally, as with the Greeks, a citizen in arms. Then he became a professional, and finally a mercenary recruited from the barbarian tribes whom it was his main duty to repel. In this last period the traditional

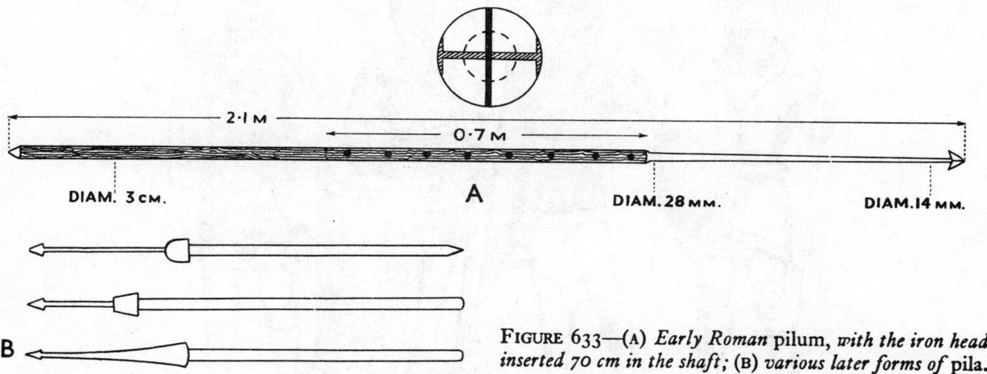

FIGURE 633—(A) *Early Roman* pilum, *with the iron head inserted 70 cm in the shaft;* (B) *various later forms of* pila.

character of the Roman legionary was lost, for he relied more and more upon weapons and methods imitated from those of his adversaries.

Rome grew by fighting. By about 300 B.C. she was mistress of southern Italy, by 270 she had incorporated the Greek cities there, and by 240 she had expelled the Carthaginians from Sicily. By the beginning of the second century she had emerged as a great power in the Mediterranean, and her forces had adopted a definite armament and a precise organization which were hardly modified before the later Imperial age.

Alexander had conquered with cavalry on Asiatic soil, but Rome's wars were to remain in Europe for some time yet, and the core of the Roman army was, and long remained, the infantry legion. This was a force of from four to six thousand men, in which the effective fighting-unit was originally a maniple of 60 (later 120) soldiers. For battle the legion was drawn up in open order, with the maniples in echelon one behind the other so that they could use their weapons to the best advantage. From the time of Marius (*c* 100 B.C.) it was found better to use a larger company-size, the cohort, nominally of 600 men.

With this change in organization came a change in weapons. The early Roman army was very much on the Greek model. The heavy infantry were armed with thrusting-spear, short Greek sword, breast-plate of bronze, helmet, greaves, and round shield. With the development of the legionary formation (*c* 400 B.C.) new tactics and weapons were adopted. The first two lines of men were armed with the *pilum* or throwing-spear, which had a wooden shaft $4\frac{1}{2}$ ft long, and an iron head of equal length of which half was a flat tang riveted into a slit in the

FIGURE 634—*Roman armour:* (A) *sixth century B.C., equipped hoplite-fashion with helmet, solid bronze cuirass, round shield, greaves,* hasta, *and short sword;* (B) *mid-second century B.C., with plumed helmet, oval shield, chain-armour,* pilum, *and longer sword;* (C) *end of first century B.C., with round helmet, scale-armour, rectangular shield,* pilum, *and sword: the Celtic leather breeches were worn in cold climates;* (D) *second century A.D., with segmented cuirass, sword, and* hasta.

shaft (figure 633). The weight of the *pilum* was 4–5 lb, and it was nicely balanced for throwing. The third rank of men was still armed with the thrusting-spear (*hasta*). All ranks carried a two-edged sword, longer than the Greek and of Spanish origin, and an oval shield about 4 ft long, made of hide-covered wood strengthened with a rim and central boss of iron. They wore crested helmets, greaves, and breast-plates of bronze, or bronze plates sewn on a leather tunic (the so-called scale-armour), or a kind of chain-armour borrowed from the Celts (figure 634 A, B, C).[1] Light troops (*velites*) carried a round buckler and short javelin. The cavalry (about 300 horsemen to each legion) had a helmet and cuirass, a double-pointed spear, round shield, and no sword.

[1] This apparently consisted of separate rings, or pieces of chain, sewn to a leather tunic.

With the reforms of Marius, the *pilum* tended to replace the *hasta* throughout the legion. Its iron head was now fitted to the shaft by means of a short tang and two pins, one made of wood which broke on impact, rendering the weapon useless to the enemy. Later the same effect was obtained by making the iron soft below the point so that it bent on striking. The familiar rectangular, semicylindrical shield of the Romans was also adopted in the first century B.C., while at the same time the solid breast-plate was universally replaced by armour made of chain or small plates sewn to a leather tunic. With Marius also the *velites* disappeared, and by the time of Caesar, half a century later, the cavalry was recruited from Celts and Germans, no longer from Roman citizens. Cretan archers and Balearic slingers were useful among the auxiliary troops.

But generalizations conceal much variation. There was no monotonous uniformity of dress and equipment in the Roman army. Particularly in the Imperial age was this true, when many different types of helmet, *pilum*, and shield are represented on monuments. During the second and third centuries A.D., the legionary was characterized, as on Trajan's column, by wearing the *lorica segmentata* (segmented cuirass), which, with many variations of detail, consisted of a number of plates and strips of iron, jointed together to give some freedom of movement, and doubtless fastened to an under-tunic of leather or stout linen to keep them in place (figure 634 D). The breast and back were covered by four plates, hinged at the back and buckled or thonged together at the front. The torso was encircled by pairs of half-hoops, similarly joined back and front, and three or more strips passed over each shoulder. All the strips were made to overlap as far as possible.

After this time, in the last two centuries of Rome, there was a great decline in military equipment. The traditional fighting-skill of the infantry legion had already disappeared when Vegetius wrote about A.D. 400. The heavy *pilum* was replaced by a lighter dart, with a short iron head. Archers were now admitted to the ranks of the legion and no longer mustered as auxiliaries. Worst of all, 'through negligence and cowardice field-exercises fell into disuse—the soldiers began to complain of the weight of their armour and besought the Emperor to let them cast it aside. And thus our soldiers, meeting the Goths with naked breasts, have been defeated' [7]. The infantry abandoned their metal cuirasses and relied at best on a tunic of leather which, it appears, was worn even by the emperors of the fifth century. Meanwhile the armour of the cavalryman increased in weight, and was extended over his mount. As in Alexander's time, the west was forced by the east to mount its finest troops on horseback, and even to adopt the characteristically eastern weapon—the composite bow.

The eclipse of the legion had been heralded 400 years before, when in 53 B.C. Crassus' army was routed by a numerically inferior Parthian force of 1000 mailed cavalry (cataphracts) and 10 000 horse-archers. In this battle the Romans assumed the defensive and waited for the Parthians to exhaust their arrows; but the Persian general Surenas had invented a new technique of carrying a reserve supply of arrows on baggage-camels. The respite expected by the Romans never came, and the sallies of their light Celtic cavalry were punished by the cataphracts.

The Parthians—who had begun to make their way towards the Fertile Crescent by way of the southern shores of the Caspian from about 200 B.C.—had perfected this new alliance of arms. The lightly protected horse-archer, with his terrible bow, was the familiar warrior of the steppe.[1] The Parthian aristocrats raised them from among their retainers. They themselves dressed in full armour—in the first century B.C. a helmet and coat of mail or scale-armour to the knee, greaves and sleeves of metal, and mail gloves (figure 635). Their horses were similarly protected—a trick known to the steppe peoples long before. In addition to the bow they carried a massive spear or lance, which became their main weapon.

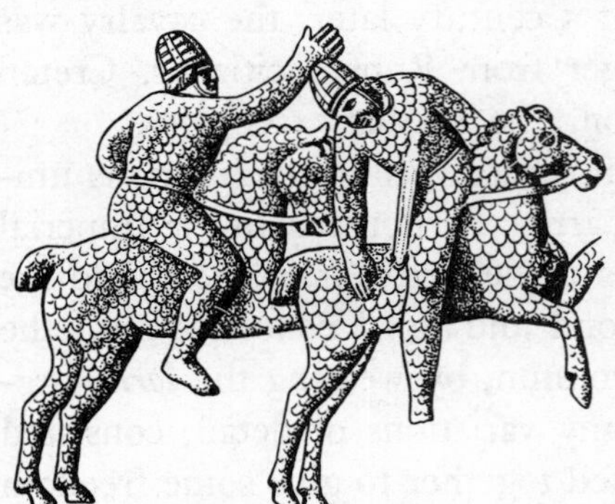

FIGURE 635—*Cataphracts. From Trajan's column. c A.D. 110.*

The cataphract was almost as well armoured as the medieval knight, and nearly as helpless when dismounted. To bear so large a weight of metal he needed, like the knight, a powerful war-horse. This the Parthians found in the Nesaean horse of Media, which they themselves improved by breeding for their purpose, producing a charger unlike any previously known. But the cataphract—like all previous horsemen in the history of equitation—had no stirrups; he rode with the grip of his knees and thighs, and was therefore more vulnerable at these points, which had to be left unprotected (p 556).

It is not known when the Roman army first raised a corps of cataphracts: perhaps not until the third century. They are referred to with enthusiasm by Procopius (sixth century), who contrasts the archers of ancient times with these modern warriors who:

go into battle equipped with cuirasses and greaves extending to the knee, with their

[1] The Red Indian horseman of the American plains, with a comparable equipment and technique, is said to have been capable of putting an arrow right through a bison.

quiver hanging from their right side and their sword on the other. Some troopers bear a lance also and, slung over their shoulders, a small handleless shield of just sufficient size to cover the face and neck. Being admirable horsemen they are trained to bend their bow without effort to either flank when riding at full gallop, and to hit a pursuing enemy in their rear [the 'Parthian shot' (vol I, figure 93)] as well as a retreating enemy to their front. They draw the bowstring to the right ear thereby charging the arrow with such an impetus as to kill whomsoever it strikes, neither shield nor cuirass having power to check its force [8].

FIGURE 636—*Auxiliary archers in the Roman army. From Trajan's column.* c *A.D. 110.*

The Romans introduced no important innovations with regard to missile weapons, learning all their technique from the Carthaginians and Greeks. The bow commonly used in war (as with the Greeks earlier) was of the curved form (figure 636)—it was said to be made from two goats' horns, fastened together at the roots in the central grip of the bow. But the straight bow was also known, since it is shown, for instance, in the hands of gods. The Romans certainly knew the cross-bow too, since it is described by Vegetius under the name *manuballista* as the same weapon as that formerly called 'scorpion' [9]. The term *arcuballista* seems to describe another kind of cross-bow. Two late-Roman representations of a simple kind of cross-bow are known from Gaul, both showing weapons of the chase rather than of war (figure 637).

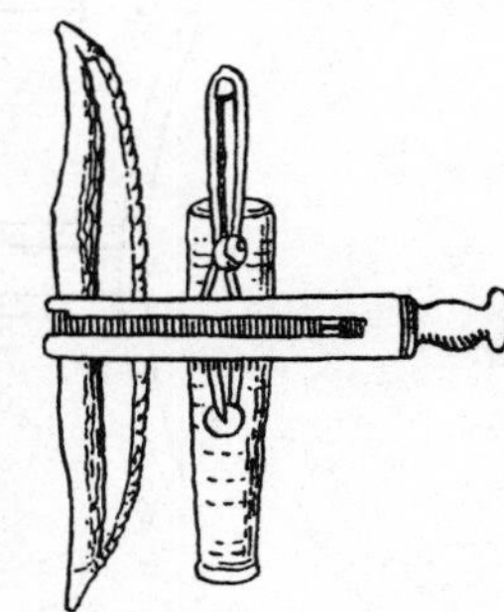

FIGURE 637—*Cross-bow and quiver depicted on a Roman tomb in Gaul. Second century A.D.*

As for torsion-artillery, this was almost as common in the Roman army, in proportion to the number of men, as heavy machine-guns and cannon are in a modern army. According to Vegetius, each century (100 men) was equipped with a *carroballista*, a small mobile catapult, operated by a crew of eleven men. These machines, which are represented on Trajan's column (figure 638), were used in the field as well as in siege-warfare. There

FIGURE 638—*A* carroballista. *From Trajan's column.* c *A.D. 110.*

were thus 55 *carroballistae* to the legion and there were in addition 10 *onagri*, one per cohort, which were reserved for siege-warfare [10]. Specialists in the construction, maintenance, and use of these and other machines marched with the legion.

We may now consider some of these engines of war in greater detail.

The *gastraphetes* is described by Hero of Alexandria [11], though it is impossible to tell whether his weapon corresponded with any in practical use.[1] It consisted of two pieces of horn fitted into a stout stock *a* (figure 639). This stock was provided with a round rest *b* at the end remote from the bow, and had a long slot *c* cut down the whole of its length. In this slot was carried the sliding piece *d*, cut to the corresponding dovetail, which was in turn grooved to take the arrow *e*. Upon the slider was mounted a lock, consisting of a double finger *f*, pivoted horizontally, with a tail extending beyond the fulcrum and resting upon the trigger *g*. Pawls *h*, engaging in racks *k*, prevented the slider from being carried

[1] *Chronological Note.* The dates of the Greek mechanicians—Ctesibius, Philo of Byzantium, and Hero of Alexandria—are all uncertain. Therefore the 'inventions' described here cannot be exactly dated either. We may assume that Ctesibius lived about the first half of the third century B.C.: that Philo flourished in the second half of the second century B.C.; and that Hero died about A.D. 75.

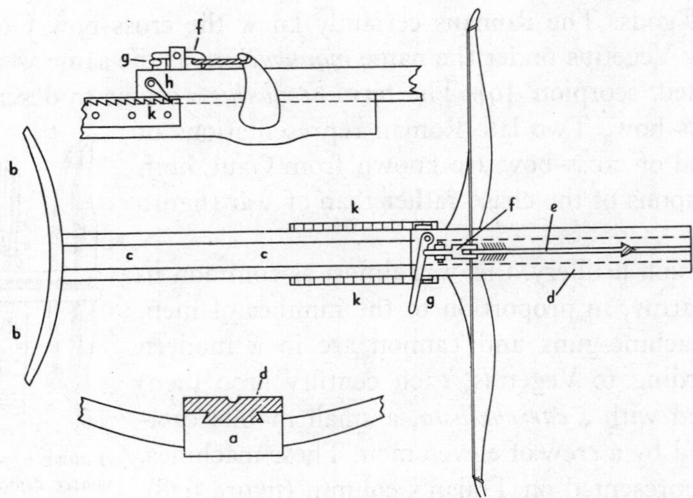

FIGURE 639—*Hero's* gastraphetes (*reconstructed*).

forward by the tension of the bow. In use, the bow-string was placed behind the fingers *f*, and locked there by the trigger. The front end of the slider was then placed against the ground or a wall, while the archer held the bow and pressed with his chest against *b*. He was thus able to use his weight and body muscles to force back the slider and bend the bow. With the slider held under tension by the

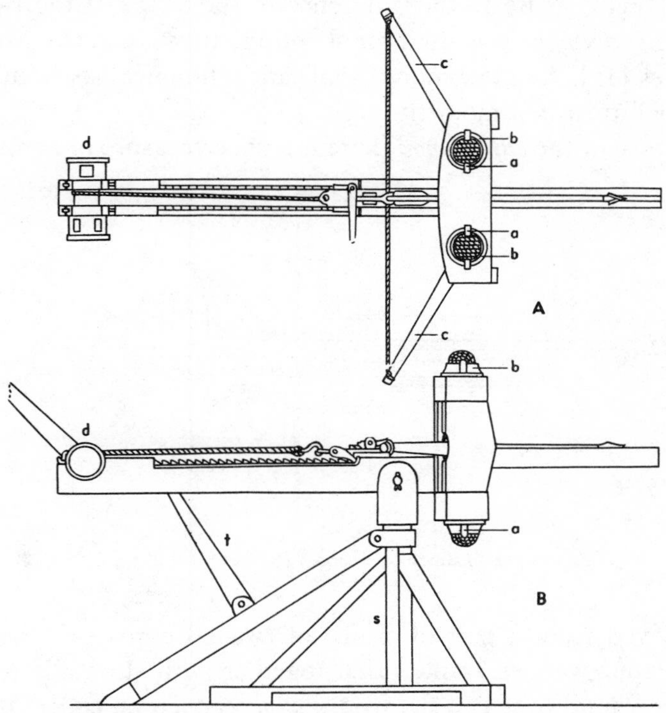

FIGURE 640—Euthytonon *described by Philo:* (A) *plan view,* (B) *elevation.*

racks and pawls, the archer placed an arrow in its groove, with the butt of the arrow between the fingers of the lock. When he pulled back the trigger *g*, the tail of the lock was free to move, releasing the bow-string from the fingers to drive forward the arrow.

From its construction this was obviously a one-man weapon, but it is not impossible that it was also used with a stand (like the seventeenth-century matchlock) or rested upon a parapet. It is said that more powerful weapons were made on the same plan, in which the bow was bent by some kind of winch, but of these there is no extant description. Nevertheless, the *gastraphetes* could have been a very effective weapon, having a tension perhaps two or three times as great as that of the ordinary bow.

As Hero himself remarked, horn proved an unsuitable material when great power was required. Such was provided by the torsion-catapult, which Greek authors describe in two forms, the *euthytonon* and *palintonon.* These are simply the Greek names for the simple and composite (double-curved) bow, respectively, and it seems plausible that some difference in the appearance of these engines of war was cognate to the difference in the shape of the two kinds of hand-bow. The *euthytonon* was used for shooting arrows, and the *palintonon* for projecting stones [12]. Another more significant difference has been suggested, to which we shall turn in a moment.

On the structure of the *euthytonon* there is a good measure of agreement. The

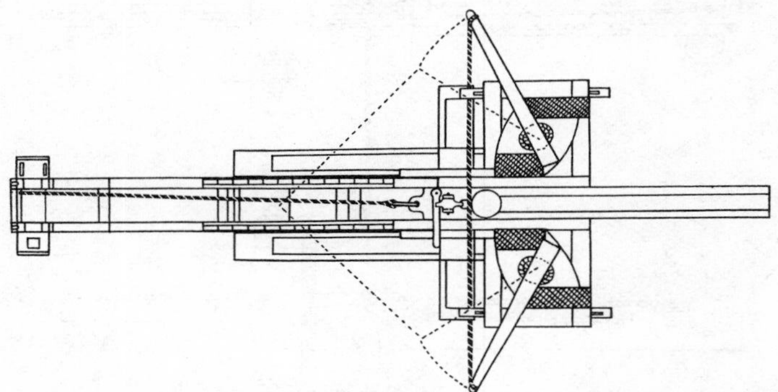

FIGURE 641—Palintonon *described by Philo (plan view).*

power was derived from a pair of skeins of twisted cords (of sinew, or hair) mounted in a stout wooden frame fitted together with dovetails (figure 640). Each skein was wound over two flat iron pins *a*, carried on collars of bronze or hard wood *b*. Two long arms of wood, *c*, were inserted into the middle of the skeins, which were then twisted in opposite directions, so that the arms were forced outwards by the torsion until they pressed strongly against the uprights of the timber frame. How this twisting was done, and how the collars were locked against the torsion, is not clear. The friction between collars and frame must have been very great, and enormous leverage must have been applied to the final twisting of the thick cords. When the two arms were joined by a cord, also of hair or sinew, the equivalent of a simple bow was produced. This could be mounted on a stock similar to that of the *gastraphetes* (figure 639), with the difference that a winch *d* was now required to strain the bow. The stock was fitted to a stand *s*, and a means provided for varying the elevation by means of the support *t*.

It will be observed that though the machine is shown with the slider and racks these were no longer essential when the winch was used. We may perhaps assume that the simplification of fitting a ratchet-wheel and pawl to the winch, and fixing the lock at the rear of the stock to take the bow-string drawn up to it (as in the medieval cross-bow), may have been made.

According to the prevailing view, the *palintonon* (*petrobolos*, *lithobolos*) (figure 641) did not differ greatly from the *euthytonon*. It was certainly a more powerful and massive weapon: the skeins of sinew were placed farther apart, each in a separate square frame. The bow-string of the *euthytonon* was replaced by a broad strap, to bear on the missile. Otherwise there was little change. A less well known interpretation of the text of Hero of Alexandria, however, would make the arms of the bow in the *palintonon* face inwards, not outwards (figure 642) [13]. In this construction the angle of twist given to the arms is increased, and the cord is made to act longer upon the projectile. Further, when strained, the arms and the cord do assume something of the shape of the compound bow. On the other hand, the framing at the front of the machine must necessarily become more clumsy to stand the strain of two widely separated points of torque.

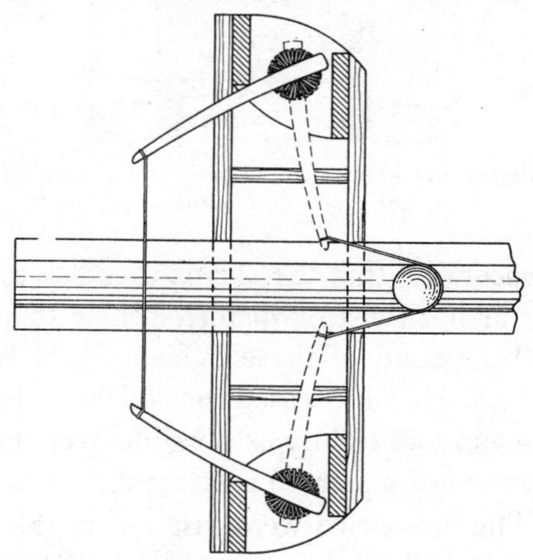

FIGURE 642—*Another reconstruction of the* palintonon, *based on a different interpretation of the classical texts. The cord is drawn to the right to strain the arms.*

The main Roman authority, Vitruvius (fl *c* 27 B.C.–A.D. 14), adds little further information save on points of minute detail [14]. He, too, described a pair of machines, one for throwing arrows and the other stones, the difference between them being only in the proportions of the different parts. Like Hero, Vitruvius stated the dimensions of the timbers making up the frames of these catapults very carefully, in terms of a modulus derived from the length of the arrow or the weight of the stone projected. In other words, the size of each machine was in a direct ratio to the size of the projectile. The dimensions of the motive elements —the skeins of twisted cords—were adjusted correspondingly.[1] No doubt it was

[1] In the arrow-shooting catapult the modulus was one-ninth the length of the arrow: this gives the diameter of the skein of sinew. In the stone-throwing catapult the modulus (as before, the diameter of the skein) was 5 digits for a projectile-weight of 2 lb, 8 digits for 10 lb, 12½ digits for 40 lb, 16 digits for 120 lb, and 18 digits for 200–360 lb.

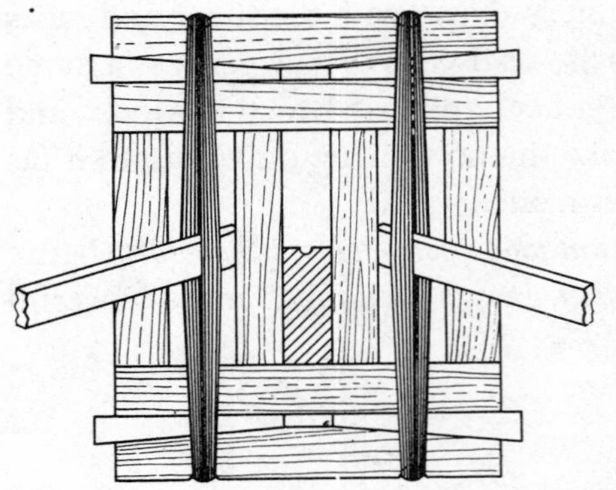

FIGURE 643—*Philo's use of wedges to tighten the skeins of cords in catapults.*

this invariable harmony of proportion which particularly interested the architect Vitruvius in the design of these machines.

Their weaknesses were obvious, and well known to the engineers concerned, such as Philo of Byzantium. The main troubles were, first, that the more powerful the skeins of sinew were made, the harder it became to contain them in a timber frame of manageable size and of strength adequate to resist the torsions and shocks set up in the machine; secondly, that the skeins themselves rapidly deteriorated and lost their tension. Further twisting distorted their shape and weakened them still further. Against the second of these defects Philo himself suggested a simple remedy (figure 643). He suppressed the collars and twisting altogether. Instead, the skeins were wound as tightly as possible over the timber framework (which did not need to have holes pierced in it), and then tightened still more by hammering in wedges. The bow-arms were inserted in the skeins as before.

In still another form of catapult, the *chalcotonon*, Philo did away with the sinew skeins in favour of springs made of bronze. This improvement had been suggested much earlier by Ctesibius. The alloy, containing 30 per cent tin, was cast into flat plates, which were shaped into curved springs and well cold-hammered. Two

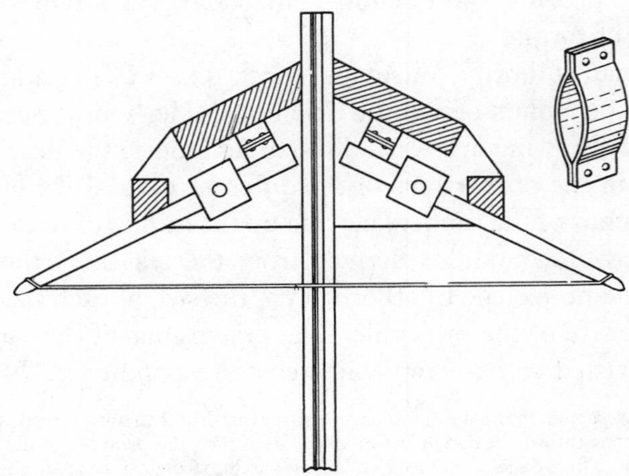

FIGURE 644—*Philo's* chalcotonon (*plan view*) *with* (*inset*) *one of the pair of elliptical bronze springs.*

elliptical springs were then assembled from four of these plates, and arranged so that they were compressed by the bow-arms on pulling back the cord (figure 644). Philo admitted that it was rather paradoxical to attribute this elasticity to a bronze plate, but thought it was due to the hammering; he seems to have confused the elastic properties of bronze and steel. The idea—if we admit the substitution of steel for bronze—was perfectly practicable, and may well have been applied to some of the smaller catapults of the latter part of this period.[1] Philo claimed that the *chalcotonon* was more powerful than other catapults, was free from the changes effected by damp in the sinews, and was as easy in its first construction as in its dismounting for transport [15].

Philo also gave a design for an automatic ballista, a kind of quick-firer, which held a number of arrows in a magazine. On turning a winch operating a chain, the bow-string was hauled back and released by a trigger when the next arrow had fallen into place, and so on till the supply of arrows was exhausted. The Chinese, in more recent times, had a magazine cross-bow of a similar but more simple construction, of which examples are often seen in museums. The most fantastic of all these designs was the *aerotonon* of Ctesibius, in which the elastic force was to be provided by the compression of air in a pair of cylinders [16]. Such inventions, though useless in practice, show that Greek engineers were fully capable of imagining complex machines.

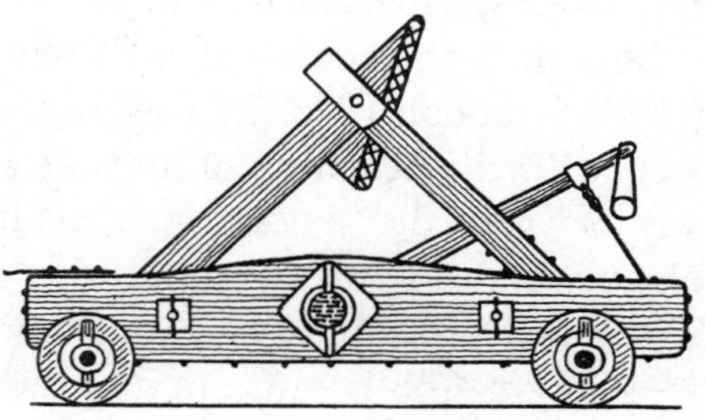

FIGURE 645—*The onager.*

That torsion-catapults, on the other hand, were formidable weapons is certain from ample evidence; and the spring-catapult, as a one-man weapon, may well have been effective, though there is no real testimony to its use.

The last of the missile weapons known to antiquity is well attested, though it was a late arrival. This was the single-armed onager described in the fourth century A.D. by Ammianus Marcellinus (figure 645) [17]. The power was provided by a single massive skein of cords passing horizontally between the two side-members of the frame of the machine, acting on a single arm which was effectively lengthened by a sling. The arm was pulled down by a winch and held in a catch which was released by the *magister* (officer) with a tap from a mallet. The impact of the arm was received upon the straw-filled cushion placed at an angle

[1] Colonel Schramm's reconstruction of this catapult very closely resembled the *carroballista* shown on Trajan's column.

of approximately 75° to the horizon. The onager was used for hurling fairly large weights, with a high trajectory and low velocity.

In this it only reinforced the tendency of all ancient catapults. The destructive effect of a projectile is proportional to mv^2, where m is the mass and v the velocity; and since the ancients could not confer a high velocity (never more than 200–220 ft/sec) they were compelled to increase the mass—hence the use of masses of rock (according to Vitruvius) weighing up to 300 lb. Further, as the initial velocity was low, to obtain an adequate range the machines had to work at a considerable elevation, of 30–45°, though we may imagine that when directed against troops catapults could be employed with a point-blank trajectory. In the main, the ancient ballistician relied for his effect on the higher trajectories, particularly as he had to aim at, or over, the tops of walls and towers.

The performance obtained from reconstructions of ancient ballistic engines (in Paris about 1860, in England and Germany about 1905–10) agrees fairly well with the accounts of ancient authors [18]. The normal effective range of any of them did not extend beyond 500 yd. Possibly with a very powerful engine,. and a light projectile, this range might have been extended by a half. The usual projectiles were an arrow or dart, about a yard long or a little more, weighing 5–6 lb because of its massive iron head, or a stone missile of 6–10 lb. It would seem that heavier projectiles must have been used but rarely, and certainly with them the range would have been much reduced. It is not impossible, as the historians relate, that missiles of one talent (about 50 lb) or more were flung on special occasions, but we need not suppose either that the ranges were long, or that the machines were capable of carrying out such a battery over a lengthy period without disintegrating.

The siege of Syracuse by the Romans (214–212 B.C.) has already been mentioned as an example of the employment of missile- and other siege-engines on the large scale. Plutarch's account of it is full of drama [19]. Marcellus, the Roman general, attacked with a vast engine raised on a raft made by lashing eight ships together. But Archimedes, persuaded to turn to military technology as a mere holiday sport for a geometer,

> opened fire from his machines, throwing upon the land forces all manner of darts and great stones, with an incredible noise and violence, which no man could withstand. Some of the [Roman] ships were seized with iron hooks, and by a counter-balance were drawn up and then plunged to the bottom. Often was seen the fearful sight of a ship lifted out of the sea into the air, to be dashed against the fortifications or dropped into the sea by the claws being let go. The great engine which Marcellus was bringing up on the raft called the 'Harp' was, while still at a distance, struck by a stone of ten talents'

weight [?], and then another and another, which fell with a terrible crash, breaking the platform on which the machine stood, loosening the bolts and tearing asunder the hulks which supported it.

In the end the Romans became so scared that 'if only a rope or small beam were seen over the wall they would turn and fly, crying out that Archimedes was bringing some engine to bear upon them'.

No doubt many of the great sieges of the ancient world could be described in similar terms, emphasizing the part played by engineering and mechanical skill in warfare. Compare, for example, Josephus's account of the sieges of Jotapata (by Vespasian) in A.D. 67 and Jerusalem (by Titus) in A.D. 70, in which Josephus himself served as a Jewish general [20]. After failing to storm Jotapata, Vespasian set engines for throwing stones and darts round about the city, 160 in all, and ordered them to dislodge the defenders of the walls. Such engines as were intended for that purpose threw lances upon the Jews with great noise; and stones weighing a talent were thrown by other engines, along with fire and a vast multitude of arrows, so that the Jews could no longer man the walls. Many were hurt by one missile; the violence of the Roman projectiles was such that they carried away the pinnacles of the walls, and dislodged the corners of the towers; no body of men could be so strong (says Josephus) that it would not be overthrown to the last rank by the force of the stones. In the end the Romans brought up a battering-ram, and made a breach which the Jews could not defend.

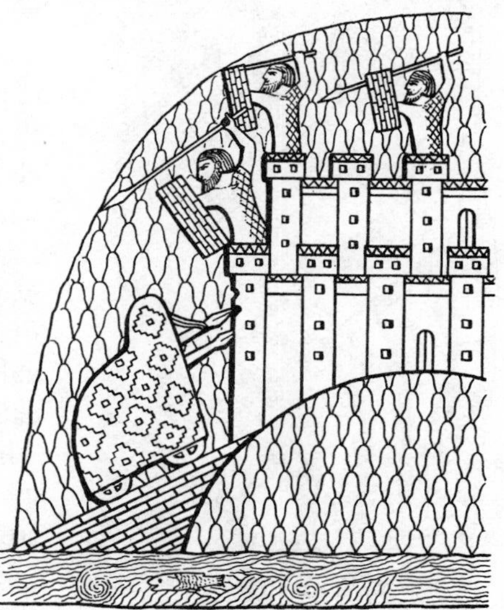

FIGURE 646—*An Assyrian relief showing a battering-ram, covered and mounted on wheels. From Nineveh. Seventh century B.C.*

Besides missile engines, many other machines, devices, and stratagems for breaking the defences of fortified places are described by Greek and Roman authors. Among the oldest and most useful of them was the battering-ram, of which illustrations have been found among the ruins of Nineveh (figure 646). Swung originally by the arms of men (as shown of Dacian tribesmen on Trajan's column), by the fifth century B.C., if not earlier, it had become a machine. In one form the ram—a long heavy beam, fitted with a heavy iron head often shaped like a ram's head—was suspended by ropes from a horizontal timber fitted in a framework built upon wheels. The framework could be pushed or pulled up to the walls of the town, and the ram swung inside it like a pendulum. In another form the ram slid on rollers mounted in a similar framework carried on wheels. The timber framework was completely covered outside by an armour of green hides to protect the men working the ram from the missiles of the besieged. This was one of the many forms of the 'tortoise'. The simplest tortoise was a party of men protected by their interlocked shields raised over their heads and shoulders (figure 647). More elaborately it consisted of a simple chassis on wheels with uprights supporting a strong roof. Beneath this shelter men could work their way to the walls to begin their operations. A line of tortoises constituted a sort of protecting gallery. The smallest of these shelters were called *musculi* (little mice); some had no wheels and were simply fixed in position to give protection, for the ancients did not dig trenches.

FIGURE 647—*The 'tortoise' formed of shields. From Trajan's column.* c *A.D. 110.*

If it were decided neither to bring down the wall by a mine, springing from beneath a tortoise, nor to use the ram, the attackers might seek to attain the level of the wall by various means. If scaling-ladders failed, the army might heap up a ramp leading to the top of the walls, as Julius Caesar did at *Avaricum* (Bourges) in 52 B.C. Alternatively, a movable tower might be built—one of the most fantastic products of ancient ingenuity (figure 648). Such towers were used on the grand scale from Macedonian times, as by Alexander and by Demetrius, who built a *helepolis* (destroyer of cities) for his attempt on Rhodes. They consisted of a chassis, which must have been supported upon many wheels. Above this rose a high timber framework, containing ten or twenty floors. Ladders enabled the troops to mount from one to another. The whole was strongly boarded and covered with green hides as a defence against fire-pots. The bottom of the tower was often occupied by a battering-ram, and the upper floors by missile engines working through loopholes. Drawbridges were fitted at appropriate levels, so that the attackers could suddenly sally forth upon the walls and towers of the besieged city. The smaller of these towers were, according to Vitruvius, about 90 ft high and 25 ft square at the base; the larger were twice as high and 33 ft square.

However absurd these dimensions seem, there is plenty of other evidence to show that towers were built, and set in motion, capable of overtopping the walls of a first-rank city. The weights suggested for them—100 tons and more—do not seem improbable; doubtless block-tackle was used to haul these monsters into position. Such tasks, apart from offensive action by the enemy, would have been comparatively light to the engineers who moved monoliths from Egypt to the forum of Rome.

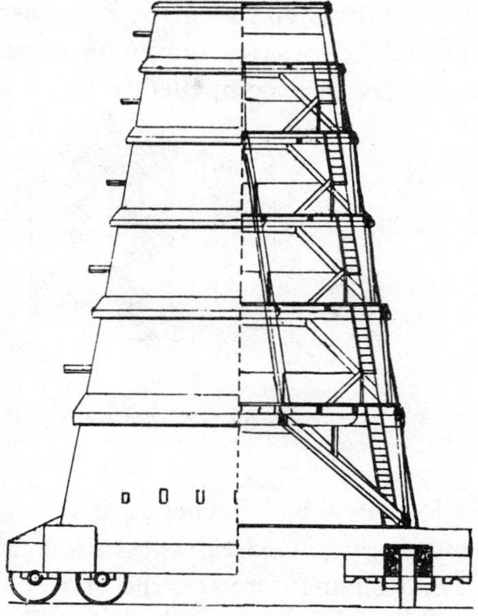

FIGURE 648—*A movable tower for assaulting fortifications (reconstructed).*

As the means of attacking fortified places did not change rapidly in the Roman world, the principles of defence also tended to remain much the same. There is not space here to describe methods of fortification in detail, though the Romans especially must have devoted a prodigious labour to their construction. We can note only the following points. (*a*) After the time of Pericles, the so-called 'polygonal' style of building—in which the blocks of stone preserved their irregular figure—had been abandoned in favour of squared stone, often in blocks of large size (plates 40 A, 41 A). (*b*) Gates were at first roughly rectangular holes in the walls, but later the Greeks, and following them the Romans, built various kinds of elaborate baffles in order to expose and delay to the greatest extent those who would force an entry. (*c*) Crenellations were adopted to improve the protection of walls; the walls themselves were sometimes built in a stepped form (figure 649), the steps being designed to facilitate the throwing of missiles upon the unshielded right sides of the assailants. (*d*) The construction of walls became architecturally more complex, so that they contained one, two, or more galleries at different levels (figure 650), elaborate staircases, chambers in towers, and the like. This is well seen in the walls of Imperial Rome. (*e*) The development of siege-engines enforced the reduplication of defences, with up to

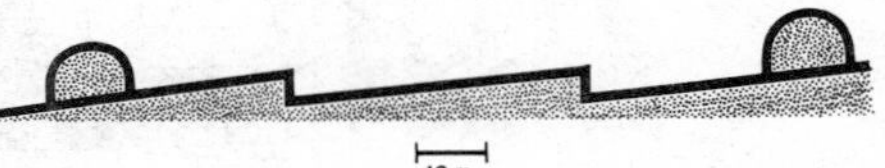

FIGURE 649—*The 'stepped' curtain-wall.*

three ditches and as many concentric walls, reinforced by towers (figure 651). The object was to dominate each wall from that interior to it, so that the besiegers were compelled to tackle each in turn.

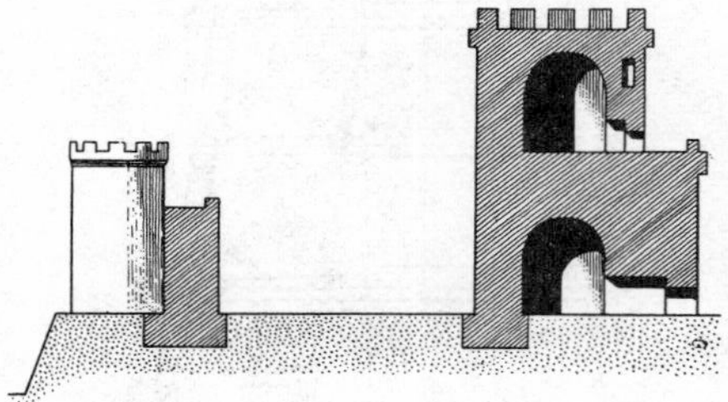

FIGURE 650—*Section through fortifications of Nicaea.*

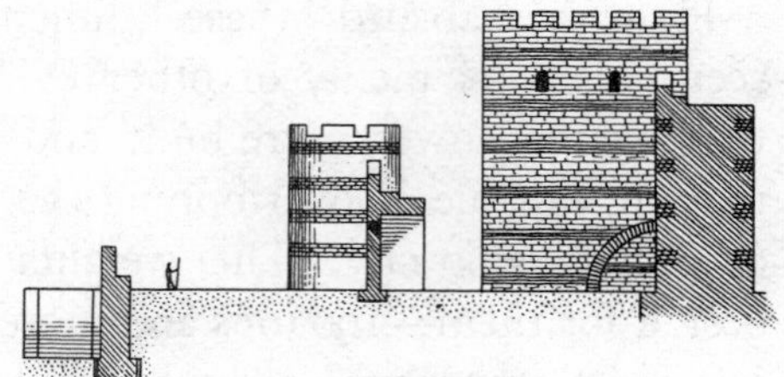

FIGURE 651—*Section through fortifications at Constantinople.*

Roman town defences are well known examples of their building-methods, and many still exist in excellent condition, as at Turin and Treves (plate 41 A, B). The Romans were also the first people to pay systematic attention to two allied problems of defence, that of the encamped expeditionary force, and that of a frontier. The methods of fortifying a camp, temporary or permanent, were laid down by military discipline. The construction of the ditch and rampart, the siting of the troops within them, and the various duties of the legionaries were all carefully defined. On Trajan's column the men are shown about their tasks,

FIGURE 652—*Legionaries building a fort (probably of turves). From Trajan's column.* c *A.D. 110.*

including the building of walls of turves (figure 652). From such camps, especially those which acquired the character of a fortified cantonment, the legions could dominate large stretches of country. From the time of Hadrian, however, it was found that the mere occupation of Britain or Germany was not enough; the occupied zone had to be protected from the incursions of the barbarians outside it. For this reason Roman strategists preferred to rest their conquests on 'natural'

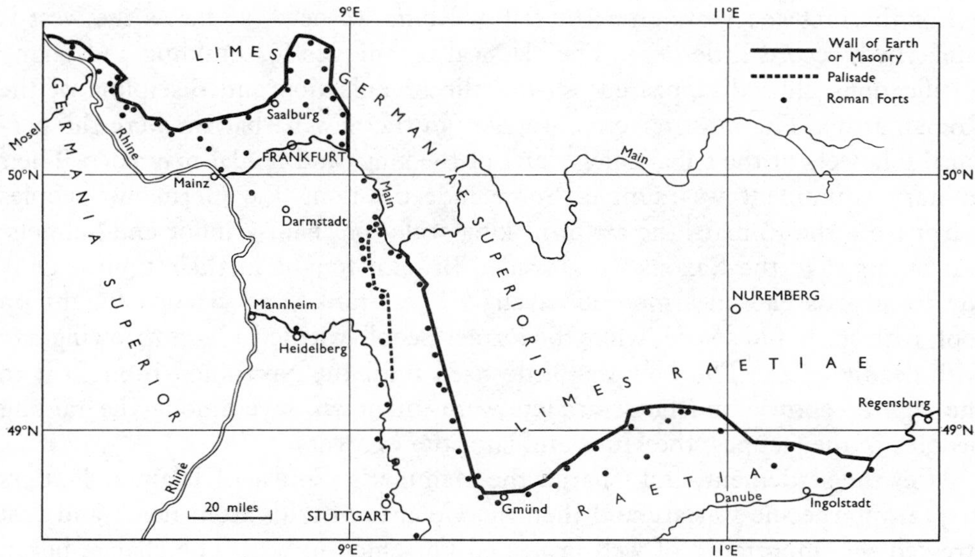

FIGURE 653—*Map showing part of the German* limes *and associated forts.*

frontiers strengthened by walls, watch-towers, guard-posts, and strong-points. The remains of these great works are still conspicuous (plate 40 B and figure 653).

It would be a pity to leave the declining Roman world without mention of its one technical inventor, the anonymous author of a work called *De rebus bellicis*. He lived about A.D. 370, and of him it has been said: 'Throughout the whole period of their greatness the Romans produced not a single mechanical invention of major importance; and [he] is the only Roman who is known to have made a conscious and serious attempt to improve the technology of his day over a wide field' [21]. He described (among other things) a paddle-wheel ship driven by oxen (figure 549), three types of frameless chariot armed with scythes, a kind of portable pontoon bridge using inflated skins, and two new *ballistae*, one mounted on a four-wheeled cart, which did not employ the principle of torsion. Unfortunately the verbal descriptions and the illustrations of these devices do not allow them to be understood in a technical sense.[1] The author's object was plain, namely, to save man-power: 'Invincible Emperor, you will double the strength of your invincible army when you have equipped it with these mechanical inventions, countering the raids of your enemies, not by sheer strength alone, but also by mechanical ingenuity' [22].

III. THE MEDIEVAL PERIOD (*c* 400–1400)

For the first 500 years after the fall of Rome we need, so far as the west is concerned, record little here. The classical techniques for building and taking fortifications alike disappeared; so did the organization and discipline of the Roman army. The most effective warriors in the new barbarism were the personal followers of the tribal chiefs, later of the kings and feudal magnates. Their military equipment was simple. Some indeed among the Germanic peoples learnt from the Romans the art of making scale- or chain-armour and helmets, but others, like the Saxons who invaded Britain, fought in their tunics. Only the Lombards had an armoured cavalry. The Franks and Saxons fought on foot, with spear and sword, while the former people wielded a light throwing-axe with deadly effect. The bow was little used until the Northmen brought it in the eighth century, and horse-archers were unknown, save among the raiding peoples of the steppes, the Huns and later the Magyars.

After the settlement, and Charles the Hammer's repulse of Islam at Poitiers in 732, internecine warfare and then the violence of raids from north and east stressed the importance of well protected horsemen in war. The change began with Charlemagne, who is depicted in the plain of Lombardy in 773 wearing an iron helm, a short coat of mail, a separate sleeve of the same on the right arm, and iron greaves with mail covering the thighs above them, and carrying spear, sword, and iron shield [23]. From his time feudalism demanded more stringently the services of well provided horsemen, and from the end of the ninth century the word *miles* was applied pre-eminently to the heavy cavalryman, the knight. A little later his armour became more complete; a kind of mail hood (hauberk) was developed to protect the neck and shoulders and the mail coat was lengthened from the hips to the knee. Meanwhile the Vikings had adopted chain-armour, and then, as they settled permanently in Normandy and the Dane-law, took to horse. They were dangerous bowmen (as the battle of Hastings proved) and they brought also the great double-headed axe as a weapon of close conflict.

In Byzantium, as might be expected, a very different state of affairs existed.

[1] The illustrations are fifteenth-century copies of a tenth-century manuscript, itself a copy of some earlier original.

The state was civilized, though hard pressed by invaders, and the spirit, if not the form, of the discipline and organization of the Roman army continued. Armies were divided into units, accompanied by engineers and even ambulances, superbly equipped by the standards of the west. The armoured horseman, wielding bow or spear, remained as before the decisive warrior. Most of the infantry also were provided with mail-shirts and steel caps; they were bowmen or swordsmen.

In the west, the growing skill of the smith is revealed in the slow perfection of armour. By the twelfth century the head was completely enclosed in a steel helm, with apertures for the eyes and nose, or swathed in fine chain-mail. The true 'mail' of the later Middle Ages, which was independent of a leather or cloth foundation, was made by forming rings of iron wire, completed by a welded or riveted joint. As they were closed they were interlaced like a net (figure 654). Mittens or gloves of mail protected the hands and joined the sleeves of the mail-coat; the feet and legs were similarly covered by cunning shaping. Underneath his mail the knight wore a quilted *gambeson*,[1] and as he was now unrecognizable he adopted armorial bearings on his shield. Sometimes a plate of iron was worn over the breast beneath the mail-coat in the thirteenth century, but the full development of plate-armour took place during the second half of the fourteenth, when it gradually extended over the whole body from plates worn over (not under) the mail to protect the breast and back. The armourer developed considerable virtuosity in providing protection at the joints between the plates, e.g. at the shoulders, but we cannot consider his methods in detail. The weight was enormous, and knights no longer young sometimes died of heart-strain without striking a blow. In beauty of design with magnificent finish and inlays of precious metals, perfection of armour was probably reached towards the end of the fifteenth century [24].

FIGURE 654—*The fabric of chain-mail.*

This encumbrance of armour was very largely due to the increased efficiency of missile weapons, particularly the English long-bow which caused such havoc at Crécy (1346). It apparently originated in south Wales, and began to make way as a national weapon about the mid-thirteenth century. It may be doubted whether the impact of its missile was ever as forceful as that of the steel cross-bow, or of the composite Turkish bow of more modern times, yet it could pierce more than one thickness of mail or thin plate. Its rate of fire was much superior to that of the cross-bow, and it was easier to train men to its use. The

[1] An alternative name, acton, is derived from the Arabic *al-qutn*, cotton.

power of a bow is proportional to both the tension of the string at the moment of release and the distance through which the string acts upon the missile; it

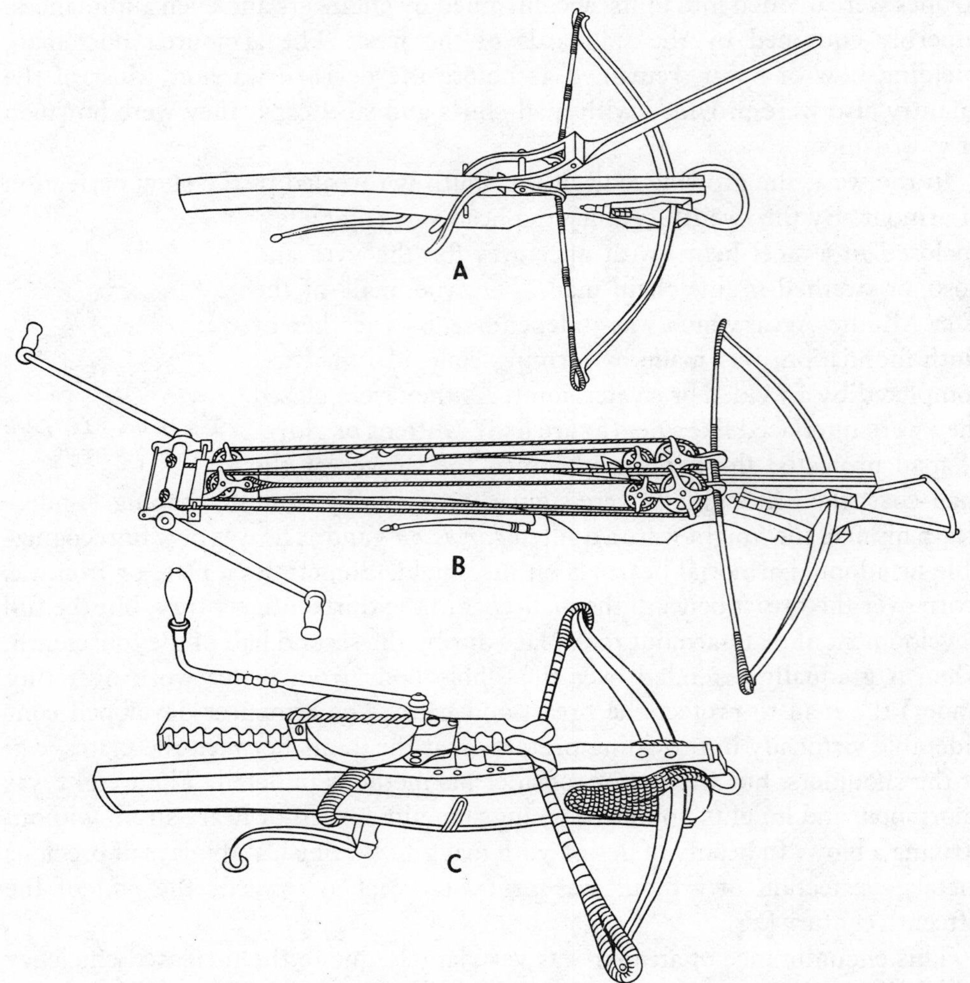

FIGURE 655—*Three different types of cross-bow; with* (A) *the goat's-foot lever,* (B) *the windlass,* (C) *the steel* cranequin (*ratchet*). *Scale 1/15.*

was the length of the arrow which could be used with the English bow, drawn from the extended left hand to the right ear, that made it formidable.

The hand cross-bow of the Middle Ages had played a larger part in war, for its history seems continuous with that of the missile engines of the ancients. Large cross-bows, firing bolts and mounted on the ramparts, are recorded at

Rome in 537 and Paris in 886 [25]; the indication would seem to be that in its last stages the 'Roman' army adopted this machine as a crude substitute for the more powerful torsion-catapult. Some sort of winch or lever must have been used to draw back the cord. The hand cross-bow is a rather later development, perhaps of the tenth century. When it was reintroduced to the east by the crusaders it was regarded by the Byzantines as an innovation.

At first the bow was made of horn, sinew, and wood; it was bent by holding it to the ground with one foot placed in a stirrup at the end of the stock, while the cord was pulled by a hook attached to a belt round the archer's waist. In this form it was of common use in war from at least 1150 to about 1370. The Italians, particularly the Genoese, were famed for skill with the weapon. It was much employed on shipboard and on horse-back. At about this latter date, 1370, the cross-bow was fitted with a stout steel bow, which could be bent only by means of a mechanical contrivance, such as the goat's-foot lever, the windlass, or the *cranequin* (figure 655). The last two of these demanded highly skilled metal-working in the manufacture of pulleys and gearing—the *cranequin* is indeed an example of heavy reduction-gearing only a little later than the wheel-work of the mechanical clock. The lock of these cross-bows was both simple and ingenious (figure 656). The steel cross-bow was a powerful weapon of war for a century until 1460–70, when it began to be displaced by the hand-gun. It lingered in hunting till the seventeenth century and indeed longer; many of the later examples intended for royal hands (such as those of Elizabeth I of England) were superb examples of workmanship and ornamentation.

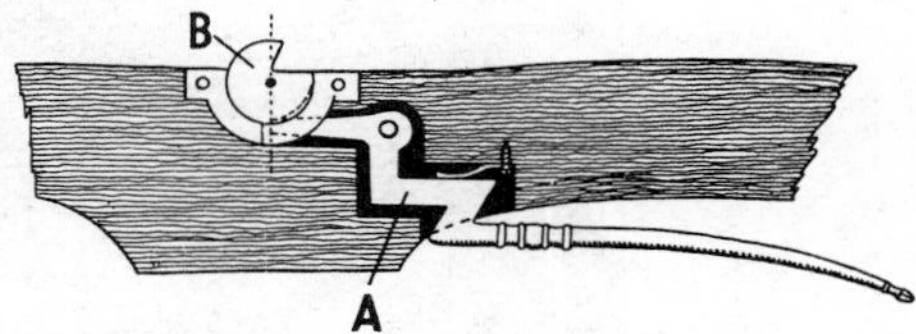

FIGURE 656—*The cross-bow lock; the piece* A *is solid, so that when it is squeezed towards the stock the roller* B *behind which the string is caught is free to turn, releasing the string.*

The ordinary military bow of the fifteenth century weighed 13–15 lb. The siege cross-bow, shot from a fixed position, was a little more weighty and powerful. With quarrels (short, square-headed arrows) 2–3 oz in weight, having steel heads, the extreme range was about 400–450 yd.

The heavier missile engines of the Middle Ages were so much copies of the Roman machines that no detailed description of them is called for. The mangonel (LL *mangonum*, a pulley) was simply the onager of Ammianus Marcellinus, but more often used with a spoon-shaped arm without the sling. Large cross-bows or *ballistae* (as they are confusingly called) for siege-warfare are fairly well attested; one having three separate bows is shown as late as 1588 (figure 657).

A medieval invention was the trebuchet (OF *trebucher*, to overturn), first heard of about 1100. Like the mangonel-onager, it had a single arm moving in the vertical plane, actuated by a heavy counterpoise hanging beyond the pivot. The

FIGURE 657—*Design for a siege cross-bow, by Ramelli, 1588.*

scheme seems fantastic but apparently it worked. No doubt ease of construction helped to render it so common in later medieval sieges (figure 583).

There is plenty of evidence, both literary and pictorial, to prove that a good deal of mechanical ingenuity was devoted to siege-warfare in the later Middle Ages. Pictures of sieges in manuscripts are indeed sufficiently stereotyped—they often show a trebuchet, knights trying to force a wall or gate, fighting with sword, mace, and axe as men strive to mount a scaling-ladder, cross-bowmen active on both sides—but the medieval soldier knew many other tricks. Most of these were familiar in the ancient world too; methods of mining, of assault with protective shelters, the use of the ram and the movable tower, had altered little in principle. There is no doubt that armies constructed these siege-works on the spot when faced with the necessity; they did not carry a siege-train proper with them. Doubtless their ranks included sufficient carpenters (figure 350) and smiths, and timber was then abundant in Europe. Perhaps metal parts were carried, but most of them could have been made at the village forge. Even a strongly fortified place would not possess an entire complement of machines standing by for use. This was obviously a serious obstacle to technical development in the days before gunpowder, after which armies had to drag their equipment with them.

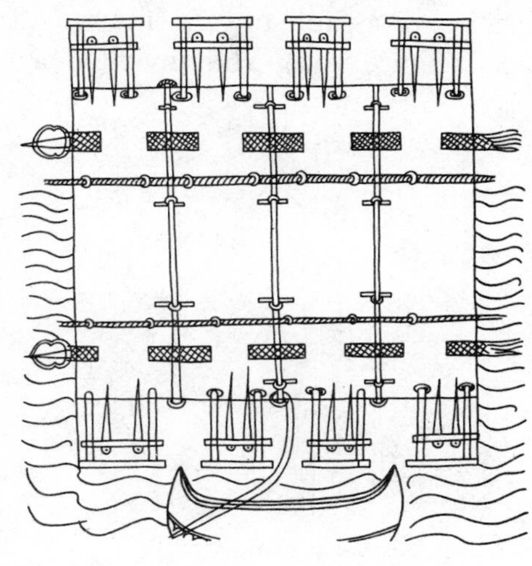

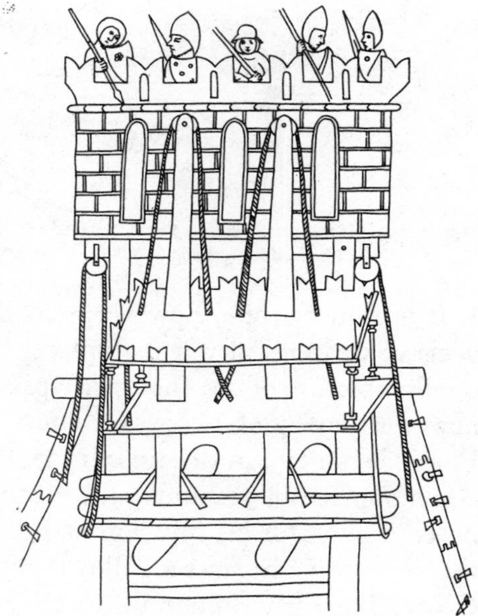

FIGURE 658—*Inventions in military engineering by Guido da Vigevano, 1335;* (above) *portable suspension-bridge,* (below) *movable tower.*

This problem was faced by a few theorists—particularly in relation to crusading campaigns, for a crusading force had to cope with new problems in the supply of food, water, fodder, timber, and munitions. It could not so easily live off the country in Syria or Palestine as in Europe, and it had to be prepared to traverse deserts.

One author who considered such problems was Marino Sanudo, called Torsellus the Elder (*c* 1260–1337), an experienced traveller in the Near East. He made some circumstantial remarks about the manufacture of *ballistae* and other war-machines, their transport by sea, and the quantity of stores required for an eastern expedition [26]. But he was not much interested in the engineering details. Another, more unusual, figure, who also hoped to further a crusade after the chance of success had passed, was Guido da Vigevano (*c* 1280–1350), an Italian physician at the French Court. His object was the lightening of the burden of engineering-material that would have to be carried, while giving it a greater flexibility in use. Clearly he was well acquainted with craft skills. He had the idea of using small interchangeable sections, transportable by pack-animals, which could be assembled to make bridges or assault-towers (figure 658). He knew of the principle of the paddle-boat (figure 594), and invented a completely enclosed fighting-car or 'tank' propelled either by men turning cranks, or by wind-power (figure 659).[1] Some of his ideas were repeated or copied by later inventors down to Leonardo da Vinci, and beyond; the figures sufficiently express the difficulties caused by absence of perspective to a medieval draughtsman, just as his text reveals the lack of an adequate technical vocabulary [27].

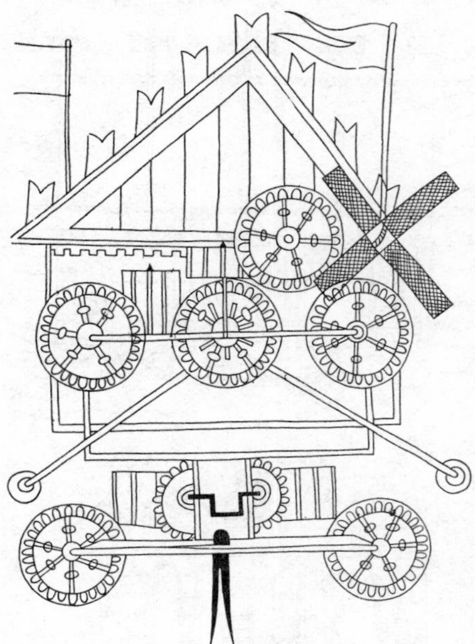

FIGURE 659—*Wind-driven fighting-car described by Guido da Vigevano.*

The future was to lie in another way, in the development of gunpowder and cannon and in the lineal descendants of these expansion-machines, the steam and internal-combustion engines. The history of pyrotechnical compounds in the Middle Ages is traced elsewhere (ch 10). Here it will suffice to note that although 'Greek fire' was a western invention, gunpowder was probably derived from the east [28].

It was in Europe, however, that the rapid development of fire-arms as the decisive weapons of war took place. Cannon were certainly in use by 1325, and the first picture of one (not, perhaps, the work of an artist who had ever seen the new weapon) is of two years later. They are mentioned at Metz in 1324, at Florence in 1326, in Britain in 1327. By mid-century they were becoming almost commonplace in accounts and inventories of military stores [29]. Edward III ordered guns for his invasion of France from 1345 onwards, and used them at the siege of Calais in the following year, though apparently not at the battle of Crécy. The early cannon were very small, sometimes no more than 20–40 lb in weight, and even late in the fourteenth century one of 150 lb could be described as a 'great cannon'. They were made, not cast, from *cuprum* or 'metal'—whether copper, bronze, or brass is uncertain—and perhaps from forged iron. Seams

[1] This is an interesting example of the use of gears and shafting for the transmission of power only a comparatively brief time after their introduction in the windmill.

would have been closed by riveting, brazing, or welding. Cannon cast from cupreous metal were introduced about the middle of the fourteenth century, and were followed at its close by the large built-up forged iron gun. The first guns were very short in proportion to their bore, and they fired darts or quarrels rather than balls. The gun was clamped directly to a stout wooden beam mounted in a fixed timber framework, without wheels. Powder at this stage was very expensive, and could be procured only in small quantities. From the middle of the century, however, the size of the gun increased rapidly: in the reign of Richard II (1367–1400) guns were bought for the Tower of London, of three, four, and even six hundred lb each. For some of these at least a new method of construction was adopted. The gun was built up from iron bars, running parallel to the axis of the bore, welded together upon a core which was later removed. One end was blocked and the barrel further strengthened by shrinking iron hoops upon it. The wrought iron of the bars was much less liable to fracture than cast metal, though the bursting of a large cannon of this type in 1460 killed James II of Scotland. Spherical projectiles had come into use by 1350, stone being substituted for iron as less expensive (pp 74 f).

The fertility of technical invention displayed in the early history of fire-arms is astonishing. Soon after the mid-fourteenth century a kind of *mitrailleuse*, the *ribaudequin*, became common; it consisted of a number of barrels mounted on a framework, so that they could be discharged in rapid succession, and was fitted with wheels. Breech-loading pieces were made too; a short iron cylinder was fixed over the open end of the breech by means of a wedge, or sometimes arranged to fit into it. But such schemes proved dangerous. Finally the small barrel mounted on a wooden stock became the hand-gun, coming into increasing prominence from the early years of the fifteenth century. The incentive provided by this new artillery towards the development of metallurgical skills was of course very great; soon a new technical literature, exemplified by the *Feuerwerkbuch* of 1420 and dealing with the manufacture and use of gunpowder and cannon, was beginning to circulate in manuscript [30].

The impact of gunpowder on the art of war was not immediate, though already before the end of the fourteenth century the catapult type of siege-engine had passed into disuse. The cross-bow, for example, had still to reach its most perfect form, and armour was by no means laid aside. Even in the seventeenth century armourers were still making body-armour impenetrable to pistol or musket-shot, despite the great weight of the plates required. The gun was for long a weapon of offence, or annoyance, demanding support from other arms; thus its role in pitched battle was far from decisive until after the close of the

Middle Ages. It was rather in siege-craft and fortification that the most rapid changes occurred, for even the strong walls and massive gates of the medieval castle were vulnerable to gunpowder. A century and a half before the invention of cannon the crude stability of a single block—*par excellence* the stone-built Norman keep, such as the Tower of London—was giving way (as long before in the ancient world) to a system of *enceintes*, which had to be forced one by one. Later castles, too, depended as much for their strength on their topographical position as on the massiveness of their construction. In the fourteenth and fifteenth centuries this later form of the feudal stronghold itself lost its tactical significance, and static defence turned on the fortified city rather than on the individual castle. For this, and against the bombards drawn in the siege-trains, new principles of castramentation were needed, gradually perfected in the polygonal fortification of the sixteenth and seventeenth centuries which was brought to its peak by the great French engineer Vauban (1635–1707).

REFERENCES

[1] TARN, SIR WILLIAM W. "Persia from Xerxes to Alexander" in 'The Cambridge Ancient History', Vol. 6, pp. 18 ff. University Press, Cambridge. 1927.

[2] *Idem.* 'Hellenistic Military and Naval Developments', p. 105. University Press, Cambridge. 1930.

[3] PLINY *Nat. hist.*, VII, lvi, 201. (Loeb ed. Vol. 2, p. 640, 1947.)

[4] KROMAYER, J., VEITH, G., *et al.* 'Heerwesen und Kriegführung der Griechen und Römer.' Handb. Altertumswiss., Abt. 4, T. III, Bd. 2, p. 234. Beck, Munich. 1928.

[5] STEIN, SIR (MARC) AUREL. 'On Alexander's Track to the Indus', pp. 136, 146 ff. Macmillan, London. 1929.

[6] TARN, SIR WILLIAM W. See ref. [2], p. 114.

[7] VEGETIUS I, xx.

[8] PROCOPIUS *Hist. bellorum*, I, i. (Loeb ed. Vol. 1, pp. 4 ff., 1914.)
TOYNBEE, A. J. 'A Study of History', Vol. 3, pp. 162 ff. Oxford University Press, London. 1934. For comment on the above passage.

[9] VEGETIUS IV, xxii.

[10] *Idem* II, xxv.

[11] HERO of Alexandria. DIELS H. and SCHRAMM, E. (Eds and Transls). 'Herons Belopoiika. Schrift vom Geschützbau.' Abh. preuss. Akad. Wiss., phil.-hist. Kl., 1918, no. 2.
PROU, V. 'La Chirobaliste de Héron d'Alexandrie.' Not. Extr. MSS. Bibl. Nat., Vol. 26, ii, p. 59. Paris. 1877. This contains a French translation of the relevant passage.
SCHRAMM, E. 'Griechisch-römische Geschütze.' Scriba, Metz. 1910.
[12] KROMAYER, J., VEITH, G., *et al.* See ref. [4], pp. 230 ff.
[13] PROU, V. See ref. [11], pp. 63 ff.
[14] VITRUVIUS X, x–xii. (Loeb ed. Vol. 2, pp. 326 ff., 1934.)
[15] PROU, V. See ref. [11], pp. 95 ff.
KROMAYER, J., VEITH, G., *et al.* See ref. [4], p. 235.
SCHRAMM, E. See ref. [11], Pl. VI.
[16] *Idem.* See ref. [11], pp. 27 ff.
[17] AMMIANUS MARCELLINUS XXIII, iv, 4–8. (Loeb ed. Vol. 2, pp. 326 ff., 1937.)
[18] PAYNE-GALLWEY, SIR RALPH W. F. 'A Summary of the History, Construction and Effects in Warfare of the Projectile-Throwing Engines of the Ancients.' Longmans, Green and Co., London. 1907. This work gives interesting and full details of the construction of mechanical artillery. The gallant knight, however, used a number of devices not attested in the ancient authors. Probably his machines performed as well as anything of equivalent size built in antiquity.
[19] PLUTARCH *Vitae parallelae: Marcellus*, xiv–xix. (Loeb ed. Vol. 5, pp. 468–86, 1942.)
[20] JOSEPHUS *De bello Judaico*, III, vii; V, vi–x. ('Works' trans. by W. WHISTON, pp. 756 ff.; 833 ff. Routledge, London. 1906.)
[21] THOMPSON, E. A. (Ed. and Trans.). 'A Roman Reformer and Inventor. Being a new text of the treatise *De rebus bellicis*', p. 44. Clarendon Press, Oxford. 1952.
[22] *Idem. Ibid.*, p. 120.
[23] OMAN, SIR CHARLES W. C. 'A History of the Art of War in the Middle Ages' (2nd rev. ed.), Vol. 1, p. 86. Methuen, London. 1924. I have drawn heavily on this work for the medieval period.
[24] FFOULKES, C. J. 'The Armourer and his Craft from the XIth to the XVIth Century.' Methuen, London. 1912.
[25] OMAN, SIR CHARLES W. C. See ref. [23], pp. 137 ff.
[26] SANUTUS, MARINUS (the Elder). *Liber secretorum fidelium crucis super Terrae Sanctae recuperatione et conservatione*, II, iv, 8, in *Gesta Dei per Francos*, ed. by J. BONGARS, Vol. 2, p. 59. Heredes Aubrii, Hanover. 1611.
[27] Two manuscripts exist: one is in the Bibliothèque Nationale in Paris (Fonds latin 11015), the other in private possession.
[28] GOODRICH, L. C. and FÊNG CHIA-SHÊNG. *Isis*, **36**, 114, 1946.
WANG LING. *Ibid.*, **37**, 160, 1947.
[29] TOUT, T. F. *Engl. Hist. Rev.*, **25**, 666, 1911.
[30] HASSENSTEIN, W. (Ed. and Trans.) 'Das Feuerwerkbuch von 1420. Neudruck des Erstdrucks. . . . 1529.' Deutsche Technik, Munich. 1941.

BIBLIOGRAPHICAL NOTE

The literature on the early history of artillery is extensive. General histories of war including chapters on this subject are:

KÖHLER, G. 'Die Entwicklung des Kriegswesens und der Kriegführung in der Ritterzeit' (3 vols). Koebner, Breslau. 1886–7.

JÄHNS, M. 'Geschichte der Kriegswissenschaften vornehmlich in Deutschland' (3 parts). 'Geschichte der Wissenschaften in Deutschland', ed. by Hist. Komm. bayr. Akad. Wiss., Vol. 21. Oldenbourg, Munich. 1889–91.

DELBRÜCK, H. *et al.* 'Geschichte der Kriegskunst' (7 parts). Stilke, Berlin. 1900–36.

Among specialized works NAPOLÉON III and FAVÉ, I. 'Études sur le passé et l'avenir de l'Artillerie' (6 vols), Dumaine, Paris, 1846–71, though old, and FFOULKES, C. J. 'The Gun Founders of England', University Press, Cambridge, 1937, are useful.

Roman soldiers crossing a bridge of boats over the Danube. From Trajan's column. c *A.D. 110.*

21

ALCHEMICAL EQUIPMENT

E. J. HOLMYARD

ALCHEMY (from the Arabic *al-kīmīa*) may be roughly defined as an art aiming at the conversion of base metals into gold and the preparation of an elixir to prolong life indefinitely. There were two sides of this art: first, a mystical and very involved body of speculative thought, and secondly, a practical technique in which chemical and metallurgical operations were carried out. Alchemical theory does not greatly concern us here, and had very little connexion with the theories of chemistry proper that came later. Alchemical practice, on the other hand, was the direct ancestor of practical chemistry and chemical technology, though other techniques contributed to various extents in the establishment of the modern pure and applied science.

The origin of alchemical theory is still far from clear, and possibly such speculation arose independently in more than one early civilization. In the west it certainly owed much to Aristotle's views on the nature of matter and to the Pythagorean theory of numbers [1], but the weight of evidence is that the fundamental ideas arose in the ancient Persian Empire, which included Mesopotamia, Asia Minor, Syria, and Egypt. Forbes suggests that three streams of thought contributed to the rise of alchemy:

(*a*) The philosophy and technology of the ancient Near East.
(*b*) The philosophical tenets of the Iranian and Indian civilizations.
(*c*) The philosophy and science of the Greeks [2].

The main alchemical belief was that the base metals, such as copper, lead, tin, and iron, were impure or unripe forms of a single metallic substance which in its pure or fully ripened state appeared as gold. It followed that if a base metal could be suitably purified or ripened it would be converted into gold, and the principal efforts of the practical alchemists were directed to the accomplishment of this transmutation. Their theorizing fellows, accepting the possibility of transmutation as a fact, wove webs of unsubstantial hypotheses to explain it. They went further, and linked their speculations on metallic metamorphoses with similar philosophies concerning the nature of man and his relation to the Deity.

This extension often led alchemy to the fringes of astrology and magic—magic squares, for instance, lay at the base of much alchemical theory. Sherwood Taylor, however, has pointed out that the alchemists in general shunned black magic, which involved the invocation of evil spirits, and confined themselves to natural magic or the search for supposed hidden relationships in objects [3]. Theoretical alchemy has a strong element of religious mysticism running throughout it; all alchemical knowledge meant 'deeper knowledge of the Cosmos as a whole' and 'another secret helping to understand the Order of Creation and maybe to master Nature' [4].

The preoccupation of alchemy with man as well as with metals accounts for the fact that practical alchemical operations were of two main kinds: those that sought the elixir of life, and those that aimed at the artificial production of gold. The former efforts, in consequence of their complete and obvious lack of success, gradually merged into iatrochemistry, the preparation of drugs for medicinal purposes; but many important remedial substances were discovered or developed in alchemical laboratories. The metallurgical operations designed to effect transmutation had a longer history, for it often happened that alloys more or less resembling gold resulted from experiments: to the alchemists success was not only always possible but frequently probable and not seldom imminent.

A general tenet was that a substance could be made which, when added to, or 'projected' upon, a properly prepared mass of base metal, would bring about the desired transmutation. This substance, the *lapis philosophorum* or philosophers' stone, was also known as the red elixir; it was credited with the power of transmuting several hundred or thousand times its own weight of copper, tin, lead, or mercury into the purest gold, 'purer than the gold of mines'. An inferior stone, the white elixir, could carry the transmutation only as far as silver, the metal next below gold in the stages of ascent to perfection. The most extravagant claims were made for these imaginary catalysts: one alchemist boasted that, were the sea composed of mercury, he could transmute it—*mare tingerem si mercurius esset*. Even more futile than the search for the elixirs was that for a universal solvent, the alkahest, which was pursued until common sense ironically inquired how such a substance could be contained.

The earliest alchemical treatises come from Hellenistic Egypt.[1] Perhaps the most ancient of them is the *Physika kai Mystika*, 'Physical and Mystical Matters', of an author writing under the name of Demokritos; he probably lived in the

[1] It has been claimed that alchemy existed in China at a much more remote date, but too little definite information on this point has so far been established. A forthcoming volume of Joseph Needham's 'Science and Civilisation in China' (the first volume was published in 1954) may settle the question, but meanwhile it should be stated that early Chinese alchemy centred on the search for a 'pill of immortality'.

second century A.D. and is not to be confused with the celebrated philosopher Demokritos of Abdera (d 376 B.C.). A large collection of alchemical material was made by Zosimos of Panopolis (*c* A.D. 300), who also wrote original works on the subject, and many commentaries on earlier authors were compiled by succeeding Graeco-Egyptians such as Olympiodorus (*c* 425). The texts of some of these

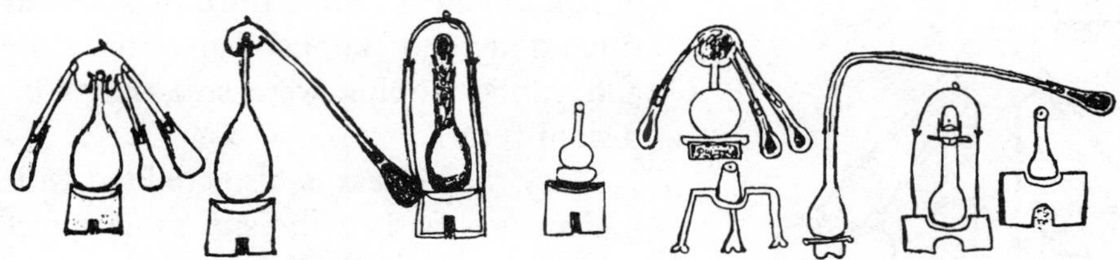

FIGURE 660—*Figures of Greek alchemical apparatus, showing a* kerotakis, *stills, flasks, water-bath, and sand-bath.*

treatises, with a translation into French, were published by Berthelot in his *Collection des anciens alchimistes grecs* (1887–8).

We may assume that the alchemists were conversant with the contemporary techniques of metal-working (chs 2, 13), glass-making (ch 9), dyeing (pp 364–9), and the like, and the assumption is indeed borne out by the content of the literature. It is, however, in these early Greek treatises that we first meet with apparatus specifically designed for alchemical purposes, typically stills and a kind of reflux device known as the *kerotakis* (p 734). Some of the different types of still employed are illustrated in figure 660, redrawn from a manuscript in the Bibliothèque Nationale, Paris. It shows flasks, stills, condensers, receivers, furnaces, a sand-bath or water-bath, a tripod, and a *kerotakis.*

One form of still consisted of a containing-vessel,[1] which could be heated, fitted with a head carrying a delivery-spout. The still-head was known as the *ambix*, but this name was often applied later to the still as a whole and has given rise, by way of the Arabic *al-anbīq*, to our word alembic. Some still-heads were provided with three spouts (figure 661), and a still of this kind was called a *tribikos.* The invention of the *tribikos* was ascribed to a woman alchemist, Mary the Jewess, traditionally supposed to be Miriam, the sister of Moses. In a work passing under her name, quoted by Zosimos, instructions for making a *tribikos* are given. Three copper or bronze tubes should be made, each a cubit and a half in length, and of a thickness of metal rather more than that of a confectioner's frying-pan. One end of each tube should be adjusted to the size of the neck of its receiving-flask, and the other should be soldered into a copper or bronze

[1] Called the cucurbit in medieval alchemy (Latin *cucurbita*, gourd).

still-head. The still-head should fit closely on to the earthenware vessel containing the substance to be distilled, and the joint should be luted with flour-paste. Similar luting was to be applied between the delivery-spouts and the receivers.

In later times, a special luting-material known as 'clay of wisdom' was highly recommended. According to one recipe, it consisted as to two-thirds of clay free of stones and as to one-third of a mixture of dried dung and chopped hair. Cork stoppers and rubber tubing were still things of the distant future.

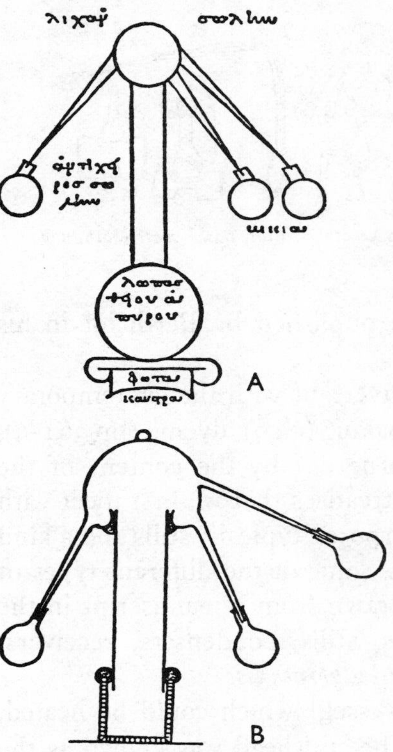

FIGURE 661—(A) *A* tribikos *or three-armed still standing on a heater.* (B) *Reconstruction of the still.*

Mary the Jewess is also credited with the invention of that indispensable piece of chemical equipment the water-bath—which is still known in France as the *bain-marie*—and of the apparatus called the *kerotakis* that figures prominently in Greek alchemical writings. In those days painting was frequently executed by the encaustic process, in which the pigments were mixed in melted wax—generally beeswax—and applied to the support by means of a brush. The *kerotakis* was the metallic plate or palette, often triangular in shape, on which artists kept their paints fluid by resting it over a small charcoal burner. In its application to alchemical operations, the *kerotakis* was placed inside a cylindrical vessel, closed at the lower end, containing mercury or sulphur or some other substance that would partly or entirely vaporize on heating. On the *kerotakis* proper[1] was placed the metal to be subjected to the action of the vapours, normally in the form of foil or powder; the upper end of the apparatus was closed with a hemispherical cover (figure 662). On heating, the volatile substance gave off its vapours, which in part attacked the metal and in part were condensed at the top of the apparatus, the liquid flowing back to the bottom. In effect, therefore, a continuous reflux action was maintained. The nature of the product varied according to the metals and vapours used; thus lead and copper on the *kerotakis*, with sulphur to provide the vapour,

[1] As with alembic for still, the name *kerotakis* was often used to denote the whole of this apparatus.

yielded a black substance that received much attention from the alchemists. Blackness or *melanosis* was supposed to represent the first stage towards transmutation; theoretically it was followed by *leukosis* or whiteness and then by *xanthosis* or yellowness, with sometimes a further stage of *iosis* or purpleness.

Altogether some eighty different kinds of apparatus are mentioned by the Greek alchemists. 'Furnaces, lamps, water baths, ash baths, dung beds, reverberatory furnaces, scorifying pans, crucibles, dishes, beakers, jars, flasks, phials, pestles and mortars, filters, strainers, ladles, stirring rods, stills, sublimatories, all make their first appearance as laboratory apparatus in their works and have persisted in somewhat modified forms to the present day' [5].

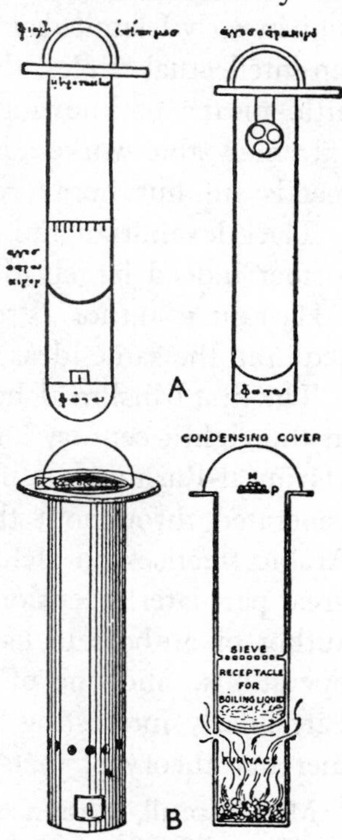

FIGURE 662—*The* kerotakis *or reflux apparatus:* (A) *as shown in Greek manuscripts;* (B) *as a conjectural restoration by F. Sherwood Taylor.*

Comparatively little is known of the fortunes of alchemy between about 400 and 700, though there is no reason to suppose that tradition underwent any striking change. With the advent of Islam, however, alchemy received a great impetus and at the same time began to show signs of a more rational outlook. The Prophet died in 632, and within a century his followers had conquered Persia, Asia Minor, Syria, Palestine, Egypt, the whole north African littoral, Gibraltar, and Spain; their victorious march west was not arrested until 732, when Charles the Hammer defeated them at Poitiers. Rather surprisingly, perhaps, the Muslims soon manifested an enthusiasm for knowledge, and having overrun all the principal centres of Greek learning they were able to indulge it to the full. Under rulers such as Hārūn al-Rashīd and Al-Ma'mūn (eighth and ninth centuries) large numbers of academies and observatories were set up and the chief Greek works on philosophy, astronomy, mathematics, medicine, and other sciences were translated into Arabic—principally by Syriac-speaking Christians. Islam was, however, soon producing scholars of her own.

Alchemy rapidly became a popular study with the Muslims; it received influential patronage, and Arabic treatises on the subject began to appear from about 700. Information concerning it was derived mainly from the Graeco-Egyptian alchemists of Alexandria on the one hand, and on the other from the

Sabians of Harran (*Carrhae*) in Syria [6]. Harran, long since destroyed, was a cosmopolitan city at which, during the Achaemenid period (*c* 650–330 B.C.), there began a remarkable intermixture of Persian, Syrian, and Greek natural philosophy. This syncretic system remained a characteristic of Harran for several centuries, and the city's reputation for catholicity of learning was still high in early Islamic days. Alchemy was among the subjects cultivated in Harranian intellectual circles, though in what proportion of theory to practice we have little means of knowing. It is at least certain that Harran possessed many extremely able workers in metals, and that it traded extensively in the precious metals, sulphur, borax, realgar, and other substances commonly used in alchemy.

The Alexandrian and Harranian alchemical tenets had much in common, the former indeed largely deriving from the latter; Zosimos, for instance, quotes a Harranian author. From both sources, therefore, the Muslims would have acquired the same ideas about the possibility of metallic transmutation.

The first Muslim alchemist of note was Jābir ibn Hayyān, who probably lived in the eighth century and appears to have been court alchemist to the Caliph Hārūn al-Rashīd. His name was westernized to Geber, and in that form was venerated throughout the course of European alchemy. The very numerous Arabic treatises on alchemy that have been ascribed to Jābir are probably in great part later recensions and expansions of his original texts. They show their author or authors to have had a wide knowledge of chemical and metallurgical operations, and one of them, 'The Book of Explanation' (*Kitāb al-Īdāh*), is particularly interesting as containing the first clear statement of the sulphur–mercury theory of metallic constitution:

> Metals are all, in essence, composed of mercury combined and solidified with sulphur. . . . They differ from one another only because of the difference of their accidental qualities, and this difference is due to the difference of their varieties of sulphur, which again is caused by a variation in the soils and in their situations with respect to the heat of the Sun.

This theory in one form or another permeated alchemy for several centuries and helps to explain many of the processes that were undertaken in the hope of bringing about transmutation. In the course of time it developed into the phlogiston theory of combustion, the overthrow of which by Lavoisier (1743–94) marked the dawn of modern chemistry.

From Jābir's books it would be possible to draw up a list of the alchemical apparatus employed by Muslim workers, but the task is unnecessary inasmuch as a full catalogue of such equipment is given by a slightly later writer, Rhazes or Rāzī. Abū Bakr Muhammad ibn Zakariyya al-Rāzī ('the man of Ray') was a famous Persian physician who lived from 866 to 925. He practised and taught medicine at his native town of Ray, the ancient *Rhagae*, and at the great hospital at Baghdad; the 'greatest clinician of Islam and the Middle Ages', he combined 'immense learning' with 'true Hippocratic wisdom'. His interests included not only medicine but alchemy, both of which he did much to rationalize. According to Al-Bīrūnī (973–1048), Rāzī wrote twenty-one books on alchemy; they have not all survived, but one of them, 'The Book of the Secret of Secrets' (*Kitāb Sirr al-Asrār* [7], gives a list of the chemicals and apparatus required in an alchemical laboratory. From this and other contemporary sources it is plain that Rāzī introduced more system and order into alchemy than it had previously possessed, even though its operations and materials were not greatly in advance of those of earlier times.

An example of Rāzī's orderly procedure is his classification of material substances into animal, vegetable, and mineral, and his further sub-classification of minerals into spirits, bodies, stones, vitriols, boraces, and salts. There were four spirits, two volatile and incombustible (mercury and sal ammoniac) and two volatile and combustible (sulphur and 'arsenic'—that is, realgar and orpiment). The bodies were the metals then known, while the stones comprised such substances as pyrites, malachite, lapis lazuli, gypsum, haematite, turquoise, galena, and stibnite, as well as glass. The five vitriols are not well defined, but included ferrous sulphate and probably alum. Natron (sodium sesquicarbonate) is classed among the boraces, and the salts include common salt, rock-salt, slaked lime, and impure sodium and potassium carbonates. Faulty though the classification was in many ways, largely owing to the lack of satisfactory methods of purification and analysis, it was based on the sound plan of grouping substances according to their properties and was therefore a long step in the right direction. Though it was gradually extended by the discovery and investigation of further substances, it was not essentially improved upon until the late eighteenth century.

In addition to the natural materials just mentioned, many derivative substances were employed. They included such compounds as cinnabar, white lead, red lead, litharge, ferric oxide, verdigris, copper oxide, and wine vinegar; according to Stapleton [8], Rāzī may also have known caustic soda and impure glycerine. A Persian writer about a century later than Rāzī, Abū Manṣūr Muwaffak, distinguished between sodium carbonate and potassium carbonate, was able to prepare a pure white specimen of arsenious oxide, and was familiar with silicic acid (*tabashīr*) obtained from the bamboo. He also recommended heated gypsum mixed with white of egg as a plaster of great service in the treatment of fractures of bones.

From the list of apparatus it may be concluded that Rāzī's laboratory was very well furnished. The items include beakers, flasks, phials, basins, glass crystallizing-dishes, jugs, casseroles, pots with lids, and porous cooking-jars. As sources of heat there were candle-lamps, naphtha-lamps, braziers, athanors (p 743), smelting-furnaces, and wind-furnaces; bellows of leather were employed when additional draught was required. Among the tools were files, spatulas, hammers, ladles, shears, tongs or forceps, and a roller and flat stone for crushing solids. Moulds and crucibles, including the double crucible or descensory[1] (*būṭ bar būṭ*) with the bottom of the upper half perforated, were used for fusing and purifying metals, and the tidy mind of Rāzī does not forget to include dusters.

Just as alchemy was transmitted to Islam from Alexandria and Harran, so Islam in turn handed it on to Latin Europe. Translation of Arabic alchemical treatises into Latin began early in the twelfth century, the initial step being taken by the English scholar Robert of Chester. Some idea of the extent of the borrowing may be gained from the following list of Arabic or Persian words to be found (usually in badly transliterated form) in European alchemical literature: abric (sulphur), alkali, alcazdir (tin), alcohol, almizadir (sal ammoniac), anticar (borax), asabon (soap), ased (lion), athanor (furnace), aludel, camphor, elixir, daeb (gold), hadid (iron), hager (stone), kamar (moon, silver), luban (resin), malek (salt), martak (litharge), matrass, naphtha, nard, noas (copper), nora (lime), realgar, tain (clay), talc, tartar, tutty, usifur (cinnabar), zaibuch (mercury), ziniar (verdigris).

There was soon a widespread circulation of alchemical manuscripts, and in 1350 a monk of Bologna possessed as many as seventy-two works on the subject. The invention of printing a century or so later increased the diffusion, though so far as is known the only fully alchemical incunabula were two works ascribed to Geber (p 736), one containing the *Summa perfectionis*, *Liber trium verborum*, and *Investigatio magisterii* (? 1485), and the other a *Flos naturarum* (1473) [9, 10]. From about the middle of the sixteenth century publication of alchemical works became prolific, and continued so until alchemy began to fall into disrepute from the date of the publication of Robert Boyle's 'Sceptical Chymist' (1661) [11].

The most celebrated treatises of this period of alchemy were those ascribed to Geber, identified with Jābir ibn Hayyān. It is true that two minor Latin works of Geber are translations of extant Arabic works of Jābir, but the major works, such as the 'Sum of Perfection', correspond to no known Arabic originals. They must therefore be considered as of doubtful authenticity, and it is possible that they

[1] The term descensory was also applied to stills heated from above instead of from below.

were written by an unknown European alchemist of the thirteenth century, when they first appeared in manuscript. Their content, as far as reference to materials and apparatus is concerned, is typical of western alchemy in general; as in Muslim and even earlier alchemy the principal emphasis is on stills and furnaces.

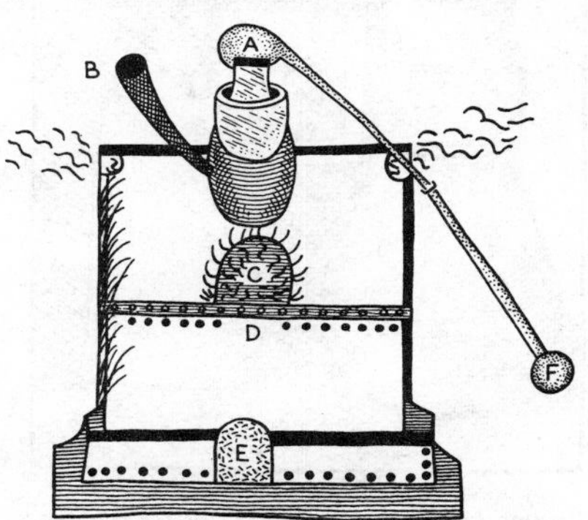

FIGURE 663—*Distillation apparatus in a furnace:* (A) *alembic;* (B) *spout to fill water-bath;* (C) *furnace;* (D) *grate;* (E) *aperture for removal of ashes;* (F) *receiver.*

Though considerable skill and ingenuity were expended on the design of elaborate stills, little improvement on the earliest models was made throughout the whole alchemical period. Figure 663 shows some of the types commonly used in the Middle Ages and later; it will be seen that there is practically no significant variation from the Alexandrian stills. Yet while the apparatus showed small advance there was certainly progress in manipulation and in realization of the results that distillation was capable of producing. Thus in the 'Sum of Perfection', translated into English by Richard Russell in 1678, Geber writes:

One method of distillation is performed in an earthen pan full of ashes; but the other with water in its vessel, with hay or wool, orderly so disposed, that the cucurbit, or distillatory alembeck, may not be broken before the work be brought to perfection. That which is made by ashes, is performed with a greater, stronger, and more acute fire; but what is made by water, with a mild and equal fire. For water admits not the acuity of ignition, as ashes doth. Therefore, by that distillation, which is made in ashes, colours, and the more gross parts of the earth, are wont to be elevated; but by that which is made in water, the parts more subtile, and without colour, and more approaching the nature of simple wateriness, are usually elevated. Therefore more subtile separation is made by distillation in water, than by distilling in ashes. This he knows to be true, who when he had distilled oyl by ashes, received his oyl scarcely altered into the recipient; but willing to separate the parts thereof, was by necessity forced to distill it by water. And then by reiterating that labour, he separated the oyl into its elemental parts [12].

FIGURE 664—*Athanor and apparatus for sublimation.*

FIGURE 665—ı *urther apparatus for sublimation. The receiver on the right fits over the top of the vessel on the left.*

It is clear from this passage that, even some 700 years ago, the idea of fractional distillation was already familiar. That it was not more fully exploited may be explained, at least in part, by the absence of a thermometric scale and the lack of close control of methods of heating. References to temperature in alchemical literature are inevitably vague; it was of course realized that a water-bath could not be made as hot as an ash-bath, but apart from such obvious distinctions the only 'fixed' points were such indefinite criteria as red-hot, white-hot, or as hot as a manure-heap or a broody hen. As late as 1622 the German alchemist J. D. Mylius recognized only four degrees of heat: that of the human body, that of June sunshine, that of a calcining fire, and that of fusion.

Sublimation was widely employed as a method of purifying such substances as sulphur and white arsenic, and emphasis was laid upon preliminary tests to discover whether it should be carried out with strong, moderate, or gentle heat. Figure 664 shows one type of sublimatory. The furnace or athanor (p 743) has an iron bar running transversely through it, about five or six inches from the bottom, and the glass vessel containing the substance to be sublimed is supported so that it nearly but not quite touches the bar. The perforated disk is arranged over the neck of the flask and helps to keep it in position, the holes in the disk acting as exits for hot gases from the furnace. The conical receiver (another shape is shown in figure 665) fits over the mouth of the flask and serves to collect the sublimate. When it was necessary to ascertain if the sublimation was complete, an earthenware rod about the diameter of the little finger, and hollow for the lower half of its length, was held above

FIGURE 666—*A worker testing purity of roll-sulphur by holding it to his ear and listening for the crackling caused by the warmth of the hand. Unless the sulphur is very pure the crackling does not occur.*

the substance under experiment; if any solid formed in the hole the sublimation was regarded as unfinished.

FIGURE 667—*Jean Béguin's apparatus for sublimation. Note the conical receiver of folded paper.*

Sulphur refined by sublimation was sometimes then fused and cast; its purity was ascertained by holding a lump of the solid to the ear (figure 666) to discover whether it crackled or not. Even very small quantities of impurity prevent the crackling. It is a matter of interest that, during the alchemical period, both benzoic acid and succinic acid were prepared by sublimation, from gum benzoin and amber respectively; benzoic acid was first described by Michel de Nostradame (Nostradamus) in 1556, and succinic acid by Agricola in 1546. An apparatus used by Jean Béguin early in the seventeenth century for the sublimation of benzoic acid from gum benzoin is illustrated in figure 667. Here the receiver consists of grey paper folded double and twisted into the shape of a cone or '*manche d'Hippocrate*'.

The variety of stills was matched or even exceeded by the variety of furnaces. This multiplicity was necessitated not only by a diversity of fuels such as charcoal, wood, peat, and dried dung, each of which required its particular form of grate, flue, damper, and so on, but by the difficulty of so regulating a furnace as to obtain different degrees of heat. Moreover the expense of fuel made it desirable to have furnaces of various sizes, so that the heating of small apparatus should not involve waste. Thomas Norton, an alchemist of Bristol (*c* 1477), claimed indeed to have invented a general-purposes alchemical furnace in which no fewer than sixty different operations, each requiring a different degree of fire, could be carried out at the same time, but he gives no details of its construction, and a drawing of his laboratory (figure 668) shows it to be as cumbered with a superfluity of furnaces as usual.

FIGURE 668—*Part of Thomas Norton's laboratory showing the processes of distillation and sublimation (see also figure 677).*

A French alchemist, Denis Zachaire or Zacaire, some fifty years later, started

FIGURE 669—*Apparatus for calcination.*

FIGURE 670—*Apparatus for fusion: note the clay crucibles.*

FIGURE 671—*Apparatus for crystallization on a water-bath.*

equipping his laboratory by building small furnaces under the guidance of a more experienced adept. From small he went to large, and soon had a room entirely furnished with them, some for distilling, others for subliming, others for calcining, others for dissolving in the water-bath, others for melting. It is scarcely surprising that his master 'died of a continued fever which seized him during the summer by reason of the soot that he breathed and swallowed . . . he hardly ever left the room, where he made scarcely less soot than there is in the Arsenal of Venice' [13].

Geber wrote a small book on furnaces [14], 'that artificers may the better attain to the compleatment of the Work'. In it he describes seven different types: the calcinatory, the sublimatory, the distillatory, the descensory, the fusory, the solutory, and the fixatory. The calcinatory furnace (figure 669) should be square, in length four feet and in breadth three, with walls six inches thick. The substances to be calcined were to be introduced into the furnace on pans made of strong clay such as that used for making crucibles, and then submitted to great heat or 'asperity of fire'. 'Be you not weary of calcination', says Geber lyrically; 'calcination is the treasure of a thing'.

The sublimatory furnace is identical with the distillatory in construction; they differ only in the degree of heat employed. Too great heat in the sublimatory may melt the sublimate in the aludel or receiver and cause it to run back, while in the distillatory the 'fire must be administered according to the exigency of things to be distilled'. The fusory furnace is for melting metals, which are placed in earthenware crucibles (figure 670), while the dissolving or solutory furnace (figure 671) consists of a pan of water supported over the fire and provided with iron clamps or other devices to hold large glass conical round-bottomed flasks.

FIGURE 672—*Stills on furnace, with feed-tube for fuel at back.*

The fixatory furnace or athanor (Arabic *al-tannūr*) 'must be made after the manner of the furnace of calcination; and in it must be set a deep pan full of sifted ashes. But the vessel, with the matter to be fixed, being first firmly sealed, must be placed in the midst of the ashes, so that the thickness of the ashes underneath, and above in the circuit of the vessel, may be answerable to the thickness of four fingers'; though this thickness may be varied according to the substance under treatment. Geber modestly adds: 'Yet if any one can more ingeniously invent the like, let not our invention retard him from so doing.'

The athanor shown in figure 669 is built of solid brickwork and apparently has no grate; another type, shown in figure 664, has a grate and perforated sides, and stands some inches from the ground. Sometimes the solid stand was replaced by a tripod, and the arrangement then resembled a brazier.

Self-feeding furnaces were known even from the time of the Greek alchemists. The fuel was stored in a container attached to the furnace and sank to the hearth by gravity; the container itself had a removable lid (figure 672) to prevent combustion from taking place in it [15]. Figure 673 shows a water-jacketed still mounted on such a furnace. The cylindrical furnace in figure 674 is provided with a damper or register, a device that first appears about 1500 and may be due to Thomas Norton. The invention of a chimney to increase the draught through a furnace was described by Glauber (p 745) in 1646 (figure 675); chimneys had been used for centuries for the removal of smoke and fumes, but Glauber was

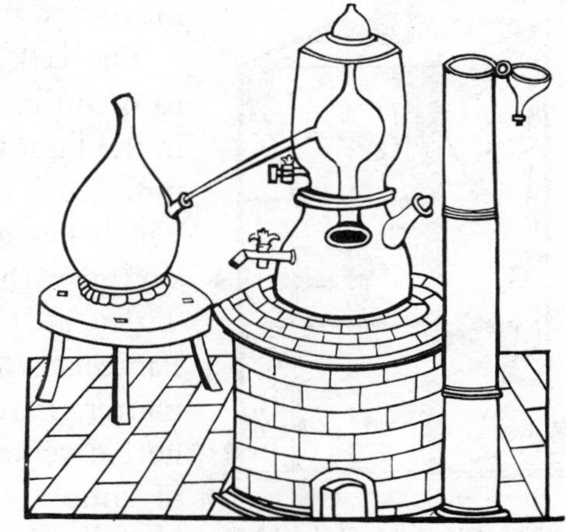

FIGURE 673—*Still of about 1500 heated in water-bath on furnace with refuelling tube at the side.*

the first to realize and apply their function in accelerating the flow of air. That Leonardo da Vinci came near to anticipating him may be deduced from the alchemical furnace shown in figure 676.

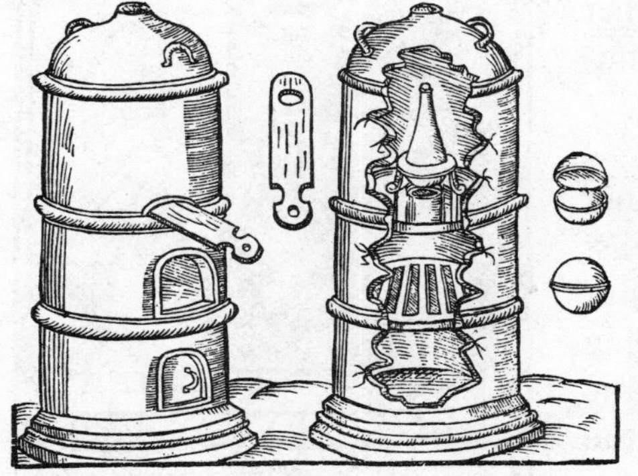

FIGURE 674—*Alchemical furnace: note damper.*

Anyone who studies the content of an alchemical laboratory cannot fail to notice the absence or unimportance of the balance. It is true that balances were used, but only for weighing out substances as a cook uses the kitchen scales; quantitative chemistry was still a science of the future, and had to wait for its inauguration until the days of Joseph Black (1728–99). That does not mean that accurate balances were unknown: as early as 780 the coiners at a Muslim mint could weigh to within a third of a milligram. 'To reach such accuracy it was needful to use the finest chemical balance, with closed case, double weigh the glass weights against each other, and read a long series of swings of the balance' [16]. To the alchemist, changes of weight had little significance and the balance was therefore in the background. There are a few exceptions, and a drawing in Thomas Norton's 'Ordinall of Alchimy' [17] (*c* 1477) shows what is probably the earliest representation of a balance in a glazed case (figure 677). Sixteenth-century assay balances are illustrated in figure 678.

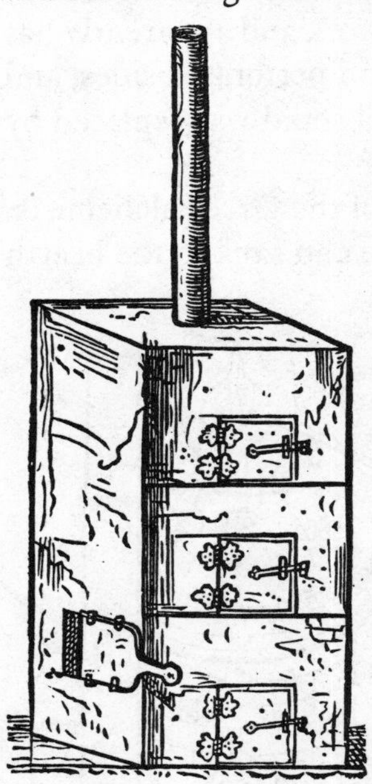

FIGURE 675—*One of Glauber's furnaces with a tall chimney to promote draught.*

The lack of importance attached by the alchemists to quantitative observations becomes less surprising in the light of the fact that the idea of a pure chemical individual had not yet been formulated. That idea, which lies at the root of scientific chemistry, was of slow growth and cannot be traced to any single origin. Titley [18] has suggested that the medical views of Paracelsus (1493–1534) have some bearing on the matter. Paracelsus rejected the complex mixtures of herbal remedies favoured by the Galenists in favour of simple extracts from plants or minerals. He and his followers believed that the freer such extracts were from contaminating substances the more efficacious they would be, and this led to the development of tests and standards of purity. The way was thus prepared for the conception of the individuality of chemical bodies, which slowly but conclusively took shape through the work of Boyle, Van Helmont, Black, Lavoisier, and Dalton.

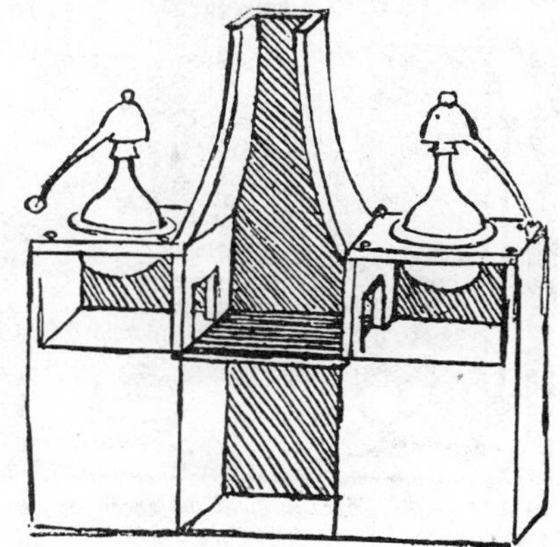

FIGURE 676—*Furnace and stills designed by Leonardo da Vinci.*

Paracelsus and his followers, the iatrochemists, were more interested in the therapeutic side of alchemy than in metallic transmutation. The materials and apparatus they used were of the established types, but two iatrochemists, J. B. Van Helmont (1577–1644), and J. R. Glauber (1604–68) deserve mention for the advances they made in general chemical knowledge. Van Helmont was the first to realize that there are many different kinds of gas, and was indeed the inventor of the word 'gas', which he derived from the Greek chaos. Thus he recognized—though he did not isolate—more or less impure forms of the gases we now call carbon dioxide, carbon monoxide, hydrogen, methane, sulphur dioxide, nitric oxide, and chlorine. Furthermore, Van Helmont had notions about ferments which in many respects are similar to our own, and believed that the digestion of food was a fermentative, or as we should say enzymatic, process.

FIGURE 677—*Part of Thomas Norton's laboratory. Note the balance: this is probably the first representation of a balance in a glass case* (*see also figure 668*).

Glauber, though a metallurgist and assayer, was not to be led astray by the will-o'-the-wisp of the philosophers' stone. For the most part his work was concerned with the preparation of substances of medicinal value, and in the course of it he made many discoveries of chemical importance. He was the first to prepare hydrochloric acid in a state of approximate purity by distilling a mixture of alum or green vitriol with salt, and later by heating salt with oil of vitriol. The sodium sulphate left in the latter operation, when crystallized from

FIGURE 678—*Metallurgical balances as used by Agricola.*

water as the decahydrate, he called *sal mirabile* and ascribed wonderful remedial powers to it; under the name of Glauber's salt it is still used in medicine. Glauber also distilled wood in closed retorts and showed that, in addition to tar, an acid and a spirit were produced. On the theoretical side, Glauber had inklings of the conception of chemical affinity, and correctly explained some cases of double decomposition. His use of the chimney to improve furnace-draught has been mentioned earlier (pp 743 f).

The cost of carrying out alchemical experiments can never have been inconsiderable and must often have been very heavy. In the 'Canterbury Tales' the Canon's Yeoman complains:

That slyding science hath me maad so bare,
That I have no good, wher that ever I fare;
And yet I am endetted so ther-by
Of gold that I have borwed, trewely,
That whyl I live, I shal it quyte never,
Let every man be war by me for ever!
What maner man that casteth him ther-to,
If he continue, I holde his thrift y-do,
So helpe me god, ther-by shal he nat winne,
But empte his purs, and make his wittes thinne.

A century and a half later an alchemist named John Damian enjoyed the patronage of King James IV of Scotland, and the Scottish Lord High Treasurer's accounts for the period 1501–8 give the prices of certain alchemical materials and apparatus used during that time [19]. They are as follows:

	£	s	d
Alembic, silver	6	19	9
Alum, per lb			7
Aqua vitae, ordinary, per quart		8	0
————, thrice-drawn, per quart		12	0
Bellows, small		1	0
Cauldron, 18 gal	1	1	0
Cinnabar, per lb		16	0

	£	s	d
Flasks, large glass		4	0
	and	6	8
—, small glass		2	0
Glass, cakes of		5	0
Gold, per oz	6	10	0
Linseed oil, per quart		8	0
Litharge, per lb		5	0
Massicot, per lb		4	0
Mortar (of unspecified metal), weighing 53 lb	3	9	0
Mortar and pestle, brass	1	0	0
Orpiment, per lb		6	0
Pitchers, earthenware			4½
	and	1	3
Pot, large earthenware		1	0
Quicksilver, per lb		4	0
Red lead, per lb		2	6
Sal ammoniac, per lb	1	15	0
Saltpetre, per oz			3
Silver, per oz		14	0
Sugar, per lb		1	6
Sulphur, per stone		8	0
Tin, per lb		1	2
Verdigris, per lb		6	0
Vermilion, per lb		6	0
Vinegar, per gal		4	0
White lead, per lb		2	0

These prices are difficult to compare with those now prevailing, but Read [19], writing in 1938, suggests that they might perhaps be multiplied by three or four to convey an idea of their modern equivalents. Expenditure is thus likely to have been high—though this did not deter the king from providing the alchemist with a damask gown at £15 16*s*, velvet short hose at £4, and scarlet hose at 30*s*. When wealthy patrons were ready to spend so lavishly, unscrupulous adventurers were not slow to impose upon them, and Thomas Norton (pp 741, 744) finds it necessary to give this warning against charlatans:

The trew men search and seeke all alone
In hope to find our delectable stone,
And for that thei would that no Man shulde have losse,
They prove and seeke all at their owne Coste;
So their owne Purses they will not spare,
They make their Coffers thereby full bare,

whereas the 'fals man' is

> *Ever searching with diligent awaite*
> *To winn his praye with some fals deceit* [20].

From about 1600 more care and thought were bestowed upon the design and arrangement of chemical laboratories—as we may judge, for example, from the works of Libavius (1540–1616). Though not disputing the central theory of alchemy, Libavius was essentially a practical chemist and several discoveries stand to his credit; thus he was the first to prepare stannic chloride and ammonium sulphate, he observed the deep blue colour imparted to solutions of copper salts by ammonia, and he invented various wet and dry methods of analysis. In the 1606 edition of his *Alchymia* [21], a work first published in 1595, he gives an elevation (figure 679 (left)) and ground-floor plan (right) of his ideal 'chemical house'. This contained a main laboratory, doubtless housing the array of furnaces that played so obsessing a part in alchemical operations, a store-room, a preparation-room, a room for sand-baths and water-baths, a room for crystallizing, a room for the laboratory assistants, a fuel-store, and a wine-cellar. The last amenity was probably intended not so much for the convivial refreshment of the staff as for storing the wine from which alcohol was distilled (tailpiece).

In the principal laboratory, apparatus was arranged round the walls; besides the furnaces there were descensories, sublimatories, stills, crucibles, mortars and pestles, phials, flasks, and basins. There was no balance-room. Altogether,

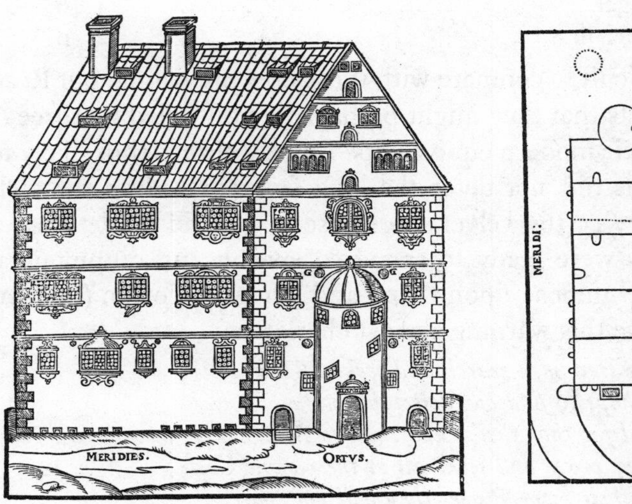

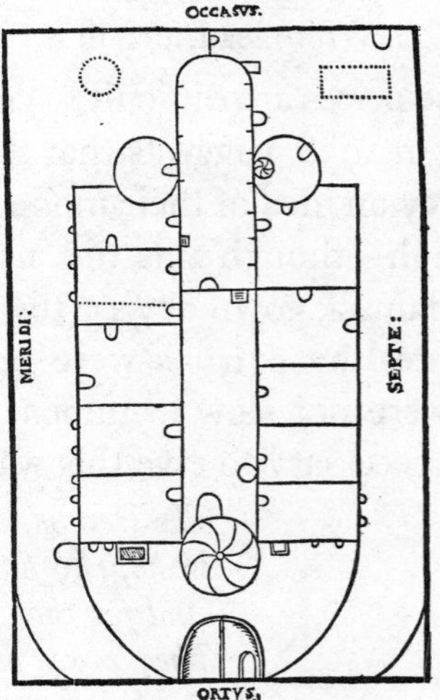

FIGURE 679—*Libavius's 'chemical house': (left) elevation; (right) plan.*

Libavius's chemical house, in its workmanlike design and orderly plan, contrasts very strongly with the usual alchemical laboratory of the time, of which a good impression can be obtained from the paintings of Pieter Brueghel (*c* 1525–69),

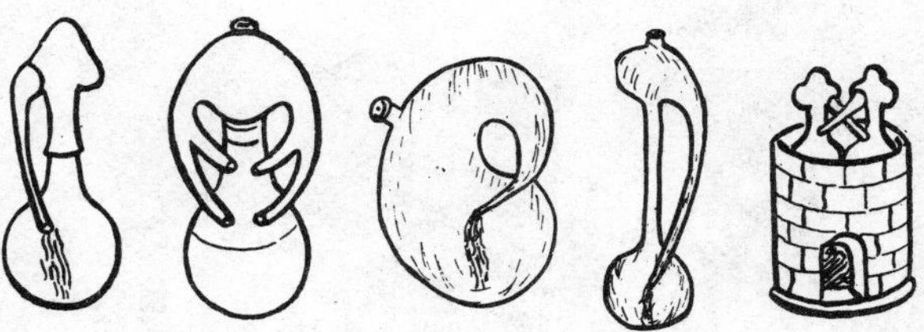

FIGURE 680—*Alchemical stills known as pelicans.*

Johannes Stradanus (*c* 1530–1605) (plate 42 A, B), J. Pinas (*c* 1600), David Teniers the younger (1610–90) (plate 43 A), and Jan Steen (1626–79). One would like to attribute Libavius's desire that everything should be ship-shape and Bristol fashion to his academic experience: he was for a time head master of Coburg Grammar School and professor of history in the University of Jena.

That the cost of fitting out a laboratory had not risen appreciably during the sixteenth century may be gathered from a manuscript preserved at Alnwick Castle and quoted by J. W. Shirley [22]. It refers to the preparation of a 'still-house' in the Tower of London in 1606–7 for the 'Wizard Earl', Henry Percy, ninth Earl of Northumberland, who was imprisoned there with Sir Walter Ralegh, Lord Cobham, and Lord Gray. Bricks, tiles, and other necessaries for making a furnace cost 12*s* 6*d*, and the carpenter was paid 22*s* 10*d* for timber for the bench, shelves, and stairs. Two lead cisterns cost 15*s* 10*d*, two stills 8*s* 6*d*, a copper vessel 7*s* 7*d*, bellows 10*d*, a steel chisel 4*d*, a pair of long compasses 4*d*, various pieces of glassware 31*s* 6*d*, and 'diverse other necessaries as Colebasketts handbasketts Syves payles potts pannes packe threed wyer Streyners treys pastbord wax paper matts Candlesticks postage and such like' 22*s* 6*d*. Further expenses in the following year included 11*s* 8*d* for a new tin still, 5*s* 11*d* for a goldsmith's balance with case, 4*s* 4*d* for weights to go with the balance, 3*s* 8*d* for phials, 2*s* 1*d* for repairs to the walls and glazed windows, and 3*s* 10*d* for an iron pestle and mortar.

The progress towards more rational design, and indeed the gradual change from alchemical to chemical laboratories, is further exemplified in the laboratory set up by Robert Boyle (1627–91) at Maiden Lane, near Covent Garden, some time after 1668 [23]. One side of it is shown in figure 681, though the drawing was made after Boyle's death and at a time when the original equipment had

FIGURE 681—*Boyle's laboratory.*

probably been augmented. The laboratory was taken over by Ambrose Godfrey, son of Boyle's laboratory assistant, and a chemical manufacturing business was carried on there for many years. Pride of place is still occupied by furnaces (bottom left, and 21–32), each destined for a particular kind of operation. One (22) was for distilling hartshorn; another (27) for making spirit of sal ammoniac; a third (31) for extracting volatile salts, 'as from vipers'; and a fourth (24) for distilling in the *bain-marie*. An apparatus for subliming is shown in (5). It consists of a small furnace from the top of which protrudes the neck of the vessel containing the substance to be sublimed; above are three receivers or aludels, the whole being surmounted by a spherical head with a conical spout. The vessel (4) with its arms akimbo is a 'pelican'; it is a reflux or circulatory apparatus and the arms are delivery-tubes. Earlier forms of pelican are shown in figure 680. In one shape or another the device is found throughout the history of alchemy.

If Boyle's laboratory is compared with that of Priestley (plate 43 B) a century later an entire change of emphasis is immediately noticeable. The chimera of transmutation has vanished; a domestic grate and a candle are ample substitutes for the regiment of furnaces; the apparatus is simple but appropriately functional; and gases are the chief object of study. Yet the Arabian bird was not altogether dead: we have witnessed it rise from its ashes as the atomic alchemy of today.

REFERENCES

[1] STAPLETON, H. E. *Ambix*, **5**, 1, 1953.
[2] FORBES, R. J. *Chymia*, **4**, 1, 1953.
[3] TAYLOR, F. SHERWOOD. 'The Alchemists: Founders of Modern Chemistry', p. 217. Schuman, New York. 1949. Reprinted: Heinemann, London. 1951.
[4] FORBES, R. J. See ref. [2], p. 4.
[5] TAYLOR, F. SHERWOOD. See ref. [3], p. 46.
[6] MARGOLIOUTH, D. S. "Harranians" in HASTINGS, J. (Ed.) 'Encyclopaedia of Religion and Ethics', Vol. 6, pp. 519–20. Clark, Edinburgh. 1913.
[7] RĀZĪ. 'Kitāb Sirr al-Asrār.' (German trans. with introd. and comm. by J. RUSKA.) Quell. Gesch. Naturw., Vol. 6. Springer, Berlin. 1937.
[8] STAPLETON, H. E. *Mem. Asiat. Soc. Beng.*, **8**, no. 6, 391–3, 1927.
[9] HIRSCH, R. *Chymia*, **3**, 115, 1950.
[10] THORNDIKE, LYNN. *Ambix*, **2**, 29, 1938.
[11] BOYLE, ROBERT. 'The Sceptical Chymist.' J. Cadwell for J. Crooke, London. 1661.
[12] GEBER. 'The Works of Geber Englished by Richard Russell, 1678' (reprinted with introd. by E. J. HOLMYARD), p. 98. Dent, London. 1928
[13] DAVIS, T. L. *Isis*, **8**, 287, 1926.
[14] GEBER. See ref. [12], pp. 227–61.
[15] GOLDSMITH, J. N. and HULME, E. W. *Trans. Newcomen Soc.*, **23**, 3, 1942 (1948).
[16] PETRIE, SIR (WILLIAM MATTHEW) FLINDERS. *Numism. Chron.*, fourth series, **18**, 115–16, 1918.
[17] NORTON, THOMAS. 'The Ordinall of Alchimy By Thomas Norton of Bristoll, Being a facsimile reproduction from *Theatrum Chemicum Britannicum*, with annotations by Elias Ashmole' (with an introd. by E. J. HOLMYARD). Arnold, London. 1928.
[18] TITLEY, A. F. *Ambix*, **1**, 182–83, 1938.
[19] READ, J. *Ibid.*, **1**, 60–65, 1938.
[20] NORTON, THOMAS. See ref. [17], chap. 1, p. 17.
[21] LIBAVIUS, ANDREAS. *Alchymia Andreae Libavii recognita, emendata, et aucta, tum dogmatibus et experimentis nonnullis* . . ., pp. 95, 97. Johannes Sauer for Peter Kopff, Frankfurt-am-Main. 1606.
[22] SHIRLEY, J. W. *Ambix*, **4**, 61, 1949.
[23] PILCHER, R. B. *Ibid.*, **2**, 17, 1938.

BIBLIOGRAPHY

ASHMOLE, ELIAS. *Theatrum Chemicum Britannicum.* Printed by J. Grismond for Nathaniel Brooke, London. 1652.
BERTHELOT, P. E. M. 'Les origines de l'alchimie.' Steinheil, Paris. 1885.
Idem. 'Collection des anciens alchimistes grecs.' Steinheil, Paris. 1887–8.
Idem. 'La chimie au Moyen-Âge' (3 vols). Imprimerie Nationale, Paris. 1893.
CARBONELLI, G. 'Sulle fonti storiche della chimia e dell'alchimia in Italia.' Istituto Nazionale Medico Farmacologico, Rome. 1925.
FERGUSON, J. *Bibliotheca Chemica* (2 vols). Maclehose, Glasgow. 1906. Reprinted: Holland Press, London. 1954–5.
HOLMYARD, E. J. 'Makers of Chemistry.' Clarendon Press, Oxford. 1931.

JĀBIR IBN ḤAYYĀN. 'The Arabic Works' ed. by E. J. HOLMYARD, Vol. 1, Part I: Arabic texts. Geuthner, Paris. 1928.

KRAUS, P. 'Jābir ibn Ḥayyān: Contribution a l'histoire des idées scientifiques dans l'Islam', Vol. 1: 'Le corpus des écrits Jābiriens'; Vol. 2: 'Jābir et la science grecque.' Mémoires présentés à l'Institut d'Égypte, Vols. 44, 45. Imprimerie de l'Institut Français d'Archéologie Orientale, Cairo. 1943, 1942.

LIPPMANN, E. O. VON. 'Die Entstehung und Ausbreitung der Alchemie' (2 vols). Springer, Berlin. 1919, 1931.

PARTINGTON, J. R. 'A Short History of Chemistry.' Macmillan, London. 1954.

READ, J. 'Prelude to Chemistry. An Outline of Alchemy, its Literature and Relationships' (2nd ed.). Bell, London. 1939.

SARTON, G. 'Introduction to the History of Science' (3 vols). Carnegie Institution of Washington Publ. 376. Baltimore. 1927–48.

TAYLOR, F. SHERWOOD. 'The Alchemists: Founders of Modern Chemistry.' Schuman, New York. 1949. Reprinted: Heinemann, London. 1951.

TEMPKIN, O. "Medicine and Graeco-Arabic Alchemy." *Bull. Hist. Med.*, **29,** 134–55, 1955.

THORNDIKE, LYNN. 'History of Magic and Experimental Science during the first Sixteen Centuries of our Era' (6 vols). Vols 1–2: Macmillan, London; Vols 3–6: Columbia University Press, New York. 1923–41.

Woman laboratory assistant. From M. Puff van Schrick's book on brandy. Strasbourg, 1512.

EPILOGUE

EAST AND WEST IN RETROSPECT

CHARLES SINGER

I. TECHNOLOGY AND CHRONOLOGY

LIKE other histories, that of technology must adopt a chronological order. But even more than most histories, such as those of politics, of art, of science, of religion, of philosophy, that of technology is resistant to any but the very broadest time-division. In volume I we parted, though far from abruptly, with the ancient empires. In this volume we have crossed the equally ill defined chronological frontiers of the great Mediterranean civilization—that of Greece and Rome—and have passed to the Middle Ages and beyond. The reader will therefore not be surprised that the techniques treated in this volume were largely continuous both with those discussed in volume I and with those that will be discussed in volume III, but they can hardly be made to fit even this very loose chronological framework. Thus, for example, the mining-methods of ancient Rome were hardly improved till the eighteenth century (ch 1) while those of the leather-worker were inherited direct from the ancient empires and remained almost static till well into the nineteenth (ch 5). Again, the methods of obtaining silk threads came to the west from China in the early Christian centuries (ch 6), while the use of horses for draught with the collar (ch 15) was a medieval improvement coming from the steppe-lands and open frontiers of eastern Europe and perhaps ultimately from east Asia.

Many such chronological misfits are inevitable in any attempt at technological period-making. Technological devices may pass from land to land, often ignoring political frontiers, cultural differences, linguistic diversities, and even the great natural barriers of mountains, seas, rivers, and deserts. The reader must therefore guard himself from thinking of the vague periods assigned to volume I or volume II, or to the subsequent volumes, as being more than matters of convenience in the composition of the narrative. He must take this chronological scheme as no more than a general suggestion that he should keep his eye on a time-indicator which, for his convenience, is roughly graduated in centuries. To get technology into perspective in the panorama of general history, he must constantly refer to the chronological tables of general history attached to these volumes.

Even when all such allowances have been made, it must be further remembered that there are certain great technological movements, within the period under review, which cannot be brought out either in time-charts or even in the narrative that makes up this volume. Discussion of the development of certain such movements must necessarily be spread over two or more of these volumes. This matter needs some further treatment, because conventional history has distorted the picture that this work seeks to convey.

Ever since the fifteenth-century revival of learning the debt of the western world to Greece and Rome has been stressed in our whole educational system. Everyone knows that this view was originally based on a time-scale vastly different from that now adopted. Geological science was hardly born till the end of the eighteenth century, and anthropological science did not come into being till the nineteenth. Before then men necessarily thought of history as packed into a few thousands of years. Best known of these time-scales is that which placed the creation at 4004 B.C., a date made conventional by Bishop Ussher in 1650. This space of time takes us, as we now know, no farther back than the late pre-dynastic cultures of Egypt and Mesopotamia (vol I, p lv). But this foreshortened history laid an undue stress on the Graeco-Roman heritage of our own civilization and necessarily ignored our even deeper debt to the great civilizations of the Near East on which those of Greece and Rome were themselves built. No-one now defends Ussher's or any similar time-scale, but it is well to remember that certain of its indirect effects remain very firmly rooted in our cultural system. The spirit of this outdated tradition has delayed the development of a philosophical and integrated view of the course of our civilization in general and of our technology in particular.

The present volume does not attempt to rewrite the history of two thousand years, but with reference to technology the reader must be reminded of certain necessary major adjustments of conventional historical perspectives. For remains of the old time-scale still linger, well enough concealed, in elementary history books and, moreover, still form a large part of the educational background in the so-called humane studies.

II. THREE NECESSARY HISTORICAL ADJUSTMENTS

(i) Whatever view be taken of the beauty and interest of the art, literature, ethics, and thought of Greece and Rome, it can no longer be held that their technology was superior to that of the ancient empires. It is true that the Greeks and Romans, in their later classical period, had a readier access to iron and to certain other metals than their predecessors. They had gained greater mastery in the treatment of glass and of several other basic substances. But, in skill in using the means at their disposal, certain of their predecessors were at least their equals and not seldom their superiors. The curve of technological expertness tends to dip rather than to rise with the advent of the classical cultures. This will become apparent if the relevant chapters of volume I be compared with the corresponding chapters in the present volume. It is a phenomenon inherent in the history of Greece and Rome, for they rose to their might by the destruction of the more ancient civilizations that they displaced. To say the same thing in other terms, the rise of the Hellenic and Roman peoples represents a 'heroic age' which, like many heroic ages, was primarily a victory of barbarians over an effete but ancient civilization.

(ii) We are accustomed to think of post-Roman history as centred on Europe and especially on north-west Europe, now for centuries its technologically most developed part. Europe, however, is but a small peninsula extending from the great land mass of Afrasia. This is indeed its geographical status and this, until at least the thirteenth century A.D., was generally also its technological status. In skill and inventiveness, during most of the period treated in this volume, the Near East was superior to the west, and the Far East perhaps superior to both.

(iii) During most of the period discussed in this volume the cultures and civilizations of several peoples of the Far and Middle East were at their highest. Many technological ideas came thence to western Europe. Some of them came direct through the open barbarian frontier of eastern Europe. Most came either through the Byzantine Empire, which for a thousand years was the wealthiest and most civilized Christian state, or through Islam which, from the ninth century till the fourteenth, was technologically far superior to western Europe (figure 683).

Of these three statements (i) is illustrated by the contents of volume I of this work. On (ii) and (iii) it is necessary to enlarge, for on their proper appreciation depends the perspective in which volume II can alone be rightly understood.

III. TECHNOLOGICAL CONTRASTS OF EAST AND WEST

During the millennium roughly extending from 500 to 1500, the relations of east and west were very different from those with which we are today familiar. Thoughtful men who have been nurtured in any of the great and ancient civilizations of the east—Chinese, Indian, Japanese, Islamic, or other—whatever they may think of us and of our ways of life, however anxious they may be for freedom from western control, nevertheless accord to the western industrial system, and especially to the technology on which it is based, the sincerest form of flattery. Though they often forget the intimate relation of that technology to the social evolution and political history which are its roots, yet they are not far wrong in supposing that industrial techniques—especially those of mass-production—form the foundations of western superiority.

During most of the period under discussion here, the boot was on the other foot. For nearly all branches of technology the best products available to the west were those of the Near East, at first those from the Byzantine Empire and later also from the Islamic Caliphate or from Persia. However much the hold of Saracens on the Holy Places might be resented by the west, the booty won from them by the crusaders or bought from them in trade was thought fit to adorn the most revered shrines in western Christendom. Thus fine fabrics of Islamic origin—sometimes even bearing invocations to Allah and the Prophet—were used to enwrap the bones of Christian saints (figure 682). Byzantine mosaics, silks, and ivories, Persian ceramics and textiles, Egyptian and Syrian glass and metal-work, as well as many of the products of Mesopotamia and Moorish Spain, were highly prized as being manifestly superior to anything that could be made in western Europe. It was largely by imitation and, in the end, sometimes by improvement of the techniques and models that had come from or through the Near East, that the products of the west ultimately rose to excellence (figure 692).

Here is needed a further adjustment of the traditional historic view. The crusades were in name, and largely in intention and in fact, an attempt to wrest the Holy Places from Islam in the name of Christianity. They were also, especially in their later stages, largely in intention and in fact an attempt to rob and disintegrate the remains of the Roman Empire that survived, and to secure trade advantage or loot from its capital, East Rome or Byzantium, by then called Constantinople (p 760). The crusades, from this point of view, may be regarded as an extension of the barbarian incursions that had once broken the power of western Rome.

Technologically, the west had little to bring to the east. The technological movement was in the other direction. Not seldom, and specially under stress of persecution and war, there were emigrations of eastern craftsmen to the west. These taught their methods to European pupils and apprentices, and so added the technical traditions of their own lands to those already being practised in

FIGURE 682—(Above) *Pattern on pallium of St Cuthbert.* (Below) *Its interpretation in Kufic–Arabic lettering. St Cuthbert died at Jarrow in 687. His remains were reinterred at Durham in 1104 in a silk wrapping of Mesopotamian workmanship. Woven into its fabric was this pattern representing the Islamic confession of faith: 'There is no god but God'. Compare figure 689.*

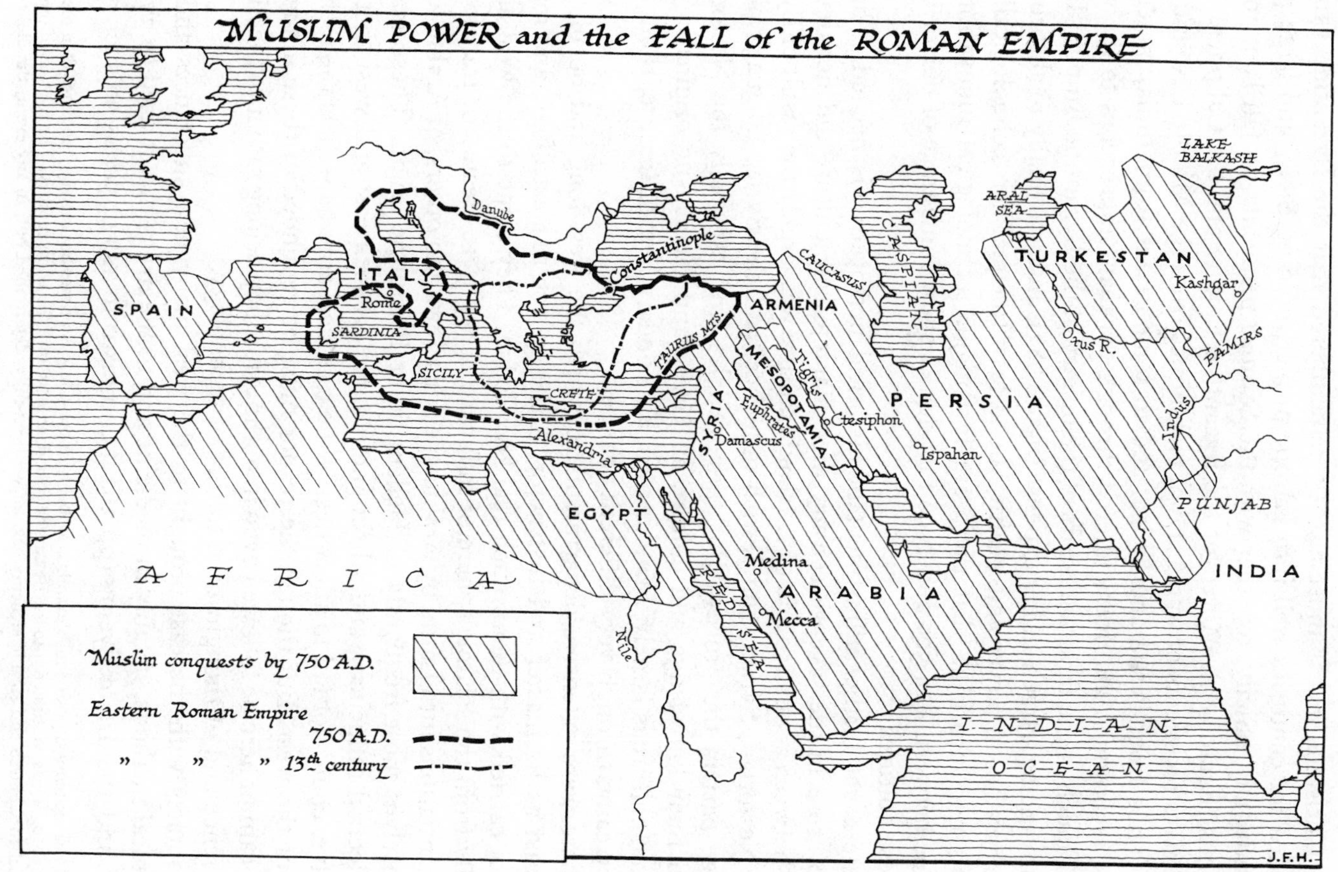
MUSLIM POWER and the FALL of the ROMAN EMPIRE
SPAIN
ITALY
Rome
SARDINIA
SICILY
Danube
Constantinople
CRETE
TAURUS MTS.
Alexandria
ARMENIA
CAUCASUS
CASPIAN
ARAL SEA
LAKE BALKASH
TURKESTAN
Kashgar
Oxus R.
PAMIRS
MESOPOTAMIA
Tigris
Euphrates
Ctesiphon
PERSIA
Ispahan
Indus
PUNJAB
INDIA
SYRIA
Damascus
EGYPT
Nile
Medina
Mecca
ARABIA
RED SEA
AFRICA
INDIAN OCEAN
Muslim conquests by 750 A.D.
Eastern Roman Empire
750 A.D.
" " " 13th century
J.F.H.

FIGURE 683.

Latin Christendom. It must also be remembered that the Byzantine Empire extended for centuries well into Europe proper, holding Sicily and large parts of south Italy—Sicily, together with the Spanish peninsula, to fall later to the Muslims (figure 683). Moreover the Byzantine Empire and the Caliphate were themselves recipients of wares and of techniques from the Middle East and east Asia. Thus from Persia and China, and to some extent from India, materials, wares, techniques, and ideas filtered through the main approaches to the west.

For much of the thousand years of the history of the Byzantine Empire, Islam was advancing, largely at her expense (figure 683), and was finally to absorb her, to surpass her in power and wealth, and at least to equal her technologically. It is thus not surprising that there are few major technological innovations in Europe between A.D. 500 and 1500 that do not show some trace of one or other of these cultures.

Here a yet further adjustment of historical perspective is needed. Though Islam arose in the seventh century, her characteristic culture did not develop until between about 850 and 950. Islamic culture was largely the result of Byzantine, Nestorian, Syrian, and Jewish influence, and the Syriac language played a large part in the process. But these peoples, and especially the Nestorians, based their knowledge and culture on Greek and Hellenistic originals. Thus Islamic culture is largely, like our own, a secondary heir of Greece, though the legacy came through very different channels.

There were extensive departments in which the west long and openly held the east as its master and instructor. Alchemy[1] (ch 21), astrology, and mathematics carried with them much real knowledge. They came to the west almost wholly from the Islamic sphere of influence. They were created or transformed by translations into Latin, by Arabic-speaking Jews and others, of Arabic works, themselves most frequently of Hellenistic or Byzantine origin. These massive intellectual influences, and all the intercourse between east and west that they involve, could not fail to affect deeply the technical aims as well as the technical arts of the west. But though the reader must be reminded of these influences, they cannot here be followed even in outline, for they belong rather to the history of science and of thought than to that of technology.

Contacts with the east brought to the west many new substances and new technical methods, notably in medicine and in the drugs that it used (p 371), in dyes and the art of dyeing (pp 364 ff), in instruments of precision such as the

[1] Greek alchemy appeared in Alexandria probably in pre-Christian times but certainly not later than the second century A.D. It was continued into the Byzantine period, at least until the tenth century. The relation of this Greek material to Chinese and Arabic alchemy has not yet been fully determined but, in any event, the alchemy that came to the west was in Arabic dress.

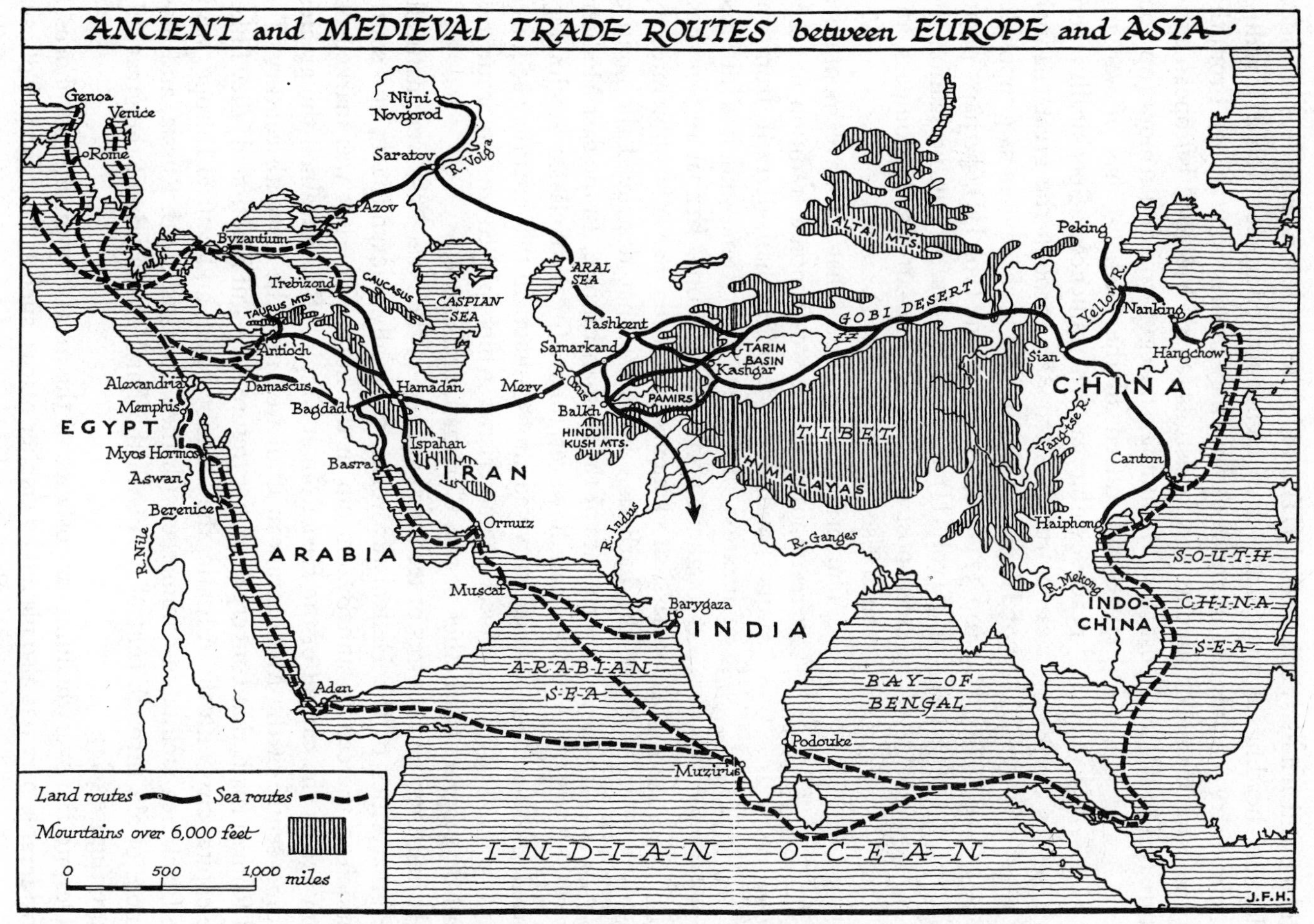

FIGURE 684—*Map showing the 'old silk-route' across Asia between China and Byzantium.*

astrolabe and armillary sphere (vol III), and in the whole range of ideas included under such headings as metallurgy (ch 2), alchemical equipment (ch 21), industrial chemistry (ch 10), and methods of building-construction (ch 12). From the east too came many condiments and foodstuffs (ch 4). Medieval European agriculture (pp 83 ff), transport (p 538), mining, strategy, arms and armour (pp 706, 715), and indeed all the medieval arts of war and peace exhibit eastern influences. Specifically there came from eastern Asia through the great channels of Islam and Byzantium such devices as gunpowder (pp 374–82), paper-making (vol III), perhaps printing (vol III) and canal-locks, and in navigation the stern-post rudder, fore-and-aft rig, the compass, and much else (table, pp 770–1).

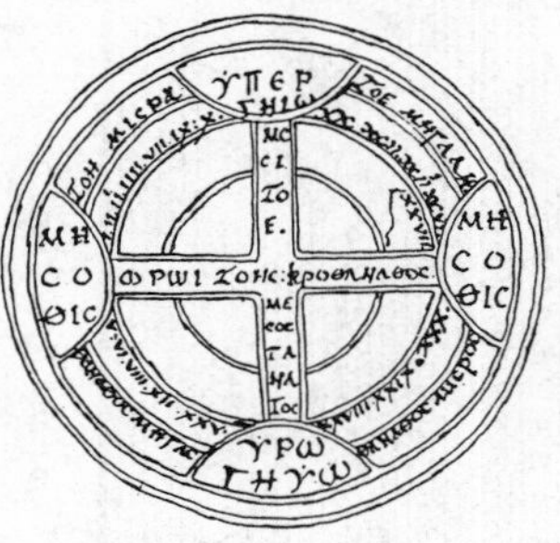

FIGURE 685—*'Circle of Pythagoras' written at Ramsey Abbey in Huntingdonshire* c *1110 by an Anglo-Saxon hand in wretched Greek. Thus between the two circles above and to the left is* ZOE MICRA *('little life') and to the right* ZOE MĒGALĒ *('big life'), while below these is* THANATHOS MĒGAS *('big death') and* THANATHOS MICROS *('little death') with the numbered days of the month. At the four points of the cross and along its limbs are words in even more corrupt Greek.*

The oriental influences that we have mentioned were imposed in the west upon a tradition of techniques and of styles that had derived from the civilization of Greece and Rome and had diffused throughout north-west Europe. In that region these classical elements had been partially obliterated by the barbarian invasions that destroyed the political structure of the Roman Empire. In the east Mediterranean, however, the traditions of classical technology had survived almost intact. In that area they acquired local characteristics which played an important role in all developments whether pagan, Christian, or Islamic.

At first the main centres of technical activity in the Near East were naturally the great cities of the Hellenistic world, especially Alexandria in the south, Antioch in the centre, and Byzantium in the north. The last was marked out from 330 as the capital of the East Roman Empire. She remained the greatest trading-centre of the Mediterranean until well into the eleventh century. The old Roman roads that converged on her (figure 459) still functioned; they linked the capital city with the Adriatic, the Danube basin, and central Europe, while the Black Sea facilitated commerce with the Caucasus, south Russia, and most remote of all; through central Asia with the Far East by way of the great silk-route (figure 684).

Despite the overwhelming importance of the great city on the Bosphorus, a considerable part was also played by smaller places, such as Damascus, Ephesus, Nicaea, Pergamum, Sinope, Smyrna, Trebizond, and Tyre. The controlling factor in development was the new faith of Christianity, which received its official recognition from the Emperor Constantine when he transferred the capital of the civilized world from Rome to Byzantium, and so took the first step in setting eastern Roman culture on its special course.

IV. PILGRIMS, CRUSADERS, AND TRADERS

With the astonishingly rapid expansion of Islam in the seventh century, much of the territory of the Byzantine Empire, including two of its most important cities, Alexandria and Antioch, passed from Christian control. The new Islamic rulers were only too eager to employ those Christian workers who were skilled in crafts which to their own rude followers were sources of admiration and wonder. Developments in the arts thus continued in the conquered territories. At first there was little to differentiate their products from those of neighbouring Christian lands. Thus the Umayyad caliphs (661–750) employed Byzantine craftsmen on their two greatest places of worship, to build the Dome of the Rock at Jerusalem and to make the mosaics of the Damascus mosque. Soon, however, as a result of contact with countries yet farther east, the craftsmen of the Caliphate became more and more affected by other and especially by Persian elements, as well as by ideas engendered by the faith of Islam itself. Thus arose distinctive arts and techniques. Islamic arts were to grow in importance, finally superseding those of eastern Christendom. The change, however, was very gradual, and Byzantine influence predominated in Islamic techniques for centuries.

FIGURE 686—*Part of the pattern and inscription of a Byzantine silk weave of the tenth century from the tomb of Charlemagne at Aix-la-Chapelle. The pattern is of stylized elephants, and the textile was deposited in the coffin probably by the Emperor Otto III when he opened it in the year 1000. Byzantine influences were strong at his Court through his mother, the Byzantine Princess Theophano. Greek words woven into the selvedge fringe give two names. One is Petros, here described in Byzantine letters as superintendent of the Zeuxippos, the district of Constantinople which included the Imperial silk-factory.*

In the tenth or eleventh century, trading and political contacts of the west with both the Christian and the Muslim east rapidly increased and, despite numerous interruptions, remained extensive. Such contacts brought to the west not only objects and materials, but a knowledge of many techniques, crafts, and decorative motifs, since the eastern craftsmen and artists were far more widely skilled than those of the west. The techniques and even the patterns and decorations of metal-working, glass-cutting, enamelling, pottery-making, and textile-weaving were copied for centuries in the west by craftsmen who perceived that there was nothing of such high quality of material and workmanship in their own lands.

FIGURE 687—*Design of tenth-century silk fabric used for relics of St Josse, legendary king of Brittany in the seventh century. In 1195 his remains were brought to the abbey dedicated to him in the Pas de Calais and there enwrapped in a shroud of Islamic workmanship bearing this pattern. The Arabic in Kufic letters may be translated 'Glory and rising fortune to the Commander Abu Mansūr Ḥaidar; may God lengthen his life'. Ḥaidar ruled Khurasan in Persia in the tenth century, which dates the weaving. In the absence of diacritical marks (which change the letter-values) the word rendered 'rising fortune' may mean 'elephants'! The ambiguity may be intended. The weaving was on a loom on which the elaborate pattern could be, and was, reversed, though the lettering is not.*

Apart from those made by eastern craftsmen migrating westward, the first contacts were those of pilgrimage. The devout visited the Holy Cities in large numbers. They often brought back souvenirs, or relics, or ornate bindings. Such things were consigned to western ecclesiastical treasuries, where they were either copied or served to suggest new ideas, methods, and techniques. The portraits of the evangelists in the famous Anglo-Saxon manuscript known as the Lindisfarne Gospels, for example, are modelled on east Christian originals, just as were the figures of the lowly herbals of Anglo-Saxon leeches,[1] whose magical formulae were sometimes spelt in broken Greek (figure 685). The sculptures of crosses in Northumbria, executed around 700, are said to follow east Christian prototypes. The ninth-century church of Germigny-des-Prés, near Orleans, follows an east Christian plan, and contains mosaics of Byzantine character (plate 44). Examples of this sort from before the year 1000 could be multiplied indefinitely.

In the seventh century, after the Holy Cities of Palestine had passed from Christian control, pilgrimage from the west began to centre on Constantinople, whither numerous sacred relics had been taken. Such contacts were supplemented by commercial intercourse with all parts of the Near East. There were, too, trading colonies of Greeks and Syrians in the west, notably at Marseilles and in Sicily.

In the sixth and early seventh centuries the imperial looms at Constantinople

[1] Both the Lindisfarne Gospels and the Anglo-Saxon herbals were derived from south Italian models, but south Italy was then part of the Byzantine realm and under full Byzantine influence.

exercised almost a monopoly for export of woven silk, though the raw material came from China by the silk-route (figure 684). With the introduction of sericulture at Constantinople under Justinian I (527–65), manufacture of the raw material as well as its weaving began there. But when the looms of Syria and Egypt, which had always been of outstanding importance, passed under Muslim control in the seventh century, their silken products also became available for

FIGURE 688—(Left) *Eagle-and-gazelle pattern of an Islamic silk made in Sicily or Spain,* c *1200.* (Right) *Copy of the pattern painted* c *1200 on the vault of the crypt in the cathedral of Clermont, Puy-de-Dôme. The painter followed his model as faithfully as he could, but he did not understand the Arabic inscriptions on the wings of the eagles, and he misinterpreted the heads of the gazelles.*

export, in competition with what had been a Constantinopolitan monopoly. Thus the fabrics used in the west as grave-wrappings for saintly or royal personages (figures 682, 687), or for the protection of relics (figure 686), or even as costumes at the courtly centres, were sometimes Byzantine and sometimes Islamic. In the tomb of St Cuthbert (d 687) at Durham, whither his body was transferred in 1104, for example, the pallium was of Islamic and probably Mesopotamian origin (figure 682). Some of the wrappings of saints now in Germany and France are Byzantine (figure 686), some Persian (figure 687), and some perhaps Syrian. Many examples of these oriental fabrics survive in the west, though many more must have perished. Their decorations often inspired local carvers as well as weavers. Such favourite Romanesque emblems as the eagle (figure 688), the winged dragon, or the gryphon, on the capitals of early medieval columns, were derived ultimately from eastern textiles, as were many floral or more formal designs. There are even cases of actual imitation by western craftsmen of ornamental Arabic texts in script (figures 688, 689).

Though textiles, as the most portable articles, formed the most usual items of trade or exchange, objects in metal, enamel, ivory, or even pottery and glass

FIGURE 689—*Arabic lettering, especially of the decorative Kufic, readily passed into ornament and was widely copied by western workmen ignorant of its meaning or origin.* (Left) *A twelfth-century wood-panel from the cathedral of Le Puy (Haute-Loire) shows how a Christian French carver, not limited by such technical restrictions as were weavers by their looms, used Arabic script decoratively. He has framed New Testament scenes with part of the Islamic confession of faith.* (Right) *The words shown in more detail at* A *are equivalent to '... but God', last part of the first phrase of the Muslim creed 'There is no god but God'. Compare* B, *from an oriental rug of* c *1300 from a church in the Tyrol, with its formularized first part of the phrase 'There is no god but ...', where the change to pure ornament goes yet further.*

were also brought. They were frequently copied. Indeed, many of the treasures of the early Middle Ages from the ninth century onwards would never have assumed the form they did but for oriental models (figures 690, 691). Sometimes, especially in the eleventh and twelfth centuries, it is hard to be sure whether certain objects were made in the Byzantine realm, or in the west in imitation of Byzantine models (figure 692). The relationships were further cemented by considerable influxes of craftsmen from the east at the time of the Islamic conquests of Syria and Egypt and when the Iconoclast Emperors (726–843) came to power in Constantinople.

These east–west relationships were accentuated in the twelfth century by the crusades. The first crusade reached Jerusalem in 1099, going by way of Constantinople. The resulting establishment of Latin colonies in Palestine and Syria could not fail to have its effects. All sorts of new techniques in the crafts, new methods in building, and new ideas in the useful arts were thus introduced to the west. An entire vocabulary was coined, or rather transmitted, and there are still very numerous words in our daily language adopted from the Arabic at that time. Two words that came to the west as a result of the textile trade are damask and muslin, the former from Damascus, the latter from Mosul, the places in which the respective types of textile were believed to be manufactured. The 'damascening' of metal (p 57) similarly derives from a method originally associated with Damascus. Other such names are considered in the chapters on industrial chemistry (pp 351 f, 354) and alchemical equipment (pp 733, 738).

FIGURE 690—*Part of a reliquary of* cloisonné *enamel made at Constantinople in 569 for a fragment of the 'True Cross'. It was presented by the Byzantine Emperor Justin II (565–78) to Queen Radegund* (b 519, d 587), *abbess of the convent of St Croix at Poitiers. (See p 452.)*

If the result of the earlier crusades was to bring the culture of the Islamic world more before western eyes, that of the fourth crusade, in 1204, brought closer acquaintance with Byzantine arts. For instead of expelling the infidels from Palestine this enterprise concentrated on capturing and sacking Christian Constantinople. Most of the accumulated wealth of art was wantonly destroyed, but many precious objects were taken to the west and presented to cathedrals and churches there. The vast treasure of articles of Byzantine origin in St Mark's at Venice was thus acquired (figure 693). This crusade was the greatest but not the only plundering exploit of its kind. It was followed by many minor acts of comparable pillage by Italian, French, and Catalonian pirate merchants in Asia Minor and Syria. The richness and excellence of their hoards served to stimulate local craftsmen. Much work was executed in the west in emulation, if not in exact imitation, of eastern models, though the craftsmanship of the imitations was generally coarser and the materials were less sumptuous.

FIGURE 691—*Byzantine cross of* c 800 *of* cloisonné *enamel on gold (description on p 466).*

The next phase of Near Eastern history, in the fourteenth century, saw less close contacts, for the Byzantine Empire was now much diminished and impoverished. Outrages by Italians and Franks, and extra-territorial demands by Venice and Genoa in the capital itself, aroused feelings of deep resentment in Constantinople against the west. Moreover, the aftermath of the crusades partly interrupted more peaceful contacts with the Muslim world. From now on the west was to become slowly but progressively the master of the east.

V. THE 'ARABIC' NUMERALS

There is one device originating from far beyond Islam that so deeply affected almost every form of western technology as to need special mention. It is the introduction of the so-called 'Arabic' system of numeration, which came ultimately from India. The importance for technology of this invention, to which we are so accustomed that we take it for granted, can hardly be exaggerated.

FIGURE 692—*The 'Alfred jewel' of similar technique to the cross of figure 691, but of English workmanship (description on p 465).*

Until the thirteenth century all but the very simplest reckoning was done with an abacus. This instrument is traceable to Greek times and, till the sixteenth century, was in general use in the west as it still is in the east. Numbers in Europe were expressed in the old Roman notation which, for higher orders, is inconvenient to read or record (figure 694). With the abacus addition and subtraction are easy, but other arithmetical processes were beyond the powers of most abacus-users, who often had to invoke professional abacists to perform what we now regard as quite elementary arithmetical operations. The difficulty of these operations was far greater then than now because of the character of the Roman notation. Both abacus and Roman numerals were ultimately displaced by another method, introduced to the west in the thirteenth century.

At the end of the twelfth century a merchant, Leonardo (Fibonacci) of Pisa (*c* 1170–1245), resided for commercial purposes in north Africa. There he learned from Arabic-speaking colleagues the use of the Indian system of numerals, in which the value of a digit, including the nought or cipher, depends on its place in a line of digits.[1] It is the ordinary method of numeration that we now employ.

[1] Thus in the number 222 the digit 2 stands successively for two hundred, twenty, and two.

It has made the elementary rules of arithmetic accessible to every schoolchild.

In 1201 Leonardo, having returned to his native city, composed there a work which was the most important western contribution to mathematics since antiquity. It is the first by a Latin Christian in which this system of numerals,

FIGURE 693—*Byzantine agate chalice with* cloisonné *enamelled panels and pearl borders in the treasury of St Mark at Venice, brought from Constantinople after its sack by the crusaders in 1204 (p 467).*

then long in use by Arabic-speaking craftsmen and merchants, was expounded for technical and commercial use in the west.[1] Very slowly this system displaced the clumsy old Roman system with its necessary use of the abacus, but by the sixteenth century the Arabic system of numeration had become almost universal in western Europe. Its adoption was a major factor in the rise of science, and was not without effect in determining the relations of science and technology in the sixteenth and seventeenth centuries.

[1] The system was used in the west a century earlier for scientific and calendarial purposes.

VI. CHINA

Having glanced at the historical processes through which the Near and Middle East came to influence European technology, it is desirable to indicate the influence of the Far East. It happens that the period with which this volume is mainly concerned corresponds fairly closely with the most flourishing time of the great civilization of China.

The history of Chinese civilization, especially in its earlier stages, has been far less subjected to scientific analysis by western scholars than has our own. Positive information being scanty, fanciful antiquity has often been ascribed to it. Nevertheless it is, in fact, the oldest surviving and most continuous civilization known to us, except our own, and our own is ancient and continuous only as an extension of the classical Mediterranean civilization, which, on the technological side at least, was mainly inherited from the ancient empires of the Near East. The Chinese is the sole technology the history of which can be reasonably and fairly contrasted with that of our own. To get the two technologies into some related historical perspective it is useful to have a few leading dates in Chinese history for comparison with the date-tables of Near Eastern and European history on pages lii, liii, and liv of volume I and on pp lv–lix of the present volume.

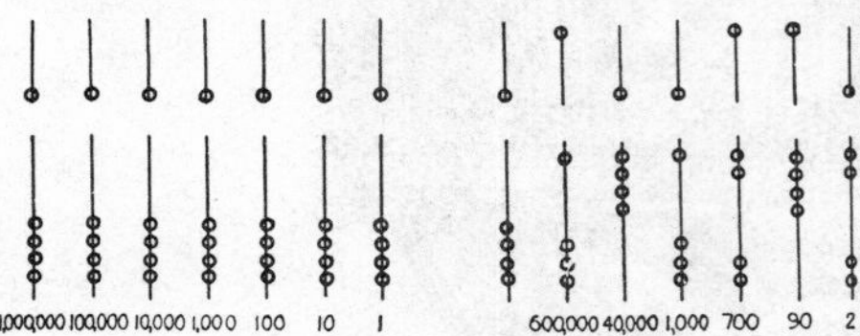

FIGURE 694—*Essentials of the Roman abacus, consisting of beads running on wires. On the left it is set for reckoning. On the right a total of 641 792 is represented. Without an abacal representation and in Roman figures this would need twenty-one elements, namely CCCCCCXLI*^{M}*VII*^{C}*LXXXXII.*

First signs of Neolithic man in China	*c* 2500 B.C.
Signs of civilization in China	from *c* 1000
First unification of China	221
Han dynasty begins	202
Partition into three kingdoms	A.D. 221
Second unification	265
Second partition (north and south)	479
Third unification	581
Third partition (north and south)	907
Fourth unification	960
Mongol dynasty	1260
Ming dynasty	1368

Joseph Needham is doing for China the things that these volumes seek to do for the west. He holds that during the period covered by this volume many devices were introduced into Europe from eastern Asia, where they had been invented. Such devices, mainly of Chinese origin, could reach the west only through Islam, or Byzantium, or, especially for agricultural methods or implements, through the steppe-peoples passing on a tradition to eastern barbarian

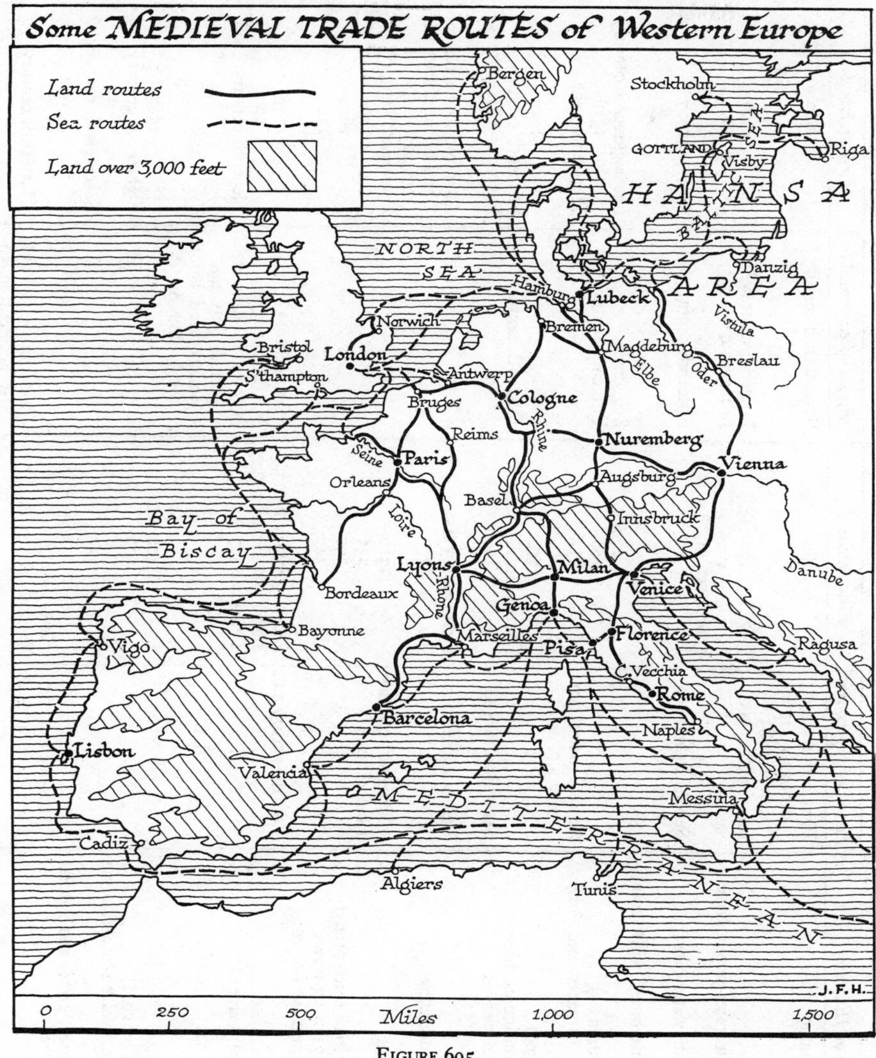

FIGURE 695.

Europe. At the same time there was naturally a certain amount of transmission in the opposite direction, that is, to or towards eastern Asia from western Asia. Here, however, we must confine ourselves to the westward movement of techniques in the period under review. To illustrate that movement, Needham has drawn up the following striking table:

TRANSMISSION OF CERTAIN TECHNIQUES FROM CHINA TO THE WEST[1]

	I. Invention or Discovery	II. China		III. Europe		IV. Approximate minimum time-lag, in centuries
		Period of experiment	First precise date	Period of experiment	First precise date	
a	Square-pallet chain-pump	First century B.C.; probably mentioned in 83	189	Seventeenth century?	1672	15
b	Edge-runner mill	First century B.C.	170	Fifteenth century?[2]	1607	13
	Edge-runner mill with water-power drive	..	400	Fifteenth century?	1607	11
c	Trip-hammer mill	..	Fourth century B.C.	..	..	..
	Trip-hammer mill, with water-power	Second to first century B.C.	20	Fifteenth century?[3]	1607	14
d	Rotary winnowing-machine, with crank-handle	..	40 B.C.	850, crank	Late eighteenth century	14
	Rotary fan for ventilation	..	180	Fifteenth century?	1556	12
	Blowing-engines for furnaces and forges, with water-power	Second to first century B.C.	31	..	Thirteenth century	11
	(*idem*., crank-drive type)	..	1310	..	1757	4
e	Piston-bellows, for continuous blast	..	Fourth century B.C.	Fifteenth century?	Sixteenth century	14
f	Draw-loom for figured weaves	Second century B.C.	c 100 B.C.	..	Fourth to fifth century	4
g	Silk-working machinery:					
	reeling machine	Before 100 B.C.	First century B.C.	..	All introduced about the end of the thirteenth century	3–13
	flyer; twisting and doubling	..	1090	..		
	water-power applied	Thirteenth century	1310	..	Fourteenth century	..
h	Wheel-barrow	First century	231	..	c 1200	9–10
i	Sailing-carriage (first high land speeds)	..	552	..	1600	11
j	Wagon-mill, grinding during travel	..	340	..	1580	12
k	Efficient draught-harness for horses:					
	breast-strap	Fourth century B.C.	Second century B.C.	Sixth century?	c 1130	8
	collar	..	Third to seventh century	Ninth century?	c 920	6
l	Cross-bow (individual weapon)	..	Third century B.C.	Known in Roman period, but not widely used before the Middle Ages[4]	Eleventh century	13
m	Kite	..	c 400 B.C.	..	1589	12
n	Helicopter top (spun by cord)	..	320	..	Eighteenth century	14
	Zoetrope (lamp-cover revolved by ascending hot air)	..	180	..	Seventeenth century	c 10
o	Deep drilling (for water, brine, and natural gas)	Second century B.C.	First century	..	1126	11
p	Iron casting	Fourth century B.C.	Second century B.C.	Cast iron yielded only by accident in ancient and medieval times	Thirteenth century	10–12
	Concave curved iron mouldboard of plough	..	Ninth century	..	c 1700	25
	Seed-drill plough, with hopper	..	85 B.C.	Known in antiquity	c 1700	14

	I. Invention or Discovery	II. China		III. Europe		IV. Approximate minimum time-lag, in centuries
		Period of experiment	First precise date	Period of experiment	First precise date	
q	Gimbals	First century B.C.	180	..	c 1200	8–9
r	Segmental arch bridges	..	610	..	1345	7
s	Cable suspension-bridges	..	First century B.C.	..	..	..
	Iron chain suspension-bridges	..	580	Proposed 1595	1741	10–13
t	Canals and rivers controlled by series of gates	..	Third century B.C.	..	1220	17
	True lock-gates and chambers	..	825	..	1452	7
u	Ship-building:					
	stern-post rudder	..	Eighth century	..	1180	3
	water-tight compartments	..	Fifth century	..	1790	12
v	Rig:					
	efficient sails (mat-and-batten principle)	..	First century B.C.	..	Nineteenth century	18
	fore-and-aft rig	..	Third century	..	Ninth century	6
w	Gunpowder:	Eighth century	c 850	..	Thirteenth century	4
	as an igniter for an incendiary weapon	..	919	[Greek Fire used in the seventh century]	..	..
	rockets and fire-lances	..	c 1100	Described c 1300	Fifteenth century	3–4
	projectile artillery	..	c 1200	..	c 1320	1
	explosive grenades and bombs	..	c 1000	..	Sixteenth century	4–5
x	Magnetism:					
	lodestone spoon rotating on bronze plate	First century B.C.	83	..	..	..
	floating magnet	..	1020	..	1190	4
	suspended magnetic needle	Tenth century	1086	..	..	..
	compass used for navigation	Eleventh century	1117	..	..	2
	knowledge of magnetic declination	..	1030	..	c 1450	4
	theory of declination discussed	..	1174	..	c 1600	4
y	Paper:	..	105	..	1150	10
	printing with wood or metal blocks	Sixth century	740	..	c 1400	6
	printing with movable type	..	1045 (earthenware) 1314 (wood)	..	..	..
	printing with movable metal type	c 1340	1392 (Korea)	..	c 1440	1
z	Porcelain	First century	Third to seventh century	..	Eighteenth century	11–13

NOTES: 1. Amplified from Table 8 (p 242) in 'Science and Civilisation in China', vol I (1954), by Joseph Needham with the research assistance of Wang Ling. By permission of the authors and the Cambridge University Press.
All dates are A.D. except where otherwise indicated. The periods in Column IV attempt to allow for considerable doubt and obscurity, and therefore are frequently less than the number obtained by subtracting the date in Column II from that in Column III.

2. The Chinese edge-runner mill had a roller revolving on a plane surface, and differed considerably from the split-ball-and-cup arrangement of the classical *trapetum*. See above, pp 112, figure 80.

3. Water-driven fulling-mills were known in Europe in the thirteenth century and perhaps earlier (p 216), but it is not clear whether these were of the Chinese tilt-hammer type, or vertical-lift stamp-mills. The tilt-hammer was however associated with the development of the blast-furnace.

4. For further details of the history of the cross-bow in the west, see pp 707 and 722–4.
Full discussion of the chronology tabulated above will be found in future volumes of 'Science and Civilisation in China'.

VII. INDIA

The other east Asian civilizations—Korean, Japanese, Indo-Chinese, Siamese, even Burmese—depend to a greater or less extent on that of China or India, much as our western civilization derives from the Graeco-Roman. None of the derivative east Asian cultures needs discussion here, but we must touch on the great Middle Eastern civilizations of India.

The early history of India is still far too obscure for treatment in continuous narrative or for presentation in chronological tables. The Indus valley civilization, beginning about 2500 B.C., is touched upon elsewhere (vol I, chs 2, 9, table F). It was destroyed *c* 1500 B.C. by invading barbarians from the north, speaking an 'Aryan' language related to Sanskrit, the earliest literary monuments of which may be of *c* 1200 B.C. To what extent the Indus culture represents that of India as a whole, or how widely it was spread over that sub-continent, and to what extent it was introduced or altered by incursions from the north, little is known with certainty. For many centuries, despite the great interlude of Buddhism, India remained divided into many states, of the history of which too little is known. The first precisely established date in Indian history is as late as 326 B.C., when Alexander invaded the country.

Some trade-contact of Europe with India existed at least as early as Homer. From the early years of the Roman Empire to long after its fall, and well into the Middle Ages, sea-traffic of Europe with India by way of the Red Sea continued with varying degrees of activity. Roman and medieval contacts are traceable even up the east coast of the Indian peninsula and beyond. Trade from India also reached Europe through the Persian Gulf and overland through Bactria (figure 684). Commerce necessarily passed in both directions along these routes. Nevertheless when we consider the great numbers and high cultural state of the Indian peoples, their technological influence on the west in the period under review has been conspicuously small. To one great Indian contribution, the so-called 'Arabic' system of numeration, we have already referred (p 766). From a very early date the west has imported from India jewels, pearls, ivory, dyes, spices, incense, drugs, lac, cotton, rice, sugar, and other raw materials. It may be that iron was cast earlier in India than elsewhere. But evidence of technological influences of India on the west is as yet extremely hard to find. Perhaps this is due to inadequate archaeological exploration, but in the present state of knowledge this negative attitude is inevitable for the history of technology.

VIII. TRADE ROUTES

Technology implies a society that needs technological products, and these it will normally obtain by the easiest channels. Thus trade-routes are a very early and natural result of the formation of stable human aggregates. They can be traced in Palaeolithic times, and are demonstrable in the economy of even the most lowly food-gathering peoples of today.

All such routes must be adapted to two-way traffic, for men cannot buy unless they also sell. Some standardization of prices long preceded money, and their recording is held to have been a major factor in the invention of writing (vol I, pp 49, 101, 745). The operation of exchange, which was once barter, became modified with the advent of money[1] from the sixth century B.C. From then on prices and consequently book-keeping became ever more important factors in commercial undertakings. As open-sea traffic developed further with the rise of the Mediterranean civilizations, the sea routes became as standardized as had long been those on land. When the Roman Empire attained her greatness both land and sea routes were most efficiently policed and kept open for regular traffic. Roads were built, regular sea-going services were established, and frontiers, passes, and river-crossings were guarded (figure 459).

Beyond the Roman imperial frontiers Roman traders or their agents penetrated far, reaching India and China by both sea and land. At the same time the eastern peoples penetrated westward. After the fall of the Roman Empire the traffic-ways thus established continued in diminished numbers and activity. They survived in outline through the early Middle Ages, to become more frequented again towards the end of the period covered by this volume. In the later Middle Ages additional routes were established, notably those across the Alps connecting the ancient civilization of Italy with the states of north-west Europe which had risen from barbarism (figure 695)

Trade-routes—like exchanges, commerce, and book-keeping—are topics of the history of economics. They cannot be discussed in these volumes, although they are so intimately linked with our subject. The reader may, however, be helped by a few charts displaying the general direction and nature of trade under the Empire (figure 459), their development for Europe during the Middle Ages (figure 695), the penetration of the Alps, and the great land and sea routes whereby the farther reaches of Asia have, through the ages, maintained some communication, however tenuous, with the west (figure 684).

This volume opens when man had been half a million years on the Earth. Certain communities had by then attained very high cultural standards,

[1] From Moneta, a title of Juno, in whose temple the Roman state mint was for a time situated.

especially in the Near East, where great empires had risen and fallen. Those of Egypt and Mesopotamia have left their marks on our own civilization, which is, in large part, their heir. Smaller peoples, the Mycenaeans, the Hebrews, the Phoenicians, the Etruscans, reached their brief terms of minor power when the great ancient empires were in decline. On our own civilization these minor powers too have left their marks, some of which are profound and far-reaching, even though technologically less significant. The main concern of this volume, however, is with technological events between *c* 500 B.C. and A.D. 1500. These have been discussed first for the great Mediterranean civilizations of Greece and Rome, and then for the technologically less understood millennium in Europe—and especially western Europe—that followed the fall of the western Roman power.

But this volume has also seen to its end a relationship between east and west that we shall not encounter again. When the Middle Ages closed, the east had almost ceased to give techniques and ideas to the west and ever since has been receiving them. This reversal of an age-old cultural current provokes two reflections. One is to marvel at the high degree of technological attainment achieved without guidance from anything worthy of the name of science. Progress in technology, though exceedingly slow, was cumulatively astonishing in its extent. Secondly, it is impossible to miss the sharp contrast between the earlier (and far longer) empirical phase of technology as described in the first two volumes of this History and its dependence on science in later times. Our next volume will deal with the frontier centuries between these two tendencies, and will treat of the early impact of a new and self-conscious science on a technology becoming progressively industrialized.

In our own time technology has become almost synonymous with the application of scientific knowledge to practical ends. To us it seems that science is the source, the parent, of technology. Up to the end of the period treated in this volume, that is up to about 1500, and perhaps much later, it would be more accurate to say that technology was the parent of science. But from the rather indeterminate period usually called the Renaissance, natural phenomena came to be more and more systematically observed. Moreover, knowledge of nature was ever-increasingly elicited by experiment, which is another word for controlled observation—a far more rapid and reliable way of extending knowledge. The rise of the use of experiment and the recession of empiricism from technology will become increasingly evident as these volumes proceed. In parting from this one, it may be well to glance back at the main technological achievements, almost wholly empirical, of the two millennia under consideration. Many different selections are possible, for each of which much is to be said, and perhaps no two persons would choose precisely the same set of examples.

Some would give first place to the process by which the forces of nature have been made to do the work previously performed by the muscles of men and animals. Thus regarded, primacy would be given to the mill, first the horizontal water-mill, then the vertical mill, then the windmill (ch 17). It is to be remembered, however, that the development of such machines owes much to a growing mastery of the principles of rotary motion, introduced long before (vol I, ch 9), and especially to their application in the construction of gearing (pp 645 ff). The fabrication and smooth working of mills was facilitated by the introduction of several tools—the screw, the plane, the lathe, the brace, the pulley, and the haulage-tongs—the arrival of all of which comes within our period (chs 11 and 18). The exploitation of new sources of power with the aid of such simple tools may be held to represent the first step in the passage from antiquity to modernity.

Or it might be said that man's prime need is for food and that therefore improvement of crops and stocks must take first place (ch 3, and vol I, chs 13 and 14). On this view the change from the dry-farming of the Mediterranean area and the irrigation-methods of the Near East to the drainage-farming and deep ploughing of the heavy soils of the west and north made possible the development of the higher forms of technology in the north-west. A very important place would then be accorded to the improvement of farming-instruments and especially the development of the plough (ch 3), which opened up northern and western Europe to the possibilities of civilization.

Yet again, since techniques can have no permanent frontiers, everything that increased means of communication, and especially of transport, must have hastened the spread of local devices into wider cultural areas. This line of thought would stress many of the ideas and inventions that have spread from east Asia to west Europe, such as the wheelbarrow (p 642), deep drilling, cast iron (pp 74–5), lock gates (p 688), gimbals, paper, printing, and so on. It would emphasize the importance of the mechanisms of the intercommunication of peoples, such as made roads, passenger-vehicles, the introduction of horses and camels, draught with the horse-collar, advances in the wheelwright's craft (chs 14, 15), and perhaps above all improvements in shipbuilding and navigation (ch 16 and vol III).

There is surely a case for judging a civilization by its use of such social surplus as is at the disposal of the favoured few for the lofty purposes of art and literature. This, in a history of technology, would lead to stress on the techniques of the fine arts, such as those of the golden age of Greece with its architecture and sculpture, and of fifteenth-century Italy with its pictorial art and its beginnings of modern science. It would emphasize too the amenities of life introduced by glazing and by blown glass (ch 9), by paper (vol III), parchment (pp 187 ff), printing (vol III), and the luxury of figured weaves (vol III). Such a judgement would give less attention to the mechanical advances of the Middle Ages and make the change to modernity rather later than is implied in these volumes.

Others, thinking backward from our time, would see in the period we have here considered the beginnings of types of manufacture characteristic of modern industry. They would attach great importance to metallurgy, especially to the introduction of case-hardened steel and to the newer alloys of copper, notably brass (ch 2). They would also lay emphasis on the preparation on an industrial scale of substances that include many of those nowadays of prime industrial importance, such as alkalis, pigments and dyes, nitre, alum, soap (ch 10), and alcohol (ch 4).

Yet others would claim the prevalence of the scientific mood as the test of modern civilization, since its diffusion has determined the course of our lives today. They would therefore give much attention to instruments of precision in antiquity, to the variety of ingenious and skilfully made measurers of time, and notably to the astrolabe. For them the rise of modern science, and, following it, of scientific technology, would be a continuation, *longo intervallo*, of the science of the ancients. On this view, modernity would perhaps begin in the seventeenth century, and therefore the consideration of these topics has been deferred until our third volume.

All these, and many other, ways of contemplating the technology of the past have had their advocates. On each of these ways many books have been written. But the work now before the reader has been put together in a spirit that takes none of these views. It seeks to present what seem to the authors and editors the salient facts arranged in the best sequence that they have been able to devise, and with that very high degree of condensation of a vast mass of material that is demanded even by these portly volumes. When their task is complete there may emerge a clearer interrelationship of the techniques to each other and to the cultures in which they have arisen. Until that broad view can be taken we must be content to know that techniques, like those who practise them, are all parts one of another, and that we live in a world the unity of which can emerge only very slowly. This may be a hard saying, but it is the way and method of science.

I. INDEX OF PERSONAL NAMES[1]

Abū Bakr Muḥammad ibn Zakariyya al-Rāzī (866–925), Persian physician, 736–8.
Abū Manṣūr Muwaffak (10th cent.), Persian writer, 737.
Abu'l-Qāsim of Kashan (*c* 1300), 302.
Adalhard (*fl* 822), abbot of Corvey, 140.
Aeschylus (525–456 B.C.), Greek dramatist, 168, 233.
Æthelwulf (d 858), Saxon king, 481.
Agnes, St (*fl c* A.D. 304), 461.
Agricola (Georg Bauer) (1494–1555), 8 n., 13–23, 34, 44, 45, 68, 290, 315, 379, 380, 655, 741, 746.
Agricola, Gnaeus Julius (A.D. 40–93), 404.
Agrippa, Marcus Vipsanius (b *c* 63 B.C.), 671.
Aḥmad ibn Mu'tasim (842–6), caliph, 635.
Alan of Walsingham (*fl* 1322), 442.
Albertus Magnus (? 1193–1280), 64.
Al-Bīrūnī (973–1048), 737.
Aldhelm, St (d 709), 32.
Al-Dimashqī (1256/7–1326/7), Syrian cosmographer, 615, 616.
Alexander III the Great (336–323 B.C.), Macedonian king, 57, 199, 697–8, 700–1, 716, 772, table I.
Alfonso X the Wise (1252–84), king of Castile, 12.
Alfred (871–99), English king, 580–1.
Al-Ḥasan al-Rammāḥ (*fl c* 1280–90), 379.
Al-Jazarī, Muḥammad ibn Ibrahīm (d 1338), historian, 614.
Al-Ma'mūn (813–33), caliph, 735.
Al-Mas'ūdī (10th cent.), geographer, 616.
Al-Qazwīnī (1203–83), geographer, 597, 614.
Amman, Jost (1539–91), 552.
Ammianus Marcellinus (b *c* A.D. 330), 713, 723.
Anna Comnena (1083–? 1148), 30, 376.
Antigonus II (*c* 320–239 B.C.), Macedonian king, 566.
Antipater of Thessalonica (1st cent. B.C.), 593.
Antoninus Pius (A.D. 138–61), Roman emperor, 30.
Antonius, Marcus (Mark Antony, *c* 82–30 B.C.), 572, 573.
Apollodorus (*fl* A.D. 100), Roman engineer, 512, 636.
Appian (2nd cent. A.D.), Alexandrian historian, 516.
Appius Claudius (*fl* 300 B.C.), Roman censor, 28, 500, 512, 670.
Arcadius (395–408), emperor of the East, 601.
Archimedes of Syracuse (*c* 287–212 B.C.), mathematician, 116, 604, 631, 632–3, 634, 658, 676–7, 699, 714.
Archytas of Tarentum (*fl* 400 B.C.), Pythagorean philosopher, 631–3, 677.
Aristobulus (*fl* 330 B.C.), Alexander's historian, 199.
Aristotle (384–322 B.C.), Greek philosopher, 528, 624, 630, 638, 658, 731.
Arnald de Villanova (? 1235–? 1311), physician and alchemist, 138, 142.
Ashur-nasir-pal I (883–859 B.C.), Assyrian king, 495.
Ashur-nasir-pal II (*fl* 850 B.C.), Assyrian king, 235.
Athenaeus of Attalia (*fl* A.D. 50), 674.
Athenaeus of Naucratis (*fl* A.D. 200), 133.
Augustus (63 B.C.–A.D. 14), Roman emperor, 28, 508, 512–13, 571, 670, table II.
Aurelius, Marcus (A.D. 121–80), Roman emperor, 530.
Ausonius (*fl* A.D. 370), Roman poet, 31, 597, 600.

Bacon, Francis (1561–1626), 604.
Bauer, Georg, *see* Agricola.
Bede, the Venerable (673–735), 7, 32, 425, 429.
Béguin, Jean (17th cent.), 741.
Bélidor, B. Forest de (? 1697–1761), 608.
Benedict Biscop (628 ?–690), abbot of Wearmouth, 32, 326.
Besson, Jacques (*fl c* 1567), 626.
Beukelszoon, William (*c* 1375), 124.
Biringuccio, Vanoccio (*fl* 1540), 475.
Black, Joseph (1728–99), chemist, 744–5.
Bodyington, Stevyn (16th cent.), grocer, 248.
Boëmus, Johannes (*fl* 1490–1520), German humanist, 126.
Boyle, Robert (1627–91), chemist, 738, 749–50.
Brendan, St (484–577), 581.
Browne, Edward (1644–1708), physician, 22–4.
Brueghel, Pieter (*c* 1525–69), 96, 175, 548, 680, 749.

Cadeby, John (16th cent.), mason, 248.
Caesar, Gaius Julius (102–44 B.C.), Roman emperor, 10, 52, 134, 168, 514, 530, 672, 705, 716.
Caligula (A.D. 37–41), Roman emperor, 518, 522, 569, 572–3, 600.
Calvisius Sabinus (*fl* 48–27 B.C.), Roman general, 513.
Campin, Robert (*fl* 1430), painter, 249, 653.
Caracalla (211–17), Roman emperor, 418, 673.

[1] The dates given for rulers are regnal dates, not years of birth and death.

Carew, Richard (1555–1620), 12.
Carvilius Pollio (*c* 80 B.C.), Roman furniture maker, 227.
Cassivellaunus (*fl* 1st cent. B.C.), British chieftain, 544.
Cato, Marcus Porcius (234–149 B.C.), 52, 86, 95, 110, 114, 134.
Catullus (*c* 87–54 B.C.), Roman poet, 201, 561.
Celsus, Aulus Cornelius (*fl* A.D. 14–37), Roman physician, 121.
Cennino Cennini (*c* 1370–*c* 1440), 352–3.
Chang Ch'ien (*fl* 128 B.C.), Chinese general, 131.
Charlemagne (Charles the Great) (768–814), emperor, king of the Franks, 65, 138, 139, 386–7, 459, 608, 761, table III.
Charles VI (1380–1422), French king, 691.
Charles of Anjou (1266–85), king of Naples, 489, 491.
Charles the Hammer (689–741), 720, 735.
Childeric (458–81), king of the Franks, 451, 453.
Chosroes II (589–628), Persian king, 451.
Cicero, Marcus Tullius (106–43 B.C.), 514.
Claudius (A.D. 41–54), Roman emperor, 29, 518, 530, 670, 671, 679, table II.
Colaeus of Samos (*fl* 650 B.C.), 568.
Columella (*fl* A.D. 36), 111, 134.
Conrad II (1024–39), king of Germany, 455.
Constantine I the Great (312–37), Roman emperor, 29, 515, 599, 600–1, 639, 761, table II.
Constantine Monomachos (1042–54), emperor of the East, 462.
Corbulo, Gnaeus Domitius (*fl* 45–63), 680.
Cornelisz, Cornelis (1562–1638), 625.
Cosmas Indicopleustes (6th cent. A.D.), 55.
Crassus (*c* 112–53 B.C.), 706.
Ctesibius of Alexandria (*c* 300–230 B.C.), mechanician, 633, 708 n., 712–13.
Cuthbert (*fl* 758), abbot of Wearmouth and Jarrow, 326.
Cuthbert, St (d 687), bishop of Lindisfarne, 169, 451, 457, 756, 763.
Cyrus (550–530 B.C.), Persian king, 495, table I.
Cyrus the Younger (424–401 B.C.), Persian king, 496–7, 697.

Damian, John (16th cent.), alchemist, 746.
Darcy, Edmund (*fl* 1593), 155.
Darius I (521–485 B.C.), Persian king, 54, 496, 666.
Demetrius I (307–283 B.C.), Macedonian king, 567, 700, 702, 716.
Demokritos (? 2nd cent. B.C.), alchemist, 732–3.
Demokritos of Abdera (d 376 B.C.), philosopher, 733.
Demosthenes (384–322 B.C.), Greek orator, 4–5, 58.
Diades (*c* 340–320), engineer, 698.
Diderot, Denis (1713–84), 152, 153.
Diocletian (A.D. 284–305), Roman emperor, 55, 639, table II.
Diodorus Siculus (*fl* 60–30 B.C.), 677.
Diogenes Laertius (*c* 3rd cent. A.D.), 632.
Dionysius I (*c* 405–367 B.C.), tyrant of Syracuse, 699.
Dioscorides Pedanius (1st cent. A.D.), Greek physician, 5–6, 370, 371.
Domitian (A.D. 81–96), Roman emperor, 134.
Drebbel, Cornelius (1572–1634), 366.
Drusus, Marcus Livius (*fl* 91 B.C.), Roman tribune, 46.
Drusus, Nero Claudius (38 B.C.–A.D. 19), 680.

Edward the Confessor (1042–66), 426.
Edward I (1272–1307), English king, 65, 532.
Edward III (1327–77), English king, 525, 726.
Edward VI (1547–53), English king, 141.
Edward, Prince of Wales, the Black Prince (1330–76), 177–8.
Egmont, Count Lamoral (1522–68), 684.
Eleanor of Aquitaine (? 1122–1204), English queen, 624.
Eleanor of Castile (d 1290), English queen, 478.
Elizabeth I (1558–1603), English queen, 13, 248–9, 723.
Eloi, St (d 660), 76, 77.
Empedocles (*c* 493–*c* 433 B.C.), 678.
Ennion, glass-maker, 322, 325, 337.
Essarhaddon (681–668 B.C.), Assyrian king, 495.
Ethelbert II (725–62), 610.
Eucratides (*c* 180–*c* 160 B.C.), Bactrian king, 487.
Eupalinus of Megara (? 6th cent. B.C.), hydraulic engineer, 668.
Euripides (? 485–? 406 B.C.), Greek dramatist, 630.
Eurysaces (1st cent. B.C.), Roman baker, 118.
Eupatrids, family of Attica, 129.

Festus (2nd cent. A.D.), 212.
Flaminius, Gaius (*fl* 223 B.C.), Roman censor, 500, 512.
Floris V (1254–96), count of Holland, 619.
Frazer, Sir James, 25.
Francke, Meister (*fl c* 1424), painter, 652.
Frederick I Barbarossa (1152–90), emperor, 75.
Frederick (1104–23), archbishop of Bremen, 683.
Frescobaldi, Florentine family, 65.
Frontinus, Sextus Julius (*c* A.D. 30–104), Roman consul, 405, 419, 671–4.
Frontinus (*fl* A.D. 300), glass-maker, 338.

Galen (? A.D. 129–99), physician, 121, 674.
Geber, *see* Jābir ibn Ḥayyān.
Giovanni da Fontana (*fl* 1400), 692.
Glauber, J. R. (1604–70), iatrochemist, 743–6.
Glaukos of Chios (7th cent. B.C.), 58.
Godfrey, Ambrose (d 1741), 750.
Gower, John (? 1325–1408), English poet, 126.
Gracchus, Gaius (*fl* 120 B.C.), 505, 507.
Gregory of Tours (*c* 540–94), 608.
Guénolé, St, of Brittany (414–504), 139.
Guido da Vigevano (*c* 1280–1350), Italian physician, 620, 725–6.

Hadrian (A.D. 117–38), Roman emperor, 9, 498, 507, 575, 598, 670.
Ḥaidar, Abu Mansūr (10th cent.), Persian king, 762.
Hajek, Václav (Wenceslaus Hagecius, d 1552), 611.
Hammurabi (*c* 1750 B.C.), Assyrian king, 495.
Hannibal (247–182 B.C.), Carthaginian general, 133.
Hanno the Carthaginian (*fl* 450 B.C.), 568.
Harrison, William (1534–93), 244.
Hārūn al-Rashīd (785–809), 735, table III.
Helena (d *c* A.D. 330), empress, 29.
van Helmont, J. B. (1577–1644), chemist, 745.
Henry II (1002–24), emperor, 468.
Henry I (1100–1135), English king, 611.
Henry III (1216–72), English king, 37, 478.
Henry V (1413–1422), English king, 588.
Henry VII (1485–1509), English king, 306.
Henry VIII (1509–1547), English king, 141, 178, 248.
Henry the Lion (1142–95), duke of Saxony, 491.
Henry de Yevele (d 1400), 385.
Heraclius (610–41), emperor of the East, 300, 315, 329, 332.
Hero of Alexandria (1st cent. A.D.), mechanician, 114, 115–17, 614–15, 618, 631–2, 634–6, 638, 643, 648, 660–2, 669, 708 *and* n., 710–11.
Herodotus (*c* 484–425 B.C.), Greek historian, 55, 197, 212, 568, 668, 674.
Herrad of Landsperg (12th cent.), 89, 115.
Hesiod (*c* 700 B.C.), Greek poet, 85, 549–50.
Hiero II (*c* 306–216 B.C.), ruler of Syracuse, 27.
Hildegard of Bingen (12th cent.), 141.
Himilco (*fl* 450 B.C.), Carthaginian navigator, 568.
Hippodamus of Miletus (5th cent. B.C.), 528.
Hokusai (1760–1849), Japanese artist, 107.
Holmes, Randle (1627–99), 254.
Homer (? 8th cent. B.C.), Greek poet, 55, 58, 148, 166–7, 193, 545, 563–7.
Honorius (395–423), emperor of the West, 601.

Ibn al-Baiṭār (d 1248), 370, 372.
Ibn al-Khatīb (1313–74), Muslim physician, 139.
Ibn Khurdādhbih (9th cent.), 524.
Isimkheb (*c* 1000 B.C.), Egyptian queen, 164.

Jābir ibn Ḥayyān (8th cent.), alchemist, 356, 736, 738–9, 742–3.
James II (1437–60), Scottish king, 727.
James IV (1488–1513), Scottish king, 746.
Jefferson, Thomas (1773–1826), president of the United States of America, 90.
Jenner, Dr Edward (1749–1823), 254.
Jerome, St (4th cent.), 97, 106, table II.
John V Palaeologus (1354–91), emperor of the East, 585.
John the Calabrian (15th cent.), 656.
John of Garland (13th cent.), 641, 653.
Josephus (*c* 37–100), Jewish historian, 715.
Josse, St (d 669), legendary king of Brittany, 762.
Julian the Apostate (361–3), Roman emperor, 140, 515, table II.
Julius Africanus (*fl.* A.D. 220–35), 376.
Justin II (565–78), emperor of the East, 452, 765.
Justinian I (527–65), Emperor of the East, 197, 664, 670, 763, table III.

Khalaf ibn ʻAbbās al-Zahrāwī (*Lat.* Albucasis, d 1013), 371.
Knocker, G. M., 294 n.
Kyeser, Konrad (*fl* 1400), 691–2.

Langland, William (? 1330–? 1400), English poet, 126 n., 214, 541.
Laurentius Vitrearius (Norman glass-maker, *fl.* 1226), 326.
Lavoisier, A.-L. (1743–94), 736, 745.
Leeghwater, J. A. (1575–*c* 1650), engineer, 688.
Leo I (457–74), emperor of the East, 599.
Leo VI (886–911), emperor of the East, 376, 516.
Leonardo (Fibonacci) of Pisa (*c* 1170–1245), merchant, 766–7.
Leonardo da Vinci (1452–1519), 205, 395, 648, 651–2, 654–7, 699, 726, 744.
Libavius, A. (1540–1616), chemist, 748–9.
Livy (59 B.C.–A.D. 17), Roman historian, 27.
Lou Shou (12th cent.), Chinese poet, 220.
Louis VII (1137–80), French king, 608.
Louis IX (St Louis) (1226–70), French king, 377, 453, 457, 585, 681.
Lucian of Samosata (*c* A.D. 160), 630.

Lucretius (? 94–? 55 B.C.), Roman poet, 9, 596.
Luke, St, evangelist, 188.
Lull, Raymond (? 1235–1315), alchemist, 142, 356.
Lycurgus (d 324 B.C.), 3.

Marcellus (*fl* 214 B.C.), Roman general, 714.
Marcus Graecus (*c* 1250), 351, 379.
Mariano, Jacopo (15th cent.), engineer, 610.
Marius, Gaius (*c* 100 B.C.), 703, 705.
Mary the Jewess, alchemist, 733–4.
Maximilian I, Emperor (1493–1519), 389, 491.
Maximus (383–8), Roman emperor, 136.
Megasthenes (*fl* 300 B.C.), 666.
Metagenes (*fl* 410–400 B.C.), 660.
Metrodorus (3rd cent. A.D.), engineer, 600.
Michelangelo (1475–1564), 28.
Michel de Nostradame (*fl* 1556), 741.
Mithridates VI the Great (1st cent. B.C.), king of Pontus, 593.
Moxon, Joseph (1627–1700), 242, 250–3.
Mylius, J. D. (*fl* 1600–40), German alchemist, 740.

Neckam, Alexander (1157–1217), 124, 532.
Needham, Joseph, 556 n. 2, 732 n., 768–71.
Norton, Thomas (*fl* 1477), alchemist, 741, 743–5, 747.
Nostradamus, *see* Michel de Nostradame.

Olaf Geirstad-Alv (d *c* 800), Norwegian king, 580.
Olaf Trygvasson (d 1000), Norwegian king, 583.
Olympiodorus (*fl* 425), 733.
Omar I (634–44), caliph, 615.
Otto the Great (936–73), emperor, 455.

Palladius (? 4th cent. A.D.), 97, 601.
Papin, Denis (1647–1712), 634.
Pappus of Alexandria (3rd or 4th cent.), 603, 636, 645.
Paracelsus (1493–1534), physician, 372, 744–5.
Pausanias (*fl c* A.D. 150), Greek geographer, 6, 25, 30, 498.
Pepys, Samuel (1633–1703), 174.
Percy, Henry (1564–1632), 9th earl of Northumberland, 749.
Peter, abbot of Gloucester (1104–13), 479–80.
Petrie, Sir W. M. Flinders, 228, 332, 336, 341.
Pettyt, Thomas (1510–58), military engineer, 536.
Pheidon (7th cent. B.C.), king of Argos, 485.
Phidias (b *c* 490 B.C.), Greek sculptor, 470.
Philip II (359–336 B.C.), Macedonian king, 697, 700.
Philip II Augustus (1180–1223), French king, 691.
Philo of Byzantium (*fl c* 150–100 B.C.), mechanician, 614, 633–4, 636, 708 n., 709–10, 712–13.
Piccolpasso, Cipriano (1524–79), 285, 289, 291, 293–4, 297–300, 303 n.
Pietro dei Crescenzi (1233–1320), 138–9.
Pinas, J. (*fl.* 1600), painter, 749.
Plato (*c* 429–347 B.C.), Greek philosopher, 27, 233, 632.
Plautus (? 254–184 B.C.), Roman playwright, 134.
Pliny the Elder (A.D. 23/4–79), Roman writer, 3, 8, 19, 27, 29, 42, 46, 86, 88, 96, 97, 104–5, 106, 113–14, 116–18, 134–5, 140–1, 151 *and* n., 189, 192–5, 197, 199, 215, 218, 226, 233, 323, 328, 350, 355, 401, 531, 681, 700.
Plot, Robert (1640–96), 35.
Plutarch (*c* A.D. 46–*c* 120), philosopher and biographer, 505, 507, 604, 633, 714.
Pollux (2nd cent. A.D.), Greek lexicographer, 199.
Polo, Marco (? 1254–? 1324), 156.
Polybius (? 205–120 B.C.), Greek historian, 5.
Priestley, Joseph (1733–1804), chemist, 750.
Priscian (*fl* 350–80), physician, 355.
Procopius (d *c* 565), Byzantine historian, 28, 375, 569, 601, 607, 706.
Pyrrhus (d 272 B.C.), king of Epirus, 670.
Pytheas (*fl* 300 B.C.), Greek seaman, 569.

Qaisar ibn Abu'l-Qāsim (d 1251), engineer, 614.
Quintus Candidius Benignus (1st cent. B.C.), 599.

Radegund, Queen (519–87), abbess of St Croix, Poitiers, 139, 452, 765.
Ramelli, A. (*fl* 1588), 626, 724.
Raud the Strong (11th cent.), 583.
Read, J., 747.
Rhazes, *see* Abū Bakr Muḥammad ibn Zakariyya al-Rāzī.
Richard I (1189–99), English king, 70.
Richard II (1377–1399), English king, 727.
Robert of Chester (12th cent.), 738.
Roger, Bishop of Sarum (1107–42), 428.
Rudborne, Thomas (*fl* 1460), 428.
Russell, Richard (*fl* 1678), 739.

Sanudo, Marino (Torsellus the Elder) (*c* 1260–1337), 725.
Sargon II (722–705 B.C.), Assyrian king, 54, 321, 336, 495.
Scaurus, Marcus Aemilius (d 90 B.C.), consul, 679.
Scipio Africanus (236–184 B.C.), Roman censor, 121.
Ségolène, Ste (*fl* 770), 139.
Seleucus II (246–226 B.C.), Seleucid king, 701.
Seneca, Lucius (55 B.C.–A.D. 39), Roman writer, 27, 211.

Servius (4th cent. A.D.), grammarian and commentator, 98.
Severus, Lucius Septimius (193–211), Roman emperor, 137, 519, table II.
Shalmeneser III (859–824 B.C.), Assyrian king, 556 n, table I.
Shirley, J. W., 749.
Smith, Adam (1723–90), political economist, 252.
Solon (*c* 640–*c* 560 B.C.), Greek statesman and poet, 1, 105, 129.
Sophocles (*c* 496–406 B.C.), Greek dramatist, 58.
Sostratus of Cnidos (*fl* 280 B.C.), engineer, 521, 566.
Stapleton, H. E., 737.
Statius (*c* A.D. 90), Roman poet, 500–1, 503.
Steen, Jan (1626–79), painter, 749.
Stowe, John (1525–1603), 244.
Strabo (? 63 B.C.–? A.D. 24), Greek geographer, 3, 28, 45, 48, 97, 135, 136, 323, 529, 593, 597, 676.
Stradanus, Johannes (*c* 1530–1605), painter, 749.
Suetonius (*c* A.D. 69–*c* 140), Roman biographer, 522.

Tacitus (*c* A.D. 55–*c* 120), Roman historian, 11, 140.
Taylor, F. Sherwood, 732, 735.
Teniers, David, the younger (1610–90), painter, 348, 749.
Tertullian (*c* A.D. 160–*c* 225), 531.
Thales of Miletus (640–548 B.C.), philosopher, 606.
Theodelinde (d 628), Lombard queen, 451.
Theodoric the Great (489–526), 457.
Theodosius I the Great (379–95), Roman emperor, 531, 515.
Theophano (958–91), wife of Otto II, 761.
Theophilus Presbyter (10th cent.), 12, 49, 63–64, 174, 189, 300, 315, 329, 332, 351, 354, 358, 458, 480.
Theophrastus of Eresus (372/369–288/285 B.C.), 85, 131, 134, 199, 226, 234, 360, 398 n.
Thothmes III (*fl* 1480 B.C.), Egyptian king, 341, 544.
Thucydides (*c* 460–*c* 400 B.C.), Greek historian, 529.
Tiberius (A.D. 14–37), Roman emperor, 52.
Tibullus (? 48–19 B.C.), Roman poet, 28.
Tiglath-Pileser I (*c* 1115–1102 B.C.), Assyrian king, 494–5.
Tillmann of Cologne (*fl* 1359), 65.
Titley, A. F., 744.
Torsellus, *see* Marino Sanudo.
Trajan (A.D. 98–117), Roman emperor, 498, 508, 512, 518, 530, 601, table II.
Tredgold, Thomas (1788–1829), engineering writer, 242.
Turner, W. E. S., 314 n., 319 n.

Ulpian (d A.D. 228), Roman jurist, 325.

Valentinian I (364–75), emperor, 515.
Valens (364–78), emperor, 664.
Valturio, Roberto (d 1484), Italian engineer, 620.
Vauban, S. le Prestre de (1633–1707), engineer, 728.
Vegetius (*c* A.D. 383–*c* 450), military writer, 121, 636, 705, 707.
Venantius Fortunatus (6th cent.), 597.
Veranzio, F. (1561–1617), 627.
Vespasian (A.D. 69–79), Roman emperor, 489, 511, 601, 715, table II.
Vierlingh, Andries (*fl* 1579), 682, 684.
Villard de Honnecourt (*c* 1250), 69, 610, 643–4, 647.
Vincent of Beauvais (d *c* 1264), encyclopaedist, 665–6.
Virgil (70–19 B.C.), Roman poet, 94, 111, 135, 195, 233.
Visconti, Filippo Maria (1402–47), duke of Milan, 657.
Vitruvius Pollio (1st cent. B.C.), Roman architect, 27, 31, 117, 396–419 *passim*, 596, 597, 614, 630, 658–60, 662, 663, 671, 674–6, 711–12, 714, 717.

Walter de Millinate (*fl c* 1300–30), 620.
William I the Conqueror (1066–87), English king, 128–9, 648.
Wren, Sir Christopher (1632–1723), architect, 34, 442.

Xenophon (*fl c* 400 B.C.), Greek historian, 496, 557, 697.

Zachaire, Denis (16th cent.), French alchemist, 741–2.
Zonca, Vittorio (*c* 1580–?), 207, 216, 645, 685–6.
Zopyrus of Tarentum (1st cent. B.C.), 699.
Zosimos of Panopolis (*c* A.D. 300), alchemist, 140, 733, 736.

II. INDEX OF PLACE NAMES

Abbeville, France, 644, 649.
Abusir, near Cairo, Egypt, 665.
Actium, Greece, 571, 573.
Admont, Styria, 69.
Agrigento (Acragas), Sicily, 401, 403, 678.
Agrilesa, Crete, 545.
Ahenny, Co. Tipperary, Eire, 544.
Aix-la-Chapelle (Aachen), Germany, 55, 423, 761.
Ajustrel, Spain, 53.
Alatri, Italy, 530, plate 40 A.
Albi, France, 440.
Alcantara, Spain, 508, 510.
Aleppo, Syria, 199.
Alexandria, Egypt, 53, 322, 350–1, 410, 521–2, 528–9, 566, 670, 735–6, 738, 761.
Aljubarrota, central Portugal, 583.
Alston Moor, Cumberland, 65.
Amalfi, Italy, 454.
Amberg, Bavaria, 75.
Amiens, France, 435.
Amsterdam, Holland, 685.
Amyklae, Greece, 499.
Antioch, Syria, 465, 504, 531, 614, 761.
Anzio, Italy, 520.
Aornos, Mount, unidentified site near Taxila, 701.
Aosta, Italy, 511.
Apamaea, Syria, 637.
Aquileia, Italy, 321, 323, 520.
Ardagh, Ireland, 464.
Arezzo, Italy, 271, 274.
Arles, France, 68, 598–9, 671, 679.
Artemisium, Euboea, 484.
Ashur, Assyria, 494, 495.
Astyra, Troad, 42.
Atchana, Turkey, 316.
Athens, 25, 27, 233, 259–62, 267, 276, 398–403, 470, 499, 529, 602, 663, 665, 667.
Aude, Gaul, 60.
Augsburg, Bavaria, 75, 611, 666.
Autun, France, 507.
Avaricum (Bourges), France, 716.

Baalbek, Syria, 30, 409–10, plate 2 A.
Babylon, 494, 496 (map).
Badari, Egypt, 315.
Baetica (Andalusia), 8.
Baghdad, Iraq, 286–7, 303, 497, 608, 614.
Bagram (Kapisa), Afghanistan, 324.
Baiae, near Naples, Italy, 665.
Balabish, Egypt, 151, 158, 161.
Ballinderry Crannog, Ireland, 255.
Barbegal, near Arles, 598–9, 602, 639.
Barygaza (Broach), India, 53.
Basle, Switzerland, 468.
Basra, Iraq, 498, 614.
Bassae, Greece, 400, 401.
Bath, Somerset, 32.
Battersea, London, 460, 462, 468.
Bayonne, France, 648.
Beaune, France, 137.
Beer, Devon, 34.
Belleville, France, 691.
Berne, Switzerland, 531.
Beth-Maré, Lebanon, 575.
Bettystown, near Drogheda, Eire, 473.
Bhita, India, 528.
Bignor, Sussex, 446.
Bir Sgaoum, Algeria, 639.
Birdlip, Gloucestershire, 460, 463.
Birka, Sweden, 313.
Blakeney, Norfolk, 439.
Blois, France, 649.
Bologna, Italy, 205, 206, 269, 738.
Bolsena, Italy, 376.
Bonn, Germany, 671.
Bons, France, 511.
Bordeaux, France, 137.
Boscoreale, near Naples, 223.
Boughton Mounthelsea, Kent, 34.
Boulogne, France, 522.
Bourges, France, 716.
Bowness, Cumberland, 31.
Box, Wiltshire, 34.
Boxmoor, Hertfordshire, 134 n.
Brampton, Cumberland, 31.
Breac Moedoc, Scotland, 169.
Brescia, Italy, 60, 75.
Briey, France, 69.
Brill, Buckinghamshire, 296.
Bristol, England, 741.
Brittany, 46.
Brixworth, Northamptonshire, 425.
Brücken, Palatinate, Germany, 76.
Bury St Edmunds, Suffolk, 426, 446.
Buxton, Derbyshire, 169.
Byzantium, 367, 368, 553, 577, 600, 720; *see also* Constantinople.

Cabeira, Pontus, 593.
Caen, Normandy, 33, 428, 577.
Cairo, Egypt, 286, 304.
Calabria, Italy, 199.
Calais, France, 527, 536, 726.
Cales, Apulia, 268.
Calydon, Greece, 403.
Cambridge, England, 34, 432–3, 436, 532, 582.

Canosa, Apulia, 268, 342, 558, 559.
Canterbury, Kent, 178, 623, 690, 691.
Capua, Italy, 53, 500, 501 (map), 531, 659.
Carchemish, Mesopotamia, 167.
Carinthia, 60, 67, 70.
Carlisle, Cumberland, 437.
Carnarvon, Wales, 237.
Carrara, Italy, 24, 28, 406.
Cartagena, Spain, 44.
Carthage, 516–17, 568, 666, 670, 699, table I.
Cassandra, Greece, 3.
Castel Durante (Urbania), Italy, 285.
Castille, 356.
Castle Eden, Durham, 340.
Castor, Northants, 429.
Catalonia, Spain, 56.
La Caunette, Languedoc, 65.
Centuripe, Sicily, 268.
Cerne, West Africa, 321.
Chalcis, Palestine, 504.
Champagne, France, 69.
Chaourse, Aisne, France, 468.
Charnwood, Leicestershire, 32.
Chartres, France, 444, 445, 644, 645, 649.
Cheam, Surrey, 297, 310.
Chedworth, near Cirencester, 446.
Chellaston, Derbyshire, 37.
Chemnitz, Saxony, 66.
Chenôve, Burgundy, 647.
Chertomlyk, south Russia, 470–1, 555, 556.
Chester, Cheshire, 532.
Chesterfield, Derbyshire, 445.
Chesterford, Essex, 218.
Chesters, Northumberland, 415.
Chichester, Sussex, 505–6.
Chiaramonte Gulfi, near Syracuse, 343.
Chiddingfold, Surrey, 443.
Chignal Smealey, Essex, 439.
Cirencester, Gloucestershire, 32.
Civitavecchia, Italy, 520.
Clairvaux, France, 650.
Clermont, Puy-de-Dôme, France, 763.
Clipsham, Rutland, 35.
Clos de Vougeot, Burgundy, 647.
Coburg, Bavaria, 749.
Colchester, Essex, 32, 323, 515.
Collingham, Yorkshire, 33.
Collyweston, Northamptonshire, 34.
Cologne, Germany, 76, 86, 287, 299, 313, 323, 339, 340, 454, 607, 620, 671.
Como, Italy, 60.
Compostella, north Spain, 526.
Conques, France, 462, 468, 474.
Constance, Baden, 195, 691.
Constantinople, 29–30, 99, 197, 288, 422–3, 452, 454, 467, 497, 556, 664, 718, 761–6.
Copais, Lake in Bœotia, Greece, 678.
Corbridge, Northumberland, 409.
Cordel, near Treves, Germany, 313.
Cordouan, France, 524.
Corfe Castle, Dorset, 36.
Corinth, 260, 288, 295, 328, 332, 498, 670.
Corvo, Azores, 568.
Cos, Aegean Sea, 566–7.
Cowdray Manor, Sussex, 522.
Crécy, France, 721, 726.
Crete, 140, 262, 317, 318, 321, 545, 618, 663.
Cromberg, 75.
Cumae, Campania, 323.
Cyprus, 7, 48, 50–1, 55, 199, 274, 276–7, 318–19, 341.

Damascus, Syria, 57, 301, 328, 760, 761, 765.
Damastion, Epirus, 5.
Damme, Oldenburg, 657.
Danzig, 76.
Dean, Forest of, 10, 61, 70.
Dejbjerg, Jutland, 538, 540, 545, 547, 548, 551.
Delft, Holland, 620.
Delos, Greece, 404, 446.
Delphi, Greece, 484, 499.
Dendra, south Greece, 481.
Dieppe, France, 141.
Djemila, Tunisia, 639.
Dodona, Greece, 557.
Dolaucothy, south Wales, 10.
Dolkheim, East Prussia, 559.
Donnerupland, Jutland, 83.
Dorchester, England, 671.
Dover, Kent, 427, 523, 648.
Dreros, Crete, 469, 470.
Dublin, 307, 475, 581.
Dugga (ancient Thugga), near Tunis, 530.
Dura-Europos, Mesopotamia, 363.
Durham, 457, 525, 756, 763.
Dürrenberg, Austria, 166.
Dursley, Gloucestershire, 32.
Dyers Cross, near Pickhurst, Surrey, 326.

Earls Barton, Northamptonshire, 237.
Easby, Yorkshire, 437.
El-Amarna, Egypt, 163, 315, 318, 320, 332, 336, 341.
Elbing, Poland, 583.
Eleusis, Greece, 499.
Elis, Greece, 499, 529.
Ely, England, 679.
Emilia, Italy, 681.
Enkomi, Cyprus, 318, 481.
Ephesus, Asia Minor, 341, 671, 760.
Epiphania (Hama), Syria, 199.
Eshnunna, Mesopotamia, 528.
Esslingen, Germany, 138.
Euboea, Greece, 52.
Exeter, Devon, 293.

Faenza, Italy, 311 n.

Fairlight, Sussex, 34.
Falmouth, Cornwall, 577.
Fano, Italy, 414.
Faversham, Kent, 339.
Fayum, Egypt, 315, 671, 676.
Fleet Ditch, London, 168.
Florence, Italy, 198, 201, 202, 207, 365, 726.
Foggia, Italy, 510.
Folkestone, Kent, 419, 446.
Fortezza, Crete, 321.
Fostat (Cairo), Egypt, 199.
Foyle, Lough, north Ireland, 582.
Frankfurt-am-Main, Germany, 75, 531.
Freiberg, Saxony, 11, 23, 65, 66.
Friesland, province, Holland, 681–3.
Fucinus, Lake, Italy, 679.
Fuente de Guarrazar, near Toledo, Spain, 451, 455.

Galgenberg, Rhineland, 295.
Gallipoli, Greece, 625.
Genoa, Italy, 327, 523, 526, 766.
Gerar, Palestine, 85–6.
Germigny-des-Prés, near Orleans, 762, plate 44.
Gerona, Spain, 641.
Ghent, Belgium, 526.
Gloucester, England, 479.
Gnathia, south Italy, 268.
Godalming, Surrey, 445.
Gokstad, Norway, 237, 579–80.
Goslar, Harz Mountains, Germany, 11, 611.
Göttingen, Germany, 532.
Gragnano, near Naples, 520.
Grand Congloue, island near Marseilles, 569.
Greenstead, Essex, 425.
Groningen, Holland, 681.
Guadalcanal, Seville, Spain, 11.
Gurob, Egypt, 318.
Gytheion, Laconia, 30.

Haarlem, Holland, 619.
Hadra, near Alexandria, Egypt, 268.
The Hague, Holland, 680.
Hallstatt, Austria, 166, 322, 455, 515.
Hama, Syria, 614.
Hamburg, Germany, 189.
Happisburgh, Norfolk, 664.
Harran (Carrhae), Syria, 736, 738.
Hastings, Sussex, 587, 720.
Hattusas, Asia Minor, 494.
Hawara, Egypt, 231.
Heeshugh, Schleswig-Holstein, 165.
Herculaneum, near Naples, Italy, 225.
Hierakonpolis, Egypt, 161, 484.
Hildesheim, Germany, 454.
Hit, Mesopotamia, 375.
Hjortspring, South Jutland, 165, 578–9.
Holt, Denbighshire, 300, 305.
Honfleur, France, 657.
Hook Norton, Oxfordshire, 184.
Hradish, Moravia, 69.
Huddleston, near Tadcaster, Yorkshire, 32, 33, 384.
Hull, Yorkshire, 305, 438, 527.
Hymettos, Mount, Greece, 25–7.

Idbury, Oxfordshire, 37.
Iglau, Czechoslovakia, 66.
Irenopolis, 57.

Jarrow, Durham, 756.
Jena, Germany, 749.
Jerusalem, 715, 761.
Joachimsthal, Bohemia, 11, 13 *and* n.
Jotapata (Tel Jefat), Palestine, 715.
Jumièges, Normandy, 426.

Kaar-el-Banaf, Egypt, 212.
Kamaresa, Crete, 545.
Karnak, Egypt, 164.
Kerch, Crimea, 545.
Kharga Oasis, Egypt, 666.
Kherbet-Agoub, Algeria, 639.
Khorsabad, Mesopotamia, 54.
Khotan, Turkestan, 197.
King's Lynn, Norfolk, 440, 462, 463, 449.
Kingston Down, near Canterbury, 456.
Kish, Mesopotamia, 328.
Klagenfurt, Austria, 546.
Knole, Kent, 256.
Knossos, Crete, 494, 528.
Kocsi, Hungary, 547.
Kongsberg, Norway, 11.
Kouklia, Cyprus, 450, 460.

Lachish, Israel, 319.
Laconia, Greece, 58.
Lambaesis, Algeria, 530.
Langres, France, 647.
Latium, Italy, 270.
Laurion, Greece, 1–4, 43–5, 110.
Layer Marney, Essex, 439, 440.
Leghorn, Italy, 523.
Leoben, Austria, 76.
Leptis Magna, north Africa, 519–20.
Lesbos, Aegean Sea, 58.
Liége, Belgium, 68, 74, 76, 655.
Limoges, France, 464, 467.
Limpsfield, Surrey, 285, 296.
Linares, Spain, 8, 9.
Lincoln, 293, 426, 446, 671, 680.
Lindisfarne, Northumberland, 429, 762.
Lipari Islands, 368.
Lisht, Egypt, 318.
Lisieux, France, 37.
Lissue Rath, Co. Antrim, north Ireland, 255.

Liverpool, England, 532.
Llanfairfechan, Carnarvonshire, 507.
Llantwit Major, Glamorgan, 415.
Llyn Cerrig, Anglesey, 539, 540, 551.
Locarno, Switzerland, 323.
London, 31, 68, 176, 202, 408, 444, 505–6, 508–9, 532–3, 559, 576.
Lübeck, Germany, 64, 76, 531.
Lucca, Italy, 206–7.
Luristan, Persia, 474.
Lusoi, Arcadia, Greece, 416.
Lyons, France, 137, 671.

Macedonia, 48.
Maidstone, Kent, 32, 34.
Mainz, Germany, 454, 671.
Majorca (Mallorca), 358.
Malabar, India, 53.
Malta, 199, 499.
Malton, Yorkshire, 32.
Marche-sur-Meuse, France, 68.
Marseilles, France, 356, 576, 762.
Mauretania, Roman province, north Africa, 227.
Mayen, near Treves, 326.
Mechlin, Belgium, 532.
Mecklenburg, 125.
Megara, Greece, 108, 268–9, 498.
Megiddo, Palestine, 86, 544.
Meissen, Saxony, 307.
Melle (Metullo), Dép. Deux-Sèvres, France, 489.
Melrose, Scotland, 551.
Memphis, Egypt, 671.
Mendip, Somerset, 3, 10, 13.
Merida, Spain, 671.
Metz, France, 726.
Middelburg, Holland, 656.
Milan, Italy, 53, 60, 75, 468, 681.
Miletus, Anatolia, 528–9, 671.
Milos, Aegean Sea, 368.
Mitanni, empire in north Mesopotamia, 541 *and* n. 2.
Mitterberg, Tyrol, 50.
Monkwearmouth, Durham, 425.
Monreale, Sicily, 392.
Mont St Michel, Brittany, 37.
Montagne Noire, France, 69.
Monte Cassino, Italy, 126, 454.
Monte Sant' Angelo, near Foggia, Italy, 454.
Monza, Italy, 451.
Mörigen, Switzerland, 95.
Mostagedda, Tasa, Egypt, 147, 159–60, 162.
Muiden, Holland, 685.
Munich, Germany, 527, 653.
Mycenae, Greece, 167, 202, 398, 413.
Myrina, Asia Minor, 277–8.

Nantwich, Cheshire, 532.
Naples, Italy, 38.
Naqada, Egypt, 161.
Narbonne, France, 69, 70.
Narni, Italy, 508, 511.
Naucratis, Egypt, 131.
Naxos, Aegean Sea, 25.
Nemi, Lake, Italy, 569, 572.
New Carthage (Cartagena, Spain), 5.
Newstead, near Nottingham, 232.
Newton Park, Somerset, 465.
Nicaea (Iznik), north-west Turkey, 718, 760.
Niedermendig, near Andernach, Germany, 31.
Nîmes, France, 404, 405, 671.
Nimrud (Calah), Mesopotamia, 336.
Nineveh, Mesopotamia, 715.
Nishapur, Persia, 302.
Niya, central Asia, 164.
Nogent-le-Roi, France, 649.
Noricum, Roman province, 56–7, 61, 456.
Northampton, England, 532.
Norwich, Norfolk, 35.
Nottingham, England, 38, 306.
Nuceria, Italy, 513.
Nuremberg, Germany, 75, 138, 144, 208, 394, 526, 531–2, 611, 641, 655, 664.
Nuzi, Mesopotamia, 319.
Nydam, south Jutland, 457, 460, 578, 579.
Nyssa, Asia Minor, 671.

Odenwald, Hesse, Germany, 30, 31.
Oeniadae, Greece, 517.
Olympia, Greece, 275, 398, 470, 499.
Olynthus, Greece, 108, 112, 445.
Oman, Arabia, 53.
Orleans, France, 608.
Orsova, Rumania, 512.
Orvieto, Italy, 58.
Oseberg, Norway, 237.
Ostia, Italy, 103, 416, 446, 518, 639, 677, plate 29 A.
Oxford, England, 36.

Pactolus, river in Asia Minor, 42.
Palestrina (Praeneste), Italy, 513, 570.
Palmyra, Syria, 497–8.
Panopolis, Egypt, 140.
Paphos, Cyprus, 459.
Paris, 532, 608, 691, 723.
Paros, Aegean Sea, 24–5, 399.
Passau, Bavaria, 75.
Patleina, Bulgaria, 288.
Pentelicon, Mount, Attica, 24–6, 398.
Perche, La, France, 74.
Pergamum, Asia Minor, 48, 52, 188, 268, 668–9, 671, 760.
Persepolis, Persia, 496.
Peterborough, England, 441, 679.
Petra, Syria, 497–8.

Petrossa, Rumania, 452.
Phaestos, Crete, 494.
Phocaea (Fogliari), near Izmir, west Turkey, 368, 486.
Pingsdorf, Rhineland, 295.
Piraeus, Greece, 402, 403, 413, 517–18, 529.
Pisa, Italy, 582.
Plataea, Greece, 529, 698.
Plattenberg, 75.
Plymouth, Devon, 691.
Poitiers, France, 452, 720, 735, 765.
Polvaccio, near Carrara, Italy, 28.
Pompeii, Italy, 110, 118, 148, 214, 217–19, 224–5, 230, 323, 363, 364, 393, 415, 416, 446, 530, 531, 570, 571, 572, 638, 631, 639, 664, 665, 677, plates 29 B, 34 B.
Pontus, 54, 58.
Populonia, Italy, 50, 52, 56, 60.
Portus, Italy, 518, 522.
Potterspury, Northamptonshire, 296.
Pozzuoli (Puteoli), Italy, 59–60, 355, 520, 530, 531.
Praeneste, *see* Palestrina.
Prety, Burgundy, 599.
Puteoli (Pozzuoli), Italy, 59–60, 355, 520, 530, 531.
Puy, Le, Haute Loire, 764.

Ramsey Abbey, Huntingdonshire, 760.
Raqqa, Syria, 328.
Ravenna, Italy, 227, 413, 421–4, 679.
Ray (Rhagae), Persia, 736–7.
Regensburg, Germany, 531.
Rhodes, 569, 625, 700–1, 702, 716.
Rimini, Italy, 513.
Rio Tinto, Spain, 7, 48, 50.
Rome (the city), 27–8, 53, 103, 107, 122, 192, 404–7, 409–14, 417, 421–2, 454, 464, 529–31, 569–70, 602, 665, 723.
Royaumont, France, 650.

Saar, 69.
St Albans, Hertfordshire, 425.
St Gall, Switzerland, 217–18.
St Georgenberg, Inn valley, Austria, 595.
Saint-Georges de Bocherville, Normandy, 490.
St Ives, Huntingdonshire, 679.
St Lucia, Istria, 322.
St Moritz, Switzerland, 664, 666.
St Osyth, near Clacton, Essex, 440.
St Rémy, Allier, France, 317.
Saintes, France, 288, 293.
Salamis, Greece, 567.
Salerno, Italy, 454.
Salisbury, Wiltshire, 428, 431, 433.
Salonika, Greece, 303.
Salzburg, Austria, 51, 689.
Samarkand, Turkestan, 307.
Samos, Greece, 48, 398, 401, 667–8.
San Cugat, Spain, 641.
Sanchi, India, 556.
Sandon Mount, Hertfordshire, 624.
Sardinia, 44, 45.
Sardis, Sardes, Asia Minor, 600, 671.
Satala, Armenia, 478.
Saxmundham, Suffolk, 440.
Schemnitz, Czechoslovakia, 11, 17.
Schleswig, 165.
Schmalkalden, Thüringia, 76.
Schoonhoven, Holland, 689.
Schwabach, Bavaria, 75.
Segovia, Spain, 671.
Seistan, Persia, 615–16.
Seleucia, Mesopotamia, 316.
Selinus, Sicily, 401, 678.
Sempringham, Lincolnshire, 429.
Semur-en-Auxois, France, 216, 218.
Sens, France, 214–15, 217.
Shetland, 590.
Sidon, Phoenicia, 367, 516.
Siegburg, Rhineland, 287.
Siegen, Prussia, 74, 76.
Siena, Italy, 620.
Silchester, Hampshire, 32, 61, 216, 414–15, 420, 671, 678.
Sinope, Asia Minor, 5, 361, 701, 760.
Siphnos, Aegean Sea, 43, 58.
Siris (Sinni), river, south Italy, 471, 472.
Smyrna, Asia Minor, 671, 760.
Snartemo, Norway, 474.
Sodermanland, Sweden, 551.
Solingen, Prussia, 75.
Soro, Denmark, 69.
Sotiel, Spain, 676.
South Shields, Durham, 455–6.
Southampton, England, 533.
Southwold, East Anglia, 436.
Spaarndam, Holland, 688.
Sparta, Greece, 398, 499, 529, 700.
Speyer (Spires), Germany, 138, 202, 620, 644.
Stade, near Hamburg, 150, 159.
Stoke-by-Clare, Suffolk, 624.
Stonesfield, Oxfordshire, 32, 34.
Stora Kopparberg, Sweden, 65.
Stradonitz, Bohemia, 559.
Strasbourg, France, 354.
Styria, 69, 70, 73, 75–6.
Susa, Persia, 496 (map), 674.
Sutton Courtenay, Berkshire, 93 n.
Sutton Hoo, Suffolk, 169, 450, 451, 454, 456, 458, 461, 464, 492, 579.
Sweffling, Suffolk, 173.
Syracuse, Sicily, 26–7, 499, 604, 637, 658, 699, 702, 714–15.

Tabriz, Persia, 302, 307.
Tanagra, Boeotia, 277.

Taplow, Buckinghamshire, 482, 483.
Tarentum, Italy, 460, 461.
Tarrant Keynston, Dorset, 37.
Tarshish (Tartessus), Spain, 568 *and* n.
Tarsus, Asia Minor, 317.
Tartessus, Spain, 568 *and* n.
Tarxdorf, Silesia, 56.
Taxila, India, 528.
Tel Halaf, Syria, 108.
Tell Defenneh, Egypt, 231.
Tène, La, Lake Neuchâtel, 166, 170, 557.
Terracina, Italy, 510, 520.
Thasos, Aegean Sea, 43, 130.
Thebes, Egypt, 150, 160, 162–4, 228–9, 319, 551, 561.
Thetford, Norfolk, 295.
Tibur (Tivoli), Italy, 27, 406.
Timgad, Algeria, 530–1.
Tiryns, Greece, 29, 398.
Tivoli (Tibur), Italy, 27, 406.
Toledo, Spain, 57, 680.
Tolfa, Italy, 368–9.
Tømmerley, Denmark, 85, 87.
Trebizond, Black Sea, 760.
Trento, Italy, 66.
Treves, Germany, 313, 323, 544, 671, 718.
Trient, Switzerland, 612.
Troy, 701–2.
Troyes, France, 609.
Turin, Italy, 718.
Tusculum, near Rome, 678.
Tutbury, Staffordshire, 37.
Tyre, Phoenicia, 367, 516, 665, 700, 702, 760.

Uitgeest, Holland, 625.
Ur, Mesopotamia, 83, 163 n., 318, 319.
Uruk, Mesopotamia, 494.
Utica, near Tunis, 408.

Vaison, south France, 553.
Valsgärde, Sweden, 313.
Vaux de Cernay, near Paris, 650.
Vebbestrup, Jutland, 84.
Velinus, Lake, Italy, 679.
Venafro, near Naples, 597–8, 602.
Vendel, Sweden, 557, 559.
Venice, 194–5, 207, 301, 326–7, 328, 356, 368, 372, 391, 393, 395, 423, 424, 454, 608, 620, 648, 765–7.
Verulamium, St Albans, Hertfordshire, 169.
Vézelay, Yonne, France, 127, 647.
Vienna, 559.
Vix, Châtillon-sur-Seine, France, 451.
Volterra, Italy, 52, 60.
Vulci, near Rome, 51.

Wadi Natrun, Egypt, 354.
Wallsend, Northumberland, 31.
Walthamstow, Essex, 577.
Waluta (Majorca), 597.
Weald, Sussex, 10, 61.
West Smithfield, London, 165.
Weston Favell, Northampton, 295.
Wilderspool, near Warrington, 56.
Wimborne Minster, Dorset, 237.
Winchelsea, Sussex, 582.
Worms, Germany, 138.
Wroxeter, England, 671.

York, England, 31–2, 293, 425, 428, 429, 444.
Ypres, France, 208, 209, 213, 438.

Zeeland, Holland, 681–4, 685.
Zlokutchene, near Sofia, 105.
Zofingen (Aargau), Switzerland, 306.
Zuider Zee, Holland, 683.
Zürich, Switzerland, 138.

III. INDEX OF SUBJECTS

Abacus, 766, 768.
Abyssinia, 55, 57.
Acids, *see* Chemical industry.
Adzes, 36, 98, 228, 231, 232, 243, 252, 390.
Aes Marianum, 52.
Afghanistan, windmills, 616–17.
Agriculture, 775.
  corn-drying kilns, 99–100.
  digging, 98–9.
  fencing, 99.
  harrows and rakes, 94.
  harvesting implements, 94–7.
  ploughs and ploughing, 81–93.
  seed sowing, 100.
  threshing and winnowing, 97–8, 105–6.
  see also Irrigation; Viticulture.
Alabaster, 37–8.
Alchemy, 351.
  alchemical furnaces, 741–5, 750, plates 42, 43 A.
  alkahest, 732.
  balances, 744–6, 749.
  cost of alchemical materials, 746–50.
  defined, 731–2.
  elixirs, 732.
  sublimation, 740–1, 750.
  theories and equipment: Hellenistic, 733–5; Muslim, 735–9.
  treatises, 736–7, 738–9.
  see also Chemical industry; Distillation.
Alchymia (Libavius), 748.
Alcohol, 141–4, 375 n., 381.

Alkalis, *see* Chemical industry.
Alum, 149–50, 151, 154–5, 350, 355, 366–9, 737, 746.
Ammonia, 355.
Anchors, 76.
Anemourion, 615.
Anglo-Saxon Chronicle, 580.
Animals, representations of, in furniture, 224, 226.
Antimony, 12, 74.
  substitute for tin, 47.
  see also Stibnite.
Anvils, 60, 61.
Apollo Belvedere, 28.
Aqua ardens (alcohol), 375 n.
Aqua regia, 356–7, 358.
Aqueducts:
  connected to water-mills, 596–7, 602.
  origin, 666–9.
  Roman, 7, 10, 47, 404, 405, 418–19, 669 (map), 670–4, 691, plate 39.
Arabs, 356, table III.
  alchemy, 735–9, 758.
  alum recipe, 368.
  camel-breeders, 497.
  glassware, 337, 341–2.
  irrigation and drainage, 680–1.
  metal-work, 453.
  numerals, 766–7, 772.
  sugar production, 372.
  technological treatises, 351–2.
Archery, 542, 705–7.
Archimedean screw, 4, 5, 7, 116, 633, 637, 657, 676–7, 689.
Architecture, 25, 27; *see further under* Building construction; *for individual buildings see* Buildings.
Ardagh chalice, 464, 474, plate 32 A.
Arkwrights, 254.
Armour, 59–60, 472.
  cataphracts: (cavalry), 706; (ships), 566, 570–1.
  Greek, 696, 698.
  medieval, 720–1, 727.
  Roman, 704–7.
Arsenic, 47–8, 74, 362, 737.
Artillery:
  cannon balls of cut stone, 34, 727.
  cannons of cast iron, 75, 726–7.
  missile engines, 698–701, 707–15.
  primitive guns, 378–9, 771.
Artisans:
  arkwrights, 254.
  carpenters, 388–95, 641, 652.
  coffer-makers, 170–1, 255–6.
  craftsmen inferior to philosophers in the ancient world, 603–4; influence of Byzantine craftsmen on the West, 761–4.
  glass-workers (q.v.), 323–5, 327, 335–7, 341.
Artisans (*cont.*):
  leather-workers, 350.
  millers, 601.
  moneyers, 490–1.
  paviours, 525–6, 532–3.
  potters, 261–2, 284–5, 288–91.
  shipwrights, 230, 573.
  shoemakers, 166–8, 170.
  smiths (q.v.), 53, 58–61, 69–70, 75, 236, 242–3, 256, 396, 472, 479, 482.
  stone-masons, 384–8, 641, 662.
  wheelwrights, 552.
Assyrians:
  battering-ram, 715.
  carpenter's tools, 229–31.
  glass, 313; recipes, 328–9.
  roads, 494–5.
Athanors, 739, 740, 743.
Augers, 231, 253, 394, 395, 631.
Automata, 635, 638.
Axes:
  carpenter's, 100, 228, 252, 389–90.
  mason's, 37.

Babylonian leather-work, 151.
Bailers, 8.
Balances, 744–6, 749.
Bapistère of S Louis, 453, 457, plate 32 B.
Barges, 576–7, 686–7.
Barrels and casks, 133, 136–7, 255, 354, plate 3 B.
Basalt, 29, 406.
Basilican churches, 421–2.
Baskets, 16, 115.
Bath stone, 32, 34.
Battering-rams, 702, 715–16.
Bayeux tapestry, 91, 128–9.
Beads, glass, 320, 322, 336.
Beer, 105, 106, 123.
  adulteration, 128.
  manufacture, 139–41.
Bellows, 5, 19, 66, 165, 643, 738, 746, 749.
  hydraulic, 68–9, 73, 348, 612–13, 655.
  piston-bellows, 770.
Bells, casting of, 64.
Benzoic acid, 741.
Bermannus (Agricola), 13, 15.
Bible, 44, 54, 59, 105, 106, 262, 495, 496, 591, 605, 700.
Birth of Venus (Botticelli), 363.
Bitumen, 374, 376, 494.
Black Death (1348–52), 65, 76–7, 128, 142, 651, 691.
Blow-pipes, 329, 331, 332.
Bogies, 548 *and* n.
Bone implements, 147–9.
Book-binding, 169, 173–4, 180–1.
'Book of Trades' (Ypres), 208, 209, 213.
Boring-apparatus, 655–6, 664.
Bossing-tools, 469.

Brace and bit, 229, 230, 253, 393–4, 652–3.
Brass, 53–5, 63, 67.
  nails of, 255–6.
Bread, *see* Flour and bread.
Bricks and tiles, 278, 304–7, 387–8, 749.
  English, 438–40.
  flue-tiles, 420, 421.
  Greek, 398, 402–3.
  Roman, 407–9, 415, 418.
  Romanesque, 429.
Bridges, 771.
  Roman, 508, 510–11, 512, 513, plate 34 A.
Bronze, 47, 53–4, 63, 327.
  casting, 476–80.
  'Corinthian bronze', 43.
  daggers, 481.
  furniture, 223–4, 227, 235–6.
  mirrors, 449, 460, 463.
  patination, 484.
  shields, 460, 462, 468.
  springs, 712–13.
  used in building construction, 403, 417–18; doors of, 454.
  water-pipes, 665.
Bronzo antico, 291.
Brotherhood of man, 605.
Buckets:
  copper, 689.
  leather, 184.
  ox-hide, 14, 16.
  wooden, 16, 100, 255, 689.
Buckwheat, 125.
Building construction (plates 29–31), 384.
  arches, 423, 432, 433–5.
  bricks and tiles, 278, 304–7, 387–8, 402–3, 415, 418, 429, 438–40.
  churches: Gothic, 430–45; Romanesque, 420–30.
  domestic building, 445–6, 641–2.
  doors and locks, 403, 415–16, 425.
  flying buttresses, 435–6.
  fortifications, 701–3, 717–19, plates 40, 41.
  heating systems, 404, 419–20.
  periods: Gothic, 430–46; Greek, 397–404; Roman, 404–20; Romanesque, 420–30.
  plaster-work, 418, 443.
  roofs, 401–3, 414–15; domed, 412–13; drainage of, 404, 444–5; trussed, 442–3; vaulted, 411–12, 427, 428–9, 433–5, 441–2.
  scaffolding, 386, 388.
  timber, 397–8, 413–15, 440–3.
  water-pipes, 418–19, 445.
  windows, 403, 416, 429, 443–4; traceried, 437–8.
Building-stone:
  English, 31–8, 384, 427–8, 436–7.
  Greek, 398–9.
  Roman, 27–32, 406–7.
Buildings (cathedrals, churches, mausoleums, temples, etc.):
  Amiens Cathedral, 435, 436.
  Arch of Constantine, 28.
  Basilica Nova of Maxentius, Rome, 412, 421.
  Blue Mosque, Tabriz, 307.
  Canterbury Cathedral, 429, 431–2, 436.
  Carlisle Cathedral, 437.
  Chartres Cathedral, 444–5.
  Chertsey Abbey, 305.
  Chichester Cathedral, 436.
  Clarendon Palace (near Salisbury), 305.
  Conservatori Palace, Rome, 227.
  Diana, temple of, at Ephesus, 660.
  Durham Cathedral, 436.
  Eltham Palace, 442.
  Ely Cathedral, 438, 442.
  Erechtheum, Athens, 399, 402, 403.
  Eton College, 384, 436.
  Geometric temple, Dreros, Crete, 469–70.
  Halesowen Priory (near Birmingham), 305.
  Jupiter, temple of, Baalbek, 410.
  King's College Chapel, Cambridge, 432–3, 436.
  Pantheon, Rome, 411, 413, 417.
  Parthenon, Athens, 399–403.
  Peterborough Cathedral, 428.
  'Pompey's Pillar', Alexandria, 410.
  Propylaea, Athens, 399, 400, 403.
  Romulus, temple of, Rome (church of SS. Cosma e Damiano), 416.
  St Denis, near Paris, 431.
  St Gervais, Paris, 438.
  St Mark's, Venice, 328, 393, 395, 423, 424, 464, 466–7.
  St Martin, Tours, 429.
  St Paul's (Old) Cathedral, London, 445.
  St Peter's, Rome, 464, 467.
  St Sophia, Constantinople, 423, 424, 454.
  St Vitale, Ravenna, 422–3.
  Sainte Chapelle, Paris, 436.
  Salisbury Cathedral, 428, 431, 433.
  Santa Maria Maggiore, Rome, 410.
  Tamerlane's mausoleum, Samarkand, 307.
  Theodoric's mausoleum, Ravenna, 413, 424.
  'Theseum', Athens, 399.
  Tower of London, 426–7, 432–3, 436, 438.
  Trajan's Column, 407, 705, 707–8, 713 n., 716, 718, 730.
  Treasury of Atreus, Mycenae, 413.
  Westminster Abbey, 305, 385, 426, 435, 438, 442, 478.
  Winchester Cathedral, 427–8.
  Windsor Castle, 34, 35, 38, 443.
  York Minster, 425, 429.
Byzantium (table III):
  architecture, 421–4.

Byzantium (*cont.*):
- ceramics, 287–8.
- enamel, 464–5, 466.
- navy, 578; invention of lateen sail, 583–4; rudders, 585.
- technological achievements, 756–8.
- weapons and armour, 720–1.

Cadmeia, 54, 55.
Calamine (zinc carbonate or silicate), 2, 13, 54.
Calcination, 742.
Calipers, 331.
Camel-transport, 497.
Cam-shafts, 635, 643.
Canals, 657, 679–80, 685–9, 771.
Cannon, *see* Firearms.
Canvas, 363.
Capitalism, rise of, 66–67, 78, 640.
Capitulare de villis (Charlemagne), 608.
Cards (for carding of wool), 193.
Carpentry:
- techniques: Greek and Roman, 233–8, 413–15; medieval, 241–3, 440–3.
- tools, 228–33, 388–95.

Carts, *see* Vehicles.
Carving in wood, 245–7.
Catapults, 34, 375, 573, 699–701, 707–8, 709, 710–15, 723–4, 727.
Celtic peoples:
- chariots, 537–40, 542–3.
- harness, 560–1.
- wheels, 550–1.

Cement, 386–7, 407.
Central heating, 404, 446.
Ceramics, origin of the word, 258; *for further entries see under* Pottery.
Chain-mail, 721.
Chalk, 34–5.
Chalybes, tribes of smiths, 54.
Champlevé work defined, 450 n.
Charcoal, 41, 48, 62, 68, 348, 368, 369, 380.
Chariots:
- harness, 163–4, 165.
- types: ceremonial, 544; fighting, 537, 539, 540–4, 696; racing, 542, 544.

Chasing-tools, 469.
Chatsworth Apollo, 476–7.
Chemical industry:
- acids, 356–7.
- alkalis, 354–5.
- cleansing agents, 355–6.
- combustibles, 369–71, 374–82.
- drugs, 371–2, 732.
- high-temperature processes, 347–8.
- low-temperature processes, 348–50.
- pigments, varnishes, grounds, binding media, 359–64.
- records of ancient techniques, 350–3.
- *see also names of separate substances.*

China:
- alchemy, 732 n.
- beer, brewing of, 140.
- ceramics, 286, 302.
- explosives, 370, 377–8.
- harness, 538, 553; saddles, 556, 557.
- magazine cross-bow, 713.
- paddle-wheels, 607.
- rotary winnowing-machine, 98.
- seric iron, 57.
- silk production, 197–8, 210.
- stern-post rudder, 583.
- stoneware, 286–7, 307–8.
- technology contrasted with that of the West, 755–60; influence on European technology, 768–71.
- viticulture, 131.
- water-mills, 614.
- wheeled vehicles, 544, 549; wheelbarrows, 642.
- windmills, 615, 617.

Chisels, 3, 5, 229, 231, 250, 252–3, 385, 391, 392, 453, 469.
Christianity, 605–6, 761–2; *see also* Bible.
Chroniques de Hainaut (Jacques de Guise), 526.
Church architecture:
- Gothic, 430–45.
- Romanesque, 420–30.
- *for specific churches see under* Buildings.

Cider and perry, 139, 637.
Cineres clavelati, 355.
'Circle of Pythagoras', 760.
Cistercian monks, 192.
Clapboard, 244–5.
'Clay of wisdom', 734.
Clipsham stone, 35–36.
Cloisonné work defined, 450 n.
Clothes:
- cleaning of, 355.
- leather, 148, 150, 156, 158, 160, 161, 165, 175–8.
- linen, 195.
- *for boots, shoes, etc., see* Footwear.

Coal, 62, 68, 347–8, 355, 369–70.
Coccid insects, 362, 366.
Cochineal, 366–7.
Codex Amiatinus, 243.
Coffer-makers, 170–1, 255–6.
Cog-wheels, 635, 648, 654, 675–6.
Coins; Coinage, 10–11, 568.
- brass, 55.
- copper, 53.
- design and manufacture, 491–2.
- die-cutting and stamping, 485–90.
- silver, 13 n., 46.

Colossus of Rhodes, 470.
Combs, 193, 194, 196.
- weaving-combs, 212, 213–14.

Compasses, 386, 391, 749.
Compositiones ad tingenda musiva (*Compositiones variae*), 63, 329, 351.
Concrete, 410–11, 428.
Conservatism of craftsmen, 350, 558, 649.
Coopers, *see* Barrels and casks.
Copper:
 'black' copper, 50.
 coins, 53.
 copper-glaze, 301.
 enamelled and gilt, 462–3, 466, 467.
 mines, 7, 48.
 production by Greeks and Romans, 51–3.
 refining and extraction: from sulphide ores, 11, 48–51; from gold, 42–3.
 solder of, 352.
 used in brass manufacture, 53–4.
 verdigris, 362.
Coppersmiths, 53.
Cordwain, 150–1, 155, 177.
'Corinthian bronze', 43.
Cotton, 199–200, 218.
Cow's milk, Greeks' abhorrence of, 119.
Craftsmen, *see* Artisans.
Cranes, 602, 607, 638, 656–7, 658–62, plate 38 B.
 see also Hoists.
Cranks, 393–4, 635, 651–5.
Crete, windmills, 618–19.
Cross-bow (*gastraphetes*), 647, 699, 707–9, 721–4.
Crow-bars, 252.
Crusades, 756, 764–6, table III.
Cuir bouilli, 171, 173, 180.

Daggers:
 bronze, 481.
 flint, 159, 164.
Damascus pigment (cobalt blue), 301.
Damascene, 57 n.
'David' of Michelangelo, 28.
De Aquis (Frontinus), 405, 419.
De coloribus et artibus Romanorum (Heraclius), 63, 351.
De materia medica (Dioscorides), 371.
De natura fossilium (Agricola), 34–5.
De re metallica (Agricola), 13–24, 290.
De rebus bellicis (*c* 370), 605, 607, 719.
Denmark, ploughs, 83–5, 87, 91.
Diffusion, 321, 379–80; *for trade-routes see* Trade and commerce.
Dikes and polders, 680–9.
Distillation:
 of alcohol, 141–4, 375 n.
 ancient techniques, 348–9.
 by medieval alchemists, 739–45.
 'pelicans', 749, 750.
Diversarum artium schedula (Theophilus), 63–4, 351.
Divining-rods, 24.
Docks, *see* Harbours and docks.
Domesday Book, 35, 91, 93, 611.
Drainage, *see* Irrigation and drainage; *for roof drainage see* Water-engineering.
Draught-animals, 90–2, 94, 514–15, 546–7, 592–3.
 see also Harness; Horses.
Drills:
 bow-drill, 230–1, 631.
 pump-drill, 231.
 scoop-drill, 230–1.
Drink:
 barley-water, 119.
 beer, 105, 106, 123; adulteration, 128; manufacture, 139–41.
 brandy, 142–3, 380.
 cider and perry, 139, 637.
 liqueurs, 142.
 milk, 119, 120, 123.
 wine, 106, 112–13, 116, 119, 123; adulteration, 128; manufacture, 131–3, 135–6, 137–8; trade in, 129–31, 134, 137.
 see also Alcohol; Viticulture.
Dyes, 151 n., 156, 168, 350, 359, 364–7.
 see also under names of dyestuffs.

Earthquakes, 423, 424, 470, 522.
Egypt, 41, 104, 212.
 aqueducts, 670–1; water-screws, 677.
 carpenter's tools, 229–31.
 chariots, 541.
 coinage, 488.
 flax cultivation, 195–6.
 furniture, 221–2, 235.
 glass-making, 312–16; frit, 317–18; glass analysis, 329; glassware, 318–20, 322, 324–5, 327; techniques, 335–6, 340.
 leather-work, 147–8, 149–50, 151, 156, 162–4, 172.
 pharos of Alexandria, 521–2.
 pigments, 359.
 ploughs, 85.
 roads, 494.
 Roman quarries in, 29.
 ships, 564.
Electrum, 485, 486.
Elixir of life, 732.
Emery, 29.
Enamels; Enamelling, 328, 342.
 bassetaille, 461–3, 465, 468, plate 33.
 champlevé, 450 *and* n., 458, 460–1, 462–8.
 cloisonné, 450 *and* n., 458–61, 462, 464–7, 765, 767.
 encrusted, 462.
 painted, 462.
England:
 aqueducts, 671, 691.
 building: Norman, 426–30; Roman, 408–9, 414–15, 419–20; Saxon, 425–6.

England (*cont.*)
- coal, first use, 68.
- cotton, 199.
- diet, medieval, 123–8.
- drainage systems, 679–80.
- glass manufacture, 323, 326, 443–4.
- iron production, 61, 70, 76–7.
- leather-work, 155, 157, 169, 182–4.
- lighthouses, 523.
- metal-work, 450, 451–2, 454–5.
- mining, 9–10, 12–13.
- ploughs and ploughing, 86–7, 89–90, 92–3.
- pottery, 289, 293; bricks and tiles, 305–6; kilns, 295–7, 300.
- quarrying, 31–38.
- roads: city streets, 532–3; medieval, 525; Roman, 503, 505–6.
- ships, 576–9.
- viticulture, 138.
- wall-linings and panels, 244.
- water-mills, 610–11.
- weapons and armour, 721–2.
- wool, 192, 193.

Epinetron, 201.
Eskimo, 148, 149, 582.
Etruscans (table I), 167, 531.
- architecture, order of, 404, 408, 410, 413.
- chariots, 542–3.
- fine metal-work, 470.
- furniture, 223–4, 235, 237.
- pottery, 269–71.
- statuary, 275–6.

Faience, origin of the term, 311 n.; *see also* Glazing.
Fan, winnowing, 97–8.
Feuerwerkbuch, 379–80, 727.
Files, 60, 230, 395.
Filters, 674.
Fire:
- fire-fighting equipment, 184–5.
- firing of bricks and tiles, 305, 307–8; of pottery, 266, 276, 294–9, 300, 307–8.
- used in glass-making, 312; in mining, 8.
- as a weapon of war, 374–82.

Firearms, 75, 378–9, 726–8.
- *see also* Artillery.

Fireworks, 377–9.
Fish-traps, 596.
Flails, 97, 98, 106, 164.
Flax:
- cultivation and preparation, 195–7.
- flax-breakers, 196.

Flint:
- book-binder's, 180.
- as building stone, 35.

Flour and bread, *see under* Food.
Flyer and spinning-wheel, 203–5.
Food:
- adulterants, use of, 128.
- butter, 127.
- diet: Greek and Roman, 118–21; medieval, 123–8.
- disintegrating techniques; milling, 107–12 (*see also* Water-mills); pounding, 106–7; pressing, 112–18; threshing, 106.
- fish, 119, 120, 123–4.
- flour and bread, 103–5, 118, 120–1, 140, 594, 599–601, 605.
- groats, 118–19.
- honey, 127, 138, 139.
- meat, 119, 120, 123–4.
- sweetening-agents, 127, 139, 372.
- vegetables, 120, 124.

Footwear:
- boots, 167, 175–7.
- buskins, 167, 175.
- coloured, 168, 177.
- *monodermon*, 167.
- sandals, 150, 151, 156, 160, 162–3, 166–8.
- shoes, 166–8, 170, 175–7.

Forel (parchment), 180.
Forks:
- agricultural, 98, 99.
- glass-making, 331.
- table, 126, 127.

Fortification, *see under* Building Construction.
Fosse Dyke, 680.
France:
- coal-mining, 68.
- coinage, 489–90.
- drainage systems, 681.
- iron production, 70, 76.
- metallurgy, 62–3; fine metal-work, 474–5.
- pottery, 293, 300.
- roads, 524–5, 526; city streets, 532.
- water-mills, 598–9, 608–10, 639.
- water-supply in cities, 691.
- weapons and armour, 720.
- wine production, 137–8.

Friction in early machines, 652–3.
Frit, 317–18.
Fulling, 193, 214–17.
- fulling-agents, 215, 355.

Furnaces:
- alchemical, 741–5, 750; athanors, 739, 740, 743.
- blast-furnaces, 62, 66, 70, 655.
- bloomery fire, 45, 50–1, 56, 71.
- Catalan furnace, 71–3.
- Corsican furnace, 71, 72.
- glass-making, 329–30, 332, 358.
- Osmund, 70, 72.
- *Saigerhütten*, 78.
- shaft-furnace, 50, 56, 73.
- *Stückofen*, 72–5.

Furnaces (*cont.*)
two types used in industrial chemistry, 347–8.
see also Fire; Kilns.
Furniture (plates 13–15):
design and taste in the Greek and Roman world, 221–6.
Gothic, 240–1, 245.
leather used for, 171, 178–9.
medieval, 241–3, plate 11.
of metal, 235, plate 10; metal fittings, 235–6.
painted, 247–9, 363.
panelling, 179–80, 243–5, 247–8, 363.
specific articles: ambries, 241, 248; arks, 254; armchairs, 223; beds, 221, 235, 248; bookcases, 243; chairs, 178–9, 221–2, 248–9, 252–4, 256–7; chests, 170–1, 180, 222, 237, 241, 247, 253; coffers, 170–1, 255–6; couches, 221, 223; cupboards, 225, 241, 246; drawers, 240–1; sideboards, 224–5, 249; stools, 221, 249, 251–2; tables, 224, 249, 250–1; theatre-seats, 234; thrones, 222–3, 226, 235.
stick-furniture, 251–3.
veneers, 227.
of wicker-work, 237, 253–4.
wood, kinds used for, 226–7, 228.
see also Carpentry; Joinery.
Fustian, origin of the word, 199.

Galena (lead sulphide), 1–2, 44, 43–4, 50, 737.
Gas, origin of the word, 745.
Gears, gear-ratios, 635, 645, 648, 654.
reduction-gearing, 647.
Gems, *see* Jewelry.
Germany:
beer production, 140–1.
coinage, 491–2.
glass-making, 325.
metallurgy, 65–8, 76–7.
mining, 11, 13–24, 65.
pottery, 295, 300.
roads, 524–5, 526; city streets, 531–2.
textiles, 199–200.
water-mills, 611–12.
wine production, 137, 138.
Gesso, 249, 363.
Gimlets, 394.
Glass; Glass-making, 64, 300, 301, plates 24–8.
decoration, 323–4, 328, 338–43.
history: ancient glass-making and glassware, 314–15, 318–22; spread of glass-making in Roman times, 322–5; medieval glass, 325–8.
inlay work, 320.
nature and analysis, 311–14, 333; materials, 328–30, 332, 354, 358–9; weathering, 333–5.
Glass (*cont.*):
opaque glass, 333.
still-bodies, 348–9.
techniques of manufacture: blowing, 322, 337; cane technique, 322, 335; cold-cutting, 322, 336; core-winding, 336–7; mould-pressing, 322, 335–6; sand-core, 321, 322, 336–7.
tools, 331, 332–3.
windows: church, 358–9, 425, 429, 443–4; domestic, 416.
see also Glazing.
Glass-workers, 323–4, 327, 337.
vitrearii and *diatretarii*, 325, 335–6, 341.
Glauber's salt (*sal mirabile*), 745–6.
Glazing:
analyses of ancient glazes, 315–17; Mesopotamian recipe, 316.
of pottery, 261–2, 263–4, 272, 274, 276, 286, 299–302, 357–8.
Gloves, 163, 177–8.
Glycerol, 356 n.
Gold, 63, 747.
alchemical theory of, 731–2.
dissolved by *aqua regia*, 356–7.
gilding of glass, 342–3.
gold-tooling, 173–4, 180, 481.
mines, 10, 12, plate I.
patination, 483–4.
pigment from, 362.
plating on furniture, 227.
refining techniques, 12, 42–3; gold-washing, 65.
repoussé work, 468, 470–4.
specific objects; book-cover at Monza, 451; cross of St Cuthbert, 451–2, 457; crown of Charlemagne, 455, 459; cups, 461–2, 465, 469; purse-mount from Sutton Hoo, 451, 456; reliquary cross from St Peter's, Rome, 464–5, 467; sword of Childeric, 451, 453; Tara brooch, 473–4.
wire, 481–3.
Granite, 30, 406.
Greece:
alchemy, 733–5, 758 n.
beer, Greeks' aversion to, 140.
building construction, 397–404; domestic, 445–6.
carpenters, 232–3.
chemical techniques, 350–1; cleaning-agents, 355; pigments, 359 361–2.
coinage, 485–7.
copper production, 51–53.
food, 103; diet, 118–21; disintegrating by hand- and roller-mills, 108–10, 112; olive-oil production, 121–3; presses, 112–13.
furniture, 221–3.
harbour-installations, 517–18.

Greece (*cont.*):
  horse-riding, 555–8.
  iron production, 58–9.
  irrigation and drainage, 678.
  leather, use of, 150; footwear, 166–8.
  liberal and mechanical arts sharply distinguished, 603–4, 638.
  machinery, 630–6.
  metallurgy, 41; casting techniques, 476–8; extraction techniques, 44–5; fine metal-work, 449–50, 469–70.
  mining techniques, 1–6.
  ploughs, 82–5.
  pottery, 259–69; techniques, 262–7.
  quarries, 24–7.
  roads, 498–9; city streets, 528–9.
  ships, 563–9.
  slave-labour, 591.
  statuary, terracotta, 275–6.
  textiles, 191–7; spinning, 200–1; weaving, 211–12.
  viticulture, 128–33.
  water-engineering, 663–4, 667–9.
  water-mills, 593–5; Vitruvian type, 599–600.
  weapons and armour, 695–703.
  wheeled vehicles, 539–40, 541–5, 549–50, 554.
Greek Anthology, 228.
Greek fire, 375–7, 700, 726.
Grimani Breviary, 192.
Grit-stone, 31–3.
Groats, 118–19.
Gruit, ingredient of medieval beer, 141.
Gunpowder, 1, 64, 377–82, 726–8, 771.
Gypsum, 737.

Hadrian's Wall, 31, 719, plate 40 B.
Hammers, 8–9, 12, 15, 66.
  carpenter's, 231, 243; claw-hammer, 252, 394–5.
  hydraulic, 69, 73, 75, 396, 645–6.
  mason's, 36, 385.
  metallurgical, 643.
Handbooks on metallurgy, 63–5.
Harbours and docks:
  Greek, 517–18.
  medieval, 536, 607, 625; dredging operations, 656–7.
  Roman, 516–17, 518–21.
Harness:
  in antiquity, 552–3.
  bridle-bits, 558–61.
  of chariots, 163–4.
  early Roman, 515, 592.
  harness-leather, 157, 165, 172–3, 183.
  horse-collars, 90, 91, 527, 538, 554, 592, 639, 770.
  rein and stirrup, 516, 538, 556–7, 559.
Harness (*cont.*):
  saddles, 182, 538, 554–7.
  shafts, 547, 553–4.
  traces, 554.
Harp, Egyptian, 149.
Harrows, 91, 94.
Hatchets, 243, 253.
Helena, Empress, sarcophagus of, 29.
Hemp, 197, 363.
Hinges, 235–6, 242 n., 429, 443.
Hippocratic Collection, 119–20.
Hispano-Moresque ware, 304.
Hittites, 151, 163, 167, 494.
Hoes, 5, 16, 98.
Hoists, 388, 602–3, 637, 641–2.
Holland:
  drainage systems, 680–9.
  windmills, 619–20, 625–6.
Hops, cultivation of, 140–1.
Hornblende-granite, 29.
Horses:
  bridle-bits, 558–61.
  cavalry, 697–8, 705–6, 720.
  draught-horses, 592–3.
  horse-cloth, 556.
  horse-collars, *see under* Harness.
  horse-shoes, 76, 77.
  introduced into the ancient empires, 495, 555–6.
  messenger-service, 495–6, 508, 524, 527.
  Nesaean horse of Media, 706.
  rein and stirrup, 516, 538, 556–7, 559.
  saddles, 182, 538, 554–7.
  spurs, 557–8, 559.
  see also Harness.
Horse-whim, 18.
Hose-pipes, 184–5.
Hundred Years War (1338–1453), 76, 651.
Hungary:
  horse-riding, 557, 559.
  vehicles, 547–8.
Hydrochloric acid, 357.
Hypocausts, 419–20.

Iatrochemistry, 745.
India:
  city-streets, 527–8.
  influence on technology of the West, 772.
  steel production, 57.
  water-engineering, 666.
Indians, North American, 148.
Indigo, 350, 361, 364–5.
Inks, 359–60.
Ireland:
  fine metal-work, 464, 471–5.
  ships, 581–2.
Iron, 2, 4, 10, 11, 301.
  cast, 6–7, 56, 72–5, 770.
  Greek and Roman production, 55–61.

Iron (*cont.*):
'hard' and 'soft' iron, 56.
medieval production, 69–78.
pig iron, 70, 73–4.
sheet iron, 75.
trade in, 76.
used in: agricultural implements, 85–6, 95; building construction, 403, 417–18, 437–8, 442; horse-shoes, 515, 561; ornamental use, 429–30, 443; tires, 551; weapons, 455–7, 696, 705, 721, 726–7; wood-workers' tools, 228–32.
wire, 74, 75.
wrought iron, 70, 73, 74.
Irrigation and drainage, 617, 637, 670, 678–81.
Islam, *see* Arabs.
Italy:
coinage, 490–1.
glass-ware, 321–2, 326–7.
pottery, 285, 287–8, 293–4.
weapons, 723.
see also Etruscans; Rome.
Ivory, 225–6, 227, 243.

Japan:
incendiary warfare, 378.
mining techniques, 3–4.
pivoted pestle, 107.
water-mills, 614.
Javelins, 698, 704.
Jewelry, 449.
beads, glass, 320, 322, 336.
cloisonné work, 450–5, 459–61.
gems, artificial, 11–12.
Joinery, 237.
carving, 245–7.
growth of, 242.
tools and techniques, 242–3.
Jupiter, temple of, at Baalbek, 30.

Keys, *see* Locks and keys.
Kilns:
brick-kilns, 388.
corn-drying, 99–100.
lime-kilns, 347–8, 355, 407.
muffle-kilns, 300, 304.
potters', 266, 276, 294–9, 307–8.
Kitāb al-Īḍāḥ ('Book of Explanation'), 736.
Kitāb Sirr al-Asrār ('Book of the Secret of Secrets'), 737.
Klismos, 222, 234.
Knives, 125, 126, 127.
draw-knives, 231, 252–3, 392.
half-moon, 166, 167, 168, 172, 178, 188, 189.

Laboratories, chemical, 748–50, plate 43 B.
Lacquer, 362.
Ladders, 100.
scaling-ladders, 716, 724.
Lamps and lanterns, 278.
cressets, 299.
illuminants, 370.
of leather, 184.
mining, 2, 3, 9, 24.
street lighting, 531.
Lapland, 209.
Lasur, 265–6.
Lathes, 222–3, 232–3, 249–51, 392, 631, 643–4.
Laws:
Drapers' Guild regulation, 202.
mining, 53, 66.
Lead, 2, 9, 43, 50, 67.
extraction processes, 44–5.
gutters and pipes, 404, 444–5, 665.
lead-glazes, 300–1, 316–17.
lead sheathing of ships' hulls, 569, 572.
red lead, 361.
roofs of, 429, 445.
use by Greeks and Romans, 46.
'white' and 'black', 47, 747.
Leather:
buff leather, 175, 182.
chamoising, 154.
colouring and stamping, 173–4.
cordwain, 150–1, 155, 177.
cuir bouilli, 171, 173, 180.
dressing and maintenance, 173.
finishing operations, 156–7.
tanning of pelts: primitive, 147–50; Greek and Roman, 150; medieval, 150–2; early modern, 152–6.
tawing, 149–50, 154–5.
techniques of leather-work, 170–4.
tools, 160, 162, 165–6, 169, 350.
used for: book-binding, 169, 173–4, 180–1, plate 7; boxes and caskets, 170–1, 172, 183, 255–6; buckets, 14, 16, 171, 184; clothes, 148, 150, 156, 158, 160, 161, 165, 175–8, plate 5 A, B; dagger-sheaths, 159; footwear, 150, 151, 156, 160, 162–3, 166–8, 175–7; furniture, 171, 178–9, 256; gloves, 163, 177–8; hangings and screens, 179–80, plate 6 A; harness, 157, 163–4, 183; lanterns, 184; miners' kit-bags, 166; musical instruments, 180, 182; saddles, 182; sedan chairs and carriages, 182–3; shields, 164, 165, 167; vessels, 150, 158–9, 165, 169–70, 171, 174; writing material, 164, plate 5 C.
Leather-workers, 350.
Levels:
spirit, 386.
water, 672.
Levers, 629–30, 632–3, 646.
Liber ignium (Marcus Graecus), 351, 379–80.

Liber sacerdotum, 352.
Liber servitoris (Khalaf ibn 'Abbās al-Zahrāwī), 371–2.
Lighthouses, 518, 521–4, 573.
  pharos of Alexandria, 521–2.
Lime, 152–3, 189, 355, 363.
Limestone, 26, 30, 31–2, 35.
  pōros stone, 398, 400.
Limonite, 71.
Linen, 195, 215–16, 363.
  bleaching, 217–18.
  smoothing, 219.
  see also Flax; Textiles.
Linseed, 195.
Liqueurs, 142.
Lock-gates, 657, 686–8, 771.
Locks and keys, 415–16, 429–30.
Long-bow, 721–2.
Looms, 770.
  horizontal, 212–14, plate 9.
  vertical two-beamed, 211–12.
  warp-weighted, 209, 211.
Lustre-painting, 303–4, 342.
Luttrell Psalter (*c* 1338), 89, 100, 124, 177, 184, 187, 203, 548–9, 596.
'Lymbrique stuff', 76.

Machines:
  developed within monasteries, 649–51.
  fundamental components, 629–30, 660.
  Greek, 630–6, 675.
  mechanization of motion, 640–5, 652–7; of power, 645–9.
  Roman, 636–8.
  transition to the Middle Ages, 638–40.
Magic, 732.
Magnesian limestone, 31, 33.
Magnetism, 771.
Malachite, 362.
Mappae clavicula de efficiendo auro, 63, 351, 352.
Marble, 24–6, 27.
  furniture, 224–5.
  Greek, 398–9, 403.
  Purbeck 'marble', 36–7.
  Roman, 406, 411.
  white Carrara marble, 28, 275, 406.
Mass-production, 603, 604.
Mathematics, 632–3, 675.
Mattocks, 5, 99.
Mechanica (pseudo-Aristotle), 630, 632, 658.
Mercury, 42, 737.
Mesopotamia, 83, 106.
  glass-ware, 319–20.
  glazing, 316–17.
  roads, 494–5; city streets, 528.
  sanitation, 665.
  ships, 566.
  water-mills, 614.
Messenger-services, 495–6, 508, 514, 524, 527.
Metal; Metallurgy:
  casting, 475–80; bell-casting, 64; *cire-perdue*, 478, 480.
  comparative smelting techniques, 70–4.
  die-cutting and stamping, 485–92.
  inlay-work, 63, 450–5.
  medieval metallurgy, characteristics of, 62–9; handbooks, 63–5.
  niello, 453, 454, 480–1.
  ornamentation of weapons, 455–8.
  patination, 483–4.
  refining and extraction: copper, 48–50; gold, 42–3; iron, 56–8; silver and lead, 43–5; tin, 47.
  repoussé work, 468–75.
  and the rise of capitalism, 66–7, 78.
  riveting, 469, 483.
  transmutation of base metals, 352, 731–2, 736.
  use of water-power, 62, 68–9, 73–4.
  see also Enamel; Smiths; Solders; Tools; Wire; *and names of individual metals*.
Milestones, 507.
Mills:
  fulling-mills, 216, 355, 610, 611.
  grinding-mills, 104, 110–11, 349.
  hammer-mills, 610, 611–12, 770.
  hand-mills, 653–4.
  roller-mills, 112, 382.
  rotary querns, 107–11, 127.
  saw-mills, 62.
  stamp-mills, 66, 69, 107, 642.
  treadmills, 16, 591, 602, 636–7, 642, 659–60.
  twisting-mills, 206–7.
  water-mills (q.v.), 69, 349, 381, 395, 593–614, 647–8; tide-mills, 610, 614, 648.
  windmills (q.v.), 349, 395, 396, 610, 614–20, 623–8; post-mills, 536, 623–4, 649.
Mining:
  Greek, 1–6.
  medieval, 10–24.
  mining laws, 53, 66.
  pumping devices, 7–8, 16–19, 654–5; drainage of mines, 19–22.
  Roman, 6–10, 41.
  tools used in, 3, 5, 8–10, 12, 14–19, 24, 110.
  ventilation, 19, 20, 23.
Minium (red lead), 361.
Mirrors, bronze, 449, 460, 463.
Missile-engines, 698–701, 707–14.
Mittelalterliche Hausbuch, Das, 66, 67, 68.
Money, origin of the word, 773 n.
Monodermon, 167.
Mordants, 350, 359, 364, 366–9.
Mortar, *see* Cement.
Mossynoeci (people), 54.
Musical instruments, 149, 180, 182, 256, 614, 634.

Nails, 76, 234, 242 *and* n., 394, 403.
  of brass, 255–6.
  nail-lifters, 231.
  not to be used in setting up cranes, 661.
  of wrought iron, 443.
Naphtha, 374–5, 376.
Natron (sodium carbonate), 354, 355, 737.
Needles, 75.
Niello, 453, 454, 480–1.
Nitric acid, 356.
Numerals, Arabic, 766–7, 772.

Olive-oil, 106, 112–13, 116, 121–3, 127, 356.
Onagers, 713–14, 723–4.
Organs:
  hydraulic, 634.
  wind, 614–15.
Orpiment, 362, 737, 747.
  see also Arsenic.
Osmund, Swedish iron, 70 *and* n., 76.

Painting:
  Cennini's handbook, 352–3.
  on enamel, 462.
  on glass, 323, 327–8, 341–2, 444.
  layering of paints, 362.
  minai technique, 306–7.
  mural painting, 363.
  pigments, varnishes and grounds, 359–64; gesso, 249, 363.
  on pottery, 264–5, 268, 301–4.
  tempera painting, 363–4.
  on wood, 242, 247–9, 363, 443.
  see also Lustre-painting.
Palestine, 106, 140, 319.
Paludina (fresh-water snail), 36.
Paper, 189, 771.
  paper-mills, 608, 611, 643.
Papyrus, 188, 197.
Parchment, 164, 180.
  manufacture, 187–8, 189–90.
  rolls and codices, 188–9, 190.
Parthians, 706.
Pavements, 28, 528–30, 531–3.
Paviours, 525–6, 532–3.
Peat, 683.
Peperino, 406, 410, 529.
Persia:
  brass production, 54.
  ceramics, 285–6, 302; ornamental tiles, 306–7.
  furniture, 224.
  metal-work, 453, 472, 481–2; gold-tooling, 174.
  roads, 495–7.
  spinning and weaving, 191.
  steel production, 57.
  water-wheels, 676.
  windmills, 615–17.
Pestle and mortar, 106–7, 380, 382, 747, 749.
Petroleum products, 374–6.
Pewter, 47.
Philosophers' stone, 732.
Phoenicians, 321, 516, 568.
Physika kai Mystika (Demokritos), 732.
Picks, 3, 5, 8–10, 12, 16, 99.
Pigments, *see* Chemical industry.
Pile-driving apparatus, 410, 436.
Pincers, 630.
Pirotechnia (Biringuccio), 475.
Piston and cylinder, 634.
Planes, 231–2, 243, 250, 253, 392–3, 641.
Plants and trees:
  Acacia nilotica, 151.
  Ammophila arenaria, 682.
  Brassica napus, 125.
  B. rapa, 125.
  Caesalpinia, 367.
  Cannabis sativa, 197.
  Carthamus tinctorius, 366.
  Chenopodium, 157.
  Dipsacus fullonum, 193.
  Fagopyrum esculentum, 125.
  Gypsophila struthium, 215.
  Indigofera tinctoria, 364–5.
  Isatis tinctoria, 364.
  Linum usitatissimum, 195.
  Morus alba, 197.
  Opuntia, 366.
  Punica granatum, 151.
  Pyrus malus, 139.
  Quercus aesculus, 413.
  Q. cerris, 413.
  Q. infectoria, 151.
  Q. robur, 413.
  Reseda luteola, 365.
  Rhamnus saxatilis, 362.
  Rhus coriaria, 151.
  R. cotinus, 366.
  Roccella tinctoria, 366.
  Rubia, 366.
  Salicornia, 682.
  Salsola soda, 354.
  Saponaria officinalis, 215.
  Stipa tenacissima, 197.
Pliers, 331.
Ploughs, 81–93, 770.
Plumb-lines, 252, 386, 672.
Pneumoconiosis, 24 n.
Polders, *see* Dikes and polders.
Polishes, 234.
Polverine (*rocchetta* or *barilla*), 354, 356, 358.
Pontil (punty), 331, 332.
Porcelain, 286–7, 307, 771.
Porphyrogenitus, origin of the title, 30.
Porphyry, 24, 29, 406.
Portland cement, 387 n.

Portland stone, 32, 34.
Portland vase, 323, 338, plate 26 A.
Portugal:
  mining, 8, 12.
  windmills, 618.
Potash, 354–5.
Potassium nitrate, 377–8, 380.
Potters, 261–2, 284–5, 288–91.
Pottery (plates 16–23), 128.
  bowls, 267; Arretine, 272–3; Megarian, 269, 272; *terra sigillata*, 274, 297.
  bucchero ware, 270–1.
  cups, 261, 263, 269, 274–5.
  decoration: *barbotine*, 273; glazes, 261–2, 263–4, 272, 274, 276, 286, 299–302, 357–8; lustre-painting, 303–4; painting, 264–5, 277, 301–2; plastic ornaments, 292–4; relief-work, 268–9, 270–1; *sgraffiato*, 303; slip, 302–3.
  Etruscan, 269–71.
  firing, 266, 276, 294–9, 300.
  Greek, 259–69; techniques, 262–7.
  impasto ware, 269–70.
  jugs, 293, 310.
  majolica ware, 300, 302, 358.
  medieval European techniques, 284–5; Byzantine influences, 287–8; drying and firing, 294–9; Islamic and Chinese influences, 285–7, 302–3; ornament, 292–4; shaping, turning and finishing, 288–92.
  pots and pitchers, 267, 268, 292.
  Roman: Arretine ware, 270, 272–3; *terra sigillata*, 273, 274.
  statuary and reliefs, 274–9.
  stilts, 274, 275.
  stoneware, 286–7.
  vases, 260, 263, 268, 269–70.
  wine-jars, 130–1, 132–3, 135–6, 263.
Power:
  prime-movers, 589–90.
  slaves and animals, 590–3, 600–1.
  steam, 1, 590, 608.
  water, *see* Water-mills; Water-power.
  wind, 614–20, 623–8.
  see also Mills.
Pozzolana (*pulvis puteolanis*), 407, 410, 412.
Presses:
  bag-press, 112.
  beam-press, 112–16.
  coupled with tread-mills, 637.
  screw-press, 113, 116–17, 218–19, 638, 647.
  wedge-press, 117–18.
Printing, 771.
Pudding-stone, 35.
Pulleys, 630, 646–7, 658–62, 723.
Pumps, 7–8, 376, 688–9, 691.
  chain, 17, 19, 22, 617, 626, 770.
  suction, 16–17, 633–4, 654.
Punches, 229, 469, 486, 490.

Qanaats, 666–7.
Quarrying:
  Greek, 24–7, 398–9.
  Roman, 27–32, 406, plate 2 A.
  Saxon and Norman quarrying in England, 32–8.
  tools used in, 36–7, 384.
Quartz, 348.

Rakes, 94, 331.
Rasps, 228–9, 230, 231.
Reaping machines, 96–7.
Reels, 208.
Reliefs of terracotta, 278–9.
Riveting, 469, 483.
Roads and streets:
  in the ancient empires, 493–8.
  city streets: ancient cities, 527–8; Greek, 528–9; medieval, 531–3, 690, plate 38 A; Roman, 529–31, plate 34 B; street lighting, 531.
  Greek, 498–9.
  maps: Britain, 506; Italy, 501; Persia and Asia Minor, 496; Roman Empire, 502.
  medieval trade-routes, 525–7, 537.
  Roman, 500–8: bridges, cuttings and tunnels for, 508–12; control of contracts and finance, 512–13; disintegration, 524; milestones, 507; straightness of, 505; traffic regulations, 514–15.
  rut-roads, 499, 511–12, 539, 548.
  specific roads; Persian Royal Road, 496; Stane Street, 505–7; *Via dell' Abundanza*, 531; *Via Aemilia*, 501, 505, 679; *Via Altinata*, 679; *Via Appia*, 28, 500–1, 504, 505, 507, 510; *Via Cassia*, 501, 510; *Via Domitiana*, 500, 507; *Via Flaminia*, 500–1, 504, 508, 511, 513; *Via Praenestina*, 28, 504, 513; *Via Sacra*, 28; *Via Traiana*, 508, 510.
  tolls and imposts, 524–5, 533.
Rome:
  aqueducts, 7, 10, 47, 404, 405, 418–19, 669 (map), 670–1; water-engineering, 663–4, 671–4.
  beer, Romans' aversion to, 140.
  brass production, 55.
  bridges, 508, 510–11.
  building construction, 404–20; domestic, 446.
  chemical techniques, 350–2; cleaning-agents, 355; pigments, 359, 361–2.
  copper production, 52–3.
  food; diet, 118–21; disintegration by mills, 108–12; flour and bread, 103–5, 600–1; olive-oil production, 121–3; presses, 114–16.
  furniture, 224–5; carpenter's tools, 230–3.
  glass-ware, 322–5, 340.

Rome (*cont.*):
harbour-installations, 516–17, 518–21.
horse-riding, 556, 560–1.
iron and steel production, 55–7, 59–61.
irrigation and drainage, 678–80.
leather, use of, 150, 166, 168.
lighthouses, 518, 521–3.
machinery, 636–8.
metals, use of, 41; extraction techniques, 42–5, 47–50; fine metal-work, 449–50.
mining techniques, 6–10; laws, 53.
mural paintings, 363.
ploughs, 86–8; harvesting implements, 95–7.
pottery, 272–4.
quarries, 27–32.
roads, 500–8, 511–12; city streets, 529–31; *for names of specific roads see* Roads.
sanitation, 530–1.
ships, 569–76.
statuary, terracotta, 276–9.
textiles, 191–7; spinning, 201–2; weaving, 211–12.
viticulture, 133–7.
water-mills, 595–600.
weapons and armour, 703–17; fortifications, 717–20.
wheeled vehicles, 538, 541, 545, 550, 554; harness, 515, 592.
for references to the city see Rome *in* Index of Place Names, *p* 786.
Rotary motion, conversion to reciprocating motion, 652–5.
Rotary querns, 107–10.
Russia, 470, 624.
'Russian' leather, 156, 177.

Saddles, 182, 538, 554–7.
Saggars, 297.
Sails, *see* Ships.
Salt, 125, 127, 381, 683.
mines, 166.
as preservative, 148–9, 155.
salt-glaze, 287, 307.
transport of, 527.
Saltpetre, 369–71, 374, 376–82, 747.
Sand, used in:
building, 407.
glass-making, 314, 320, 323, 328.
metal-casting, 475.
Sandals, 150, 151, 156, 160, 162–3, 166–8.
Sanitation, 527–32, 689–90, 691.
see also Water-engineering.
Sarsen, 35.
Saws:
carpenter's, 228–30, 390–2.
frame-saws, 391–2.
joiner's, 253.
machine-saws, 643–4.
Saws (*cont.*):
two-handled, 230, 389, 391.
used in quarrying and stone-masonry, 29, 30–1, 385.
Saxons, 32–3, 93.
Scientists in the ancient world, 603–4, 606, 638.
Scissors, 100.
Screws, 242 n., 394–5, 631–2.
screw-cutters, 631–2, 655.
screw-jacks, 647.
screw-stoppers, 293–4.
see also Archimedean screw.
Sculpture, *see* Statuary.
Scythes, 75, 95–6, 396.
Seedlip, 100.
Serpentine quarries, 30.
Set-squares, 253, 386–7.
Shagreen, 157.
Sharpening-stones, 97.
Shears, 100, 193, 215, 217–18, 331.
Shellac, *see* Lacquer.
Shields:
bronze, 460, 462, 468, 696.
leather, 164, 165, 167.
'tortoise', 716.
wood, 169.
Ship-sheds, 517–18.
Ships, 390.
barges, 576–7, 686–7.
dredgers, 656–7.
Greek: Homeric warships, 563–7; cataphracts, 566–7; methods of rowing, 565–6.
Irish, 581–2.
medieval, 582–8; seamanship, 583; stern-post rudder, 583–5; three-masted carracks, 586–7.
merchantmen, 567–9, 573–6, 585–6.
paddle-driven, 606, 607, 651, 719.
Roman, 569–76; pleasure-ships, 572; warships, 570–2.
sail, development of, 567, 571, 574, 583–4, 771.
Scandinavian, 578–81.
ship-burials, 578–80, 581; *see also* Sutton Hoo *in* Index of Place Names, *p* 786.
Shipworm (*Teredo navalis*), 569.
Shoemakers, 166–8, 170.
Shoes, *see* Footwear.
Siamese spoke-reel, 206.
Sickles, 94–5.
Siege-engines, 718–19, 724–5, 727–8.
battering-rams, 702, 715–16.
scaling-ladders, 716, 724.
towers, 716–17, 725–6.
see also Missile-engines.
Sieves, 5, 104.
Silex, 28, 529.

Silk, 756, 761.
  bleaching and dyeing, 365.
  early sericulture, 197–9.
  old silk-route, 759, 763.
  spinning techniques, 205–7, 644–5, plate 8.
  warping, 210.
Silver, 63, 356–7, 747.
  assay by touchstone, 45–6.
  coins, 46, 485.
  cups, 449–50, 463.
  mines, 1–4, 5, 65, 110.
  plating on furniture, 227.
  refining techniques, 42 n., 43–5, 65–7.
  repoussé work, 468, 470–2.
  wire, 452–3.
Sinopia, 361.
Siphons, 668–9, 671.
Slates, roofing, 34.
Slave-labour, 590–2, 600–1, 638–9, 671–2.
  and the rise of Christianity, 605–6.
Sleds, 19.
Smalt, 361.
Smiths:
  blacksmiths, 58–61, 69–70, 236, 242, 256, 396; London Blacksmiths' Company, 243.
  bronzesmiths, 472.
  coppersmiths, 53.
  locksmiths, 60.
  needlesmiths, 75.
  silversmiths, 482.
  swordsmiths, 479–80.
Soap, 355–6.
Soda, 354, 737.
Solders, 46, 47, 352, 453, 665–6.
Spades and shovels, 9, 12, 16, 98–99.
Spain:
  aqueducts, 47, 670, 671.
  cotton, 199.
  flax, 195.
  glassware, 323.
  gold production, 42.
  iron production, 61.
  leather, 150–1, 180.
  mining, 5, 11–12, 13, 46–8.
  ornamental tiles, 307.
  viticulture, 134.
  wool, 192.
'Spanish' leather, 150–1, 179.
Spears, 696, 698.
  Roman *pila*, 703–5.
Sphyrelaton work, 469–70.
Spindle-whorls, 201–2.
Spinning:
  silk '205–7, 644–5.
  wool and vegetable fibres, 200–5.
Spinning-wheels, 201–5, 644–5, 656.
Spoons, 125, 127.
Springs, spring mechanism, 629–30, 643–4, 712–13.
Stamps, stamping-mills, 66, 69.
Statuary:
  Colossus of Rhodes, 470.
  Greek temple figures, 470.
  terracotta, 274–9.
Steel, 75.
  Chalybean, 5, 59.
  crucible and damask steel, 57.
  Greek and Roman production, 60.
  tools of, 453.
  trade in, 76.
  weapons of, 63.
Stibnite (antimony trisulphide), 12, 47, 74, 737.
Stills, *see* Distillation.
Streets, *see* Roads.
Sublimation, 740–1, 742, 750.
Succinic acid, 741.
Sugar, *see* Sweetening-agents.
Sulphur, 42 n., 47, 352, 371, 376, 380–2, 737, 740–1, 747.
Sulphuric acid, 357.
Sumac, tanning-agent, 151, 155–6.
Sumerians, 164.
Summa perfectionis (Geber), 356, 738–9.
Sweetening-agents, 127, 139, 372.
Switzerland, 219.
  relief-tiles, 306.
Swords, 57, 60, 69, 75, 651, 704.
  ornamented, 451, 453, 455–8, 474.
  sword-moulds, 479–80.
Syria:
  glass industry, 322, 324–5, 327–8, 337, 340.
  merchant ship model, 575–6.
  Roman roads, 504.
  water-wheels, 614.

Talmud, 117.
Tannin, 151, 155.
Tanning, *see* Leather.
Tartar (argol), 354–5, 358.
Tawing of leather, 149–50, 154–5.
Taynton stones, 35.
Teazel, fuller's, 193, 217.
Temperature, alchemical classification of, 740.
Templets, 289.
Teredo navalis (shipworm), 569.
Terracotta, 262, 269, 274–9.
Textiles:
  fibres: cotton, 199–200; flax and hemp, 195–7; silk, 197–9; wool, 192–5.
  finishing, 214–19.
  reeling, winding, and warping, 208–10.
  spinning techniques: silk, 205–7; wool and vegetable fibres, 200–5.
  weaving, 210–14.
Théâtre des Instrumens Mathématiques (Besson), 626.

Theodosian Code (A.D. 438), 63, 539–40.
Threshing-sled, 106.
Tickhill Psalter, 177.
Tiles, *see* Bricks and tiles.
Time-scales, 753–4.
Tin, 63, 747.
  alloys, 47.
  mines, 10, 12–13, 46, 67.
Tongs, hinged, 58, 60, 61, 630.
Tools, used for:
  agriculture (q.v.), 81–100.
  carpentry, 228–33, 252, 389–95, 641–4, 652, plate 12.
  coin-stamping, 488.
  cooking, 127–8.
  food disintegrating techniques, 106–18.
  glass-making, 331, 332–3.
  joinery, 242–3, 250, 253.
  leather-work, 160, 162, 165–6, 169, 350.
  metallurgy and fine metal-work, 58–61, 66, 67, 396, 453, 469.
  mining, 3, 5, 8–10, 12, 14–19, 24.
  pottery, 274, 289–90, 294 (*see also* Wheels).
  quarrying and stone-masonry, 36–7, 384–6.
  tanning, 147–9, 152–4.
Tortoise-shell as a veneer, 227.
Touchstones, 45–6.
Toys, 6–7.
Trade and commerce, 321, 493.
  in grain, 103.
  Greek trade-routes, 567–9.
  influence on the spread of technology, 772–3.
  medieval trade-routes, 525–7, 537, 759 (map), 760, 769 (map).
  Roman trade, 573–4.
  in textiles, 191, 197, 527, 763–5.
  tolls and imposts, 524–5, 533.
  in wine, 129–31, 136–7.
Transmutation of base metals, 352, 731–2, 736.
Transport:
  by camel and dromedary, 497.
  of grain, 103, 498.
  litters and sedan chairs, 182–3.
  of metals and ores, 568 n.
  of salt, 527.
  of stone blocks, 401, 545, 660, 662.
  of wine, 129–31, 136–7, 527, 545.
  of wool, 527, 537.
  see also Roads; Vehicles.
Travertine, 27, 406.
Treadmills, 16, 591, 602, 636–7, 642, 659–60.
Trebuchets, 643, 644, 724.
Trees, *see* Plants and trees.
Tribulum, 97.
Triticum, 103 n.
Tufa, 406, 410.
Tunnels, 511.
Turnery, 222–3, 232–3, 234, 237, 249–51.
Tutankhamen's tomb, 163, 221–2, 235, 319–20, 336, 460, 484.
Tutty, 54.
  recipe, 352.

Utrecht Psalter, 95–6, 651.

Varnishes, 248, 249, 362.
  see also Lacquer.
Vehicles:
  carts and wagons: *caretta*, 541; Greek, 544–5, 549–50; *hamaxa*, 498, 539; medieval, 546–7, 552, plate 37 A; Roman, 545, 550.
  coaches, 547–8.
  see also Chariots.
Vellum, 180, 188.
Veneering, 227, 240.
Verjuice, 138 *and* n., 357.
Vermilion, 5–6, 361, 747.
Vikings, 237, 474, 479, 557, 562, 579–81, 720.
Villanovan culture, 269, table I.
Vinegar, 8 n., 128, 133, 357, 360, 362, 747.
Virgula divina, *see* Divining-rods.
Viticulture:
  Greek, 128–33.
  medieval, 137–9.
  Roman, 133–7.
  Spanish, 134.

Wainscot, 244.
Warping, 209–10.
Water-budgets, 184, 187.
Water-carriers, 187, 689.
Water-engineering, 404–5.
  dikes and polders, 681–9.
  irrigation and drainage, 617, 637, 670, 678–85.
  medieval water-supply, 689–92.
  pipes and sewers, 418–19, 527–31, 664–6.
  roof drainage, 404, 444–5.
  water-installations of Rome, 671–4.
  water-lifting, 675–8.
  wells and cisterns, 663–4, 666, 689.
  see also Aqueducts.
Water-mills, 69, 349, 381, 395, 770, 775.
  floating mills, 607–8, 647.
  Greek or Norse mill, 593–5.
  introduced into Rome, 600–1; obstacles to its introduction and spread, 601–6; rapid spread in Middle Ages, 608–14.
  specialized functions of, 609–11.
  tide-mills, 610, 614, 648.
  Vitruvian, 595–600.
Water-power:
  fulling-mills, 216.
  hydraulic hammers, 69, 73, 75.
  twisting-mills, 207.
  use in metallurgy, 62, 68–9, 73–4, 75, 78, 348.

Water-power (*cont.*):
  water-wheels, 7, 10, 20–1, 23, 66, 68, 78, 605, 675–6, 677–8, 681.
  see also Aqueducts; Water-mills.
Water-wheels, *see* Water-power.
Weapons:
  ornamented, 451, 453, 455–8.
  pyrotechnic, 374–82, 771.
  of steel, 63, 456.
  see also Firearms *and names of individual weapons*.
Weaving, 198, 210–14.
  see also Looms.
Wedges, 3, 8, 12, 15, 30, 36, 117–18, 242 n., 252, 630, 712.
Weights and measures, 128.
Welding, 58.
Wells and cisterns, 663–4, 666, 689.
Wheat, 103–4.
Wheelbarrows, 17, 546, 547, 641, 642, 770.
Wheels:
  on carts, 545, 548–50.
  on chariots, 539, 542, 549, plate 5 D.
  cog-wheels, 635, 648, 654, 675–6.
  'dished', 552.
  evolution of the spoked wheel, 548–52.
  paddle-wheels, 606, 607, 647–8, 651, plate 37 B.
  on ploughs, 88–9.
  potter's, 261–3; kick-wheels, 288–90; Piccolpasso's diagram, 291.
  scoop-wheels, 685, 688.
  solid, 545, 548, 550.
  spinning-wheels, 201–5, 644–5, 656.
  tires, 163, 549, 551–2.
Wicker-work, 237, 253–4.
Winches, 630, 642, 646.
  horse-winch, 685.
Windlasses, 14, 16, 658–9, 686.
Windmills, 349, 395, 396, 610, 689, 775.
  development in Europe, 617–20, 625–8.
  post-mills, 536, 623–4, 649, 689.
Wind-organ, 614–15.
Wine, 106, 112–13, 116, 119.
  adulteration, 128.
  dregs, 354–5.
  manufacture, 131–3, 135–6.
  trade in, 129–31, 134, 137.
  see also Viticulture.
Wine-skins, 160, 161, 167.
'Winged Victory', 566–7, plate 35.
Wire, 336.
  gold, 481–3.
  hydraulic wire-drawing machine, 655.
  iron, 74, 75.
  silver, 452–3.
Woad, 349, 361, 365.
Wood:
  ashes, 354, 358.
  barrels and casks, 133, 136–7, 255, 354.
  bending, 234.
  in building construction, 397–8, 413–15, 440–3.
  carvings, 245–7.
  doors and locks, 415–16, 425, 429–30, 443.
  as fuel, 347–8, 369.
  ploughs, 85, 89.
  furniture (q.v.), 221–56 (especially 226–8).
  seasoning, 233.
  shields, 169.
  vessels, 100, 127–8.
  wall-linings and panels, 243–5, 247–8, 363, 443.
  water-pipes, 664.
  wheels, 549–52.
  for carpenters see under Artisans.
Wool, 192–5, 218.
  transport of, 527.
Writing-materials:
  ink, 359–60.
  leather, 164.
  paper, 189, 771.
  papyrus, 188, 197.
  parchment, 164, 180, 187–90.
  vellum, 180, 188.

Zinc, 2, 53–4, 67.

PLATES

Interior of a Japanese gold-mine at Sado of about 1850. The appearance of the mine closely resembles that of the ancient Greek mines of Laurion. (p 3)

A. *The big squared limestone block in the quarry south of Baalbek, near Damascus. It was blocked out, but not used, for the Temple of Jupiter. Second century A.D.* (p 30)

B. *Marks of cross-ploughing under a barrow at Vesterlund, Jutland. They are about one inch in cross section and depth. Bronze Age.* (p 92)

A. *Peasants' wedding-feast. The menu seems to be simple. Pottery dishes containing the food, probably a cereal, are carried on a coarse tray. A knife and slices of bread are lying on the edge of the long table. Note* (*left corner*) *the size of the mugs being filled, with beer perhaps, to celebrate the occasion* (p 216). *The rural stick-furniture is also worth noting* (p 252). *Pieter Brueghel*, c *1568*

B. *Twelfth-century north Italian sculpture of a cooper. He is pushing the hoops of his stave-built cask into place by knocking with his right mallet on the left one.* (pp 136, 245, 255)

C. *A migrant philosopher wearing a wallet hanging from his girdle. English sculpture. Sixteenth century.* (p 183)

PLATE 4

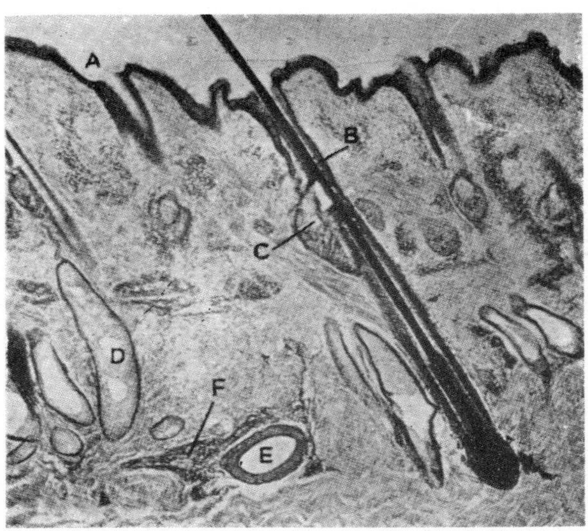

A. *Photomicrograph of epidermis and superficial layers of an ox-hide, showing at* (A) *epidermis, at* (B) *hair-sheath or follicle with hair, at* (C) *sebaceous or lubricating gland, at* (D) *sudoriferous or sweat-gland, at* (E) *artery, at* (F) *vein.* (pp 148, 187)

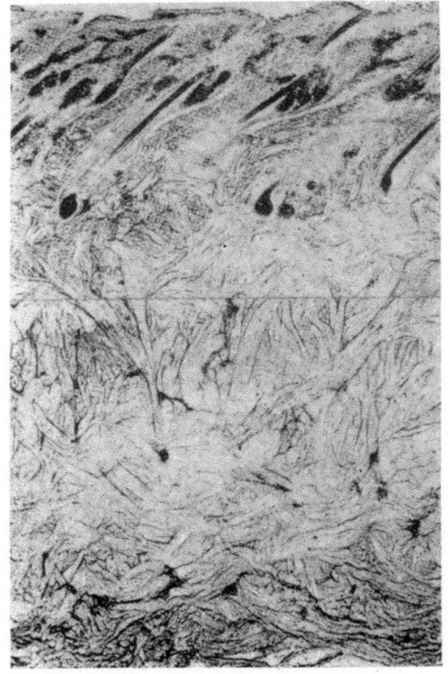

B. *Photomicrograph of ox-hide after de-hairing but before tanning, showing outer or epidermal layer and inner dermis or corium. The inner layer is composed of long fine collagen fibrils forming fibres, which are grouped into bundles and bound together by reticular tissue.* (pp 148, 187, 189)

C. *Photomicrograph of tanned sole-leather, showing the grain side compressed by heavy rolling.* (pp 148, 149, 189)

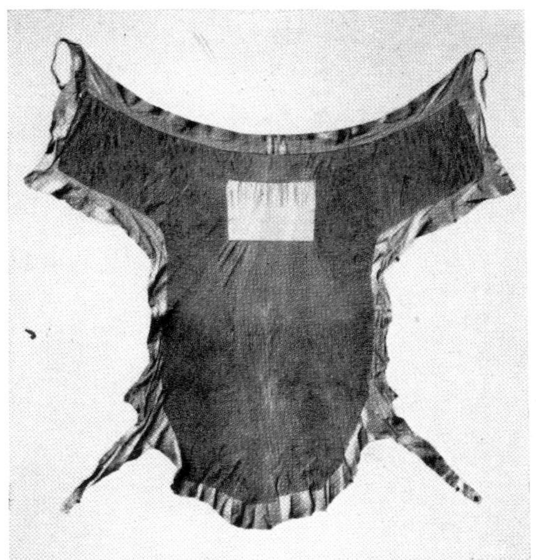

A. *Egyptian loin-cloth made of a gazelle-skin. By fine slitting all but the border and an oblong patch has been converted into a net.* c *1500 B.C.* (p 172)

B. *Enlarged detail showing the slits cut into the skin which allowed it to stretch.* (p 172)

C. *Section of the 'Book of the Dead'. The writing in hieratic, on leather, is made visible by an infra-red photograph.* c *3000 B.C.* (p 164)

D. *Chariot-wheels from the tomb of Tutankhamen, showing the lashing of all joints with raw-hide, and the leather tires.* c *1350 B.C.* (p 163)

PLATE 6

A. *Portion of leather wall-hanging covered with silver foil, richly embossed, and painted with flower scrolls and birds.* (p 179)

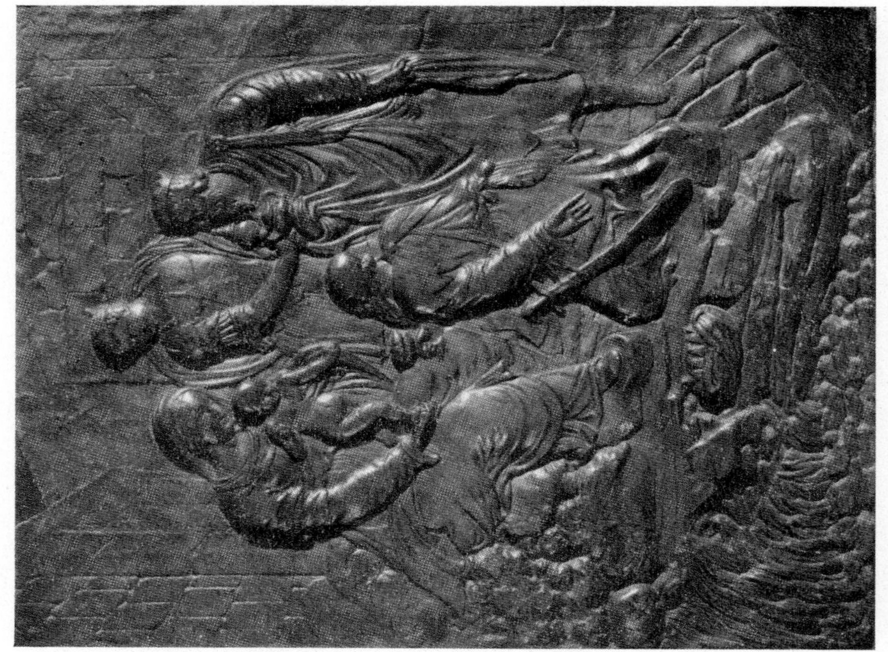

B. *Leather panel embossed in high relief and tooled. Seventeenth century.* (p 180)

A. *Red morocco binding blind- and gold-stamped, thought to have been done by Neapolitan craftsmen,* c *1480–90, for Matthias Corvinus, King of Hungary. Gold-impressing with hot iron is thought to have spread from Moorish Spain and Naples about this date all over the continent, coming to England during the reign of Henry VIII.* (p 180)

B. *Gold-tooled binding by Roger Payne (1739–97) in diced brown Russia calf.* (p 180)

PLATE 8

Reeling silk in sixteenth-century Florence. The method is similar to the Chinese one represented in figure 160. The cocoons were placed in a trough of hot water to soften the gum, and the filaments wound into four hanks on a long reel. The reel was fitted with a crank and was slowly rotated by means of a lever. The crank is clearly seen in the second reel which is being examined. The furnace to heat the trough and the chimney leading to the open air are technical improvements on the Chinese system. c 1510. (p 198)

'The suitors surprising Penelope', a fresco by Pinturicchio (1454–1513). The artist has depicted her working at a contemporary two-heddle Italian loom of the type shown in figure 182 but with a high frame. The heddles, treadles, reed, cloth-beam, and shuttle are all clearly shown (p 213). *A maidservant (lower left) winds a shuttle-spool from a ball of wool with which the cat is playing. Through the window is seen the ship on which Ulysses has returned—a three-masted vessel of the carrack type.* (p 586)

A. *Headboard of a bronze couch from Amiterno, near Rome, showing the upper finial in the form of a mule's head, and a scene inlaid in silver showing satyrs and maenads gathering grapes. First century* A.D. (p 227)

B. *Bronze table or stand from a private house in Herculaneum, a richer and more ornate version of the type of tripod shown in figure 198. First century* A.D. (p 224)

Chair of St Maximin at Ravenna, of wood (renewed and restored in several places) and covered with ivories in the forceful Syro-Hellenistic style of the sixth century A.D. *Probably made at Antioch in the eastern Byzantine Empire.* (pp 225, 227, 243)

St Joseph, the carpenter, making mousetraps in his workshop. He is boring holes with a brace. This is one of the earliest representations of this tool (*see p 394*)*; note also the auger, chisel, hammer, pincers, chopper, one-handled saw, and squaring-axe spread out over his work-bench and stool, both of stick-furniture type. The right-hand wing of an altarpiece by Robert Campin,* c *1420–38.*

(pp 242, 243, 252, 653)

A

B

C

D

A. *An ambry or food-cupboard of panelled construction. The oak shows signs of being adzed. The earliest extant drawers in English furniture are to be found in ambries such as this example. First half of the sixteenth century.* (pp 241, 243, 249)

B. *Court cupboard with open shelves of an unusual design with heraldic animal supports in upper tier. English, late Elizabethan.* (p 249)

C. *A court cupboard with small cupboard or ambry fitted in upper tier. Such a court cupboard may have been a livery-cupboard. English, early seventeenth century.* (p 249)

D. *An oak joined press with decoration of various types: low relief carving, gouge and punch work. Late Elizabethan.* (pp 246–7, 249)

A. *Oak trestle table. The trestles are held rigid by the central stretcher; so that the table can be easily dismantled the stretcher is fixed by wedges. Sixteenth century.*
(See detail figure 229 and pp 249, 251)

B. *Long table with turned trestle supports, the top formed of a single oak board. English, late sixteenth century.*
(pp 249, 251)

C. *Oak draw-leaf table with turned and carved bulbous legs. English, Elizabethan.* (pp 249, 251)

D. *Walnut draw-leaf table of French or continental provenance. Early seventeenth century.* (pp 249, 251)

A. *Small oak box decorated with chip-carved roundels. Late sixteenth century.* (See detail figure 215)

B. *Oak table-box, the front carved in high relief with a male and female figure representing the owner and his wife. Below are their initials, and on the right-hand side of the box is the date, 1597, probably that of their marriage.* (See detail figure 211)

C. *Chest constructed of six boards. Front decorated with three rows of Gothic ornament, the centre being a trail of conventionalized leaves. English, first half of the sixteenth century.* (p 242)

D. *Oak chest with linenfold panels. English, mid-sixteenth century.* (pp 245, 249)

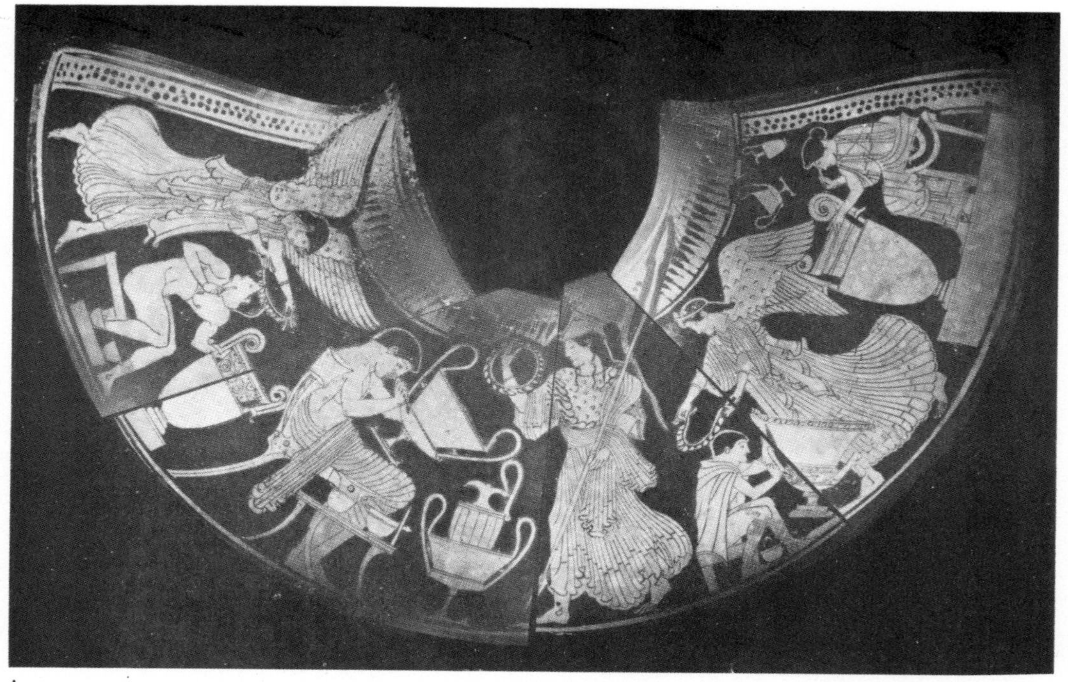

A

B

C

Athena and Victories crowning potters at work

A. *Unrolled painting from an Athenian hydria found at Ruvo in the grave of a woman.* B, C. *Two views of the hydria. Fifth century* B.C. (p 261)

A. *Potters at work. Painting on an Attic krater of the fifth century* B.C. (p. 261)

B. *White-ground Athenian krater. Fifth century* B C. (p 265)

PLATE 18

A

C

D

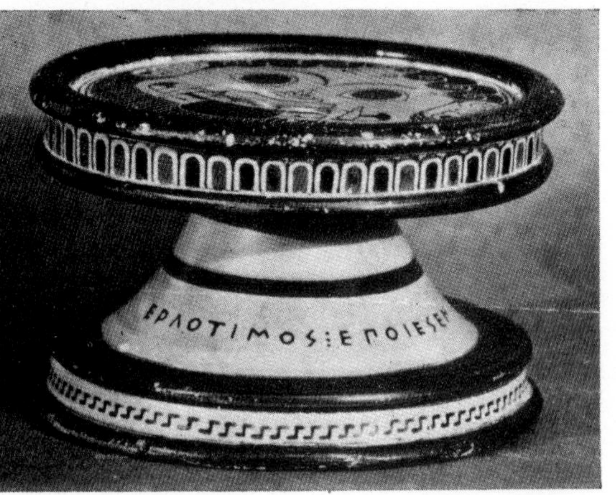

B

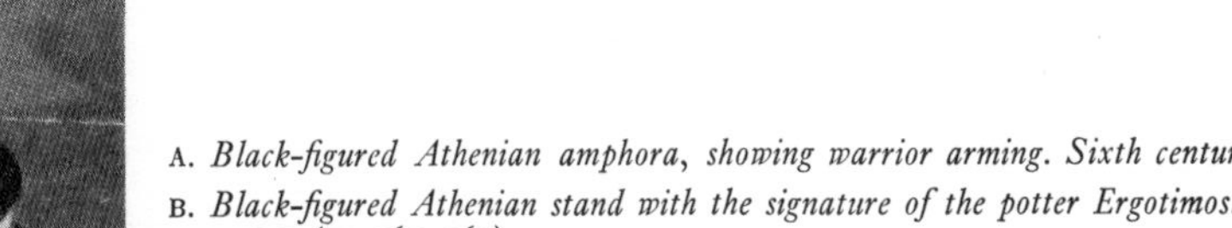
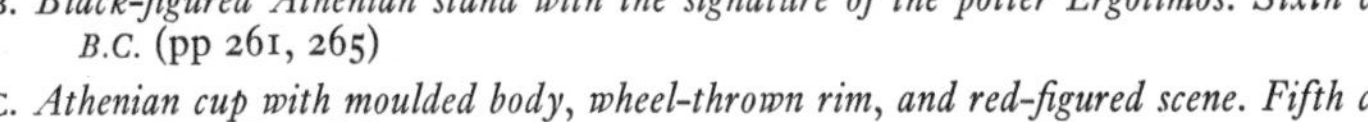

A. *Black-figured Athenian amphora, showing warrior arming. Sixth century* B.C. (p 265)

B. *Black-figured Athenian stand with the signature of the potter Ergotimos. Sixth century* B.C. (pp 261, 265)

C. *Athenian cup with moulded body, wheel-thrown rim, and red-figured scene. Fifth century* B.C. (pp 265, 267)

D. *Black-figured Etruscan hydria. Sixth to fifth century* B.C. (p 271)

Greek terracotta group of Zeus and Ganymede, under life-size, from Olympia, front and rear views. 475–470 B.C. (pp 275, 276)

A

C

B

A. *Head of a Greek terracotta statue of Athena, height 22·4 cm., from Olympia.* c *490* B.C. (p 275)

B. *Head of the terracotta statue of the Apollo from Veii. Sixth to fifth century* B.C. (p 275)

C. *Terracotta antefix with satyr and maenad from Satricum. Sixth to fifth century* B.C. (p 278)

A

B

C

A, B. *Terracotta statuettes of cupids. Fourth to third century* B.C. (p 277)
C. *Melian relief with representation of Bellerophon. Fifth century* B.C. (p 278)

A

B

C

D

E

A. *Dish of* sgraffiato *ware with mottled green and bronze lead-glaze. The colour, initially put into the scratched lines, may be seen running down in streaks through the lead-glaze. Nishapur, Persia. Ninth century. Diameter* $10\frac{1}{4}$ *in.* (pp 302, 303)

B, C. *Drug-jars* (albarelli). B. *Painted in colours (blue, turquoise, and black. Height 13 in. From Sultanabad, Persia. Fourteenth century.* C. *Painted dark blue. Height* $15\frac{3}{8}$ *in. From Faenza, Italy. Fifteenth century. This* albarello *shape is of oriental origin, and became widely adopted in the west from the later Middle Ages onwards, a good example of the profound influence of the orient on European ceramics.* (p 286)

D. *Painted screw-topped bottle decorated with plastic ornament, made at Urbino in the workshop of Orazio Fontana. Height* $17\frac{1}{2}$ *in.* c *1560.* (p. 293)

E. *A tin-glazed plate painted in purple-black, pale blue, and yellow ('Protomajolica' ware). Diameter 8 in. From Corinth, late twelfth or thirteenth century. Fishes of this style are found on such pottery from Palestine (Athlith), and Apulia, south Italy, and illustrate the Levantine element in the development of painted tin-glazed wares in Italy.* (p 287)

A

B

C

A. *A large dish, made at Manises, south Spain, in the fifteenth century, painted in yellow (copper lustre) and blue. Diameter* $16\frac{7}{8}$ *in. This shows a combination of European heraldic devices with arabesque patterns of Islamic inspiration. European heraldic devices themselves developed with some Islamic influence, and Islamic pottery sometimes bears Islamic heraldic devices.* (pp 286, 304)

B. *Majolica plate from Caffagiolo, near Florence; blue, yellow, orange, green, red, and purple. Diameter* $9\frac{1}{4}$ *in. It probably represents the potter Jacopo Fattorini painting pottery.* c *1515*. (p 302)

C. *Majolica dish from Deruta, central Italy, showing a potter's wheel. Diameter* $14\frac{1}{2}$ *in.* c *1530*. (p 288)

PLATE 24

A

B

C

D

E

F

G

A. *Goblet of blue-green faience, from Medûm, Egypt,* c 1550–1350 B.C. *Scale* c $\frac{1}{3}$ (pp 311, 315). B. *Kohl-pot of green glazed steatite, from Riqqeh, Egypt,* c 1550–1350 B.C. *Scale* c $\frac{7}{21}$ (p 315). C. *Green-glazed pottery jar, from Kish, Iraq. Neo-Babylonian. Scale* c $\frac{3}{7}$ (p 316). D. *Blue glass vase with yellow painted decoration and cartouche of Thothmes III.* c 1500 B.C. *Scale* $\frac{3}{5}$ (pp 320, 341). E. *Cane inlay fragment. Second–first century* B.C. *Scale* $\frac{1}{1}$ (pp 322, 338). F. *Jar, green 'frit' ware, from Kahun, Egypt.* c 1450–1350 B.C. *Scale* c $\frac{2}{5}$ (p 317). G. *Cup in lead-glazed pottery, green. From Kuklia, Cyprus. First century* A.D. *Scale* $\frac{1}{2}$. (p 317)

PLATE 25

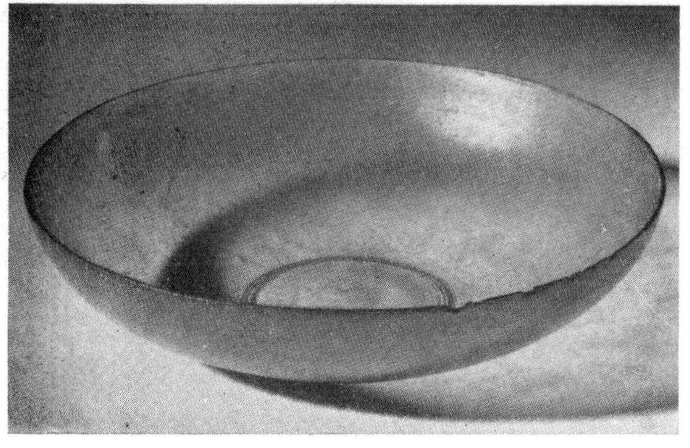

A

B

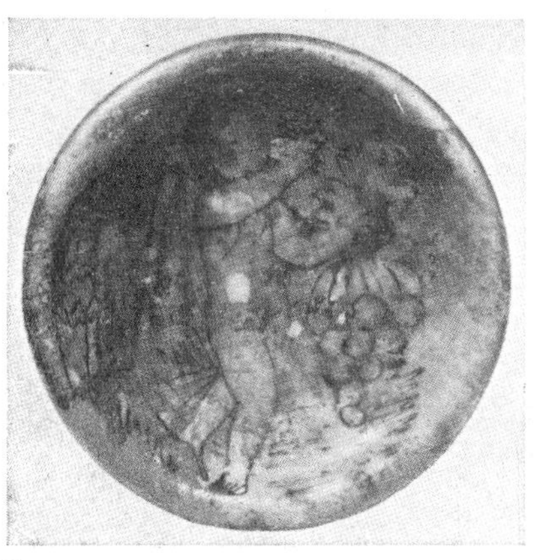

C

D

E

A. *Greenish bowl, mould-pressed, ground and polished. From the Aegean archipelago. First century* B.C. *to first century* A.D. *Scale* $\frac{1}{3}$ (p 335). B. *Bowl with gilt decoration between two layers of colourless glass, from Canosa, Italy. Hellenistic period. Scale* $\frac{1}{3}$ (p 342). C. *Colourless lid, with painted design on its under side. From Cyprus. Second century* A.D. *Scale* $\frac{2}{3}$ (p 341). D. *Green bowl, with brown lustre decoration. From Atfih, Egypt. Eighth to tenth centuries* A.D. *Scale* $\frac{1}{3}$ (pp 327, 342). E. *Greenish bowl with mould-blown corrugations, Frankish. Sixth century* A.D. *Scale* $\frac{1}{2}$. (p 339)

PLATE 26

A

B

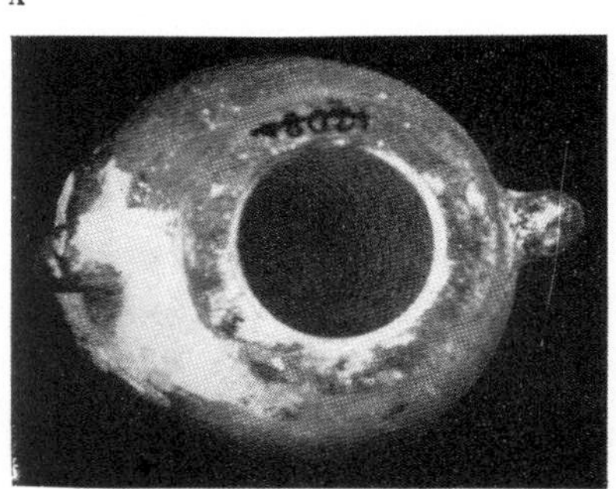

C

D

E

A. *The 'Portland' cameo vase, white figures on blue ground. From Rome. First century A.D. Scale* $\frac{1}{3}$ (pp 323, 338, 341)

B. *Colourless beaker, with fine facet-cut figures. Second century A.D. Scale* $\frac{1}{2}$. (pp 324, 341)

C, D. *Top and side views of Sargon II vase, green, cold cut and ground. From Nimrud.* c *720 B.C.* C, *scale* $\frac{2}{3}$, D, *scale* $\frac{1}{2}$. (See fig. 299, pp 321, 336)

E. *Colourless beaker with engraved figures, from Worringen, near Cologne. Late third or fourth century A.D. Scale* $\frac{1}{2}$. (p 341)

A

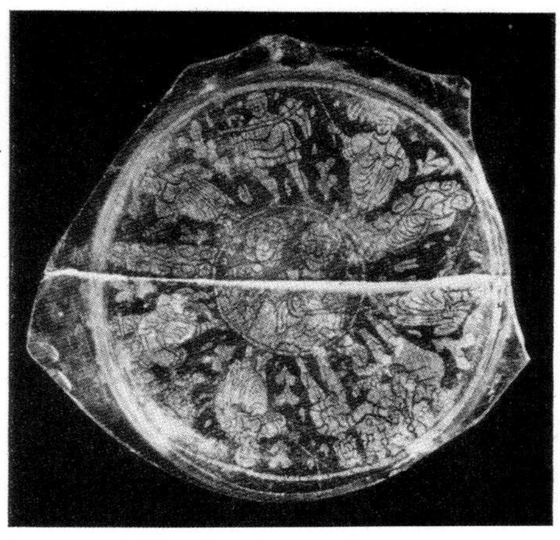

B

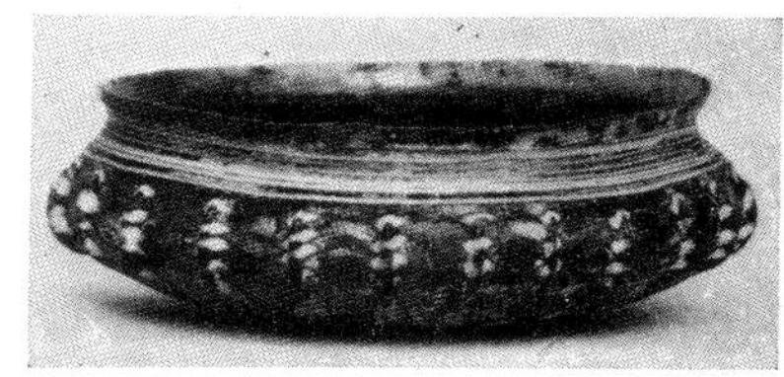

C

D

E

F

A. *Opaque white jug, with painted scene of Apollo and Daphne, from south Russia. Late second or early third century* A.D. *Scale* $\frac{1}{2}$ (pp 324, 341, 343). B. *Bottom of a bowl with gilt decoration between two layers of colourless glass, from Italy. Fourth century* A.D. *Scale* $\frac{1}{2}$ (p 342). C. *Ribbed bowl, blue with opaque white trails. First century* A.D. *Scale* $\frac{1}{2}$ (p 340). D. *Mould-blown beaker, brown, with knopped pattern. First century* A.D. *Scale* $\frac{1}{3}$ (p 339). E. *Colourless jug, with nipped ribbing and trellis handle. From Colchester. Third century* A.D. *Scale* $\frac{1}{4}$ (p 340). F. *Amber jug with opaque white marvered blobs. From the Aegean archipelago. First century* A.D. *Scale* $\frac{1}{4}$. (p 340)

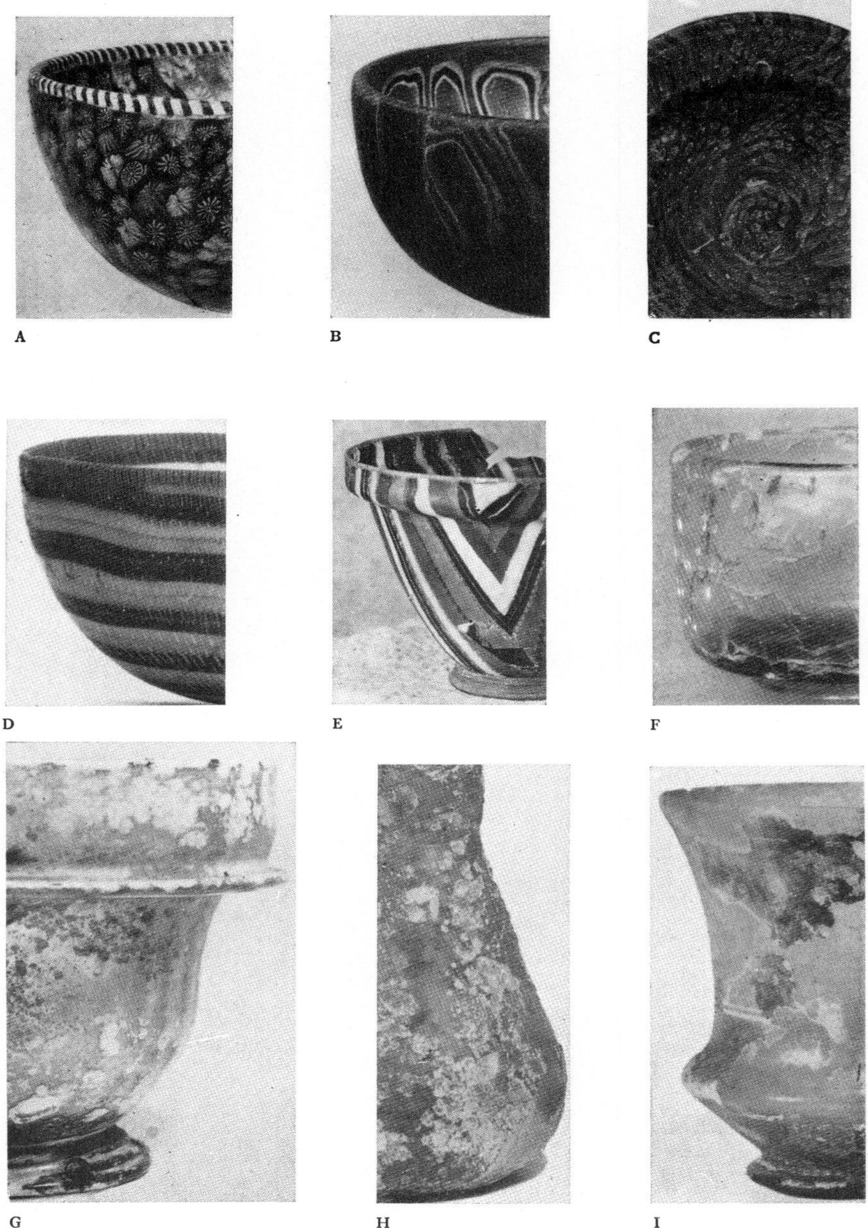

A B C

D E F

G H I

A–E. *Portions of bowls showing the five main varieties of mosaic glass:* A, *floral;* B, *marbled;* C, *mottled;* D, *laced;* E, *striped. First century* B.C. *to first century* A.D. *Scale* $\frac{1}{2}$. (pp 322, 338)

F–I. *Portions of vessels showing the four main varieties of weathering on ancient glass:* F, *strain-cracking;* G, *pitting;* H, *pitted;* I, *flaking. Second to fourth centuries* A.D. *Scale* $\frac{2}{3}$ (pp 333–4)

A. *'Street of Balconies' at Ostia showing the support to the second-floor balconies. These* insulae *or apartment-blocks had shops on the ground floor, and some of the buildings originally rose to four or five storeys.* (p 446)

B. *A 'peristyle' or colonnaded courtyard in a typical example of a Greco-Roman patrician dwelling at Pompeii. The low wall is painted with plants and hunting scenes. The columns, painted red and black, enclose a garden. Note the cistern* (impluvium) *in the centre, a feature which the peristyle has in common with the atrium.* (pp 446, 530)

A. *A medieval artist's conception of Noah building the Ark. He depicts carpenters constructing a timber-framed building of three floors, the roof and sides of which are boarded. The posts and rails of the timber framing are mortise and tenoned and held together by wooden pegs. The carpenter drilling the holes for the pegs with an auger is to be seen on the right-hand side. The boarded roof is nailed; a box of nails can be seen on the part already finished. Other wood-working operations—planing, drilling, trimming with an axe, and sawing—are carefully depicted in the foreground. Among the tools lying on the ground are a brace and a frame-saw. From the 'Bedford Book of Hours', Flemish, early fifteenth century.* (pp 241, 389, 391, 393–5, 440)

B. *A nobleman discusses the progress of the builders with his master-mason. The mason carries a plumb-bob square for checking horizontal levels. Behind him carpenters use the trimming axe and frame-saw. To the left, one man takes a new hod of mortar on his shoulder, while another prepares to ascend the ladder, possibly with liquid refreshment. Loads of stone are sent up to the workers with a simple hoist. Small figures in the background dig a ditch and hammer home stakes for a fence. From a late fifteenth-century French manuscript.* (pp 385–6, 392)

A. *Building the Tower of Babel. Masons with scabbling axes, toothed bolster, square, and compasses carve the stone. This is transported in two stages with crank-turned hoists. The rope on the lower ones is doubled to halve the tension in it. There is a wooden shed where water from a large keg is mixed with the sand and lime. Mortar is carried up on foot. Scaffolding is rudimentary and affords inadequate support to builders attacked in their impious work by angels. The tower is of exotic architecture, and does in fact bear some resemblance in its square diminishing storeys to a Mesopotamian ziggurat. Otherwise the only concession to the orient is the camel loaded with stone on the right. There is a tower-mill on one of the hills. From a French manuscript of* c *1425.* (pp 385–6, 388, 394)

B. *A remarkable picture of churches in various stages of completion. The scaffolding is elaborate, and material may be drawn up by rope to the highest level. Mortar is mixed in the foreground, and two freemasons are engaged on sections of fluted columns. The building behind them is of brick, but its bonding cannot be determined. The bricklayers are supplied with mortar in hods and bricks in wooden buckets. Roofs are tiled. Flemish, fifteenth century.* (pp 385, 388, 440)

A. *The Ardagh chalice of silver, gold, enamel, crystal, filigree-work, and jewels. Irish. Ninth century.* (pp 464, 474)

B. *The* baptistère *of St Louis, made of bronze with gold and silver inlays. 1290–1310.* (see figure of detail 416, and p 453)

The King's Lynn cup. It is made of silver gilt and decorated with bassetaile *enamel.* c *1325.* (pp 461, 463, 469)

A. *Roman bridge built over the Marecchia in the times of Augustus and Tiberius. First century A.D.* (p 510)

B. Via di Nola *at Pompeii showing raised side-walks. Before A.D.* 79 (p 530)

Winged Victory erected at Samothrace probably to commemorate the naval battle of Cos, c *258 B.C. The statue is standing on the prow of a cataphract* (cf figure 519). c *250–180 B.C.* (p 566)

Richly ornamented harness of the fifteenth century. The knights hold only the plain snaffle-reins, while the decorated curb-reins lie free on the horses' necks. The saddle of the first knight has a large upright pommel and a heavy breast-band, richly ornamented like all these harness-straps. Notice the iron ring-stirrup and the rowel-spur with very long shank (p 560). *The forelock of the white horse is plaited. From Van Eyck's Altar of Ghent. 1432*

A. *Remains of one of the Dejbjerg wagons which was probably used as a funeral vehicle. Celtic work, probably made in south Germany or north-east France. First century B.C.* (figures 481, 504 C, pp 540, 545, 547)

B. *Galatea rides a wheel-propelled shell in her triumphal procession. The device, an invention of late Roman times and illustrated in the fifteenth century* (figure 549), *must have been quite popular. Raphael painted it for the wealthy Chigi, c 1514, and a Pesaro potter adapted the painting, including the technical feature, for the decoration of a cistern in the late sixteenth century. The bowl, 19·7 inches in length and 8·7 inches in height, is painted in yellow, green, and blue, and illustrates also the tradition of plastic and polychrome achievements in ceramic technique maintained in the Renaissance.* (pp. 306, 607)

A. *Street scene in a French fifteenth-century city. The street has no gutters but the pavement is well laid in large slabs. Manuscript illumination by Foucquet, 1458.* (p 531)

B. *The large crane at Bruges still worked as in Roman times by a cage-wheel. The casks have been hoisted from a barge. The primitive sledge is apparently waiting to cart one of them away over the cobblestone pavement of the quayside into the narrow street in the background.* (pp 245, 388, 531, 638)

Intersection of five aqueducts, south-east of Rome (see figure 614). *The arches to the right carry Marcia, Julia, and Tepula. Those to the left Claudia and Anio Novus. A fiscal tower is built over the main crossing. Modern pictorial reconstruction* (pp 670, 672)

A. *Section of polygonal masonry wall of the citadel at Alatri, south Italy, built by the local tribe. Fourth century* B.C. (p 717)

B. *Aerial photograph of a section of Hadrian's wall running along the summit of the ridge, with a fort in the foreground. The line of the track to the left of the wall is clearly visible.* c *A.D. 122.* (p 719)

A. *The Roman* Porta Nigra *at Treves.* c *A.D. 300.* (p 718)

B. *The* Porta Palatina *at Turin, built by the Emperor Augustus. 27 B.C.–A.D. 14.* (pp 717, 718)

A. *Alchemical laboratory represented by Brueghel. Note the great variety of apparatus depicted. As usual, stills and crucibles form the major portion of the equipment. The crudity of the scales reflects the unimportance attached to quantitative work.* (p 749)

B. *Though this picture by Stradanus shows many of the pieces of apparatus commonly used by alchemists, the absence of crucibles, ladles, files, and so on seems to indicate that the numerous workers are engaged upon the preparation of herbal remedies. The figure in the alcove on the right may be carrying out alchemical work proper.* (p 749)

A. *A stylized representation by Teniers of an alchemist at work. The average alchemical work-room was far from being so tidy and uncrowded, and normal operations would not have left the tablecloth in such a good state of repair. The charcoal furnace in the foreground is apparently unprovided with a damper.* (p 749)

B. *Priestley's laboratory. The simplicity of the apparatus is in striking contrast to the complex heterogeneity of typical alchemical equipment. The collection and investigation of gases was no part of alchemy.* (p 750)

Mosaic of the figure of an angel in the vault of the church of Germigny-des-Prés, near Orleans. Ninth century. Medium, workmanship, and design are all under strong Byzantine influence. (p 762)